# Handbuch der experimentellen Pharmakologie

# Vol. 43 Heffter-Heubner New Series

# Handbook of Experimental Pharmacology

# Anabolic-Androgenic Steroids

Contributors

N. Arimasa · A. Arnold · D. L. Bennett · A. L. Beyler
W. A. Daniel · J. A. Friedland · G. S. Gordan
C. W. Gurney · R. Hilf · A. B. Houssay · C. D. Kochakian
H. Kopera · R. L. Landau · I. Lewin · T. W. Perry
G. O. Potts · A. J. Ryan · H. Spencer

Editor

Charles D. Kochakian

With 38 Figures

Springer-Verlag Berlin Heidelberg New York 1976

CHARLES D. KOCHAKIAN, Ph. D., Director and Professor, Experimental Endocrinology, The University of Alabama in Birmingham, Medical Center, 1919 Seventh Avenue South, Birmingham, AL 35294/USA

ISBN-13: 978-3-642-66355-0 e-ISBN-13: 978-3-642-66353-6
DOI: 10.1007/978-3-642-66353-6

Library of Congress Cataloging in Publication Data. Main entry under title: Anabolic-androgenic steroids. (Handbook of experimental pharmacology: New series; v. 43). Includes bibliographies and indexes. 1. Androgens. 2. Anabolic steroids. 3. Androgens—Therapeutic use. 4. Anabolic steroids—Therapeutic use. I. Arimasa, N. II. Kochakian, Charles D., 1908— III. Series: Handbuch der experimentellen Pharmakologie: New series; v. 43. QP905.H3 vol. 43 [QP572.A5] 615'.1'08s [599'.01'927] 76-12989



Softcover reprint of the hardcover 1st edition 1976

# Preface

This volume was planned to provide a comprehensive survey of the role of the anabolic-androgenic steroids in the vital economy exclusive of the androgenic (sexual) functions. It seemed appropriate to bring together all of this information in an organized fashion in one volume at this time not only to serve as a source of information but also to indicate and suggest areas that need further exploration. The anabolic action of the steroid hormones has gone through a period of great activity in both basic and clinical research. A complete understanding of the manifold anabolic effects still remains to be elucidated and the art of clinical application is only gradually becoming apparent.

This volume should be useful not only to the experienced investigator in both basic and clinical research but also for the novice. Furthermore, it should serve as a source of information for the careful use of these steroids in certain metabolic diseases. These steroids have had wide clinical application with variable results. In many instances further careful exploration is suggested. Other instances have demonstrated varying degrees of usefulness.

The editor has chosen to write most of the chapters related to the basic aspects of the action of the anabolic-androgenic steroids and for the clinical areas has selected contributors who have been active not only in research but in the clinical application of these steroids. Since the material for these chapters could readily overlap with consequent duplication and also possible serious omissions, a conference of these contributors was held on January 27, 1975. A detailed review was presented by each contributor with active informal discussion. The conference proved of great value. Changes were made and hopefully overlap and omissions have been minimized.

I would like to thank the many contributors for their enthusiastic interest and cooperation. I wish to especially thank Richard Landau for his many constructive suggestions and Allan Kenyon for his interest and suggestions. Finally, I wish to express my appreciation to my secretary, Frances Youngblood, for her dedicated assistance in the many details of editing such a volume and the extensive typing and literature search required in the preparation of my chapters. To my wife, I owe my special gratitude for her continuous support and laboratory assistance during the initiation of the renotrophic-androgenic and enzyme studies.

Autumn, 1976 CHARLES D. KOCHAKIAN

# Table of Contents

## III. Sites of Tissue Formation

### A. Body and Organ Weights and Composition. C. D. Kochakian. With 13 Figures

**B. Effect of Anabolic-Androgenic Steroids on the Skin, Including Hair and Sebaceous Glands.** A. B. HOUSSAY. With 2 Figures



**C. Anabolic-Androgenic Steroids and Experimental Mammary Tumors.** R. HILF

## IV. Mechanism of Action

### A. Regulation of Protein Synthesis by Anabolic-Androgenic Steroids.

C. D. KOCHAKIAN



### B. Regulation of Nucleic Acid Synthesis by Anabolic-Androgenic Steroids.

C. D. KOCHAKIAN

## VII. Clinical Applications

### A. Effect of Androgens on Bone, Calcium, and Phosphorus Metabolism.
H. SPENCER, J. A. FRIEDLAND, and I. LEWIN. With 2 Figures



### B. The Use of Anabolic-Androgenic Steroids in Childhood and Adolescence.
W. A. DANIEL, JR. and D. L. BENNETT

# List of Contributors


N. ARIMASA, M. D., The First Department of International Medicine, Okayama University Medical School. 5-1 2-chome, Shikata-cho, Okayama 700, Japan

A. ARNOLD, Ph. D., Sterling-Winthrop Research Institute, Rensselaer, NY 12144, USA

D. L. BENNETT, M. D., Instructor, Adolescent Medicine, University of Alabama in Birmingham, Medical Center, Birmingham, AL 35294, USA

A. L. BEYLER, Ph. D., Sterling-Winthrop Research Institute, Rensselaer, NY 12144, USA

W. A. DANIEL, J.R. M.D., Professor of Pediatrics, Director, Adolescent Unit, University of Alabama in Birmingham, Medical Center, Birmingham, AL 35294, USA

JOAN A. FRIEDLAND, M. D., Veterans Administration Hospital, Houston, TX 77031, USA

G. S. GORDAN, M. D., Department of Medicine, University of California, San Francisco, CA, USA

C. W. GURNEY, M. D., Deputy Dean/Clinical Sciences, Department of Medicine, University of Chicago, Chicago, IL 60637, USA

R. HILF, Ph. D., Associate Professor of Biochemistry, University of Rochester, School of Medicine and Dentistry, Rochester, NY 14642, USA

A. B. HOUSSAY, M. D., Professor of Physiology, 2a. Catédra de Fisiologia, Facultad de Medicina, Universidad de Buenos Aires, Buenos Aires, Argentina

C. D. KOCHAKIAN, Ph. D., Director and Professor, Experimental Endocrinology, The University of Alabama in Birmingham, Medical Center, Birmingham, AL 35294, USA

H. KOPERA, M. D., Lecturer, Head Clinical Pharmacology Unit, Dept. of Exper. and Clin. Pharmacology, University of Graz, A-8010 Graz, Austria

R. L. LANDAU, M. D., Department of Medicine, University of Chicago, Chicago, IL 60637, USA

I. LEWIN, M. D., Cancer Cooperative, Springfield, MA 01103, USA

T. W. PERRY, Ph. D., Purdue University, School of Agriculture, Agricultural Experimental Station, Cooperative Extension Service, West Lafayette, IN 47907, USA

G. O. Potts, Ph. D., Head, Department of Endocrinology, Sterling-Winthrop Research Institute, Rensselaer, NY 12144, USA

A. J. Ryan, M. D., University Health Service, University of Wisconsin, Madison, WI 52706, USA

Hérta Spencer, M. D., Chief, Metabolic Section, Edward Hines, Jr. Hospital, Hines, IL 60141, USA

# I. Historical Review of Anabolic-Androgenic Steroids

C. D. Kochakian

Initial interest in the role of the testis in the vital economy was directed towards the development and the regulation of the functions of the accessory sex organs. Indeed the classical experiments of Berthold (1849) on the regulation of the growth of the comb and wattles of fowl anticipated endocrinology by 50 years. Certain differences in physical development and metabolism between the male and female organism suggested that the testis regulated other events in the development of the male animal. Thus, the growth rate, especially the prepubertal spurt in the male, the greater rate of energy metabolism, and size of the male were obvious. Korenchevsky (1925) was one of the first to attempt to correlate these changes with the testis by direct experimentation with testicular extracts. He injected crude preparations of testis from various sources into rabbits and dogs but was not able to demonstrate any effective changes in energy and nitrogen metabolism. More dramatic claims based on subjective observations had been made earlier (1889) by Brown-Sequard, who administered crude testis extracts to himself and others with reports of a marked increase in vigor.

The first clear demonstration of an effect of the products of the testis was not produced until 1935. Kochakian (1935) and Kochakian and Murlin (1935) reported a marked decrease in protein metabolism by the nitrogen balance technique and noted a slight or no increase in energy metabolism in castrated dogs after the injection of a potent androgen extract from human male urine. Shortly thereafter, testosterone was characterized (David et al., 1935) as the internal secretion of the testis, and methods for its synthesis from cholosterol were developed (Butenandt and Hanisch, 1935; Ruzicka and Wettstein, 1935). Kochakian and Murlin (1936) prepared androst-4-ene-3,17-dione, administered it to the castrated dogs, and reproduced the effect on nitrogen metabolism produced by the urinary extract. Shortly, testosterone became available and also proved to be effective (Kochakian, 1937), thus establishing a protein anabolic action for androgens.

The rapid commercial availability of testosterone permitted Kenyon and his associates (1938) to establish, by careful balance studies, effects in the human parallel to those in the dog. These studies opened up many potential clinical applications of the protein anabolic action of the androgens. Many of the early explorations are recorded in the historically recognized Josiah Macy, Jr., Conferences on Convalescences, under the chairmanship of Fuller Albright and the guidance of Frank Fremont-Smith.

I have been asked on many occasions what motivated the study on protein metabolism. I am, therefore, recording the events that led to these classical studies because they illustrate an unconscious operation of the prepared mind. John R.

MURLIN'S career had been devoted to nutrition and metabolism with extensive studies on energy and protein metabolism. It was his custom to devote the last 3 weeks of laboratory work in the course on Vital Economics (Nutrition and Endocrinology) at the University of Rochester School of Medicine to small research projects by the medical students. One of the projects in 1932 was the effect of a male hormone containing urine extract on the energy metabolism of the castrated dog. The presence of male hormone-active substances (growth of the capon's comb) in the urine of human males and an effective method of extraction had only recently been reported by FUNK et al. (1930). The students prepared the extract and, in the short time available to them, observed by the difficult (closed circuit) calorimeter type of study an apparently clear calorigenic effect of the extract. These results stimulated JOHN R. MURLIN to obtain a grant ($ 500) to hire an assistant to confirm and extend these studies. He felt that, since extracts had to be prepared, a person with an interest and training in organic chemistry would be required. I was selected as the graduate research assistant. Extracts from thousands of gallons of urine from medical students were prepared. Administration of these extracts at various doses and periods of time did not elicit the anticipated calorigenic action. Murlin was disappointed but suggested the study of protein metabolism. This, I feel sure, was prompted unconsciously because of his earlier detailed studies on protein metabolism in the pregnant dog (MURLIN, 1910, 1911). The initial experiment gave a marked decrease in urinary nitrogen excretion and a concomitant increase in body weight. The calculation of the nitrogen retained as new tissue agreed with the increase in body weight. Furthermore, it was obvious that the stimulation of growth in the accessory sex tissues could account for only a fraction of these changes. Obviously other tissues were stimulated to grow. Therefore, KOCHAKIAN and many others extended the studies to obtain a more detailed analysis of this phenomenon. These studies progressed from the whole animal to the molecular level (KOCHAKIAN, 1965) and opened new parameters of endocrine research.

The initial reaction to the protein anabolic action of the androgen was that it was induced via the growth hormone of the anterior pituitary. This was not only dispelled, but also the anabolic effect was established as direct effect of the androgen by systematic nitrogen balance studies in rats and dogs surgically deprived separately of each of the endocrine organs (KOCHAKIAN, 1964). Studies by KENYON and others in patients with endocrine deficiencies gave the same conclusions.

Extension of the studies revealed that the growth of almost every tissue was influenced to some extent by the androgens. The degree of the effect varied widely among the different animals. The marked responsiveness of the mouse kidney to the level of androgen suggested to KOCHAKIAN that this organ might be representative of the protein anabolic action of the C-19 steroid hormones. Modifications of chemical structure indicated partial dichotomy in the response of the kidney and the seminal vesicles and prostate. Thus, androstane-3$\alpha$,17$\beta$-diol was preferentially renotrophic while androstenedione was androgenic. Similar results were obtained in a comparison of the response of the temporal muscle of the guinea pig, which had been shown by PAPANICOLAOU and FALK (1938) to be greatly dependent on androgen. These results, first reported at the JOSIAH MACY, JR., Conferences on Convalescence, stimulated interest and raised hopes for the development of a steroid hormone that would be exclusively protein anabolic for use in many debilitating conditions. A further impetus for the search of such a compound was the report that nortestosterone had a preferential effect on the levator ani muscle, first introduced as a test object by EISENBERG and GORDAN (1950) and refined as an assay method by HERSHBERGER et al. (1953). The simplicity of the assay prompted many investigators and pharmaceutical companies to exploit it as a screening procedure. Over a thousand compounds have been synthesized and screened (VIDA, 1969). Many of these compounds now are marketed as anabolic agents and have been tried in a wide variety of metabolic disturbances and in animal husbandry with varying success. Since a complete separation of the anabolic from the androgenic properties of these compounds has not been achieved, they should be more properly designated as anabolic-androgenic steroids rather than anabolic steroids.

It was recognized very early that elucidation of the mechanism of action of these hormones would not only provide a better understanding of their metabolic effects but would also make their clinical application more feasible. KOCHAKIAN initiated two approaches for the elucidation of the mechanism of action. Since these hormones regulated growth processes, it seemed reasonable to expect that certain enzymes were involved. The initial studies demonstrated striking changes in kidney arginase activity and subsequent studies by a number of investigators have revealed the involvement of a wide variety of enzymes. These enzymes may by classified into at least three categories: (1) constitutive; (2) regulatory; and (3) RNA and protein synthetic (KOCHAKIAN, 1962, 1964). The other approach to the main problem was the question of the involvement of the hormone at the site of action. The first studies (CLARK and KOCHAKIAN, 1947) demonstrated that a number of enzymes were present in tissues that converted the hormone to several metabolites. Subsequent studies (KOCHAKIAN, 1959) indicated that the type and intensity of the activity of these enzymes not only varied among the tissues but also among animal species and between males and females. These studies provided supplementary support to the studies on the relationship of chemical structure of the steroid and response of a tissue. Thus, it seemed that the purpose of the C-19 steroid oxidoreductases present in a tissue was to provide the specific tissue with the proper steroid hormone or spectrum of steroid hormones. More recent studies have indicated that the various tissues probably produce specific proteins to serve as receptors for specific steroids, which in turn induce the necessary metabolic changes.

## References


BERTHOLD, A.: Transplantation der Hoden. Arch. Anat. Physiol. Wissensch. **1849**, 42—46.

BROWN-SEQUARD, C. E.: The effects produced on man by subcutaneous injections of liquid obtained from the testicles of animals. Lancet **1889 II**, 105—107.

BUTENANDT, A., HANISCH, G.: Über Testosterone. Umwandlung des Dehydro-Androsterons in Androstendiol und Testosteron; ein Weg zur Darstellung des Testosterons aus Cholesterin. Z. physiol. Chem. **237**, 89—97 (1935).

CLARK, L. C., JR., KOCHAKIAN, C. D.: The in vitro metabolism of testosterone to $\Delta^4$-androstenedione-3,17; cis-testosterone and other steroids by rabbit liver slices. J. biol. Chem. **170**, 23—33 (1947).

DAVID, K., DINGEMANSE, E., FREUD, J., LACQUEUR, E.: Über krystallinisches männliches Hormon aus Hoden (Testosteron), wirksamer als aus Harn oder aus Cholesterin bereitetes Androsteron. Z. physiol. Chem. **233**, 281—282 (1935).

EISENBERG, E., GORDAN, G. S.: The levator ani muscle of the rat as an index of myotrophic activity of steroidal hormones. J. exp. Therap. **99**, 38—44 (1950).
FUNK, C., HARROW, B., LEJWA, A.: The male hormone. Amer. J. Physiol. **92**, 440—449 (1930).
HERSHBERGER, L. G., SHIPLEY, E. G., MEYER, R. K.: Myotrophic activity of 19-nortestosterone and other steroids determined by levator ani muscle method. Proc. Soc. exp. Biol. (N.Y.) **83**, 175—180 (1953).
KENYON, A. T., SANDIFORD, I., BRYAN, A. H., KNOWLTON, K., KOCH, F. C.: The effect of testosterone propionate on nitrogen, electrolyte, water and energy metabolism in eunuchoidism. Endocrinology **23**, 135—153 (1938).
KOCHAKIAN, C. D.: The effect of male hormone on protein metabolism of castrate dogs. Proc. Soc. exp. Biol. (N.Y.) **32**, 1064—1065 (1935).
KOCHAKIAN, C. D.: Testosterone and testosterone acetate and the protein and energy metabolism of castrate dogs. Endocrinology **21**, 750—755 (1937).
KOCHAKIAN, C. D.: Metabolism of androgens by tissue enzymes. In: Biochemistry of Steroids. Fourth International Congress of Biochemistry **4**, 196—207 (1959).
KOCHAKIAN, C. D.: Intracellular regulations in the kidney by androgens. Amer. Zoologist **2**, 361—366 (1962).
KOCHAKIAN, C. D.: Protein anabolic property of androgens. Ala. J. Med. Sci. **1**, 24—37 (1964).
KOCHAKIAN, C. D.: Mechanisms of anabolic action of androgens: In: Mechanism of Hormone Action, pp. 191—210. New York: Academic Press 1965.
KOCHAKIAN, C. D., MURLIN, J. R.: The effect of male hormone on the protein and energy metabolism of castrate dogs. J. Nutr. **10**, 437—459 (1935).
KOCHAKIAN, C. D., MURLIN, J. R.: The relationship of the synthetic male hormone, androstenedione, to the protein and energy metabolism of castrate dogs and the protein metabolism of a normal dog. Amer. J. Physiol. **117**, 642—657 (1936).
KORENCHEVSKY, V.: The sexual glands and metabolism. III. The influence of testicular or ovarian emulsions upon the nitrogen and gaseous metabolism of dogs and rabbits. Brit. J. exp. Path. **6**, 158—172 (1925).
MURLIN, J. R.: Metabolism of development. II. Nitrogen balance during pregnancy and menstruation of the dog. Amer. J. Physiol. **27**, 177—205 (1910).
MURLIN, J. R.: Metabolism of development. III. Qualitative effects of pregnancy on the protein metabolism of the dog. Amer. J. Physiol. **28**, 422—454 (1911).
PAPANICOLAOU, G. N., FALK, E. A.: General muscular hypertrophy induced by androgenic hormone. Science **87**, 238—239 (1938).
RUZICKA, L., WETTSTEIN, A.: Sexualhormone VII. Über die künstliche Herstellung des Testikelhormons. Testosteron (Androsten-3-on-17-ol). Helv. chim. Acta **18**, 1264—1275 (1935).
VIDA, J. A.: Androgens and Anabolic agents. Chemistry and Pharmacology. New York-London: Academic Press 1969.

# II. Metabolic Effects

## A. Metabolic Effects of Anabolic-Androgenic Steroids in Experimental Animals

C. D. Kochakian

### I. Introduction

#### A. Early Studies

The influence of the testis on metabolic processes in tissues other than the accessory sex organs had been recognized for a long time. The effect of castration on the size and behavior of domestic animals, the difference in BMR, skeleton, body weight, and shape, etc., between males and females was known. The dramatic effects on vigor and energy reported by Brown-Sequard (1889) after the self-administration of testis glycerol extracts focused attention on energy metabolism. Early attempts to confirm these results were conflicting and unconvincing. Later several investigators administered extracts to experimental animals, with small and questionable results on energy and nitrogen metabolism (Koch,1939; Moore, 1939).

#### B. Establishment of Anabolic Action

##### 1. Dog

*a) Urinary Extract*

Kochakian (1935), Kochakian and Murlin (1935) reported that a lipid-soluble extract from human male urine (Funk and Harrow, 1930), which contained the ability to stimulate comb growth in the capon and the accessory sex organs in the rat, induced consistent nitrogen retention in the castrated dog previously adjusted to nitrogen and caloric equilibrium (Fig. 1). The positive nitrogen balance was due entirely to a decrease in urinary nitrogen excretion. The fecal nitrogen was not significantly changed. A maximum retention was attained within 5 days, and further injection of the extract maintained the urinary nitrogen excretion at the new level. The response was dose-related but once a maximum dose was attained, no further increase in nitrogen retention could be induced by the use of a larger dose. The positive nitrogen balance was accompanied by an increase in body weight. Cessation of the injections resulted in a transitory increase in nitrogen excretion and a loss in body weight. The loss in nitrogen was only a fraction of that retained under the stimulus of the androgenic extract.

The injection of a single large dose of the urinary extract into the castrated dogs induced a definite positive nitrogen balance on the day after the injection and a maximum effect on the 3rd day that was maintained for 2 more days. The urinary nitrogen then gradually returned to normal by the seventh day, without the extra

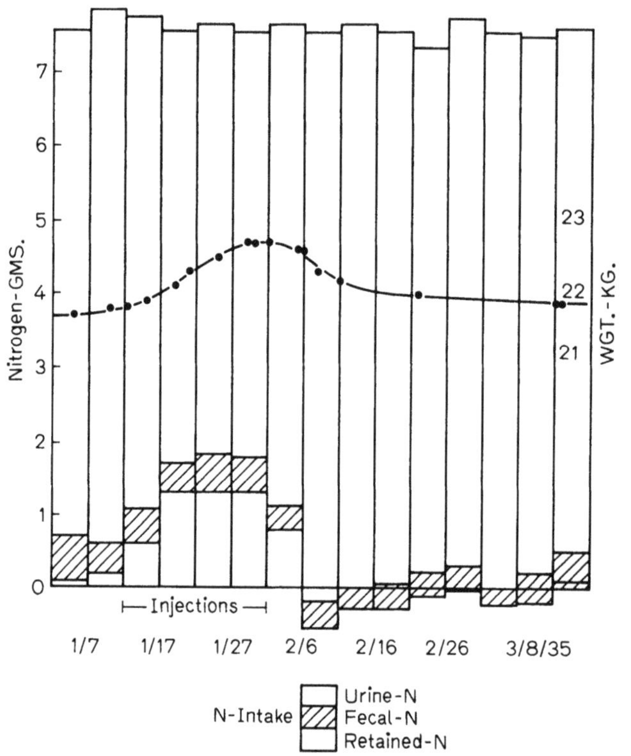


Fig. 1. Effect of a "male hormone" urine extract on nitrogen metabolism and body weight of castrated ("fat") dog. The dose, 133 I.U./day, has been converted to international units (1 I.U. = 0.1 mg androsterone) on the basis of subsequent assays by standardized methods. (Redrawn from KOCHAKIAN and MURLIN, 1935)

excretion of nitrogen noted after prolonged injections of the extract at the maximum effective dose.

The presence of anabolic stimulating compounds was later also demonstrated in pregnant mare's urine (KOCHAKIAN, 1950a). A large proportion of the activity was in the α-ketonic fraction of a neutral fat-soluble extract. The maximal response in terms of nitrogen retention and body weight exhibited by the castrated dogs was essentially identical to that exhibited in response to a crystalline testosterone suspension, and the maximum daily nitrogen retained per kilogram per day was the same as that following administration of male urine extract and androst-4-ene-3,17-dione (KOCHAKIAN and MURLIN, 1936).

### *b) Pure Steroids*

Identification of the product of the testis by DAVID et al. (1935) and development of the partial synthesis from cholesterol of related compounds permitted extension of the studies to pure compounds. The first compound studied was androst-4-ene-3,17-dione, which was prepared by the essentially identical methods reported almost simultaneously by BUTENANDT and KUDSZUZ (1935) and RUZICKA and WETTSTEIN

(1935). The injection of the pure androstenedione for 3 days into the castrated dogs produced a decrease in urinary nitrogen similar to that observed with the urinary extract (KOCHAKIAN and MURLIN, 1936). Indeed, the maximum nitrogen retained per kg body weight was identical for the two preparations. The blood NPN and urea exhibited a variable small decrease, indicating that the decrease in urinary nitrogen excretion was not due to a retention of nonprotein constituents.

When testosterone and its acetate became available (BUTENANDT and HANISCH, 1935), these two compounds also stimulated nitrogen retention upon injection into castrated dogs (KOCHAKIAN, 1937). In later studies, the injection of a crystalline suspension of testosterone (KOCHAKIAN, 1950a) and an oil solution of testosterone propionate (KOCHAKIAN and COSTA, 1959) into castrated dogs produced a marked positive nitrogen balance with a concomitant increase in body weight.

The studies with the pure compounds indicated that the effect of the androgenic urinary extract was due to its androgen content and established the protein anabolic property of androgens.

## 2. Rat

### *a) Nature of the Protein Anabolic Effect*

The studies on dogs were extended to rats to provide a more practical and feasible model for elucidation of the nature of the anabolic action of the androgens. The attainment of nitrogen and caloric balance in the rat proved to be a delicate and sensitive matter. A satisfactory balance was not reached until after several weeks, and frequently months, of careful adjustment of the food intake. Adjustment of food intake resulted in the establishment of a new equilibrium in an exponential fashion (FORBES, 1973).

The castrated rat responded to the testosterone propionate injection in a manner similar to that of the castrated dog. Testosterone propionate produced an almost immediate decrease in urinary nitrogen (Fig. 2) (KOCHAKIAN, 1950b). The decrease was definitely evident on the second day after the beginning of injections and possibly within the first 24 h and continued to decrease for 3 days to a minimum value, where it was maintained for approximately a week. Continuation of the injections was not able to maintain the new level. The urinary nitrogen gradually returned to the basal level despite continued injections of the androgen. This unexpected secondary phenomenon was designated as "the wearing-off effect." Upon cessation of the androgen injections, a small negative nitrogen balance was observed for approximately one week, with a subsequent return to the basal level. This effect was termed "the rebound phenomenon."

Body weight reflected the changes in urinary nitrogen excretion (Fig. 2). A rapid increase in body weight was induced during the initial period of positive nitrogen balance followed by a decrease in the rate of increase during the wearing-off period, when the rat had returned to nitrogen equilibrium despite continued androgen injections. Cessation of the injections resulted in a further decrease in body weight to a point slightly below the starting weight. The overall loss in body weight, however, was not due to loss of protoplasmic mass. The rat was not in negative nitrogen balance during the 56 days of injection and exhibited only a small negative nitrogen

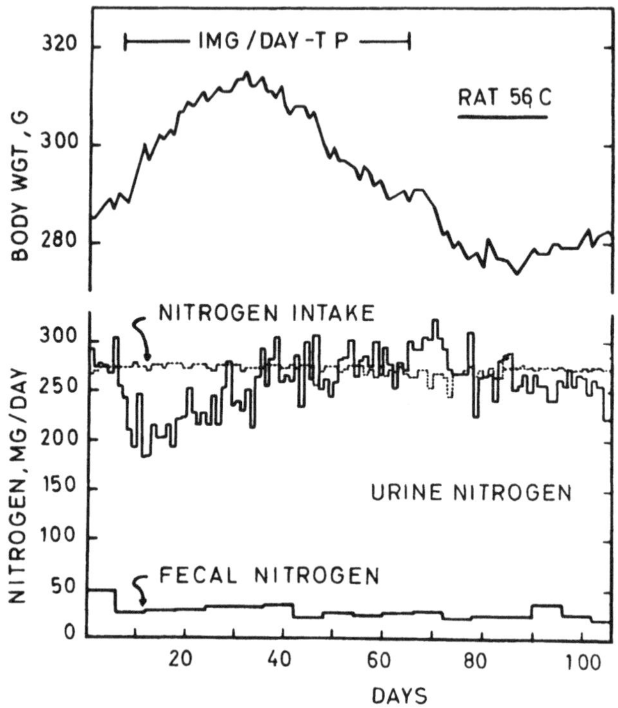


Fig. 2. Effect of testosterone propionate (TP) on body weight and nitrogen balance of castrated rat (KOCHAKIAN, 1950b)

balance on cessation of injections. If the retained nitrogen were converted to protoplasm by the factor proposed by REIFENSTEIN et al. (1945), the observed gain in body weight during the initial phase of nitrogen retention (the first 2 weeks) would be due to the formation of protoplasm. The latter, however, increased at a slower rate during the wearing-off period and was maintained during the rest of the injection period; therefore the observed loss in body weight was due to constituents other than protein, which were shown by carcass analyses (KOCHAKIAN et al., 1950a) to be fat.

The effects on nitrogen metabolism and body weight were produced repeatedly and consecutively in the same rat. The wearing-off effect, therefore, was not due to resistance of the rat to continued injection. Furthermore, an increase in the dose during the wearing-off period reversed the disappearance of the nitrogen retention and the loss in body weight (KOCHAKIAN et al., 1950b).

As in the case of the castrated dogs, fecal nitrogen excretion was not changed by testosterone propionate injections.

These studies provided not only a smaller animal in which to study the protein anabolic action of androgens but also further information concerning the nature of the protein anabolic effect. The wearing-off effect was not observed in the dog but would have been apparent if the injection period had been extended (KOCHAKIAN and COSTA, 1959). In some instances, it was difficult to bring the rats into body weight and nitrogen equilibrium. When the rats persisted in positive nitrogen balance and injections were made, the decrease in nitrogen excretion was only that between the existing positive balance and the anticipated maximum effect due to the injection. The rest of the response pattern—the wearing-off effect and the return to nitrogen equilibrium—proceeded in the usual manner. If the body weight was increasing prior to hormone administration, then the induced increase in body weight was superimposed and gave a false maximum increase. The same was observed in growing rats (SAUNDERS and DRILL, 1958). These phenomena were not appreciated by some later investigators, as is indicated below. Furthermore, some of the studies did not provide adequate time for adjustment to a new equilibrium (FORBES, 1973) after placing the rats on a new, fixed food intake.

*b) Steroids with Protein Anabolic Activity*

Testosterone propionate and several other steroids, studied at various dose levels for prolonged periods, produced the same response pattern in terms of nitrogen excretion and body weight (KOCHAKIAN, 1950b). The injection of testosterone propionate at 0.4 mg/day gave the same results as 1.0 mg/day, indicating that the maximal effective dose was between 0.4 and the less effective 0.125 mg/day dose. Urea excretion changed in parallel with the total urinary nitrogen. Testosterone, 17α-methyltestosterone, and 17β-hydroxy-5α-androstan-3-one were equally effective but less effective than testosterone propionate. Testosterone administered as a subcutaneously implanted pellet proved to be as effective as testosterone propionate. The other compounds induced the anabolic effect but to lesser degrees. Androsterone was the least effective. Several of the more active compounds were compared simultaneously at two or three dose levels and were quantitated on a log-dose basis (cf. Table 1).

Two strains of rats, Wistar and Holtzman, were used in these studies, and the Carworth-Sherman (KOCHAKIAN and VAN DER MARK, 1952; KOCHAKIAN et al., 1950a) and Sprague-Dawley (KOCHAKIAN, 1950c) strains were used in later studies, but no qualitative or quantitative differences in response were detected. Subsequent studies (see below) by many other laboratories with other strains of rats did not reveal any strain differences.

## II. Comparison of Protein Anabolic Activity of Natural and Synthetic Steroids

### A. Assay Procedure in Rat

The balance method in the rat has been adopted by many laboratories to compare the anabolic action of many synthetic steroids. In most instances the procedure of KOCHAKIAN (1950b) has been used, with a limitation of the steroid administration to the initial period of nitrogen retention; in a few instances other modifications also have been made. ARNOLD et al. (1959) and POTTS et al. (1960) substituted dextrin for sucrose and changed the main source of protein from casein (low in cysteine) to the more complete protein, lactalbumin, and also decreased the quantity of protein to ½ that used by KOCHAKIAN, presumably based on the principle that the body makes more economical use of protein at the lower level. They found that the rats more readily equilibrated their nitrogen excretion when the protein of the diet was de-

Table 1. Estimated relative nitrogen retention activities of subcutaneously injected steroids in castrated rats[a]

| Steroid | Dose g/day | Rel. activity Max. N Ret'd/day | References |
|---|---|---|---|
| Testosterone propionate | * | 100 | |
| 4-Chlorotestosterone acetate | 1 | 100 | SALA et al. (1957) |
| 4-Chloro-19-nortestosterone acetate | 1 | 100 | SALA et al. (1957) |
| 17β-Hydroxy-(DL)-13β,17α-Diethylgon-4-en-3-one | * | 100 | TOMARELLI and BERNHART (1964) |
| 17α-Ethyl-19-nortestosterone | 5 | 100 | ASCHKENASY (1959) |
| 17α-Ethyl-19-nortestosterone | 0.02, 0.05, 0.25, 1 | 100 | SAUNDERS and DRILL (1958) |
| 19-Nortestosteronecyclopentylpropionate | 0.25, 0.5, 2 | 75 | STAFFORD et al. (1954) |
| 17β-Hydroxy-17α-ethylnorandst-4-en-3-one | 0.25, 1 | 50 | SAUNDERS and DRILL (1958) |
| 17β-Hydroxy-17α-ethylnorandst-4-en-3-one | 1, 2, 5 | 50 | LENNON and SAUNDERS (1964) |
| 17β-Hydroxy-17α-ethylnorandst-4-en-3-one | 5 | (100) | ASCHKENASY (1959) |
| 17β-Hydroxy-17α-methyl-2-oxa-5α-androstan-3-one | 1, 2, 5 | 30 | LENNON and SAUNDERS (1964) |
| 17β-Hydroxy-2α,17α-dimethyl-5α-androstan-3-3'-azine (Dimethazine) | 0,25, 1 | 20 | MATSCHER et al. (1962) |
| | | | DERUGGIERI et al. (1963) |
| 17α-Methyltestosterone | * | 17 | POTTS et al. (1960) |
| | | | TOMARELLI and BERNHART (1964) |
| 17αMethyltestosterone | * | 7 | KOCHAKIAN (1950) |
| 17β-Hydroxy-17α-methyl-5α-androstan-3-one | * | 17 | BEYLER et al. (1968) |
| 17β-Hydroxy-17α-methyl-5α-androstan-3-one | 1, 2, 5 | (50) | SAUNDERS and DRILL (1958) |
| | | | LENNON and SAUNDERS (1964) |
| Testosterone | * | 15 | KOCHAKIAN (1950) |
| Testosterone | 1 | (35) | SAUNDERS and DRILL (1958) |
| 17β-Hydroxy-5α-androstan-3-one | * | 15 | KOCHAKIAN (1950) |
| 17β-Hydroxy-5α-androstan-3-one | 2 | 18 | STAFFORD et al. (1954) |
| 17β-Hydroxy-17α-methylandrosta-4,6-dien-3-one | * | 12 | POTTS et al. (1960) |
| Androst-4-ene-3,17-dione | * | 7 | KOCHAKIAN (1950) |
| 3β-Hydroxy-5α-androstane-6,17-dione | 2.5 | 5 | KASSENAAR et al. (1953) |
| 17β-Hydroxy-17α-methyl-5α-androstane-(3,2c)-pyrazole | * | 5 | ARNOLD et al. (1959) |
| | | | BEYLER et al. (1961) |
| 17α-Methylandrost-5-ene-3β,17β-diol | 2.0 | 4 | STAFFORD et al. (1954) |
| 17α-Methylandrost-5-ene-3β,17β-diol | * | 2.5 | POTTS et al. (1960) |
| 5α-Androstane-3.17-dione | * | 4 | KOCHAKIAN (1950) |
| 5α-Androstane-3α,17β-diol-17-acetate | 6 | 2 | KOCHAKIAN (1950) |
| 17α-Methyl-5α-androstane-3α,17β-diol | 2.0 | 2 | KOCHAKIN (1950) |
| Testololactone | 10 | 1 | SHEMANO et al. (1951) |
| 5α-Androstane-3α,17β-diol | 0.5, 1 | 1 | KOCHAKIAN (1950) |
| Androsterone acetate | 20, 30 | 1 | KOCHAKIAN (1950) |

| | | | |
|---|---|---|---|
| 17β-Hydroxy-17α-methyl-19-norandrost-5(10)-en-3-one | * | 0 | POTTS et al. (1960) |
| 17β-Hydroxy-17α-methylandrost-4-ene-(3,2c)-pyrazole | * | 0 | POTTS et al. (1960) |
| 4-Hydroxy-17α-methyltestosterone | 1 | + | BALDRATTI and ARCARI (1962) |

[a] Values are referred to testosterone propionate as 100. Asterisk (*) indicates that unknown and testosterone propionate standard were determined at several dose levels to obtain log-dose response curves per ARNOLD et al. (1959). Values in parentheses are crude estimates as are values obtained at one or two dose levels. See Appendix I for trivial, systematic, and trade names.

Table 2. Estimated relative nitrogen retention activities of orally administered steroids in castrated rats[a]

| Steroid | Dose mg/day | Rel. activity Max. N ret'd/day | References |
|---|---|---|---|
| 17α-Methyltestosterone | * | 100 | |
| 17β-Hydroxy-(DL)-13β,17α-diethylgon-5-en-3-one | * | 16,300 | TOMARELLI and BERNHART (1964) |
| 17β-Hydroxy-7α-dimethylandrost-4-en-3-one | * | 4,200 | ARNOLD et al. (1963b) |
| 9αF-11β,17β-Dihydroxy-17α-methylandrost-4-ene-3-one | * | 3,800 | ARNOLD et al. (1963b) |
| 17β-Hydroxy-17α-methyl-5α-androstane-(3,2c) pyrazole | * | 3,000 | ARNOLD et al. (1959) |
| 17β-Hydroxy-17α-methyl-5α-androstane-(3,2c) pyrazole | * | 3,400 | BEYLER et al. (1961) |
| 17β-Hydroxy-17α-methyl-5α-androstane-(3,2c) pyrazole | * | 1,000 | ARNOLD et al. (1963c) |
| 17β-Hydroxy-17α-methyl-5α-androstane-(3,2c) pyrazole | * | 4,000 | TOMARELLI and BERNHART (1964) |
| 4,17β-Dihydroxy-17α-methylandrost-4-en-3-one | * | 1,800 | ARNOLD et al. (1963b) |
| 17β-Hydroxy-2α,17α-dimethyl-5α-androstan-3,3′-azine | 0.5, 2.0 | 1,000 | MATSCHER et al. (1962) DERUGGIERI et al. (1963) |
| 17β-Hydroxy-17α-methyl-5α-androstan-(2,3d)-isoxozol | * | 970 | ARNOLD et al. (1963b) |
| 17α-Methylnortestosterone | * | 460 | ARNOLD and Potts (1966) |
| 17α-Ethyl-19-norestra-4-en-17β-ol | * | 390 | ARNOLD et al. (1963c) |
| 17α-Ethyl-19-norestra-4-en-17β-ol | * | 170 | ARNOLD et al. (1963a) |
| 17β-Hydroxy-17α-methyl-5α-androstan-2-[hydroxy-methylen]-3-one | * | 175 | ARNOLD et al. (1963c) |
| 17β-Hydroxy-17α-methyl-5α-androstan-(3,2c)-isoxazol | * | 155 | ARNOLD et al. (1963a) |
| 17α-Ethyl-19-nortestosterone | 5 | 150 | ASCHKENASY (1959) |
| 17α-Ethyl-19-nortestosterone | 5 | 150 | SAUNDERS and DRILL (1958) |
| 17β-Hydroxy-17α-methylandrost-4-en-(2,3d)-isoxazole | * | 120 | ARNOLD et al. (1963a) |
| 17β-Hydroxy-17α-methylandrosta-1,4-dien-3-one | * | 120 | ARNOLD et al. (1963c) |
| 17β-Hydroxy-17α-methyl-5α-androstan-3-one | * | 80 | ARNOLD et al. (1963b) |
| 17β-Hydroxy-17α-methylandrosta-4,6-dien-3-one | * | 70 | BEYLER et al. (1961) |
| 17α-Methylandrost-5-ene-3β,17β-diol | * | 70 | ARNOLD and POTTS (1966) |
| 17α-Methyl-4-chlorotestosterone | * | 65 | ARNOLD and POTTS (1966) |
| Testosterone | | 38 | ARNOLD et al. (1963a) |
| 3-Oxo-1-methylandost-1-en-17β-yl-acetate | * | 6 | ARNOLD and POTTS (1966) |

[a] Values are referred to 17α-methyltestosterone as 100. See Table 1 for further details.

creased from 18% to approximately 9%. The daily nitrogen intake was 150–175 mg/day. DE RUGGIERI et al. (1963) further reduced the protein content of the diet to 3.5% so that the rats received 58–61 mg nitrogen per day. STAFFORD et al. (1954), SAUNDERS and DRILL (1958), LENNON and SAUNDERS (1964), and TOMARELLI and BERNHART (1964) force-fed a liquid diet (STAFFORD et al., 1954), which had essentially the same composition as that of KOCHAKIAN (1950b). As in the initial studies of KOCHAKIAN, the castrated rats were brought into nitrogen equilibrium by gradual adjustment of the food intake over a long period of time (2–3 months).

SALA et al. (1957) made a more drastic change in the procedure. The rats were castrated at 150 g and after 10 days were placed on a protein-free diet for 7 days. Then they were fed 8 g of a semisynthetic diet containing mixed vegetable and animal proteins (2.3% nitrogen) and were simultaneously injected for 7 days with the test steroid. In later studies (BALDRATTI and ARCARI, 1962), a liquid diet containing hydrolyzed casein as the source of protein was fed at a daily intake of 150 mg nitrogen, the experimental periods were 10 days each with the injections beginning on the 4th day. The urinary nitrogen excretion of the test rats was compared with that of control rats. Other modifications in nitrogen balance studies were made in studies with normal rats and rabbits (see below).

ARNOLD et al. (1959) observed a log-dose response to the steroids (POTTS et al., 1960), which was utilized in all of their studies (Tables 1, 2). TOMARELLI and BERNHART (1964) adopted this method of evaluation. The many other reports, however, were at only one or two dose levels in comparison with either testosterone propionate for injection or methyltestosterone for oral studies, usually at the same dose levels. Frequently these comparisons were with testosterone propionate or methyltestosterone at or greater than the maximum effective dose. Therefore, the estimation of the quantitative relationships are hazardous. Nevertheless, a useful rough comparison has been attempted and some conclusions may be made. The results are presented together with full caution in regard to their limitations (Tables 2, 3). A further complication in the evaluation of the data was the failure of some investigators to recognize that it is not possible to detect the complete ability of the steroid to produce a positive nitrogen balance in rats already exhibiting a strong positive nitrogen balance.

Another complication is the effect of the steroids on appetite. Most of the steroids at the higher dose levels will inhibit the food intake of the rat, but this was encountered usually if the diet was not limited to maintain the rats in both body weight and in nitrogen equilibrium (e.g. ASCHKENASY, 1959). The effect on appetite is specially strong if the steroid is contaminated with or has estrogenic activity (POTTS et al., 1960) and also is dependent on the endocrine status of the rat (KOCHAKIAN, 1950b; KOCHAKIAN and DOLPHIN, 1955). Moreover, estrogens and other compounds with estrogenic activity impair the utilization of the ingested protein, resulting in an increase in fecal nitrogen (BEYLER et al., 1961). The C-19 steroids (KOCHAKIAN, 1950b) and other steroids with nitrogen-retaining properties without estrogenic activity (BEYLER et al., 1961) do not affect fecal nitrogen excretion.

## B. Comparison of Steroids

The relative activities of the many steroids were determined both by subcutaneous injection (Table 1) and by oral administration (Table 2).

#### 1. Subcutaneous Administration (Table 1)

None of the compounds were more effective than testosterone propionate. A few compounds were as or nearly as effective as, but most were much less (20%) effective than testosterone propionate. Esterification definitely enhanced the potency of the steroids. This probably was primarily due to the retardation in the rate of absorption. The subcutaneous implantation of a pellet of testosterone enhanced its potency to that of testosterone propionate (KOCHAKIAN, 1950b). Esterification also prolonged the activity of the steroid, and the duration of the prolongation was dependent on the nature of the ester group. Thus, a single injection of 19-nortestosterone-cyclopentylpropionate exhibited its nitrogen-retaining effect longer than a similar dose of testosterone propionate (STAFFORD et al., 1954).

#### 2. Oral Administration (Table 2)

Oral administration revealed many compounds with activities greater than that of 17α-methyltestosterone (Table 2). This could be at least in part due to the rate, degree, and route of absorption from the gastrointestinal tract and metabolism by the liver. 17α-Methyl-5α-androstan-17β-ol was completely ineffective in stimulating the kidney and accessory sex organs of the castrated mouse when implanted subcutaneously as a pellet because of its lack of absorption. When it was administered orally, it was more effective than 17α-methyltestosterone in inducing growth of the kidney and less effective on the accessory sex organs (KOCHAKIAN, 1952). There is one unexplained major inconsistency. 17β-Hydroxy-17α-methyl-5α-androstan-(3,2c)pyrazole in two studies from the same laboratory (ARNOLD et al., 1959; BEYLER et al., 1961) was 30 times as effective as methyltestosterone, but in a later study was only 10 times as effective (ARNOLD et al., 1963c). No explanation could be provided[1]. Another laboratory (TOMARELLI and BERNHART, 1964), however, was in agreement with the two earlier studies. The relative effectiveness of orally administered 17α-methyltestosterone and subcutaneously injected testosterone propionate can be estimated from the report of TOMARELLI and BERNHART (1964). 17β-Hydroxy-(DL)-13β,17α-diethylgon-4-en-3-one was reported to have the same potency as testosterone propionate by subcutaneous injection, to be 16.3 times as effective as 17α-methyltestosterone by the oral route, and 6 times more effective subcutaneously than orally. Thus, 17α-methyltestosterone is only about 1% as effective by the oral route as testosterone propionate by subcutaneous injection, and even the most active of the new oral compounds is still much less effective than testosterone propionate by injection.

## III. Effect of Castration on Protein and Energy Metabolism

### A. Dog

Although castration has been practiced in animal husbandry for centuries to produce desirable modifications in behavior and metabolism, careful analysis of the

[1] The authors suggest (personal communication) that the discrepancy probably was due to a difference in administration of the hormones. The lower value was obtained when the steroids were incorporated in the daily portion of food while the higher value was obtained by stomach tube administration.

changes by metabolic studies have been made only in a few instances. KORENCHEVSKY (1925) reviewed the earlier reports on the effect of gonadectomy in several species of male and female animals and humans. In an attempt to resolve the apparently conflicting reports, he studied 11 dogs (4 males, 7 females) and 5 rabbits (4 males, 1 female), before and after castration. These animals were acclimated to the laboratory, and the food intake was adjusted to a minimum level for maintenance of body weight. Some of the dogs gained over 5% in body weight, and others either lost a little or gained less than 5%. They were designated "fat" and "thin" dogs, respectively. Nitrogen excretion in the fat dogs was decreased in some instances in the 1st week and always after about 2 months after the operation. The decrease was surprisingly high, −32%, and presumably constant. In the thin dogs, a small decrease in nitrogen excretion was occasionally seen at first but after a few weeks was at the normal level. The gaseous metabolism was lower (−18 to −34%) during 2–8 months after the operation in the fat dogs and in the thin dogs showed a small decrease or a small increase. Similar results were obtained in the rabbits. KORENCHEVSKY attributed the changes in metabolism to secondary changes in the other endocrine glands and emphasized the different responses of the fat and thin types.

The results in the fat dogs were in agreement with the earlier report of LOEWY and RICHTER (1899, 1902) of a 10% decrease in the energy metabolism of one dog accompanied by an increase in fat over a period of 2 years after castration. The results in the thin dogs were consistent with the very careful studies by LUTHJE (1903) on 2 littermate dogs. One of the dogs was castrated at 113 days of age, and the nitrogen balance and BMR were determined at intervals over a 2-year period. The dogs received identical food intake, and the nitrogen balance was determined daily for periods of 7 days to 2 months. A comparison of the average nitrogen metabolism and BMR revealed no difference between the castrated dog and the littermate control. The simultaneous study of two female dogs from the same litter showed no effect of ovariectomy.

A more dramatic effect was reported by PTASZEK (1928), in one dog. The BMR decreased from 657 to 431 calories/24 h after castration. Injection of a testis preparation restored the BMR to normal, which dropped to 564 calories after 4 weeks. In the light of our present knowledge, these results are difficult to accept and cannot be due to the possible known androgens of the testis.

## B. Rat

SANDBERG et al. (1939) reported the results of castration at 7 weeks and 6 months of age. They did not distinguish between male and female rats. The adult rats were in very strong positive nitrogen balance (+70%), and castration had little or no effect on nitrogen excretion. The young rats were at an even higher positive balance (+87% of intake), and castration gradually reduced the positive balance to +60% after 7 months. The fecal nitrogen was not changed. Parallel results were obtained with sulfur, chloride, sodium, and potassium excretion. The extremely strong positive balances in the young and specially in the adult rats are disturbing.

When adult male rats were brought into nitrogen balance on a constant food intake prior to castration, a small, temporary (4 days) increase in nitrogen excretion occurred, which then dropped to slightly below the level of that of the control animals and gradually returned to the preoperative level (KOCHAKIAN, 1950c).

## C. Guinea Pig

Maignon (1913) found no difference in the BMR of castrated as compared to normal guinea pigs studied frequently over a period of 1 year except for a small increase in the normal but not in the castrated guinea pigs in January and February and a slightly higher value for the normal guinea pigs during October. These differences were attributed to adaptation of the normal guinea pigs to the temperature changes and an imperviousness to temperature changes by the castrated animals. The differences, however, were very small, and the figures rather indicate a better consistency in determinations throughout the year for the castrated than the normal guinea pigs.

## D. Fowl

Heymans (1921) detailed studies in the rooster, which always becomes fat after castration, provided an explanation for the occurrence of a decrease in BMR in the "fat" dogs. Castration produced a 20–30% decrease in BMR after 1 month. The castrated roosters (capons) ate less food but gained more weight, were less active, contained more body fat, and better tolerated starvation because of their greater reserve of body fat. Later Aude (1927) also observed a lower rate of energy metabolism by cockerals after castration.

# IV. Effect of Endocrine Status on Anabolic Activity of Steroids

## A. Testis

### 1. Dog

The initial study in normal animals was performed on a young adult dog. The administration of 3 daily doses of androst-4-ene-3,17-dione at a dose greater than that required to produce a significant decrease in nitrogen metabolism in the castrated dog (Kochakian and Murlin, 1936) was ineffective. Testosterone propionate also was ineffective in the normal male dog at a single dose of 25 mg and produced a slight decrease after a large single dose of 125 mg (Thorn and Engel, 1938). A remarkable decrease in the nitrogen excretion of a normal male dog was obtained after the injection of 25 mg/day for 7 successive days (Thorn and Engel, 1938). Curiously, the decrease occurred in the first 3 days of the postinjection period and was 7 times that previously observed in castrated dogs (Kochakian and Murlin, 1935, 1936; Kochakian, 1937).

### 2. Rat

Studies in the normal rat (and also man) demonstrated that a much higher dose of androgen is needed in the normal than in the castrated animal (Kochakian et al., 1950b). The response pattern in both nitrogen excretion and body weight, however, was similar to that of the castrated rat. Comparable results were obtained in young adult male rats adapted to tube-feeding and injected with 2 mg/day of testosterone propionate (BRAASCH et al., 1950). 17α-Methyltestosterone included in the diet at 0.04 or 0.14 mg/day did not affect the very strong positive nitrogen balance or the body weight of growing (initial weight 160–170 g) male rats (BAVETTA et al., 1962). This was not surprising in view of the strong positive nitrogen balance (50% of the intake) and the low dose of the steroid (KOCHAKIAN, 1952; ARNOLD et al., 1963c). BOURLIERE et al. (1958) also observed a decrease in nitrogen retention in normal old adult male rats, 23 and 32 months of age, after the injection of 0.25, 0.50, or 1.0 mg/day of testosterone propionate but did not carry out simultaneous studies in castrated rats of the same age. The same response was obtained with each of the three doses. In contrast, 17β-hydroxy-5α-andostan-3-one, at doses of 0.25, 1.0, or 2.0 mg/day, increased the urinary nitrogen excretion; 19-nortestosterone and 17α-methylandrost-5-ene-3β,17β-diol at 1.0 mg/day were ineffective. The balance studies were not conducted in the conventional manner. The rats received the customary semisynthetic diet containing 18% casein. When the rats were in equilibrium, the casein was removed from the diet for 6 days, then replaced until a new equilibrium was reached. At this point the casein again was removed, and androgen was administered daily for the 6-day nonprotein period. Nitrogen excretion during the two nonprotein periods was compared to determine the effect of the steroids.

The positive nitrogen balance (about 20% of intake) of adult normal male rats (450–500 g) fed *ad libitum* became slightly less positive upon the injection of an excessive dose (2.5 mg/day) of testosterone propionate with a concomitant decrease in appetite (nitrogen intake) (KOCHAKIAN and WEBSTER, 1958). Cessation of the injections resulted in an increase in appetite and a return of the positive nitrogen balance to the control level. The changes in the positive nitrogen balance were proportional to the changes in the nitrogen intake and are consistent with the relationship between nitrogen balance and the absorbed nitrogen (BROWN, 1949).

### 3. Rabbit

KRUZE, et al. (1966) administered 1.5 mg/kg of Dianabol (17β-hydroxy-17α-methylandrosta-1,4-dien-3-one) every second day for 2 weeks to male and female rabbits (8–10 months old) fed *ad libitum* and in two other studies, a total dose of 50 mg of the long-acting steroid, Deca-Durabolin (19-nortestosterone-17β-decanoate), and Sternabol (4-chloro-17β-hydroxyandrost-4-en-3-one) for 2 weeks. In every instance a significant increase in body weight was observed. The increase was apparent almost immediately with Dianabol but was delayed with the long-acting steroids. The urinary nitrogen excretion increased during the injection period and was further increased in the 4-weeks postinjection period. Parallel changes occurred in urea and α-amino nitrogen excretion, but these constituents of the blood were not significantly altered. A comparison between male and female rabbits was not presented. The authors interpreted the increase in nitrogen excretion as a reflection of an increased turnover and pool of metabolized nitrogen compounds with an associated increase in protein metabolism. These results on nitrogen metabolism seem to be peculiar to the rabbit and are contrary to results in other animals and man. The rabbits were fed "standard food" *ad libitum*. No measurement of intake was made nor were control rabbits simultaneously studied.

## B. Ovary

### 1. Rat

The studies on normal and castrated male rats were extended to the female to determine whether the ovary influenced the nitrogen-retaining effect of androgens (KOCHAKIAN and BEALL, 1950). Testosterone propionate was only one-third to one-fifth as effective in inducing nitrogen retention, and the wearing-off effect was more gradual and prolonged in the normal female rat, 175–265 g body weight, than in the castrated male rat, 288 g body weight. Furthermore, the observed increase in body weight was greater than that calculated from the nitrogen retained and did not reach a plateau as quickly as in the castrated male rats. This effect was more pronounced in the younger rats. Comparable results were obtained in young adult female rats adapted to tube-feeding (BRAASCH et al., 1950). Testosterone propionate injected into ovariectomized rats produced changes in both weight and nitrogen excretion intermediate between those of the normal females and the castrated males (KOCHAKIAN and BEALL, 1950).

GORDAN et al. (1947a) noted that normal adult female rats required a high dose of testosterone propionate to produce a significant decrease in nitrogen excretion and increase in body weight. The injection of 0.5 mg/day of testosterone propionate for 5 days produced a small decrease in urinary nitrogen excretion. This dose is almost twice that required to produce a maximum effect in castrated rats (KOCHAKIAN, 1950b). An increase of the daily dose to 2.5 mg/day produced a further decrease in nitrogen excretion and increase in body weight. The increase in body weight at both dose levels was greater than that calculated from nitrogen retained and also persisted after cessation of injections, as in the studies of KOCHAKIAN and BEALL (1950).

ROOKS et al. (1971) gave a single subcutaneous injection of 6 mg of a new long-acting synthetic steroid [19-nortestosterone-17-(4′-methylbicyclo-[2.2.2.]-oct-2′-ene-1′-carboxylate)] to adult female rats on an isocaloric diet of powdered laboratory chow. Urea excretion was decreased in the following 2 weeks and in the subsequent second 2-week period was partially returned towards normal. In a cross-over experiment on the same rats, nitrogen retention was observed only during the first 2 weeks. Nitrogen excretion for the second 2 weeks was the same as that for the controls. No rebound phenomenon was observed. The urinary nitrogen values were the same as those of the controls during the third 2-week period. Body weight increased in both experiments and remained so even after cessation of injections.

### 2. Dog

GAEBLER and TARNOWSKI (1943) reported a decrease in nitrogen excretion in normal female dogs after injection of 25 mg/day of testosterone propionate, comparable to that obtained in the castrated male dog (KOCHAKIAN, 1937), but nitrogen retention was not evident until after 2–3 days, and was not accompanied by any changes in water intake and urine volume. Similar results were obtained with 50 mg/day of testosterone propionate for 4 days (GAEBLER et al., 1953). Apparently the induction of nitrogen retention by testosterone propionate is induced later in the normal female than in the castrated male dog. Blood glucose, NPN, and fat were not affected in the above experiments. SIREK and BEST (1953) also did not detect an effect on blood glucose of normal female dogs after injection of testosterone propionate.

### 3. Monkey

STUCKI et al. (1960) developed a quantitative bioassay method using metabolic balance studies in the ovariectomized young female monkey *(Macaca mulatta)*. Although considerable scatter was obtained in the individual data, the average values gave a log-dose response not only for nitrogen retained, but also for potassium retention and increase in body weight. This bioassay method has been utilized to compare the anabolic activity of a number of steroids (Table 3) (STUCKI et al., 1965; LYSTER and DUNCAN, 1963). Maximum nitrogen retention was induced in a few days and was maintained during the 14 days of injection. The injection period was not extended to determine whether a wearing-off effect occurred as in rats. The positive nitrogen balance, however, decreased almost immediately after cessation of injection

Table 3. Comparison of protein anabolic activities of several orally active steroids in ovariectomized female monkeys. Values refer to 17α-methyltestosterone as 100. Data were recalculated from effect on nitrogen and potassium balance and increase in body weight reported by STUCKI et al. (1965)

| Steroid | Relative activity |
|---|---|
| 17α-Methyltestosterone | 100 |
| 9α-Fluoro-11β,17β-dihydroxy-17α-methylandrost-4-en-3-one | 500 |
| 17β-Hydroxy-7α,17α-dimethylandrost-4-en-3-one | 575 |
| 17β-Hydroxy-17α-methylandrosta-1,4-dien-3-one | 116 |
| 17β-Hydroxy-2(hydroxymethylene)-17α-methyl-5α-androstan-3-one | 115 |
| 17β-Hydroxy-17α-methylandrostano[3,2c]pyrazole | 102 |
| 11β,17β-Dihydroxy-17α-methylandrost-4-en-3-one | 18 |

## C. Pituitary

### 1. Dog

Subcutaneous injection of 25 mg/day of testosterone propionate into the castrated-hypophysectomized adult dog (KOCHAKIAN, 1943, 1946) decreased urinary nitrogen excretion in a manner similar to that in the castrated dog.

### 2. Rat

Hypophysectomy in the adult male rat induced a rapid loss in body weight, a strong negative nitrogen balance that gradually decreased but was still evident after 3 weeks, and a decrease in appetite. The rats reached body weight and nitrogen equilibrium after several weeks at a food intake of 50–70% of normal (KOCHAKIAN, 1950c). Castration at the time of hypophysectomy did not detectably alter the changes.

Injection of testosterone propionate produced the characteristic changes in urinary nitrogen excretion. Body weight increased to a plateau where it was maintained with continued injections and decreased only after cessation of injections. The increase in body weight and the decrease in nitrogen excretion after testosterone propionate injection were less than those of the nonhypophysectomized rats. RUPP and PASCHKIS (1953) tube-fed (INGLE et al., 1943) hypophysectomized male rats twice daily, after 6 weeks, they injected 2.5 mg/day of testosterone propionate for 9 days to obtain a twofold greater daily nitrogen retention and increase in body weight than observed by KOCHAKIAN (1950c). The greater effect was due apparently to the greater protein and caloric intake, which prevented the usual loss in body weight and decrease in food intake following hypophysectomy (KOCHAKIAN, 1950c). The observed weight gain was greater than the theoretical.

Administration of a growth hormone preparation to the hypophysectomized rats produced a response pattern similar to that of the steroid, but the decrease in nitrogen excretion was greater, the wearing-off period was longer, and the increase in body weight was greater (KOCHAKIAN, 1950c). The loss of body weight and the rebound in nitrogen excretion on cessation of injections of the two hormones were comparable. The presence of the testis did not have a detectable effect on any of these changes. The fecal nitrogen was small and was not affected by either hormone. The appetite of several of the hypophysectomized rats receiving testosterone propionate was inhibited for varying periods of time, usually beginning after about 2 weeks.

Administration of a maximum dose of each of the hormones to separate groups of castrated rats induced practically identical responses in urinary nitrogen excretion. The initial increase of body weight also was identical for both hormones but with continued injections that of the androgen-treated rats began to decrease while that of the growth hormone-treated rats continued to increase but at a gradually slower rate.

Simultaneous administration of testosterone propionate and growth hormone, at their respective maximum effective doses, to castrated rats produced a summation in both nitrogen retention and increase in body weight (KOCHAKIAN, 1960a). ELSAIR et al. (1973) also reported a summation effect following simultaneous administration of norethandrolone and growth hormone to normal male rats. The results would have been more convincing if the authors had been more careful in the experimental design. The rats were not only the less responsive normal males but also were fed *ad libitum* and were in strong positive nitrogen balance when the experiments were initiated. These two factors undoubtedly accounted for the inability to observe an effect of the norethandrolone despite the massive dose of 20 mg/day and the small effect with growth hormone alone. Furthermore, the injection periods of only 4 days were followed by only a 3-day postinjection period, which served as the control for the next 4-day injection period.

Hypophysectomized female rats showed a small variable decrease in urinary nitrogen excretion and a slight increase in body weight after daily injection of 0.5 mg testosterone propionate for 5 days (GORDAN et al., 1948). A dose increase to 2.5 mg/day for another 5 days produced a significant drop in urinary nitrogen excretion with a concomitant increase in body weight that was only equal to about one-half the value calculated from the nitrogen retained. The normal female was more responsive (see above) than the hypophysectomized female rat, but both were less sensitive than the castrated male. It is noteworthy that the food (nitrogen) intake of the female hypophysectomized rats was the same as that of the normal female rats and nearly twice that of the hypophysectomized castrated rats (KOCHAKIAN 1950c). This difference probably is due to the fact that the period after the operation was longer (1–6 months) for the castrated rats and permitted an adjustment to the lowest basal level from the lack of the anterior pituitary hormones.

## D. Thyroid

### Rat

#### *a) Thyroidectomy*

The thyroid is essential for normal growth and development. Thyroidectomy of adult male and female rats maintained on a constant food intake by tube-feeding produced an increase in nitrogen excretion in the 1st week after the operation (RUPP et al., 1949). A similar effect was not produced by thyroidectomy of castrated male rats previously brought to nitrogen and weight equilibrium by regulation of the food intake (KOCHAKIAN and DOLPHIN, 1955) and was not evident in the 2nd week after thyroidectomy of male rats castrated 1 week prior to thyroidectomy (NOACH et al., 1952). This difference might be due to the difference in the feeding procedure.

The administration of 5 or 10 μg of thyroxine for 7 days to thyroidectomized male and female adult rats on a tube-fed constant intake decreased the urinary nitrogen excretion without a change in body weight (RUPP et al., 1949). Normal rats did not show the same response. The anabolic and catabolic effects of thyroxine were not evident until the 3rd day of injections. The nitrogen metabolism of the adult castrated rat apparently is more sensitive than that of the normal rat to thyroxine. The injection of 10 μg/day of DL-thyroxine produced an immediate increase in nitrogen excretion that returned to the basal level after 6 days, but the body weight progressively decreased for the duration (21 days) of the injections (KOCHAKIAN, 1960a). Thyroxine apparently initially stimulated the catabolism of labile protein and then shifted to the catabolism of fat and spared protein.

Injection of 2.5 mg/day of testosterone propionate into male rats tube-fed a constant food intake, 2 months after thyroidectomy, induced the characteristic changes in nitrogen excretion and body weight (RUPP and PASCHKIS, 1954). Injection of 0.5 mg/day of testosterone propionate to castrated male rats 2 (NOACH et al., 1952) and 3 weeks (KOCHAKIAN and DOLPHIN, 1955) after thyroidectomy produced changes in nitrogen excretion and body weight qualitatively and quantitatively, practically identical to those in castrated nonthyroidectomized rats except that initiation of the effects was slightly delayed.

FUJII (1959) did not observe any difference in nitrogen excretion of normal or thyroidectomized male and female rats fed *ad libitum* and injected with 0.4 or 4 mg/kg/48 h of 17α-methylandrost-5-ene-3β,17β-diol. Thyroidectomy was performed at 4 weeks and castration at 2 weeks of age, and the steroid injections were begun at about 90 days of age. The nitrogen excretion studies were done at 60, 80, and 100 days of age. Thus, it is not surprising that no effect on nitrogen excretion was detected. The rats were still in the growth phase, and in strong positive nitrogen balance, the steroid has weak protein anabolic porperties, and the rats obviously were far into the wearing-off phase.

*b) Thiouracil*

The production of hypothyroidism by the addition of 0.1% thiouracil to the diet did not change the qualitative or quantitative effect of testosterone propionate on urinary nitrogen excretion (KOCHAKIAN and DOLPHIN, 1955). Furthermore, an increase in the dose of testosterone propionate from 0.25 to 2.5 mg/day during the wearing-off period produced an identical reversal and delay of the wearing-off period in the hypothyroid and nonhypothyroid castrated rats. The maximum increase in body weight appeared to be greater in the hypothyroid rats but this effect was not real. The loss in body weight during the postinjection period was the same for both groups of rats.

Testosterone propionate had a variable inhibitory effect on the appetite of the thyroidectomized but not the thiouracil-fed rats. Also, thyroidectomy produced a small decrease in fecal nitrogen that was not evident in the thiouracil-fed rats. Testosterone propionate did not change the fecal nitrogen excretion in either group of rats (KOCHAKIAN and DOLPHIN, 1955).

*c) Hyperthyroid*

Testosterone propionate abolished the protein catabolic effect of 10 µg/day of DL-thyroxine (KOCHAKIAN, 1960b). The androgen produced an identical response on nitrogen excretion in the hyperthyroid-castrated rats as in the castrated rats. The effect of testosterone propionate on body weight was superimposed upon the decreasing weight of the hyperthyroid rats. The loss in body weight must be due to the enhancement by thyroxine of the utilization of fat, which continues and does not interfere with the protein anabolic action of testosterone propionate.

## E. Adrenals

### Rat

*a) Adrenalectomy*

Adrenalectomy of adult castrated rats in nitrogen and body weight equilibrium produced a slight transient increase in urinary nitrogen excretion. The rats were injected with 0.50, 0.75, or 1.0 mg/day of desoxycorticosterone acetate or implanted subcutaneously with a 15-mg pellet (rate of absorption 0.1 mg/day) of the steroid at the time of the operation. Injection of testosterone propionate at 0.5 mg/day produced changes in nitrogen excretion essentially identical to those produced in simultaneously treated nonadrenalectomized castrated rats (KOCHAKIAN et al., 1950b). Parallel changes occurred in the urinary excretion of $K^+$, $PO_4^{-3}$, and $Cl^-$. Changes in body weight also were identical in both groups of rats. Increasing the dose from 0.5–1.0 and 2.0 mg/day did not produce any further increase in the nitrogen retained. KASSENAAR (1953) obtained a similar positive nitrogen balance in adrenalectomized-castrated rats maintained on 0.9% NaCl by injection of either 0.1 mg/day desoxycorticosterone acetate or 1.25 mg/day testosterone propionate. Simultaneous injection of the two steroids resulted in a summation of their effects in not only the pattern and degree of nitrogen excretion but also of the changes in body weight. Thus, the two steroids were acting independently.

*b) Anticatabolic*

The protein anabolic action of testosterone and related compounds has been studied in catabolic conditions produced by glucocorticoids and various types of stress. The prior injection for 5 days of testosterone propionate into young adult male adrenalectomized rats maintained on 1 mg/day of desoxycorticosterone acetate did not produce a decrease of the increase in nitrogen excretion produced by hourly injections of adrenal cortical extract during the 24th to 32nd h of the fasting period (KOCHAKIAN and VAIL, 1947). The marked increases of nitrogen, $PO_4^{-3}$, $K^+$ excretion, and urine volume induced by the injection of 12 mg/kg for 5 days of cortisone acetate in fasted, castrated male rats were progressively decreased by the simultaneous injection on the 3rd, 4th, and 5th days of 0.2, 1.0, and 5 mg/kg of testosterone propionate or the oral administration of 5, 20, and 80 mg/kg of stanozolone or norethandrolone. Methandrostenolone, on the other hand, at 20, 40, and 80 mg/kg was ineffective and possibly added to the cortisone action (BEYLER et al., 1968; TAINTER et al., 1963a, b). The increase in urinary nitrogen excretion produced by a third degree burn in young adult male and female rats, tube-fed a constant intake, was markedly depressed by the injection of 2.0 mg/day of testosterone propionate (BRAASCH et al., 1950).

The addition of medrol at 2.0 mg/100 g diet (0.24 mg/day), but not at 0.5 mg/100 g diet, partially reduced the strong positive nitrogen balance in growing young adult male rats (160–170 g initial weight). The simultaneous inclusion of 17α-methyltestosterone in the diet (0.03 or 0.13 mg/day) did not influence the catabolic effect of the Medrol (BAVETTA et al., 1962). The failure could be due to the inadequate dose of the androgen (cf. KOCHAKIAN, 1950b; ARNOLD et al., 1963c).

## F. Pancreas

### 1. Dog

Uncontrolled pancreatic diabetes is accompanied by wasting and an intense protein catabolism that is brought to equilibrium by the injection of insulin (CHAIKOFF and FORKER, 1950). GAEBLER and TARNOWSKI (1943) injected 25 mg/day of testosterone propionate for 6 days into two female depancreatized dogs whose diabetes was nearly controlled by the twice daily injection of insulin. Urinary nitrogen and glucose excretion were decreased without a significant effect on water intake, urine volume, or body weight. On the first day after cessation of injections, there was a great increase in glucose excretion and a negative nitrogen balance.

KOCHAKIAN and COSTA (1959), in a series of experiments, made a detailed study of the effect of testosterone propionate on the metabolism of depancreatized-castrated dogs with different types of insulin and with different doses of the steroid injected over different time periods. Maintenance of the most uniform blood glucose level throughout the day was obtained when NPH insulin was used to partially control the diabetes.

Testosterone propionate always produced a positive nitrogen balance, a decrease in urine and blood glucose, and an increase in body weight. The continued injection of the steroid did not prolong the positive nitrogen balance. Nitrogen excretion gradually returned to or toward the basal level. On cessation of injections, it grad-

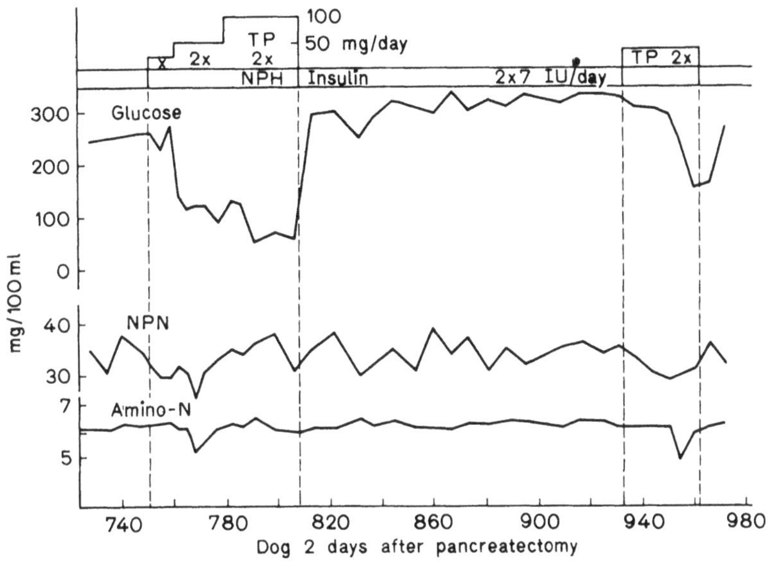


Fig. 3. Effect of testosterone propionate on basal blood constituents of depancreatized-castrated dog. NPN and amino-N are serum values. Testosterone propionate dose was divided (2×) and injected at 8 a.m. and 5 p.m. feeding times (KOCHAKIAN and COSTA, 1959)

ually (4 weeks) returned to the basal level, but at a slower rate than the urinary nitrogen. As expected, the fecal nitrogen was much higher than normal and was decreased after about 3 weeks of testosterone propionate; on cessation of injections, it slowly returned to the control level. The effect of testosterone propionate was even more pronounced when the protein content of the diet and the resultant fecal nitrogen were higher. The urine volume did not show any remarkable changes, and there was no indication of hemodilution or concentration at any time. The hematocrit, RBC, and WBC values were constant.

Testosterone propionate produced a response in the postabsorptive blood glucose level (Fig. 3) similar to that in the urinary glucose level except that it returned to normal almost immediately after cessation of injections. In some instances it was decreased below the basal level by the steroid. The availability of extra glucose could have diverted the testosterone propionate from the mobilization of body fat to compensate for the decrease in energy from protein (KOCHAKIAN and MURLIN, 1935) and at the same time enhanced the sparing effect on protein. The nitrogen retained and body weight gain in the depancreatized-castrated dog were greater than in the nondepancreatized dog. The blood glucose levels of the nondepancreatized-castrated dog (KOCHAKIAN and COSTA, 1959), the normal female dog (SIREK and BEST, 1953), and the partially depancreatized dog with normal blood sugar (SUOTO-MAIOR and FOGLIA, 1944) were not influenced by the hormone. Blood and urinary ketone body levels were higher in the depancreatized than the nondepancreatized castrated dogs. The absolute increases were small and were not indicative of excessive fat metabolism (KOCHAKIAN and COSTA, 1959).

PELLEGRINI (1943a, b, c), in brief communications, reported results in glucose metabolism similar to the above. Testosterone propionate at 50–75 mg/day, but not at 5–25 mg/day, produced a marked decrease in blood sugar and glucosuria of totally depancreatized dogs. The drop in blood sugar was evident in 1–5 h. Insulin was 4–5 times more effective in depancreatized dogs whose blood glucose had been lowered by previous treatment with testosterone propionate. These data are in apparent agreement with those of KOCHAKIAN and COSTA (1959), but unfortunately the paucity of experimental details prevents an accurate evaluation. Especially disturbing is the inference that the totally depancreatized dog can be maintained with testosterone propionate and no insulin. These animals are notorious for their rapid deterioration and death after total pancreatectomy.

In acute experiments, CORNIL and ROSANOFF (1938) and ZUNZ and LABARRE (1939) reported variable hyperglycemic and hypoglycemic effects within a few hours after the injection of testosterone propionate and acetate intravenously or intramuscularly in normal male and female dogs. ZUNZ and LABARRE attempted to correlate these changes with the anesthesia (chloral), thyroid, adrenal, and the splanchnic nerves. GASTALDI (1947) could not detect any immediate (30–180 min) change in the blood glucose of normal or depancreatized (basal blood glucose approximately 3.0 g/100 ml) dogs after doses of 5–80 mg of testosterone propionate.

## 2. Rat

### *a) Alloxan*

DE JONGH et al. (1950) reported that testosterone propionate at 0.25 mg twice daily for 7 days produced nitrogen retention in adult ovariectomized rats but not in alloxan-diabetic ovariectomized rats after a preliminary control period of 6 days. The diabetic rats excreted twice as much nitrogen as the nondiabetic rats and approximately 4 g/day of glucose. It is not clear whether the extra nitrogen excretion was due to an increase in protein catabolism or an increase in food intake (protein) to compensate for the loss of energy (glucose) in the urine. The authors indicated that the food intake of the nondiabetic rats was limited to 18 g/day but did not give the actual nitrogen intake and presumably did not make any adjustments for the diabetic rats. In contrast, WRIGHT and KOCHAKIAN (1953) observed the characteristic changes in nitrogen excretion and body weight not only in castrated but also in normal alloxan-diabetic male rats. The food intake of these animals was increased to compensate for the calories lost in the urine and to maintain nitrogen and body weight equilibrium. The diabetic normal rats responded to the testosterone propionate with a slight increase in body weight and a subsequent loss on cessation of injections. The nitrogen retention also was small but significant. Both of these effects were less than usually observed in the nondiabetic normal rat. The diabetic castrated rats were brought to weight and nitrogen equilibrium on a 43% protein diet. They showed the characteristic changes in body weight and urinary nitrogen excretion except that the maximum daily urinary nitrogen retention was approximately twice that observed in the nondiabetic castrated rats, possibly due to the approximately twofold increase in protein intake.

The urinary glucose level of the diabetic-castrated rats exhibited a possible small decrease during the first week of hormone administration and then rapidly increased during the wearing-off period to above the basal level, increased further on cessation of injections, and then gradually decreased to or below the basal level (WRIGHT and KOCHAKIAN, 1953). Thus, the induced glucosuria was not permanent. The changes in glucose excretion preceded that of the nitrogen during the wearing-off and rebound periods and may be independent of the effect of the hormone on protein metabolism. The increased excretion of glucose during the wearing-off period could be an expression of the effect noted by LEWIS et al. (1950) in castrated male and female rats subtotally depancreatized at 80–100 g body weight. Daily injection of these rats with 0.5 mg of 17α-methyltestosterone or testosterone propionate, but interestingly *not* the ring-saturated 17α-ethyldihydrotestosterone, increased over a 6-month period the incidence of and accelerated the onset of diabetes (increase in blood sugar). These effects were apparent sooner in the castrated males and required 6 months in the castrated females. Furthermore, the onset of diabetes was faster in the castrated males than in the female controls. The growth rate was slower in the treated rats, especially in the castrated males. The enhancement in diabetes presumably was permanent. The authors, however, did not make a postinjection study.

CAVALLERO and MALANDRA (1956) obtained results similar to those of WRIGHT and KOCHAKIAN (1953) following injection of alloxan-diabetic rats with 1 or 5 mg/day testosterone propionate or 4-chlorotestosterone acetate for 8 days. The blood glucose was not changed. CHIAPPINO (1960) extended the studies of CAVALLERO and MALANDRA (1956) to the use of 4-hydroxy-17α-methyltestosterone. Young adult male rats (150–180 g) were made diabetic by the injection of alloxan. After a 7-day observation period, the rats excreting 3.5–4.0 g glucose per day were divided into a control and a treated group. The 4-hydroxy-17α-methyltestosterone was fed in the diet at 10 mg/day for 7 days. Urinary glucose and nitrogen decreased slightly (540 and 14 mg/day respectively) during treatment and increased (257 and 45 mg/day) in the postinjection period. The rats received a constant food intake, but both groups lost 18 g body weight.

*b) Phlorhizin*

The characteristic effects of testosterone propionate also were observed in castrated rats made diabetic by the daily injection of phlorhizin and brought to nitrogen and body weight equilibrium by an upward adjustment of the food intake (WRIGHT and KOCHAKIAN, 1953) to compensate for the loss in calories (glucose) in the urine. The hormone, however, did not significantly affect the glucosuria.

*c) Glucagon*

HAZELWOOD and O'BRIEN (1961) explored the modification of glucagon-induced hyperglycemia in rats by the glucagon tolerance test. The rats were injected daily for 5 days with 0.15 mg of 17α-ethyl-19-nortestosterone (Nilevar) before the test. Nilevar markedly decreased the increase in blood glucose induced by the glucagon in the intact and in the 20-day previously hypophysectomized rats. In addition, the blood urea level was decreased and the liver glycogen increased in the intact but not in the hypophysectomized rats. Testosterone at the same dose, however, was not effective, possibly due to the inadequate dose (cf. Table 2 and KOCHAKIAN, 1950b; KOCHAKIAN et al., 1950b). Castrated rats were hypersensitive to the glucagon tolerance test at 2 and 7 days but were normal at 14 days and refractory at 30 days after castration.

*d) Excess Sucrose Intake*

Rats receiving 10% sucrose as drinking water showed an increase in fasting blood cholesterol and triglycerides 38 days after castration. The injection of 1 mg/day of testosterone propionate between the 8th and 38th days prevented the increase (LEDERER et al., 1973). These studies may be interpreted to indicate that testosterone propionate can direct fat formed from excess carbohydrate to reserve stores or that the excess carbohydrate is utilized to spare protein or fat. The situation may be analogous to the effect of the steroid in the partially regulated depancreatized dog and may explain the greater nitrogen retention observed in the alloxan-diabetic rat and the depancreatized dog.

## V. Effect of Nutritive State on Anabolic Activity of Steroids

The protein anabolic action of testosterone has been studied in rats and dogs on diets quantitatively and qualitatively different in protein.

### A. Rat

#### 1. Quantity of Protein

KOCHAKIAN and VAN DER MARK (1952) compared the effect of 0.25 and 0.50 mg/day of testosterone propionate in castrated male rats receiving isocaloric diets containing 18, 28, and 43% protein as casein. The response patterns in nitrogen excretion and body weight were the same for the three groups of rat. The maximum daily and total nitrogen retained were slightly increased at the higher levels of protein, but the differences were not statistically significant. In alloxan-diabetic castrated rats, however, on a 43% protein (casein) diet and an approximately twofold intake to compensate for the loss in calories (glucose) in the urine, the nitrogen retained on testosterone propionate injection was approximately twice that of nondiabetic rats on the 18% protein (casein) diet (WRIGHT and KOCHAKIAN, 1953). On the other hand, an even higher protein intake (twice that of the alloxan-diabetic rats) by phlorhizin-diabetic castrated rats did not give any greater nitrogen retention than that of castrated rats on the 18% protein (casein) diet (WRIGHT and KOCHAKIAN, 1953). Body weight responses to testosterone propionate were not significantly altered by any of the above alterations in protein intake.

WIJNAN and DE GROOT (1953) castrated adult male rats and, after 1 week feeding *ad libitum* of a normal diet, changed the diet to a protein-free diet and injected 0.5 mg of testosterone propionate twice daily. Nitrogen excretion rapidly decreased to the low endogenous level. The steroid accelerated the rate of reduction in the urinary nitrogen excretion during the first week. When testosterone propionate was injected 2 weeks after beginning the ingestion of the nonprotein diet and when the urinary nitrogen excretion was at its minimum, no nitrogen retention was observed, but a definite increase in urinary nitrogen excretion (rebound) occurred on cessation of the injections. Similar results were obtained by MANUNTA (1955). Since a complete fast is reported (USUELLI et al., 1949) not to impair the ability of 1 mg/day of testosterone propionate for 10 days to increase the weight of the seminal vesicle and prostates and to decrease the loss in weight of the kidney, liver, and heart of young adult rats, the hormone apparently utilized the amino acids produced by the catabolism of the protein of other tissues to induce protein synthesis in these organs; on cessation of hormone injections, the atrophy of the newly formed tissues provided the extra nitrogen excretion. The daily loss in body weight was less in the treated than in the nontreated fasted rats.

## 2. Quality of Protein

NIMNI and GEIGER (1957) altered the quality of the protein in the diet and observed the effect of testosterone propionate. Injection of 2.5 mg/day for 2 days produced a positive nitrogen balance in the 2nd, 3rd, and 4th postinjection days in castrated but not in normal adult male rats on a diet containing zein (deficient in tryptophan and lysine) as the sole source of protein. The lack of effect in the normal rats was attributed to their lesser sensitivity to testosterone propionate (KOCHAKIAN and DOLPHIN, 1955). When gelatin (deficient in tryptophan and phenylalanine) was the sole source of protein, the castrated rats exhibited a positive nitrogen balance in the first and second postinjection days. A more marked effect was obtained on extension of the injections to 7 days. In a third experiment, the nonessential amino acids, L-glycine, L-glutamate, and L-alanine, served as the sole source of nitrogen; the testosterone propionate effect on nitrogen retention was still evident but was less effective. The authors concluded that at least one part of the nitrogen retention under hormone stimulation cannot be for protein synthesis and is due to a decrease in renal excretion of nitrogen-containing metabolites. This is contrary to the many studies on blood nonprotein nitrogen. A more plausible explanation is that testosterone propionate induced a substitution of the amino acids deficient in the diet with those obtained by catabolism of the protein of low priority tissues (USUELLI et al., 1949; KOCHAKIAN et al., 1950a). The body weight of the rats on the amino acid-deficient diets continued to decrease even during the period of testosterone propionate treatment. The short period (2 days) was not sufficient to allow a significant manifestation of an increase in body weight.

## 3. Depletion and Repletion

Testosterone propionate also is effective in replenishing protein stores depleted by a 12-day fast (KOCHAKIAN et al., 1948). The rats were fed the same constant food intake before and after the fast. The initiation of testosterone propionate injections, 1.0 mg/day, simultaneously with refeeding, produced a more rapid increase in body weight 10 g greater than that of the controls after about 3 weeks. Cessation of injections resulted in a gradual loss, to a weight less than that of the controls, followed by a gradually increasing rate of gain so that at about the 130th day the body weight of both groups of rats were essentially identical. The nitrogen retained by the treated rats during the initial period of recovery was greater than that retained by control rats. On cessation of injections, the treated rats excreted more nitrogen than the controls for several days and then both groups excreted equal amounts. Practically identical results were obtained in another study, which was carried only through an injection period of 50 days at 2.5 mg/day. Fecal nitrogen was not altered by the hormone. Chloride excretion in the urine was less in the treated rats during the period of nitrogen retention.

Administration of phlorhizin to normal or castrated adult rats on constant food intake produced a depletion of calories by a marked glucosuria and an increase in nitrogen and phosphate excretion with a concomitant sharp continuous loss in body weight (WRIGHT and KOCHAKIAN, 1953). Simultaneous administration of testosterone propionate (1 mg/day, normal; 0.5 mg/day, castrated rats) slightly retarded the loss in body weight and the initial increase in urinary nitrogen and phosphate excretion. The wearing-off effect, however, intervened in about one week, and the excretion of these two constituents became the same as in the nonhormone-treated phlorhizinized rats. Furthermore, on cessation of hormone injections, an extra-urinary excretion (rebound) of both nitrogen and phosphate occurred with an acclera-tion in the rate of loss in body weight. Testosterone propionate enhanced the glucosuria induced by the phlorhizin; the enhancement was greater in the normal than in the castrated rats.

GEIGER and EL RAWI (1952) placed adult (238 g) male rats on a qualitatively balanced diet from which the protein (casein) was removed. Injection on alternate days of 2 mg/day of testosterone propionate did not affect the loss in body weight during the 15 days of the experiment. Replacement of the protein in the diet and continuation of the hormone treatment gave an initial, more rapid rate of increase in body weight with a subsequent retardation, so that after 16 days the body weight of the treated rats was less than that of the controls. The initiation of injections at the time of restoration of the protein to the diet also enhanced and prolonged the initial increase in body weight. These results are comparable and consistent with the repletion studies of KOCHAKIAN et al. (1948).

QUIMBY (1951) placed young (50 g, 30 days of age) male rats on a qualitatively balanced diet just sufficient to maintain body weight. After 90 days, the rats were allowed to eat *ad libitum*, and one group received 0.5 mg/day of testosterone propionate. In the following 60 days, the two groups of rats had ingested the same quantity of food, but the treated rats had gained less body weight (238 vs. 268 g). The enhanced energy metabolism during the early recovery period was prevented by the testosterone propionate and also by an anterior pituitary growth hormone preparation (QUIMBY et al., 1948).

ASCHKENASY and DRAY (1954) depleted adult (200–250 g) male rats by feeding a protein-free diet for 60 days. The diet then was altered to contain only 7% casein to provide a slow repletion, and one group of rats received 2.5 mg of testosterone propionate twice per week. After 30 days of repletion, 2 of 8 control (oil) and 8 of 16 testosterone propionate-treated rats were dead. The surviving treated rats increased their body weights slightly more than that of the controls and also ingested slightly more food. The rats of neither group attained the weight of rats on an 18% casein diet. The high mortality of the treated rats might be due to aggravation of the sulfur amino acid deficiency of the 7% casein diet by the protein-anabolic demands of the testosterone propionate.

A comparison of the effectiveness of the body weight stimulating activity of testosterone in the various conditions of inanition is interesting. Loss of body weight due to enhancement of energy metabolism by thyroxine did not alter the manifestations of body weight-stimulating or nitrogen-retaining properties of testosterone propionate (KOCHAKIAN and DOLPHIN, 1955). On the other hand, the withdrawal of calories by the several procedures or withdrawal of protein with the subsequent decrease in energy metabolism due to undernutrition (QUIMBY et al., 1948) greatly diminished the manifestation of the body stimulating activity of the hormone.

## B. Guinea Pig

Guinea pigs were castrated at 500 g body weight (89 days of age) and 42 days later were deprived of food but not water for 9 days. The loss in body weight was the same in the normal controls (–32%) as in the castrated (–28%) guinea pigs (KOCHAKIAN et al., 1957). Subcutaneous implantation of a 15-mg pellet of testosterone or the injection of 10 mg/day of testosterone propionate during the fast had no detectable effect on the rate or total loss (–31%) of body weight.

## C. Dog

PERLMAN and CASSIDY (1953) compared the effect of 25 mg/day testosterone propionate in two dogs placed first on a diet containing 23.5% and then on 11.8% casein as the source of protein. On the "high" protein diet a positive nitrogen balance was induced in two experiments in one dog and in one experiment in the other dog. In the second experiment of the latter dog, no change in the nitrogen balance during the 21 days of testosterone propionate was observed, but surprisingly a positive balance appeared during the 7th to 16th postinjection days. When this same dog was placed on the low protein diet, a negative nitrogen balance appeared during the 10th to 16th days of testosterone propionate treatment with a prompt return to equilibrium on cessation of injections. The nitrogen balance of the other dog was not changed during the 15 days of treatment or during the subsequent 15 days. These results are difficult to interpret. There is also the question of whether the low protein diet should have been supplemented either with cystine or methionine (ARNOLD and SCHAD, 1954). GORDAN et al. (1948) reported that nitrogen retention and increase in body weight induced in rats by growth hormone were linear from 6–24% casein in the diet and showed no further increase at 48% casein. The addition of 66 mg/day of DL-methionine to the 6% casein diet greatly enhanced the body weight-stimulating and nitrogen-retaining properties of the growth hormone preparation (GORDAN et al., 1947b).

# VI. Kinetics of Protein Synthesis

## [$^{15}$N]-Glycine Studies

BARTLETT (1953) utilized [$^{15}$N]-glycine to explore the effect of testosterone propionate and also growth hormone alone and in combination on the storage of protein in two female dogs. When the dogs were in weight and nitrogen equilibrium, testosterone propionate at 50 mg/day for 5 days or growth hormone at 5 mg/day or a combination of the two hormones was injected subcutaneously for 5 days. [$^{15}$N]-glycine was administered by mouth on the fourth day, and urinary excretion of $^{15}$N was determined for the subsequent 24 and 48 h. In general, testosterone propionate induced a higher rate of protein synthesis, a lower rate of amino acid catabolism, and no significant change in the size of the nitrogen pool. Growth hormone, in contrast, increased the nitrogen pool. The combined hormones did not appreciably affect the rate of protein synthesis and decreased the rate of amino acid catabolism and the size of the nitrogen pool. The need for data on protein degradation was emphasized for a more complete understanding of the kinetics of protein storage under the stimulus of the hormones.

BARTLETT and STEVENSON (1954) studied the effect of testosterone propionate alone and in combination with growth hormone on the rate of protein degradation and loss in fasting female dogs. When the dogs were in body weight and nitrogen equilibrium, the food was removed for 6 days. The [$^{15}N$]-glycine was administered intravenously on the second day of the fast, and the excretion of the $^{15}N$ was determined for the next 5 days. In general, testosterone propionate, growth hormone, and a combination of the two hormones diminished the percentage of $^{15}N$ excreted during the 2nd to 5th days of the hormonal stimulation to give a significant decrease in protein degradation and loss. These data and previous observations (BARTLETT and GAEBLER, 1952; BARTLETT, 1953) indicate that an increase in protein synthesis and a decrease in amino acid catabolism are the major shifts in protein metabolism after both testosterone propionate or growth hormone induction of protein anabolism in the female dog.

## VII. Nitrogen Constituents of Urine and Blood

### A. NPN, Urea, Amino-N and Protein

#### 1. Urine

KOCHAKIAN and MURLIN (1935) in their initial studies demonstrated that the changes in urinary nitrogen during and after injection of the male hormone urinary extracts into the castrated dogs were accompanied by parallel changes in urinary excretion of urea. Excretion of ammonia was not changed. Identical parallel changes in urinary urea excretion were observed after the injection of androst-4-ene-3,17-dione into the castrated dog and of testosterone propionate in the depancreatized-castrated dog (KOCHAKIAN and COSTA, 1959) and in the castrated rats (KOCHAKIAN, 1950b). ROOKS et al. (1971) utilized this parallelism in the excretion of urea with total nitrogen to study the effect of a long-acting steroid on nitrogen retention in normal female rats.

KRUZE et al. (1966) observed a parallel increase in the excretion of urea and amino-nitrogen with total nitrogen by male and female rabbits injected with 17β-hydroxy-17α-methylandrosta-1,4-dien-3-one; 19-nortestosterone-17β-decanoate or 4-chloro-17β-hydroxyandrost-4-en-3-one. The rabbits were fed *ad libitum*.

#### 2. Blood

*a) Dog*

Since a decrease in urinary nitrogen excretion could reflect kidney damage and retention of nonprotein nitrogen constituents, KOCHAKIAN and MURLIN (1936) studied the blood NPN and urea of the castrated dogs in the experiments with androst-4-ene-3,17-dione. These constituents were not increased but were slightly decreased. Testosterone propionate stimulated an even greater decrease in the blood NPN of the depancreatized-castrated dog whose diabetes was partially regulated by insulin (Fig. 3) (KOCHAKIAN and COSTA, 1959). Indeed, the changes in blood NPN were in parallel with those of urinary nitrogen. SIREK and BEST (1953) suggested that the presence of insulin was necessary to obtain a decrease in blood NPN after testosterone propionate administration. Depancreatized female dogs maintained on insulin and then injected with 50 mg/day of testosterone propionate for 2 days with a further dose of 25 mg on the afternoon of the 2nd day, did not exhibit a decrease in blood NPN on the 3rd day after 24 h without food and insulin or during the following 10 h without insulin and food. These results would be consistent with the failure of GAEBLER and TARNAWSKI (1943) to detect a decrease in blood NPN of the normal female dog similarly treated with testosterone propionate except that SIREK and BEST (1953) observed a remarkable decrease (50%) in the NPN of their normal female dogs treated with testosterone propionate. The decrease was even greater than that observed in the castrated (KOCHAKIAN and MURLIN, 1936) or in the depancreatized-castrated dog ( Fig. 3) (KOCHAKIAN and COSTA, 1959).

The blood amino-N of the castrated-depancreatized dog exhibited a transitory decrease after about 2 weeks of testosterone propionate treatment (KOCHAKIAN and COSTA, 1959).

*b) Rat*

Rats castrated at 150–155 g body weight and fed a 33% protein diet showed no change in blood NPN 20 days later (LEATHAM, 1948a). Injection of testosterone propionate at 0.1, 0.25, and 1.25 mg/day beginning the day after castration slightly but not significantly increased the blood NPN at the lowest and highest doses and unexplainably greatly increased the NPN after 0.25 mg/day. Normal rats receiving 0.5, 1.25, and 2.5 mg/day of testosterone propionate showed nonsignificant changes. Plasma globulin but not albumin was increased after castration. Administration of testosterone propionate prevented the rise in the globulin and had a slight variable effect on the concentration of albumin and globulin of the normal rats. The hematocrit was not changed in any of the experiments.

Elevated blood NPN produced by feeding a high protein (78% casein or lactalbumin) diet for 20 days to adult male rats was not changed by injection of 1.25 mg testosterone propionate 3 times weekly (LEATHAM, 1951). The concentrations of plasma albumin and globulin also were not significantly altered by the high protein diet or by the testosterone propionate.

Simultaneous injection of testosterone propionate at 0.1 or 0.5 mg/day prevented the slight elevation in plasma NPN concentration in adult rats made hypothyroid by feeding thiourea or thiouracil, but the elevated concentrations of total protein and globulin were not affected (LEATHAM, 1948b). The hormone did not change the plasma NPN of the nonhypothyroid normal rats.

Injection of testosterone propionate, 4-chlorotestosterone acetate, or 4-chloro-19-nortestosterone acetate for 7 days did not change the serum NPN of adult castrated rats on constant food intake (SALA et al., 1957). Injection of testosterone or

Nilevar at 0.15 mg/day for 5 days significantly decreased the concentration of blood urea of adult rats fed *ad libitum* (HAZELWOOD and O'BRIEN, 1961). Similar injections begun 20 days after hypophysectomy were ineffective. The concentration of blood urea 25 days after hypophysectomy was increased approximately 50%, which was probably a reflection of the enhanced protein catabolism (KOCHAKIAN, 1950c).

## B. Creatine-Creatinine

The early recognition of creatinuria in children and its disappearance in boys and decrease in girls after puberty (cf. WANG, 1939) prompted further studies when the gonadal hormones became available.

### 1. Dog

KOCHAKIAN and MURLIN (1935) observed a creatinuria in castrated dogs when the main source of protein of the diet was beef heart. Since beef heart is relatively rich in creatine, it seemed that the creatinuria was probably due to this exogenous source rather than being endogenous in nature. Substitution of casein for beef heart in the diet eliminated the creatinuria. NITZESCU and GONTZEA (1937) also noted that the creatinuria of dogs disappeared upon removal of meat from the diet, and the intravenous administration of 0.5 g creatine resulted in a prompt creatinuria.

Daily administration of a potent male hormone urinary extract produced a variable decrease in the creatinuria of castrated dogs (KOCHAKIAN and MURLIN, 1935). The variation was probably due to the variation in the creatine content of the various batches of beef heart used to prepare the daily diet. A more consistent decrease in the creatinuria was noted in a short-term experiment with daily urine analysis and a single batch of beef heart for the duration of the experiment. Creatinine excretion was not altered in any of the studies. BUHLER (1935) reported a marked increase in urinary creatine excretion of dogs $1^1/_2$ months after castration that returned to the precastration level 5–6 months later. A normal littermate control, however, showed the same changes. Creatinuria obviously was due to the variations in the creatine content of the diet. When Proviron, a male hormone urine extract, was administered, the creatinuria of both the castrated and normal dogs was diminished.

### 2. Rat

COFFMAN and KOCH (1940) also did not detect any creatine in the urine of normal or castrated adult (2 months) rats on a creatine-free diet. Oral administration of creatine, however, produced an intense creatinuria of the same intensity in both groups of rats. Administration of testosterone propionate together with continued feeding of creatine produced a sharp decrease in the creatinuria accompanied by an increase in body weight. After about 8 days, no further increase in body weight occurred, and the creatinuria rose sharply (wearing-off effect) despite the continued injection of testosterone propionate. The response pattern was identical for both the normal and castrated rats, but the decrease in creatinuria was less in the normal than in the castrated rats. The response pattern of exogenous creatinuria to testosterone propionate injection was identical to that observed for urinary nitrogen (see above). These results clarified and extended the earlier report of KUN and PECZENIK (1935) that "male sex hormone" injected into rats castrated at 3–8 months of age abolished the creatinuria induced by feeding creatine in milk, and normal rats did not exhibit creatinuria.

CROXATTO et al. (1941, 1942) reported that creatine excretion, including the enhanced excretion after glycine feeding, was decreased to a greater degree in normal than in castrated rats by injection of testosterone propionate. TEDESCHI and MANGIANTINI (1957) also noted that testosterone decreased the creatinuria produced by feeding creatine not only to normal rats but also to rats that were fed diets deficient in riboflavin, pyridoxine, nicotinic acid, pantothenic acid, or ascorbic acid. The creatinuria induced by injection of thyroglobulin was reduced by the simultaneous administration of testosterone.

The importance of the diet was recognized by TERROINE et al. (1941), who noted that the effects of castration on creatine and creatinine excretion of adult male and female rats fed various diets were very irregular.

### 3. Monkey

BUHLER (1935) explored the role of several hormones in creatine and creatinine metabolism in not only the dog but also the rabbit and a monkey *(Java macaque)*. Proviron decreased creatine excretion by the male monkey when administered $2^1/_2$ months after castration but not when administered before or shortly after castration. JAILER (1940) studied monkeys *(macaca mulatta)* on a diet practically free of creatine. Immature monkeys exhibited as a rule a slight creatinuria and a low ability to retain exogenous creatine (227 mg creatine hydrate injected intraperitoneally). Subcutaneous injection of 5 mg/day of testosterone propionate for 15 days resulted in a gradual disappearance of the creatinuria. Repetition of the creatine retention test now indicated a 70% retention of the creatine in contrast to 30% before administration of the hormone. Mature monkeys exhibited variable amounts of creatine in the urine—two were entirely free, one had creatinuria one-third of the time, and two excreted creatine about one-half the time. All of the monkeys exhibited a strong retention of exogenous creatine. Upon castration, the creatinuria was enhanced, and the ability to retain exogenous creatine was decreased. Subcutaneous injection of 2 or 5 mg/day of testosterone propionate abolished or nearly abolished the creatinuria within a few days and enhanced the ability to retain exogenous creatine. Creatinuria reappeared within a few days after cessation of the hormone injections. Creatinine excretion was not affected either by castration or hormone administration. While these studies suggest that immature and castrated monkeys do not retain exogenous creatine as well as mature or testosterone propionate-treated monkeys, the wide variability of the creatinuria suggests the need of stricter dietary control of creatine intake.

In a later study (JAILER, 1941), the intense creatinuria produced by injection of 1 mg/day of thyroxine in two castrated and one normal male and three ovariectomized adult monkeys was abolished or nearly abolished by the simultaneous injection of 10 mg/day of testosterone propionate.

## 4. Rabbit

SCHRIRE and ZWARENSTEIN (1932a) reported that castration of adult male rabbits induced a 25–50% increase in creatinine excretion that returned to normal levels upon transplantation of testes and that reappeared upon removal of the transplants. No creatine was detected in the urine. Later these same authors reported that the increased excretion of creatinine produced by castration was decreased by the injection of a suspension of testes or testicular extracts (SHRIRE and ZWARENSTEIN, 1932b). Curiously, the lipoid extract induced an increased excretion of creatinine in normal and recently castrated rabbits. (CHEETHAM and ZWARENSTEIN, 1938).

BUHLER (1935) reported increased excretion of both creatinine and creatine after castration of adult rabbits. The creatinuria but not the creatininuria was decreased in normal and castrated rabbits by the injection of an androgen (Proviron) or an estrogen (Progynon). SEGHINI (1937) reported that the intense creatinuria of recently castrated rabbits spontaneously disappeared 1 month after castration and was decreased by the injection of a testicular extract. WILLIAMSON and GULICK (1941) showed that 2–5 mg/day of testosterone propionate decreased within 4 days the creatinuria of normal male rabbits with or without administration of creatine. The muscle creatine was increased without any change in the blood values. The blood and urinary creatinine was strikingly constant.

The above studies indicate that the rabbit, too, will retain creatine under the stimulus of testosterone but the creatinuria after castration needs further clarification because of the lack of control of dietary intake of creatine.

## 5. Guinea Pig

Castration or thymectomy of male or female guinea pigs is reported (COMSA, 1945a, b) to increase the excretion of creatine during the subsequent 15–30 days. The effect was less marked in immature than in mature animals (COMSA, 1947a). A larger and more prolonged creatinuria occurred if both organs were removed simultaneously (COMSA, 1945a). Enhancement of creatinuria after castration was further increased by thyroidectomy. If the guinea pigs were thyroidectomized, the resulting creatinuria was not further enhanced by castration (COMSA, 1947b). Creatinuria subsequent to the single and combined endocrine extirpations was abolished within 24–48 h by injection of testosterone propionate at 0.25 mg/day (COMSA, 1947c). Simultaneous removal of the three organs had no effect on creatine excretion (COMSA, 1947d).

Changes in creatine excretion may be related to the synthesis of new protein in the skeletal muscle under the influence of testosterone. Muscle is known to be relatively rich in creatine and contains 98% of the body creatine. Recently, creatine was reported (INGWALL et al., 1974) to stimulate selectively the rate of synthesis of actin and the myosin heavy chain in cultures of differentiating skeletal muscle. The rate of degradation was not affected. Furthermore, the inclusion of creatine in the diet inhibits transamidinase activity in the rat and the chick (WALKER, 1960), thus presumably sparing arginine, glycine, and methionine for protein synthesis. Addition of creatine to the diet, however, did not enhance the nitrogen-retaining and body weight-stimulating effect of testosterone propionate (KOCHAKIAN, 1950a) or the creatine content of the gastrocnemius muscle of rats (COFFMAN and KOCH, 1940).

The increase of creatine in the rabbit muscle (WILLIAMSON and GULICK, 1941), however, suggests that the analysis should be made at various time intervals during hormone administration.

The effect of the steroid hormones on creatine excretion has been studied extensively in man (see Chap. II.B).

### C. Coproporphyrin

Male rats excreted more coproporphyrin in the feces than female rats. Castration and ovariectomy produced a slight increase. Testosterone propionate increased the level of urinary coproporphyrin excretion by ovariectomized rats to that of the normal males, and estradiol-17β decreased that of the castrated males to the female level (HOFFBAUER et al., 1953).

## VIII. Electrolytes and Water

### A. Dog

The similar chemical structures of the sex steroids and the adrenal cortical steroids prompted THORN and ENGEL (1938) to study the effect of testosterone propionate and other steroids on the urinary excretion of electrolytes and water. They found testosterone propionate to be approximately one-fourth as effective as an adrenal cortical extract in the retention of $Na^+$ in a 24-h test in normal male and female dogs. Androst-4-ene-3,17-dione and androst-5-ene-3β,17β-diol at 40 mg were ineffective. GAEBLER and TARNOWSKI (1943) did not observe any changes in water intake or excretion by normal or depancreatized female dogs injected with 25 mg/day of testosterone propionate for 3–5 days. The urine volume of depancreatized-castrated dogs partially regulated by insulin was not remarkably altered during or after the long-term injection of testosterone propionate (KOCHAKIAN and COSTA, 1959).

WELSH et al. (1942) injected young adult female dogs on a constant food intake with 100 mg/day of testosterone propionate for 12 days after seven control periods of 7 days each and observed a decrease in diodrast Tm of as much as 100%. The rates of glomerular filtration (creatinine clearance) and effective renal blood flow (diodrast clearance) were unaffected. The proportion of potassium to sodium retained was too high to assume that an increase in extracellular fluid occurred.

Substitution of 25 mg/day of testosterone propionate for a maintenance dose of adrenal cortical extract in an adrenalectomized dog did not alter the symptoms of adrenal cortical deficiency (THORN and ENGEL, 1938). The typical sodium and chloride diuresis was not influenced, the dog lost weight and strength, and the blood NPN was elevated.

DEMUYLDER and BORGHGRAEF (1957) concluded that the endocrine secretion of the testis did not play a major role in electrolyte and water balance in the adult dog. These authors found no significant change in the sodium space measured with radioactive sodium after 30 days of castration. Injections of testosterone propionate at 1.0 or 1.5 mg/kg/day for 54–63 days and 61–63 days, respectively, produced an inconsistent augmentation and, upon cessation of injections, a consistent temporary shrinkage of the sodium space.

## B. Rat

SELYE and BASSETT (1940) were unable to detect any changes in water or chloride metabolism in rats treated with testosterone. In nitrogen balance studies, however, not only chloride but also potassium and phosphates changed in parallel with the changes in nitrogen excretion stimulated by testosterone propionate in castrated (KOCHAKIAN et al., 1950) and adrenalectomized (KOCHAKIAN et al., 1950) rats. Similar results were obtained for $PO_4^{-3}$ in phlorhizinized rats (WRIGHT and KOCHAKIAN, 1953) and for chloride in castrated rats recovering from starvation (KOCHAKIAN et al., 1948). 4-Chloro-19-nortestosterone acetate, 4-chlorotestosterone acetate, and testosterone propionate induced retention of $PO_4^{-3}$ and of $K^+$ in parallel with the nitrogen retention (SALA et al., 1957). A similar effect was indicated for $Na^+$ but the results were too variable to be conclusive. KIM (1961), however, reported changes in urinary $Na^+$ and $K^+$ in parallel with the changes (retention and wearing-off effect) in urinary nitrogen excretion after injection of testosterone propionate, Dianabol, and Durabolin into adult female rats on constant food intake and attempted to correlate these changes with the decrease in histamine excretion.

SELYE and DOSNE (1940) reported that testosterone propionate at 1 mg/day for 10 days decreased the blood chloride of rats. MILLER (1943), however, found no changes in the chloride, sodium, or potassium concentrations of the serum of rats after 14–26 days of testosterone propionate at 1 mg/day. Continuation of injections for 6 months resulted in a decrease in the concentrations of serum and muscle potassium, but those of sodium and chloride were not changed. SALA et al. (1957) also did not detect any change in blood chloride on short-term (7 days) injection of testosterone propionate, 4-chlorotestosterone acetate, or 4-chloro-19-nortestosterone acetate.

Oral 17α-methyltestosterone and oral and subcutaneous injection of dimethazine (17β-hydroxy-2α,17α-dimethyl-5α-androstan-3,3′-azine) changed the urinary $PO_4^{-3}$ excretion in parallel with the change in urinary nitrogen with a concomitant change in both urinary and fecal $Ca^{2+}$ (DERUGGIERI et al., 1963). In an earlier study MANDEL et al. (1954) found that the calcium balance of rats became negative after castration and was restored to normal by the injection of testosterone. The changes occurred in the feces. The urinary calcium was not affected.

## C. Mouse

The concentration of the blood potassium of mice was slightly decreased by castration and restored to normal by the subcutaneous implantation of a pellet of testosterone propionate (KOCHAKIAN, 1941).

## D. Rabbit

Testosterone at 1 mg/kg body weight was reported (FICHERA and CATANIA, 1938) to increase the blood potassium of adult rabbits approximately 12%.

Retention of these elements in parallel with nitrogen is consistent with the stimulation by the steroids of the synthesis of new protoplasm. The lack of or at best small changes in the concentration of blood electrolytes is in agreement with the failure to observe edema in experimental animals in contrast to certain instances in man as a result of the administration of testosterone or related steroids.

# IX. Energy Metabolism

## A. Dog

Although the effects of the androgenic urinary extracts on nitrogen metabolism were essentially the same in a "thin" and a "fat" lethargic castrated dog, the effect on energy metabolism was different (KOCHAKIAN and MURLIN, 1935). The thin dog showed no changes in BMR after single large doses or after daily injections at different doses of the male hormone urine extract for 15–25 days. The fat dog also showed no changes in BMR after single doses of various quantities, but on continued daily injections, it responded with a gradual small (10%), but consistent, increase. Partition of the energy metabolism demonstrated a decrease in protein metabolism and a compensatory increase in fat catabolism. Injection of androst-4-ene-3,17-dione (KOCHAKIAN and MURLIN, 1936) and testosterone or testosterone acetate (KOCHAKIAN, 1937) for 3 days did not alter the BMR of either the thin or fat dog. The results with the pure hormones are consistent with the single large doses of the urinary extracts.

## B. Rat

Testosterone propionate did not alter the BMR of rats (KOCH, 1937). Thyroidectomy produced an approximately 30% decrease in the BMR of normal adult male rats (MEYER and DANOW, 1942). Castration 20 days before or 25 days after thyroidectomy did not produce a further decrease, but castration alone decreased the BMR about 10%. The effect of castration is consistent with the earlier studies of KOJIMA (1917), who observed a decrease of 27% in the energy metabolism of rats 1 and 2 months after castration with a concomitant increase in appetite and body weight. The daily injection of 0.5 mg 17α-methyltestosterone into thyroidectomized, thyroidectomized-castrated, or castrated rats produced a slight increase in the BMR of only the doubly-operated rats (MEYER and DANOW, 1942). These rats also were more sensitive to thyroxine.

QUIMBY et al. (1948) reported that the enhanced energy metabolism of rats during the initial replenishment period after chronic undernourishment was prevented by the injection of 0.5 mg/day testosterone propionate. They attributed the action of the steroid to inhibition of the production of growth hormone from the anterior pituitary.

These studies, though limited, indicate that testosterone is not a calorigenic agent but rather stimulates a selective utilization of the various caloric foodstuffs.

# X. Summary

The protein anabolic effect of testosterone as indicated by the induction of a positive nitrogen balance and an increase in body weight is well established. The positive balance is due to a decrease in urea production and excretion. Fecal nitrogen is not changed. The fact that this property of the internal secretion of the testis is not mediated through any of the other endocrine organs is evident when the various other endocrine organs are removed. The protein anabolic effect is due to better utilization of the ingested protein, but it does not continue indefinitely. After a period of time, which varies with the species of animal under study, the nitrogen balance gradually returns to the basal level despite the continued injection of the steroid. This phenomenon has been termed the wearing-off phase.

A large number of testosterone related compounds have been found to possess the same protein anabolic effect. Esterification greatly enhances the effect of testosterone and other compounds. Testosterone propionate, administered parenterally, is the most potent compound. Many compounds are also effective by oral administration but are much less effective than by subcutaneous or intramuscular injection.

The steroids also influence carbohydrate and fat metabolism. Energy metabolism is not greatly changed, but prolonged hormone treatment will produce a small (10%) increase in the BMR of a fat but not of a thin dog. In both instances, however, analysis of the distribution of the calories indicates that not only protein but also carbohydrate is spared, and the utilization of fat is enhanced to compensate for the loss of the calories from protein and carbohydrate and to provide the extra energy in the fat dog. The sparing effect on carbohydrate is very evident in the depancreatized-castrated dog. Testosterone propionate markedly reduces blood and urinary sugar to or below normal levels upon injection into dogs whose diabetes is partially regulated with insulin. The effect on blood sugar levels is not evident in the nondepancreatized dog or rat.

The nutritive state of the animal will modify but not abolish the protein anabolic effect of the steroids. The quantity of protein in the diet will influence the magnitude of nitrogen retention. Increase of the protein (casein) above 18% in the castrated rat will result in a smaller enhancement in the castrated rat but in the alloxan-diabetic rat, the result will be a nearly proportional increase in nitrogen retention. The fasting animal will show a nitrogen-retaining effect in the first few days of the fast, but once the nitrogen excretion has reached a minimal level a positive nitrogen balance cannot be detected; however, withdrawal of hormone treatment will result in an extra excretion of nitrogen in the urine, presumably due to atrophy of newly formed tissue under the stimulus of the hormone (USUELLI et al., 1949). The simultaneous administration of testosterone during repletion from a fast will hasten and enhance the recovery of protein and body weight. The quality of the dietary protein also is important. A deficiency in any of the essential amino acids will decrease but not abolish the protein anabolic effect of the hormones. If the animal is in positive nitrogen balance due to feeding *ad libitum* or has not reached equilibrium under the new constant intake of food (FORBES, 1973), the hormone-induced positive nitrogen balance is partially or completely undetected.

Testosterone propionate enhances the utilization of exogenous creatine in parallel with the changes in urinary nitrogen excretion. Furthermore, the degree of utilization is greater in the castrated than in the normal animal. Retention of creatine presumably is associated at least in part with the newly formed muscle.

Induction of nitrogen retention by the steroids is accompanied by a parallel change in $Na^{+}$, $K^{+}$, $Cl^{-}$, and $PO_4^{-3}$, indicating the new formation of protoplasm. The urinary urea and blood NPN and urea also change in parallel with the changes in urinary nitrogen excretion.

Castration produces a transitory small postoperational increase in nitrogen excretion and possibly a small decrease in energy metabolism in the rat. The careful studies in the dog do not reveal any change in nitrogen or energy metabolism, although decreases were indicated on long-term castration if the dogs tended to become fat. Support for the latter possibility is provided by a decrease in both BMR and nitrogen metabolism in the rooster after castration. The rooster always becomes fat and inactive after castration.

## References


ARNOLD, A., BEYLER, A. L., POTTS, G. O.: Androstanazole, a new orally active anabolic steroid. Proc. Soc. exp. Biol. (N.Y.) **102**, 184—187 (1959).

ARNOLD, A., POTTS, G. O.: Oral anabolic-androgenic evaluations of four anabolic steroids. Acta endocr. (Kbh.) **52**, 489—496 (1966).

ARNOLD, A., POTTS, G. O., BEYLER, A. L.: Relative oral anabolic to androgenic activity ratios of androisoxazole, ethylestrenol, methylandrostenolisoxazole and testosterone. Acta endocr. (Kbh.) **44**, 490—498 (1963a).

ARNOLD, A., POTTS, G. O., BEYLER, A. L.: The ratio of anabolic to androgenic activity of 7:17-dimethyltestosterone, oxymesterone, mestanolone, and fluoxymesterone. J. Endocr. **28**, 87—92 (1963b).

ARNOLD, A., POTTS, G. O., BEYLER, A. L.: Evaluation of the protein anabolic properties of certain orally active anabolic agents based on nitrogen balance studies in rats. Endocrinology **72**, 408—417 (1963c).

ARNOLD, A., SCHAD, J. S.: Nitrogen balance studies with dogs on casein or methionine-supplemented casein. J. Nutr. **53**, 265—273 (1954).

ASCHKENASY, A.: Etude expérimentale de l'action protidoanabolisante ainsi que de la toxicité aigue et chronique de la 17-ethyl-19-nortestosterone. Therapie **14**, 332—349 (1959).

ASCHKENASY, A., DRAY, F.: Action de la thyroxine, des hormones corticotrope et somatotrope et des hormones genitales sur la reprise du poids et sur la regeneration des organes apres inanition proteique. Ann. endocr. (Paris) **15**, 441—460 (1954).

AUDE, D.: Influence de l'hormone testiculaire sur les echanges respiratoires chez les gallinaces. Rev. franç. Endocr. clin. **5**, 81—115 (1927).

BALDRATTI, G., ARCARI, G.: The anabolising effect of 4-hydroxy-17α-methyltestosterone. Panminerva med. **4**, 137—139 (1962).

BARTLETT, P. D.: Rates of protein synthesis, amino acid catabolism and size of nitrogen pool during nitrogen storage induced with testosterone propionate, and testosterone propionate combined with growth hormone. Endocrinology **52**, 272—276 (1953).

BARTLETT, P. D., GAEBLER, O. H.: Studies on the mechanism of nitrogen storage. VI. Rate of protein synthesis and size of the nitrogen pool. J. biol. Chem. **196**, 1—9 (1952).

BARTLETT, P. D., STEVENSON, A.: Effects of protein anabolic hormones on rates of protein degradation and protein loss in the fasting dog. Endocrinology **55**, 200—204 (1954).

BAVETTA, L. A., BEKHOR, I., SHAH, R., O'DAY, P., NIMNI, M. E.: Metabolic and anti-inflammatory properties of 6-methylprednisolone alone and in combination with anabolic hormones. Endocrinology **71**, 221—226 (1962).

BEYLER, A. L., ARNOLD, A., POTTS, G. O.: Methods for evaluating anabolic and catabolic agents in laboratory animals. J. Amer. med. Wom. Ass. **23**, 708—721 (1968).

BEYLER, A. L., POTTS, G. O., ARNOLD, A.: Influence of molecular unsaturation on hormonal activity pattern of certain heterocyclic steroids. Endocrinology **68**, 987—995 (1961).

BOULIERE, F., CENDRON, H., TESNIERE, B.: Comparaison de l'action frénatrice de quelques stéroides sur la perte d'azote, chez le Rat adulte il age. C. R. Soc. Biol. (Paris) **152**, 1636—1640 (1958).

BRAASCH, J. W., WAKERLIN, G. E., BELL, J. H., LEVENSON, S. M.: Effect of testosterone propionate on total urinary nitrogen excretion of the rat following burns. Proc. Soc. exp. Biol. (N.Y.) **75**, 183—189 (1950).

BROWN, J. H. U.: Influence of methionine and thiouracil on nitrogen balance index and organ weights of adult rats. Proc. Soc. exp. Biol. (N.Y.) **71**, 612—616 (1949).

BROWN-SEQUARD, C. E.: The effects produced on man by subcutaneous injections of a liquid obtained from the testicles of animals. Lancet **1889 II**, 105—107.
BÜHLER, F.: Tierexperimentelle Untersuchungen über den Einfluß von verschiedenen Hormonen auf die Ausscheidung von Kreatin und Kreatinin in Urin. Z. ges. exp. Med. **96**, 821—844 (1935).
BUTENANDT, A., HANISCH, G.: Über Testosteron. Umwandlung des Dehydro-androsterons in Androstendiol und Testosteron; ein Weg zur Darstellung des Testosterons aus Cholesterin. Z. physiol. Chem. **237**, 89—97 (1935).
BUTENANDT, A., KUDSZUS, H.: Über Androstendion, einen hochwirksamen männlichen Prägungsstoff. Z. physiol. Chem. **237**, 75—88 (1935).
CAVALLERO, C., MALANDRA, B.: Sul'effeto anabolico del testosterone e del 4-chlorotestosterone nel ratto con diabete allossanico. Boll. Soc. ital. Biol. sper. **32**, 748—750 (1956).
CHAIKOFF, I. L., FORKER, L. L.: The antidiabetic action of insulin on nitrogen metabolism. Endocrinology **46**, 319—326 (1950).
CHEETHAM, R. W. S., ZWARENSTEIN, H.: The influence of the gonads on protein metabolism. V. The effect of testicular, urinary and anterior pituitary preparations on the urinary creatinine of normal and castrated rabbits. Biochem. J. **32**, 871—877 (1938).
CHIAPPINO, G.: Azione del 4-idrossi-17α-metiltestosterone sul bilancio azotato nel diabete allossanico del ratto. Arch. Stud. Fisiopat. Clin. Ricambio **24**, 1—5 (1960).
COFFMAN, J. R., KOCH, F. C.: The effect of testosterone propionate on induced creatinuria in rats. J. biol. Chem. **135**, 519—530 (1940).
COMSA, J.: Action de l'oestrone sur la créatine urinaire du Cobaye. C. R. Soc. Biol. (Paris) **139**, 1033—1034 (1945a).
COMSA, J.: Conséquences de la castration sur la créatinurie du Cobaye thymiprive. C. R. Soc. Biol. (Paris) **139**, 391—392 (1945b).
COMSA, J.: Nouvelles recherches sur la créatinurie chez le Cobaye. C. R. Soc. Biol. (Paris) **141**, 413—415 (1947a).
COMSA, J.: Evolution de la créatinurie chez le Cobaye thyréoprive castré. C. R. Soc. Biol. (Paris) **141**, 511—513 (1947b).
COMSA, J.: Nouvelles recherches sur l'action de la testosterone sur la créatinurie du Cobaye. C. R. Soc. Biol. (Paris) **141**, 917—921 (1947c).
COMSA, J.: Nouvelles recherches sur la creatinurie du Cobaye thymithyreoprive castré. C. R. Soc. Biol. (Paris) **141**, 513—514 (1947d).
CORNÍL, L., ROSANOFF, H.: Influence du propionate de testosterone sur la glycémie. C. R. Soc. Biol. (Paris) **129**, 983—986 (1938).
CROXATTO, H., MUERMANN, J., DE LA PARRA, R.: Influence of gelatin or glycine administration on the excretion of creatine by the rat. Bol. Soc. Biol. Santiago **4**, 337—339 (1941, 1942).
DAVID, K., DINGEMANSE, E., FREUD, J., LAQUEUR, E.: Über krystallinisches männliches Hormon aus Hoden (Testosteron), wirksamer als aus Harn oder aus Cholesterin bereitetes Androsteron. Z. physiol. Chem. **233**, 281—282 (1935).
DE JONGH, S. E., SCHIERNECKER, J., VAN WIERINGEN, G.: Remarks on the anabolic effect of testosterone. Acta physiol. pharmacol. neerl. **1**, 294—297 (1950).
DE MUYLDER, E., BORGHGRAEF, R.: Influence des oestrogenes et des androgenes sur l'espece extracellulaire mesure au sodium radioactif. Bull. Soc. roy. belge. Gynéç. Obstét. **27**, 426—439 (1957).
DE RUGGIERI, D. P., MATSCHER, R., GANDOLFI, C., CHIARAMONTI, D., LUPO, C., PIETRA, E., CAVALLI, R.: Un nuovo steroide aventa azione proteoanabolica: La dimetazina. Arch. Sci. biol. **47**, 1—19 (1963).
ELSAIR, J., POEY, J., NACEUR, J., DEVINE, R., MRABET, N.: Effets de l'hormone de croissance, de la norethandrolone et une association des deux sus le bilan d'azote du rat. Ann. Biol. clin. **31**, 225—230 (1973).
FICHERA, G., CATANIA, C.: Metabolismo del Calcio e del Potassio sotto l'azione dell' Ormone sessuale maschile. Rev. Pat. sper. **21**, 187—192 (1938).
FORBES, G. B.: Another source of error in the metabolic balance method. Nutr. Rev. **31**, 297—300 (1973).
FUJII, T.: Influence of androgen on hypothyroidism in rats. I. Effects of methylandrostenediol on thyroidectomized rats. Endocr. Jap. **6**, 47—58 (1959).
FUNK, C., HARROW, B.: The male hormone. Biochem. J. **24**, 1678—1680 (1930).

Gaebler, O. H., Beher, W. T., Crigger, E. M.: Conjugation of benzoic and phenylacetic acids during nitrogen storage induced with growth hormone or testosterone propionate. Amer. J. Physiol. **172**, 152—156 (1953).

Gaebler, O. H., Tarnowski, S. M.: Effects of estrone, ascorbic acid and testosterone propionate on nitrogen storage and insulin requirement in dogs. Endocrinology **33**, 317—324 (1943).

Gastaldi, F.: Modificazioni glicemiche immediate indotte dal propionato di testosterone in cani normali e pancreasectomizzati. Studi sassaresi **25**, 601—606 (1947).

Geiger, E., El Rawi, I.: Lack of effect of testosterone propionate on body weight of rats during protein depletion and repletion. Metabolism **1**, 145—148 (1952).

Gordan, G. S., Bennett, L. L., Li, C. H., Evans, H. M.: Effect of hypophyseal growth hormone on rats fed low protein diets. Proc. Soc. exp. Biol. (N.Y.) **65**, 317—319 (1947 b).

Gordan, G. S., Bennett, L. L., Li, C. H., Evans, H. M.: The effect of dietary protein content upon the nitrogen retention and weight gain produced by the hypophyseal growth hormone. Endocrinology **42**, 153—160 (1948).

Gordan, G. S., Evans, H. M., Simpson, M. E.: Effects of testosterone propionate on body weight and urinary nitrogen excretion of normal and hypophysectomized rats. Endocrinology **40**, 375—380 (1947 a).

Hazelwood, R. L., O'Brien, K. D.: Modification of glucagon-induced hyperglycemia in rats by 17-ethyl-19-nortestosterone. Proc. Soc. exp. Biol. (N.Y.) **106**, 851—854 (1961).

Heymans, C.: Influence de la castration sur les echanges respiratoires, la nutrition et le jeune. J. Physiol. Path. Gén. **19**, 323—331 (1921).

Hoffbauer, F. W., Watson, C. J., Schwartz, S.: Urinary and fecal coproporphyrin excretion in rats. I. Results in normals and castrates. Proc. Soc. exp. Biol. (N.Y.) **83**, 228—232 (1953).

Ingle, D. J., Ginther, G. B., Nezamis, J.: Effect of diet in rats on adrenal weights and on survival following adrenalectomy. Endocrinology **32**, 410—414 (1943).

Ingwall, J. S., Weiner, C. D., Morales, M. F., Davis, E., Stockdale, F. E.: Specificity of creatine in the control of muscle protein synthesis. J. Cell. Biol. **62**, 145—151 (1974).

Jailer, J. W.: The effect of castration and androgen therapy on creatine and creatinine excretion in monkeys. Amer. J. Physiol. **130**, 503—511 (1940).

Jailer, J. W.: Effect of testosterone propionate on creatinuria of experimental hyperthyroidism in male and female monkeys. Endocrinology **29**, 89—92 (1941).

Kassenaar, A. A. H.: On the metabolic action of testosterone and related compounds. V. The effect of DOCA and testosterone on androgen metabolism in the adrenalectomized castrated male rat. Acta endocr. (Kbh.) **14**, 130—136 (1953).

Kassenaar, A. A. H., van Bekkum, D. W., Querido, A.: On the metabolic action of testosterone and related compounds. Acta endocr. (Kbh.) **12**, 153—158 (1953).

Kim, K. S.: Anabolic agents and histamine metabolism. Nature (Lond.) **191**, 1368—1370 (1961).

Koch, F. C.: The biochemistry of androgens. In: Sex and Internal Secretions, pp. 807—845. Baltimore: Williams and Wilkins Co. 1939.

Kochakian, C. D.: Effect of male hormone on protein metabolism of castrate dogs. Proc. Soc. exp. Biol. (N.Y.) **32**, 1064—1065 (1935).

Kochakian, C. D.: Testosterone and testosterone acetate and the protein and energy metabolism of castrate dogs. Endocrinology **21**, 750—755 (1937).

Kochakian, C. D.: The rate of absorption and effects of testosterone propionate pellets on mice. Endocrinology **28**, 478—484 (1941).

Kochakian, C. D.: Conference on metabolic aspects of convalescence. Josiah Macy, Jr., Foundation, p. 125 (1943).

Kochakian, C. D.: The protein anabolic effects of steroid hormones. Vitam. and Horm. **4**, 255—310 (1946).

Kochakian, C. D.: Protein anabolic activity of pregnant mare's urine. Proc. Soc. exp. Biol. (N.Y.) **75**, 543—545 (1950 a).

Kochakian, C. D.: Comparison of protein anabolic property of various androgens in the castrated rat. Amer. J. Physiol. **160**, 53—61 (1950 b).

Kochakian, C. D.: Comparison of protein anabolic properties of testosterone propionate and growth hormone in the rat. Amer. J. Physiol. **160**, 66—74 (1950 c).

Kochakian, C. D.: Renotrophic-androgenic properties of orally administered androgens. Proc. Soc. exp. Biol. (N.Y.) **80**, 386—388 (1952).

KOCHAKIAN, C. D.: Summation of protein anabolic effects of testosterone propionate and growth hormone. Proc. Soc. exp. Biol. (N.Y.) **103**, 196—197 (1960a).

KOCHAKIAN, C. D.: Protein anabolic action of testosterone propionate in hyperthyroid-castrated rats. Endocrinology **66**, 786—788 (1960b).

KOCHAKIAN, C. D., BEALL, B.: Comparison of the protein anabolic property of testosterone propionate in the male and female rat. Amer. J. Physiol. **160**, 62—65 (1950).

KOCHAKIAN, C. D., COHN, L., QUIGLEY, E., TRYBALSKI, E.: Effect of testosterone propionate on nitrogen and chloride excretion and body weight of castrated rats during recovery from fasting. Amer. J. Physiol. **155**, 272—277 (1948).

KOCHAKIAN, C. D., COSTA, G.: The effect of testosterone propionate on the protein and carbohydrate metabolism in the depancreatized castrated dog. Endocrinology **65**, 298—309 (1959).

KOCHAKIAN, C. D., DOLPHIN, J.: The protein-anabolic effect of testosterone propionate in the hypothyroid rat. Amer. J. Physiol. **180**, 317—320 (1955).

KOCHAKIAN, C. D., MOE, J. G., DOLPHIN, J.: Protein anabolic effect of testosterone propionate in adrenalectomized and normal rats. Amer. J. Physiol. **162**, 581—589 (1950b).

KOCHAKIAN, C. D., MURLIN, J. R.: The effect of male hormone on the protein and energy metabolism of castrate dogs. J. Nutr. **10**, 437—459 (1935).

KOCHAKIAN, C. D., MURLIN, J. R.: The relationship of the synthetic male hormone, androtenedione, to the protein and energy metabolism of castrate dogs and the protein metabolism of a normal dog. Amer. J. Physiol. **117**, 642—657 (1936).

KOCHAKIAN, C. D., ROBERTSON, E., BARTLETT, M. N.: Sites and nature of protein anabolism stimulated by testosterone propionate in the rat. Amer. J. Physiol. **163**, 332—346 (1950a).

KOCHAKIAN, C. D., TILLOTSON, C., AUSTIN, J.: A comparison of the effect of inanition, castration and testosterone on the muscles of the male guinea pig. Endocrinology **60**, 144—152 (1957).

KOCHAKIAN, C. D., VAIL, V. N.: The effect of adrenalectomy, adrenal cortical hormones, and testosterone propionate plus adrenal cortical extract on the arginase activity of the liver and kidney of the rat. J. biol. Chem. **169**, 1—6 (1947).

KOCHAKIAN, C. D., VAN DER MARK, W.: Relationship of protein intake to protein anabolic activity of testosterone propionate. Proc. Soc. exp. Biol. (N.Y.) **79**, 74—75 (1952).

KOCHAKIAN, C. D., WEBSTER, J. A.: Effect of testosterone propionate on the appetite, body weight and composition of the normal rat. Endocrinology **63**, 737—742 (1958).

KOJIMA, M.: Studies on the endocrine glands V. Effects upon metabolism of castration, of thyroidectomy, of parathyroidectomy, and of thyroid and parathyroid feeding. Quart. J. exp. Physiol. **1**, 351—360 (1917).

KORENCHEVSKY, V.: Sex glands and metabolism. I. Influence of castration on nitrogenous and gaseous metabolism. Brit. J. exp. Path. **6**, 21—35 (1925).

KRUZE, D., DUBILOWSKI, P., WIEROZCHOWSKI, P., WYSOKOWSKI, J.: The effect of anabolic steroids (Dianabol, Steranabol, and Decadurabolin) on nitrogen metabolism in the rabbit. Endokr. pol. **17**, 368—371 (1966).

KUN, H., PECZENIK, O.: Geschlechtsspezifische Wirkung der Sexualhormone auf den Kreatinstoffwechsel. Pflügers Arch. ges. Physiol. **236**, 471—480 (1935).

LEATHAM, J. H.: Plasma protein concentrations and organ weights of castrated and testosterone propionate treated rats. Amer. J. Physiol. **154**, 459—464 (1948a).

LEATHAM, J. H.: Influence of testosterone propionate on the plasma and liver proteins of hyprothyroid rats. J. biol. Chem. **176**, 1285—1289 (1948b).

LEATHAM, J. H.: Plasma and liver proteins of rats following high protein diets and testosterone propionate. Amer. J. Physiol. **165**, 73—78 (1951).

LEDERER, J., DELHAYE, M., POTTIER-ARNOULD, A. M., NIETHALS, E.: Influence de la testosterone et de ses antagonistes, acetate de cyproterone et oestradiol sur l'action lipidogene du saccharose chez le rat. Ann. endocr. **34**, 336—340 (1973).

LENNON, H. D., SAUNDERS, F. J.: Anabolic activity of 2-oxa-17α-methylidihydrotestosterone (oxandrolone) in castrated rats. Steroids **4**, 689—697 (1964).

LEWIS, J. T., FOGLIA, V. G., RODRIGUEZ, R. R.: The effects of steroids on the incidence of diabetes in rats after subtotal pancreatectomy. Endocrinology **46**, 111—121 (1950).

LOEWY, A., RICHTER, P. F.: Sexualfunction und Stoffwechsel. Arch. Anat. Physiol. Suppl. B, **1899**, 174—198.

LOEWY, A., RICHTER, P. F.: Zur Frage nach dem Einfluß der Castration auf den Stoffwechsel. Centralbl. Physiol. **16**, 449—452 (1902).

LUTHJE, H.: Über die Castration und ihre Folgen. Arch. exp. Path. **48**, 184—222 (1903).

LYSTER, S. C., DUNCAN, G. W.: Anabolic, androgenic and myotropic activities of derivatives of 7α-methyl-19-nortestosterone. Acta endocr. (Kbh.) **43**, 399—411 (1963).
MAIGNON, M. F.: Influence des saisons et des glandes génitales, sur les combustions respiratoires chez le cobaye. C. R. Soc. Biol. (Paris) **156**, 347—349 (1913).
MANDEL, P., EBEL, A., FONTAINE, R.: Effet de la testosterone sur le bilan calcique chez le rat castre. C. R. Soc. Biol. (Paris) **148**, 713—715 (1954).
MANUNTA, G.: Influenza della castrazione e della somministrazione di testosterone sull'escrezione di azoto di ratti in carenza proteica. Studi Sassaresi **33**, 1—8 (1955).
MATSCHER, R., LUPO, C., DE RUGGIERI, P.: Attivita biologica della Dimetazina in senso proteoanabolica. Boll. Soc. ital. Biol. sper. **38**, 1—3 (1962).
MEYER, A. E., DANOW, H.: Influence of methyltestosterone on metabolism of normal, castrate and thyroidectomized rats. Proc. Soc. exp. Biol. **49**, 598—601 (1942).
MILLER, H. C.: Effect of the injection of sex hormones on serum and muscle electrolytes of rats. Endocrinology **32**, 443—448 (1943).
MOORE, C. R.: Biology of the testes. In: Sex and Internal Secretions, pp. 353—451. Baltimore: Williams and Wilkins Co. 1939.
NIMNI, M. E., GEIGER, E.: Testosterone-induced N-retention in castrated rats on diets containing incomplete proteins. Endocrinology **61**, 753—757 (1957).
NITZESCU, I., GONTZEA, I.: La foie et la créatinurie. C. R. Soc. Biol. (Paris) **125**, 77—80 (1937).
NOACH, E. L., WIJNANS, M., DE JONGH, S. E.: The influence of the thyroid gland on the anabolic effect of testosterone. Proc. kon. ned. Akad. Wet., Series C **55(5)**, 562—566 (1952).
PELLEGRINI, G.: Modification immediate dell'iperglicemia provasabili dal propionato de testosterone nel diabete mellito sperimentale. Boll. Soc. ital. Biol. sper. **18**, 176—177 (1943a).
PELLEGRINI, G.: Azione del propionato di testosterone sul ricambio idrocarbonato nel diabete mellito sperimentale. Arch. Sci. Med. **21**, 177 (1943b).
PELLEGRINI, G.: Azione del propionato di testosterone sul ricambio idrocarbonato nel diabete mellito sperimentale. Boll. Soc. ital. Biol. sper. **18**, 174—176 (1943c).
PERLMAN, P. L., CASSIDY, J. W.: Influence of nitrogen intake on nitrogen-retaining action of testosterone propionate. Proc. Soc. exp. Biol. (N.Y.) **83**, 674—675 (1953).
POTTS, G. O., ARNOLD, A., BEYLER, A. L.: Comparative myotrophic and nitrogen retaining effects of several steroids. Endocrinology **67**, 849—854 (1960).
PTASZEK, L.: Influence des hormones sexuelles (testiculaires) sur le metabolisme basal chez les animaux. C. R. Soc. Biol. (Paris) **99**, 929—931 (1928).
QUIMBY, F. H.: Effects of hormone, vitamin and liver supplements on the appetite and growth of the young rat during recovery from chronic starvation. Amer. J. Physiol. **166**, 566—571 (1951).
QUIMBY, F. H., PHILLIPS, N. E., WHITE, I. U.: Chronic inanition, recovery, and metabolic rate of young rats. Amer. J. Physiol. **154**, 188—192 (1948).
REIFENSTEIN, E. C., JR., ALBRIGHT, F., WELLS, S. L.: The accumulation, interpretation and presentation of data pertaining to metabolic balances, notably those of calcium, phosphorus and nitrogen. J. clin. Endocr. **5**, 367—395 (1945).
ROOKS, W. H. II., LLOYD, E., DORFMAN, R. I.: The nitrogen retaining activity in rats of 19-nortestosterone-17(4'-methylbicyclo[2.2.2.])-oct-2'-ene-1'-carboxylate. Proc. Soc. exp. Biol. **136**, 681—683 (1971).
RUPP, J. J., PASCHKIS, K. E.: Influence of testosterone propionate on the protein metabolism of hypophysectomized rats. Metabolism **2**, 268—270 (1953).
RUPP, J. J., PASCHKIS, K. E.: Influence of testosterone propionate on protein metabolism of thyroidectomized rats. Proc. Soc. exp. Biol. **86**, 399—401 (1954).
RUPP, J. J., PASCHKIS, K. E., CANTAROW, A.: Influence of thyroxine on protein metabolism. Endocrinology **44**, 449—453 (1949).
RUZICKA, L., WETTSTEIN, A.: Sexualhormone. V. Künstliche Herstellung des männlichen Sexualhormons trans-Dehydro-androsterone und des Androsten-3,17-dions. Helv. chim. Acta **18**, 986—994 (1935).
SALA, G., BALDRATTI, G., RONCHI, R.: Influenza del 4-chlorotestosterone, del 4-chloro-19-nortestosterone e del testosterone sul bilancio azotato del ratto. Folia endocr. **10**, 729—735 (1957).
SANDBERG, M., PERLA, D., HOLLY, O. M.: The effect of castration in albino rats on nitrogen, sulfur, chloride, sodium and potassium metabolism, and on changes in weight and food intake. Endocrinology **24**, 503—509 (1939).
SAUNDERS, F. J., DRILL, V. A.: Nitrogen-retaining properties of 17-ethyl-19-nortestosterone. Metabolism **7**, 315—321 (1958).

SCHRIRE, I., ZWARENSTEIN, H.: Influence of the gonads on protein metabolism. I. Urinary creatinine after castration. Biochem. J. **26**, 118—121 (1932a).
SCHRIRE, I., ZWARENSTEIN, H.: Influence of gonads on protein metabolism. II. Effect of injections of saline suspension of testes and testicular extracts on urinary creatinine after castration. Biochem. J. **26**, 1886—1889 (1932b).
SEGHINI, C.: La creatinuria in rapporto alla funzione endocrina della ghiandole sessuali. Pathologica **29**, 53—57 (1937).
SELYE, H., BASSETT, L.: Effect of desoxycorticosterone and testosterone on water and chloride metabolism. Proc. Soc. exp. Biol. (N.Y.) **45**, 272—277 (1940).
SELYE, H., DOSNE, C.: Influence of steroid hormones on metabolism of chlorides. Annales Acad. biol. (Univ. Chile) **3**, 93—103 (1940).
SHEMANO, I., GORDAN, G. S., EISENBERG, E.: The singularly brief anabolic activity of testololactone, a compound lacking androgenicity. Proc. Soc. exp. Biol. (N.Y.) **78**, 612—613 (1951).
SIREK, O. V., BEST, C. H.: The protein anabolic effect of testosterone propionate and its relationship to insulin. Endocrinology **52**, 390—395 (1953).
STAFFORD, R. O., BOWMAN, B. J., OLSON, K. J.: Influence of 19-nortestosterone cyclopentylpropionate on urinary nitrogen of castrate male rat. Proc. Soc. exp. Biol. (N.Y.) **86**, 322—326 (1954).
STUCKI, J. C., DUNCAN, G. W., LYSTER, S. C.: Anabolic and androgenic activities of 7α,17α-dimethyltestosterone (U-19763), a new anabolic steroid. In: Proc. First Intern. Cong. Hormonal Steroids, Vol. II, pp. 119—132. New York: Academic Press 1965.
STUCKI, J. C., FORBES, A. D., NORTHAM, J. I., CLARK, J. J.: An assay for anabolic steroids employing metabolic balance in the monkey: The anabolic activity of fluoxymesterone and its 11-keto analogue. Endocrinology **66**, 585—598 (1960).
SUOTO-MAIOR, M. C., FOGLIA, V. G.: Accion de la testosterone sobre la diabetes pancreatica. Rev. soc. argentina Biol. **20**, 79—87 (1944).
TAINTER, M. L., ARNOLD, A., BEYLER, A. L., BURNHAM, D. F., POTTS, G. O., ROTH, C. H.: Synthetic anabolic steroids in geriatrics. Medical Times **91**, 747—759 (1963b).
TAINTER, M. L., ARNOLD, A., BEYLER, A. L., DROBECK, H. P., POTTS, G. O., ROTH, C. H.: Anabolic steroid therapy in the management of cancer: A summary. J. Amer. Geriat. Soc. **11**, 1140—1151 (1963a).
TEDESCHI, G. G., MANGIANTINI, M. T.: Attivita di alcuni ormoni steroidi sulla regalozione del metabolismo creatinica. Boll. Soc. ital. Biol. sper. **33**, 1619—1623 (1957).
TERROINE, E. F., LELU, P., DEVRIENT, T., DE LA BERNARDIE, A.-M., CHEDRU, J.: Control of creatine and creatinine metabolism by sex hormones. II. Influence of castration on the excretion of creatine and creatinine by adult rats under various conditions of nitrogen metabolism. Trav. membres. soc. chim. biol. **23**, 1247—1272 (1941).
THORN, G. W., ENGEL, L. L.: The effect of sex hormones on the renal excretion of electrolytes. J. exp. Med. **68**, 299—312 (1938).
TOMARELLI, R. M., BERNHART, F. W.: Oral anabolic activity of (±)13β,17α-diethyl-17β-hydroxygon-4-en-3-one (norbolethone) in a nitrogen retention assay. Steroids **4**, 451—456 (1964).
USUELLI, F., PIANA, G., MAINARDI, B.: Testosterone e digiuno. Folia endocr. **2**, 31—37 (1949).
WALKER, J. B.: Metabolic control of creatine biosynthesis. I. Effect of dietary creatine. J. biol. Chem. **235**, 2357—2361 (1960).
WANG, E.: Clinical and experimental investigations on creatine metabolism. Acta med. scand. Suppl. **105**, 1—338 (1939).
WELSH, C. A., ROSENTHAL, A., DUNCAN, M. T., TAYLOR, H. C., JR.: The effects of testosterone propionate on renal function in the dog, as measured by the creatinine and diodrast clearance and diodrast Tm. Amer. J. Physiol. **137**, 338—347 (1942).
WIJNANS, M., DE GROOT, C. A.: Contribution to the knowledge of the protein anabolic effect of testosterone propionate. Acta physiol. pharmacol. neerl. **3**, 85—94 (1953).
WILLIAMSON, M., GULICK, A.: Influence of testosterone on distribution and excretion of creatine. Endocrinology **28**, 654—658 (1941).
WRIGHT, P. M., KOCHAKIAN, C. D.: Metabolic effects of testosterone propionate in experimental diabetes. Amer. J. Physiol. **173**, 217—222 (1953).
ZUNZ, E., LABARRE, J.: Contributions a l'étude des effets des hormones sexuelles sur la régulation de la glycémie. III. Action de l'acetate de testosterone. Arch. int. Physiol. **49**, 474—489 (1939).

# B. The Metabolic Effects of Anabolic Steroids in Man

R. L. LANDAU

Several clinical observations anticipated the experimental demonstration that male hormone is a growth promoting agent. In 1895 Sacchi described a 9-year-old boy who started growing very rapidly at the age of 5. He was 4 ft 9 in tall and weighed 97 lb, exceeding the height and weight of the average American 9-year-old by 6 in and 30 lb. He also exhibited a deep beard and masculine body hair. These were the clinical expressions of a 289-g interstitial cell tumor of the testis. Ten months after removal of the tumor the virilizing process had subsided and the physical state was reported to be essentially unchanged. In 1929 ROWLANDS and NICHOLSON, described another virilized 9-year-old boy who had grown unusually rapidly for 3 years. He was 5 ft tall and quite muscular. A large interstitial cell tumor was also removed in this case and rapid growth stopped after the surgery. A third case of this sort, a 5-year-old, was described by STEWART et al. (1936). This patient was 51 in tall, 5 in taller than the American average, and was beginning to virilize. The interstitial cell tumor in this case measured 1 cm in diameter. These case reports comprise the evidence available indicating the anabolic influence of testicular androgen when testosterone became available for clinical study. Since then there have, of course, been other reports.

Advancement of secondary sex characters in boys or virilization in girls due to the adrenogenital syndrome provides another natural demonstration of the capacity of the virilizing hormones to induce somatic growth. Precocious puberty due to hypothalamic disease is also associated with accelerated somatic growth. However, neither of these syndromes offers the uncomplicated experiment presented by the functioning interstitial cell tumor of the testis. The completion of Koch's postulates for this metabolic phenomenon required the demonstration that the most potent androgen isolated from testicular tissue, testosterone, would enhance protein anabolism.

## I. The Anabolic Effect of Testosterone

The comprehensive description of the metabolic influence of testosterone in man was carried out by ALLAN KENYON and his collaborators in a series of studies extending from 1938 to 1944. More or less simultaneous and confirming observations were also reported, but the studies and reports of the Chicago group were a model for physiologic investigation in man and remarkable for their perceptive exposition of the significance of experiments. KENYON clearly defined the potential of the male hormone as a stimulus of somatic growth and development, and the possibilities for analysis and employment of these somatotropic effects of androgen in clinical medicine. During the subsequent 35 years, we have been filling the interstices in his original outline.

Testosterone was identified as the most potent testicular androgen and probable male hormone in 1935 (DAVID et al.). Its synthesis was described the same year (BUTENANDT and HANISCH, 1935).

Androgen bioassays in experimental animals indicated that the actions of the steroid could be substantially enhanced by esterification with short-chain fatty acids such as acetate, formate, propionate, and butyrate. The propionate was shown to be about four times as effective as the free steroid in the bioassay system (MIESCHER et al., 1936). It was the metabolic effect of this ester of testosterone which has been studied in greatest detail. For many years testosterone propionate, administered intramuscularly in an oil medium, was the most frequently employed testosterone preparation for humans.

Fortunately, an appropriately effective dose of testosterone propionate was arrived at quite promptly. Initially studies with 2.5 mg/day of the free testosterone in sesame oil demonstrated not much more than a transient sexual arousal influence in eunuchoids. Five mg/day of testosterone propionate had more definite effects in eunuchoids. Twenty-five mg/day was selected as the dosage for most of the original metabolic studies because this quantity exhibited a distinct and prompt influence. This dosage, if continued indefinitely, is clearly in excess of the average daily output of testosterone by the normal testis, but it is not outrageously beyond the physiologic range.

Along with other investigators (HAMILTON, 1937; MCCULLOUGH, 1939; KEARNS, 1939; VEST and HOWARD, 1938), KENYON (1938) demonstrated that this dosage of the testicular androgen would virilize adult eunuchoids or castrates more promptly than occurs in normal puberty. Penile and prostatic growth, the development and growth of pubic, axillary, and beard hair, deepening of the voice due to laryngeal enlargement, sharply increased sexual drive, and frequent penile erections were noted by all observers. Slight breast growth occurred in several of the subjects. All four of the adult eunuchoids studied by KENYON experienced a marked increase in body weight; two developed pitting edema, and two experienced a notable increase in appetite while eating ad lib. The weight gains (Fig. 1) amounted to as much as 8.5 kg and were at least partially sustained for months despite some diminution of testosterone dosage. Clearly such weight gains could not be accounted for on the basis of linear growth or just the accessory sex apparatus, beard, and larynx.

The series of metabolic balance studies by KENYON, KNOWLTON, and their associates (1938, 1940, 1944) which followed the clinical observations demonstrated that testosterone induced a prompt decline in urinary nitrogen, urea, inorganic phosphorus, creatine, potassium, sodium, and chloride (Figs. 2 and 3). Similar observations had just been made by KOCHAKIAN with respect to nitrogen and urea (1937) and by THORN and HARROP (1937) regarding sodium and chloride in dogs. The reduced urea output without accompanying azotemia in castrate dogs was correctly interpreted by KOCHAKIAN as reflecting protein deposition, although it was initially assumed that the protein was deposited solely in the growing accessory sex tissues. In the castrate dog, nitrogen retention ranged from about 20–60 mg/kg body weight. On a weight basis the nitrogen retention induced in eunuchoids corresponded closely. Translated into protein the nitrogen retention observed in the eunuchoids treated for 8–15 days

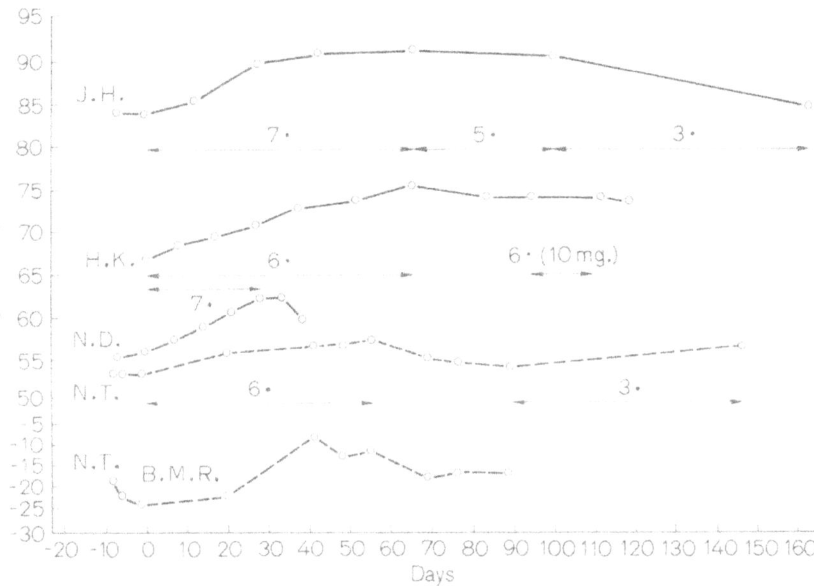


Fig. 1. Effect of testosterone propionate on body weight of 4 eunuchoids and on BMR of one, diet and activity being uncontrolled. Numbers followed by X indicate number of times 25 mg of testosterone propionate given each week. (From KENYON, 1938)

totaled from 108.8–394.6 g. Assuming a normal prostate weight of about 20 g, it quickly became obvious that protein must have been deposited in a number of somatic tissues. Thus the impression gained from clinical observations in treating hypogonadal men was confirmed and extended in a more explicit and quantitative manner.

The assumption that urinary nitrogen retention reflects protein deposition was strongly supported by other data. Almost all the drop in urinary nitrogen consisted of urea, the end product of protein catabolism. The blood urea concentration did not rise, and fecal nitrogen was unaffected. Other constituents of tissue protein were also retained under the influence of testosterone. Thus urinary inorganic phosphate excretion declined in rough parallel with that of nitrogen without there being any increase in plasma phosphate concentration or fecal phosphate excretion. Urinary potassium and sulfate excretion were also suppressed. In the usual short studies, none of these substances was retained in a stoichiometric relationship with its content in tissue protein, but the general correspondence between the patterns of their retention and that of total nitrogen or urea is quite persuasive. Urinary creatinine, the indicator of completeness of 24-h urine collection, was unaffected. Creatinuria whether occurring spontaneously or induced by high meat protein in the diet or creatine hydrate feeding was promptly reduced by testosterone administration. The critical association of creatine with energy metabolism in striated muscle suggests that its retention, induced by testosterone, specifically reflects anabolism within voluntary muscle.

The studies of others, who have confirmed the androgen induced anabolic process, have on occasion indicated a decline in the serum potassium concentration

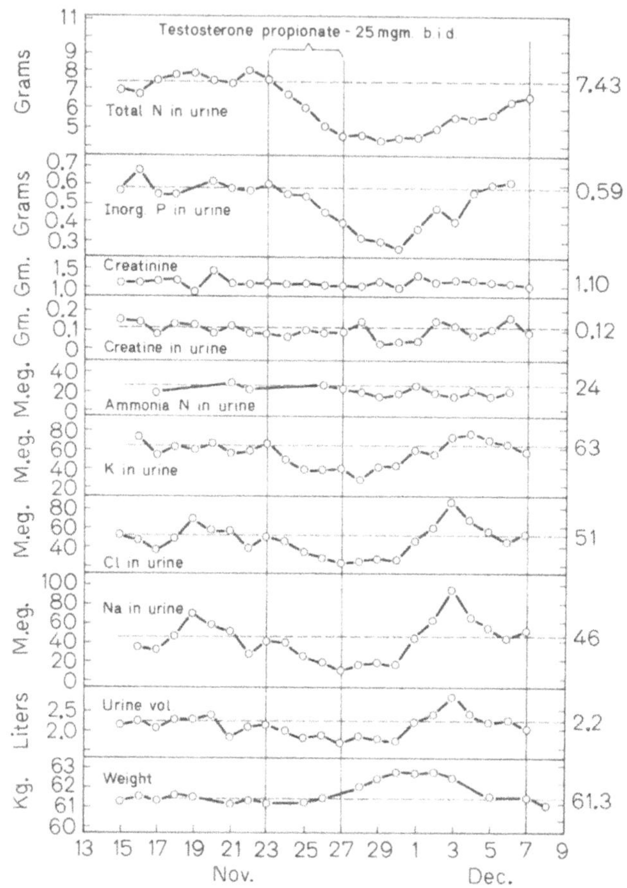


Fig. 2. Effect of intramuscular injections of testosterone propionate on excretion of several urinary constituents and body weight in eunuchoid on constant adequate diet. (From KENYON et al., 1940)

(BRAY et al., 1971; BUTLER et al., 1942) and the serum inorganic phosphate concentration in prepubertal children (KEITEL, 1957). Both are intracellular tissue constituents which are excreted in lesser quantities during protein anabolism (see above). This lowering of serum phosphate and potassium concentration, even though not a uniform development, strengthens the evidence that androgens induce tissue anabolism.

Figures 2 and 3 are typical of the results obtained in metabolic experiments. Nitrogen retention begins on day 1–3 reaching a maximum in about 5–8 days. This reduced level of nitrogen excretion may then be maintained for at least a month or two depending upon the subject's nutritional status and diet (see below). Obviously, however, such anabolism cannot continue indefinitely, and compensatory mechanisms are brought into play as they must be in a normal man whose testes continue to secrete testosterone at a considerable rate well after he has matured. Typically, when testosterone in the eunuchoid is discontinued, urinary nitrogen rises slowly to

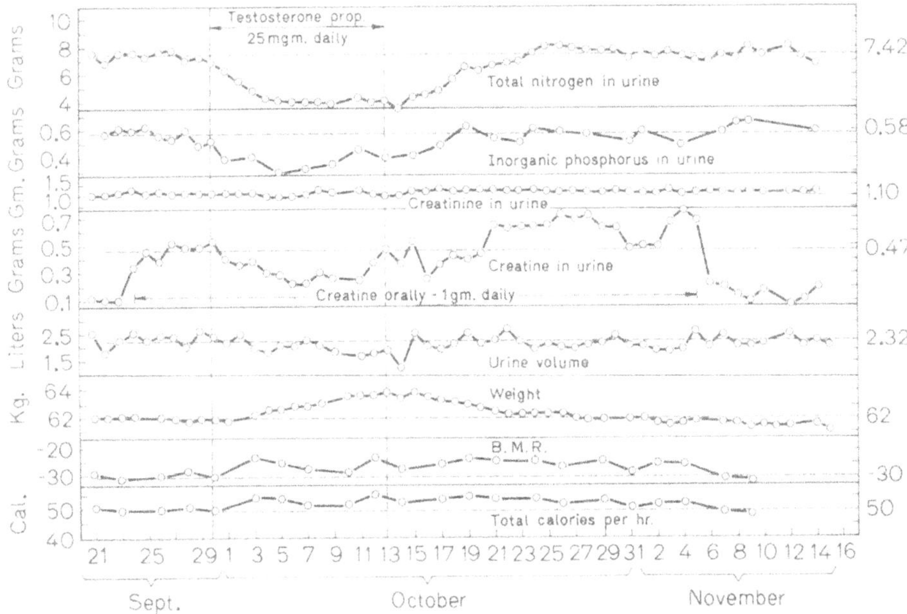


Fig. 3. Effect of testosterone propionate given intramuscularly on several urinary constituents, body weight, and BMR in eunuchoid. (From KENYON et al., 1940)

the pretreatment baseline over a period of 6–12 days. It is important to appreciate that despite the absence of testicular androgen in these hypogonadal men the protein appears to be retained more or less permanently. More likely it was dissipated, but at a rate too slow to detect in a short term by balance methodology, perhaps in part because the amount of protein synthesized under the influence of a short course of testosterone was quite limited.

In one study the total daily caloric expenditure, as determined indirectly by insensible weight loss measurement, was shown to be uninfluenced by testosterone administration. This experiment (Fig. 4) not only proved that the anabolism could not be explained by diminished heat production, it provided also a clear illustration of the several components in the metabolic processes induced by the addition of androgen to this subject. The patient was consuming a diet mixture anticipated to result in a slight gain in weight. This expectation was exceeded substantially when testosterone was given. The excess over the calculated gain was accounted for in part by the protein storage and resultant shift in the metabolic mixture of protein to fat, assuming that 3 g of water is deposited with each gram of protein. The remainder of the induced gain in weight was the result of the retention of sodium chloride and water held in extracellular intravascular and extravascular spaces.

A general comment on long-term metabolic balance observations is desirable. Shifts of up to 1 g per day in the baseline output of urinary nitrogen, while subjects are on constant intake and activity, occur from time to time in most subjects and are readily detected when short pools (24 or 48 h) are measured. The nitrogen balance thus moves about the null point when equilibrium is being maintained; at one time or another balance may be slightly negative or slightly positive. The net effect over

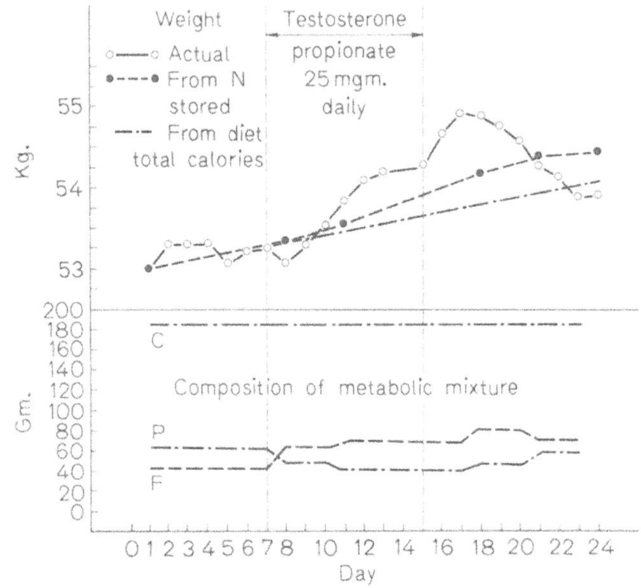


Fig. 4. Comparison of actual weight gain in eunuchoid with that expected from nitrogen stored and with that predicted from excess of diet calories over total calories expended. Curve of weight gain from nitrogen stored is calculated with the assumption that 3 g of water is deposited with each g of protein stored. Total calories expended were derived from insensible loss of water (calculated from insensible weight loss, urinary nitrogen, and carbohydrate of diet). The third curve represents gain in body weight which might be expected from slight disproportion between diet calories and total calories, had not testosterone been given, and had the condition of control period been maintained throughout. Lower half of figure illustrates shift in metabolic mixture resulting from administration of testosterone. It is assumed that carbohydrate oxidized equals carbohydrate ingested. (From KENYON et al., 1938)

the long pull in healthy adults is to maintain a steady state. The manner by which these shifts about an equilibrium level are accomplished is obscure. However, the endocrine system is almost certainly involved. For example, the loss of testosterone, as by discontinuance of the hormone in a chronically treated hypogonadal man would seem to be a factor tending to diminish the protein content of the body. The results noted above indicate that in this particular circumstance the retained protein is not dissipated rapidly, but may be lost undetectably in the manner suggested.

A positive calcium balance was not observed when testosterone propionate was administered for a short time. However, positive calcium balances have been observed in long-term observations (ALBRIGHT et al., 1941); bone growth and maturation are stimulated by the male hormone (see Chapters on calcium metabolism and pediatrics). It may be that significant protein deposition must occur in bone before significant calcium retention can be observed.

The anabolic influence of testosterone is qualitatively the same in eunuchoids, normal young and aged men, and women (KENYON et al., 1940, 1942). The intensity of the effect is usually significantly greater in eunuchoids than in persons of either sex with functioning gonads. Based on body weight, men with an androgen deficiency alone retain at a maximal level an average of about 0.06 mg/kg of nitrogen daily, while normal men and women retain about half this much (comparisons while on maintenance diets). In prepubertal dwarfs the rate of nitrogen retention was almost double that in the eunuchoids (WILKINS and FLEISCHMANN, 1942). It appears that in normal men the discontinuance of testosterone is frequently followed by a distinct rebound loss of previously retained nitrogen. The induction of nitrogen retention in men with normal plasma testosterone concentrations must be interpreted as indicating that somatic tissues are capable of greater androgen induced anabolism despite the fact that tissue protein is being physiologically maintained by secreted testosterone. Looked at in this manner, it is no surprise that the intensity of protein storage induced in short experiments is less in normal men than in those not continuously subjected to the effects of a potent androgen. Although there are no comparative studies on the duration of effects in eunuchoids and normals, it seems reasonable to presume that intrinsic compensation would terminate anabolism sooner in normal men.

Another factor to be considered as being partially responsible for the more modest anabolism induced in normal men is the expectation that exogenous testosterone will suppress testosterone secretion by Leydig cells. LH secretion is suppressed by testosterone administered by intravenous infusion or intramuscularly in amounts which elevated the plasma testosterone levels well above control values. LH suppression averaged about 40%, probably levels which supported some Leydig cell function (LEE et al., 1972; SHERINS and LORIAUX, 1973). Since these studies and others indicate that 25 mg per day of testosterone would elevate the plasma level of normal men much higher than control concentrations, endogenous testosterone suppression cannot be regarded as a significant modifying factor. Careful evaluation of the possibility would require careful studies with 5–10 mg of testosterone propionate daily, measuring nitrogen balance and hormone titers. Testosterone analogues can be expected to have some qualitatively different actions, and hence results with them cannot be readily extrapolated to the native hormone. It has been shown, for example, that fluoxymesterone will suppress testosterone output promptly and quite effectively without influencing the circulating gonadotropin concentrations (JONES et al., 1974).

The situation in women is not strictly analogous. Estradiol is also generally anabolic (see below), but less so than androgen. It is not known whether the process stimulated in somatic tissues is identical with that induced by testosterone. Nonetheless, it is fair to assume that in the normal reproductive age woman responsive tissues are subjected to the mild anabolic stimulus of estradiol when testosterone is administered. In addition, progesterone which is secreted only during the postovulatory phase is distinctly catabolic, a process which would diminish the anabolic effect of added androgen (LANDAU and LUGIBIHL, 1961).

From the beginning, KENYON expressed concern that the somatotropic influence of testosterone could have been mediated by another hormone or perhaps required the presence of other internal secretions for full expression. The full effects were elicited in two patients with Addison's disease treated only with desoxycorticosterone and added salt (KENYON et al., 1943), and this was confirmed by TALBOT et al. (1943). Thus, cortisol is not required, even permissively, and the effect is not mediated by the adrenal cortex. Ovaries and testes are obviously not required. A full effect has also been noted in treated myxedema (CRISPELL et al., 1957). One of the first four hypogonadal men studied by KENYON was apparently severely hypopituitary, and the response in this man was substantially less than that in the other hypoandrogenic subjects. Furthermore, in this subject some of the retained nitrogen was lost during the immediate 10 days to 2 weeks following testosterone therapy in contradistinction to the observations in the patients with only a testicular deficit. This diminished effect has been confirmed in subsequent observations made in growth hormone deficient children. It appears that the maximal response to testosterone under any particular conditions requires the presence of pituitary secretions.

ZACHMANN and PRADER (1970) determined that a standardized dosage of a mixture of long-lasting testosterone esters administered to boys with sexual immaturity induced a maximum increase in growth velocity during the first 6 months of treatment. This was analogous to "catch-up" growth. Although a sharp increase in growth was also induced in boys with a growth hormone deficiency, it was less marked than in boys whose pituitaries secreted growth hormone. The development of secondary sexual characteristics such as pubic, axillary, and beard hair, penile growth, and voice deepening was also slowed in the growth hormone deficient patients. It thus appears that in man pituitary principles exert some sort of permissive action with respect to the full response of the male hormone. Prolactin may also be important in this regard as illustrated by the studies in the hypophysectomized rat in which prolactin is required in addition to testosterone in order for prostatic growth to be fully restored (GRAYHACK et al., 1955).

The basal metabolic rate is characteristically somewhat lower than normal in eunuchoids, whether the androgen deficiency is due to primary testicular failure or is secondary to a gonadotropin deficiency. There is no evidence that this moderate depression of basal heat production is related in any way to a thyroid hormone deficiency. Carefully determined rates in hypogonadal men usually ranged from −25 to −15 (Mayo Foundation standard), and the treatment with full replacement quantities of testosterone propionate resulted in a slow progressive rise in the metabolic rate totaling about 10% (SANDIFORD et al., 1941). The mechanism by which heat production is increased remains obscure, but it is likely that the hormone acts directly on responsive tissues to increase oxygen consumption. Testosterone is not one of the steroidal compounds which increases the basal body temperature (MAGALLON and MASTERS, 1950). Thyroid function is not affected by androgen administration. However, the plasma concentration of thyroxine binding globulin is significantly lowered by androgen. This explains the lowered total plasma thyroxine concentration which has been observed (FEDERMAN et al., 1958).

The proof that endogenous testosterone of testicular origin could stimulate substantial anabolism physiologically required the demonstration that stimulation of Leydig cells would induce a metabolic response like that of the injected hormone. This was accomplished first in an undersized prepubertal boy who was treated with chorionic gonadotropin, a preparation which possesses mostly luteinizing or interstitial cell stimulating activity (KENYON et al., 1942). A dosage of 750–1500 U of highly purified chorionic gonadotropin induced nitrogen retention of 65 mg/kg/day and parallel suppression of urinary inorganic phosphates, sulphate, potassium, and creatine. A small amount of sodium was also retained. A 5 mg daily dosage of testosterone propionate induced a qualitatively identical process but the magnitude of the effect was slightly less. It is of interest that both responses were brought about by amounts of testosterone insufficient to cause an elevation of urinary 17-ketosteroids or androgen. This attests to the lack of both specificity and sensitivity of these procedures as indices of testicular androgen secretion (ENGSTROM et al., 1951); it is

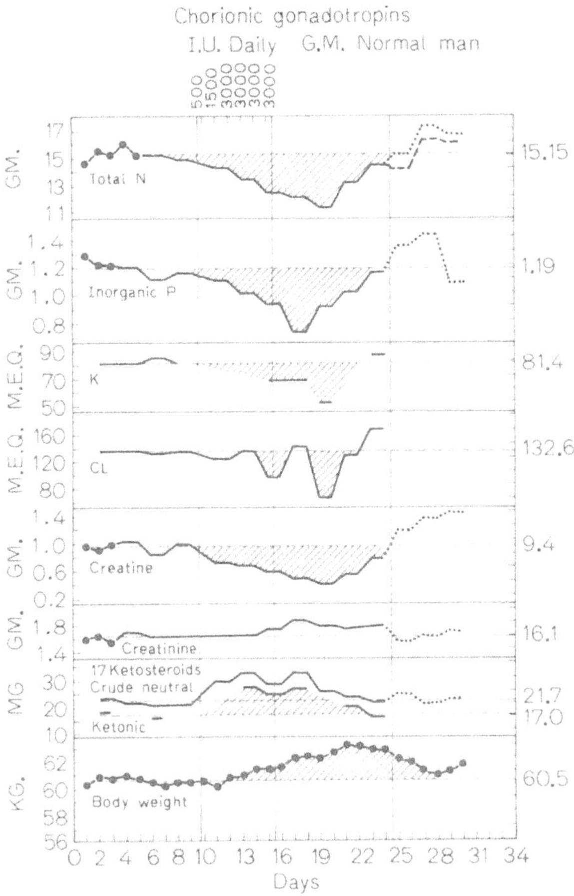


Fig. 5. Effects of chorionic gonadotropin on several urinary constituents and on body weight in a normal 23-year-old man. (From LANDAU et al., 1950)

now known that this dosage of chorionic gonadotropin will greatly increase plasma testosterone.

In normal virilized men in the third decade a standardized dosage of chorionic gonadotropin (500, 1500, and 3000 X 4 U/day) induced an anabolic process at least as great as that resulting from 25 mg/day of testosterone propionate (Fig. 5) (LANDAU et al., 1950). It was qualitatively identical to that resulting from testosterone propionate. Urinary 17-ketosteroids rose to about 8 mg/day, an amount approximating the urinary 17-ketosteroids derived from 20 mg of testosterone. Urinary inorganic phosphate and potassium dropped slightly; salt was retained. In a eunuchoid who responded satisfactorily to just 5 mg of testosterone propionate per day, the standard course of chorionic gonadotropin had no distinct anabolic influence demonstrating that this interstitial cell stimulating preparation had no direct influence and acted only through the testis. Subsequent studies in women during their reproductive years (LANDAU and LUGIBIHL, 1961) failed to disclose any anabolic influence of large amounts of chorionic gonadotropin further supporting these points.

The ease by which the adult testis can be stimulated to secrete more testosterone raises the question of whether alterations in the rate of secretion of testosterone may be invoked to actively regulate protein metabolism on a day to day basis. Until recently the technology for studying this possibility was not available, and none of the early reports are strictly applicable. One can only speculate at this time. Such an active process would require a regulating sensor, presumably in the hypothalamus, which would detect the need for more or less protein anabolism. If there were such a center, stimulated testosterone secretion via the gonadotropin releasing hormone and pituitary LH secretion is a clear possibility. The dissipation of tissue protein could conceivably be enhanced by the reversal of this system. The minimal nitrogen loss induced by estradiol benzoate and stilbesterol administration in men with poliomyelitis and a normal man (WHEDON and SHORR, 1957; KENYON, A. T. and LANDAU, R. L., unpublished study) may be explained by the resulting suppression of testosterone secretion (SHERINS and LORIAUX, 1973) and thus may provide an example of this phenomenon. Such testosterone suppression, if brought about by physical stress or serious injury could facilitate the catabolic phase which follows (see below). Whatever the participation of the testis in the regulation of protein metabolism, it is not vital for survival. Other endocrine and metabolic processes handle this quite well in children, women patients with the feminizing testis syndrome, and castrates. However, if the testis participates, it should be assumed that its milieu is the masculine muscular physique, and its minimal role is to sustain that status.

## II. Salt-Retaining Action of Testosterone

The demonstration that testosterone and estradiol, like desoxycorticosterone, were salt-retaining steroids led to an initial interpretation now known to be incorrect. At that time the salt-retaining action of adrenocorticoids had just been shown to be a life-preserving influence of the hormones. It was suggested that the structurally similar sex hormone possessed a degree of adrenocortical-like activity (THORN and HARROP, 1937). The induction of sodium retention in desoxycorticosterone-treated addisonians proved that neither the adrenal cortex or cortisol were required for this influence of testosterone (KENYON et al., 1943; TALBOT et al., 1943). The subsequent demonstration that mineralocorticoid activity involves the stimulated exchange of sodium ions for hydrogen and potassium ions in the distal renal tubule and that this results in a kaluresis accompanying sodium retention (GIEBISH, 1969) clearly delineates this process from the salt-retaining effect of testosterone. Potassium excretion is not increased by testosterone. As previously noted, urinary potassium usually declines in parallel with nitrogen. A mineralocorticoid type of effect would result in a prompt rise in urinary potassium, and this would not be masked by the slowly progressive retention of potassium which accompanies anabolism. The means by which sodium and chloride retention are induced by testosterone is not known.

## III. Anabolic Effects of Estrogen

In planning these reviews it was generally assumed that anabolic steroids referred to the male hormone, testosterone, its biologically active metabolites, adrenocortically derived androgens, and synthetic analogues of testosterone which are strongly anabolic and in addition possess androgenic and sometimes progestational activity (see below). However, in addition to a selective growth stimulating effect on the uterus and breasts, estrogens are, at least in man and dogs, a somatic growth-promoting agent. The growth-promoting estrogenic steroids cannot be neglected just because they are female hormones, even though they are not as effective as the male variety. Endocrinologists have long felt rather comfortable with the assumption that the superior muscular development of men as compared with women could be explained on the basis of the somatotropic effect of androgen. This interpretation is certainly supported by the physical transition of eunuchoids with testosterone treatment, the muscular development which accompanies puberty in boys and the growth promoting influence of testosterone in dwarfs. What evokes the pubertal growth spurt in girls? Could some of the somatic differences between the two sexes be genetically determined?

In evaluating the metabolic effects of the administered estradiol (usually given intramuscularly as the benzoate ester), it should be noted that the physiologic replacement dosage of estradiol in women is less than 1/10 of the testosterone secretion rate of normal men. The Chicago group (KNOWLTON et al., 1942) reported careful comparisons between the response to 5 mg of estradiol benzoate/day and 5 mg testosterone propionate in several subjects (Fig. 6). Nitrogen, inorganic phosphorus, sulfate, and sodium retention were induced by the estrogen in eunuchoid men,

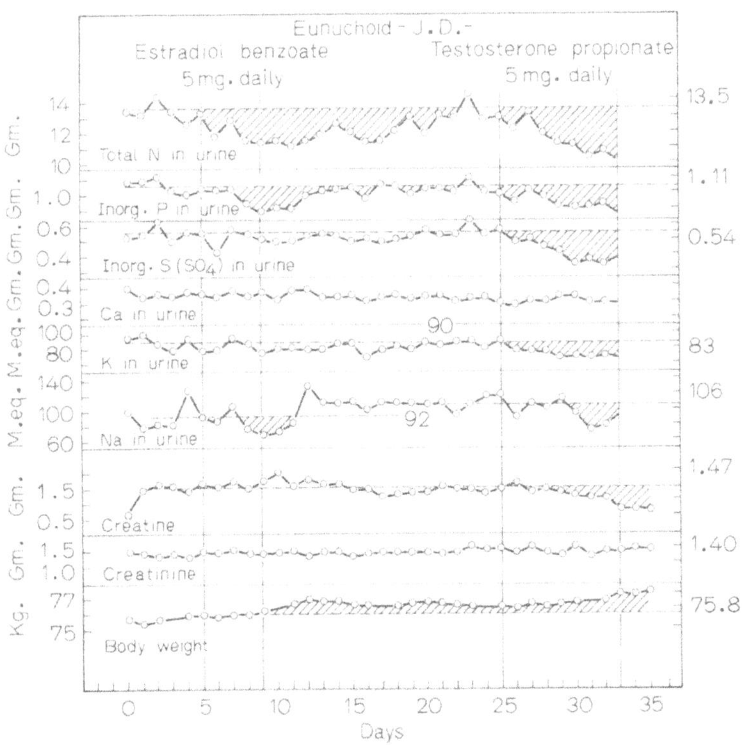


Fig. 6. Effects of estradiol benzoate and testosterone propionate on several urinary constituents in eunuchoid. 1.32 g creatine hydrate taken daily. (From KNOWLTON et al., 1942)

hypogonadal and normal female subjects. Although the effects were truly significant, it was apparent, by employing identical comparative dosages, that estradiol was less anabolic than testosterone. While this dosage of male hormone approached the daily maintenance level, 5 mg/day of estradiol is well beyond physiologic range except in the pregnant state. By employing this dosage of estrogen the weak to modest protein anabolic potential of the hormone was exposed. Subsequent investigations have indicated that dosages of estradiol benzoate as small as 0.5 mg per day may induce a modest anabolic response in women without ovarian function, pointing to the likelihood that this influence has physiologic significance (LANDAU et al., 1957).

The relative mildness of the anabolic force of estradiol is also indicated by several other studies. No nitrogen retention was induced by estradiol benzoate in a virilized woman with the adrenogenital syndrome (KNOWLTON et al., 1942), a patient with Cushing's syndrome (ALBRIGHT et al., 1941), a man with osteoporosis (ALBRIGHT and REIFENSTEIN, 1948), and two normal young men recovering from acute anterior poliomyelitis (WHEDON and SHORR, 1957). As a matter of fact, the estrogen was actually slightly catabolic in the last two young men, as previously noted. The lack of effect in Cushing's syndrome suggests that the female hormone was unable to overcome the catabolic influence of a glucocorticoid excess as does testosterone (see below). In the face of virilizing levels of testosterone it may be assumed that the somatic tissues stimulatable by estradiol were already strongly affected by the circulating androgen.

There are, in addition, qualitative as well as quantitative differences between the effects of estrogen and androgen. Potassium retention was not observed in any of the estradiol studies and in one normal woman potassium excretion was actually increased slightly through the 8 day period of estrogen therapy. In addition, whereas creatine retention has been uniformly observed as a very sensitive response to testosterone (when creatinuria was present or induced by creatine feeding), creatine excretion in the urine was either enhanced or unaffected by short courses of estradiol. This phenomenon provides at least a partial explanation for the fact that creatinuria of up to 0.5 g per day may be seen in normal women. Whether this creatinuria is the result of increased synthesis in the liver (see below), and/or a failure of muscle to respond as it does to testosterone, remains to be determined. The fact that suppression of creatinuria has not been observed could be a reflection of the fact that although estrogen is anabolic with respect to some somatic tissues, voluntary muscle is not responsive. This concept would explain the more limited anabolic force of estradiol as well as the fact that although estrogen is growth-promoting in girls and anabolic with respect to bone (ALBRIGHT and REIFENSTEIN, 1948), it obviously fails, to result in muscular growth, as does testosterone. There is no reason to assume that a somatotropic agent must affect all tissues in the same manner, i.e., testosterone does not stimulate albumin synthesis and apparently suppresses the synthesis of thyroxine binding globulin by the liver (GROSSMAN and YALLOW, 1965; FEDERMAN et al., 1958).

## IV. Effect of Estrogen on Salt Metabolism

Sodium and chloride retention were evident in the first studies in man employing 5 mg per day of estradiol benzoate (KNOWLTON et al., 1942), and this was confirmed using 8 mg per day in estrogen deficient women (DIGNAM et al., 1956). A decline in urinary sodium excretion has also been observed in estrogen deficient women using doses of estradiol within the physiologic range (LANDAU et al., 1957). Retention was made particularly evident in these cases by the sharp rebound loss of sodium several days after estrogen discontinuance. PREEDY and AITKEN (1956a, b, c) studied the effect of 10 mg per day in normal women (ages not recorded), some patients with hepatic disease, and others with congestive heart failure. The results are in general agreement with those of the other investigators, although they are not always recorded as being so (CHRISTY and SHAVER, 1974). The point of difference is only the transient and relatively slight nature of the sodium and chloride retention noted in the normal women. This might well be explained by the very large dosage and the complexity of the several interacting targets of estrogen action discussed briefly below. Salt retention was more distinct in those with hepatic and cardiac disease.

Like androgen, the mechanism of salt-retaining action is obscure but is probably at least partially a direct renal influence. The lack of the sort of rise in urinary potassium output, observed when mineralocorticoid acts (GIEBISH, 1969), indicates that increased aldosterone secretion is not the major factor. However, the fact that estrogen increases plasma renin substrate concentration and thus transiently the plasma renin activity and the plasma angiotensin level (RIEGER et al., 1972) indicates that to a minor degree the renin-angiotensin-aldosterone system may play a supporting role in estrogen-induced salt retention. Thus a slight increase in aldosterone secretion effected in this manner would be expected to forestall the decline in aldosterone secretion anticipated as a part of the usual homeostatic mechanism for plasma volume control when salt and water are retained. Two other possibilities must also be mentioned in explanation of the salt retention. Some of the suppression of salt and water excretion may be extrarenal in origin, resulting from estrogen modified connective tissue. Such tissues may become more hydrophilic in a manner analogous to the estrogen-induced change in the sex skin of some monkeys (AYKROYD and ZUCKERMAN, 1938). Estradiol also stimulates prolactin secretion (FRANTZ et al., 1972) and this hormone exerts a distinct salt-retaining influence (HORROBIN et al., 1971).

The salt-retaining action of estrogen and the interaction of the hormone with prolactin secretion and the renin-angiotensin-aldosterone system are of importance in the regulation of salt and water metabolism during the menstrual cycle and in the pregnant state. The physiologic regulation of electrolyte and fluid metabolism in women during their reproductive years and the disturbances in the homeostatic mechanisms which lead to disease states can be the subject of several reviews and are not proper subjects for detailed exposition here. It should be sufficient to mention that contraceptive pill hypertension, periodic and premenstrual edema, excessive fluid retention, and toxemia of pregnancy are relevant.

## V. The Use of Androgen in Debilitated States

KENYON suggested, as the basis for a military research proposal in 1942, that the administration of testosterone propionate might be an effective adjunct for the treatment of some war casualties. It seemed reasonable for him to assume that in the recuperation from debilitated states in which body protein had been depleted, a potent anabolic agent might hasten protein restitution and recovery of the patient to full action. This seemingly simple and important proposal has been neither fully proven nor disproven after 30 years of exposure. In part this is because the military and social pressure for an answer has diminished, but probably more importantly because the problem demands such a large volume of carefully controlled and difficult research, taking into consideration such variables as diet, types of injury, and diseases. The simple question has no straightforward uncomplicated answer. Nonetheless, the idea persists. It has stimulated an immense amount of research including the development of an array of steroidal compounds with potential therapeutic advantages—oral effectiveness and relatively greater anabolic than androgenic activity.

Before reviewing these studies designed to evaluate the clinical efficacy of anabolic agents, it would be useful to briefly survey the essentials of the influence of injury and the effect of dietary intake on protein metabolism. The studies of CUTHBERTSON (1932, 1942), confirmed by a number of other workers, clearly demonstrated that the normal healthy body responds to acute significant physical stress or injury by a sharp and quite intense catabolic phase which may last for 20–60 days and is often associated with a poor appetite. This negative nitrogen balance was noted to coincide with the stress-evoked adrenocortical activation, and indeed is qualitatively very similar to the protein catabolism induced in man by an excess of glucocorticoids alone (FORSHAM et al., 1948). This similarity and the coincident outpouring of large amounts of cortisol led many observers to postulate that the accelerated protein catabolism of severe physical stress was a manifestation of hyperadrenocorticism (ALBRIGHT, 1942). The studies of INGLE and BAKER (1953), however, led to the presently accepted view that increased protein catabolism induced by stress requires the presence of cortisol but is probably not mediated directly by this hormone.

The loss of 60 g or more of protein per day in an injured or seriously ill patient is repugnant to some physicians concerned with the maintenance of nutrition through a serious ailment. However, as a normal phenomenon in well-nourished individuals subjected to acute injury this loss of what has sometimes been termed "deposit protein" should probably be regarded as serving a useful purpose. "Deposit protein" does not imply naively that there exists within some cells a circumscribed locus of protein which is not functionally active or in equilibrium with circulating amino acids. The fact is that in a healthy state some of the protein in cells from a number of organs and tissues is very readily mobilized as amino acids for gluconeogenic and other needs of the body as a whole. The protein must thus be regarded as a reserve which can be utilized without apparently impairing cellular function. When this portion of the body protein has been dissipated, conservatory mechanisms are set into play. At this stage, despite continuing illness or debility, nitrogen equilibrium can be readily achieved if caloric and protein intake is maintained at an adequate level. A recurrence of the original injury or a new stress occurring at this time does not evoke a catabolic process, and this may relate to the weakened resistance to stress of protein-depleted patients (HOWARD, 1944). Recovery from a severe or debilitating ailment or injury requires protein restitution. This recovery process readily evolves if the person consumes a caloric surfeit and adequate protein.

The subtlety and precision with which the body reconciles energy expenditure, tissue requirements for growth, and restitution after loss with protein and caloric intake is remarkable indeed. This incredibly complex regulatory system which doubtless involves the endocrine system in a major way is the one upon which it has been proposed that the anabolic influence of administered androgen can on occasion be added with benefit. The detailed questions which have been asked are the following[1].

1. Can the catabolic response to injury be significantly modified by androgen, and if so, is there an advantage to be gained by its administration?

2. In debilitated protein-depleted states, is there a place for treatment with anabolic androgen?

3. When the depleted patient is able to consume enough protein and calories to induce protein sparing, or anabolism, is the anabolic effect of androgen additive?

4. Can androgen counteract the catabolic influence of other hormones in excess, i.e., glucocorticoid and thyroid?

5. Can the appetite-stimulating effect of androgen, first noted in the treatment of eunuchoids, be employed clinically?

6. Are there other applications of this anabolic effect?

In studies carried out in attempts to answer these questions, investigators have employed testosterone esters and synthetic steroids with formulae similar to that of testosterone. The chemical maneuvers to provide pharmacologic advantages such as protracted action, ease of administration, and accentuated anabolicity with relatively less androgenicity have been summarized elsewhere in this volume. The unusual metabolic effects of some of these synthetic steroids in man will be summarized below. However, from the viewpoint of the application of anabolic steroids in disease, the results will be organized here on the basis of the metabolic and clinical circumstances in which any anabolic steroid has been tested. The assumptions made are that the anabolic influence of all of the androgenic compounds is exerted by the same biologic mechanism and that in regard to this particular response they differ from one another in a quantitative sense only.

The acute protein wasting which follows a fracture of the lower extremity without much additional injury may be ameliorated or perhaps even abolished by the feeding of a high caloric, high protein diet (PEDEN et al., 1960; WEBB et al., 1960; KENNEDY et al., 1962). The simultaneous administration of an androgenic steroid provides additional anabolic force. It does thus appear that when good nutrition can be maintained following such injuries, the anabolic steroids may help to restore nitrogen

---

[1] In this portion of the review many more specific references could be made to the JOSIAH MACY JR. Conferences on the Metabolic Aspects of Convalesence. Instead the interested reader is referred to the entire series.

Dr. FRANK FREMONT-SMITH initiated the Macy Foundation Conference Series when he called this one together in September 1942 as an aid to World War II medical research. Dr. SMITH, and the Conference Chairman, FULLER ALBRIGHT, brought together a remarkably able and articulate group of investigators who discussed freely and stimulated each other mightily. The proceedings, ably edited by E. C. REIFENSTEIN, JR. convey their spirit and contain details of a number of experiments not published elsewhere.

The principal thread of the 17 Conferences which continued through 1948 was maintained by the regular members which included: W. D. ARMSTRONG, JOHN S. L. BROWNE, KONRAD DOBRINER, JOHN E. HOWARD, ALLAN T. KENYON, CHARLES KOCHAKIAN, EPHRAIM SHORR, and LAWSON WILKENS. A number of equally distinguished "visitors" were invited on occasion to present relevant research and join the discussions.

equilibrium. What is not certain is whether in such injuries the minimization of protein loss is of any clinical value. In other patients in whom the injury or illness is much more severe and good nutrition cannot be maintained because the patient cannot eat, the combined nitrogen loss from stress and from undernutrition may be immense. In this circumstance, the administration of androgen may lessen somewhat the daily loss of tissue protein but whether this is significant in the overall economy, and influential in recovery has not been established.

The original suggestion that administered androgen would increase anabolism during recovery from injuries and debilitating illnesses has been confirmed by a number of observers. Patients recovering from gunshot wounds, pyogenic infections, major compound and simple fractures, burns, common bile duct obstruction, and uncontrolled diabetes mellitus are representative of the array of disease states which have been studied. When the conditions for study have been firmly met, testosterone or one of the steroids related to it has induced nitrogen retention ranging from 1–5 g per day with great uniformity (FORSYTH, 1954). The results indicate that the nutritional status of the subject at the time the study begins and the diet he is given or able to consume are crucial factors which condition the magnitude of response. The food intake is obviously dependent in part on whether the disease is continuing to smolder or whether the patient is in a recovery phase of his illness.

Persons with substantial inanition who require only time and food for full recovery possess a remarkable "spontaneous" anabolic stimulus, and often an equally remarkable appetite. When provided with a caloric surfeit and adequate protein they may easily retain as much as 5 g of nitrogen per day. Surprisingly, testosterone will impose additional anabolism in these cases (FORSYTH, 1954). On the other hand, when *well-nourished* chronically ill patients are made anabolic by a caloric excess regimen, testosterone induced little or no additional nitrogen retention (FORSYTH and PLOUGH, 1955) (Fig. 7). The underlying difference in these two situations has not been studied. It appears likely that the degree of protein depletion at the time the investigation was carried out is the key factor, i.e., if protein is needed, testosterone will be more effective. From the cold, practical point of view, it is important to question whether anabolic steroid therapy is indicated as an aid in the recovery of such patients. Presumably in a situation in which time is crucial the attempt should be made. It could then be determined whether the key factor in total clinical recovery is protein restitution.

The medical need for a tissue protein conserving agent is certainly clear in circumstances in which food intake must be limited and/or a disease state which may be protein wasting in and of itself continues for a long time. In many such individuals, prescribed or voluntary food restriction impairs the anabolic activity of administered androgen. Anabolism has been demonstrated, however, during strict weight reduction regimens in obese persons (KINSELL et al., 1951; BRAY et al., 1971; CRISPELL et al., 1957). It is more surprising that minimal protein sparing has been observed during a six day fast in a normal man and in fasts of the same duration supplemented only with 50–300 g of carbohydrate daily (BUTLER et al., 1945) (Fig. 8). In patients with acute renal failure given no food or 100 g of carbohydrate daily or 100 g of carbohydrate and very limited protein, a retardation in the rate of rise of the blood urea nitrogen has been observed; and this may well be clinically useful in these acute problems (BLAGG et al., 1962). In one study in a normal man the nitrogen

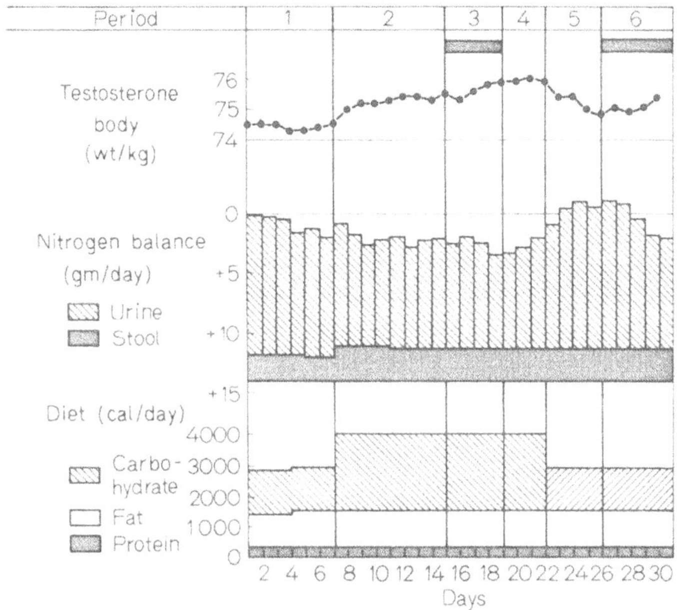


Fig. 7. Effect of testosterone propionate in 24-year-old well-nourished patient with rheumatoid arthritis. Note negligible anabolic effect when androgen administered during diet-induced protein sparing. (From FORSYTH and PLOUGH, 1955)

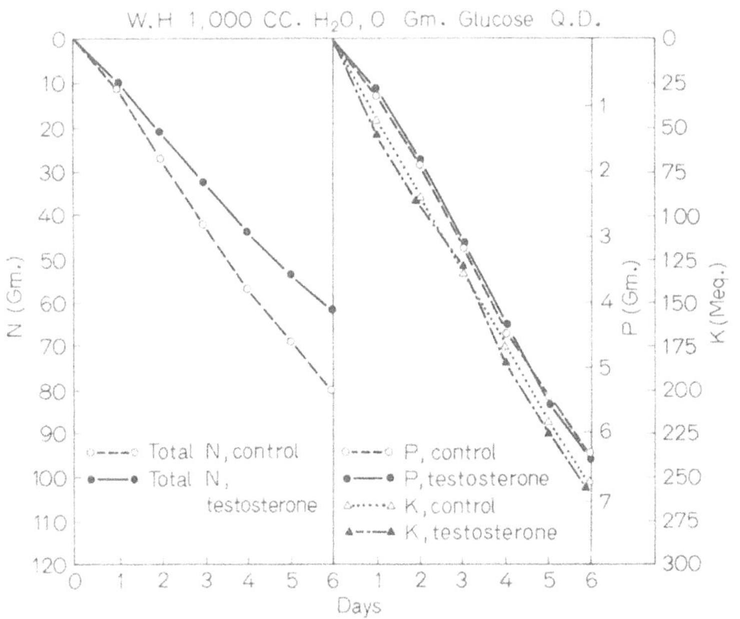


Fig. 8. Effect of testosterone propionate (25 mg/daily) on daily urinary nitrogen of totally fasted man *(left hand section)*. Values are cumulative. On right hand portion are depicted cumulative losses of potassium and inorganic phosphate. (From BUTLER et al., 1945)

balance was determined on three diets, one with very adequate calories and protein, one with adequate calories and restricted protein and one with caloric and very significant protein restriction; then the entire program was repeated while the subject was given testosterone propionate each day (see Macy Conference footnote). The nitrogen sparing influence of the androgen ranged from 3 g per day on the full diet to slightly less than 1 g per day as a nitrogen minimum excretion was approached. Similar effects have been reported by SAARNE et al., 1965. Failure of anabolic steroids to induce anabolism on a deficient calorie-protein diet has also been reported and this is disconcerting in view of the effect noted above in a fasting man (SAARNE et al., 1965). In general it may be assumed that the androgens will exert some anabolic influence or protein sparing under virtually all dietary restricted states, but that it may sometimes be trivial, quantitatively speaking.

There is full agreement that if the calorie-protein intake can be held at a minimal adequate level or somewhat better, the androgens will induce significant protein anabolism. The major problem which has not been fully answered is whether the anabolic effect will continue as long as the drug is administered. This influence of testosterone propionate has been demonstrated in several patients with Cushing's syndrome (ALBRIGHT et al., 1941; DEAKINS et al., 1944; KEUTMANN et al., 1948), a catabolic state (although it has been rare indeed for a patient with hyperadrenocorticism to be studied during the active catabolic phase). Testosterone is not proposed as the treatment for Cushing's syndrome although in addition to the nitrogen retention induced, the hypochloremic alkylosis was also corrected in one patient by testosterone (TEABEAUT et al., 1950). The clinical importance of the capacity of testosterone to counteract the catabolic influence of glucocorticoids occurs when ACTH or adrenocorticosteroids must be given in greater than replacement quantities and the protein wasting itself becomes a problem (BROCHNER-MORTENSEN et al., 1959; SKOOG et al., 1962). Such cases involving underlying disease are usually not fully controlled by the hyperadrenocorticism, and in such instances the anabolic steroids are appropriately used. This form of hyperadrenocorticism must be continued for the welfare of the patient; the anabolic agent may paliate toxicity. The androgens have also been shown to have the capacity to nullify the protein wasting of induced hyperthyroidism and Graves' disease (BRAY et al., 1971; KINSELL et al., 1944; CRISPELL et al., 1957).

The anabolic and appetite-stimulating influence of these compounds has been applied to a variety of other wasting conditions in which food intake or assimilation are limited. From the scientific viewpoint this treatment is rational. From the same aspect measurement of therapeutic efficacy is lacking. Efforts have been made to quantitate the relative nutritional efficacy of several of the anabolic steroids in man (ALBANESE, 1965), but no practical means of quantitating their efficacy as adjuncts in chronic debilitative disease has been concocted. The best evidence of effectiveness is that the physicians continue to prescribe, and the collective wisdom of the profession should not be lightly discounted.

In aging men and women, anabolic androgens have been shown in a double-blind switchover study to have induced both an increase in appetite and weight (BEW, 1965). In psychiatrically involved states such as anorexia nervosa, the efficacy of androgens is virtually nil. The failure of androgens to increase weight gain in premature infants suggests that testosterone may not possess the capacity to promote anabolism and growth at this early age (HARDY and WILKENS, 1949; FOSS, 1949).

The anabolic influence of androgens, in particular their effect on skeletal muscle, suggested that pharmacologic quantities of steroids might be therapeutically useful in patients with muscle wasting diseases. This has not been shown to be true (DOWBEN and PERLSTEIN, 1961).

No compound has been shown to be more anabolic than testosterone administered as the propionic acid ester. However, other esters such as the enanthate and cypionate (JUNKMANN, 1957; REIFENSTEIN and HOWARD, 1958) have a greater capacity to enhance and prolong the activity of testosterone and hence they are more useful products for long-term testosterone administration. For full metabolic activity, injections need to be given intramuscularly no more than every 10–14 days with these products. Idiosyncratic toxicity is nil. No ill effects have been reported from the use of excessive testosterone in men. These long-acting esters would therefore be the treatment of choice for stimulating anabolism in men when no significant objection is raised to repeated intramuscular injects. It must be presumed that the possibility of permanent spermatogenic damage from androgen treatment is remote.

## VI. Effects of Androgens on Carbohydrate and Fat Metabolism

In the total energy balance study (Fig. 4), it was assumed that since caloric intake and utilization were not modified, the positive nitrogen balance necessitated a compensatory increase in caloric expenditure from another source. It was assumed that the major source of these calories would be fat. The studies of BUTLER et al. (1945) provide experimental evidence for this supposition (Fig. 9). It can be seen that in the fasting man, sparing some nitrogen under the influence of testosterone (see above), the acetone body excretion in the urine was very substantially increased. These fasting studies have also shown that the plasma glucose was slightly lower when testosterone was administered, presumably because less protein was available for gluconeogenesis. The effect on plasma glucose concentration has not been confirmed by all studies, perhaps because the conditions in other experiments were less rigorous (ZIMMERMAN et al., 1951; LANDON et al., 1962a). The blood glucose response to insulin and epinepherine was not affected by androgen, but the hyperglycemic effect of glucagon was blunted in one study (LANDON et al., 1962b). It is apparent that the effect of androgens on carbohydrate metabolism is trivial. It is entirely likely that the influence of androgenic steroids on the utilization and availability of both fat and carbohydrate is secondary to the protein anabolic process which makes amino acids less available for other purposes.

The influence of androgens on circulating levels of cholesterol and lipoproteins appears to be a phenomenon unrelated, at least directly, to the anabolic process. The results reported are sometimes contradictory and sometimes, to this reviewer, confusing. The weakly androgenic metabolite of dehydroepiandrosterone and of testosterone, androsterone, will lower the serum cholesterol concentration when administered in very large dosage (80–240 mg/day) parenterally (HELLMAN et al., 1959). It is effective orally if combined with parachlor-phenoxyisobutarate (HELLMAN et al., 1962). The effects of testosterone, methyltestosterone, and other anabolic androgens

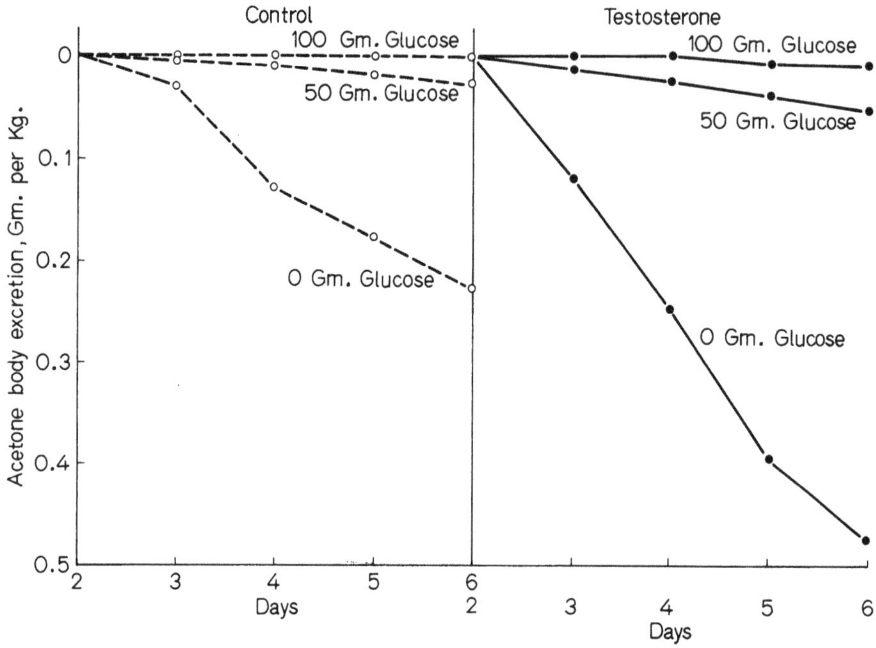


Fig. 9. Effect of testosterone and of diet upon urinary output of acetone bodies. Subjects had been either fasted completely (except water) or given just the amounts of glucose indicated each day. (From BUTLER et al., 1945)

on the total serum cholesterol concentration is erratic with some reports of little or no effect. Quite uniformly, however, especially in patients with atherosclerosis or hypercholesterolemia, the androgens bring about a redistribution of the cholesterol, a lowering of the alpha lipoprotein and an increase in the beta lipoprotein fraction. Estrogens have the opposite influence (EDER, 1958, and HOWARD and FURMAN, 1962a). Testololactone, a nonandrogenic, nonanabolic derivative of testosterone, has the same effect on lipoprotein concentrations indicating that this effect can be dissociated from the anabolic process (HOWARD and FURMAN, 1962b). It has also been reported that androgens increase hepatic triglyceride mobilization through accelerated lipoprotein release and/or synthesis (MENDENHALL, 1974).

## VII. Synthetic Testosterone Analogues

The desirability of having orally effective products and compounds which would have minimal androgenicity while preserving anabolicity have led to the development of a sizable array of competing products. Some of the latter must be administered parenterally. It is not in the interest of this article to review in detail the metabolic activity of each of these, comment on relative potency, progestational activity if present, special therapeutic indications, or toxicity. See Table 1 for references to metabolic studies with these products. All of them have nitrogen-retaining capacity but vary somewhat in the relative intensity of their salt-retaining potency.

Table 1. Selected references for demonstration of anabolic effects of synthetic testosterone analogues

| Compound | Route of administration | Reference |
|---|---|---|
| Methyltestosterone | Oral | Wilkins and Fleischmann, 1945 |
| | | Kinsell et al., 1944 |
| | | Johnston, 1947 |
| | | Samuels et al., 1942 |
| | | Schedl et al., 1959 |
| Fluoxymesterone | Oral | |
| Nandrolone | Intramuscular | Nowakowski, 1961 |
| | | Lambrechts and Jadoul, 1961 |
| | | Brøchner-Mortensen, 1961 |
| Mestanolone | Oral | Harris, 1961 |
| | | Anderson, 1961 |
| Methylandrostenediol | Oral | Anderson, 1961 |
| | | Homburger et al., 1953 |
| | | Warren and Hays, 1952 |
| | | Henderson and Weinberg, 1951 |
| Methandrostenolone | Oral | Queritdo et al., 1952 |
| | | Bray et al., 1971 |
| Stanozolol | Oral | Bray et al., 1971 |
| | | Bew, 1965 |
| | | Howard and Furman, 1962 |
| Norethandrolone | Oral | Stoker et al., 1959 |
| | | Brøchner-Mortensen et al., 1959 |
| | | Spencer et al., 1959 |
| | | Schedl et al., 1959 |

Those selected for lower androgenicity have passed that test, but can be virilizing if sufficient material is given over a long enough period.

The 17a-alkylated products, the ones which are orally effective, deserve some special comment as a group. Methyltestosterone, the first of these, has been the subject of most of the studies to be reported here, but newer compounds with an alkyl substitution exhibit similar unique actions. Although methyl testosterone has shown distinct nitrogen retention with no elevation in plasma urea, the anabolic response is always associated with the progressive increase in urinary creatine which may reach values in excess of 1.0 g per day (Wilkins et al., 1942). Creatininuria is not significantly affected. A number of days are required for the creatininuria to reach a plateau. When the drug is discontinued, the creatine excretion usually rises sharply before dropping back to pretreatment control levels (Fig. 10). The time course of the progressive rise in the plasma creatine and creatinuria have indicated that the effect is due to enhanced hepatic and renal creatine synthesis (Wilkins and Fleischmann, 1945; Samuels et al., 1946; Fleischmann et al., 1960; Marquardt et al., 1961). The posttreatment secondary rise noted above has been interpreted as reflecting the previous deposition of some of the preformed creatine in muscle, the creatine retention which was shown to occur when testosterone propionate is given.

A second unexpected influence of methyl testosterone is the exceptional elevation of basal metabolic rates which was induced. Rates as high as + 38 have been noted (McCullough and Rossmiller, 1941). This appears to be a substantially greater

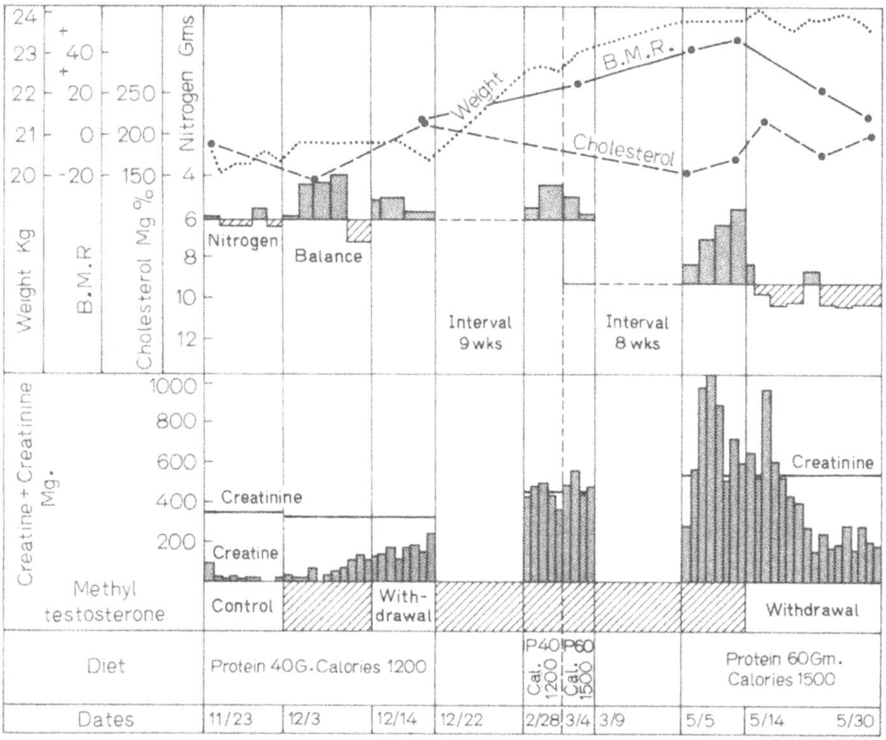


Fig. 10. Effect of methyltestosterone on creatine and nitrogen metabolism in 16-year-old boy. Black columns in nitrogen category indicate positive balance. (From WILKINS et al., 1941)

capacity for increasing oxygen consumption than was brought about by the male hormone itself, but actual comparisons of the two products in the same subject have not been made. No evidence of hyperthyroidism has been associated with the administration of either methyl testosterone or testosterone propionate. The only known effect of the androgens on thyroid activity is the suppression of the level of thyroxine binding globulin capacity in plasma with an accompanying rise in the thyroxine binding pre-albumin previously noted. No sustained changes in free thyroxine have been observed, although the drop in thyroxine binding globulin may lead to a lower PBI or total plasma thyroxine (FEDERMAN et al., 1958).

Two potential toxic reactions of the 17-alkylated steroids must be mentioned. These compounds have a variable potential for slightly increasing bromsulphalein retention without clinically impairing liver function (MARQUARDT et al., 1961). Cholestatic jaundice is, however, a relatively rare toxic effect (FOSS and SIMPSON, 1959; BRICK and KYLE, 1952). Fortunately, this untoward reaction has always reversed itself with liver function returning to normal after discontinuance of the drug. A more dangerous effect is the recently reported development of hepatic carcinoma which has been reported in five patients who received very large doses of a 17-alkylated steroid to treat unusual refractory anemia (JOHNSON et al., 1972).

Since compounds with anabolic activity have therapeutic potential, a nonsteroidal, nonandrogenic drug will be mentioned. The very effective antipyretic, aminopyrine, is anabolic in normal men without lowering their temperature (EHRLICH et al., 1964).

## VIII. Conclusions

The dominant metabolic action of testosterone is protein anabolism. This somatotrophic action of the male hormone is exerted on most of the tissues of the body. It appears to be a direct action of the hormone on the organs and tissues affected, not requiring the mediation of any other endocrine gland or secretions. This action is growth promoting and probably accounts at least in part for the large stature and greater muscular development of men as compared with women. It is possible that the secretion of testosterone by the testes plays a significant role in the regulation of protein metabolism through adult life. The hormone probably affects the net metabolism of carbohydrate and fat indirectly, as a result of its protein anabolic influence. It also possesses a modest salt-retaining activity and has a minor influence on plasma lipoprotein concentrations.

Estradiol is also anabolic although this effect is substantially weaker than that of the male hormone. It probably has little or no effect on skeletal muscle, but stimulates skeletal growth and may be anabolic in other tissues in addition to the sexual apparatus. Estradiol also has a salt-retaining influence which may be conditioned to a limited extent by the fact that it also induces an increased concentration of renin substrate, and thus may enhance the action of the renin-angiotension-aldosterone system.

A number of modified androgens, most of which are alkylated at the 17a position and some of which are 19 nor compounds, have been developed as orally effective anabolic agents. Most are relatively more anabolic than androgenic and hence are more appropriate for use in women than testosterone. The 17-alkylated steroids have metabolic and endocrine effects and rare toxic reactions which distinguish them from that of testosterone. Esterification of testosterone at the $17\beta$ position enhances and prolongs the action of the hormone without causing qualitative changes in the response. A number of similar esters of steroids similar to testosterone have been synthesized and made available for human use because bioassays indicate that they retain the anabolic force of testosterone but less androgenicity. There are few if any contraindications for employing any of these esters for anabolic influence when there is medical need (the less androgenic compounds for women).

When the caloric and protein intake are adequate, the anabolic force of the androgens may significantly hasten protein restitution in debilitated states or diminish excessive protein catabolism when it is present and thus have a favorable influence on the ultimate recovery of a patient.

A number of questions of importance have not been fully answered:

1. Does the testis participate actively in the regulation of protein metabolism by altering the rate of testosterone secretion?

2. Can the difference between the musculature in men and women be based solely on the different hormonal milieus in the two sexes?

3. What is the significance of the creatinuria induced by 17 alkylated androgens and by estradiol?

4. Can a more quantitative evaluation be made of the clinical benefit accruing from androgen administration in debilitated or catabolic states?

## References


ALBANESE, A.: Newer methodology in the clinical investigation of anabolic steroids. J. New Drugs **5**, 208–224 (1965).

ALBRIGHT, F.: Cushing's syndrome. Its pathological physiology, its relation to the adrenogenital syndrome and its connection with the problem of the reaction of the body to injurious agents ("Alarm Reaction of Selye"). Harvey Lectures **38**, 123–186 (1942).

ALBRIGHT, F., PARSONS, W., BLOOMBERG, E.: Cushing's syndrome interpreted as hyperadrenocorticism leading to hypergluconeogenesis. Results of treatment with testosterone propionate. J. clin. Endocrinol. **1**, 375–386 (1941).

ALBRIGHT, F., REIFENSTEIN, E. C.: The Parathyroid Glands and Metabolic Bone Disease. Baltimore: Williams and Wilkins Co 1948.

ANDERSON, I. A.: The metabolic effects of certain anabolic steroids. Acta endocr. Suppl. **63**, 54–74 (1961).

AYKROYD, O. E., ZUCKERMAN, S.: Factors in sexual skin oedema. J. Physiol. (Lond.) **94**, 13–25 (1938).

BEW, K.: A double blind trial of an anabolic steroid in 18 elderly women. Practitioner **194**, 530–534 (1965).

BLAGG, C. R., PARSONS, F. M., YOUNG, G. A.: Effect of dietary glucose and protein in acute renal failure. Lancet **1962 I**, 609–612.

BRAY, G. A., RABEN, M. S., LONDONO, J., GALLAGHER, T. F. JR.: Effects of triiodothyronine, growth hormone, and anabolic steroids on nitrogen excretion and oxygen consumption of obese subjects. J. clin. Endocr. **33**, 293–300 (1971).

BRICK, I. B., KYLE, L. H.: Jaundice of hepatic origin during the course of methyltestosterone therapy. New Engl. J. Med. **246**, 176–179 (1952).

BROCHNER-MORTENSEN, K., GJORUP, S., THAYSEN, J. H.: The metabolic effect of a new anabolic 19-nor steroid. Acta med. scand. **165**, 197–205 (1959).

BUTENANDT, A., HANISCH, G.: Über Testosterone. Z. physiol. Chem. **237**, 89–97 (1935).

BUTLER, A. M., TALBOT, N. B., MACLACHLAN, E. A.: Effect of testosterone therapy on concentration of potassium in serum. Proc. Soc. exp. Biol. (N.Y.) **51**, 378–380 (1942).

BUTLER, A. M., TALBOT, N. B., MACLACHLAN, E. A., APPLETIN, J. E., LINTON, M. A.: Effect of testosterone propionate on losses incident to inadequate dietary intake. J. clin. Endocr. **5**, 327–336 (1945).

CHRISTY, N. P., SHAVER, J. C.: Estrogens and the kidney. Kidney International **6**, 366–376 (1974).

CRISPELL, K. R., WILLIAMS, G. A., PARSONS, W., HOLLIFIELD, G.: Metabolic studies in myxedema following the administration of 1-triiodothyronine. J. clin. Endocr. **17**, 221–231 (1957).

CUTHBERTSON, D. P.: Observations on the disturbance of metabolism produced by injury to the limbs. Quart. J. Med. **25**, 233–246 (1932).

CUTHBERTSON, D. P.: Post-shock metabolic response. Lancet **1942 I**, 433–437.

DAVID, K., DINGEMANSE, E., FREUD, J., LAQUEUR, E.: Über Krystallinisches Männliches Hormon aus Hoden (Testosteron) Wirksamer als aus Harn oder aus Cholesterin Berlititer Androsteron. Z. physiol. Chem. **233**, 281–282 (1935).

DEAKINS, M. L., FRIEDGOOD, H. B., FERRIBEE, J. W.: Some effects of testosterone, testosterone propionate, methyltestosterone, stilbesterol, and x-ray therapy in a patient with Cushing's syndrome. J. clin. Endocr. **4**, 376–384 (1944).

DIGNAM, W. S., VOSKIAN, J., ASSALI, N. S.: Effects of estrogens on renal hemodynamics and excretion of electrolytes in human subjects. J. clin. Endocr. **16**, 1032–1042 (1956).

DOWBEN, R. M., PERLSTEIN, M. A.: Muscular dystrophy treated with norethandrolone. Arch. intern. Med. **107**, 245–251 (1961).

EDER, H. A.: Recent Progr. Hormone Res. XIV, 405–416 (1958).
EHRLICH, E. N., LUGIBIHL, K., LANDAU, R. L.: The anabolic influence of aminopyrine. Metabolism **13**, 799–807 (1964).
ENGSTROM, W. W., MUNSON, P. L., COOK, L., COSTASTED, P. J.: Studies on the conversion of testosterone to the urinary 17-ketosteroids in normal human males and males with hepatic and other diseases. J. clin. Endocr. **11**, 416–426 (1951).
FEDERMAN, D. D., ROBBINS, J., RALL, J. E.: Effects of methyltestosterone on thyroid function, thyroxine metabolism, and thyroxine-binding protein. J. clin. Invest. **37**, 1024–1030 (1958).
FLEISCHMANN, W., FERRILL, C., FISCHER, R. L., ETTELDORF, J. M.: Effect of certain nor-steroids on creatine excretion. J. Pediat. **57**, 370–375 (1960).
FORSHAM, P. H., THORN, G. W., PRUNTY, F. T. G., HILLS, A. G.: Clinical studies with pituitary adrenocorticotropin. J. clin. Endocr. **8**, 15–66 (1948).
FORSYTH, B. T.: The effect of testosterone propionate at various protein calorie intakes in malnutrition after trauma. J. Lab. clin. Med. **43**, 732–740 (1954).
FORSYTH, B. T., PLOUGH, I. C.: The protein sparing effect of carbohydrate with and without testosterone. J. Lab. clin. Med. **46**, 840–847 (1955).
FOSS, G. L.: Some experiences with a new anabolic steroid, methyltestosterone and the treatment of premature infants. J. Pediat. **34**, 439–442 (1949).
FOSS, G. L., SIMPSON, S. L.: Oral methyltestosterone in jaundice. Brit. med. J. **1959 I**, 259–263.
FRANTZ, A. G., KLEINBERG, D. L., NOEL, G. L.: Studies on prolactin in man. Recent Progr. Hormone Res. **28**, 527–590 (1972).
GIEBISH, G.: Functional organization of proximal and distal tubular electrolyte transport. Nephron **6**, 260–281 (1969).
GRAYHACK, J. T., BUNCE, P. L., KEARNS, J. W., SCOTT, W. W.: Influence of the pituitary on prostatic response to androgen in the rat. J. Hopkins Hosp. Bull. **96**, 154–163 (1955).
GROSSMAN, J., YALLOW, A. A.: Effects of anabolic steroids on albumin metabolism. J. clin. Endocr. **25**, 698–707 (1965).
HAMILTON, J. B.: Treatment of sexual underdevelopment with synthetic male hormone substance. Endocrinology **21**, 649–654 (1937).
HARDY, J., WILKINS, L.: Methyltestosterone and the treatment of premature infants. J. Pediat. **34**, 439–442 (1949).
HARRIS, L. H.: The protein anabolic action of mestanolone. J. clin. Endocr. **21**, 1099–1105 (1961).
HELLMAN, L., BRADLOW, H. L., ZUMOFF, B., FUKUSHIMA, D. K., GALLAGHER, T. F.: Thyroid androgen interrelations and the hypocholesterolemic effect of androsterone. J. clin. Endocr. **19**, 936–943 (1959).
HELLMAN, L., ZUMOFF, B., KESSLER, G., ROSENFELD, R. S., GALLAGHER, T. F.: Reduction of serum cholesterol in man by an oral androsterone preparation. J. clin. Invest. **41**, 1364 (1962).
HENDERSON, E., WEINBERG, M.: Methylandrostenediol. J. clin. Endocr. **11**, 641–652 (1951).
HOMBURGER, F., DART, R. M., BONNER, C. D., BRANCHE, G., KASDON, F. E., FISHMAN, W. H.: Some metabolic and biochemical effects of methylandrostenediol. J. clin. Endocr. **13**, 704–711 (1953).
HORROBIN, D. F., LLOYD, I. J., LIPTON, A., BURSTYN, P. G., DURKIN, N., MUIRURI, K. L.: Actions of prolactin on human renal function. Lancet **1971 II**, 352–354.
HOWARD, J. E.: Metabolic observations on patients convalescent from fracture. Trans. Ass. Amer. Phycns. **58**, 162–170 (1944).
HOWARD, R. P., FURMAN, R. H.: Metabolic and serum lipid effects of methylandrostene pyrazoles. J. clin. Endocr. **22**, 43–51 (1962a).
HOWARD, R. P., FURMAN, R. H.: Metabolic and serum lipid effects of *Δ'* testololactone. Proc. Soc. exp. Biol. (N.Y.) **110**, 227–229 (1962b).
INGLE, D. J., BAKER, B. L.: A consideration of the relationship of experimentally produced and naturally occurring pathologic changes in the rat to the adaptation diseases. Recent Progr. Hormone Res. **8**, 143–169 (1953).
JOHNSON, F. L., LERNER, K. G., SIGEL, M. C. U., FAGLER, J. R., MAJERUS, P. W., HARTMANN, J. R., THOMAS, E. D.: Association of androgenic-anabolic steroid therapy with the development of hepatocellular carcinoma. Lancet **1972 II**, 1273–1276.
JOHNSTON, J. A.: Factors influencing retention of nitrogen and calcium in period of growth VII. Effect of methyltestosterone. Amer. J. Dis. Child. **74**, 52–57 (1947).

JONES, T. M., LANDAU, R. L., FANG, V. S.: Effects of administration of synthetic oral androgens on testicular function in young adult men. Proc. 56th Ann. Meet. Endocrine Soc., June 12–14, 1974.

JUNKMANN, K.: Long-acting steroids in reproduction. Recent Progr. Hormone Res. **13**, 389–419 (1957).

KEARNS, W. M.: The clinical application of testosterone. J. Amer. med. Ass. **112**, 2255–2258 (1939).

KEITEL, H. G.: Plasma inorganic phosphate and glucose in prepubertal children taking testosterone. J. Pediat. **34**, 439–442 (1957).

KENNEDY, J. H., PETERS, H. G., PILSON, M.: The anabolic effect of SC 7294 upon the nitrogen balance in patients undergoing pulmonary resection for tuberculosis. Amer. J. med. Sci. **243**, 193–195 (1962).

KENYON, A. T.: The effect of testosterone propionate on the genitalia, prostate, secondary sex characters and body weight in eunuchoidism. Endocrinology **23**, 121–134 (1938).

KENYON, A. T.: The First Josiah Macy Jr. Conference on Bone and Wound Healing. Sept. 1942.

KENYON, A. T., KNOWLTON, K., LOTWIN, G., MUNSON, C. D., JOHNSTON, C. D., KOCH, F. C.: Comparison of metabolic effects of testosterone propionate with those of chorionic gonadotropin. J. clin. Endocr. **2**, 685–689 (1942a).

KENYON, A. T., KNOWLTON, K., LOTWIN, G., SANDIFORD, I.: Metabolic response of aged men to testosterone propionate. J. clin. Endocr. **2**, 690–695 (1942b).

KENYON, A. T., KNOWLTON, K., SANDIFORD, I.: The anabolic effects of the androgens and somatic growth in man. Ann. intern. Med. **20**, 632–654 (1944).

KENYON, A. T., KNOWLTON, K., SANDIFORD, I., FRICKES, L.: Metabolic effects of testosterone propionate in Addison's disease. J. clin. Endocr. **3**, 131–136 (1943).

KENYON, A. T., KNOWLTON, K., SANDIFORD, I., KOCH, F. C., LOTWIN, G.: A comparative study of the metabolic effects of testosterone propionate in normal men and women and in eunuchoidism. Endocrinology **26**, 26–45 (1940).

KENYON, A. T., LANDAU, R. L.: Unpublished study.

KENYON, A. T., SANDIFORD, I., BRYAN, A. H., KNOWLTON, K., KOCH, F. C.: The effect of testosterone propionate on nitrogen, electrolyte, water, and energy metabolism in eunuchoidism. Endocrinology **23**, 135–153 (1938).

KEUTMANN, E. H., FRIEDMAN, H. A., BASSETT, S. H., KOCHAKIAN, C. D., RUSS, E. N.: Metabolic studies in Cushings syndrome. Amer. J. Med. **5**, 518–531 (1948).

KINSELL, L., HERTZ, S., REIFENSTEIN, E. C. JR.: The effect of testosterone compounds on the nitrogen balance and creatine excretion in patients with thyrotoxicosis. J. clin. Invest. **23**, 880–890 (1944).

KINSELL, L. W., MARGEN, S., MICHAELS, G. D., MCCALLIE, D. P.: Studies in fat metabolism. II. Evaluation of the effect of testosterone propionate upon ketone, carbohydrate and protein metabolism in a patient with diabetes mellitus complicated by thyrotoxicosis. J. clin. Invest. **30**, 1486–1490 (1951).

KNOWLTON, K., KENYON, A. T., SANDIFORD, I., LOTWIN, G., FRICKES, R.: Comparative study of metabolic effects of estradiol benzoate and testosterone propionate in man. J. clin. Endocr. **2**, 671–684 (1942).

KOCHAKIAN, C. D.: Testosterone and testosterone acetate and the protein and energy metabolism of castrate dogs. Endocrinology **21**, 750–755 (1937).

LAMBRECHTS, A., JADOUL, R.: Influence of norandrostenolone phenylpropionate on growth and balance experiments in children. Acta endocr. Suppl. **63**, 111–120 (1961).

LANDAU, R. L., BERGENSTAL, D. M., LUGIBIHL, K., DIMICK, D. F., RASHID, E.: The relationship of estrogen and of pituitary hormones to the metabolic effects of progesterone. J. clin. Endocr. **17**, 177–185 (1957).

LANDAU, R. L., KNOWLTON, K., LUGIBIHL, K., BRANDT, M., KENYON, A. T.: The anabolic effects of chorionic gonadotropin in normal young men. J. clin. Invest. **29**, 619–629 (1950).

LANDAU, R. L., LUGIBIHL, K.: The catabolic and natriuretic effects of progesterone in man. Recent Progr. Hormone Res. **17**, 249–292 (1961).

LANDON, J., WYNN, D., COOKE, J. N. C., KENNEDY, A.: Effects of the anabolic steroid methanedione on carbohydrate metabolism in man. Metabolism **11**, 501–512 (1962a).

LANDON, J., WYNN, D., HOUGHTON, B.J., COOKE, J.N.C.: The effects of anabolic steroid methandione on carbohydrate metabolism in man. Metabolism **11**, 513–523 (1962b).

LEE, P.A., JAFF, R.B., MIDGLEY JR., A.R., KOHEN, F., NISWENDER, G.D.: Regulation of human gonadotropins VIII. Suppression of serum LH and FSH in adult males following exogenous testosterone administration. J. clin. Endocr. **35**, 636–641 (1972).

MAGALLON, D.T., MASTERS, W.H.: Basal temperature studies in the aged female: Influence of estrogen, progesterone and androgen. J. clin. Endocr. **10**, 511–518 (1950).

MARQUARDT, G.H., FISCHER, C.I., LEVY, P., DOWBEN, R.M.: The effect of anabolic steroids on liver function tests and creatine excretion. J. Amer. Med. Ass. **175**, 851–853 (1961).

MCCULLOUGH, E.P.: Treatment of testicular deficiency with testosterone propionate. J. Amer. med. Ass. **112**, 1037–1043 (1939).

MCCULLOUGH, E.P., ROSSMILLER, H.R.: Methyltestosterone II. Calorigenic activity. J. clin. Endocr. **1**, 503–506 (1941).

MENDENHALL, C.L.: Augmented release of hepatic triglycerides with anabolic steroids in patients with fatty liver. Amer. J. dig. Dis. **19**, 122–126 (1974).

MIESCHER, K., WETTSTEIN, A., TSCHOPP, E.: The activation of the male sex hormone II. Biochem. J. **30**, 1977–1990 (1936).

NOWAKOWSKI, H.: Metabolic studies with anabolic steroids. Acta Endocr. Suppl. **63**, 37–48 (1961).

PEDEN, J.C., OHIN, A.O., WILLIAMS, P.T.: The influence of an anabolic agent on the nitrogen balance of patients with major extremity fractures. Arch. Surg. **80**, 1036–1042 (1960).

PREEDY, J.R.K., AITKEN, E.H.: The effect of estrogen on water and electrolyte metabolism I. The normal. J. clin. Invest. **35**, 423–429 (1956a).

PREEDY, J.R.K., AITKEN, E.H.: The effect of estrogen on water and electrolyte metabolism II. Hepatic disease. J. clin. Invest. **35**, 430–442 (1956b).

PREEDY, J.R.K., AITKEN, E.H.: The effect of estrogen on water and electrolyte metabolism III. Cardiac and renal disease. J. clin. Invest. **35**, 443–451 (1956c).

QUERITDO, A., KASSENAAR, A., SCHUURS, M.A.M., SELDENRATH, H.C.: A balance study of methylandrostenediol in a patient with hypogonadism. J. clin. Endocr. **12**, 1077–1082 (1952).

REIFENSTEIN, E.C. JR., HOWARD, R.P.: Protein anabolic effectiveness in post-menopausal and senile osteoporosis of a single injection of the long acting steroid ester, testosterone enanthate. Metabolism **7**, 364–373 (1958).

RIEGER, D., ROMERO, J.L., LAZAR, J., HOOBLER, S.W.: Definition and use of renin reaction velocity in the study of human hypertension. J. Lab. clin. Med. **80**, 342–350 (1972).

ROWLANDS, R.P., NICHOLSON, G.W.: Growth of left testicle with precocious sexual and bodily development (macro-genitosomia). Guy's Hosp. Rep. **79**, 401–408 (1929).

SAARNE, A., BJERSTAF, L., EKMAN, B.: Studies on the nitrogen balance in the human during long term treatment with different anabolic agents under strictly standardized conditions. Acta med. scand. **177**, 199–211 (1965).

SACCHI, E.: A case of infantile gigantism (pedomacrosomia) with a tumor of the testicle. Riv. sper. Freniat. **21**, 149–161 (1895).

SAMUELS, L.T., HENSCHEL, A.F., KEYS, A.: The influence of methyltestosterone on muscular work and creatine metabolism in normal young men. J. clin. Endocr. **2**, 649–654 (1942).

SAMUELS, L.T., SELLERS, D.M., MCCAULEY, C.J.: The source of excess creatine following methyltestosterone. J. clin. Endocr. **6**, 655–663 (1946).

SANDIFORD, I., KNOWLTON, K., KENYON, A.T.: Basal heat production in hypogonadism in men and its increase by protracted treatment with testosterone propionate. J. clin. Endocr. **1**, 931–939 (1941).

SCHEDL, H.P., CELEA, C., BARTTER, F.C.: Structure-activity relationships of anabolic steroids—role of the 19 methyl group. J. clin. Endocr. **19**, 921–935 (1959).

SHERINS, R.J., LORIAUX, D.L.: Studies on the role of sex steroids in the feedback control of FSH concentrations in men. J. clin. Endocr. **36**, 886–893 (1973).

SKOOG, W.A., ADAMS, W.S.: Metabolic balance studies of a patient with multiple myeloma given combined ACTH-androgen therapy. Metabolism **11**, 611–625 (1962).

SPENCER, H., BERGER, E., CHARLES, M.L., GOTTESMAN, E.D., LASZLO, D.: Metabolic effects of 17 ethyl-19-nor testosterone in man. J. clin. Endocr. **17**, 975–984 (1957).

STEWART, C. A., BELL, E. T., ROEHLKE, A. B.: Interstitial-cell tumor of testis with hypergenitalism in a child of 5 years. Amer. J. Cancer **26**, 144–150 (1936).

STOKER, P. E., HORWITH, M., PENNINGTON, T. G., CLARKSON, B.: 17a ethyl-19-nortestosterone as an anabolic agent and its effect on creatine metabolism and vaginal cytology in humans. Metabolism **8**, 709–721 (1959).

TALBOT, N. B., BUTLER, A. M., MACLACHLAN, E. A.: The effect of testosterone and allied compounds on the mineral, nitrogen, and carbohydrate metabolism of a girl with Addison's disease. J. clin. Invest. **22**, 583–593 (1943).

TEABEAUT, R., ENGEL, F. L., TAYLOR, H.: Hypokalemic hypochloremic alkalosis in Cushing's syndrome, observations on the effects of treatment with potassium chloride and testosterone. J. clin. Endocr. **10**, 399–409 (1950).

THORN, G. W., HARROP, G. A.: The sodium retaining effect of sex hormones. Science **86**, 40–41 (1937).

VEST, S. A., HOWARD, J. E.: Clinical experiments with the use of male sex hormones. J. Urol. **40**, 154–183 (1938).

WARREN, W. D., HAYS, M. A.: Comparative study of anabolic effects of three androgenic steroids. Proc. Soc. exp. Biol. (N. Y.) **79**, 503–505 (1952).

WEBB, W. R., DOYLE, R. S., HOWARD, H. S.: Relative metabolic effects of calories, protein and an anabolic hormone, 19 nor testosterone in early post-operative period. Metabolism **9**, 1047–1057 (1960).

WHEDON, G. D., SHORR, E.: Metabolic studies in paralytic acute anterior poliomyelitis IV. Effects of testosterone propionate and estradiol benzoate on calcium, phosphorus, nitrogen, creatine and electrolyte metabolism. J. clin. Invest. **36**, 995–1033 (1957).

WILKINS, L., FLEISCHMANN, W.: The protein anabolic activity of steroid compounds. REIFENSTEIN, E. C. JR. (Ed.) Josiah Macy Jr. Conference on Metabolic Aspects of Convalescence. 1942.

WILKINS, L., FLEISCHMANN, W.: Studies on the creatinuria due to methylated steroids. J. clin. Invest. **24**, 21–32 (1945).

WILKINS, L., FLEISCHMANN, W., HOWARD, J. E.: Creatinuria induced by methyltestosterone in the treatment of dwarfed boys and girls. Johns Hopkins Hosp. Bull. **69**, 493–503 (1941).

ZACHMANN, M., PRADER, A.: Anabolic and androgenic effect of testosterone in sexually immature boys and its dependency on growth hormone. J. clin. Endocr. **30**, 719–726 (1970).

ZIMMERMAN, H. J., PARRISH, A. E., ALPERT, L. K.: Failure of testosterone to alter responsiveness to insulin in diabetes mellitus. Proc. Soc. exp. Biol. **77**, 646–648 (1951).

# III. Sites of Tissue Formation

## A. Body and Organ Weights and Composition

C. D. KOCHAKIAN

### I. Introduction

The early studies on the effect of androgenic preparations and steroid hormones on the body and organ weights resulted in apparent conflicts. Further studies, however, resolved the apparent inconsistencies and indicated the dependence of the effect of the hormones upon age, sex, and species of the experimental animals, the dose and duration of treatment, and the endocrine and nutritional status of the animals. In addition, the effect on body weight could be misinterpreted due to the lack of sufficiently frequent determinations. Finally, the degree of handling of the animals was also a possible factor in some instances (BERNSTEIN and ELRICK, 1957).

The effect of the hormones on the body weight and nitrogen equilibrium of adult animals was clear-cut, but was also modified by the status of the animal and the treatment (Chap. II.A.).

### II. Body Weight and Composition

#### A. Effect of Castration

Castration has been practiced since early times in domestic animals with observed gross changes in body weight and composition. KORENCHEVSKY (1925) reported that dogs, after castration, became either fat or lean.

##### 1. Rat

Numerous experiments in rats (DONALDSON and HATAI, 1911; HATAI, 1915; EVANS and SIMPSON, 1927; KORENCHEVSKY, 1930; COMMINS, 1932; LAWLESS, 1936; RUBINSTEIN et al., 1939a; DEANESLEY and PARKES, 1941) demonstrated a retardation in the rate of growth of male rats after castration. The difference did not begin to become apparent until after about 7 weeks of age (RUBINSTEIN et al., 1939a). Several studies (STOTSENBURG, 1909; MOORE, 1922, 1939; OVERHOLSER, 1929; ANDERSON and KENNEDY, 1933; HOLT et al., 1936) failed to note consistent and reliable differences. Inconsistency was noted even within the same study (KORENCHEVSKY, 1930). It should be recognized that the differences between the normal and castrated rats was never very great (cf. Fig. 1), and that careful control of the many variables are essential for definite results (HOLT et al., 1936).

A clear retardation in growth rate was reported for rats castrated at 20–23 days of age (Fig. 1) (SCOW, 1952), at 22–35 days (KORENCHEVSKY et al., 1953a), and 36–40 days (KOCHAKIAN et al., 1956a). SCOW and HAGAN (1957), in a later, identical study to the above, observed no effect on body weight after castration. SWARTZ

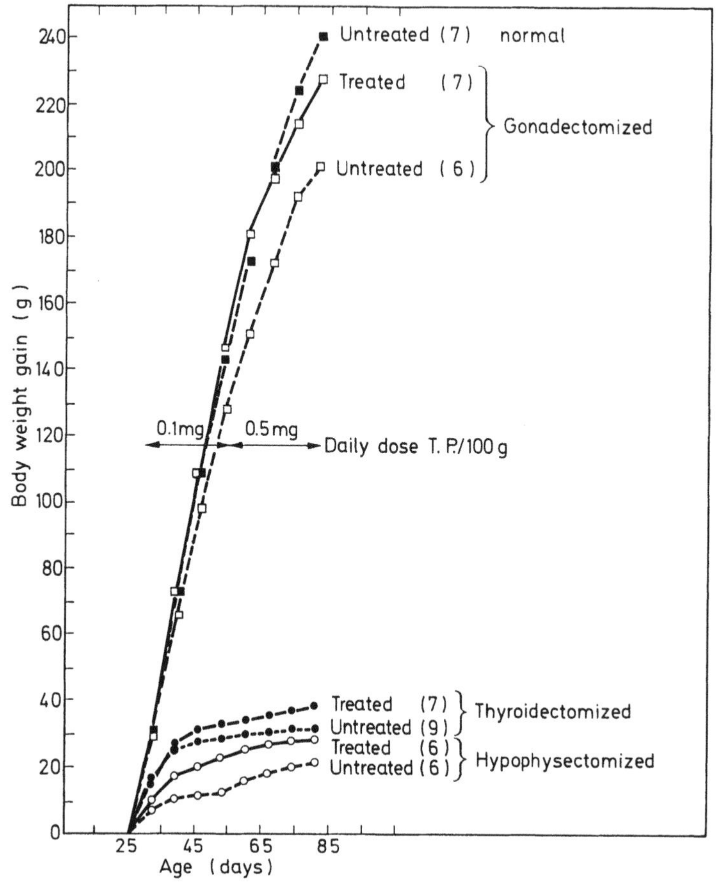


Fig. 1. Effect of testosterone propionate and endocrinectomies on body weight gain in rats. Injections were begun on the fourth post-operative day. Numbers in parentheses indicate the number of animals in each group (SCOW, 1952)

(1962) also noted a small difference in body weight of rats castrated at weaning when compared with their litter mate controls, but a much greater difference was obtained when the rats were castrated at birth. In both instances the difference between normal and castrated rats did not begin to appear until the time of sexual maturity (about 50 days of age).

LEATHAM (1948a) reported that rats castrated at 150–155 days of age (268 g) gained only 5 g in body weight while the normal controls gained 17 g in the following 25 days. No difference in appetite was noted. Castration produced a slight increase in total plasma protein due to a rise in plasma globulin. Plasma albumin, nonprotein nitrogen and hematocrit levels were not significantly altered.

## 2. Mouse

Castration of the mouse at 26–28 days of age produced a small but consistent retardation in the initial rate of growth (MILLER, 1950; KOCHAKIAN, 1959), followed by a

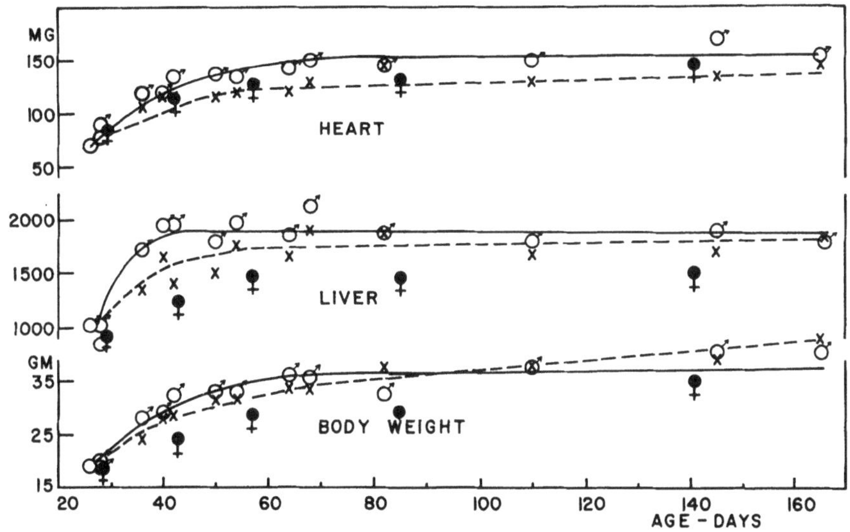


Fig. 2. Effect of castration on body, liver, and heart weight. Mice of the Holtzman-Swiss strain were castrated at 26—28 days of age and autopsied at the indicated intervals. Normal males: O—O, Castrated males: x — x, females: ♀--♀ (KOCHAKIAN, 1959; 1964)

more rapid rate (Fig. 2). At about 90 days, the body weight of the castrated mice had reached that of the unoperated controls and thereafter exceeded that of the controls (KOCHAKIAN, 1959, 1964). The lack of difference at 90 days is in agreement with the earlier study of MASUI and TAMURA (1926), who castrated mice at 20–22 days and killed them at 90 days. The secondary acceleration in the increase in body weight probably represents the deposition of extra fat (KOCHAKIAN and STETTNER, 1948) rather than true growth.

### 3. Guinea Pig

MOORE (1922, 1939) castrated guinea pigs before 30 days of age and noted a decrease in the rate of growth beginning at about 90 days so that at 1 year the body weight of the castrated guinea pigs was about 10% less than that of the controls. Castration of the guinea pig at 1–2 months (KOCHAKIAN et al., 1948a) or 3–4 months (KOCHAKIAN et al., 1956b) did not produce an immediate change in the body weight. A statistically insignificant decrease appeared after 6 weeks, which gradually increased to a small but significant difference after another 3 (KOCHAKIAN et al., 1948b) or more weeks (KOCHAKIAN et al., 1956b). A later experiment (KOCHAKIAN and TILLOTSON, 1957b) also showed a small but consistent retardation in growth within 6 weeks. Castration at an older age (7–8 months) produced a significant decrease in body weight within 6 weeks (KOCHAKIAN et al., 1956b). Almost identical results were reported by SCOW. Castration of male guinea pigs at 25 days resulted in a reduction in body weight gain during the subsequent 60–90 days (SCOW and ROE, 1953; SCOW and HAGAN, 1955). Castration of adult (600–700 g) guinea pigs resulted in a small decrease in body weight (SCOW and ROE, 1953).

### 4. Hamster

Castration of the hamster at 45–75 g body weight did not affect the growth rate during the next 60 days: thereafter, the weight of the castrated hamsters definitely surpassed that of the normal animals (KOCHAKIAN et al., 1948b; KEYES, 1948). Upon autopsy at 140 days after castration, an extensive deposit of fat was grossly noted in the abdomen of the castrated hamsters. The castrated hamsters consumed more food and drank more water than the normal males, but less on a per 100 g body weight basis after their weight exceeded 100 g due to the then much heavier (50%) body weight (KEYES, 1948).

## B. Steroid Administration

### 1. Rat

*a) Normal and Castrated*

α) Young

KORENCHEVSKY et al. (1935) administered androsterone and 5α-androstane-3α,17β-diol in oil solution and as water solutions of their lithium succinates in several doses to rats (2–3 per group) beginning 2–84 days after castration for 17–53 days. In general the large doses of the 5α-androstane-3α-17β-diol depressed appetite and body weight while the medium and low doses and the oil-soluble androsterone slightly increased the body weight. In earlier experiments (KORENCHEVSKY et al., 1933; KORENCHEVSKY and DENNISON, 1934a), male hormone extracts of urine stimulated a slight increase in body weight, appetite and retroperitoneal fat. The rats were castrated at 21 days and injected at 50 days for 21 days.

RUBINSTEIN et al. (1939b) and RUBINSTEIN (1944) found that 1 mg/day of testosterone propionate injected intraperitoneally into 26-day-old male rats began to inhibit the growth rate after 4 to 5 weeks. Similar results were obtained by SAINTON et al. (1938). The injection of 0.05 mg/day of the steroid, on the other hand, accelerated the growth rate of normal (RUBINSTEIN and SOLOMON, 1940a) and castrated (RUBINSTEIN and SOLOMON, 1941; RUBINSTEIN, 1944) rats. The stimulation of growth in the normal rats varied with the time of year in an irregular manner (RUBINSTEIN and SOLOMON, 1940b). The inability of MCEUEN et al. (1937) to detect an inhibitory effect in a similar study was probably due to the early termination (23 days) of the treatment and the use of the less potent testosterone. In a subsequent study (SELYE, 1941b), the alternate subcutaneous and intraperitoneal injection for 20 days of 0.1, 0.5, or 1.0 mg/day of a saline suspension of testosterone did not affect the body weight of male rats (5 weeks old) but 10 and 20 mg/day produced a marked inhibition.

MEYER (1949) could not confirm the body weight stimulatory effect in male rats of small doses of testosterone propionate varying from 2 to 200 μg/day for 176 days beginning at 24 days of age. The injections of 2 mg/day, however, began after 28 days gradually to decrease the rate of growth so that after 154 days the experimental animals weighed 92 g less than the controls.

CLAUSEN and FREUDENBERGER (1939) observed a 9.5% increase in the body weight of female rats injected with 1 mg/day of testosterone propionate on alternate days from the 21st to the 29th day of age. On the other hand, SHAY et al. (1941) injected 1 mg testosterone propionate 3 times a week, beginning at birth for 20–40 weeks and did not observe any change in the growth rate of the male but an acceleration in that of the females beginning after about 7 weeks; both the control and treated groups attained a plateau about the 15th or 16th week. The spacing of the hormone injections greatly diminishes its effectiveness (KOCHAKIAN, 1938) and may explain its lack of effect in the male animals. In any event, the female rats responded differently. TURNER et al. (1941) also found no effect of testosterone propionate, up to 2 mg/day, on the increase in body weight or bone growth in normal or castrated rats even with prolonged injections beginning at the day of birth. JOSS et al. (1963) found that both testosterone propionate and 17α-methyltestosterone either accelerated, was ineffective, or inhibited the growth rate of female rats depending on the dose of the hormone.

The injection of lactating rats at the time of delivery with 5 mg/day of methandrostenolone accelerated the growth rate of the pups in both small and large litters during the 21-day period of lactation (KOWALEWSKI, 1965). The inhibition of growth by 5 mg/day of cortisone was counteracted by the simultaneous injection of the methandrostenolone.

Male rats castrated at 21–23 days and injected with 0.1 mg of testosterone propionate per 100 g body weight for 28 days, beginning 4 days after the operation and for another 28 days at 0.5 mg/100 g body weight, exhibited a greater growth rate than the castrated controls (SCOW, 1952). Indeed, the growth rate was practically identical with that of the normal rats except for the fact that a retardation began to appear during the last two weeks. In a later, almost identical study, SCOW and HAGAN (1957) reported that testosterone propionate was ineffective. They attributed this inability to confirm the earlier study to the small effect of castration.

SMITH and ALLISON (1965), in the course of their studies on skin, also observed the effect of 10 mg/kg initial weight per day of testosterone propionate for 21 days on the body weight of weanling (68; 87 g), young adult (271; 218 g), and adult (565; 360 g) male and female rats respectively. The testosterone propionate had no effect on the body weights of the weanling or adult male rats but produced an inhibition in the young adult rats beginning at 7 days which was not significant until after 21 days. In contrast, the body weights of the female rats were increased but to a small degree and at varying periods for the three age groups. The changes were significant after 41 and 21 days in male and female weanling rats, respectively; after 14 days in the young adult rats; and after 21 days in the adult rats. These results are consistent with the dose of the testosterone propionate. Testosterone propionate at 1 mg/day inhibited the food intake and body weight of castrated rats but increased them in adrenalectomized castrated rats (ASCHKENASY, 1959a).

Many of the newer steroids also result in stimulation or inhibition of growth in rats depending on the dose, differing between males and females. LIGHT and TORNABEN (1953a, b) applied by inunction 200 μg/day of 17α-methyltestosterone to weanling rats for 200 days. The growth curve of the treated male rats showed a small and insignificant inhibition beginning at about 120 days, while that of the female rats was definitely accelerated after a few weeks with a concomitant increase in food consumption. KAMSTRA and THURSTON (1961) also noted a difference in response of male and female weanling rats which were fed 17α-ethyl-19-nortestosterone at graded doses up to 0.044% of the diet (commercial laboratory chow). The body weights of the male rats were increased at the lower doses and inhibited at the higher doses, while body weight of the female rats progressively increased with an increase in dose. The injection of 10 mg/kg of testosterone propionate, 19-nortestosterone phenylpropionate, or 4-chlorotestosterone at 48-hour intervals from the 7–100 days of age inhibited the growth rate of male rats but accelerated that of female rats (Fujii et al., 1962). The total nitrogen content of the thigh muscles was not altered by the steroids.

Edgren (1963) adapted the body weight stimulation by means of lower steroid doses to a semiquantitative assay method and compared the growth-stimulating activity of a number of the synthetic steroids with that of testosterone propionate at selected one- or two-dose levels. Immature, 4-week-old, castrated rats were injected for 14 days with the designated steroid. Significant accelerated growth was evident after about 1 week and continued or gradually disappeared at varying degrees depending on the potency and the dose of the steroid. Barnes et al. (1954) injected rats castrated at 45–55 g body weight for 20 days with daily doses of 19-nortestosterone or testosterone propionate, varying from 10 to 500 μg/day. The maximum increase was obtained at about 50 μg/day with no further increase at the higher doses. 19-Nortestosterone propionate and 19-nortestosterone cyclopentylpropionate also produced maximal increases in body weight at a dose of about 50 μg/day.

The single injection of 30 mg/kg of 3-oxo-1-methyl-5α-androst-1-en-17β-yl-enanthate immediately stimulated an increase in body weight of the adult (200 g) female rat and accelerated that of the young (80 g) female rat only after 4–5 weeks. The periodic (every 14 days) injection of the steroid ester at 15 or 50 mg/kg did not affect the time of appearance or degree of the growth-accelerating effect in young female rats (60–70 g) (Jochle and Langecker, 1962). A similar effect was stimulated by injection of 3 mg/kg of acetate but 19-nortestosterone phenylpropionate at 12 mg/kg every 2 days for 14 weeks did not affect the growth curve. The administration of these steroids was ineffective in normal male rats.

Vargas and Abeliuk (1972) administered orally 1 mg/day (except Sundays) 17α-methyltestosterone or stanozolol to 26-day-old female rats for 3 months. After 20 days, stanozolol produced a significant increase while 17α-methyltestosterone required 30 days. The increase continued for the duration of the experiment. The intramuscular injection of testosterone in corn oil or stanozolol in either corn oil or as an aqueous suspension enhanced the growth rate of 4-day-old male and female rats over the 90 days of treatment.

β) Adult

The dose and duration of testosterone propionate treatment, the age, sex, and presence or absence of the testes influenced the response in body weight and appetite also of the adult rat.

The injection of testosterone propionate at 0.1 and 0.25 mg/day for 20–25 days enhanced (while 1.25 mg/day depressed) the increase in body weight of castrated rats of 150–155 days of age; and the injection of 0.5, 1.25, and 2.5 mg/day had a progressively depressing effect on the body weight of normal rats (Leatham, 1948a). The different doses of the hormone prevented the small rise in plasma protein following castration but had no definite effect in the normal rats. The plasma nonprotein

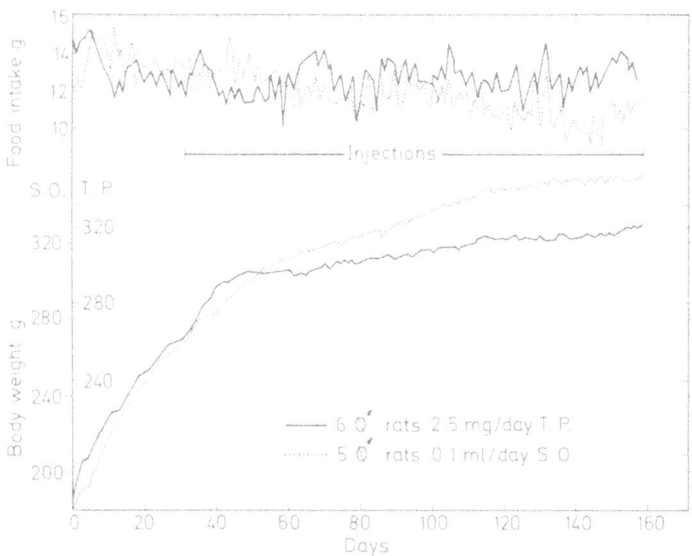


Fig. 3. Effect of testosterone propionate on the food intake and body weight of castrated rats. Castration was performed at zero day (KOCHAKIAN et al., 1950)

nitrogen was too variable to warrant the suggestive increase noted in the normal and castrated rats after testosterone propionate. In subsequent studies (LEATHAM, 1948b), the plasma NPN of normal rats (300 g) was not altered by 0.1 or 0.5 mg/day of testosterone propionate for 20–27 days but the slightly elevated values in the hypothyroid (thiourea or thiouracil) was not restored to normal by the hormone. The elevated plasma NPN produced in young adult male rats (150–160 days of age) by the feeding of a 78% protein (casein or lactalbumin) diet was not influenced by injection of 12.5 mg testosterone propionate 3 times weekly for 20 days (LEATHAM, 1951). The body weight also was not affected. Testosterone propionate for 30 days in a similar study in 50-day-old male rats also was ineffective on both plasma NPN and body weight (LEATHAM, 1951).

Testosterone propionate at 2.5 mg/day, begun 30 days after castration, produced an immediate sharp increase followed by a decrease in the rate of gain in body weight (Fig. 3) (KOCHAKIAN et al., 1950). The initial body weight affected the degree and duration of these effects. The larger rats (ca. 200 g) attained a maximum increase in body weight in 8–10 days, after which the rate decreased in 20 days to less than that of the control rats while the younger rats (ca. 150 g) attained these levels in half the time, and the increase in body weight was barely detectable. After about 60 days, the injected rats began to consume more food but their rate of gain in body weight was less than that of the controls (Fig. 3) with a resulting smaller final body weight. Analyses of these animals demonstrated an interesting reshuffling of protein, water, and fat (Table 1). The prolonged injection, 125 days, of 2.5 mg/day of testosterone propionate decreased the gain in weight of the carcass due to a storage of less water and protein and especially of fat. The kidney, liver, and accessory sex organs, however, were larger and contained more protein and water. The increase in protein and

Table 1. Effect of 125 days of injections of testosterone propionate (T.P.) on the composition of gain in body weight of castrated rats fed ad libitum[a]

| | Carcass gm | Liver gm | Kidney gm | Sem. ves. and pros. gm | Total gm |
|---|---|---|---|---|---|
| Weight | −34.30 | +1.00 | +1.30 | +7.52 | −24.48 |
| Water | − 1.00 | +0.54 | +0.99 | +5.41 | + 5.94 |
| Protein | − 2.92 | +0.33 | +0.67 | +3.41 | + 1.73 |
| Fat | −30.6 | −0.13 | +0.05 | +0.19 | −30.49 |

[a] Values are the differences between the changes in the androgen treated and control rats during the 125 days (Fig. 3). (KOCHAKIAN et al., 1950).

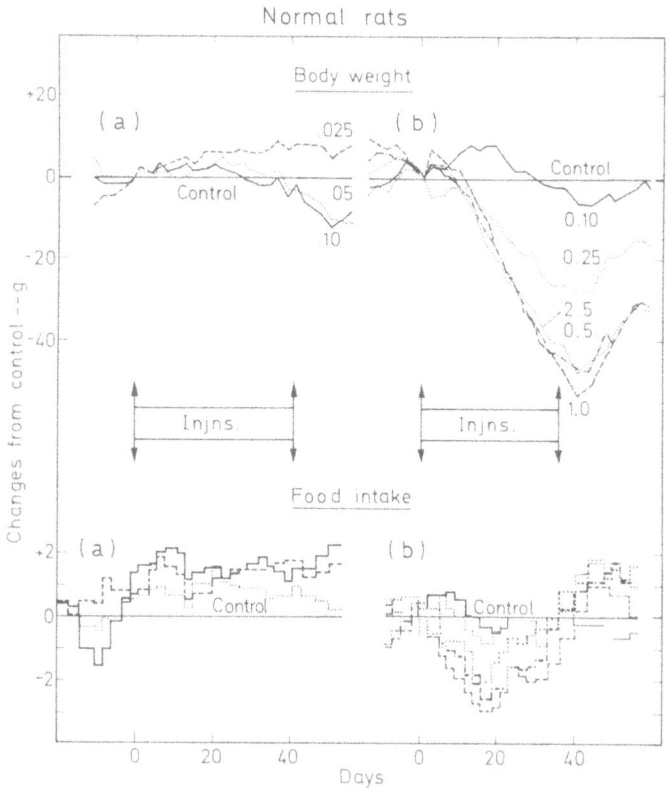


Fig. 4. Changes in body weight and food intake in normal rats injected with various doses of testosterone propionate. Changes are from control rats which during the injection period gained 107 g in series A and 90 g in Series B. Average body weights at beginning of injections were: Series A 301–305 g and series B 276–283 g per group. Dose of testosterone propionate (mg/day) is indicated beside body weight curves of respective groups (KOCHAKIAN and ENDAHL, 1959)

water in the liver, kidney, and especially the seminal vesicles and prostate was sufficient to more than counterbalance the loss of these constituents by the carcass. Thus, the hormone stimulated an inhibition of protein synthesis and/or an increase

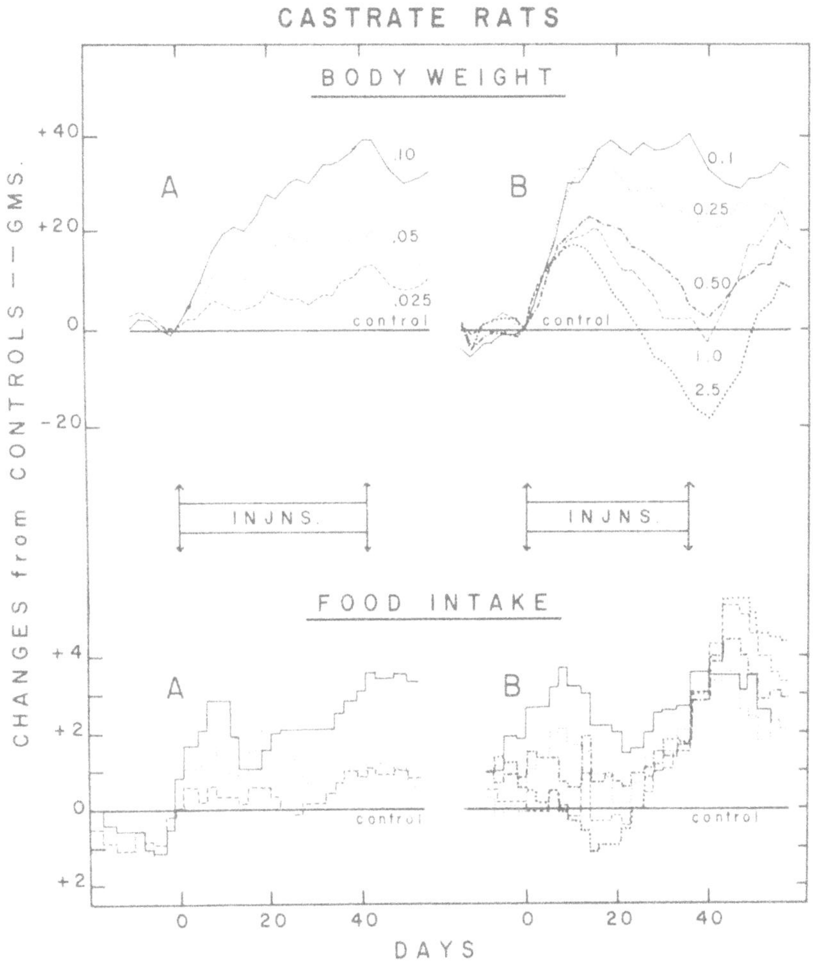


Fig. 5. Changes in body weight and food intake in castrated rats injected with various doses of testosterone propionate. Changes are from control rats which during injection period gained 107 g in series A and 77 g in series B. Average body weights at beginning of injections were: Series A 254–260 g/group and series B 274–294 g/group. Series A contained 10 rats/group and series B 5 to 7 rats/group. Dose of testosterone propionate (mg/day) is indicated beside body weight curves of respective groups (KOCHAKIAN and ENDAHL, 1959)

in protein catabolism in the carcass which was accompanied by a concomitant increase in protein synthesis in the internal organs.

The inhibitory effect of the testosterone propionate occurred at a lower dose in the normal (Fig. 4) than the castrated (Fig. 5) rat (KOCHAKIAN and ENDAHL, 1959). Castration was at 110 g (40 days) and the injections were begun 35 days later and continued daily for 40 days. Testosterone propionate initially produced a small, greater gain in body weight of the normal rats than in the controls (Fig. 4a, b) which was maintained only at the lowest dose (0.025 mg/day). Increases of the daily dose produced a progressive decrease in the rate of gain in body weight. The maximum inhibitory effect was reached at 0.5 mg/day. Cessation of the injections resulted in a gradual restoration of the rate of gain. The food intake was increased at the lower doses but was sharply decreased at the higher doses for about 3 weeks and then gradually returned toward that of the controls. On cessation of injections a further and marked increase in food intake occurred in those rats receiving more than 0.1 mg/day of the steroid. The castrated rats responded in the same pattern as the normal rats except that they were not as sensitive to the inhibitory effects. Every dose of testosterone propionate produced an increase in the gain in body weight above that of the controls. The maximum, a substantial increase, was obtained at a dose of 0.1 mg/day (Fig. 5a, b). The higher doses produced progressively smaller increases followed by an earlier and more rapid decrease. On cessation of injections the rate of gain in body weight increased. The amount and rate was greater after the halting of the higher doses of testosterone propionate.

The inhibitory effect of testosterone propionate is more severe in the heavier (fatter) rats (KOCHAKIAN and WEBSTER, 1958). A significant inhibitory effect on body weight was observed in the initial experiment only after 1.0 and 2.5 mg/day of testosterone propionate. In the subsequent experiment when the same rats were 130 g heavier, the inhibition was evident at 0.25 mg/day and increased in proportion to the dose of the steroid. An analysis of the body indicated that the loss in body weight was due primarily to a loss in fat and also a small amount of protein. The body fat but not the protein was regained within 21 days after cessation of injections. The changes in body fat were grossly visible and were calorically equal to the decrease in food intake during treatment and the increase during the recovery period. The loss in protein was equal to the decrease in the positive nitrogen balance induced by the inhibition of food intake by the hormone. The losses and restorations in weight and fat occurred in about equal amounts in the carcass, skin and hair, and viscera. The greatest loss in protein occurred in the skin and hair. The seminal vesicles and prostates and the kidneys responded to the hormone with a gain in all of their constituents (fat, water, protein) while the liver showed small decreases. These constituents were returned to normal 21 days after cessation of injections. The decrease in body weight was apparent before the decrease in food intake. Thus, it seems that the testosterone propionate mobilized the fat reserves for energy and induced a compensatory decrease in energy from the diet. The greater effectiveness of the hormone in the older animals may be due to the greater quantity and more labile lipid reserves.

KORNER and YOUNG (1955) also noted that age was an important factor in the response of both male and female rats on a constant maximum food intake to 0.5 or 0.3 mg/day of 17$\alpha$-methylandrost-5-ene-3$\beta$,17$\beta$-diol. Male rats at $3^1/_2$ months and female rats at 4 months gained more weight than control rats after 20 and 23 days of treatment. Older or younger rats exhibited a smaller final gain in weight and a smaller gain in protein but the loss in fat increased with the age of the female rats; male rats were not analyzed. The younger (3 months) and older (5 months) male, but not the female, rats gained less body weight than the controls. These data are consistent with the above and provide further support to the importance of the age of the rat and the quantity (and quality?) of the endogenous fat of the rat in its response to steroid treatment.

Further evidence for the fat-mobilizing property of testosterone is provided by studies on blood lipids. The unesterified fatty acids of the blood of young (111–116 g) female rats was further increased on the 2nd day of starvation by the simultaneous injection of 10 mg/day of testosterone propionate. The effect was not evident after 1 day of starvation and one injection of testosterone propionate, but if the dose was increased to 20 mg/day, a very variable increase was observed after 24 h (LARON and KOWADLO, 1963). The blood glucose and NPN were not changed by the testosterone propionate. Testosterone propionate and also dihydrotestosterone, but not epiandrosterone, at 5 mg/kg is reported (FISUM and SHARKEVICH, 1971) to increase the unesterified fatty acid levels of starved rabbits.

KOCHAKIAN et al. (1950) extended their studies to rats which were brought into weight and nitrogen equilibrium to obtain further insight into the nature of the changes in nitrogen balance and body weight induced by testosterone propionate. The rats were castrated at 270 g and the injections were started 5 months after castration when the body weights had reached an equilibrium at 292–298 g for the respective groups. The controls at the end of the experiment lost an average of one gram. The rats receiving 0.125 and 1.0 mg/day of testosterone propionate had gained 15 g by the 14th and 15th day and maintained the gain to the end (21 days) and the rats receiving 1.0 mg/day for 46 days attained a maximum increase of 17 g after 12 days which was decreased to 9 g at the end of the treatment. The main site of increase in weight was the carcass but as the dosage and duration of treatment were increased, there was a shift from the carcass to the accessory sex organs and, to a smaller degree, the kidney. These changes were accompanied by proportionate changes in protein and water. The rats treated for 46 days lost fat while those in the other two groups (lower dose and shorter period of treatment) gained fat; especially in the carcass and viscera. The weight of the liver was slightly increased with a proportionate increase in protein and water.

In another study the rats were castrated at a younger age (180–200 g body weight) and the injection of testosterone propionate at 1 mg/day was begun much later, at 7 and 11 months. The increase in body weight, 13 g, was essentially the same as in the previous experiment and agreed with that calculated from the nitrogen retained. Castration at the younger age and the longer period after castration resulted in a larger accumulation of body fat which was sensitive to the fat-mobilizing action of the testosterone propionate. These rats, after only 13 days of injections, lost 13.49 g of fat from the carcass and gut but gained almost twice as much protoplasm (protein plus the accompanying water) to give a net increase in body weight of 13.62 g. The major increase was in the carcass (and gut) but the seminal vesicles and prostates, liver, and kidney also contributed to the gain. The greater increase in the carcass protein of these rats over that of the "lean" rats may be due to a sparing action of the greater and more readily available fat stores.

### *b) Hypophysectomized*

#### α) Young

The early studies (see above) suggested that the acceleration of the growth rate by low doses of testosterone propionate and the inhibition by high doses was a reflection of the stimulation or inhibition of growth hormone of the anterior pituitary by testosterone propionate. TYSLOWITZ (1943) reported an acceleration of growth in young (27–50 g), but not in older, hypophysectomized rats upon the injection of testosterone propionate. LEONARD (1943), in a study of the influence of testosterone on the development of the mammary gland, recorded that the injection of 1.25 mg/day testosterone propionate for 10 days into male rats hypophysectomized at 22–36 days of age retarded the loss in body weight due to hypophysectomy. SIMPSON et al. (1944) conducted an extensive series of studies at different doses, 0.05–4.0 mg/day, in different age groups (females hypophysectomized at 26 days, males at 30, 40, 60, and 70 days), injections at different postoperative intervals (0–100 days after hypophysectomy), and for different lengths of time (4–20 days). Testosterone propionate did not produce a significant acceleration in growth but in some instances appeared to retard the loss in weight and augmented the growth-stimulating effect of pituitary growth hormone. PAESI et al. (1956) observed a retardation of the growth-inhibiting effect of 0.5 μg but not 2.5 μg estradiol-17β-benzoate for 12 days by the simultaneous injection of 40 μg/day testosterone propionate in young, 30–35 g, hypophysectomized female rats.

SCOW (1952) hypophysectomized male rats at 21–23 days of age and 4 days later injected them daily with 0.1 mg testosterone propionate for 28 days and at 0.5 mg/100 g body weight for another 28 days (Fig. 1). The slow rate of growth was accelerated within a few days so that after 21 days the body weight of the treated rats was 9 g greater than that of the control hypophysectomized rats but much less than that of the normal rats. Continuation of the injections diminished the difference so that at the end of the experiment the difference was reduced to 6.8 g. The increase in weight of the accessory sex tissues accounted for 58% of the increase in body weight and the increase in weight of the "levator ani" muscle was completely out of proportion with the increase in body weight. The hormone treatment had no appreciable effect on the composition (protein, water, fat) of the total carcass and the thigh muscles. THYBERG and LYONS (1948), in a previous, similar study, had observed a maximum difference in body weight of 12 g between treated and nontreated hypophysectomized rats.

STEINETZ et al. (1972) hypophysectomized male rats at 90–100 g body weight and 10 days later injected 0.04 mg testosterone propionate or 1 mg methandrostenolone (17β-hydroxy-17α-methylandrosta-1,4-dien-3-one) alone or simultaneously with pituitary growth hormone for 7 days. The steroids had no influence on the body weight but augmented the growth-stimulating effect of the pituitary preparations. The growth hormone, in turn, enhanced the growth-stimulating effect of testosterone propionate on the seminal vesicles, ventral prostate and "levator ani." Furthermore, the methandrostenolone at the used dose was ineffective on these organs but produced significant increases when the growth hormone preparation was administered simultaneously.

#### β) Adult

KORNER and YOUNG (1955) found that 0.3 mg/day of 17α-methylandrost-5-ene-3β,17β-diol for 30 days retarded the loss in body weight of $3^1/_2$ month old, hypophysectomized female rats. The body of the treated rats had more protein and water and less fat than the controls. The decanoate ester of DL-17β-hydroxy-13β-ethylgon-4-en-

3-one also prevented the weight loss in hypophysectomized male rats (REISS et al., 1965). Human chorionic gonadotropin produced the same effect as the steroid in hypophysectomized-normal but not hypophysectomized-castrated rats indicating that the endogenous production of androgens had an effect similar to that of the exogenous hormone.

Adult female hypophysectomized rats on constant food intake had a small and questionable increase in body weight after 5 days of 0.5 mg/day of testosterone propionate and a definite increase during the next 5 days with an increase of the dose to 2.5 mg/day (GORDAN et al., 1947). When adult castrated and noncastrated hypophysectomized male rats were allowed to adjust their food intake to their new low basal level, they responded to testosterone propionate with a significant increase in body weight to a maximum within about 2 weeks, where it was maintained on continued injections for as long as 2 months (KOCHAKIAN, 1950a). A maximum effect was attained at 1.0 mg/day; increase of the dose to 1.25, 2.0, or 5.0 mg/day did not produce a significant further increase. Simultaneous injection of such rats with an anterior pituitary growth hormone preparation produced a summation of their respective effects on body weight (KOCHAKIAN, 1960).

*c) Adrenalectomized*

The injection of testosterone propionate or acetate, androst-4-ene-3,17-dione or urinary androgen extracts decreased the survival time of immature (40–60 g) adrenalectomized male rats in direct relation with the physiologic dose (capon test) of the hormones. No beneficial effects were noted at the lower doses. The survival time of adrenalectomized female rats also was decreased (8.7–3.8 days) by injection of 5 mg/day of testosterone propionate (SPURR and KOCHAKIAN, 1939). ASCHKENASY (1959b) reported that injection of 1 mg/day of testosterone propionate during the 8th to 20th postoperative days increased the body weight of adrenalectomized-castrated rats (160–190 g) but decreased that of the nonadrenalectomized-castrated rats. DUCOMMUN et al. (1959) adrenalectomized female rats at approximately 150 g body weight and substituted 1% NaCl solution for the drinking water. The injection of an aqueous crystalline suspension of 17α-ethyl-19-nortestosterone for 3 weeks produced a greater increase (50 g) in body weight than in the controls (30 g).

*d) Thyroidectomized*

Metabolic studies in adult castrated rats demonstrated that surgical or chemical (thiouracil) thyroidectomy did not interfere with the body weight-stimulating and nitrogen-retaining properties of the testosterone propionate (see Chap. II.A). The body weight-stimulating effect of testosterone propionate was also apparent in immature thyroidectomized rats. SCOW (1952) thyroidectomized male rats at 21–23 days of age and 4 days later injected them daily with 0.1 mg of testosterone propionate per 100 g body weight for 28 days and increased the dose to 0.5 mg/100 g for another 28 days (Fig. 1). The body weights of the treated and untreated thyroidectomized rats showed a divergence after about 2 weeks and at the end of the experiment the treated rats were about 7 g heavier than the controls; more than 50% of the increase was due to the growth of the accessory sex tissues. The steroid produced a 25% reduction in the fat percentage and a small increase in the protein and water percentage of the carcass. The composition of the thigh muscles was not altered by the hormone.

CARRIERE (1966) thyroidectomized and castrated rats at 49 g body weight and observed 80 days later the expected marked retardation in growth after thyroidectomy and a further retardation in the doubly operated rats. The injection of 0.25 mg/day for 10 days and 0.5 mg/day of testosterone propionate for the last 32 days decreased the loss in weight in an amount equal to that lost after castration.

## 2. Mouse

### a) Normal

The injection of 0.2 or 0.4 mg/day of testosterone propionate for 140 days progressively retarded the growth rate of young male mice (23–48 days of age) but had no effect on simultaneously and similarly treated female mice (KOCHAKIAN, 1940). The males also showed loss of hair while the females did not. The toxic effects of 0.10 mg/day of estradiol dipropionate was partially offset by the simultaneous injection of 0.4 mg/day of testosterone propionate. A complete cessation of growth was produced in males after 7 weeks and in females after 2–3 weeks by the estrogen, and alleviation of both the effect on body weight and mortality by testosterone propionate was more pronounced in the females than in males. The toxic effects (weight loss) produced by total body irradiation of mice was enhanced by the injection of 0.1 mg/day of testosterone propionate but was reduced when 0.1 mg of testosterone cyclopentylpropionate was injected at 14-day intervals (SPELLMAN et al., 1954).

### b) Castrated

CHAI (1956) castrated mice at 33 days of age (16.8 g) and 1 day later injected testosterone propionate at 1–254 μg/day. Each dose level was twice that of the preceding dose. Body weight increased progressively with dosage and reached that of normal mice at the dose of 32 μg/day. It was not decreased or further increased by the several higher doses. The weight of the seminal vesicles also attained the normal size at the same time as body weight but continued to increase to double normal size with the increase in daily dose of the hormone.

Mice castrated at 1 month of age reached a plateau in body weight after about another month (KOCHAKIAN, unpublished). Injection of the mice at any time after this period with the various steroids produced an immediate and marked increase in body weight. For example, testosterone propionate for 6 days produced an increase of 2 g at 5 μg, 4 g at 10 μg, and 5 g at 25 μg/day. At 500 and 1000 μg/day, a maximum increase of 5 to 6 g was obtained after 6 days, and contination of injections for 10 days did not produce any further change. A log-dose response was obtained in the dose range of 0–200 μg/day for 7 days beginning the day after castration (28 days of age) or 28 days later. A wide variety of steroids stimulated an increase in body weight of the castrated mouse (KOCHAKIAN, 1949). Furthermore, the somatotrophic activity did not parallel either the renotrophic or androgenic activities. Compounds which had weak androgenic and/or renotrophic activity, e.g., androsterone, dehydroepiandrosterone, were nearly or as effective as the most potent compounds, e.g., 17$\beta$-hydroxy-5$\alpha$-androstan-3-one, testosterone, but required a much higher dose level. Thus, the body weight of the castrated mouse readily responds to the administration of testosterone propionate over a wide range of doses with an appreciable increase.

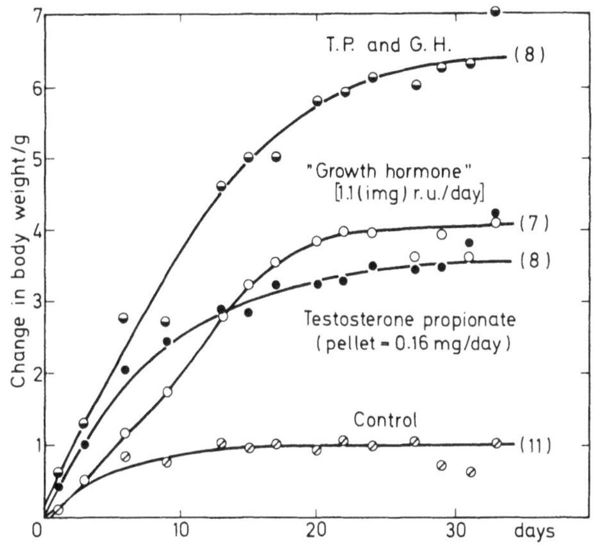


Fig. 6. Effect of testosterone propionate (TP) and growth hormone (GH) on body weight of castrated mice. Castration was at 17 to 19 g body weight and hormone treatment was begun 30 days later (KOCHAKIAN and STETTNER, 1948)

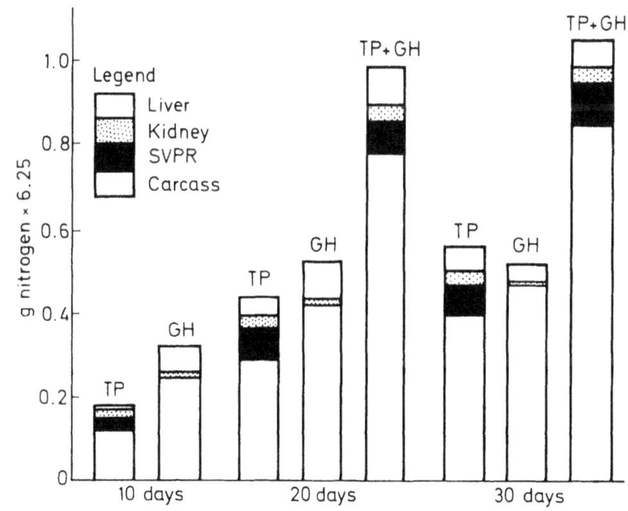


Fig. 7. Composition of the increase in body weight of castrated mice treated with testosterone propionate (TP) and growth hormone (GH) over that of control mice. See Figure 6 for further details (KOCHAKIAN and STETTNER, 1948)

The analysis of the mouse indicated that the increase in body weight is true growth. KOCHAKIAN and STETTNER (1948) castrated mice at 17–19 g body weight, and 1 month later one group of mice was implanted subcutaneously with a 14-mg pellet of testosterone propionate, another group received daily injections of a growth hormone preparation, and a third group was treated with both hormones. Mice from the treated groups and an untreated group were autopsied after 10, 20, and 34 days of treatment. The two hormones stimulated similar maximum increases in body weight and the combination gave a nearly exact summation of the effect of the two hormones (Fig. 6). Since both hormones were administered at the maximal dose, the stimulatory effects of the two hormones were independent and supplementary. An analysis of the carcass and the various organs indicated a progressive increase in protein synthesis with a proportionate increase in water (Fig. 7). The growth hormone was more effective than the testosterone propionate and the combined treatment gave a summation of the protein anabolic effect of the two hormones. Both hormones decreased the amount of fat but the effect of the growth hormone occurred sooner and was greater than that of the steroid hormone. The combined hormone treatment produced a summation of the fat-mobilizing activity of the two hormones. The greatest increase in protein synthesis and the loss in fat occurred in the carcass. The seminal vesicles, prostates, kidneys, and livers contributed to the newly formed protein with a proportionate increase in fat. JOCHLE and LANGEBECKER (1962) observed similar results 30 days after the injection of 10 mg of 3-oxo-1-methyl-5α-androst-1-en-17β-yl-enanthate in female mice (15–20 g). The body weight of the injected mice was 3.5 g greater than that of the controls. A proportional increase in calcium, phosphorus, and DNA and an actual percentage increase in nitrogen in the fat-free carcass supported the conclusion that true growth had occurred.

*c) Hypophysectomized*

FERGUSON and VISSCHER (1953) administered graded doses of testosterone propionate for 14 days into castrated-hypophysectomized mice and found that 20 μg/day was the minimum dose which would produce the maximum growth in the seminal vesicles. The minimum dose and twice the minimum dose were injected into hypophysectomized-castrated male and female mice respectively for 14 days, beginning 21–35 days after the operation at 3–4 months of age. The operated mice lost 13–32% of their body weight. The final weight of the treated male mice was practically identical to that of the controls but that of the treated female mice was slightly greater. Furthermore, the testosterone propionate was not able to counteract the weight depressing effect of stilbestrol or estradiol-17β. Studies with graded doses of testosterone propionate are needed to assess the role of the pituitary in the body weight-stimulating effect of the steroid in mice.

### 3. Guinea Pig

Guinea pigs castrated at 250 g (KOCHAKIAN et al., 1948b) or 350 g (KOCHAKIAN and TILLOTSON, 1957a) were implanted subcutaneously approximately 1 month later with pellets of various steroids for 30 days. At the end of the experiment the slightly depressed body weight due to castration was restored to normal. The restoration was attained with a relatively small dose and a severalfold increase in the dose of the steroids did not produce a further increase in body weight. SCOW and HAGAN (1955) also noted a restoration of the body weight of guinea pigs castrated at 20 days of age and injected with 2 mg/day of testosterone propionate from the 50th to 95th day of age. BOTTOMLEY and FOLLEY (1938) gave daily injections of 2 mg/day of a number of steroids to immature guinea pigs (170–190 g) for 28 days. Only 17α-methyltestosterone stimulated a significant increase in body weight.

### 4. Hamster

The implantation of a 24–26 mg pellet (0.15 mg/day) of testosterone propionate at the time of castration (48 g body weight) prevented the large increase in body fat and weight noted after 140 days of castration (KOCHAKIAN et al., 1948a). The implantation of 3 pellets of 14–15 mg each of 17α-methyltestosterone (0.5 mg/day) was more effective than testosterone pellets (0.72 mg/day for 20 days) in decreasing the body weight. The injection of testosterone propionate at 1 mg/day for 20 or 60 days had no significant effect on the final body weight of the castrated hamsters.

### 5. Monkey

The ovariectomized monkey has been demonstrated to respond to various steroids with an increase in body weight and a positive nitrogen balance (see Chap. I).

An immediate and remarkable increase in body weight of 5-month-old female monkeys *(Macaca mulatta)* was obtained upon injection of 7.5 mg/kg/week of testosterone propionate; 7 months later the body weights and lengths were equal to that of normal monkeys 2 years old (VAN WAGENEN, 1949). There was a simultaneous acceleration of menarche. An identical acceleration in growth and maturity of male monkeys was reported previously (VAN WAGENEN, 1947). The testosterone propionate injections were begun at 7 months of age, and at 2 years the body weights were 50% greater than comparable untreated males. The monkeys exhibited by X-ray a massive increase in musculature. After 10 months the monkeys showed full maturity, which is normally attained only after 6 years.

PHOENIX (1974) castrated male rhesus monkeys and 3 years later injected them with 1 mg/kg of dihydrotestosterone propionate for 56 days. The body weights of the monkeys were immediately and progressively increased during the 56 days of treatment and into 2 weeks of the postinjection period. The final increase was approximately 20% of the initial body weight. In the following 7 weeks the body weights gradually returned to nearly the original values.

The monkey is very sensitive to the protein anabolic action of testosterone and deserves further study.

### 6. Rabbit

The injection of 25 mg testosterone propionate or 10 mg methandrostenolone 3 times weekly for 5 weeks enhanced the increase in body weight of male rabbits (2.3 kg) (RASMUSSEN, 1968). The increase with methandrostenolone (850 g) was greater than with testosterone propionate (750 g); the increase in the controls was 650 g. These results are not in agreement with the earlier report of WRETE (1945). Rabbits (6 months old, 2.27–3.30 kg) were injected every other day for 30 days with 0.5, 10, and 25 mg of testosterone propionate. The lowest dose produced a small but not significant decrease in gain in weight but the 10 mg dose produced a definite inhibition, which was further enhanced by the 25 mg dose.

## III. Muscle

### Introduction

The changes in body weight after castration and testosterone administration were early associated (KORENCHEVSKY and DENNISON, 1934a, b; KOCHAKIAN, 1941) with changes in not only the weight of the carcass and the accessory sex organs but also other specific organs. The initial studies on nitrogen balance in castrated dogs indicated that the newly synthesized protein and increase in body weight could not be accounted for by the stimulation of growth of only the accessory sex organs (KOCHAKIAN and MURLIN, 1935). The observations of PAPANICOLAOU and FALK (1938) that the temporal and "other" muscles were greatly dependent on the internal secretion of the testes for growth emphasized the skeletal muscles as a prime site of testosterone action.

### 1. Guinea Pig

*Skeletal Muscle*

The most striking effects of castration and hormone administration is observed in certain muscles of the guinea pig. PAPANICOLAOU and FALK (1938) noted a marked increase in the normally flat and small temporal muscle of the female guinea pig injected with gonadotropic hormone (Follutein) in their studies on the estrous cycle. The increase was such that it was obvious by simple palpation and attained the size of that of the normal male guinea pigs. Castration of the immature male guinea pig prevented the growth of the muscle; it remained at the same size of the female. Injection of gonadotropic hormone into ovariectomized females or castrated males was ineffective. The injection of testosterone propionate but not progesterone or an estrogen (Amniotin) into castrated male and normal adult or ovariectomized female guinea pigs stimulated growth of the temporal and "other" muscles. The simultaneous administration of progesterone or the estrogen did not prevent the effect of the testosterone propionate. These results have been confirmed (Fig. 8) (KOCHAKIAN and TILLOTSON, 1956; KOCHAKIAN, 1964, 1966) and extended.

α) Castration

Castration produced a decrease in the weight of not only the temporal muscle (KOCHAKIAN et al., 1948a; SCOW and ROE, 1953; SCOW and HAGAN, 1955) but also the rate of growth of 47 other skeletal muscles. The retractor penis responded in the same fashion as the seminal vesicle and prostate (KOCHAKIAN et al., 1956b). The degree of retardation of growth varied widely among the muscles. In general, the degree of decrease was from the head to the hind legs. Some of the muscles did not show a retardation until about 12 weeks or longer of castration. Castration of adult male guinea pigs resulted in various degrees of atrophy of the responsive muscles within 6 weeks. The decrease in total muscle mass was consistent with the lower body weight of the castrated guinea pigs. The muscle atrophy was due to a diminution in the diameter of the myofibrils with a proportionate decrease in protein.

When castrated guinea pigs were studied over a long period of time, a small gradual increase in weight of the atrophied muscles was observed so that after

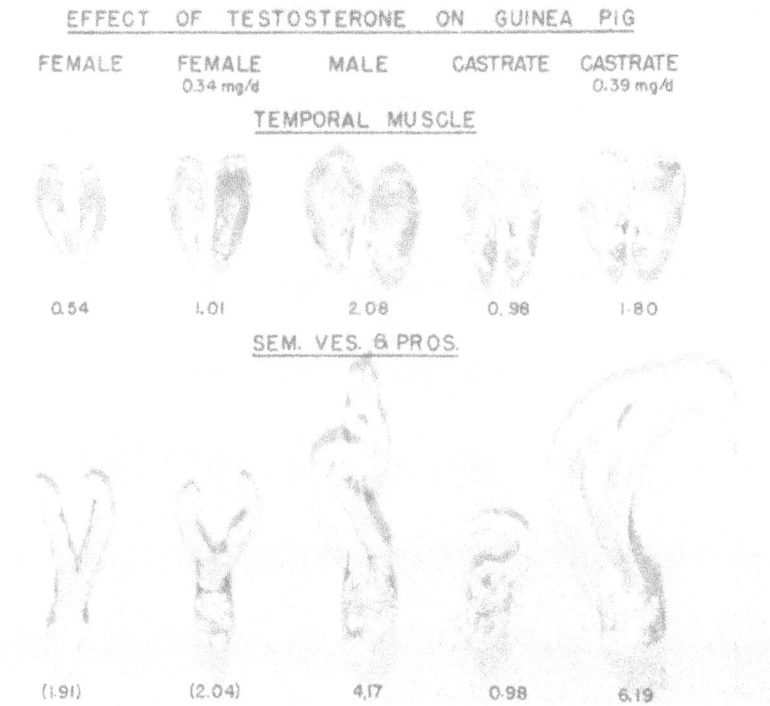


Fig. 8. Comparison of the size of the temporal muscle of the male and female adult guinea pig and the effect of testosterone propionate. Castration was performed at 400g body weight. Two pellets of testosterone propionate (15±1mg each) were implanted subcutaneously at the time of castration. Autopsy was 54 days later. Total testosterone propionate absorbed was 13 to 14 mg. The weights of the tissues in grams are presented below each tissue (KOCHAKIAN and TILLOTSON, 1956). SCOW and ROE (1953) also observed a smaller weight for the temporal and also the masseter and digastric muscles of the adult female guinea pig

230 days a significant increase in the temporal and masseter muscles was apparent (KOCHAKIAN et al., 1964a).

β) Steroids

*i) Dose and Duration of Treatment.* The administration of testosterone as a subcutaneously implanted pellet restored the weight of nearly all of the muscles, especially the temporal and retractor penis muscles (Fig. 9) (KOCHAKIAN and TILLOTSON, 1957a). A fivefold increase in the dose produced a further increase in the seminal vesicles and prostates to the normal level but did not produce a consistent further increase in the growth of even those muscles that gave the greatest responses. The previous study (KOCHAKIAN et al., 1948a) also showed a lack of parallelism between the response of the temporal muscle and the accessory sex organs. The subcutaneous implantation of two pellets of testosterone propionate to provide a dose of 0.24 mg/day restored the weight of the seminal vesicles and prostates nearly to normal after 90 days but the temporal muscles were only partly restored (40% of normal). In a

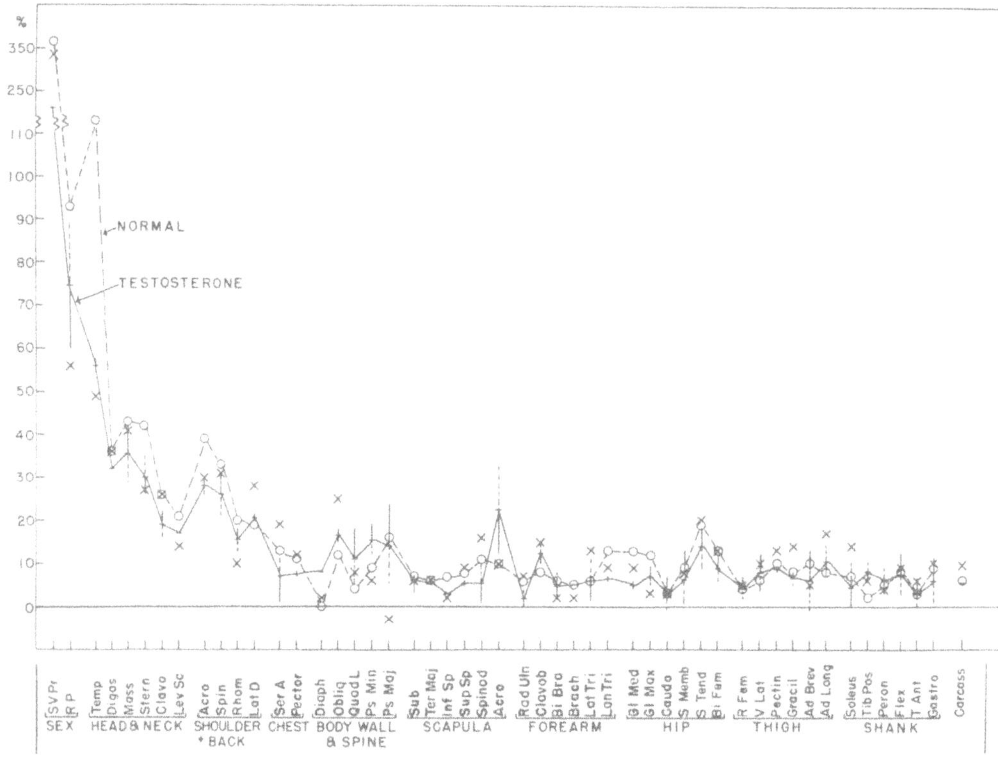


Fig. 9. The response of the individual muscles of the castrated male guinea pig to testosterone stimulation. The values are the averages of two separate experiments at approximately the same dose level (5.9 and 5.4 mg testosterone/21 days) 8 and 7 guinea pigs per group. The difference from the average for the first experiment is indicated by the short solid line and for the second experiment by the broken line. The third experiment (×) was at a five-fold increase in dose (28.4 mg/21 days). The normal values (○—○) are the averages of 12 guinea pigs. The values for the castrated control guinea pigs (the zero line) are the averages of 30 animals. Castration was at 510 g body weight. The testosterone pellets were implanted subcutaneously 30 days later (Kochakian and Tillotson, 1957)

further study, the guinea pigs were castrated at approximately 250 g and treatment was begun 35 days later. A large dose—12.5 mg/day—of testosterone propionate for 14 days returned the weights of both the temporal muscle and the seminal vesicles and prostates of the castrated guinea pigs to only approximately one-half that of the normal animals. If the guinea pigs were castrated 35 days later and three 15-mg pellets (0.71 mg/day) were implanted at the same time, the weight of the temporal muscles of the castrated guinea pigs after 30 days was the same as that of the normals but the weight of the seminal vesicles and prostates now was 43% greater than normal. An increase of dosage to 1.1 mg/day did not enhance the response of the temporal muscle but the seminal vesicles and prostates then were 73% greater than normal. As a further attempt to stimulate the growth of the muscles above normal, guinea pigs were castrated at 68 days of age (549–698 g) and injections of testosterone propionate were begun immediately at 5, 10, or 25 mg/day for 65 days (Kochakian and COCKRELL, 1958a). The 5 and 10 mg/day doses maintained all of the muscles except the temporal and the digastric muscles at or within the normal level. No significant differences between the dose levels were detected. The 25 mg/day of testosterone propionate not only maintained most of the muscles, including the temporal and digastric within the normal level, but actually increased some 20–30% above normal. The rate of growth of the body was maintained at the normal level by all three doses of the testosterone propionate. It would appear that the guinea pig requires a relatively large dose of testosterone to restore or even maintain its muscles to or at the normal level but part of this requirement is due to the method of administration. When testosterone was injected subcutaneously as an aqueous crystalline suspension at 0.25 mg/day for 14 days, it was twice as effective on the temporal muscle and 8 times as effective on the seminal vesicles and prostates of the castrated guinea pigs as a similar dose in oil (KOCHAKIAN and ROBERTSON, 1950a). An increase of the dose to 0.50 mg/day enhanced the differential to 3 and 16 times respectively and made the potency essentially the same as that of a similar dose of testosterone propionate.

*ii) Comparison of Various Steroids.* KOCHAKIAN et al. (1948a) noted that 7 active steroids stimulated a log-dose response in the weight of the temporal muscles of the castrated guinea pig and a dichotomy between the response of the muscle and the seminal vesicles and prostate on modification of the chemical structure of testosterone. This study was extended to the 46 other muscles and the internal organs with 13 different steroids (KOCHAKIAN and TILLOTSON, 1957a). A comparison of the relative potency of the active steroids on the more responsive muscles revealed that 17β-hydroxy-5α-androstan-3-one was the most potent steroid (Table 2). It was 50% more

Table 2. Relative myotrophic and androgenic potencies of several $C_{19}$ steroids. The respective values are compared to those of testosterone set at 100 (KOCHAKIAN and TILLOTSON, 1957)

| | Sem. ves. and pros. | Digastric | Masseter | Clavo-trapezius | Temporal | Sterno-mastoid | Retractor penis |
|---|---|---|---|---|---|---|---|
| 17β-Hydroxy-5α-androstan-3-one | 150 | 380 | 320 | 280 | 230 | 190 | 150 |
| Testosterone | 100 | 100 | 100 | 100 | 100 | 100 | 100 |
| 17β-Hydroxy-17-methylandrostan-3-one | 100 | 140 | 120 | 80 | 90 | 100 | 80 |
| 5α-Androstane-3α, 17β-diol | 50 | 60 | 70 | 280 | 70 | 90 | 30 |
| 17α-Methyltestosterone | 50 | 60 | 50 | 40 | 30 | 20 | 100 |
| 17α-Methyl-5α-androstane-3α, 17β-diol | 16 | 30 | 20 | 9 | 20 | — | 4 |
| Androst-4-ene-3,17-dione | 17 | 10 | 10 | — | 10 | 3 | 20 |
| 5α-Androstane-3,17-dione | 16 | 20 | 10 | 20 | 10 | 8 | 8 |
| Dehydroepiandrosterone | 15 | 10 | 4 | 20 | 10 | 3 | 4 |
| Epiandrosterone | 5 | 10 | 6 | — | 4 | — | 30 |

Table 3. Ratio of myotrophic to androgenic (seminal vesicles and prostates) activity of several $C_{19}$ steroids. The respective values are compared to those of testosterone set at 1.0 (KOCHAKIAN and TILLOTSON, 1957)

| | Digastric | Masseter | Clavo-trapezius | Temporal | Sterno-mastoid | Retractor penis |
|---|---|---|---|---|---|---|
| 17β-Hydroxy-5α-androstan-3-one | 2.5 | 2.1 | 1.9 | 1.5 | 1.3 | 1.0 |
| 5α-Androstane-3α,17β-diol | 1.2 | 1.4 | 5.7 | 1.4 | 1.8 | 0.6 |
| 17β-Hydroxy-17-methyl-5α-androstan-3-one | 1.4 | 1.2 | 0.8 | 0.9 | 1.0 | 0.8 |
| 17α-Methyl-5α-androstane-3α, 17β-diol | 1.8 | 1.2 | 0.6 | 1.3 | — | 0.3 |
| Testosterone | 1.0 | 1.0 | 1.0 | 1.0 | 1.0 | 1.0 |
| 17α-Methyltestosterone | 1.2 | 1.0 | 0.7 | 0.5 | 0.4 | 1.9 |
| 5α-Androstane-3,17-dione | 1.3 | 0.8 | 1.4 | 1.3 | 0.5 | 0.5 |
| Epiandrosterone | 2.6 | 1.2 | — | 0.8 | — | 5.8 |
| Androst-4-ene-3,17-dione | 0.8 | 0.8 | — | 0.6 | 0.2 | 1.3 |
| Dehydroepiandrosterone | 0.7 | 0.3 | 1.5 | 0.7 | 0.2 | 0.3 |

effective than testosterone in the stimulation of growth of the seminal vesicles and prostates and the retractor penis and was twice to three times more effective in stimulating the growth of the muscles. The introduction of a 17α-methyl group greatly reduced the efficacy of both 17β-hydroxy-5α-androstan-3-one and testosterone. The 3α,17β-diols were more effective than the diones and the 3β-hydroxy steroids were the least effective. 17α-Methyl-5α-androstane-3β,17β-diol and 17α-methylandrost-5-ene-3β,17β-diol were inactive in the doses used (3.5 and 6.5 mg/21 days, respectively). Testosterone propionate was studied at only a single dose (3.2 mg/21 days).

The stimulation of growth of the more responsive muscles did not parallel the response in the seminal vesicles and prostates (Table 3). The dichotomy in response was related to the chemical nature of the steroids. The reduced compounds had a more favorable effect on the muscles while the more oxidized compounds were relatively more effective on the accessory sex organs. Four compounds were more effective than testosterone in the relative stimulation of growth in the various muscles and 17β-hydroxy-5α-androstan-3-one was the most effective. The introduction of the 17α-methyl group decreased the favorable myotrophic effect of the steroids.

*iii) Thyroidectomized.* Since the thyroid has a profound effect on body weight, KOCHAKIAN and TILLOTSON (1957b) studied the effect of thyroidectomy and testosterone propionate on weights of the various muscles and organs. Guinea pigs were castrated and thyroidectomized at 470–549 g body weight. Thyroidectomy did not change the rate of growth of the body, carcass, internal organs, or the 48 muscles during the following 51 days. The subcutaneous implantation of a 15-mg pellet of testosterone into the thyroidectomized-castrated guinea pigs stimulated the usual increase in the weight of the body, muscles, and internal organs after 21 days. Apparently, the absence of the thyroid was without influence. The negative results seemed anomolous and prompted a search for aberrant thyroid tissue (KOCHAKIAN and COCKRELL, 1958b). Guinea pigs thyroidectomized at 60–100 days showed 37–47 days later by $^{131}$I uptake several aberrant nodules in various sites. The nodules were verified by histologic section. Only one of the nonthyroidectomized guinea pigs exhibited an apparent nodule. Thus, the negative results on body and muscle weight after thyroidectomy were due to maintenance of the animals in a euthyroid state by the activation of the aberrant nodules.

*iv) Starvation.* Since starvation results in a loss of protein from the body and internal organs as well as a loss of fat, KOCHAKIAN et al. (1957a) compared the effects of starvation with castration on the weights of the individual muscles and the organs with the simultaneous administration of testosterone. The guinea pigs were castrated at 485–542 g body weight (83–92 days of age); food was removed 42 days later from three groups: 1. normal, 2. castrated, and 3. castrated and implanted subcutaneously with a 15-mg pellet of testosterone. Nine days later all the animals were killed, including a group of normal and a group of castrated guinea pigs which were not fasted. The deprivation of food for 9 days decreased the weight of the individual muscles to approximately the same degree with some exceptions. An identical response occurred in the castrated guinea pigs. The lack of food caused only a small further decrease in those muscles which already had decreased as a result of castration. Thus, most of the labile protein of these muscles already was lost due to castration. On the other hand, the protein of the muscles which were not or only slightly responsive to castration readily lost their labile protein on fasting. Castration for 9 days was more effective in reducing the weight of the seminal vesicles and prostate than a 9-day fast but less effective on the weight of the individual muscles.

The administration of a pellet of testosterone at the beginning of the fast had a slight stimulatory effect only on the seminal vesicles, prostates, and retractor penis. Muscles were not affected. When a large dose—10 mg/day—of testosterone propionate was injected during the fasting period, the accessory sex tissues and several of the muscles were stimulated to increase in weight. Thus, testosterone propionate was able to mobilize protein from other parts of the body and to resynthesize the endogenous amino acids into the newly formed muscle and sex tissues. The internal organs decreased in weight and protein during the fasting period.

### γ) Growth Hormone

A growth hormone preparation of the anterior pituitary was injected daily for 14 days into castrated guinea pigs 30–35 days after castration (KOCHAKIAN and ROBERTSON, 1950b). The initial dose produced a slight increase in the weights of the temporal muscle and seminal vesicles and prostates. A twofold increase in the dose enhanced the increase in the weights of both tissues but a fourfold increase in dose did not stimulate a further increase in either tissue. No change in body weight was observed. The weight of the adrenals and thyroids were slightly but insignificantly increased.

### δ) Muscle Composition

The decrease in weight of the individual muscles after castration and the increase upon hormonal stimulation were accompanied by proportional changes in protein (nitrogen) and water (KOCHAKIAN et al., 1948a; KOCHAKIAN et al., 1956b; KOCHAKIAN and TILLOTSON, 1957a; KOCHAKIAN et al., 1964a). The concentration of nonprotein nitrogen was decreased 65 days after castration (KOCHAKIAN et al., 1948a) and after 75 but not after 30 days (SCOW and HAGAN, 1955). Testosterone administration restored the values toward normal. Myoglobin concentration was increased in the temporal muscle of the castrated guinea pigs 60 and 90 days after castration and was restored to normal by the injection of 2.5 mg/day of testosterone propionate from the 85th to 115th day of age (SCOW and ROE, 1953). The calf muscles were not affected by either castration or hormone administration. The concentration of myosin was decreased and that of collagen was increased 75 days after castration (SCOW and HAGAN, 1955). The injection of 2 mg testosterone propionate per 100 g body weight daily from the 50th to 95th day of age restored the various values to or near normal. The concentration of myoglobin in the small temporal muscle of the female guinea pig was higher than that of the normal male and similar to that of the castrated male guinea pig. The high concentration of myoglobin in the female and castrated male guinea pigs imparted a dark red hue to the temporal muscle. The amino acid composition of the protein (perchloric or trichloracetic acid precipitable) of the temporal, masseter, and six other muscles was not detectably changed by castration and testosterone or testosterone propionate administration (KOCHAKIAN et al., 1964b).

## 2. Rat

### *a) Skeletal Muscle*

Castration of rats at 36–40 days of age resulted in the expected slight retardation in the rate of increase in body weight (KOCHAKIAN et al., 1956a). Autopsy of the rats after 56, 94, and 132 days indicated that the growth of 29 individual skeletal muscles and the heart and diaphragm occurred at the same rate as those in the normal controls and the increase in body weight. The percent of nitrogen (protein) was unaltered. SCOW (1952) also noted that the weight and protein of the thigh muscle of rats castrated at 21–23 days increased in proportion to the increase in body weight and the enhancement of the rate of increase in body weight by testosterone propionate was accompanied by a proportionate increase in the weight and protein of the thigh muscles.

KORNER and YOUNG (1955) studied the effect of 1 mg/day of 17α-methylandrost-5-ene-3β,17β-diol for 35 days on weight and protein of seven skeletal muscles and the heart and diaphragm of $3^1/_2$-month-old female rats. The acromiotrapezius and digastric muscles were slightly (10%) increased and the plantarius and soleus were decreased; the other muscles changed in proportion to the increase in body weight. The same muscles were increased, but the masseter decreased in 3-month-old male rats injected with 0.5 mg/day of the steroid for 28 days. The response in the muscles was modified by the age of the rats and the dose of the steroids as well as body weight (KORNER and YOUNG, 1955). Adrenalectomized or hypophysectomized male and female rats responded with isometric increases in all of the muscles. The injection of 4-month-old female rats for 20 days with 0.4 mg of the above steroid caused no significant change in the protein content of the myofibrillar, sarcoplasmic, or collagen fractions of the acromiotrapezius, masseter, quadriceps, heart, or diaphragm muscles (KORNER, 1955). No change was found in the granular, but the myofibrillar adenosine triphosphatase activity of the acromiotrapezius and quadriceps muscles was significantly reduced by the steroid. No change was obtained in the cytochrome C concentrations of the gastrocnemius muscle of rats castrated at 30 days of age and injected 5 months later with 12 mg of a crystalline suspension of testosterone propionate at 10-day intervals for 40 days (TISSIERES, 1947). Testosterone accelerated creatine metabolism in the skeletal muscle of the young (150–170 g) female rat (TEDESCHI et al., 1960). The testosterone was injected at 5 mg/day for 3 days, on the 4th day 10 μCi (275 μg) [methyl-$^{14}$C]-methionine was injected and 6 h later, the creatine of the skeletal muscle of the treated rats contained more radioactivity than that of the controls.

WORTMAN and LEONARD (1953) injected female rats (180–220 g) for 3 days with 1 mg/day of testosterone propionate and observed a significant increase in the concentration of glycogen in the rectus femoris, abdominal, and diaphragm muscles. The injection of testosterone propionate did not influence the glycogenolytic effect of epinephrine. BERGAMINI (1975) observed a small linear increase in the glycogen content of five different skeletal muscles of male rats castrated at 200 g and injected with a large dose, 5 mg/day, of testosterone propionate (see p. 100).

Testosterone propionate injected at 1 mg/day prior to and for 30 days after severance of the sciatic nerve of adult male (300 g) and female (236 g) rats did not modify the marked atrophy of the tibial muscles in spite of a marked retardation in the loss in body weight (GORDAN et al., 1949). There was no difference in length-tension relationships between the treated and untreated rats but fibrillation of the denervated anterior tibial muscle was increased by the steroid.

### *b) Perineal Muscle*

#### α) Initial Observations

WAINMAN and SHIPOUNOFF (1941), in the course of other studies, observed that castration of the rat caused a marked decrease in the mass of the striated perineal musculature—the bulbocavernosus, ischiocavernosus, and levator ani. Since comparable changes were not noted in other muscles, they investigated the hormonal regulation of the growth of these muscles. Sixty-day-old rats were castrated, and a group was injected for 35 days with 0.2 mg/day of testosterone propionate. At the same time two groups of normal rats of the same age were injected daily with 0.2 and 0.5 mg of hormone respectively. At autopsy the perineal complex of the castrated control rats weighed only 19.3% of the normal. The injection of testosterone propionate at 0.2 mg/day increased the weight of the perineal complex of both the castrated and normal rats to approximately 50% above normal and the 0.5 mg/day dose produced a further small increase to 59% above normal. After castration the muscle fibers were greatly decreased in width to varying degrees. The nuclei appeared to be increased in number due to the decrease in fiber width and the infiltration of connective tissue. No increase in mitotic figures of the nuclei were apparent. The injection of testosterone propionate restored the normal histological appearance of the muscles with an increase above normal in fiber width.

The perineal complex in the normal female was extremely small and barely dissectable but the injection of 1 mg/day to ovariectomized, 25-day-old female rats for 25 days greatly increased the weight of the perineal complex (WAINMAN and SHIPOUNOFF, 1941). GUTMANN et al. (1967) injected female rats with 1 mg testosterone propionate twice per week beginning at birth and observed that after 1 month the "levator ani" was the same size as that of an untreated male of the same age, but after 2 months of treatment it was only about two-thirds the weight of that of the normal male. This difference could be due to the dose regimen being less efficient than the endogenous secretion by the normal male. It was of interest that 1 month after cessation of injections, the weight of the muscle had only partially regressed.

EISENBERG et al. (1949) confirmed the decrease in weight and the quantitative effects of testosterone propionate on the "levator ani" muscle, and EISENBERG and GORDAN (1950) proposed the use of this muscle for the bioassay of protein anabolic activity (see Chap. V). The validity of the use of this muscle to distinguish the protein anabolic from androgenic activity has been questioned (NIMNI and GEIGER, 1957; SCOW, 1952; SCOW and HAGAN, 1957; AHREN et al., 1962; APOSTOLAKIS et al., 1963; BERANGER et al., 1963; BERGAMINI, 1973; BRESLOFF et al., 1974). HAYES (1965) pointed out that the muscle of the perineal complex designated as the "levator ani" is in fact the dorsal bulbocavernosus and is peculiar to rodents. CIHAK et al. (1967), in a careful embryologic study of the development of this muscle, have substantiated HAYES' criticism. (However, since the designation levator ani has had widespread use in the literature, it will continue to be used in this paper.)

### β) Weight and Composition

The "levator ani" muscle has been used extensively not only for bioassay of hormones (see Chap. V), but also has undergone intensive study with respect to atrophy after castration and hypertrophy after hormone stimulation. The following will be devoted to the characterization of the growth of the muscle after castration and androgen treatment.

*i) Castration.* EISENBERG and GORDAN (1950) noted that the "levator ani" of the rat castrated at 30 days of age continued to grow during the subsequent 30 days but at a much slower rate than in the normal rat. SAUNDERS (1957) and BURESOVA et al. (1972) also observed a continued growth for 30 days after castration of rats at 23 and 40 days of age respectively. No further growth, however, was noted after the subsequent 20 and 30 days. In contrast, the seminal vesicles and the prostates of the rats in these studies ceased to grow immediately after castration.

GORI et al. (1967) castrated mature (240–350 g body weight) rats and found a rapid decrease in the weight of the "levator ani" during the first 3–4 weeks and a further slow decrease up to 8 months after castration. A more detailed study of the decrease in weight during the first week after castration was made by BERGAMINI et al. (1969). The diameter of the fibers decreased rapidly but to varying degrees during the first 15 days and decreased more slowly during the next 45 days, followed by a slow progressive small increase up to 8 months after castration. Careful and detailed electron microscopic studies demonstrated that myofilaments were lost from the periphery of the myofibrils. A marked decrease in glycogen also was found in confirmation of biochemical data (see below). The mitochondria were reduced in number and volume roughly in parallel with the myofibrils. The lysosomes were distributed more widely and showed an increase in size and number in the atrophic muscle. GALAVAZI and SZIRMAI (1971) confirmed the decrease in fiber width after castration and also found that the number of fibers was not changed in 119-day-old rats castrated at 42 days. They did not observe, however, indications of disorganization or other changes upon examination by electron microscopy which could be interpreted as signs of loss of contractile material as observed by GORI et al. (1967) in the older rats. GALAVAZI and SZIRMAI (1971) attributed this difference to age at castration. Myonuclear multiplication was suppressed in the castrated rats and the quantity of contractile material per myonucleus was lower than in the normal rats of the same age.

*ii) Steroids.* A detailed analysis of the growth and composition of the levator ani muscle under steroid stimulation has been reported. GORI et al. (1969) castrated rats at 200–300 g body weight and, 2 months later, testosterone propionate was injected at 2.5 mg/day and autopsy was performed after 24, 48, 72, 96, 120, and 168 h. The weight of the muscle increased rapidly to 75% of the controls after 48 h, where it remained until the 5th day and then exhibited a second growth phase which attained the normal level after 15 days without any further increase during the following 15 days. A similar diphasic increase was noted in rats treated only 15 days after castration (BERGAMINI et al., 1965), but the first increase was less evident and earlier. The early increase may be due to imbibition of water (EISENBERG and GORDAN, 1950). Testosterone propionate reversed the changes detected by light and electron microscopy after castration. The diameter of the muscle fibers and myofibrils was rapidly and markedly enlarged. The volume of the nucleoli was returned to normal after 24 h and continued to increase to more than twice normal but after the 4th day began to decrease and returned to normal by the 15th day of treatment. GALAVAZI and SZIRMAI (1971) carried out similar studies but castrated the rats at immaturity (42 days of age) and used a lower but still high dose of testosterone propionate, 1.25 mg/day, for 15 days beginning 77 days after castration. The diameter of the muscle fibers was rapidly increased but the length and number was unchanged. A marked increase in the number of myonuclei occurred between the 4th and 7th day of treatment. The frequency distribution of the width of the individual fibers indicated that the hormonal influence was markedly different for the individual fibers.

### γ) Effect of Thyroidectomy and Hypophysectomy

The "levator ani" muscle of the rat thyroidectomized or hypophysectomized at 21–23 days of age was only one third that of rats castrated at the same age. The daily injection of testosterone propionate at 0.1 mg/100 g for 28 days and 0.5 mg/100 g body weight for the next 28 days produced a six- to seven-fold increase in the weight of the muscle (SCOW, 1952). LINET and BUMBA (1969) hypophysectomized male rats at 150–170 g body weight and immediately gave daily injections of testosterone propionate at 1.7 mg/100 g or 19-nortestosterone phenylpropionate at 4 mg/100 g. The wet and dry weight and nitrogen content of the treated rats was much greater than that of the controls after the 5th, 6th, and 7th days of treatment. LEONARD (1952) hypophysectomized adult (190–240 g body weight) rats and noted an approximate 100% increase in the concentration of glycogen 7 days after injection of 2 mg/day of testosterone propionate, or 1 mg/day begun 7 days after the operation, or begun 4 days after the operation and continued for only 3 days. A progressive increase in the concentration of glycogen was observed after 1, 2, and 3 days of injection of 1 mg/day of testosterone propionate, begun 7 days after hypophysectomy.

δ) Effect of Denervation

BASS et al. (1969) compared the atrophy of the "levator ani" after castration with that after denervation. They bilaterally sectioned the pudendal nerve in one group of rats and castrated another group at 1 month of age. In agreement with an earlier report by LEONARD (1957), the atrophy of the muscle was greater in the denervated than in the castrated rats 1, 2, and 3 months after the operations. The contraction time of the muscle from the castrated rats was slightly but significantly increased after 1 month, but no further increase was detected after 2 and 3 months. The relaxation time showed a similar response. The contraction and relaxation times of the denervated muscle were more profoundly affected than those of the castrated rat; they steadily increased. The percent of contracttile and sarcoplasmic proteins in the muscle after 6 months of castration was unchanged. In contrast, the denervated muscle showed a relative decrease of the contractile and an increase of the sarcoplasmic proteins.

The injection of testosterone propionate, 1 mg twice weekly, beginning 2 weeks after castration or denervation of 1-month-old rats, produced the expected increase in the weight of the "levator ani" muscle of the castrated rat but was without effect on the denervated muscle (BURESOVA et al., 1972). On the other hand, denervation of the female rat at birth followed by injection of testosterone propionate for 30, 40, and 75 days prevented the normal complete involution of the "levator ani" muscle (BURESOVA et al., 1972).

HANZLIKOVA and GUTMANN (1972) utilized the cross-union of nerves, which is known to change the histochemical pattern of muscle, to help clarify differential effects of hormonal and neurotrophic agents. The cross-union of the tibial and pudendal nerves in 2-month-old rats reversed the homogeneous to a heterogenous population of muscle type in the "levator ani" muscle as indicated by ATP-ase and succinic dehydrogenase activities. Castration of the rats 2 months after cross-union of the nerves resulted in the usual atrophy of the "levator ani" muscle but the heterogeneous pattern was maintained. The injection of testosterone propionate for 30 days produced the usual marked increase in weight and a maintenance of the heterogeneous pattern. The same effect was obtained in long-term castrated rats. The authors concluded that the muscle fiber pattern is neurally regulated and is not affected by testosterone. In a subsequent study, HANZLIKOVA and GUTMANN (1974) transplanted the soleus muscle into the site of the "levator ani" muscle with innervation by the pudendal nerve. The soleus muscle acquired the histochemical properties and fast contraction time of the "levator ani" muscle but showed no changes in weight, contraction, or histochemical properties after castration or the injection of testosterone propionate for 2 months.

ε) Glycogen

The glycogen concentration of the "levator ani" muscle is rapidly changed after castration and testosterone propionate treatment. LEONARD (1952) castrated mature (approximately 300 g) rats and observed a 15% decrease after 1 day, and 27% and 32% after 2 and 7 days, respectively. After 15–96 days, however, no difference between the normal and castrated rats was detected. The injection of testosterone propionate at 1 mg/day for 3 days greatly increased the concentration of glycogen in the normal and the 42–50-day and 7-day castrated rats. Essentially the same result was produced 24 h after a single injection of the hormone in rats castrated 4 days previously. BERGAMINI et al. (1969) noted a greater decrease in glycogen content than the weight of the muscle 2 days after castration of 250–300 g rats and a parallel slow decrease in the subsequent 28 days of study. The injection of testosterone propionate at 1 mg/day into rats castrated 15 days previously at 100-g body weight produced a 100% increase after 12 h, and a 400% increase after 24 h. Subsequently the total glycogen content remained unchanged while the weight of the muscle increased so that the concentration of the glycogen after 8 days was essentially the same as that in the normal rat. The lag in the appearance of the glycogen was reduced to 6 h by the intravenous injection of an aqueous alcohol solution of 0.8 mg of testosterone or 1 mg of testosterone propionate. The increase in glycogen induced by testosterone propionate was not altered by adrenalectomy or by alloxan diabetes. The formation of glycogen from [$^{14}$C]-glucose injected into 15-day castrated rats was diminished (-40%). Treatment with testosterone propionate produced an increase of 300% in glycogen-specific radioactivity after 12 h and a further increase after 24 and 48 h. Experiments in vitro with muscle from immature rats injected with 1 mg/day of testosterone propionate duplicated and further substantiated the experiments in vivo.

BERGAMINI and his associates conducted a number of experiments to delineate the effect of testosterone on glycogen synthesis. BERGAMINI (1969) studied xylose transport in immature 50-g rats injected with a maximal dose (5 mg) of testosterone propionate 24 h before the experiment. The hormone treatment increased the penetration rate of xylose and distribution space which was inhibited by phlorizin. The penetration of L-glucose, a nonactively transported sugar, was not changed. Adrenalectomy or hypophysectomy 2 days previously and the production of alloxan diabetes 3 days before sacrifice did not alter the effect of the hormone.

Enzyme studies (BERGAMINI et al., 1969), the above studies, and further transport studies (PAGNI et al., 1973) suggested that the increased glycogen synthesis under testosterone stimulation could be attributed to greater penetration and more rapid phosphorylation of glucose and to increased glucose-6-P-independent glycogen synthetase activity. Incubation of the muscle with D-glucose produced a rapid accumulation of D-glucose-6-phosphate in the muscle within 12 h and a further increase in 24 h. The free glucose content of the muscle was not significantly increased. [$^{14}$C]-xylose showed an accumulation in the muscle fibers parallel with that of glucose in agreement with an earlier report by ARVILL (1967) and consistent with the transport of 2-deoxyglucose and α-amino-isobutyric acid reported by MILLS and SPAZIANI (1968). The content of glucose-6-P and glucose-1-P of the "levator ani" muscle 15 days after castration was restored to normal 12 h after injection of 1 mg of testosterone propionate. The related enzymes also showed changes. Hexokinase activity was increased 40% after 12 h of testosterone propionate. Glucose-6-P-independent glycogen-synthetase was also significantly (+100%) increased. Both of these enzymes decreased slightly after the maximum at 12 h. Total glycogen-phospharylase activity also showed a significant increase but at a later time, 24 h. Phosphoglucomutase and UDPG pyrophosphorylase activities were not changed within the period of study 48 h.

BERGAMINI (1975) compared the effect of glycogen synthesis in the "levator ani" muscle with that in five skeletal muscles—diaphragm, extensor digitorum longus, soleus, quadriceps, and obliquus externus. Castration of 200 g rats after 15 days decreased the glycogen only in the "levator ani" muscle. The injection of 5 mg/day of testosterone propionate produced a small linear increase in the glycogen of the skeletal muscles but a very rapid and larger increase to a maximum within 16 h in the "levator ani" muscle. Testosterone propionate at 5 mg/day for 2 days had a greater stimulatory effect in hypophysectomized than normal castrated rats. In adrenalectomized and alloxan-diabetic rats, the effect of the hormone was practically null in all of the muscles except the "levator ani." The administration of hydrocortisone or insulin to the respective groups provided a permissive effect and restored the activity of the testosterone propionate. The various enzymes and xylose uptake increased by testosterone propionate in the "levator ani" muscle (see above) were not affected in the skeletal muscles.

## 3. Mice

### *a) Perineal Muscles*

#### α) Weight

*i) Castration.* Castration of the immature (26–28 days old) mouse resulted in an immediate cessation of growth of not only the "levator ani" but also the bulbocavernosus muscle (KOCHAKIAN, 1959). No change in weight of the muscles was noted over the following 4 months in contrast to that observed in the rat. VENABLE (1966a) castrated adult (43-day-old) mice and observed a decrease in weight to one third of normal after 99 days of age and no further change at 227 days of age.

*ii) Steroids.* The injection of testosterone propionate at doses of 25–200 μg/day for 7 days immediately after castration at 28 days (18–21 g) or 28 days later stimulated a linear log dose response of increase in weight (KOCHAKIAN, unpublished). As in the case of the rat (HERSHBERGER et al., 1953), the response of the "levator ani" muscle was at a slower rate than that of the seminal vesicles and prostates but that of the bulbocavernosus was parallel with that of the accessory sex organs.

In contrast to the rat (EISENBERG and GORDAN, 1950; HERSHBERGER et al., 1953, 1966), estrogens also stimulated growth in the complete perineal complex (HERSHBERGER et al., 1966). The mice were castrated at 22–23 days; 3 weeks later, they were injected daily for 10 days. Stilbestrol, estradiol, and Tace (chlorotrianisine) in a wide range of doses produced significant increases in the weight of the perineal complex. MER-25 [1-(p-2-dimethylamino-ethoxyphenyl)-1-phenyl-2-p-methoxyphenylethanol], an estrogen antagonist without estrogenic activity, was ineffective. Stilbestrol also induced increases in the weight of the perineal complex of mice hypophysectomized 1 to 2 days before treatment or adrenalectomized and maintained on saline or 0.02 mg/day cortisone acetate and injections begun immediately.

#### β) Composition

VENABLE (1966a) has performed careful histologic studies to determine the nature of the change in mass of the "levator ani" muscle after castration and stimulation with testosterone. The mice were castrated at 40 days and studied 45 days later. The treated mice were injected with 0.5 mg Depo-Testosterone (19-nortestosterone cyclohexylpropionate) 9 days previous to sacrifice. The weight changes in the muscle were accounted for solely by changes in the size of the muscle fibers. No mitotic activity (determined by microscopic examination and $[^3H]$-thymidine incorporation in nuclei) accompanied the atrophy or hypertrophy. The number of cell nuclei and the amount of DNA in the muscle was not changed in the castrated or treated mice. VENABLE (1966b) at the same time made a careful study of the morphology of the cells by light and electron microscopy. The distribution of the cell types in the muscle bundle was determined by counting the respective nuclei and was found to be unchanged after castration or testosterone treatment. The percentages of each type of nucleus were: muscle, 57%; endothelial, 20%; fibroblast, 14%; pericyte, 7%; and satellite cell, 2%. The atrophy of the "levator ani" muscle after castration was accompanied by a loss of myofilaments, but no detectable degeneration of muscle tissue was observed. Electron microscopy demonstrated a decrease in the average diameters of the muscle fibers. Injection of 0.5 mg depotestosterone 8 days previous to sacrifice restored the average diameter of the muscle fibers. The observations indicated the formation of new myofilaments along the myofibers. The components of the sarcoplasm—mitochondria, sarcoplasmic reticulum, and sarcoplasmic granules—changed in proportion with the changes in weight.

*b) Inguinal*

Hernias were produced in estrogen-treated male and testosterone-treated intact and ovariectomized female mice (HAZARY and GARDNER, 1960). Estrogen administered to castrated male mice did not produce hernias. The development of the hernias was accompanied by an increase in the diameter of the muscle fibers in the inguinal and scrotal regions.

### 4. Rabbit

The tibial muscle of rabbits (1 kg) was increased 56% compared to 41% for controls by the oral administration of 0.2 mg/kg for 21 days of formylandrostadiendiol (DEFILIPPI and ZACCHE, 1968). The muscle size was visualized by a new radiographic technique and confirmed by removal and determination of the actual weights of the muscle.

SCHUMANN (1940) analyzed muscles obtained by biopsy from 2–2.5-kg male rabbits and found a decrease (-52%) in the concentration of glycogen, an increase in inorganic phosphate with no change in ATP and creatine phosphate in the quadriceps muscle 3–5 days after castration. The injection of 5 mg/day of testosterone propionate for 3 days produced an increase in the concentration of glycogen. A small decrease in the inorganic phosphate and no change in ATP and creatine phosphate of the gastrocnemius muscle in normal and 35-day castrated rabbits, except that the inorganic phosphate in the castrated animals was increased.

### 5. Other Species

The guinea pig and rat represent two different categories with respect to regulation of muscle growth by the anabolic steroids. While specific muscles of the guinea pig are very dependent on the anabolic-androgenic steroids for their full development, all of the skeletal muscles of the rat show a uniform dependency. The hamster skeletal muscles (Kochakian, unpublished), and very likely that of the mouse, behave like those of the rat. On the other hand, the grossly obvious greater muscular development of the head, neck, chest, and shoulders of the male as compared to the female in many species, e.g., bos, moose, elk, buffalo, seal, cat, etc., and even man, suggest that their muscle development is similar to that of the guinea pig. Jasienski (1929) made a histologic study of the dorsoscapular muscle of the bull, steer, and cow and found that the diameter of the myofibrils of the muscle of the cow was much smaller than that of the bull and similar to that of the steer.

The frog probably presents the earliest recognition of sexual dimorphism in the skeletal muscles and many amphibian species also have been noted to show dimorphism (Szirmai, 1962). The musculature of the forelimbs in the male frog undergoes hypertrophy during the breeding season and then atrophies. The muscles mainly involved are the flexors of the ventral arm and shoulders and some of the pectoral muscles. The myofibrils have a larger diameter (Jasienski, 1929). Castration prevents the usual enlargement of these muscles during the breeding season. Gutmann and associates (Melichna et al., 1972) confirmed the larger diameter of the myofibrils in the flexor carpi radialis muscle of the male frog during the mating season. All of the fibers in the female muscle contained a higher concentration of ATP-ase activity determined histochemically while the male contained a mixture of high and low ATP-ase active fibers. The difference in ATP-ase activity was associated with the slower contraction properties of the male muscle.

It is of interest that the presumably nonsexually-related gastrocnemius muscle of the frog (Gaule, 1901) and the hind limbs of the alligator and crocodile are larger in the male than the female.

## IV. Urinary Bladder

### A. Mouse

Castration retarded the growth of the urinary bladder in the mouse to that of the female (Kochakian et al., 1957b; Kochakian, 1964) (Fig. 10). The implantation of a pellet (15 mg) of testosterone at the time of castration (30 days) or 28 days later produced a progressively larger urinary bladder than that of the controls (Kochakian et al., 1957b). A proportionate increase in protein and water indicated a true growth of the bladder. Many other steroids also stimulated an increase in urinary bladder weight (Kochakian, unpublished see Table 6).

### B. Rat

Aschkenasy (1959c) did not observe a change in the weight of the urinary bladder of rats castrated at approximately 210 g body weight and receiving orally 5 mg/day of 17α-methyltestosterone or 17α-ethyltestosterone for 29 days beginning 1 week after castration. The incorporation of 17α-ethyltestosterone in the diet at 25 mg/100 g of diet produced after 4 months a significant increase in the weight of the urinary bladder of normal male (125 g), but not of female (135 g), rats. An increase of the steroid to 100 mg/100 g diet did not produce a further increase in the bladder weight of the male but did produce an even greater effect in the female rats. Aschkenasy

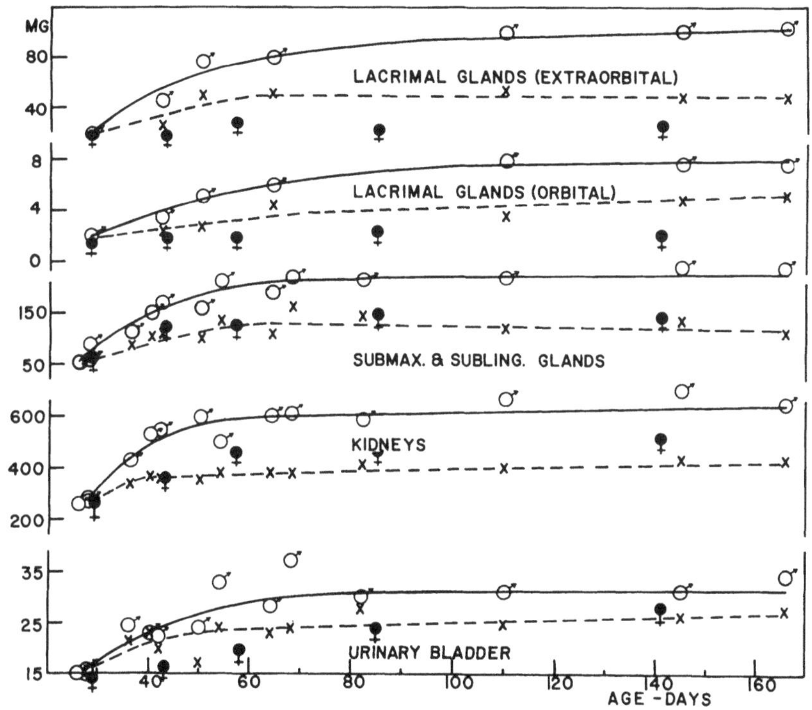


Fig. 10. The effect of castration and sex on the weight of mouse organs which show a partial dependence on the internal secretion of the testis. Castration was at 26—28 days of age. Normal male: ○—○, Castrated: ×----×. Normal female: ♀—♀ (KOCHAKIAN, 1959, 1964)

(1959a), in another study, noted a small 5% and 9%, increase in the urinary bladder weight respectively of male rats castrated or castrated and adrenalectomized at 160–190 g body weight and injected from the 8th to 21st postoperative days with 1 mg/day of testosterone propionate.

## C. Dog

The urinary bladder of growing puppies, $3^1/_2$ months old, was increased 16–97% over that of controls after 6–14 weeks of injection with testosterone propionate. The most marked enlargement was observed in two female puppies treated for 14 weeks (BLACKMAN et al., 1944).

# V. Internal Organs

Castration and administration of the anabolic-androgenic steroids affect the weight of several internal organs. In most instances, the changes in these organs were determined simultaneously with that of the accessory sex organs.

## A. Kidney

### 1. Rat

#### a) Castration

KORENCHEVSKY (1930) and KORENCHEVSKY and DENNISON (1934b) reported that castration of immature (4-week) rats resulted in a slight retardation of approximately 15% in the growth of the kidney after about 70 days. These results were further confirmed with respect to degree and the requirement of the duration of castration by KORENCHEVSKY and ROSS (1940), KORENCHEVSKY et al. (1953b), LAWLESS (1936), LEATHAM (1948a), KOCHAKIAN et al. (1950, 1954, 1956a, 1959), and ENDAHL and KOCHAKIAN (1956).

These results emphasized the small change and the long duration of castration required to significantly detect the change. Thus, no detectable difference (approx. 5%) between normal and castrated rats was observed after 30 days of castration of immature rats (KOCHAKIAN et al., 1950a; KORENCHEVSKY et al., 1953b). Castration of adult rats (250–280 g), however, is reported (BASSINGER and GITTES, 1974) to produce a significant decrease (18%) in kidney weight after 7 days and no further decrease (16%) after 30 days. On the other hand, castration of aged rats (510–515 g) decreased the kidney weights 35 and 23% only after 3 and 4 months respectively (KORENCHEVSKY et al., 1953b).

#### b) Steroids

α) Castrated Male

KORENCHEVSKY et al. (1933) castrated rats at 21–24 days of age and after at least 65 days, injected them subcutaneously with 1.5 capon units of two commercial male hormone preparations from human urine for 21 days to produce an increase of 6–24% in kidney weight. In a later study (KORENCHEVSKY and DENNISON, 1934b), the difference was more consistent (10–15%) but of the same small degree. The authors emphasized the necessity of at least 21 days of injection for a significant difference to become apparent. The simultaneous injection of an estrogenic preparation slightly enhanced the effect of the male hormone preparations. Further studies by KORENCHEVSKY et al. (1935) demonstrated that androsterone, 5α-androstane-3α,17β-diol, and their succinates injected immediately after castration into immature rats for 21–59 days prevented the usual atrophy of the kidney and when injected for 17–32 days beginning 20–84 days after castration restored the kidney weights to normal. In a subsequent series of papers, (KORENCHEVSKY et al., 1953a) similar results were noted with testosterone propionate and other crystalline steroids. KORENCHEVSKY and ROSS (1940) recommended that androsterone be used as a kidney stimulant because of its weak androgenic but strong renotrophic activity.

SELYE and ALBERT (1942a, b) did not observe a significant increase in kidney weight of immature castrated rats injected for 10 days with relatively large doses, 2 × 1 or 2 × 5 mg/day, of testosterone, 17α-methyltestosterone, or other steroids. The injection of the higher doses in mature castrated rats also was ineffective. The failure to note a significant response was probably due to the short duration of hormone treatment. LEATHAM (1948b) observed that 0.1 mg/day of testosterone propionate injected into 150–155 g rats for 20–25 days, beginning the day after castration, maintained the kidney weight at the normal level. Increases of the dose to 0.25 or 1.25 mg/day did not enhance the effect of the hormone.

ENDAHL and KOCHAKIAN (1956) castrated immature rats and two to three months later implanted subcutaneous pellets of testosterone, testosterone propionate, or 17α-methyltestosterone. A kidney weight-stimulating effect of the steroids was not detectable after 6 days and was barely detectable after 12 days. After 22–26 days, kidney weights were restored to normal. An increase of the dosage did not produce a further increase in kidney weight. The injection of 0.5 mg/day of testosterone propionate for 60 days (from 140–200 days of age) also restored the kidney weights only to, or possibly slightly above, normal. In a subsequent study (KOCHAKIAN et al., 1959), these same steroids implanted at the time of castration (48 days of age) prevented the atrophy of the kidneys. When they were implanted 25–34 days after castration, they restored the atrophied kidneys to normal within 28 days. The injection of a larger dose, 2.5 mg/day, of testosterone propionate to rats castrated at about 170 g body weight produced an increase after 8 days of about 20% in kidney weight which was not increased further after 16 days but after 125 days was greatly increased, 1.30 g or + 70%, over the castrated control and 0.83 g greater than that of the normal rat (KOCHAKIAN et al., 1950). ASCHKENASY (1971) injected 1 mg/day of testosterone propionate or applied percutaneously 1 mg/day of testosterone in alcohol or gelatin for a long period—(3 months) to rats castrated at 50 days. The kidneys of the testosterone-in-alcohol-treated rats were increased approximately 50% and those of the other two groups 25–30%.

KOCHAKIAN (1950b,c, 1964) compared the renotrophic and androgenic activity of four active steroids injected at 1 mg/day for 21 days in adult castrated rats under nitrogen balance study. 5α-Androstane-3α,17β-diol stimulated a 45% increase in kidney weight, 17β-hydroxy-5α-androstan-3-one exhibited a 25% increase, testosterone propionate 11%, and androst-4-ene-3,17-dione 8% (Table 4). The androgenic activity of the steroids did not parallel the renotrophic activities and 5α-androstane-3α,17β-diol exhibited a greater preferential renotrophic activity while testosterone propionate showed the lowest.

Table 4. Comparison of the renotrophic-androgenic action of various androgens at 1 mg/day for 21 days in the rat

| Treatment | Rats | Increase | | | |
|---|---|---|---|---|---|
| | | Body Wgt. | Kidney | Sem. ves. and pros. | K/Sv. Pr. |
| | (No) | g | g | g | g/g |
| Castrated | 12 | 2 | (2.00) | (0.21) | |
| Normal | 6 | 0 | — | 2.27 | |
| 5α-Androstane-3α,17β-diol | 2 | 25 | 0.94 | 3.30 | 0.29 |
| 17β-Hydroxy-5α-androstan-3-one | 2 | 14 | 0.50 | 2.83 | 0.18 |
| Testosterone propionate | 8 | 21 | 0.22 | 4.27 | 0.05 |
| Androst-4-ene-3,17-dione | 2 | 10 | 0.15 | 1.70 | 0.09 |

Rats on constant food intake for nitrogen balance studies; body weights approx 300 g. The figures in parentheses are the absolute values. (From KOCHAKIAN, C. D., 1964).

MARCHETTI and BALLESIO (1961) castrated rats at 30 days of age and on the following day injected 1 mg of either the phenylpropionate or the decanoate of 17β-hydroxy-19-norandrost-4-en-3-one. The kidney weights of the rats treated with the decanoate were slightly increased after 21 days and that of both groups of rats after 42 days.

The oral administration of 1 mg/day (except Sunday) of 17α-methyltestosterone or stanozolol to 26-day-old female rats increased the kidney weights only 15 and 10% after 3 months (VARGUS and ABELIUK, 1972).

BASSINGER and GITTES (1974) found that 0.2 mg/day of testosterone propionate injected into adult (250–280 g) rats, beginning at castration, prevented the decrease in kidney weight due to castration for 7 or 30 days. A greater dose, 2.5 mg/day, gave the same response as the smaller.

### β) Normal Male

The early studies of KORENCHEVSKY et al. (1933, 1935, 1937 b, 1939) indicated that the kidney weight of normal rats of different ages showed little or no change after treatment with testosterone propionate or other hormones. In a later experiment (KORENCHEVSKY et al., 1953 a), testosterone propionate implanted subcutaneously as a 25-mg pellet in 18-month-old rats stimulated after 91 days a 16% increase in kidney weight accompanied with an enlargement of the cells of the convoluted tubules. SELYE (1940 a) reported a 20% increase in the kidney weights of young adult (154 g) rats injected for 21 days with 2 mg/day of testosterone. SELYE (1941 a) also observed a small increase in immature rats but was not able to confirm it in a subsequent study (1941 b). Rats 5 weeks old were injected alternately subcutaneously and intraperitoneally for 20 days with an aqueous suspension of testosterone at 0.1, 0.5, 1.0, 10, and 25 mg/day without any detectable differences in the kidney weights.

LECOMPTE (1944) injected adult male rats with 0.5 mg of testosterone propionate 3 times per week for 1 month and then raised the dose to 1 mg without any effect on the weight of the kidney after 4, 14, or 27 weeks. PIERRE and CAHN (1956) reported a questionable 10% increase in the kidney weights of young normal rats (120 g) injected for 8 days with 1 mg/kg/day of 17β-hydroxy-5α-androstan-3-one or androst-5-ene-3β,17β-diol dipropionate. A sixfold increase in the dose, however, was without any effect. COBET et al. (1967) found a questionable 10% increase in the kidney weight of adult (200–250 g) rats 14 days after injection of a single dose (3 mg/100 g) of 19-nortestosterone-17β-decanoate. BASINGER and GITTES (1974) injected adult male rats (250–280 g) for 7 days with 0.2, 1.0, or 2.5 mg/day of testosterone propionate without any effect on kidney weights. In a more prolonged study, NATHAN and GARDNER (1965) were unable to detect any change in kidney weight of adult (300 g) male rats after 6 weeks of testosterone enanthate injected at 4 mg twice per week. The long-term (136 days) injection of adult normal rats with 2.5 mg/day of testosterone propionate produced only a 10% increase in kidney weight (KOCHAKIAN et al., 1950). The final weight, 2.62 g, moreover, was much less than that (3.20 g) of similarly treated castrated rats. In both instances the seminal vesicles and prostates were stimulated to identical supernormal weights, 7.70 and 7.60 g respectively.

In contrast to all of the above studies, LUDDEN et al. (1941) injected male rats (120–250 g) with testosterone propionate for 21 days and observed that 0.5 mg/day stimulated a 16% increase and 1.0 mg/day a large (40%) increase which was not further increased by increasing the dose to 10 mg/day or by extending the injections to 42 days. Indeed, there was a possible smaller effect; the increase now was 32%. These results would be more convincing if LUDDEN et al. had used a greater number of rats with more uniform body weights, e.g., the mean body weights of the rats at 1.0 mg/day was 174 g, at 10 mg/day was 265 g, and the controls 215 g. These differences in body weight cannot be attributed to the hormone treatment. A more convincing renotrophic response was provided by a different study. LIGHT and TORNABEN (1953a) observed a 35% increase in the kidney weights of male rats inuncted from weaning (36–42 g) with 400 μg of 17α-methyltestosterone in 200 mg of ointment for 200 days. No change was detected in the histology of the kidneys.

The accumulated evidence indicates that the presence of the testis even in the immature rat somehow deters the manifestation of the renotrophic activity of these steroids. This protective action is selective because the responsiveness of the accessory sex organs is apparently not affected. Very long-term administration of the steroids, however, seems to produce a breakthrough, and apparently the effectiveness is related to the early initiation of treatment, the route of administration and subsequent metabolism of the steroid, or some other factor. The mechanism of this protective action and breakthrough demands elucidation.

### γ) Female

The kidney of the female rat is smaller than that of the male (MACKAY and MACKAY, 1927; KORENCHEVSKY et al., 1937a; KORENCHEVSKY and ROSS, 1940) and is increased 10 days after the ovariectomy of young adults (about 4 months) (PIERRE and CAHN, 1956). The injection of 0.5 and 1.5 mg/day of testosterone propionate for 21 days increased the kidney weight of young adult (77 days old) female rats 17 and 28% respectively (KORENCHEVSKY et al., 1937a). LUDDEN et al. (1941) noted a 16% increase and a 24% increase in the kidney weights of normal (120–250 g) female rats injected for 21 days at 1.0 and 10 mg/day of testosterone propionate. KORENCHEVSKY and ROSS (1940) reviewed previous studies and in new experiments reported enlargement of the kidneys of normal and ovariectomized female rats after the injection of 7.5 mg/day of androsterone or esters of testosterone at 0.75 and 7.5 mg/day for 21 days. Dehydroisoandrosterone at 7.5 mg/day was ineffective. The effect of androsterone on the kidney weight of ovariectomized rats was confirmed in a later study (KORENCHEVSKY and JONES, 1948) and also an enlargement of the tubules, but no mitoses were observed.

SELYE (1940a) observed a 10% increase in the kidney weights of young (109 g) adult female rats after the injection of 2.0 mg/day of testosterone propionate for 21 days. LECOMPTE (1944) injected normal adult (200 g) female rats with testosterone propionate, 0.5 mg 3 times per week, for the first month and thereafter (14 and 27 weeks) with 1 mg 3 times per week. The kidney weights were not detectably different from those of the sesame oil controls. PIERRE and CAHN (1956) report that androst-5-ene-3β,17β-diol dipropionate and 17β-hydroxy-5α-androstan-3-one injected into 4-month-old female rats at 1 mg/day for 8 days beginning 10 days after ovariectomy did not enhance the increase in kidney weight caused by castration. The 1 mg and 6 mg/day dose of 17β-hydroxy-5α-androstan-3-one and the 1 mg/day, but not the 6 mg/day, of androst-5-ene-3$\beta$,17$\beta$-diol dipropionate increased the kidney weight of the normal female rats.

NATHAN and GARDNER (1965) injected adult (300–350 g) female rats twice per week with 4.0 mg of testosterone enanthate for 6 weeks. The kidney weights were substantially increased, from 2.26 to 2.97 g but still were much less than those (3.60 g) of comparably aged male rats.

Weanling female rats inuncted for 186 days with 200 mg of an ointment containing 400 µg of 17$\alpha$-methyltestosterone showed an increase in kidney weight of 35% (LIGHT and TORNABEN, 1953 b).

SELYE (1940 b) injected 1 mg/day of testosterone propionate, beginning the day after birth, and thereafter for 30 days. One month after the last injection, the kidney weights of the treated rats were 44% larger than those of the controls. SELYE interpreted this to indicate that the increase in kidney weight caused by androgens may persist for a long time.

$\delta$) Infantile

SHAY et al. (1941) injected 5 mg/day of testosterone propionate into female rats, beginning the first day of life. The rats were killed at various periods between 30 and 60 days of age. The kidneys per 100 g body weight of the treated rats were approximately 15% greater than those of the control rats. Male rats similarly treated showed a slight decrease in kidney weight.

ZUFAROV et al. (1974) reported that the intraperitoneal injection of 1 mg/day of testosterone propionate during the first 7 days of life increased both kidney and body weights. The increase in kidney weight involved an increase in the number of formed structures and hypertrophy and hyperplasia of the cells.

$\varepsilon$) Hypophysectomized

SELYE (1941 a) reported that 4 mg/day of testosterone for 20 days produced a 25% increase in the kidney weights of female rats hypophysectomized at 90 g body weight. The injections were begun on the day of hypophysectomy. KOCHAKIAN et al. (1950) found that the marked atrophy of the kidney of the adult (300 g) hypophysectomized rat was not detectably changed 13 days after the subcutaneous implantation of a 14–15 mg pellet of testosterone. 17$\beta$-Hydroxy-5$\alpha$-androstan-3-one and androst-5-ene-3$\beta$,17$\beta$-diol at 40 mg/kg for 8 days were ineffective in recently (4 days) hypophysectomized rats (PIERRE and CAHN, 1956). JAHN (1955) observed a slight but statistically significant enlargement of the nuclei of the cells of the kidney convoluted tubules of 120–150 g male rats injected with 1.0 mg/day of testosterone propionate for 14 days beginning 28 days after hypophysectomy.

The period of time after hypophysectomy for initiation of hormone administration may possibly be a factor in the determination of the effectiveness of the steroid. The atrophy of the anterior pituitary-dependent endocrine organs, especially the thyroid, is a slow, gradual process. The duration of hormone treatment also must be considered.

$\zeta$) Thyroidectomized

KOCHAKIAN et al. (1954) castrated rats at 82–119 g body weight and thyroidectomized them 35 and 50 days later. Thyroidectomy within 25 days produced a further and even greater decrease in kidney weight than castration. The decrease was not further enhanced 40 days after thyroidectomy. The decrease in kidney weight was much less when adult (325–350 g) rats were thyroidectomized 25 days after castration and the body weight was maintained by a constant food intake, even though the period after thyroidectomy was 328 days. The injection of 0.5 or 1.0 mg/day of testosterone propionate for the last 14 days was unable to change the kidney weights of these rats. Longer periods of treatment were not done.

MASSON and ROMANCHUK (1945) reported that the renotrophic activities of testosterone and thyroxine are additive in the castrated, normal, and hypophysectomized immature (47–51 g) rat. The testosterone (60 μg/day) and thyroxine (10 μg/day) were injected for 10 days, beginning 24 h after the operation. Unfortunately, untreated controls were not studied except in the normal series. If the stimulation of the respective hormones is deduced by difference, a small but questionably significant effect was produced by testosterone. In addition, testosterone produced only an approximately 10% increase in the kidney weights of the normal rats. These small increases, however, summated with the greater increase induced by the thyroxine.

It seems that either the thyroid or its internal secretions are required for the renotrophic action of the anabolic-androgenic steroids in the rat. The atrophy or decrease in thyroid activity after hypophysectomy (BRAUN-MENENDEZ, 1950) may be responsible for the ineffectiveness of the steroid hormones in the posthypophysectomized rat, possibly secondarily through the slowing of metabolic processes.

#### η) Repletion from Chronic Starvation

QUIMBY (1948) fed 30-day-old male rats only enough of a balanced diet to maintain body weight. After 90 days, they were allowed full rations for 60 days and simultaneously injected with 0.5 mg/day of testosterone propionate. After 60 days, the kidney weights were no different than those of saline controls but significantly greater than normal controls of the same age. ASCHKENASY (1971) placed 180-g male rats on a protein-free diet for 55 days and then for 28 days on a low-protein (7% casein) diet to provide an incomplete and slow restoration. During the recovery period, the rats were injected with 1 mg/day of testosterone propionate or applied percutaneously with 1 mg/day of testosterone in alcohol or suspended in a gel medium. The kidney weights of the treated rats were slightly greater than those of the controls. The hemoglobin and RBC, especially after the percutaneous treatment, also were increased. The authors suggested that this correlation supports the hypothesis that testosterone augments the production of the renal erythropoietic factor.

#### ϑ) Starvation

USUELLI et al. (1949) deprived young adult male rats (129–240 g body weight) of food and injected 1 mg/day of testosterone propionate. After 10 days, the kidneys and also the liver and heart had lost less weight than those of the nontreated fasted rats. The weights of the seminal vesicles and prostates were increased to the same degree (3 times normal) as in the injected nonfasted rats.

### *c) Kidney Composition*

LUDDEN et al. (1941) reported an increase in water content in proportion to the increase in kidney weight in both normal male and female rats after injection of testosterone propionate.

COXON et al. (1956) noted proportionate changes in nitrogen (protein) and water with changes in kidney weight after castration or steroid treatment. Castration was performed at about 4 months of age. The testosterone propionate was implanted as a 25-mg pellet 3 months after castration and the rats were killed 54 days later. An additional 2 mg of testosterone propionate was injected as a crystalline suspension 2, 7, and 9 days before autopsy. KOCHAKIAN et al. (1950) also observed proportionate changes not only in nitrogen and water but also lipids. Proportionate changes in water, nitrogen, potassium, and phosphorus content were reported (HORKY and GANGOLLI, 1973) for the kidneys of male rats injected every other day for 3 weeks with 0.2 mg/100 g body weight of 19-nortestosterone-17β-phenylpropionate, 17β-hydroxy-17α-methylandrosta-1,4-diene-3-one, 3-oxo-1-methyl-5α-androst-1-ene-17β-yl-acetate, or 17β-hydroxy-17α-ethyl-19-norandrost-4-en-3-one.

## 2. Mouse

### a) Castration

MASUI and TAMURA (1926) castrated mice at 20 days of age and noted at 90 days of age that the kidney weight of the normal controls was 53% greater than that of the castrated mice. Thus the kidney weight of the castrated mouse was comparable to that of the female mouse which was 45% smaller than that of the normal male mouse. Ovariectomy of the female had a slight, questionable effect on kidney weight.

KOCHAKIAN (1941) confirmed the effect of castration, and in a later, detailed study (Fig. 10), KOCHAKIAN (1964) castrated mice at 26–28 days of age and noted an immediate retardation in the rate of growth of the kidney. The kidney reached maximum size at about 40 days of age while that of the normal mice was reached about 2 weeks later, after which the kidney reached maximum size at about 40 days of age while that of the females was slightly greater than those of the castrated mice. Castration of adult mice produced a maximum decrease of 30–35% in kidney weight within 2 weeks (KOCHAKIAN and HARRISON, 1962; KOCHAKIAN, 1969; KOCHAKIAN et al., 1969; KOTHS et al., 1972).

### b) Steroids

#### α) Normal Female

SELYE (1939b), in the course of studies with testosterone, noted that the kidneys of mice treated with testosterone were enlarged. To establish this observation, he injected adult female albino mice subcutaneously with 0.5 mg/day of testosterone propionate for 10 days and 1.0 mg/day for another 6 days. The kidney weights were increased approximately 50%. A dose of 5 mg/day for 20 days gave a smaller (30%) increase in another strain of mice (dba Bar Harbor). Another experiment in younger female (albino?) mice gave a 55% increase in kidney weight after 14 days of 3 mg/day of testosterone propionate. The increases in weight were unquestionable and large in contrast to those obtained in rats. These results in the female mouse were confirmed almost immediately by FEYEL (1942, 1943), WRETE (1943), and later by HOMBURGER et al. (1950), VAN BEKKUM and KASSENAAR (1951), and subsequently by several other investigators in the course of exploration of the various aspects of the nature of the hypertrophy stimulated by testosterone. TRUCKENBRODT (1969) and HEIM et al.

(1969) reported that 0.5 mg/g of methenolone (17$\beta$-hydroxy-1-methyl-5$\alpha$-androst-1-en-3-one) injected as a single subcutaneous dose produced a more pronounced growth of the kidneys of adult (6-month-old) than of growing (4-day-old) mice. A maximum increase of about 60% was observed in the adult male and female mice while an increase of slightly more than 20% was obtained in the infantile mice after 14–21 days.

### $\beta$) Normal Male

Almost simultaneously with SELYE'S observations in female mice, PFEIFFER et al. (1940) reported similar results in the male mouse. Immature (30 days) and young adult (100 days) mice were injected weekly with 1.25 mg testosterone propionate, 16.6 μg estradiol-17$\beta$-benzoate (immature mice), 50 μg estradiol-17$\beta$-dipropionate (young adult mice), or a combination of the two hormones. After 30 days, the weight of the kidneys of the estrogen-treated mice was slightly above that of the controls while that of the testosterone propionate-treated mice was 35% greater. The effect of the two hormones summated when given simultaneously. No difference was noted among the 3 different strains of mice used in the study. Even greater increases (80%) were obtained by extending the duration of the testosterone propionate injections to 60 days. The substitution of testosterone for testosterone propionate produced only a small increase in kidney weight, due undoubtedly to the long interval between injections which was partially counterbalanced in the case of testosterone propionate by the more prolonged action of the ester (KOCHAKIAN, 1938). FEYEL (1942, 1943) and CLARK et al. (1943) confirmed the above results in normal male mice.

### $\gamma$) Castrated Male

KOCHAKIAN (1941) castrated mice at 5 weeks and 1 month later implanted subcutaneously a pellet of testosterone propionate weighing 11 or 22 mg or two of 11 mg each. The treated mice and controls were killed at fixed intervals over a 96-day period. The rate of increase of the kidneys was uniform, rapid, and the same at both dose levels. The maximum effective dose apparently was already provided by the lower (11-mg pellet) dose. Kidney weights were restored to the normal levels within about 1 week and continued to increase at a gradually decreasing rate and were more than double the normal weight after 96 days. MASSON and BELAND (1943) castrated mice at 23 g body weight and at the same time implanted subcutaneously two 10-mg tablets of 17$\alpha$-methyltestosterone. The kidney weights were increased 84% after 30 days and the substitution of 1.5% KCl for the drinking water did not alter (82%) the effect of the steroids.

### $\delta$) Comparison of Various Steroids

The consistent and striking response of the kidney to the administration of testosterone propionate suggested to KOCHAKIAN (1944, 1946) that this phenomenon might be representative of the extra sexual activity of testosterone. Therefore, a comparison of the effect of a variety of steroids on kidney and seminal vesicle and prostate weights was undertaken (KOCHAKIAN, 1944, 1946, 1949). The mice were castrated at 16.0–19.5 g body weight and 30 days later they were implanted subcutaneously with a 14-mg pellet of the various steroids. A comparison of their relative effects on the

Table 5. Comparison of effect of chemical structure on the renotrophic-androgenic activities of various steroids in the castrated mouse (KOCHAKIAN, 1944)

| Steroid | Increase | | | |
|---|---|---|---|---|
| | Steroid Abs'd mg/30 days | Kidney | Sem. ves. and pros. mg/ | K/Sv. Pr. mg/mg |
| Androsterone | 13.0 | 65 | 20 | 3.3 |
| 17α-Methyl-5α-androstane-3α,17β-diol | 2.6 | 261 | 157 | 1.7 |
| 5α-Androstane-3α,17β-diol | 1.7 | 202 | 137 | 1.5 |
| 17β-Hydroxy-17-methyl-5α-androstan-3-one | 0.2 | 193 | 148 | 1.3 |
| Testosterone | 6 0.14 | 125 | 144 | 0.9 |
| Normal mice | — | 162 | 216 | 0.7 |
| 5α-Androstane-3,17-dione | 32.3 | 60 | 95 | 0.6 |
| Androst-4-ene-3,17-dione | 10.6 | 84 | 173 | 0.5 |

two different types of tissue revealed a definite relationship between chemical structure and preferential stimulation of the respective tissues (Table 5). The saturated diols had a preferential effect on the kidney; androst-4-ene-3,17-dione on the seminal vesicles and prostates and the 17β-hydroxy-3-oxo compounds had an intermediate effect. These studies were extended to determine whether the dose of the steroids influenced the relative growth-stimulating properties of the steroids. The relative activity of testosterone and related compounds was not influenced by dose (Fig. 11) but that of the diols was markedly affected. The preferential renotrophic activity was

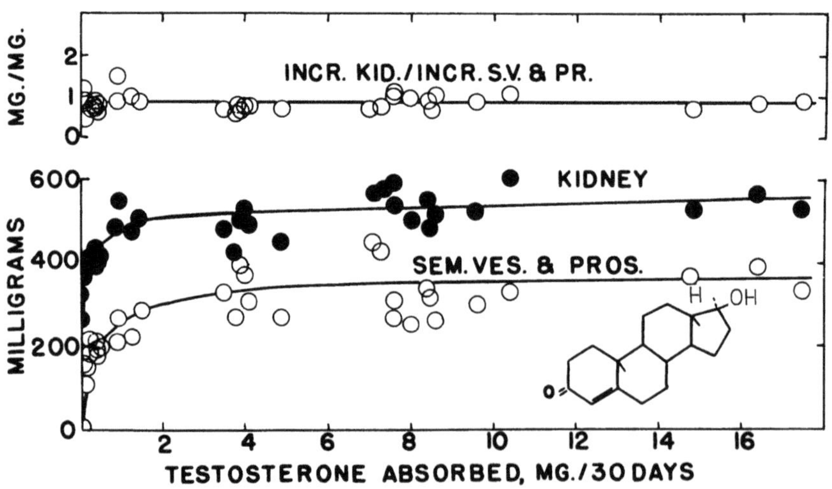


Fig. 11. A comparison of the renotrophic and androgenic activities in castrated mice stimulated by different doses of testosterone. The mice were castrated at 16 to 19.5 g body weight. The pellets of testosterone were implanted subcutaneously 30 days after castration. The average values for normal mice were: Kidneys, 414 mg; seminal vesicles and prostate 222; incr. Kid/incr. SVPr = 0.74 (KOCHAKIAN, 1946)

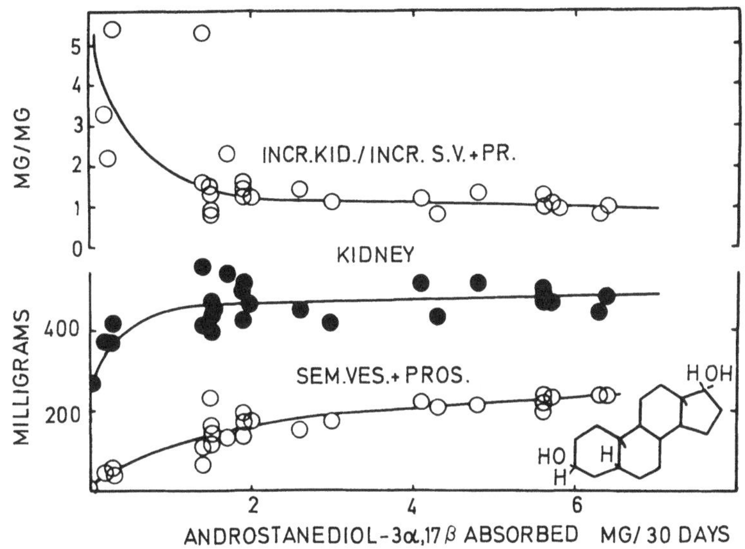


Fig. 12. A comparison of the renotrophic and androgenic activities in castrated mice stimulated by different doses of 5α-androstane-3α,17β-diol. See Fig. 11 for further details

greatest at the lowest dose and gradually decreased to that of the testosterone as the dose was increased (Fig. 12). The change was not due to an increase in renotrophic activity but to a lower rate of increase in the weight of the seminal vesicles and prostate. The preferential renotrophic activity occurred at the dose level which stimulated a normal or lower than normal response in the kidney. Once both the kidney and the seminal vesicles and prostate were stimulated to their maximum size, a further, severalfold increase in the dose did not produce a further increase in the weight of these organs.

The maximum renotrophic and androgenic activities were different for the various steroids. A manyfold increase of the dose was not able to increase the characteristic maximum for the respective compounds (Kochakian, 1946, 1949).

The simultaneous administration of estrogens (Kochakian, 1944; Kochakian et al., 1948c) resulted in a summation of the renotrophic activities but an inhibition of the androgenic activity, giving a preferential renotrophic/androgenic ratio similar to that of the diols.

Since the presence of the 17β-hydroxyl seemed to be essential for maximum activity, and the reduction of the α,β-unsaturated group of ring A reduced androgenic but not renotrophic activity, Kochakian (1949) studied several mono-substituted steroids for a further possible separation of the two biological effects. The tissue fluid, however, was not able to absorb any of these compounds from the implanted pellets. Kochakian (1952), therefore, administered the one available compound, 17α-methyl-5α-androstan-17β-ol, by mixture in the food and compared its renotrophic-androgenic activity with 17α-methyltestosterone and several other methylated and nonmethylated compounds. The 17α-methyltestosterone had the most potent renotrophic-androgenic activity of the several compounds. The 17α-methyl-5α-androstan-17β-ol, however, had essentially the same renotrophic activity but only a fraction of the androgenic activity of 17α-methyltestosterone. The oral route of administration proved to be much less efficient than the subcutaneous implantation of the very active steroids but more efficient for the less soluble and less active steroids.

SEGALOFF (1963) and SEGALOFF and GABBARD (1963) studied the renotrophic activity of a wide variety of the newer steroids by subcutaneous injection at 10, 100, or 1000 μg/day for 7 days in immature castrated mice. The responses were too small and too variable for satisfactory comparisons.

ε) Hypophysectomized

FERGUSON and VISSCHER (1953) castrated and hypophysectomized male and female mice 3–4 months old and 31–35 days later injected them subcutaneously for 14 days. The males received 20 μg/day of testosterone, 0.1 μg/day of estradiol-17β, or 0.1 μg/day stilbestrol and combinations of the testosterone with each estrogen. The females received 40 μg/day of testosterone, 0.08 μg stilbestrol, 0.1 μg estradiol-17β, and combinations of testosterone with each of the estrogens. The doses were determined after studies with graded doses of each steroid to find the minimum dose which produced the maximum effect on the appropriate target tissues. Hypophysectomy reduced the kidney weights 55% in the female and 61% in the male. The testosterone produced a significant increase in the kidney weight but did not restore the kidneys to the normal level. The estrogens were ineffective and the combination of the estrogens with testosterone did not enhance or inhibit the effect of the testosterone. QUINTARELLI and KOCHAKIAN (unpublished) obtained comparable decreases in the kidney weight of normal female (−63%) and male (−75%) mice after hypophysectomy. Injection of 100 μg/day of testosterone propionate for 5 days into the male hypophysectomized mice increased the kidney weight 27% above that of hypophysectomized control mice.

As in the rat (KOCHAKIAN et al., 1950), the decrease in kidney weight after hypophysectomy was greater than that after castration presumably because of the thyroid deficiency (BRAUN-MENENDEZ, 1950). Thyroidectomy produced a greater decrease in kidney weight in the rat than castration (KOCHAKIAN et al., 1954). The stimulation by testosterone in the hypophysectomized mouse was less than that observed in the normal or castrated mouse and may be due to at least two or more factors, e.g., the smaller initial size of the kidney and/or the much lower metabolic status of the hypophysectomized mice. In any event, the renotrophic effect of testosterone is not mediated through the pituitary.

A growth hormone preparation produced an increase of about 30% in the kidney weight of adult castrated mice after daily injections for 10 days but did not produce any further increase on extension of the injections to 20 and 34 days (KOCHAKIAN and STETTNER, 1948). The increase in kidney weight was approximately one-third of that produced by a subcutaneously implanted pellet (15 mg) of testosterone propionate and partially summated with that of testosterone propionate when both hormones were administered simultaneously.

ζ) Nutritional Status

In the course of studies on the comparison of dose and renotrophic-androgenic response of various steroids, KOCHAKIAN (1946) encountered a group of mice which had not received proper care and were undernourished. The kidneys of these mice showed a much lower response than normal but the response of the seminal vesicles and prostates were not detectably altered. HOMBURGER and PETTENGILL (1955) found that the partial restriction of the intake of a balanced diet decreased the renotrophic activity of testosterone propionate in immature female mice. On the other hand, a high protein (46%) diet did not enhance the renotrophic activity of the hormone and even possibly caused a decreased response.

### η) Kidney Composition

The kidneys in many of the studies were analyzed to determine the nature of the increase in weight stimulated by the hormones. Moisture content was changed in proportion to the changes in weight (PFEIFFER et al., 1940; KOCHAKIAN, 1941; WRETE, 1943; FEYEL, 1943). A very high dose of testosterone propionate, 50 mg for 2 weeks in divided daily injections, however, was reported (FEYEL, 1943) to produce edema in the kidneys of both male and female mice.

The nitrogen (protein) and total lipid content of the kidneys also changed in direct proportion with the changes in kidney weight (KOCHAKIAN and STETTNER, 1948). HAY and GRAY (1970) observed differences in the glycosphingolipid composition after the injection of testosterone but not progesterone or estradiol-17β. The injection of 1 mg/day of testosterone for 2 weeks doubled the concentration of the glycosphingolipids in the kidneys of female mice compared to that of the normal male. The increase was in the diglycosylceramide fraction. Testosterone did not increase the glycosphingolipids in the kidneys of the male mice.

### ϑ) Kidney Histology

SELYE (1939 a, b), in his initial studies on the effect of testosterone on the weight of the kidney of the female mouse, reported a marked hypertrophy of the epithelial cells of the proximal and distal convoluted tubules. The glomeruli showed no significant changes but the normally very flat layer of epithelium which lines the parietal lamina of Bowman's capsule was markedly hypertrophied after testosterone treatment and took on the appearance, including a brush border, similar to that seen in the proximal convoluted tubule. No indication of a toxic effect of the testosterone was observed. PFEIFFER et al. (1940) also noted an enlargement of the epithelial cells and also the tubules of the kidneys of the male mouse after testosterone propionate treatment. Mitotic activity was not observed. The normal squamous epithelium of Bowman's capsule was replaced in most instances by cuboidal cells cytologically indistinguishable from those in the proximal convoluted tubules. CRABTREE (1941) resolved an apparent discrepancy between the observations of SELYE and PFEIFFER et al. concerning the size of the cells in Bowman's capsule. She compared the size of these cells with age and noted that immature mice had few cuboidal cells but after puberty an increasing number of the capsules had cuboidal cells in the male but not the female kidney. Castration caused a reversion to the immature or female pattern. The administration of testosterone propionate increased the size of these cells in the castrated male and the female to that of the normal male. The doses of testosterone propionate, 2.5, 5, and 10 mg/day for 3 weeks, were excessive and produced hydronephrosis with some urinary bladder obstruction. The cells of the convoluted tubules were so enlarged that they protruded into and narrowed the lumen.

The effect of castration and testosterone propionate on the size of the cells has been further confirmed (FEYEL, 1943; WRETE, 1943; KOCHAKIAN, 1945). The size of the cells can be actually stimulated to greater than normal not only by testosterone but also other steroids (KOCHAKIAN, 1945). FERGUSON and VISSCHER (1953) did not observe an increase in the size of the cells in the parietal layer of Bowman's capsule after the injection of testosterone propionate into hypophysectomized-castrated male and female mice, although they were able to confirm the stimulating effect of the androgen in normal male mice. The failure to detect a change in cell size in the hypophysectomized mice was very likely due to the relatively small, albeit significant, effect on the size of the kidney.

Deleterious effects on the kidney were noted only by CRABTREE (1941), and suggestive degenerative changes again by high doses of testosterone propionate by FEYEL (1943). VAN BEKKUM and KASSENAAR (1951), however, reported marked degenerative changes on prolonged administration of testosterone propionate to gonadectomized adult female mice. The injection of testosterone propionate, 2.5 mg twice per week, was begun 24 days after ovariectomy. The histologic examination of the kidneys of the female ovariectomized mice 22 and 30 days after testosterone propionate injections showed the previously reported hypertrophy of the epithelial cells and in addition a definite shrinkage of most of the glomeruli and patches of infiltration with cells of a predominantly mononuclear type in various parts of the cortical tissue. These changes were also observed in another series of studies but only after 30 days of testosterone propionate treatment. Similar changes were observed in adrenalectomized-ovariectomized mice. The changes were accompanied by widely variable but consistent increases in blood urea beginning 10 days after hormone treatment. The kidneys of castrated male mice were not studied histologically but the testosterone propionate did not produce an increase in blood urea until after 35 days of treatment and the increase was slight. The surface of the enlarged kidneys was found to be granular. KOCHAKIAN (unpublished) also has observed a granular or orange-peel-like surface with scattered blebs of blood vessels in the kidneys of mice induced to supernormal (twice or more normal) weight. Histologic studies, however, were not made of these kidneys.

RABINOVITCH and VALERI (1952) extended the histologic studies to the nuclei. Adult male mice were castrated and 30 days later 10 mg of powdered testosterone propionate was implanted subcutaneously. Noncastrated mice were similarly treated. Controls and treated mice were killed 4 and 8 days later. The usual increase in kidney weight and size of the cells of the convoluted tubules were observed. Measurement of the volume of the nuclei of the convoluted tubules were made. Castration produced a decrease in nuclear volume and testosterone propionate produced an increase in both the castrated and normal mice after 4 days and no further increase after 8 days. The tubular diameter of the kidneys of these same animals was doubled and the number of nuclei per volume of cortex was decreased 40% after testosterone administration. Castration increased the concentration of nuclei (RABINOVITCH, 1952). BERN and ALFERT (1954) confirmed these results by studying two control and two adult male mice which had been implanted subcutaneously with testosterone 23 days previously. The nuclear volume was increased 50% after testosterone treatment. They further noted a correlation in nuclear volume size with differences in nonhistone protein content. The cytophotometric studies did not indicate any change in DNA or histone content of the nuclei after castration or testosterone propionate treatment.

TESSMAN (1968) carried out a careful light and electron microscopic study of the kidneys of 25–30 g male and female mice after the injection of 0.5 mg/day of testosterone propionate for 3, 10, and 16 days. The hypertrophy of the renal proximal tubules and the cells in Bowman's capsule was confirmed. The electron microscopic studies revealed that the enlargement of the nuclei of these cells was due to increased karyoplasmic hydration and by transmutation of heterochromatin into euchromatin. Also, there was a folding of the nuclear membrane, structural loosening of the nucleoli, and a granule flow towards the nuclear membrane. The cytoplasm of the stimulated cell showed a moderate increase in hydration of the hyaloplasm. The changes in the nuclei were evident after 3 days of testosterone propionate and no or only slight further enhancement was observed after 10 or 16 days which suggested to the author the initiation of the regulating mechanism for the "wearing off" phenomenon (KOCHAKIAN, 1950a, b, 1964).

### c) *Genetic*

LYON and HAWKES (1970) described a strain of mice with a X-linked testicular feminization (Tfm) mutation. The injection of as much as 20 mg of testosterone, or 17$\beta$-hydroxy-5$\alpha$-androstan-3-one or 5$\alpha$-androstane-3$\alpha$. 17$\beta$-diol into male mice with this defect (Tfm/Y) did not stimulate an increase in the weight of the kidneys (DOFOKU et al., 1971). Furthermore, the weight of the kidneys was smaller than that of the normal male and similar to that in the female and castration was without effect.

It is also of interest that the weight of the kidney varies with different strains of mice. BARTKE and SHIRE (1972) compared the kidney weights of male mice of the C57Bl/10 and DBA/2 inbred strains and also of their reciprocal $F_1$ hybrids. The kidney weights of the DBA/2 mice were consistently and considerably heavier over the period from 7–16 weeks of age. A comparison of 7 week old males with female mice demonstrated only a small (3.8%) difference in the kidney weight of the C57Bl/10J strain and a much bigger (15.2%) difference in the DBA/2J strain. The submandibular glands curiously showed the reverse effect. The absolute weights of the glands were heavier in the males than the females but on the basis of body weight, the glands of the females of the DBA/2J strain were heavier than those of the male mice and those of the C57Bl/10J strain were slightly heavier in the male mice. The authors on the basis of cholesterol analysis of the testis and the response after various hormone injections suggested that the strain difference is due to differences in the production and utilization of testosterone.

### d) *Antiandrogen*

OHNO and LYON (1970) stated that the injection of castrated male mice with 10 mg cyproterone acetate 24 h before the injection of 5 mg of depotestosterone cypionate completely suppressed the hypertrophy of the kidney during the following 8 days. BROULIK et al. (1975), however, found that 2.5 mg/day of cyproterone acetate fed for 21 days to normal mice or to castrated mice receiving 2.5 mg of testosterone isobutyrate twice weekly decreased the kidney weight in proportion to the decrease in body weight. Thus, there was no suppression of the normal kidney or the androgen stimulated hypertrophy in the castrated mice on a body weight basis. The seminal vesicles exhibited the expected inhibition of testosterone stimulation (NEUMANN et al., 1967). These data provide further evidence that the mechanism of action of testosterone is different for these tissues.

### 3. Guinea Pig

#### *a) Castration*

CIRILLO and GUARDAVACCARO (1933) reported a decrease of about 10% in the kidney weight of guinea pigs 4 months after castration at 30 days of age. KOCHAKIAN et al. (1956b) also noted a small but statistically insignificant decrease in the kidney weight of guinea pigs castrated at 6 months of age. Guinea pigs castrated at approximately 2 months (500 g), moreover, did not show a significant difference from that of normal controls within 2 months (KOCHAKIAN et al., 1948a; HUMM et al., 1948; KOCHAKIAN and TILLOTSON, 1957a) or over the subsequent 27 weeks (KOCHAKIAN et al., 1956b).

#### *b) Steroids*

Small increases (10%) in kidney weight were induced by several androgens administered 1 month after castration as subcutaneous pellets for 30 days. An increase of the dosage as much as fivefold did not produce any further increase in kidney weight. Furthermore, a large dose (12.5 mg/day) of testosterone propionate injected subcutaneously for 14 days was also ineffective (KOCHAKIAN et al., 1948a; HUMM et al., 1948). In a later study (KOCHAKIAN et al., 1956b), guinea pigs were castrated at 500 g body weight and 1 month later were implanted subcutaneously with pellets of steroids to provide a wide range of dosage. None of the steroids (17β-hydroxy-5α-androstan-3-one; testosterone; 17β-hydroxy-17α-methyl-5α-androstan-3-one; 17α-methyltestosterone; 5α-androstane-3α,17β-diol; 17α-methyl-5α-androstane-3α,17β-diol; androst-4-ene-3,17-dione; 5α-androstane-3,17-dione; 3β-hydroxyandrost-5-en-17-one; 3β-hydroxy-5α-androstan-17-one; 17α-methyl-5α-androstane-3β,17β-diol; and 17α-methylandrost-5-ene-3β,17β-diol) significantly influenced the weight of the kidneys after 21 days.

### 4. Hamster

The hamster kidney was completely unresponsive to androgen manipulation (KOCHAKIAN et al., 1948b). Castration and administration of testosterone, 17α-methyltestosterone or testosterone propionate even at doses which produced supernormal weights in the seminal vesicles and prostates did not produce significant changes in the kidney weights. Prolongation of steroid administration up to 140 days and the use of three different diets also did not alter the results. LI et al. (1969) also did not detect any change in kidney weight after castration or the administration of testosterone to gonadectomized male and female hamsters for 3 weeks, beginning 3 weeks after gonadectomy at 60–77 days of age.

### 5. Rabbit

WRETE (1943) injected 0.5 mg testosterone propionate every other day for 30 days into 6-month-old rabbits without detecting any significant change in kidney or body weights. An increase of the dose to 10 and 25 mg also was ineffective. Microscopic examination did not indicate any toxic or other changes in the kidneys.

## 6. Dog

### *a) Female*

LATTIMER (1942) injected 3 female dogs, 7 years old, with 10 mg/day of testosterone propionate for 32 days. The kidney weight/body weight ratios were compared with those of two controls and also those of 7 and 22 dogs from two other laboratories. The treated dogs showed heavier kidney weight/body weight ratios than LATTIMER'S controls and those of the other laboratories. An overlap, however, between the treated dogs and the group of 22 dogs prompted LATTIMER to cautiously conclude that the renotrophic effect of the testosterone propionate should be considered as suggestive. LATTIMER found no change in the blood pressure as a result of testosterone propionate treatment in the above experiment.

### *b) Puppies*

BLACKMAN et al. (1944) found no difference in the kidney weight of two male puppies ($3^1/_2$ weeks old) injected with testosterone propionate for 6 weeks as follows: 25 mg 3 times weekly for 4 weeks; two 25-mg and one 10-mg injections during the 5th week and six 10-mg doses during the 6th week. In another experiment, one male and two female puppies were given 50–75 mg of testosterone propionate in divided (2–6) doses each week for 14 weeks. The kidneys of the treated puppies weighed 11–17% more than those of one male and one female untreated dog. The hormone did not produce a consistent change in blood pressure.

## 7. Human

No studies on the effect of castration or androgen treatment on the kidney weight of humans are available. Comparisons, however, have been made between males and females from autopsy data. WALD (1937) made a careful statistical study of kidney weights after traumatic deaths in increments of 10 years from the ages of 20–80 years of 307 males and 107 females. The mean kidney weights were essentially the same between 20 and 70 years of age in the males and 20–60 years in the females with a decrease in later years. The male kidneys were 40–75 g heavier than those of the females. LATTIMER (1942) made an exhaustive study of 1600 autopsy protocols from the archives of the department of pathology of the New York Hospital and selected 50 males and 50 females in the active sexual years of 20–40 years and without indication of obesity, emaciation, or kidney pathology. The males had an average 11.5% greater kidney weight/body weight ratio. A similar study on children below 11 years of age showed no sex difference to provide support for the sex differences in the adults.

## 8. Compensatory Renal Hypertrophy

### *a) Rat*

The surgical removal of one kidney of the rat results in a rapid increase in the weight of the remaining kidney (ARATAKI, 1926) during the first 48 h and then continues at a slower rate (ADDIS and LEW, 1940; ROLLASON, 1949). This phenomenon has been studied extensively by numerous investigators and will not be reviewed here. The influence of androgens on the surviving kidney, however, has received attention. MACKAY (1940) performed unilateral nephrectomy and castration at the same time on rats of about 300 g body weight and injected part of the rats with 12.5 mg of testosterone propionate on the 1st, 4th, and 6th postoperative days, and 25 mg on the 14th and 29th postoperative days. On the 34th day, the remaining kidney of the treated rats was 30% greater than that of the nontreated rats. The hormone stimulated a nonsignificant increase of about 10% in the kidneys of the nonnephrectomized-castrated rats.

HALPERN et al. (1951) and HALPERN and COURNOT (1953) made a detailed study on compensatory renal hypertrophy in rats (150 g) killed at 2-day intervals in groups of 10 for 16 days after heminephrectomy and injection of 10 mg/day of testosterone propionate. The weight of the kidneys of the untreated rats increased rapidly for 6 days and thereafter at a very slow rate. The kidney weight of the treated rats increased more rapidly reaching a maximum after 4 days where it was maintained for 6 days when a secondary spurt began and continued at a rapid rate until the end of the study (16 days). In an attempt to explain the effect of the testosterone propionate, the excretion of urine and chloride and the body weight were studied for 25 days in another set of rats. The testosterone propionate decreased the daily excretion of urine and chloride to a minimum after 5 days where it was maintained until the 8th day and then rapidly increased to a maximum above normal by the 16th day and gradually returned in the next 4 days to the preinjection level, where it was maintained for the 5-day postinjection period. The body weight showed changes in parallel with the changes in urine volume and chloride excretion. The testosterone propionate initially stimulated an increase in body weight reaching a maximum after 5 days which was maintained until the 8th day, then decreasing to slightly above normal by the 16th day, remaining at that level during the final 4 days of injection and the 5-day postinjection period. The investigators interpreted the results to indicate a retention and excretion of water by the tissues of the body. These changes parallel the changes in appetite and body weight reported by KOCHAKIAN and WEBSTER (1958) and KOCHAKIAN et al. (1959). Unfortunately, HALPERN and COURNOT (1953) did not determine the water and chloride intake. In a third series of experiments, the rats were killed 7 days after heminephrectomy and the water concentration in the kidney of the treated animals was not significantly different from that of the nontreated rats, indicating that the enhancement of the renal compensatory hypertrophy by testosterone propionate was not due to edema. In a final experiment, the concentrations of nitrogen and RNA of the kidney also were unchanged, thus leading to the conclusion that the additional kidney weight under testosterone propionate stimulation was true growth or hypertrophy and not hyperplasia which was further substantiated by the lack of a difference in the number of renal cell mitoses.

COBET et al. (1967) unilaterally nephrectomized male rats of 200–250 g body weight and injected them with a single dose of 3 mg 19-nortestosterone-17$\beta$-decanoate. Fourteen days later, the weight of the kidneys was compared to those of nontreated heminephrectomized, nontreated, and treated normal rats. Heminephrectomy produced a 27% increase, the hormone alone 12%, and the two together a summation of 44%. LATTIMER (1942) hypophysectomized young adult female rats, and after a "recovery period" removed the right kidney and injected 2 mg/day of testosterone propionate for 25 days. The left kidney of 3 of the rats was enlarged to

71% of the predicted weight of both kidneys. Removal of the pituitary prevented the usual renal compensatory hypertrophy.

GROSS and STRICKER (1951) placed 150 g male rats on either a protein-poor (8% casein) or a protein-rich (48% casein) diet, removed the right kidney and injected 5 mg/kg/day of testosterone propionate for 40 days. The weight of the remaining kidney of the rats on the protein-rich diet was increased more than that of the rats on the protein-poor diet but neither was further enhanced by the hormone.

### b) Dog

In an attempt to observe a more positive renotrophic effect in the dog, LATTIMER (1942) removed the left kidney of two female dogs (5 years old) and 58 days later injected one dog with 10 mg/day of testosterone propionate for 32 days. The right kidney showed the expected compensatory hypertrophy in the control dog and an even greater increase in the treated dog. The left kidneys weighed 20.1 and 19.8 g and the right kidneys 23.7 and 27.6 g for control and treated dogs respectively. The kidney of the treated dog showed greater tubular excretory activity (diodrast) and glomerular inulin clearance.

### c) Mouse

WRETE (1943) unilaterally nephrectomized 15-month-old male and female mice which had been gonadectomized at 4–7 weeks of age. The compensatory renal hypertrophy was enhanced by the injection of 0.5 mg/day of testosterone dipropionate and also by 5 μg/day of estradiol-17$\beta$-benzoate. The weight increase induced by a combination of kidney removal and hormone treatment was greater than the sum of the individual effects. There was a proportionate increase in the size of the epithelial cells and water and dry matter content of the stimulated kidneys.

JELINEK et al. (1964) castrated adult (17–22 g) mice and 19–24 days later performed a unilateral nephrectomy on part of these mice and a comparable age group of normal mice. Part of the mice were immediately injected subcutaneously with 3.2 mg of 19-nortestosterone phenylpropionate. Significant compensatory hypertrophy was already evident after 48 and 72 h but there was no difference between the controls and the treated mice, both castrated and noncastrated. After 96 h, however, the steroid had produced a significant and similar enhancement in the renal compensatory hypertrophy in both castrated and normal mice.

HOMBURGER et al. (1950) produced compensatory renal hypertrophy by ligating the left ureter just below the renal pelvis of adult female mice. The resulting hydronephrosis induced contralateral compensatory hypertrophy which was further increased by the injection of doses of testosterone propionate varying from 0.1–5 mg every other day for 14 days beginning on the day after ureteral ligation.

## B. Liver

### 1. Rat

### a) Castration

KORENCHEVSKY and DENNISON (1934a, b, 1935) and KORENCHEVSKY et al. (1935, 1941) reported slight atrophy in the liver of rats which required 70 days after castration to become apparent. In a later study (KORENCHEVSKY et al., 1953b), the liver of rats castrated at 22–35 days of age were 13% smaller than those of controls after 64 days and no further decrease was noted up to 150 days after castration. An identical response was obtained by KOCHAKIAN et al. (1956a). The liver weight of rats castrated at 40 days of age was 12% smaller than that of control rats after 96 days. Comparable decreases in liver weight were apparent 70 or more days after castration in two other studies (ENDAHL and KOCHAKIAN, 1956; KOCHAKIAN et al., 1954). Male rats castrated at 150–155 days of age did not exhibit a decrease in liver weight 20–25 days later (LEATHAM, 1948a) and ASCHKENASY and BLANPIN (1955) were unable to observe a significant decrease in the liver weight of 150 g male rats 8 days after castration.

Thus the effect of castration on liver weight is small but consistent and the difference from normals only appears gradually over a long castration period. On the other hand, if the rats were castrated at birth, the liver weights were approximately 35% less after 120 days than that of litter-mate controls while that of males castrated at weaning was only about 14% lower (SWARTZ, 1962) or similar to that of the earlier reports. These differences were smaller on a body weight basis.

*b) Steroids*

α) Castrated

The slight decrease in liver weight after castration was restored to or near normal by the injection for at least 21 days of a commercially prepared urinary extract (KORENCHEVSKY et al., 1933; KORENCHEVSKY and DENNISON, 1934a, b, 1935), androsterone, 5α-androstane-3α,17β-diol, and their succinates (KORENCHEVSKY and DENNISON, 1934a; KORENCHEVSKY et al., 1941), dehydroepiandrosterone, testosterone, testosterone propionate and dipropionate, and androst-4-ene-3,17-dione (KORENCHEVSKY et al., 1941). ENDAHL and KOCHAKIAN (1956) also noted that the liver weight of castrated rats was restored to or near normal only after about 3 weeks of administration of testosterone, testosterone propionate, or 17α-methyltestosterone, implanted subcutaneously after 70 or more days of castration of immature rats. No detectable increases were noted after 6, 12 (ENDAHL and KOCHAKIAN, 1956), or 14 days (KOCHAKIAN et al., 1954) of hormone administration. An increase in the dose or duration of treatment did not produce further increases in liver weight. KOCHAKIAN et al. (1950) observed the usual, approximately 13% increase (restoration to normal) in liver weight of adult castrated rats after 21 days of 0.125 mg/day of testosterone propionate. An increase of the dose to 1.0 mg/day decreased the effectiveness of the steroid even after 46 days. A further increase in dose to 2.5 mg/day produced a 15% increase in eight days but only 4% after 16 days, and about 10% after 125 days. These variable effects with dose and duration of treatment are consistent with the effects on food intake and body weight.

LEATHAM (1948a) found that 0.1 mg testosterone propionate for 20–25 days, beginning the day after castration, increased the liver weight 26% above normal. The testosterone propionate was ineffective at 1.25 mg/day and had an intermediate effect at 0.25 mg/day.

β) Normal Male

KORENCHEVSKY et al. (1933) injected a wide range of doses of a male hormone extract from urine for 21 days into immature normal male rats and did not find a significant change in the weight of the liver. The extract, however, was not too effective, especially at the lower doses, in the induction of an increase in the seminal vesicles and prostates. SELYE (1940a) injected young (154 g) male rats for 21 days with 2 mg/day of testosterone and found no significant increase when injected alone or in combination with estradiol-17$\beta$ (0.3 mg/day), progesterone (2 mg/day), or desoxycorticosterone acetate (2 mg/day). In a subsequent study, SELYE and BELAND (1943) carried out a long-term study in which immature (44–55 g) rats were injected for the 1st month with 0.5 mg/day, the 2nd month with 1.0 mg/day, and the 3rd month with 2.0 mg/day of testosterone. They ignored the changes in liver weight but the data indicated a substantial increase (approx. 50%) after 2 months, was maintained after 3 months, and surprisingly, further increased 2 months after cessation of injections. LEATHAM (1948a) found no change in the liver weight of 150–155 g rats injected for 20–25 days with 0.5, 1.25, or 2.5 mg/day of testosterone propionate. The feeding of a high protein diet (78% lactalbumin) did not modify the effect of the hormone (LEATHAM, 1951). KORENCHEVSKY et al. (1953a) implanted subcutaneously a 25 mg tablet of testosterone propionate in senescent (about 18 months old) male rats and 91 days later observed no significant change in the weight, size of hepatic cells, or number of mitoses.

A body weight depressing dose of testosterone propionate, 2.5 mg/day, injected for 28 days into adult (150 g) male rats decreased the liver weight about 14% with a proportional decrease in protein, water, and fat (KOCHAKIAN and WEBSTER, 1958). Extension of the treatment to 125 days did not give a further decrease (KOCHAKIAN et al., 1950) and the decrease was no longer evident when calculated according to body weight. LIGHT and TORNABEN (1953a) found a similar effect after the percutaneous inunction of 0.4 mg/day of 17$\alpha$-methyltestosterone for 201 days to immature (36–42 g) rats. On the other hand, ASCHKENASY (1959c) reported a 14% and 28% increase in the liver weight of young adult (125 g) male rats fed 25 mg and 100 mg of 17$\alpha$-ethylnortestosterone per 100 g diet respectively for 120 days. Apparently the orally administered compounds must have a direct and probably different effect on the liver.

### $\gamma$) Female

KORENCHEVSKY et al. (1935) obtained small and variable changes in the weight of the liver of rats 30–81 days after ovariectomy at 30 days of age. These results were confirmed in a later study (KORENCHEVSKY et al., 1941). Ovariectomy of female rats either at birth or at weaning produced a 10% increase in liver weight after 120 days but a small decrease when calculated according to body weight (SWARTZ, 1962).

The injection of 0.9–3.6 mg/day of androsterone, 0.49 or 0.62 mg/day of 5$\alpha$-androstane-3$\alpha$,17$\beta$-diol, or 2.29 mg/day of its succinate for 22–27 days (KORENCHEVSKY et al., 1935) or dehydroepiandrosterone (KORENCHEVSKY and DENNISON, 1936) for 21 days into rats, begun 10–82 days and 14–34 days after ovariectomy respectively, produced increases in liver weight similar to that in castrated male rats except for the highest dose (3.44 mg/day) of 5$\alpha$-androstane-3$\alpha$,17$\beta$-diol disuccinate, due apparently to the inhibition of appetite and the consequent loss in body and liver weight. The rats were immature at ovariectomy.

In a further study, KORENCHEVSKY et al. (1941) injected immature ovariectomized rats for 21 days, beginning at various times after ovariectomy, with various doses of several hormones. The daily injection of androsterone, dehydroepiandrosterone, and testosterone propionate or dipropionate, at 7.5 mg/week for 2–4 months produced significant increases in liver weight. Testosterone propionate at 2.25 mg/week was as effective as 7.5 mg/week. The lower doses and shorter period of injection were ineffective. KORENCHEVSKY et al. (1941) noted that the liver weights of the female rats were stimulated to above normal and approached that of the normal males while the hormones only restored the weight to the normal level in castrated males. Androsterone was also effective when injected as a crystalline suspension of 10 mg every 9 or 10 days for 43 days beginning at 132 days of age (KORENCHEVSKY and JONES, 1948). SELYE (1940a) and SELYE and BELAND (1943) did not report any change in the liver weight of young (109 g) and immature (44–55 g) female rats injected with 2 mg/day for 21 days and 0.5–2.0 mg/day up to 90 days respectively.

ASCHKENASY (1959c) fed 17α-ethylnortestosterone at 25 or 100 mg/100 g of diet to young (135 g) female rats for 120 days and observed an increase in liver weight of 18 and 33% respectively. LIGHT and TORNABEN (1953b) applied percutaneously 0.5 mg/day of 17α-methyltestosterone for 186 days and obtained a 17% increase in liver weight on a body weight basis.

KORENCHEVSKY et al. (1950), in an attempt to ameliorate some of the effects of senescence, did not observe a change in the liver weight of 22–24-month-old female rats after the injection for 60 days of either androsterone benzoate in a crystalline suspension at 15 mg 5 times a week, or testosterone propionate also a crystalline suspension at 2 mg twice per week. A small but questionable increase in liver weight was noted in 27-month-old rats after the injection of 10 mg of a crystalline suspension of androsterone once per week or the implantation of a 20 mg pellet of testosterone propionate.

## 2. Mouse

### *a) Castration*

KOCHAKIAN (1941) noted a small (7%), insignificant decrease in the weight of the liver of male mice 65 days after castration at 5 weeks of age, and no difference after 95 days. In a later temporal study (KOCHAKIAN, 1959, 1964), a clearer picture of the effect of castration was obtained. Castration at 26–28 days retarded the rate of growth of the liver so that a maximum and significant difference was evident at about 40 days (Fig. 2). The difference then began to gradually diminish so that by about 60 days of age the liver weight of the castrated mice was slightly but not significantly smaller than that of the normal controls. Thus, the livers of the normal mice attained a plateau in weight at about 40 days while the castrated mice required another 20–30 days to attain the same plateau. The weight of the liver of the female mouse was less than that of the castrated male as was the body weight.

WOLFF (1970, 1971) reported a genetic dependence for the decrease in liver weight after castration. Mice of the following inbred strains and $F_1$ hybrid crosses were used: YS/ChWf; VY/Wf; BALB/cAnNIcr; $C_{57}$BL/6JNIcr; $C_3$H/HENIcr; $C_3$H/HeNIcr A YS-$A^y$a and Vy-$A^{vy}$/a. Castration was at 28 days and autopsy occurred at 28-day intervals for 112 days. Castration had no effect on liver weight in the $C_3$H strain nor in any YS $F_1$ hybrids. Among the yellow VY $F_1$ hybrid males, castration was without effect except in the (BALB/C x VY) category. The liver weight in all castrated non-yellow VY $F_1$ hybrids was less than in the intact siblings. In contrast, the liver weight of ovariectomized yellow females was significantly greater than that of intact $A^y/a$ females. Ovariectomy had no effect on the liver weight of non-yellow HS/ChWf females. WOLFF suggested that the effect on liver weight is not directly due to a decreased level of sex hormones but rather by an alteration in the balance of a number of hormones.

*b) Steroids*

The subcutaneous implantation of a 22-mg pellet of testosterone propionate into 65-day-old mice for 30 days after castration had no effect on liver weight after 5, 11, or 13 days, but after 27–64 days produced a consistent increase which was no longer evident after 90 and 96 days (KOCHAKIAN, 1941). In a later, similar experiment by KOCHAKIAN and STETTNER (1948), a single 15-mg pellet of testosterone propionate was implanted subcutaneously and an insignificant 4% increase in liver weight was obtained after 10 days, and a 20% and 15% increase after 20 and 34 days, respectively. A growth hormone preparation was more effective after 10 (+29%) and 20 (+35%) days but possibly less effective (+11%) after 34 days. The effect of the two hormones did not summate.

FERGUSON and VISSCHER (1953) observed a very marked decrease in the liver weight of 3–4-month-old mice 45–49 days after hypophysectomy and gonadectomy. The decrease was greater in the male (−43%) than the female (−31%) mice. The injection of testosterone propionate at 20 μg/day for males and 40 μg/day for females for the last 14 days produced a small, partial restoration, an increase of approximately 20%, in both groups of mice.

### 3. Guinea Pig

*a) Castration*

Castration of guinea pigs at 300 g (KOCHAKIAN et al., 1948a), or 500 g body weight (KOCHAKIAN et al., 1956b) did not significantly affect the rate of growth of the liver during the subsequent 27 weeks. Furthermore, castration of guinea pigs at approximately 30 weeks did not change the liver weight 6 weeks later.

*b) Steroids*

The administration of testosterone propionate by pellet implantation or subcutaneous injection and several other steroids by pellet implantation at several dose levels for 21 or 30 days did not change the weights of the liver of castrated guinea pigs (KOCHAKIAN et al., 1948b, 1956b).

### 4. Hamster

KOCHAKIAN et al. (1948b) castrated hamsters at 45–75 g body weight and 1 month later injected 1 mg/day of testosterone propionate for 20 and 60 days or subcutaneously implanted 14–15 mg pellets of testosterone or 17α-methyltestosterone for 20 days; or a 24–26 mg pellet of these steroids was implanted at the time of castration for 140 days. The liver weights of the normal, castrated, and treated hamsters were essentially the same for the respective experiments.

### 5. Dog

BLACKMAN et al. (1944), in the course of their studies on the effect of testosterone propionate on the kidney of growing male and female puppies, found that the liver of the treated dogs weighed 14.1–17.9 g more than their calculated expected weight.

## C. Heart

### 1. Rat

KORENCHEVSKY and associates in their early studies also examined the effect of castration and administration of androgenic hormones on the weight of the heart simultaneously with that of the other organs of the rat.

*a) Castration*

The weight of the heart was slightly decreased in most instances after long periods of castration but these changes were not consistent in individual rats of the castrated group (KORENCHEVSKY and DENNISON, 1934a) which was confirmed (KORENCHEVSKY et al., 1935; KORENCHEVSKY and DENNISON, 1936). KORENCHEVSKY et al. (1941) pooled all of their previous studies and some new experiments and made a statistical analysis of the combined data. They concluded that castration slightly but significantly decreased the weight of the heart. KOCHAKIAN et al. (1954) confirmed the slight effect of long-term castration. Rats were castrated at 72–105 g body weight and 136 days later the heart weighed 12% less than that of the normal rats. KORENCHEVSKY et al. (1953a), in a later study, castrated rats at 22–35 days of age and in contrast to their earlier conclusion found no difference in the weight of the heart between normal and castrated rats killed at four different intervals over a period of 109 days.

*b) Steroids*

α) Castrated Rat

The small decrease in heart weight after long-term castration observed by KORENCHEVSKY et al. (1941) was restored only to, or toward, normal by the administration of a male hormone urine extract (KORENCHEVSKY et al., 1933; KORENCHEVSKY and DENNISON, 1934b), androsterone; 5α-androstane-3α,17β-diol and their succinates (KORENCHEVSKY et al., 1935), dehydroepiandrosterone (KORENCHEVSKY and DENNISON, 1936) and testosterone propionate (KORENCHEVSKY et al., 1941). Testosterone propionate was the most effective. KOCHAKIAN et al. (1954) confirmed these results. Subcutaneous implantation of two 15 mg pellets into rats castrated at 72–105 g body weight for the period of 123–136 days after castration restored the small decrease in heart weight to normal. ASCHKENASY (1959a) found no change in the heart weight of castrated or adrenalectomized castrated rats (166–190 g) after the injection of 1 mg/day of testosterone propionate for 21 days beginning 8 days after the operation. The period of castration was probably too short to exhibit a decrease in weight.

β) Normal Male Rat

Normal immature (30–35 days) male rats responded to the injection of a male hormone urine extract for 21 days with small and irregular changes in heart weight

(KORENCHEVSKY et al., 1933). SELYE (1940a) also observed irregular responses in heart weight of young (154 g) male rats after the daily injection of 2 mg of testosterone for 21 days. Later, SELYE and BELAND (1943) injected immature (44–55 g) rats daily with increasing doses of testosterone for 1, 2, or 3 months and did not observe a definite change in heart weight. In contrast to the young male rats, senescent (18-month-old) male rats implanted subcutaneously with a 25 mg tablet of testosterone propionate showed after 68 and 91 days a small consistent increase (11%) in heart weight (KORENCHEVSKY et al., 1953b). These changes were considered to be a reversal of the hypofunction of the testes due to senescence.

#### γ) Female Rat

KORENCHEVSKY and DENNISON (1936) found no definite changes in the heart weight of rats ovariectomized before sexual maturity and injected 14–34 days later for 21 days with 1, 2, or 3 mg of dehydroepiandrosterone. An increase in the dose to 7.5 mg/day and extension of the duration of treatment to 3 months resulted in an increase in heart weight towards that of the normal male; and androsterone, testosterone propionate and testosterone dipropionate at the same dose level increased the heart weight to that of the normal male (KORENCHEVSKY et al., 1941). Testosterone propionate produced the same effect at a smaller dose, 2.25 mg/day, and was effective even at shorter periods of treatment. KORENCHEVSKY and JONES (1948) confirmed the small effect of androsterone in ovariectomized rats. The androsterone was injected as the more efficacious aqueous crystalline suspension from age 132–175 days and produced only a 9% increase in heart weight.

SELYE (1940b) injected 2 mg/day of testosterone for 21 days into young (109 g) female rats and recorded a 10% decrease in heart weight. SELYE and BELAND (1943) administered increasing (0.5–1.0–2.0 mg/day) doses for 3 months to immature rats (45–55 g). The recorded heart weights did not indicate any definite changes. VARGAS and ABELIUK (1972), however, observed a 10% increase in the heart weight after the oral administration of 1.0 mg/day of either 17α-methyltestosterone or stanozolol for 3 months to 26-day-old female rats.

Senescent female rats showed a 20% increase in heart weight on the injection of 10 mg of an aqueous crystalline suspension of androsterone at 1-day intervals or the subcutaneous implantation of a 20 mg pellet of testosterone propionate for 86 days into 24-month-old female rats. The effect was greater than in the younger rats and was considered to be a reversal of the effects of senescence.

#### δ) Thyroidectomized Rat

Thyroidectomy of castrated rats results in a further and marked decrease in heart weight to a minimum level (– 30 to – 35%) within 25 days. The injection of 0.5 or 1.0 mg/day of testosterone propionate for 14 days was not able to influence the weight of the heart (KOCHAKIAN et al., 1954).

#### ε) Repletion After Underfeeding

QUIMBY (1948) found that the injection of 0.5 mg/day of testosterone to male rats for 60 days after repletion (feeding ad lib) from 20 days of underfeeding was of no value in the restoration of the heart weight to normal.

## 2. Mouse

### a) Castration

KOCHAKIAN (1941) castrated mice at 5 weeks and observed a small decrease in heart weight after 2 months. In a later, more detailed study (Fig. 2) (KOCHAKIAN, 1959, 1964), mice castrated at 26–28 days showed a slower rate of growth and a smaller maximum weight of the heart similar to that of the female.

### b) Steroids

The implantation of a 22 mg pellet of testosterone propionate 30 days after castration produced a consistent, approximately 15% increase within 5–13 days which gradually increased to about 25% during the next 83 days (KOCHAKIAN, 1941).

#### α) Hypophysectomized Mouse

Hypophysectomy produced a marked decrease in the heart weight of male (– 55%) and female (– 50%) mice which was not affected by 20 μg/day and 40 μg/day respectively of testosterone propionate for 14 days (FERGUSON and VISSCHER, 1953).

#### β) Castrated Mouse

BROULIK et al. (1973) found a parallel decrease in the cardiac output and renal blood flow of young adult mice 14 days after castration without a further decrease after another 7 days. The subcutaneous implantation of a pellet of testosterone propionate 21 days after castration restored the values to normal within 7 days after a delay of at least 2 days. Further hormone treatment did not produce a further increase in cardiac output or renal blood flow.

# D. Thymus and Spleen

## 1. Rat

### a) Male

#### α) Castration

KORENCHEVSKY (1930) castrated rats between 28 and 33 days of age, and 182 days later found that the adrenals on the average were 23.5 (8.6–41.6%) larger than those of litter mate controls. These results were confirmed in subsequent experiments (KORENCHEYSKY and DENNISON, 1934b; KORENCHEVSKY et al., 1953b) with parallel changes in the anterior pituitary and the thymus but indefinite changes in the spleen. The larger weight of the thymus after castration was not due to hypertrophy but a delay in the normal physiological process of involution (KORENCHEVSKY and DENNISON, 1936). The inability to restore the weights of these organs to normal was due to the small size of the dose.

#### β) Steroids

ROSS and KORENCHEVSKY (1941) pooled their data on the effect of various steroid hormones on the involution of the thymus and reported that the weakest male hormones, dehydroepiandrosterone and androst-5-ene-3β,17β-diol, had little or no effect. The greatest effect was produced by the strongest hormone, testosterone propionate, while androsterone, testosterone and androst-4-ene-3,17-dione produced intermediate effects. The administration of testosterone propionate as a subcutaneously implanted pellet of 25 mg accelerated the atrophy of the thymus in senescent (18-month-old) normal and castrated (1 month previously) rats after 68 or 91 days. A depressing effect on adrenal weight was small and inconsistent and no significant change in the weight of the spleen was detected.

SELYE (1940) confirmed the acceleration of involution of the thymus by the injection of 2 mg/day of testosterone for 21 days in immature male and female rats but the weights of the adrenals and spleen were not changed. SELYE and BELAND (1943) extended the duration of testosterone injections to 3 months with monthly doubling of the dose from 0.5–2.0 mg/day and noted a marked acceleration of the involution of the thymus of both the male and female rats but no significant changes in the weights of the spleen or adrenals.

ASCHKENASY (1959a) castrated or adrenalectomized and castrated rats at 160–190 g and 8 days later injected 1 mg/day of testosterone propionate for 21 days. The weight of the thymus in both groups of treated rats was less than half that of the nontreated rats. The spleen was not significantly affected in the castrated rats but was decreased about 20% in the doubly operated rats. ASCHKENASY (1959c) extended these studies to the oral administration of 17α-methyltestosterone and 17α-ethylnortestosterone at 5 mg/day for 28 days beginning 7 days after castration. The 17α-ethylnortestosterone had a slight depressing effect on the thymus and neither steroid affected the weight of the spleen. ASCHKENASY (1959c) fed 25 or 100 mg 17α-ethylnortestosterone per 100 g diet to young (about 130 g) normal male and female rats for 4 months as part of a toxicity study. The thymus weight was markedly depressed at the lower dose and only a slight further decrease was produced by the higher dose in both the male and female rats. The spleen was not weighed. LIGHT and TORNABEN (1953a) found no change in the spleen or adrenal weights of rats inuncted on the skin with 0.4 mg/day of 17α-methyltestosterone in 200 mg of ointment for 201 days beginning at weaning (36–42 g). The thymus was not weighed separately.

*b) Female*

α) Ovariectomized

ROSS and KORENCHEVSKY (1941) reviewed the earlier literature which established that gonadectomy resulted in a retardation of the involution of the thymus in the female as well as the male of a variety of animals and provided further experimental support for this phenomenon in the rat. Ovariectomy at 23–27 days of age resulted after 33–88 days in a thymus weight 40–70% greater than that of controls (KORENCHEVSKY et al., 1935).

β) Steroids

Androsterone and 5α-androstane-3α,17β-diol and its succinate at different doses for 22 to 27 days beginning 10–82 days after ovariectomy decreased the weight of the thymus of ovariectomized (23–27 days) rats (KORENCHEVSKY et al., 1935). The effect of androsterone was confirmed in a later report (KORENCHEVSKY and JONES, 1948).

Testosterone propionate, however, was even more effective (KORENCHEVSKY et al., 1937a, b; ROSS and KORENCHEVSKY, 1941). The injection of 0.5 mg/day for 21 days beginning about 50 days after ovariectomy decreased the weight of the thymus to about one third that of the control rats and 1.5 mg/day produced only a small further decrease. SELYE (1940) obtained similar results by the injection of 2 mg/day of testosterone for 21 days in normal female rats weighing 109 g. In a later study no difference was observed between female rats injected for 1 month with 0.5 mg/day of testosterone, but a definite effect was observed upon extension of the injections at double the dose for another month, and a further decrease on doubling the dose again for the 3rd month.

The already partly involuted thymus of senescent (20–24 months old) female rats was further decreased by the injection of a crystalline suspension of testosterone propionate for 60 days at 2 mg twice per week or the subcutaneous implantation of a 20 mg pellet for 86 days (KORENCHEVSKY et al., 1950).

LIGHT and TORNABEN (1953b) found no change in the weight of the spleen but a marked decrease in the weight of the adrenal of normal female rats inuncted on the skin with 0.4 mg of 17α-methyltestosterone in 200 mg of ointment for 186 days. The thymus was weighed with the heart and lungs and the group weight was unchanged. VARGAS and ABELIUK (1972) found no significant difference in the thymus weight of 26-day-old female rats after the oral administration of 1.0 mg/day for 3 months of 17α-methyltestosterone or stanozolol. These results, seen superficially, seem contradictory with the above but they more likely indicate that the dose of these steroids was not sufficient to exceed the normal involution over the 3-month period.

## 2. Mouse

The changes in weight of the thymus, spleen and also the adrenal after castration and androgen administration are readily detected in the mouse.

### *a) Castration*

KOCHAKIAN (1941) castrated mice at 5 weeks and 1 and 2 months later the thymus was approximately 50% heavier than that of the controls and the spleen and adrenals about 30% heavier. Ninety days after castration, the difference in weight of the thymus had almost disappeared due to the normal involution of the organ in both the normal and castrated mice. The difference was still apparent in the spleen and was enhanced in the adrenal. KOCHAKIAN (1959) demonstrated that the time after castration was an important factor in the degree of change in these three organs after castration. Mice were castrated at 26–28 days of age and organ weights were determined at intervals to 170 days (Fig. 13). The thymus and spleen of the castrated mice initially increased in weight more rapidly than those of the normal controls. After about 2 weeks, the increase in weight ceased and thereafter the thymus weight of the castrated mouse was only slightly greater than that of the normal. The spleen, however, showed only a slight regression in weight with age and the weight of the adrenals increased rapidly until about 70 days of age and then continued to increase but at a very slow rate. There was no indication of involution of this gland with age in either the castrated or the normal mice. The adrenals of the female mouse were

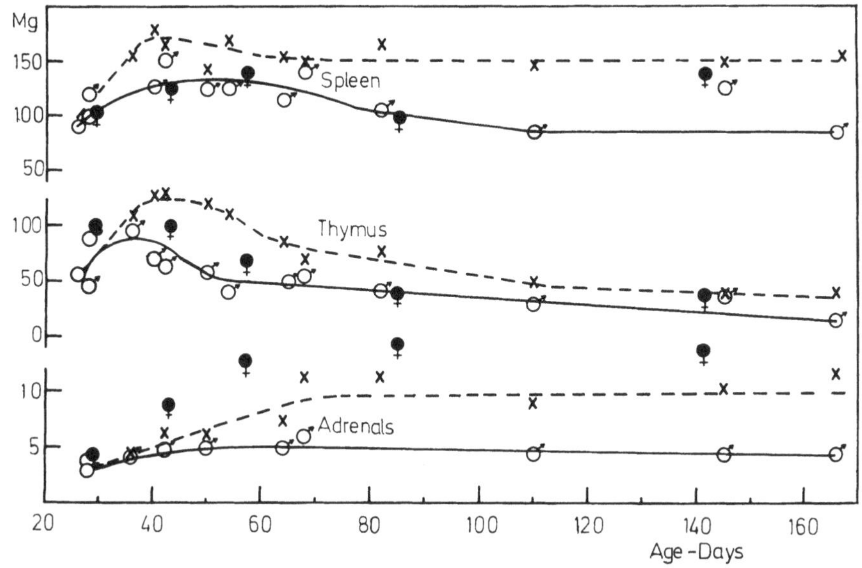


Fig. 13. The effect of castration and sex on the weights of the spleen, thymus and adrenals of mice. Mice of the Holtzman-Swiss strain were castrated at 26–28 days of age. Normal males: ○—○, castrated: ×---×, normal females: ♀—♀ (KOCHAKIAN, 1959, 1964)

consistently heavier than those of the male. The thymus and spleen of the female mouse showed essentially the same changes as those of the male with age.

### b) *Steroids*

#### α) Castrated

The implantation of a pellet of testosterone propionate one month after castration in immature (5-week-old) mice produced a definite decrease in the hypertrophied thymus within 5 days but required 11–13 days for the spleen and adrenals (KOCHAKIAN, 1941). After 20 days of hormone treatment, the weight of the thymus was only a fraction of that of the normal controls and those of the spleen and adrenal were reduced to those of the normal mice. The ability to cause regression of the thymus was demonstrated by a wide variety of steroids (KOCHAKIAN, 1944), including estrogens (KOCHAKIAN, 1944, 1947; KOCHAKIAN et al., 1948c) and adrenal cortical steroids (KOCHAKIAN, 1944). The ability to decrease thymus weight was roughly proportional to androgenic or estrogenic activity of the respective compounds. A high dose of testosterone, for example, nearly obliterated the thymus completely. A dose relationship was evident also for the adrenal cortical compounds but the C-11 steroids were more effective than the 11-desoxysteroids.

#### β) Hypophysectomized

In contrast to castration, hypophysectomy at 3–4 months of age resulted in a definite atrophy of the adrenals and spleen and a questionable small decrease in the thymus of both male and female mice (FERGUSON and VISSCHER, 1953). The injection of testosterone propionate (20 μg/day for males and 40 μg/day for females) during the 31st to 45th days after hypophysectomy produced a small decrease in the thymus of the female mice but no significant change in that of the male mice or the adrenals and spleens of both male and female mice.

### 3. Guinea Pig

*a) Castration*

Castration of guinea pigs at 457–573 g body weight resulted in a slightly greater weight of the thymus, spleen, and adrenal (KOCHAKIAN et al., 1956). The difference was evident at 6, 12, and 27 weeks after castration for the adrenals, but an unexplained larger weight of both the thymus and spleen of the normal guinea pigs after 12 weeks masked the difference at this time interval for these two organs. Castration at 21 weeks of age resulted in a definite hypertrophy of the thymus after 6 weeks but no difference in the adrenals or spleen was detected. Castration of guinea pigs at 350 g (approximately 8 weeks of age) produced no definite change in thymus weight 9 and 51 days later (KOCHAKIAN et al., 1957a).

*b) Steroids*

The administration of testosterone and several other steroids at a wide range of doses for 21 days (KOCHAKIAN and TILLOTSON, 1957a) did not affect the thymus, spleen or adrenal weights of guinea pigs castrated at 510 g body weight. The steroids were implanted subcutaneously as 14-mg pellets 30 days after castration.

## E. Lacrimal Glands

### 1. Rat

A clear-cut sexual dimorphism of the lacrimal glands was recognized by histologic studies in the rat by TEIR (1949), GABE (1955), and PARHON et al. (1955), and confirmed by WALKER (1958) and BAQUICHE (1959) in careful systematic studies. TEIR (1949), WALKER (1958), BAQUICHE (1959), and BENSON (1964) demonstrated transformations in the cytology of the cells with age and sexual development. The difference between male and female rats appeared after 3 months of age.

*a) Gonadectomy (Male vs. Female)*

On the basis of the greater size of the exorbital glands in the male than the female rat, GABE (1955) castrated rats (age not given) and observed no change in the morphology of the gland 30 days later. Ovariectomy of female rats, on the other hand, converted the cells to the male type. BAQUICHE (1959), in his careful studies, compared rats of a known same age. The male and female rats were gonadectomized at 50–55 days of age and killed 40 days later. The exorbital glands of the castrated male appeared to have no significant weight change, and histologic examination revealed that all of the cellular elements were transformed toward, but not yet into, the female type. Ovariectomy of the female also resulted only in modifications of the cellular elements towards the male type so that the histology of the cells from the castrated male was not distinguishable from that of the ovariectomized female rat. CAVAL-

LERO and MORERA (1960) confirmed the decrease in the number of large nuclei in the lacrimal glands 80 days after castration of rats at 150–200 g body weight. Detailed cytological changes similar to those described by BAQUICHE (1959) were also noted by CAVALLERO (1967).

YAMADA (1965) and MORII et al. (1966) observed that the exorbital lacrimal glands of the young adult female (190 g) rat was only slightly larger (10%) than that of the equivalent age male rat (222 g) in contrast to the report of BAQUICHE (1959) and GABE (1955). Castration did not change the weight of the lacrimal glands in the male rat but ovariectomy reduced that of the female to that of the male.

### *b) Steroids*

#### α) Morphologic and Cytologic Changes

GABE (1955) injected castrated male rats for 6 days (time after castration was not given) with 0.5 mg/day of testosterone propionate and did not detect any changes in the histology of the lacrimal glands. The injection of 100 μg/day of estradiol-17β-benzoate for 5 days into the ovariectomized rats, however, reverted the cells to the female type. BAQUICHE (1959) injected the gonadectomized rats with 0.5 mg/day of testosterone propionate or 100 μg/day of estradiol-17β-benzoate from the 30th to 40th days after gonadectomy. The testosterone propionate had no effect on the weight of the lacrimal glands of either the male or female rats. In the male rat, the acini were enlarged; the heterogeneous and small cytoplasm of the castrated rat now was dense, homogeneous and strongly eosinophilic; the protoplasm indicated secretory activity; the cellular polymorphism was accentuated to more than that of the normal males; and the nuclei were enlarged with a strong network of chromatin and prominent nucleoli. The ovariectomized female rat responded to testosterone propionate in a manner similar to the male. The estradiol-17β-benzoate had no significant effect on the weight of the lacrimal glands of either the male or female rat but converted the histological picture to that of the normal female.

BAQUICHE (1959) extended the studies to normal rats with testosterone propionate at a higher dose and a longer period of injection. The administration of 1 mg/day of testosterone propionate to 5-week-old male rats for 10 days had no effect on the weight of the lacrimal glands. The acini, however, were enlarged and many were strongly vacuolized, the nuclei were enlarged and displaced towards the periphery of the cell. In another experiment, 6-week-old male and female rats were injected with 1 mg/day of testosterone propionate for 42 days. The weight of the lacrimal glands (293 mg) of the control males was definitely greater than that (185 mg) of the females but the testosterone propionate did not produce a significant change in the weight of the lacrimal glands of either group of rats. The histological picture, however, again indicated an accentuation of the male type of morphology. Thus the testosterone propionate enhanced the functional capacity of the lacrimal glands in both the male and female rats. In another study from the same laboratory (DUCOMMUN et al., 1959), the effects of testosterone propionate was confirmed in male and female rats of 125 g body weight by the injection of testosterone propionate at 1 mg/day for 6 weeks. Similar but less pronounced effects were produced by 17α-ethyl-19-nortestosterone at 1 mg/day as a crystalline aqueous suspension. CAVALLERO and MORERA (1960) observed a definite shift toward higher nuclear volumes in the lacrimal cells of normal and especially in the castrated male rat after the injection of 1 mg/day of testosterone propionate between the 60th and 80th days after castration.

The injection of 2 mg/day of testosterone propionate for 4 weeks begun 2 days after gonadectomy had no effect on the weight of the lacrimal glands in the castrated male rat but increased it in the ovariectomized rat on a body weight basis (YAMADA, 1965; MORII et al., 1966). The authors also noted marked histological changes. The terminal portion of the gland in the male rat had more secretory granules, the nuclei of the glandular cells were larger and more vesicular and contained more phosphatase and DPN-diaphorase than that of the female rat. Castration of the male did not alter the histological picture. The prolonged administration of testosterone propionate, however, produced profound histological changes in the exorbital gland of both the castrated and ovariectomized rat, similar to those observed by the previous investigators.

β) Quantitative Changes

The morphologic changes in the exorbital lacrimal gland prompted CAVALLERO (1967) to devise a quantitative assay procedure based on the histological vesicular-mucous change in the glandular acini. Rats (150–180 g) castrated 1 week previously were injected for 20 days with the steroids suspended in an aqueous solution. The percentage of glandular acini showing vesicular change and containing PAS positive material in their lumen was determined. Using this technique, CAVALLERO (1967) compared the activity of 21 $\Delta^4$-3-oxo- and $\Delta^5$-3β-hydroxy steroids and CAVALLERO and OFNER (1967) 14 compounds of the 5α-androstane series with testosterone. The effect of the various modifications in chemical structure on activity was compared. In many instances, but not always, the relative activities were comparable to those obtained by other reported assay procedures.

γ) Biosynthesis of Mucopolysaccharides

CAVALLERO et al. (1960) studied the uptake of $^{35}S$ by adult castrated and newborn male rats. The older rats were injected for 10 days with 1 mg/day of testosterone propionate and 18 h before the end of the experiment with 0.3 μCi/g body weight of $^{35}S$-labeled $Na_2SO_4$. The $^{35}S$ uptake by the lacrimal glands was 2.5 times higher in the treated than the control rats. The newborn (2-day-old rats) were injected with testosterone propionate for 3 days and 1 μCi/g body weight of $^{35}S$-labeled $Na_2SO_4$. No difference was noted in the uptake of radioactivity between the treated and control rats.

*c) Hypophysectomized*

ARAGAKI et al. (cited by YAMADA, 1965) reported that the histologic changes induced by injection of testosterone propionate were detected in the hypophysectomized rats.

## 2. Mouse

*a) Male vs. Female*

KOCHAKIAN (1959) (Fig. 10) observed only a slight further growth of the orbital and exorbital lacrimal glands of the female mouse after 26–28 days of age. The glands of the male mouse, on the other hand, continued to grow rapidly until 60–70 days and then at a much slower rate. The final weights of the lacrimal glands of the male were approximately 4 times those of the female mouse.

MORII et al. (1966) found only a small difference in the weights of the exorbital lacrimal gland of the 3-month-old female mouse from that of the male mouse (29 vs. 34 mg). The histologic appearance of the glands of the male mouse showed differences from the female similar to those observed for the rat.

*b) Castration*

In contrast to the rat, the weight of the lacrimal glands of the mouse is very sensitive to castration. KOCHAKIAN (1959) castrated mice at 26–28 days of age and noted an immediate retardation in the rate of growth of both the exorbital and the smaller orbital lacrimal glands. The maximum final weights of these glands were only about 50% of those of the normal controls but almost twice those of the female mouse. The effect of castration was comparable to that obtained for the submaxillary and sublingual glands and also the kidneys and the urinary bladder (Fig. 10).

*c) Steroids*

α) Weight Changes

KOCHAKIAN (unpublished) studied the effect of a wide variety of steroids on the weight of the lacrimal glands with a simultaneous determination of the effect on other target tissues for comparative purposes. The mice were castrated at 18–21 g (approximately 4 weeks of age) and 1 month later were implanted subcutaneously with a 15-mg pellet of the respective steroids. The surprising activity exhibited by the 11-oxy substituted 5β-androstane compounds could be due to the minor contamination by the respective 5α-epimers which showed considerable activity (Table 6). The testosterone and nortestosterone compounds exhibited marked activity except that 17α-methyltestosterone surprisingly was ineffective. Several of the more active and less soluble compounds were studied for 10-, 20-, and 28-day periods. The weights of the lacrimal glands were further increased to as much as double the normal weight after 28 days of steroid treatment. 17α-Hydroxy-5α-androstan-3-one stimulated the lacrimal glands to 4 times that of the castrated controls, testosterone and 17α-ethyl-19-nortestosterone to 3 times and androstane-3α,17β-diol to twice that of the castrated controls. The treatment for 10 and 20 days with several of the steroids exhibited effects intermediate to the results after 7 and 28 days of treatment. The changes in the weight of the lacrimal glands seemed to parallel those of the salivary glands and kidneys with occasional exceptions.

MORII et al. (1966) reported that the weekly injection of 0.1 mg Enarmon (testosterone enanthate 11.6 mg and testosterone propionate 2 mg) for 3 weeks into 2-month-old male mice increased the weight of the lacrimal gland approximately 55%. The histologic difference between the male and female mice and the changes induced by the steroid were similar to those reported for the rat.

β) Histologic Changes

QUINTARELLI and DELLOVO (1965) made a histologic study of the effect of testosterone propionate on the exorbital lacrimal gland of the mouse. Immature, 4-week-old mice, were injected for 50 days with 1 mg/day of testosterone propionate. The glands,

Table 6. Changes in organ weights of castrated mice produced by various steroids implanted as pellets for 7 days

| Steroid | Mice | Steroid Abs'd | body Wgt. | Change from control values | | | | | | | | |
|---|---|---|---|---|---|---|---|---|---|---|---|---|
| | | | | Sal. | Lac. | Kidney | Sv.Pr. | BC. | LA. | Blad. | Thy. | Adr. |
| | no. | mg/7 d. | gm | gm | mg | mg | mg | mg | mg | mg | mg | mg |
| *Control* | 17 | — | (0.7) | (125) | (14.5) | 358) | (14) | (7.4) | (8.2) | (21.2) | (70) | (7.8) |
| *5β-Androstane* | | | | | | | | | | | | |
| 17β-Hydroxy-5β-androstan-3-one | 3 | 1.6 | −0.1 | 21 | −0.6 | 14 | 1 | −1 | 1 | 1.6 | − 9 | −0.3 |
| 5β-Androstane-3,17-dione | 3 | 12.6 | 0.8 | 10 | 1.9 | 14 | 2 | −1 | 0 | 1.4 | 5 | 0.9 |
| 9α-F1-11β,17β-Dihydroxy-17-methyl-5β-androstan-3-one[b] | 7 | 10.5 | 1.6 | 21 | 4.9 | 71 | 10 | 14 | 9 | 0 | 8 | −1.2 |
| 11α,17β-Dihydroxy-17-methyl-5β-androstan-3-one[b] | 4 | 10.8 | 2.3 | 22 | 4.8 | 91 | 7 | 7 | 7 | 9.1 | − 7 | −0.3 |
| 17β-Hydroxy-17-methyl-5β-androstane-3,11-dione[c] | 6 | 14.7 | 4.4 | 29 | 5.7 | 119 | 18 | 14 | 11 | 5.5 | − 5 | −2.3 |
| *5α-Androstane* | | | | | | | | | | | | |
| 17β-Hydroxy-5α-andristan-3-one | 7 | 0.4 | 3.6 | 75 | 8.2 | 207 | 82 | 26 | 14 | 1.3 | −46 | −3.2 |
| 5α-Androstane-3α,17β-diol | 7 | 0.6 | 3.2 | 65 | 10.1 | 196 | 48 | 22 | 15 | 3.1 | −32 | −2.4 |
| 9αF1,11β,17β-Dihydroxy-17-methyl-5α-androstan-3-one | 7 | 0.33 | 3.3 | 48 | 4.9 | 101 | 38 | 22 | 11 | 2.7 | −10 | −0.9 |
| 11α,17β-Dihydroxy-17-methyl-5α-androstan-3-one | 5 | 13.3 | 4.2 | 50 | 8.5 | 174 | 68 | 20 | 13 | 8.5 | −38 | −1.8 |
| 17β-Hydroxy-17-methyl-5α-androstane-3,11-dione | 6 | 2.2 | 3.9 | 75 | 12.3 | 223 | 94 | 23 | 13 | 4.7 | −34 | −1.6 |

Table 6 (continued)

| | | | | | | | | | | | | |
|---|---|---|---|---|---|---|---|---|---|---|---|---|
| 9α-F1-17β-Hydroxy-17-methyl-5α-androstane-3,11-dione | 5 | 1.5 | 1.5 | 12 | 5.8 | 52 | 5 | 7 | 4 | 5.1 | 12 | −0.6 |
| *Androstene* | | | | | | | | | | | | |
| Testosterone | 6 | 3.9 | 2.5 | 76 | 4.0 | 217 | 113 | 30 | 16 | 2.4 | −57 | −3.6 |
| 17α-Methyltestosterone | 3 | 2.2 | 4.4 | 45 | −4.6 | 225 | 101 | 26 | 14 | 2.7 | −47 | −2.1 |
| 11-oxo-17α-Methyltestosterone | 5 | 13.9 | 1.9 | 66 | 11.8 | 227 | 84 | 18 | 14 | 8.1 | −37 | −1.7 |
| 11α-Hydroxy-17α-methyltestosterone | 4 | 12.7 | 3.8 | 54 | 9.2 | 183 | 74 | 22 | 17 | 5.7 | −27 | −2.4 |
| 9α-F1-11β-Hydroxy-17α-methyltestosterone[d] | 2 | 0.2 | 1.1 | 37 | 2.4 | 55 | 2 | 2 | −1 | −0.6 | − 7 | +1.5 |
| *Norandrostene* | | | | | | | | | | | | |
| 19-Nortestosterone | 2 | 11.8 | 3.7 | 69 | 7.9 | 227 | 90 | 34 | 21 | 10.7 | −58 | −1.4 |
| 19-Nortestosterone prop. | 3 | 3.6 | 5.8 | 63 | 9.7 | 244 | 93 | 31 | 14 | 10.5 | −42 | −1.6 |
| 17α-Ethyl-19-nortestosterone[e] | 6 | 2.2 | 4.2 | 97 | 8.1 | 191 | 67 | 25 | 14 | 3.8 | −55 | −3.4 |

[a] The results are differences from the control values which are in parentheses. The abbreviated headings are: Lac (exorbital lacrimal glands), Sal (salivary-submaxillary and sublingual glands), SVPr (seminal vesicles and prostates, BC (bulbocavernosus), LA (levator ani). Blad (urinary bladder), Thy (thymus), Adr (adrenals). The weight of the orbital lacrimal glands changed in parallel with that of the exorbital gland and is not presented.

[b] Contain 2—3% of 5α epimer as indicated by rotatory dispersion studies (Upjohn). The activity of these two compounds may be due to the 5α impurities.

[c] Rotatory dispersion studies indicated greater than 99% purity (Upjohn).

[d] Halotestin.

[e] Nilevar.

acinar cytoplasm, and nuclei of the treated mice were enormously enlarged—to 4 to 5 times that of the control mice. The acini were filled with acidophilic material. Nearly all the inter- and intra-lobular ducts were filled with an intense periodic acid-Schiff-positive material. DNA-RNA staining was enhanced in the nuclei of the orbital glands. The RNA was largely distributed around the nuclei. KRIESTEN (1969) also observed a fourfold increase in the nuclear volume of the exorbital glands of mice after the injection of 1 mg/day of testosterone propionate for 50 days. The increase in nuclear volume was accompanied by two- to threefold increase in DNA and an increase in the number of nuclei.

*d) Dwarf Mouse*

The exorbital lacrimal gland of dwarf mice 43 days old did not exhibit the increase in weight or the sexual dimorphism of normal mice of the same age (MARTINAZZI and BARONE, 1963). The injection of gonadotropin, and to a lesser degree thyrotropin, increased the weight of the glands and induced the morphologic change in the male dwarf mouse; in the female, the changes were induced only by prolactin. Somatotropin increased the weight of the glands but was not able to change the morphology.

### 3. Rabbit

The observation that secretion of the lacrimal glands in patients with kerato-conjunctivitis was enhanced by the administration of testosterone prompted RADNOT and NEMETH (1955) to perform studies in female and castrated male adult rabbits. The testosterone propionate was injected at 25 mg twice per week for 4–14 weeks with a progressive increase in the weight of the lacrimal glands to approximately 3 times that of the controls in both groups of rabbits. There was hyperactivity of the parenchyma with indication of accelerated secretory activity. NOVER (1957) was also stimulated by the beneficial clinical effects of testosterone propionate in patients with kerato-conjunctivitis to conduct experimental studies. The injection of normal adult male and female rabbits with 25 mg of testosterone propionate or testosterone phenylpropionate 3 times weekly for a total dose of 500 mg increased the rate of secretion by the lacrimal glands.

## F. Gastrointestinal Tract

No studies on the size of these organs have been made after castration and administration of anabolic-androgenic steroids. Microscopic studies, however, indicate that the epithelial cells of the small intestine are influenced by castration and testosterone administration.

### 1. Rat

*a) Intestinal Epithelium*

α) Castration, Thyroidectomy, and Steroids

GIORDANO et al. (1962) injected rats with testosterone propionate and found an increase in the height of the villi; the crypts of Lieberkühn had more goblet cells, and more cellular infiltration and the blood vessels were dilated. Furthermore, the number of mitoses and the RNA of the cytoplasm were markedly increased. CARRIERE (1966) castrated immature (49 g) rats and observed no change in the increase in crypt cells of the duodenum 80 days later. Thyroidectomized or thyroidectomized-castrated rats showed a reduction in these cells. Injection of the doubly operated rats with 250 μg/day of testosterone for 10 days and 500 μg for 32 days before the end of the experiment significantly increased the cell number in the crypt section towards normal. Thyroxine was more effective; the cell number was increased to above normal. A similar study on castrated rats would have been of interest in view of GIORDANO et al. results. The increase in mitotic index of the cells with age was not affected by castration but was reduced by thyroidectomy and restored to normal by the injection of thyroxine. KSIAZKIEWICZ-SZAPIRO (1970) corroborated and extended the observations of GIORDANO et al. (1962) and CARRIERE (1966). She castrated young adult (190–210 g) rats and 2 days later began injection of testosterone propionate at 40 mg/kg for 42 days. Castration did not produce any changes in the morphologic studies. Hematoxylin and eosin staining and PAS staining showed no difference in the epithelial cells. Testosterone propionate, however, induced slightly taller villi of the mucosa with occasional distended apices. The stroma contained an increased quantity of histiocytic, lymphocytelike and polynuclear cells together with dilated vessels. The superficial epithelium seemed taller and an augmented number of goblet cells was detected. The electron microscopic studies showed many young epithelial cells and young goblet cells in various developmental states. Furthermore, the epithelial cells contained a very abundant granular endoplasmic reticulum in the cytoplasm which was interpreted to be consistent with an increase in the essential RNA for protein synthesis—the initiation of the anabolic action of testosterone (KOCHAKIAN, 1964).

*b) Gastric Ulcers*

KOWALEWSKI (1959) compared the possible protective action of a primarily androgenic (testosterone enanthate) and a primarily anabolic (17α-ethyl-19-nortestosterone) steroid on the development of gastric ulcers in fasting cortisone-treated male rats of 195–220 g body weight. The rats received orally 5 mg/day of 17α-ethyl-19-nortestosterone for 10 days or a single 10 mg intramuscular injection of testosterone enanthate before a 4-day fast. Cortisone acetate was injected at 10 mg/day during the fasting period. The rats were injected subcutaneously with 1 mCi of $^{35}S$ in $H_2SO_4$ 24 h before the fast to study modifications in mucopolysaccharride production. The rats lost 27–32% of their body weight during the fasting period. The testosterone enanthate increased the occurrence of ulcers in the glandular part of the stomach due to fasting only but provided some protection against the high incidence of the ulcers induced by cortisone. The 17α-ethyl-19-nortestosterone had a higher protective action against the cortisone-induced ulcers and had no influence on ulcers due only to fasting. Analysis of the stomachs for incorporation of $^{35}S$ demonstrated a parallel partial protective action against the inhibition of $^{35}S$ incorporation by cortisone. The changes were in the mucopolysacharrides. Since the protective action of 17α-ethyl-19-nortestosterone was greater than that of testosterone enanthate, KOWALEWSKI postulated that the protective action was due to the anabolic properties of the steroids.

## 2. Mouse

### a) *Intestinal Epithelium*

#### α) Castration and Steroids

TUOHIMAA and NIEMI (1968) studied the effect of testosterone on cell renewal and mitotic cycles in the small intestine of the mouse for comparison with the seminal vesicles and prostate. The mice were castrated at 3 months; 2 weeks later they were injected with 0.25 mg/day of testosterone propionate for 3 days and were killed at 1- or 2-h intervals for 24 h. Simultaneously with the last hormone injection, 50 μCi of [$^3$H]-thymidine was injected. The labeling index was increased from $44 \pm 3.0$–$57 \pm 2.7$. The mitotic index was not changed. The hormone markedly shortened the time of all phases of the mitotic cell cycle of the accessory sex organs but only that of the postmitotic presynthetic period in the intestine.

WRIGHT and MORLEY (1971) and WRIGHT et al. (1972) provided results on the epithelial cells of the mouse small intestine which are consistent with the observation in the rat and an elaboration of that reported by TUOHIMAA and NIEMI (1968). Male mice were castrated at 3 months. Fourteen days later, 250 μg/day of testosterone propionate was injected for 3 days. [$^3$H]-Thymidine, 1 μCi, was injected on the 3rd day at the same time as the third injection of testosterone propionate. The mice were killed serially from 0.5–58 h later. The mean crypt height and the cell cycle time were not changed but the labeling index was increased about 50% and the mitotic index about 80% in the crypt cells of the upper jejunum of the treated mice. The growth factor, which is defined as the fraction of a particular population that consists of cells progressing round the cell cycle, was increased in the treated mice to indicate that the upper limit for labeling was raised about four cell positions resulting in an increase in the proliferative compartment.

#### β) Steroid Protection Against Toxic Drugs

BLOCH et al. (1974) observed that 3-deazauridine increased the life span of male DBA/2Ha mice bearing leukemia L1210 cells by 30–65% but was rather toxic to female mice. To evaluate this differential sex-related toxicity, they gonadectomized 6–8-week-old male and female mice and 3 days later injected intraperitoneally 250 mg/kg/day of deazauridine for 6 days. The toxicity (body weight loss) in the females was decreased and that in the males increased so that the effect in the two gonadectomized groups was intermediate between that of the normal male and female. The subcutaneous injection of 1 mg/day testosterone enanthate completely prevented the weight loss in female mice concomitantly receiving either 125 or 250 mg/kg/day of the antitumor agent and partially reduced that produced at 500 mg/kg/day. The small loss in body weight that occurred in the male mice at the two higher doses was prevented by the testosterone enanthate. The reduction in weight loss was accompanied by a reduction in mortality. In an attempt to elucidate the beneficial effect of testosterone, BLOCH et al. conducted histophathology studies. Decreased hematopoietic activity was noted in the bone marrow and spleen after drug treatment but the most striking effect was in the intestinal epithelium. The duodenal crypts of female mice treated intraperitoneally with 250 mg/kg of deazauridine 24 h prior to autopsy revealed a reduction in the number of mitotic figures and numerous pyknotic cells which were reversed by a simultaneous injection of 1 mg of testosterone enanthate. Extension of the drug treatment to 4 days resulted in a shortening of the intestinal villi, abnormal appearance of the lining cells and dilation of the crypts. The concomitant injection of 1 mg/day of testosterone enanthate prevented these changes.

In none of the above studies in the rat or mouse was the steroid administered by mouth to observe the effect of direct application of the hormone.

## VI. Summary and Conclusions

The effect of castration and testosterone and related compounds on the body weight of several animal species has been studied at different ages, duration of castration, dose and duration of hormone treatment and during different endocrine and nutritional conditions. These factors had variable modifying influences. An increase in body weight was best elicited by the steroids after the body weight had attained a plateau. The castrated adult mouse responded readily with an increase in body weight after steroid treatment.

The decrease in body weight after castration and the increase after hormone treatment was accompanied by changes in the weight of several organs as well as the carcass (muscle). These changes were widely different for the different animals. Analyses of the growth stimulated carcass and organs gave proportionate increases in protein, water and fat indicating true growth. Of the many animals studied, the mouse presents the widest spectrum of responsive tissues.

## References

ADDIS, T., LEW, W.: The restoration of lost organ tissue. The rate and degree of restoration. J. exp. Med. **71**, 325–363 (1940).

AHREN, K., ARVILL, A., HJALMARSON, A.: Effect of testosterone phenylpropionate and 19-nortestosteronephenylpropionate on the seminal vesicles, the levator ani muscle and the mammary glands of castrated male rats. Acta endocr. (Kbh.) **39**, 584–598 (1962).

ANDERSEN, D. H., KENNEDY, H. S.: The effect of gonadectomy on the adrenal, thyroid and pituitary glands. J. Physiol. (Lond.) **79**, 1–30 (1933).

APOSTOLAKIS, M., MATZELT, D., VOIGHT, K. D.: Der Einfluß von Testosteronpropionat auf glykolytische und transaminierende Enzymaktivaten in Leber, Musculus biceps femoris und Musculus levator ani der Ratte. Biochem. Z. **337**, 414–424 (1963).

ARAGAKI, M.: Experimental researches on the compensatory enlargement of the surviving kidney after unilateral nephrectomy (albino rat). Amer. J. Anat. **36**, 437–450 (1926).

ARVILL, A.: Effects of testosterone on the metabolism of the isolated levator ani muscle of the rat. Acta endocr. (Kbh.) **56**, (Suppl. 122) 1–14 (1967).

ASCHKENASY, A.: Comparaison des effets du propionate de testosterone sur le poids de diverse organes et tissue chez des rats males castrés et castrés-surrénalectomises. Ann. Endocr. (Paris) **20**, 158–171 (1959a).

ASCHKENASY, A.: Mise en évidence d'un contrôle surrenalien de certains effets tissulaires des androgenes. J. Physiol. (Lond.) **51**, 388–389 (1959b).

ASCHKENASY, A.: Etude expérimentale de l'action protidoanabolisante ainsi que de la toxicité aigue et chronique de la 17-ethyl-19-nortestosterone. Thérapie **14**, 332–349 (1959c).

ASCHKENASY, A.: Erythropoietic effect of testosterone administered to male rats by the percutaneous or subcutaneous routes. Israel. J. med. Sci. **7**, 901–905 (1971).

ASCHKENASY, A., BLANPIN, O.: Effets de la surrénalectomie, de l'orchidectomie, et des deux interventions associées sur le poids et la composition chimique du foie chez le rat. Ann. Endocr. (Paris) **16**, 789–799 (1955).

BAQUICHE, M.: Le dimorphisme sexuel de la glande de Loewenthal chez le rat albinos. Acta anat. (Basel) **36**, 247–280 (1959).

BARNES, L.E., STAFFORD, R.O., GUILD, M.E., THOLE, L.C., OLSON, K.J.: A comparison of myotrophic and androgenic activities of testosterone propionate with 19-nortestosterone and its esters. Endocrinology **55**, 77–84 (1954).

BARTKE, A., SHIRE, J.G.M.: Differences between mouse strains in testicular cholesterol levels and androgen target organs. J. Endocr. **55**, 173–184 (1972).

BASINGER, G.T., GITTES, R.F.: Effect of testosterone propionate on compensatory renal hypertrophy. Endocrinology **94**, 599–601 (1974).

BASS, A., GUTMANN, E., HANZLIKOVÁ, V., HAJEK, I., SYROVY, I.: The effect of castration and denervation upon the contraction properties and metabolism of the levator ani muscle of the rat. Physiol. bohemoslov. **18**, 177–194 (1969).

BENSON, W.R.: Acceleration of aging changes in the exorbital lacrimal gland of the female rat. Amer. J. Path. **45**, 587–593 (1964).

BERANGER, A., CZYBA, J.C., DUMONT, L., PINATEL, M.C.: Les réponse ponderale du muscle releveur de l'anus du Rat Utilisée comme test d'activite anabolique de certains steroides. Etude critique. C. R. Soc. Biol. (Paris) **157**, 831–834 (1963).

BERGAMINI, E.: Testosterone and sugar transport in levator ani muscle of rat. Biochim. biophys. Acta (Amst.) **193**, 193–202 (1969).

BERGAMINI, E.: Criticism about the use of muscles provided with a testosterone-binding protein to evaluate the miotrophic activity of androgens. Steroids Lipids Res. **4**, 119–121 (1973).

BERGAMINI, E.: Different mechanisms in testosterone action on glycogen metabolism in rat perineal and skeletal muscles. Endocrinology **96**, 77–84 (1975).

BERGAMINI, E., BOMBARA, G., PELLEGRINO, C.: The effect of testosterone on glycogen content of levator ani muscle of castrated rats. Biochem. J. **97**, 11 P.(1965).

BERGAMINI, E., BOMBARA, G., PELLEGRINO, C.: The effect of testosterone on glycogen metabolism in rat levator ani muscle. Biochim. biophys. Acta (Amst.) **177**, 220–234 (1969).

BERN, H.A., ALFERT, M.: Hormonal influence on nuclear synthesis. II. Volume, deoxyribonucleic acid and protein content of renal nuclei in castrated and testosterone treated mice. Rev. brasil biol. **14**, 25–29 (1954).

BERNSTEIN, L., ELRICK, H.: The handling of experimental animals as a control factor in animal research—a review. Metabolism **6**, 479–482 (1957).

BLACKMAN, S.S., JR., THOMAS, C.B., HOWARD, J.E.: Effect of testosterone propionate on arterial blood pressure, kidneys, urinary bladder and livers of growing dogs. Bull. Johns Hopk. Hosp. **74**, 321–334 (1944).

BLOCH, A., DUTSCHMAN, G., GRINDEY, G., SIMPSON, C.L.: Prevention by testosterone of the intestinal toxicity caused by the antitumor agent 3-deazauridine. Cancer Res. **34**, 1299–1303 (1974).

BOTTOMLEY, A.C., FOLLEY, S.J.: The effect of high doses of androgenic substances on the weights of the testes, accessory reproductive organs and endocrine glands of young male guinea pigs. J. Physiol. (Lond.) **94**, 26–39 (1938).

BRAUN-MENENDEZ, E.: Mecanismo de la regression renal en las ratas hipofisoprivas. Rev. Soc. argent. Biol. **26**, 65–76 (1950).

BRESLOFF, P., FOX, P.K., SIM, A.W., VAN DER VIESS, J.: Effect of nandrolone phenylpropionate on $^{14}$C-leucine incorporation into muscle protein in the rat and rabbit *in vivo*. Acta endocr. (Kbh.) **76**, 403–416 (1974).

BROULIK, P.D., KOCHAKIAN, C.D., DUBOVSKY, J.: Influence of castration and testosterone propionate on cardiac output, renal blood flow, and blood volume in mice. Proc. Soc. exp. Biol. (N.Y.) **144**, 671–673 (1973).

BROULIK, P., STARKA, L., STON, J.: Effect of antiandrogen cyproterone acetate on the action of testosterone on mouse kidney. Physiol. bohemoslov. **24**, 221–227 (1975).

BURESOVA, M., GUTMANN, E., HANZLIKOVÁ, V.: Differential effects of castration and denervation on protein synthesis in the levator ani muscle of the rat. J. Endocr. **54**, 3–14 (1972).

CARRIERE, R. M.: The influence of thyroid and testicular hormones on the epithelium of crypts of Lieberkuhn in the rats intestine. Anat. Rec. **156**, 423–432 (1966).

CAVALLERO, C.: Relative effectiveness of various steroids in an androgen assay using the exorbital lacrimal gland of the castrated rat. I. $\Delta^4$-3-ketones and $\Delta^5$-3$\beta$-hydroxy-steroids. Acta endocr. (Kbh.) **55**, 119–130 (1967).

CAVALLERO, C., CHIAPPINO, G., MILANI, F., CASELLA, E.: Uptake of $^{35}S$ labeled sulfate in the exorbital lacrimal glands of adult and newborn rats under different hormonal treatment. Experientia (Basel) **16**, 429 (1960).

CAVALLERO, C., MORERA, P.: Effect of testosterone on the nuclear volume of exorbital lacrimal glands of the white rat (glandula lacrimalis praeparotidea) Experientia (Basel) **16**, 285–287 (1960).

CAVALLERO, C., OFNER, P.: Relative effectiveness of various steroids in an androgen assay using the exorbital lacrimal gland of the castrated rat. II. $C_{19}$ steroids of the 5$\alpha$-androstane series. Acta endocr. (Kbh.) **55**, 131–135 (1967).

CHAI, C. K.: Seminal vesicle growth as a function of androgen stimulation in mice. Amer. J. Physiol. **186**, 463–467 (1956).

CIHAK, V. R., GUTMANN, E., HANZLIKOVÁ, V.: Morphologische, physiologische Merkmale, Entwicklung und Homologie des M. „levator ani“ der Ratte. Anat. Anz. **120**, 492–506 (1967).

CIRILLO, N., GUARDAVACCARO, G.: Il peso dei reni, dei surreni, delle vescicole seminali e delle milza in seguito a gonadectomia in Cavia cobaya. Scritti biologici **8**, 253–265 (1933).

CLARK, L. C., KOCHAKIAN, C. D., FOX, R. P.: The effect of castration and testosterone propionate on D-amino acid oxidase activity in the mouse. Science **98**, 89 (1943).

CLAUSEN, F. W., FREUDENBERGER, C. B.: A comparison of the effects of male and female sex hormones on immature female rats. Endocrinology **25**, 585–592 (1939).

COBET, G., DIECKHOFF, J., SEIPELT, H.: Über den Einfluß von 19-nor-Testosteron-17$\beta$-decanoat und einseitiger Nephrektomie auf einige Fermente der verbliebenen Rattenniere. Z. ges. exp. Med. **142**, 180–190 (1967).

COMMINS, W. D.: The effect of castration at various ages upon the adult weight of male albino rats. J. exp. Zool. **63**, 573–579 (1932).

COXON, R. V., KORENCHEVSKY, V., LAWRENCE, A.: Chemical changes in the heart, liver and kidneys of rats after administration of sex hormones. J. Path. Bact. **72**, 613–621 (1956).

CRABTREE, C. E.: The structure of Bowman's capsule in castrate and testosterone treated male mice as an index of hormonal effects on the renal cortex. Endocrinology **29**, 197–203 (1941).

DEANESLY, R., PARKES, A. S.: Quantitative study of the effects of implanting tablets of oestrogens and androgens in rats. J. Endocr. **2**, 487–495 (1941).

DEFILIPPI, G., ZACCHE, E.: Studio della attivata anabolizzante proteica mediante una speciale tecnica radiografica. Arch. int. Pharmacodyn. **172**, 85–90 (1968).

DOFUKU, R., TETTENBORN, U., OHNO, S.: Testosterone-“Regulon” in the mouse kidney. Nature (Lond.) New Biol. **232**, 5–7 (1971).

DONALDSON, H. H., HATAI, S.: Note on the influence of castration on the weight of the brain and spinal cord in the albino rat and on the percentage of water in them. J. comp. Neurol. **21**, 155–160 (1911).

DUCOMMUN, P., DUCOMMUN, S., BAQUICHE, M.: Comparison entre l'action du 17-ethyl-19-nortestosterone et du propionate de testosterone chez le rat adulte et immature. Acta endocr. **30**, 78–92 (1959).

EDGREN, R. A.: A comparative study of the anabolic and androgenic effects of various steroids. Acta endocr. **44**, Suppl. 187; 1–27 (1963).

EISENBERG, E., GORDAN, G. S.: The levator ani muscle of the rat as an index of myotrophic activity of steroidal hormones. J. Pharmacol. exp. Ther. **99**, 38–44 (1950).

EISENBERG, E., GORDAN, G. S., ELLIOTT, H. W.: Testosterone and tissue respiration of the castrate male rat with a possible test for myotrophic activity. Endocrinology **45**, 113–119 (1949).

ENDAHL, B. R., KOCHAKIAN, C. D.: Role of castration and androgens in D-amino acid oxidase activity of tissues. Amer. J. Physiol. **185**, 250–256 (1956).

EVANS, H. M., SIMPSON, M. E.: Resemblance of litter-mates in the degree of response to anterior hypophyseal fluid. Anat. Rec. **35**, 36 (1927).

FERGUSON, D. J., VISSCHER, M. B.: The effects of androgen and estrogen in castrated mice and hypophysectomized $C_3H$ mice. Endocrinology **52**, 463–473 (1953).

FEYEL, P.: Action des hormones derivees des sterols sur le rein chez la souris. C.R. Acad. Sci. (Paris) **214**, 718–720 (1942).

FEYEL, P.: L'action trophique des hormones sexuelles sur le rein chez la souris. Ann. Endocr. (Paris) **4**, 93–110 (1943).

FISUM, S. D., SHARKEVICH, I. N.: Fat mobilizing activity of testosterone propionate, dihydrotestosterone and epiandrosterone under starvation conditions. Fisiol Biokhim. Patol. Endokr. Gist. Mater. Resp. Konf. **1971**, 119–120.

FUJII, T., MORITA, Y., KOYAMA, R.: Effect of some anabolic steroids on the growth in male and female rats. Endocr. jap. **9**, 93–102 (1962).

GABE, M.: Conditionnement hormonal de la morphologie des glandes sus parotidiennes chez le Rat albinos. C.R. Soc. Biol. (Paris) **149**, 223–225 (1955).

GALAVAZI, G., SZIRMAI, J. A.: Cytomorphometry of skeletal muscle: The influence of age and testosterone on the rat M. levator ani. Z. Zellforsch. **121**, 507–530 (1971).

GAULE, J.: Über die geschlechtliche Differenz der Muskeln bei Fröschen. Pflügers Arch. ges. Physiol. **83**, 83–88 (1901).

GIORDANO, G., FOPPIANI, E., ROMANO, R.: Le influenze ormonale sull'apparato digerente. Arch. Marogliano Pat. Clin. **18**, 51–86 (1962).

GORDAN, G. S., EVANS, H. M., SIMPSON, M. E.: Effects of testosterone propionate on body weight and urinary nitrogen excretion of normal and hypophysectomized rats. Endocrinology **40**, 375–380 (1947).

GORDAN, G. S., FEINSTEIN, B., RALSTON, H. J.: Effect of testosterone upon atrophy of denervated skeletal muscle. Exp. med. Surg. **7**, 327–334 (1949).

GORI, Z., PELLEGRINO, C., POLLERA, M.: The castration and atrophy of the dorsal bulbocavernosus muscle of rat: an electron microscopic study. Exp. molec. Path. **6**, 172–198 (1967).

GORI, Z., PELLEGRINO, C., POLLERA, M.: The hypertrophy of levator ani muscle of rat induced by testosterone: an electron microscope study. Exp. molec. Pathol. **10**, 199–218 (1969).

GROSS, F., STRICKER, E.: Die Abhängigkeit der durch Testosteron und Thyroxin bedingten Nierenhypertrophie vom Eiweißgehalt der Nahrung. Arch. int. Pharmacodyn. **88**, 127–141 (1951).

GUTMANN, E., HANZLIKOVÁ, V., CIHÁK, R.: Persistence of the levator ani muscle in female rats. Experientia (Basel) **23**, 852–853 (1967).

HALPERN, B. N., COURNOT, L.: L'action de l'hormone male sur l'hypertrophic renal compensatrice considere comme un effet anabolique des androgenes sur le metabolisme des protides et la croissance tissulure. Arch. Sci. Physiol. **7**, 111—128 (1953).

HALPERN, B. N., COURNOT, L., CAMU, J.: Action du propionate de testosterone sur l'evolution ponderale de l'hypertrophic rénale compensatrice chez le rat et la teneur de l'organ en azote total et nucleique. C.R. Soc. Biol. (Paris) **145**, 1080–1084 (1951).

HANZLIKOVÁ, V., GUTMANN, E.: Effect of foreign innervation on the androgen-sensitive levator ani muscle of the rat. Z. Zellforsch. **135**, 165–174 (1972).

HANZLIKOVÁ, V., GUTMANN, E.: The absence of androgen-sensitivity in the grafted soleus muscle innervated by the pudendal nerve. Cell Tiss. Res. **154**, 121–129 (1974).

HATAI, S.: The growth of organs in the albino rat as affected by gonadectomy. J. exp. Zool. **18**, 1–67 (1915).

HAY, J. B., GRAY, G. M.: The effect of testosterone on the glycosphingolipid composition of mouse kidney. Biochim. biophys. Acta (Amst.) **202**, 566–568 (1970).

HAYES, K. J.: The so-called "levator ani" of the rat. Acta endocr. **48**, 337–347 (1965).

HAZARY, S., GARDNER, W. U.: Influence of sex hormones on abdominal musculature and the formation of inguinal and scrotal hernias in mice. Anat. Rec. **136**, 437–443 (1960).

HEIM, F., SCHWARZLOSE, W., TRUCKENBRODT, H., MITZNEGG, P.: Effect of anabolic and antianabolic agents on desoxyribonucleic acid, ribonucleic acid and protein contents, de novo cell formation and cell size in relation to age, sex, and nutrition. Z. Gerontol. **2**, 25–47 (1969).

HERSHBERGER, L. G., SHIPLEY, E. G., MEYER, R. K.: Myotrophic activity of 19-nortestosterone and other steroids determined by modified levator ani muscle method. Proc. Soc. exp. Biol. (N.Y.) **83**, 175–180 (1953).

HERSHBERGER, L. G., THOMPSON, C. R., CLEGG, M. T.: Myotrophic effects of estrogen in mice and lack of effect in rats. Proc. Soc. exp. Biol. (N.Y.) **121**, 785–788 (1966).

HOLT, H., KEETON, R.W., VENNESLAND, B.: The effect of gonadectomy on body structure and body weight in albino rats. Amer. J. Physiol. **114**, 515–525 (1936).
HOMBURGER, F., FORBES, I., DESJARDINS, R.: Renotropic effects of some androgens upon experimental hydronephrosis and upon the clitoris in the mouse. Endocrinology **47**, 19–25 (1950).
HOMBURGER, F., PETTINGILL, O.: The protein anabolic and other effects of testosterone propionate in mice: effects of nutrition and interrelationship of biologic activities of the hormone. Endocrinology **57**, 296–301 (1955).
HORKY, J., GANGOLLI, S. D.: Changes in the weight, nitrogen, potassium and phosphorus of some organs of rats after administration of anabolic steroids. Čs. Gastroent. Výž. **27**, 151–157 (1973).
HUMM, J. H., KOCHAKIAN, C. D., BARTLETT, M. N.: Effect of castration and steroids on arginase and phosphatases of organs of the guinea pig. Amer. J. Physiol. **155**, 251–254 (1948).
JAHN, U.: Karyometrische Untersuchungen der Wirkung von Testosteronpropionat, Oestradioldipropionat und Thyroxin auf die Kerne der Tubuli contorti. I. Ordnung in Rattennieren. Endokrinologie **32**, 302–309 (1955).
JASIENSKI, J.: Influence de la castration sur la musculature. C.R. Soc. Biol. (Paris) **101**, 533–537 (1929).
JELÍNEK, J., VESELÁ, H., VALOVÁ, B.: The effect of nortestosterone phenylpropionate on compensatory hypertrophy of the remaining kidney after unilateral nephrectomy. Acta endocr. (Kbh.) **46**, 352–360 (1964).
JOCHLE, W., LANGECKER, H.: Biologische Wirkungen des 1-methyl-$\Delta^1$-androsten-17$\beta$-ol-3-on-17-acetats und 17-onanthats (Methenolon-acetat und Methenolon-Onanthat). Arzneimittel-Forsch. **12**, 218–223 (1962).
JOSS, E. E., ZUPPINGER, K. A., SOBEL, E. H.: Effect of testosterone propionate and methyltestosterone on growth and skeletal maturation in rats. Endocrinology **72**, 123–130 (1963).
KAMSTRA, L. D., THURSTON, M. M.: The effect of feeding 17$\alpha$-ethyl-19-nortestosterone to weanling rats. Proc. South Dakota Acad. Sci. **40**, 98–102 (1961).
KEYES, P. H.: Dental caries in the Syrian hamster. III. The effect of gonadectomy on caries activity and body growth. N.Y. J. Dent. **18**, 172–179 (1948).
KOCHAKIAN, C. D.: The comparative efficacy of various androgens as determined by the rat assay method. Endocrinology **22**, 181–192 (1938).
KOCHAKIAN, C. D.: The tolerance of male and female mice respectively to estrogens and androgens. Endocrinology **26**, 54–60 (1940).
KOCHAKIAN, C. D.: The rate of absorption and effects of testosterone propionate pellets on mice. Endocrinology **28**, 478–484 (1941).
KOCHAKIAN, C. D.: A comparison of the renotrophic with the androgenic activity of various steroids. Amer. J. Physiol. **142**, 315–325 (1944).
KOCHAKIAN, C. D.: The effect of various steroid hormones on the alkaline and acid phosphatases of the kidney of the mouse. Amer. J. Physiol. **145**, 118–122 (1945).
KOCHAKIAN, C. D.: The effect of dose and nutritive state on the renotrophic and androgenic activities of various steroids. Amer. J. Physiol. **145**, 549–556 (1946).
KOCHAKIAN, C. D.: Effect of estrogens on the body and organ weights and the arginase and "alkaline" and "acid" phosphatases of the liver and kidney of castrated male mice. Amer. J. Physiol. **151**, 126–129 (1947).
KOCHAKIAN, C. D.: Renotrophic-androgenic and somatotrophic properties of further steroids. Amer. J. Physiol. **158**, 51–56 (1949).
KOCHAKIAN, C. D.: Comparison of protein anabolic properties of testosterone propionate and growth hormone in the rat. Amer. J. Physiol. **160**, 66–74 (1950a).
KOCHAKIAN, C. D.: Comparison of protein anabolic property of various androgens in the castrated rat. Amer. J. Physiol. **160**, 53–61 (1950b).
KOCHAKIAN, C. D.: The mechanism of the protein anabolic action of testosterone propionate. In: Symposium on Steroid Hormones, pp. 113–149, Madison. Wisconsin: Univ. Wisconsin Press 1950c.
KOCHAKIAN, C. D.: Renotrophic-androgenic properties of orally administered androgens. Proc. Soc. exp. Biol. (N.Y.) **80**, 386–388 (1952).
KOCHAKIAN, C. D.: Mechanisms of androgen actions. Lab. Invest. **8**, 538–556 (1959).

KOCHAKIAN, C. D.: Summation of protein anabolic effects of testosterone propionate and growth hormone. Proc. Soc. exp. Biol. (N.Y.) **103**, 196–197 (1960).

KOCHAKIAN, C. D.: Protein anabolic property of androgens. Ala. J. med. Sci. **1**, 24–37 (1964).

KOCHAKIAN, C. D.: Regulation of muscle growth by androgens. In: The Physiology and Biochemistry of Muscle as a Food, pp. 81–112, Madison. Wisconsin: Univ. Wisconsin Press 1966.

KOCHAKIAN, C. D.: Early effects of androgens on the biosynthesis of protein and RNA by the postmitochondrial fraction of the mouse kidney. Steroids **14**, 77–90 (1969).

KOCHAKIAN, C. D., BARTLETT, M. N., GONGORA, J.: Effect of castration and androgens on body weight and organ weights, and the arginase and phosphatases of kidney and liver of the male Syrian hamster. Amer. J. Physiol. **153**, 210–214 (1948 b).

KOCHAKIAN, C. D., COCKRELL, D.: Response of individual muscles of the guinea pig to "large" doses of testosterone propionate. Proc. Soc. exp. Biol. (N.Y.) **97**, 148–149 (1958 a).

KOCHAKIAN, C. D., COCKRELL, D.: Presence of aberrant thyroid tissues in the guinea pig. Endocrinology **63**, 385–388 (1958 b).

KOCHAKIAN, C. D., ENDAHL, B. R.: Changes in body weight of normal and castrated rats by different doses of testosterone propionate. Proc. Soc. exp. Biol. (N.Y.) **100**, 520–522 (1959).

KOCHAKIAN, C. D., ENDAHL, G., AUSTIN, J.: The influence of testosterone on the growth of the urinary bladder. Endocrinology **60**, 80–82 (1957 b).

KOCHAKIAN, C. D., ENDAHL, B. R., ENDAHL, G. L.: Influence of androgens on the transaminases and glutamic dehydrogenase of tissue. Amer. J. Physiol. **197**, 129–134 (1959).

KOCHAKIAN, C. D., GARBER, E. E., BARTLETT, M. N.: Effect of estrogen alone and in combination with testosterone on the body and organ weights and the arginase and phosphatases of the organs of the mouse. Amer. J. Physiol. **155**, 265–271 (1948 c).

KOCHAKIAN, C. D., HARRISON, D. G.: Regulation of nucleic acid synthesis by androgens. Endocrinology **70**, 99–108 (1962).

KOCHAKIAN, C. D., HILL, J., COSTA, G.: Amino acid composition of the proteins of the muscles and organs of the normal, castrated and testosterone-treated guinea pig. Acta endocr. **45**, 613–622 (1964 b).

KOCHAKIAN, C. D., HILL, J., HARRISON, D. G.: Regulation of nucleic acids of muscles and accessory sex organs of guinea pigs by androgens. Endocrinology **74**, 635–642 (1964 a).

KOCHAKIAN, C. D., HUMM, J. H., BARTLETT, M. N.: Effect of steroids on the body weight, temporal muscle and organs of the guinea pig. Amer. J. Physiol. **155**, 242–250 (1948 a).

KOCHAKIAN, C. D., MURLIN, J. R.: The effect of male hormone on the protein and energy metabolism of castrate dogs. J. Nutr. **10**, 437–459 (1935).

KOCHAKIAN, C. D., NISHIDA, M., HIRONE, T.: Ribosomes of mouse kidney: Regulation by androgens. Amer. J. Physiol. **217**, 383–391 (1969).

KOCHAKIAN, C. D., REED, B. A., EISCHEID, A. M.: Changes in tissue weights and enzymes produced by thyroidectomy and testosterone. Amer. J. Physiol. **177**, 413–417 (1954).

KOCHAKIAN, C. D., ROBERTSON, E.: Enhancement of myotrophic and androgenic responses by aqueous suspension of steroids. Proc. Soc. exp. Biol. (N. Y.) **75**, 542–543 (1950 a).

KOCHAKIAN, C. D., ROBERTSON, E.: Effect of growth hormone on temporal muscle of the guinea pig. Proc. Soc. exp. Biol. (N. Y.) **73**, 388–389 (1950 b).

KOCHAKIAN, C. D., ROBERTSON, E., BARTLETT, M. N.: Sites and nature of protein anabolism stimulated by testosterone propionate in the rat. Amer. J. Physiol. **163**, 332–346 (1950).

KOCHAKIAN, C. D., STETTNER, C. E.: Effect of testosterone propionate and growth hormone on the weights and composition of the body and organs of the mouse. Amer. J. Physiol. **155**, 255–261 (1948).

KOCHAKIAN, C. D., TILLOTSON, C.: Hormonal regulation of muscle development. In: Hormones and the aging process, pp. 63–78, New York: Academic Press, 1956.

KOCHAKIAN, C. D., TILLOTSON, C.: Myotrophic effects of castration and testosterone in thyroidectomized guinea pigs. Amer. J. Physiol. **189**, 425–427 (1957 a).

KOCHAKIAN, C. D., TILLOTSON, C.: Influence of several $C_{19}$ steroids on the growth of individual muscles of the guinea pig. Endocrinology **60**, 607–618 (1957 b).

KOCHAKIAN, C. D., TILLOTSON, C., AUSTIN, J.: A comparison of the effect of inanition, castration and testosterone on the muscles of the male guinea pig. Endocrinology **60**, 144–152 (1957 a).

KOCHAKIAN, C. D., TILLOTSON, C., AUSTIN, J., DOUGHERTY, E., HAAG, V., COALSON, R.: The effect of castration on the weight and composition of the muscles of the guinea pig. Endocrinology **58**, 315–326 (1956b).
KOCHAKIAN, C. D., TILLOTSON, C., ENDAHL, G. L.: Castration and the growth of muscles in the rat. Endocrinology **52**, 226–231 (1956a).
KOCHAKIAN, C. D., WEBSTER, J. A.: Effect of testosterone propionate on the appetite, body weight and composition of the normal rat. Endocrinology **63**, 737–742 (1958).
KORENCHEVSKY, V.: The sexual glands and metabolism. I. Influence of castration on nitrogen and gaseous metabolism. Brit. J. exp. Path. **6**, 21–35 (1925).
KORENCHEVSKY, V.: The influence of cryptorchidism and of castration on body weight, fat deposition, the sexual and endocrine organs of male rats. J. Path. Bact. **33**, 607–636 (1930).
KORENCHEVSKY, V., DENNISON, M.: The effect on male rats of the simultaneous administration of male and female sexual hormones and the relation to the assay of the hormones. Biochem. J. **28**, 1486–1499 (1934a).
KORENCHEVSKY, V., DENNISON, M.: The manifold effects of castration in male rats. J. Path. Bact. **38**, 231–246 (1934b).
KORENCHEVSKY, V., DENNISON, M.: The manifold effects of testicular hormones (as extracted from human urine or synthetically prepared) and of oestrone on the male, as judged by experiments on rats. Proc. roy. Soc. Med. **28**, 71–73 (1935).
KORENCHEVSKY, V., DENNISON, M.: The assay of transdehydroandrosterone and its effects on male and female gonadectomized rats. Biochem. J. **30**, 1514–1522 (1936).
KORENCHEVSKY, V., DENNISON, M., HALL, K.: The action of testosterone propionate on normal adult female rats. Biochem. J. **31**, 780–785 (1937a).
KORENCHEVSKY, V., DENNISON, M., HALL, K.: The effects of testosterone and testosterone propionate on adult male rats compared with those on female rats. Biochem. J. **31**, 1434–1437 (1937b).
KORENCHEVSKY, V., DENNISON, M., KOHN-SPEYER, A.: Changes produced by testicular hormone in normal and castrated rats. Biochem. J. **27**, 557–579 (1933).
KORENCHEVSKY, V., DENNISON, M., SIMPSON, S.: The prolonged treatment of male and female rats with androsterone and its derivatives, alone or together with oestrone. Biochem. J. **29**, 2534–2552 (1935).
KORENCHEVSKY, V., HALL, K., BURBANK, R. C., COHEN, J.: Hepatotrophic and cardiotrophic properties of sex hormones. Brit. med. J. **1941**, 396–399.
KORENCHEVSKY, V., HALL, K., ROSS, M. A.: Prolonged injections of male sex hormones into normal and senile male rats. Brit. Med. J. **1939 I**, 4–8.
KORENCHEVSKY, V., JONES, V. E.: The effects of progesterone, oestradiol, thyroid hormone, and androsterone on the artificial premature "climacteric" of pure gonadal origin produced by ovariectomy in rats. J. Geront. **3**, 21–39 (1948).
KORENCHEVSKY, V., PARIS, S. K., BENJAMIN, B.: Treatment of senescence in female rats with sex and thyroid hormones. J. Geront. **5**, 120–157 (1950).
KORENCHEVSKY, V., PARIS, S. K., BENJAMIN, B.: Treatment of senescence in male rats with sex and thyroid hormones and desoxycorticosterone acetate. J. Geront. **8**, 415–434 (1953a).
KORENCHEVSKY, V., PARIS, S. K., BENJAMIN, B.: Effects of castration and the processes of aging in male rats and man. J. Geront. **8**, 6–32 (1953b).
KORENCHEVSKY, V., ROSS, M. A.: The kidneys and sex hormones. Brit. med. J. **1940**, 645–648.
KORNER, A.: The influence of methylandrostenediol on the protein fractions and adenosinetriphosphatase activity of muscles of the rat. J. Endocr. **13**, 90–93 (1955).
KORNER, A., YOUNG, F. G.: The effect of methylandrostenediol on the weight and protein content of the muscles and organs of the rat. J. Endocr. **13**, 84–89 (1955).
KOTHS, K. E., GODCHAUX, W., III., DOEG, L. H., DOEG, K. A.: The effect of testosterone on the synthesis of mouse kidney mitochondrial and microsomal proteins *in vivo* and *in vitro*. Endocrinology **91**, 125–134 (1972).
KOWALEWSKI, K.: Effect of certain androgenic steroids and cortisone on gastric ulcerogenesis in fasting rats. Proc. Soc. exp. Biol. (N.Y.) **101**, 147–150 (1959).
KOWALEWSKI, K.: Changes in the growth and body composition of suckling rats caused by the treatment of lactating mothers with steroid hormones: Its relation to litter size. Endocrinology **76**, 1089–1092 (1965).

KRIESTEN, K.: Influence of testosterone propionate on the cell nuclei of the exorbital glands of the mouse. Zbl. Vet.-Med. **16**, 557–562 [Reihe, A.] (1969).

KSIAZKIEWICZ-SZAPIRO, M. D.: Experimental investigations of the rat small intestine following castration and testosterone administration. Folia Histochem. Cytochem. **8**, 369–388 (1970).

LARON, Z., KOWADLO, A.: Fat-mobilizing effect of testosterone. Metabolism **12**, 588–591 (1963).

LATTIMER, J. K.: The action of testosterone propionate upon the kidneys of rats, dogs and men. J. Urol. **48**, 778–794 (1942).

LAWLESS, J. J.: Castration in the rat with and without removal of the epididymes. Anat. Rec. **66**, 455–473 (1936).

LEATHAM, J. H.: Plasma protein concentrations and organ weights of castrated and testosterone propionate treated rats. Amer. J. Physiol. **154**, 459–464 (1948 a).

LEATHAM, J. H.: Influence of testosterone propionate on the plasma and liver proteins of hypothyroid rats. J. biol. Chem. **176**, 1285–1289 (1948 b).

LEATHAM, J. H.: Plasma and liver proteins of rats following high protein diets and testosterone propionate. Amer. J. Physiol. **165**, 73–78 (1951).

LECOMPTE, P. M.: Effect of sex hormones on nephrotoxic nephritis in the rat. Yale J. Biol. Med. **16**, 627–638 (1944).

LEONARD, S. L.: Stimulation of mammary glands in hypophysectomized rats by estrogen and testosterone. Endocrinology **32**, 229–237 (1943).

LEONARD, S. L.: A glycostatic effect of testosterone on the perineal muscles of the rat. Endocrinology **50**, 199–205 (1952).

LEONARD, S. L.: Comparison of atrophy and glycogen storage in some muscles after castration and denervation. Proc. Soc. exp. Biol. (N.Y.) **94**, 458–460 (1957).

LI, J. J., KIRKMAN, H., HUNTER, R. L.: Sex difference and gonadal hormone influence on Syrian hamster kidney esterase isozymes. J. Histochem. Cytochem. **17**, 386–393 (1969).

LIGHT, A. E., TORNABEN, J. A.: Effects of prolonged percutaneous administration of methyltestosterone and estradiol on growing male rats. J. Nutr. **49**, 51–63 (1953 a).

LIGHT, A. E., TORNABEN, J. A.: Effects of prolonged percutaneous administration of methyltestosterone and estradiol on growing female rats. J. Nutr. **51**, 365–379 (1953 b).

LINET, O., BUMBA, V.: Effect of androgenic-anabolic steroids on adrenals and M. levator ani in hypophysectomized rats. Endocr. Exp. **3**, 89–94 (1969).

LUDDEN, J. B., KRÜGER, E., WRIGHT, I. S.: Effect of testosterone propionate, estradiol benzoate, and desoxycorticosterone acetate on the kidneys of adult rats. Endocrinology **28**, 619–623 (1941).

LYON, M. F., HAWKES, S. G.: X-linked gene for testicular feminization in the mouse. Nature (Lond.) **227**, 1217–1219 (1970).

MACKAY, E. M.: Degree of compensatory renal hypertrophy following unilateral nephrectomy. III. Influence of testosterone propionate. Proc. Soc. exp. Biol. (N.Y.) **45**, 216–217 (1940).

MACKAY, L. L., MACKAY, E. M.: Factors which determine renal weight; Sex. Amer. J. Physiol. **83**, 196–201 (1927).

MARCHETTI, M., BALLESIO, P. L.: Studio degle effetti Biologici di un nuovo estere del 19-norandrostenolone: il 19-nor-androstenolone decanato. Polyclinico **68**, 1761–1768 (1961).

MARTINAZZI, M., BARONE, C.: Hormonal control of exorbital lacrimal gland in mice with pituitary dwarfism. Folia Endocr. **16**, 123–132 (1963).

MASSON, G., BELAND, E.: Influence du potassium sur l'action renotropique des substances steroides. Rev. Canad. Biol. **2**, 487–494 (1943).

MASSON, G., ROMANCHUK, M.: Observations sur certains changements morphologiques produits par la thyroxine et la testosterone. Rev. Canad. Biol. **4**, 206–218 (1945).

MASUI, K., TAMURA, Y.: The effect of gonadectomy on the weight of the kidney, thymus, and spleen of mice. Brit. J. exp. Biol. **3**, 207–223 (1926).

MCEUEN, C. S., SELYE, H., COLLIP, J. B.: Effect of testosterone on somatic growth. Proc. Soc. exp. Biol. (N.Y.) **36**, 390–394 (1937).

MELICHNA, J., GUTMANN, E., HERBRYCHOVA, A., STICHOVA, J.: Sexual dimorphism in contraction properties and fibre pattern of the flexor carpi radialis muscle of the frog (Rana temporaria L). Experientia (Basel) **28**, 89–91 (1972).

MEYER, B. J.: The influence of testosterone propionate and stilboestrol on spermatogenesis, the prostate gland, the endocrine system, and growth of the male albino rat. S. Afr. J. med. Sci. **14**, 89–133 (1949).
MILLER, R. B.: Gonadectomy and ovarian (spleen) grafts and growth of normal male and female mice. Proc. Soc. exp. Biol. (N.Y.) **73**, 654–656 (1950).
MILLS, T. M., SPAZIANI, E.: The influence of testosterone on penetration of α-aminoisobutyric acid and 2-deoxyglucose in male rat sex accessory tissues. Biochim. biophys. Acta (Amst.) **150**, 435–445 (1968).
MOORE, C. R.: On the physiological properties of the gonads as controllers of somatic and psychical characteristics. V. The effect of gonadectomy in the guinea pig on growth, bone lengths, and weight of organs of internal secretions. Biol. Bull. **43**, 285–312 (1922).
MOORE, C. R.: Biology of the Testes. In: ALLEN, E., DANFORTH, C. H., DOISY, E. A. (Eds.): Sex and Internal Secretions, Chapter VII, 343—451. Baltimore: Williams and Wilkins Co. 1939.
MORII, S., OKAMURA, H., YAMADA, H.: Effects of male sex hormones upon exlacrymal glands of rodents. J. Kansai Medical School **18**, 29—33 (1966).
NATHAN, D. G., GARDNER, F. H.: Effects of large doses of androgen on rodent erythropoiesis and body composition. Blood **26**, 411—420 (1965).
NEUMANN, F., VON BERSWORDT-WALLRABE, R., ELGER, W., STEINBECK, H.: Hormonhemmer-Untersuchungen mit testosteron-antagonisten. Colloq. Gesell. physiol. Chem. pp. 218—248. Berlin-Heidelberg-New York: Springer 1967.
NIMNI, M. E., GEIGER, E.: Testosterone induced nitrogen retention in castrated rats on diets containing incomplete proteins. Endocrinology **61**, 753—757 (1957).
NOVER, A.: Die Wirkung von Testosteron und Hypophysin auf die Tränen-Sekretion. Arzneimittel-Forsch. **7**, 277—278 (1957).
OHNO, S., LYON, M. F.: X-linked testicular feminization in the mouse as a non-inducible regulatory mutation of the Jacob-Manod type. Clin. Genet. **1**, 121—127 (1970).
OVERHOLSER, M. D.: The effect of castration upon the size of the parathyroid glands and upon the susceptibility to tetania parathyreopriva in the albino rat. Anat. Rec. **41**, 303—321 (1929).
PAESI, F. J. A., HELLINGA, G., HOOGSTRA, M. J., VAN REES, G. P.: Antagonistic action of oestradiol and testosterone on the growth of hypophysectomized immature rats. Acta endocr. **21**, 177—184 (1956).
PAGNI, R., BERGAMINI, E., PELLEGRINO, C.: The inhibition by denervation of the effect of testosterone on glycogen metabolism in the levator ani muscle. Endocrinology **92**, 667—673 (1973).
PAPANICOLAOU, G. N., FALK, E. A.: General muscular hypertrophy induced by androgenic hormone. Science **87**, 238—239 (1938).
PARHON, C. I., BABES, A., PETREA, I., ISTRATI, F., BURGHER, E.: Structura si Dimorfismul sexual al glandelor parotide la Sobol-anul. Alb. Bul. Stiint. Sect. de Stunte Med. VII (3) Universite de Bucarest, 1955. Cited by Baquiche, M. (1959).
PFEIFFER, C. A., EMMEL, V. M., GARDNER, W. U.: Renal hypertrophy in mice receiving estrogens and androgens. Yale J. biol. Med. **12**, 493—501 (1940).
PHOENIX, C. H.: Effects of dihydrotestosterone on sexual behavior of castrated male rhesus monkeys. Physiol. Behav. **12**, 1045—1055 (1974).
PIERRE, R., CAHN, J.: Etude experimentale di l'action renotrope comparee du dipropionate di'androstenediol et de l'androstanolone. Rev. Path. gén. **56**, 735—740 (1956).
QUIMBY, F. H.: Organ weights of rats receiving hormone supplements during recovery from chronic starvation. Endocrinology **42**, 263—272 (1948).
QUINTARELLI, G., DELLOVO, M. C.: Activation of glycoprotein biosynthesis by testosterone propionate on mouse exorbital glands. J. Histochem. Cytochem. **13**, 361—364 (1965).
RABINOVITCH, M.: Quantitative histological studies on the effects of testosterone propionate and castration on the kidney cortex of the mouse. Folia. clin. biol. **18**, 109—114 (1952).
RABINOVITCH, M., VALERI, V.: Effects of castration and of testosterone propionate administration on nuclear volumes of the kidney cortex cells of the mouse. Rev. bras. biol. **12**, 417—420 (1952).
RADNOT, M., NEMETH, B.: Wirkung der Testosteronpräparate auf die Tränendrüse. Ophthalmologica (Basel) **129**, 376—380 (1955).
RASMUSSEN, F.: Effect of androgens and anabolic steroids on rabbit urinary bladder and ureter. Acta endocr. **57**, 414—426 (1968).

REISS, M., SIDEMAN, M. B., PLICHTA, E. S.: Comparison of the action of growth hormone, human chorionic gonadotrophin and an anabolic steroid in hypophysectomized rats. Acta endocr. **49**, 366—369 (1965).

ROLLASON, H. D.: Compensatory hypertrophy of the kidney of the young rat with special emphasis on the role of cellular hyperplasia. Anat. Rec. **104**, 263—283 (1949).

ROSS, M. A., KORENCHEVSKY, V.: The thymus of the rat, and the sex hormones. J. Path. and Bact. **52**, 349–360 (1941).

RUBINSTEIN, H. S.: The response of the male albino rat to the intraperitoneal administration of testosterone propionate. J. Urol. **51**, 88—91 (1944).

RUBINSTEIN, H. S., ABARBANEL, A. R., KURKLAND, A. A.: The effect of castration on body weight and length of the male albino rat. Endocrinology **25**, 397—400 (1939a).

RUBINSTEIN, H. S., KURKLAND, A. A., GOODWIN, M.: The somatic growth depressing effect of testosterone propionate. Endocrinology **25**, 724—728 (1939b).

RUBINSTEIN, H. S., SOLOMON, M. L.: Growth stimulating effect of testosterone propionate. Proc. Soc. exp. Biol. (N.Y.) **44**, 442—443 (1940a).

RUBINSTEIN, H. S., SOLOMON, M. L.: Effect on body growth of small doses of testosterone propionate administered at different seasons. Proc. Soc. exp. Biol. (N.Y.) **45**, 745—748 (1940b).

RUBINSTEIN, H. S., SOLOMON, M. L.: The growth stimulating effect of small doses of testosterone propionate in the castrate albino rat. Endocrinology **28**, 229—232 (1941).

SAINTON, P., SIMMONET, H., PETRESCU, J.: Influence de l'hormone mâle (propionate de testosterone) sur le développment de rat mâle impubère. C. R. Soc. Biol. (Paris) **128**, 698—700 (1938).

SAUNDERS, F. J.: The duration of action of several androgenic-anabolic steroids. Acta endocr. **26**, 345—352 (1957).

SCHUMANN, H.: Weitere Untersuchungen über den Einfluß der Sexualhormone auf den Glykogen-, Phosphagen- und Adenylpyrophosphatgehalt der Herz- und Skeletmuskulatur. Klin. Wschr. **19**, 364—366 (1940).

SCOW, R. O.: Effect of testosterone on muscle and other tissues and on carcass composition in hypophysectomized, thyroidectomized, and gonadectomized male rats. Endocrinology **51**, 42—51 (1952).

SCOW, R. O., HAGAN, S. N.: Effect of testosterone propionate on myosin, collagen and other protein fractions in striated muscles of gonadectomized male guinea pigs. Amer. J. Physiol. **180**, 31—36 (1955).

SCOW, R. O., HAGAN, S. N.: The effect of testosterone propionate on myosin, collagen and other protein fractions in striated muscle of gonadectomized rats. Endocrinology **60**, 273—276 (1957).

SCOW, R. O., ROE, J. H., JR.: Effect of testosterone propionate on the weight and myoglobin content of striated muscles in gonadectomized guinea pigs. Amer. J. Physiol. **173**, 22—28 (1953).

SEGALOFF, A.: The enhanced local androgenic activity of 19-nor steroids and stabilization of their structure by 17α- and 17β-methyl substituents to highly potent androgens by any route of administration. Steroids **1**, 299—315 (1963).

SEGALOFF, A., GABBARD, R. B.: 14-Dehydrogenation as a means of increasing androgenicity when measured by local applications on the chick comb. Steroids **1**, 77—90 (1963).

SELYE, H.: The effect of testosterone on the kidney. J. Urol. **42**, 637—641 (1939a).

SELYE, H.: Morphological changes in female mice receiving large doses of testosterone. J. Endocr. **1**, 208—215 (1939b).

SELYE, H.: Interactions between various steroid hormones. Canad. med. Ass. J. **42**, 113—116 (1940a).

SELYE, H.: Production of persistent changes in the genital organs of immature female rats treated with testosterone. Endocrinology **27**, 657—660 (1940b).

SELYE, H.: Effect of hypophysectomy on morphological appearance of kidney and on renotropic action of steroid hormones. J. Urol. **46**, 110—131 (1941a).

SELYE, H.: Effect of dosage on the morphogenetic actions of testosterone. Proc. Soc. exp. Biol. (N.Y.) **46**, 142—146 (1941b).

SELYE, H., ALBERT, S.: Morphogenetic actions of various steroids in the castrate male rat. J. Pharmacol. exp. Ther. **76**, 137—148 (1942a).

SELYE, H., ALBERT, S.: The effect of various steroids in intact male rats. Amer. J. med. Sci. **204**, 876—884 (1942b).

SELYE, H., BELAND, E.: The development and repair of organ changes induced by steroid compounds. Rev. Canad. biol. **2**, 271—289 (1943).
SHAY, H., GERSHON-COHEN, J., PASCHKIS, K., FELS, S.S.: Influence of testosterone propionate on somatic growth in the white rat. Endocrinology **28**, 877—884 (1941).
SIMPSON, M. E., MARX, W., BECKS, H., EVANS, H. M.: Effect of testosterone propionate on the body weight and skeletal system of hypophysectomized rats synergism with pituitary growth hormone. Endocrinology **35**, 309—316 (1944).
SMITH, Q. T., ALLISON, D. J.: Cutaneous collagen and hexosamine and femur collagen of testosterone propionate-treated rats of various ages. Biochem. Pharmacol. **14**, 709—720 (1965).
SPELLMAN, M. W., CARLSON, J. C., LILLEHEI, C. W.: The effects of testosterone propionate and testosterone cyclopentylpropionate upon morbidity and mortality in mice following lethal total body irradiation. Cancer **7**, 617—621 (1954).
SPURR, C. L., KOCHAKIAN, C. D.: The effect of androgens on the survival of adrenalectomized rats. Endocrinology **25**, 782—786 (1939).
STEINETZ, B. G., GIANNINA, T., BUTLER, M., POPICK, F.: The role of growth hormone in the anabolic action of methandrostenolone. Endocrinology **90**, 1396—1398 (1972).
STOTSENBURG, J. M.: On the growth of the albino rat (Mus Norvegicus Var Albus) after castration. Anat. Rec. **3**, 233—244 (1909).
SWARTZ, F. J.: Chemical and cytological aspects of rat liver growth. Growth **26**, 167—180 (1962).
SZIRMAI, J. A.: Histological aspects of the action of androgens and oestrogens. In: Protein Metabolism, pp. 45—77. Berlin-Heidelberg-New York: Springer 1962.
TEDESCHI, G. G., JANNAKOPULU, G., PETRELLI, F.: Interazione fra testosterone e cortisone nella regolazione del metabolism creatinico. Rend. 1st Sci. Camerino **1**, 34—38 (1960).
TEIR, H.: On the sizes of the nuclei in the glandula infraorbitalis of the white rat. Acta path. microbiol. scand. **26**, 620—635 (1949).
TESSMAN, D.: Über die Wirkung von Testosteron auf die normale Mäuseniere. Z. Zellforsch. **86**, 61—73 (1968).
THYBERG, W. G., LYONS, W. R.: Androgen-induced growth of the os penis of hypophysectomized-gonadectomized rats. Proc. Soc. exp. Biol. (N.Y.) **69**, 158—161 (1948).
TISSIERES, A.: Le taux du cytochrome C dans le muscle squelettique du rat après traitement au dinitrophénol, au desoxycorticosterone et au testostérone. Arch. int. Physiol. **55**, 252—257 (1947).
TRUCKENBRODT, H.: Tierexperimentelle Untersuchungen zur Regulation der DNS-, RNS- und Eiweiß-Synthese während des physiologischen Wachstums und deren Beeinflussung durch anabole und antianabole Pharmaca. Ärztl. Forsch. **23**, 121—138 (1969).
TUOHIMAA, P., NIEMI, M.: The effect of testosterone on cell renewal and mitotic cycles in sex accessory glands of castrated mice. Acta endocr. **58**, 696—704 (1968).
TURNER, H. H., LACHMANN, E., HELLBAUM, A. A.: Effect of testosterone propionate on bone growth and skeletal maturation of normal and castrated male rats. Endocrinology **29**, 425—429 (1941).
TYSLOWITZ, R.: Effect of testosterone propionate on growth in hypophysectomized rats. Conference on Metabolic Aspects of Convalescence Including Bone and Wound Healing. Fourth Meeting, Josiah Macy, Jr. Foundation, New York, N.Y. (1943), pp. 123.
USUELLI, F., PIANA, G., MAINARDI, B.: Testosterone e Digiuno Folia endocrinologica **2**, 31—39 (1949).
VAN BEKKUM, D. W., KASSENAAR, A. A. H.: Observations on the effect of testosterone propionate on the kidneys of the mouse. Acta endocr. **8**, 155—164 (1951).
VAN WAGENEN, G.: Maturity induced by testosterone in the young male monkey. Fed. Proc. **6**, 219 (1947).
VAN WAGENEN, G.: Accelerated growth with sexual precocity in female monkeys receiving testosterone propionate. Endocrinology **45**, 544—546 (1949).
VARGAS, L., ABELIUK, J.: An integrated bioassay for anabolic-androgenic activity. Steroids Lipids Res. **3**, 353—362 (1972).
VENABLE, J. H.: Morphology of the cells of normal, testosterone-deprived and testosterone-stimulated levator ani muscles. Amer. J. Anat. **119**, 271—301 (1966a).
VENABLE, J. H.: Constant cell populations in normal, testosterone-deprived, and testosterone-stimulated levator ani muscles. Amer. J. Anat. **119**, 263—270 (1966b).

WAINMAN, P., SHIPOUNOFF, G. C.: The effects of castration and testosterone propionate on the striated perineal musculature in the rat. Endocrinology **29**, 975—978 (1941).
WALD, H.: The weight of normal adult human kidneys and its variability. Arch. Path. **23**, 493—500 (1937).
WALKER, R.: Age changes in the rat's exorbital lacrimal gland. Anat. Rec. **132**, 49—69 (1958).
WOLFF, G. L.: Genetic influences on response to castration of liver growth and hepatoma formation. Cancer Res. **30**, 1726—1730 (1970).
WOLFF, G. L.: Genetic modification of homeostatic regulation in the mouse. Amer. Naturalist **105**, 241—252 (1971).
WORTMAN, L. C., LEONARD, S. L.: The effect of injections of testosterone propionate and cortisone acetate on glycogen depletion induced by epinephrine in different skeletal muscles. Endocrinology **53**, 480—486 (1953).
WRETE, M.: Experimentelle Untersuchungen über die Wirkung von Geschlechtshormonen auf die kompensatorische Nierenhypertrophie bei kastrierten weißen Mäusen. Upsala Läk.-Fören. Förh. **48**, 393—414 (1943).
WRETE, M.: Experiments into the action of testosterone propionate on the kidneys of male rabbits. Acta Anat. **1**, 214—223 (1945).
WRIGHT, N. A., MORLEY, A. R.: The effect of testosterone on the growth fraction of the mouse small intestine. J. Endocr. **50**, 351—352 (1971).
WRIGHT, N. A., MORLEY, A. R., APPLETON, D.: The effect of testosterone on cell proliferation and differentiation in the small bowel. J. Endocr. **52**, 161—175 (1972).
YAMADA, H.: Influences of sex hormones upon extraorbital lacrymal gland of rats and its tumorigenesis. Acta Soc. Ophthal. Jap. **69**, 139—149 (1965).
ZUFAROV, K. A., GONTMAKHER, V. M., KHIDOYATOV, B. A., SAGDULLAEV, Z. Z.: Effect of testosterone propionate on the early postnatal development of rat kidney and cell differentiation in it. Tsitologiya **16**, 564—568 (1974).

# B. Effect of Anabolic-Androgenic Steroids on the Skin, Including Hair and Sebaceous Glands

A. B. HOUSSAY

Human skin is a target organ for androgenic hormones, the skin of men presenting characteristics that distinguish it from that of women and children. In prepubertal castrated men, the androgenic characteristics of the skin do not develop, but they appear when these men are treated with testosterone or other androgenic steroids. Administration of androgenic steroids causes the development of androgenic skin characteristics in women and children. The physiologic and pharmacologic effects of these anabolic-androgenic steroids upon the skin have been studied in man and experimental animals. The aim of this review is to describe the principal changes produced in the skin and its appendages, especially the sebaceous glands and the different types of hair, by administration of these steroids.

## I. Effects upon the General Properties of the Skin

The anabolic-androgenic steroids influence the structure of the skin, its physical properties, and chemical composition. These changes may perhaps explain the so-called anabolic properties upon the skin, which are the basis of some therapeutic applications.

### A. Mechanical Properties

Some studies have been made on flaps of excised skin in rats to determine the influence of several hormones upon the mechanical properties of the skin. Measurements have been made of the load required to rupture the skin strip (g), the tensile strength (g/mm$^2$), and the stress relaxation of the stretched skin after several loads (VOGEL, 1969a, b, c, d, 1973). A sexual difference in the tensile strength of the skin was found, being higher in females (VOGEL et al., 1970). Administration of testosterone produced a biphasic effect—short-term treatments decrease, and long-term treatments increase the tensile strength of rat skin (HERRICK, 1945; VOGEL, 1969a, 1970). An anabolic steroid, methandrostenolone, increased the tensile strength of granulation tissue in skin wounds in rats. The latter phenomenon was associated with an increased hydroxyproline concentration (VILJANTO et al., 1962). The endocrine factors produce opposite changes in the tensile strength and in the stress relaxation of the rat skin, and both have been attributed to alterations in the binding forces within the connective tissue fibers (VOGEL, 1973). No sexual differences or changes due to endocrine factors have been detected in the mechanical properties of human skin (HOLZMANN et al., 1971).

## B. Collagen Content

Histologic and histochemical studies have shown that administration of testosterone increases the bundles of collagen fibers in the skin of rats, guinea pig (DE BRUX and DU BOISTESSELIN, 1953), and hypogonadal men (GABRILOVE et al., 1962). Administration of testosterone does not change (CERIMELE et al., 1971), and the administration of estrogens does not change (MORGAN, 1963) or decrease (TAKACS, 1971), the hydroxyproline content of the skin in rats. Methyltestosterone or stilbestrol, whether given alone or in combination, have not been able to counteract the inhibitory effect of glucocorticoids on collagen synthesis in subcutaneously implanted polyvinyl sponges in the rat (BAVETTA et al., 1962). On the other hand, administration of methandrostenolone increases the concentration of hydroxyproline in the repair tissue of skin wounds in the rat (VILJANTO et al., 1962) and decreases the inhibitory effect of cortisone upon the hydroxyproline content in rat skin (KOWALEWSKI, 1966).

In human skin there is a sexual difference in the collagen, measured as the total hydroxyproline concentration, which is greater in men than in women (SHUSTER and BOTTOMS, 1963). On the other hand the skin collagen (total hydroxyproline concentration) is increased in women with hirsutism (SHUSTER et al., 1970) and in women patients with osteoporosis, after treatment with androgens (BLACK et al., 1970).

## C. Skin Mucopolysaccharides

No sex difference has been found in the concentration of total acid mucopolysaccharides (glycosaminoglycans) in rat skin, but administration of testosterone propionate increases its concentration (ALLALOUF and BER, 1961; KOFOED et al., 1970; CERIMELE et al., 1971), while administration of estrogens decreases it (ALLALOUF and BER, 1961; CERIMELE et al., 1971). The increased concentration of glycosaminoglycans in the rat skin after administration of testosterone is due to a great increase in the hyaluronic acid fraction while the sulphated glycosaminoglycans were not affected (DAVIDSON and SMALL, 1963; KOFOED et al., 1970). These changes are similar to those obtained in the cock's comb in castrated animals receiving testosterone (LUDWIG and BOAS, 1950; SCHILLER et al., 1952; DOYLE et al., 1964).

A comparative study of the effects of some anabolic-androgenic steroids upon the skin glycosaminoglycans was performed in castrated male $C_3H$/Ep mice. No difference was found between normal and castrated males. Administration of testosterone propionate or androstanolone for 20 days to castrated male mice increased skin glycosaminoglycans concentration; this increase is due only to the increase in the hyaluronic acid and chondroitin-4-sulfate fractions, as shown in Figure 1. On the other hand, administration or nandrolone phenylpropionate of nandrolone decanoate decreased skin glycosaminoglycans concentration, as shown in Table 1, this decrease being due to the same two fractions, hyaluronic acid and chondroitin-4-sulfate as seen in Figure 2 (KOFOED and HOUSSAY, 1974).

Nandrolone phenylpropionate and decanoate have a stronger inhibitory effect upon hair growth in mice than testosterone propionate (HOUSSAY and GAMPER, 1974). The hair folicles of the inhibited areas are mostly in telogen. It has been shown in the hair cycle of rats that glycosaminoglycans content is lower in the telogen phase (MORETTI et al., 1968). This may explain the depressing effect upon the glycosamino-

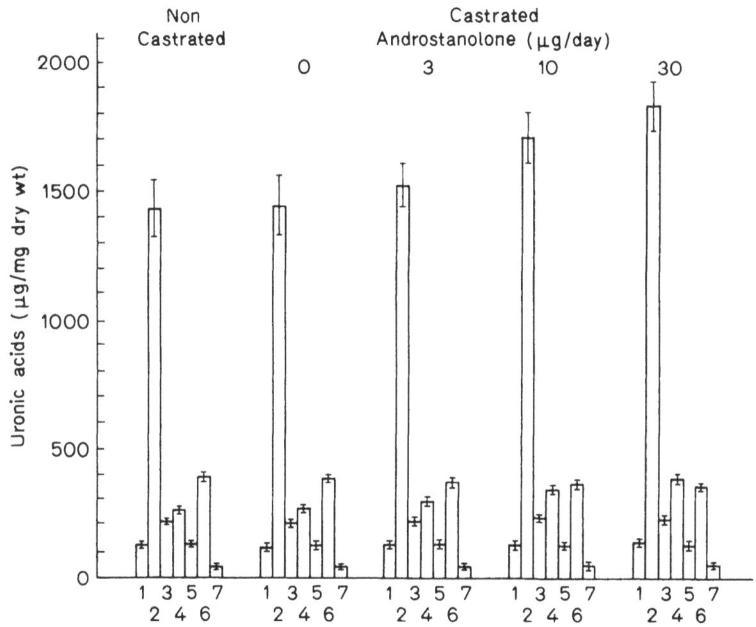


Fig. 1. Uronic acid chromatographic fractions from the skin of control and castrated rats treated with androstanolone: *(1)* glycoproteins, *(2)* hyaluronic acid, *(3)* heparitin-sulfate, *(4)* chondroitin-4-sulfate, *(5)* chondroitin-6-sulfate, *(6)* dermatansulfate, and *(7)* heparin

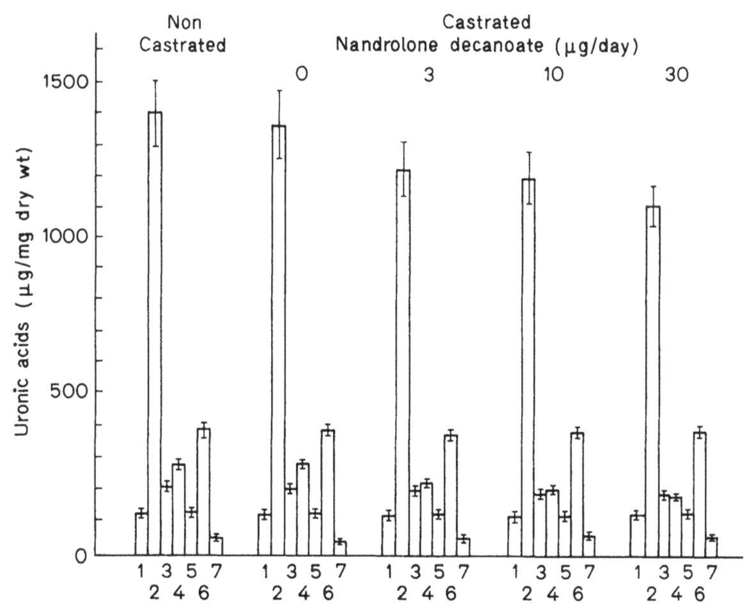


Fig. 2. Uronic acid chromatographic fractions from the skin of control and castrated rats treated with nandrolone decanoate. References as in Figure 1

Table 1. Effects of androgenic anabolic steroids upon glycosaminoglycans content of skin in castrated $C_3H/Ep$ mice (μg/Uronic Acid/mg dry tissue)

| Daily dose (μg/day) | Non-castrated | Castrated | | | | |
|---|---|---|---|---|---|---|
| | | 0 | 3 | 10 | 30 | 100 |
| Nandrolone phenylpropionate | 2.18±0.21 | 2.14±0.17 | | 2.03±0.16 | 1.64±0.09 | 1.11±0.08 — |
| Nandrolone decanoate | 2.14±0.21 | 2.17±0.23 | 2.00±0.13 | 1.68±0.12 | 1.33±0.20 | — |
| Androstanolone | 2.18±0.18 | 2.13±0.21 | 2.15±0.17 | 2.96±0.23 | 3.45±0.16 | — |
| Testosterone propionate | 2.19±0.17 | 2.22±0.19 | 2.36±0.17 | 3.69±0.19 | 3.76±0.18 | — |

glycans of those anabolic steroids, which produce greater hair growth inhibition. A similar decrease in scalp glycosaminoglycans has been observed in the telogenic loss of hair, which characterizes alopecia areata and pattern alopecia in the human (REBORA et al., 1968).

## D. Anabolic Effects

Administration of testosterone induces proliferation of the epidermis in female rats or castrated male rats, which is inhibited by antiandrogens (LÄUPPI and STUDER, 1959; SCHELL, 1973). Skin thickness is greater in men and hirsute women than in normal women (SHUSTER et al., 1970). In studies performed on the scalp, it has been shown that while the epidermis is thicker in men, the dermis and hypodermis are thicker in women (MORETTI, 1971).

In the submaxillary glands of mice, a polypeptide called epidermal growth factor (EPG) has been found that promotes epidermal growth and keratinization if given to immature mice. Administration of androgens increases this factor in the submaxillary glands, but not in plasma and urine. EPG is liberated to plasma by sympathetic stimulation or adrenergic agents (BYYNY et al., 1974). The submaxillary increase in epidermal growth factor after administration of androgens has been used as an assay method for androgenic steroids. The effect of 5α-dihydrotestosterone is 10 times, and that of 3α-androstanediol 5 times greater than that of testosterone, while 3β-androstanediol has only a slight effect (BARTHE et al., 1974).

The anabolic-androgenic steroids have been shown to antagonize the antianabolic but not the anti-inflammatory effects of glucocorticoids. Testosterone propionate and methylandrostenediol impede adrenal atrophy by cortisone, but they do not interfere with the inhibitory effect of cortisone on granuloma formation by subcutaneous cotton pellets in rats (WINTER et al., 1953). Methandrostenolone partially antagonizes the growth-inhibiting effects of glucocorticoids in rats, but the anti-inflammatory effect on experimental granulomas is not modified or is somewhat increased (RENZI and CHART, 1962). Methandione opposes the inhibitory effects of cortisone upon the repair of crushed muscle in mice (SLOPER and PEGRUM, 1967). In studies performed in tissue cultures, it has been shown that nandrolone decanoate prevents completely the mitotic inhibition induced by dexamethasone in mouse ear epidermis in vitro (BULLOUGH et al., 1968).

These anabolic properties have been used to improve the healing of skin wounds and ulcers experimentally and in human patients. Parenteral administration of testosterone propionate has been shown to improve the healing of experimental skin wounds in rats, with increase in tensile strength and collagen formation (LODDI and MOGGI, 1953; PEARCE et al., 1960). Treatment with anabolic steroids stimulated a rise in glycosaminoglycans in wound tissue (SORENSEN, 1966). A careful study of the healing of punch wounds in an X-rayed area of rabitt's ears did not show any improvement following the topical application of anabolic steroids nor any modification in the inhibitory effect of fluoroformylon (HEITE and IRRO, 1973). Treatment with methandrostenolone improved the healing of experimental punch wounds in the forearm of human subjects (CACCIALANZA and TOMMASINI, 1961). Topical application of a cream containing dexamethasone and nandrolone decanoate ensures better progress of experimental punch wounds in the forearm of human volunteer subjects than the same cream without the anabolic steroid (STEVANOVIC, 1971). Good results were obtained in the treatment of skin wounds and ulcers with topical treatment with methandrostenolone (VIVARELLI and TOMASSINI, 1964), 4-chlorotestosterone acetate combined with neomycin sulphate (BARBERIS and AMBROGGIO, 1970; ESPOSITO, 1970), and 4-chlorostestosterone acetate combined with fluoroformylon and antibiotics (BRÄUER, 1971). Some anabolic steroids have been successfully used in the treatment of burn patients (ROMAN-CALDERÓN, 1968).

The successful treatment of several skin diseases with combination ointments containing anabolic steroids, glucocorticoids, and antibiotics has been reported (BRÄUER, 1971). In studies using these combined ointments on experimental acanthosis of the guinea pig, addition of 4-chlortestosterone decreased the antiacantosis effect of fluoroformylon (HEITE and CARDONA, 1973). These studies with combined ointments were also made on Pyrifer erythema of human skin; fluoroformylon inhibited the erythema and 4-chlortestosterone had no influence upon this effect (HEITE and MEYER, 1973).

## II. Metabolism of Androgens by the Skin

The gonads and the adrenals produce testosterone and preandrogens, the most important of which are dehydroepiandrosterone and $\Delta^4$-androstenedione (see review by TAMM and VOIGT, 1971). While 35% of blood testosterone is produced directly by the gonads and adrenals, 65% is produced by the transformation of preandrogens in peripheral tissues (KIRSCHNER and BARDIN, 1972). Plasma testosterone is principally bound to the testosterone-estradiol binding globulin (Te BG), the level of this protein being higher in women than in men. The Te BG-active androgen link reduces the metabolism of androgens at the level of the target organs. The testosterone binding with this globulin is decreased in patients with hirsutism, but increases when estrogens are administered (CORVOL and MAHOUDEAU, 1971; WIELAND et al., 1973).

In vitro studies have shown that human skin is able to metabolize dehydroepiandrosterone and $\Delta^4$-androstenedione into testosterone (GALLEGOS and BERLINER, 1967; KIM and HEERMANN, 1968; BERLINER et al., 1968; HAY and HODGINS, 1971, 1973). Other metabolites of dehydroepiandrosterone are 5α-androstane-3,17-dione; androst-4-ene-3,17-dione; androsterone and smaller amounts of 5α-dihydrotesterone and 5α-androstane-3β,17β-diol (BERLINER et al., 1968; FAREDIN et al., 1969; CHAKRABORTY et al., 1970; JULESZ et al., 1971; HODGINS, 1971; HAY and HODGINS, 1971, 1973; THOMAS and OAKE, 1974). Testosterone is transformed in the human skin to several metabolites, the most important being 17β-hydroxysteroids: 5α-dihydrotestosterone; 5α-androstane-3α-17β-diol; 5α-androstane-3β-17β-diol, and androst-5-ene-3β-17β-diol but also to 17-oxosteroids: 5α-androstane-3,17-dione; androst-4-ene-3,17-dione and androsterone (WOTIZ et al., 1956; RONGONE, 1966; GOMEZ and HSIA, 1968; WILSON and WALKER, 1969; FLAMIGNI et al., 1971; SOMMERVILLE et al., 1971; HAY and HODGINS, 1971; ENGELHARDT and KARL, 1971; GOMEZ et al., 1972; JENKINS and ASH, 1973; THOMAS and OAKE, 1974). Testosterone is also metabolized in cultured human skin fibroblasts (SHANIES et al., 1972; MULAY et al., 1972; PINSKY et al., 1974). Several hydroxysteroid dehydrogenases have been localized in the skin by histochemical methods (BAILLIE et al., 1965, 1966; MUIR et al., 1968a, 1970; CALMAN et al., 1970). 17β-Hydroxysteroid dehydrogenase has been found in microsomal preparations of rat skin, and their properties and the chemical characteristics of their inhibitors have been studied (DAVIS et al., 1972).

Because it has been proposed that testosterone has to be converted into 5α-dihydrotestosterone to be able to act upon the nuclei of the target organs in order to produce its metabolic effects (WILSON and GLOYNA, 1970), 5α-reductase activity was studied in microsomal preparations of human natal foreskin (VOIGT et al., 1970). The structural characteristics of the enzyme inhibitors have been determined, the most active being progesterone, desoxycorticosterone, and a steroid without hormonal effects, 3-oxo-androst-4-en-17β-carboxylic acid (VOIGT and HSIA, 1973b). In animal trials upon the flank organ of female hamsters, topical application of this substance has been shown to inhibit the topical effect of testosterone, but not that of 5α-dihydrotestosterone (HSIA and VOIGT, 1974). Topical application of progesterone upon the skin of normal men reduces the in vitro conversion of testosterone to 5α-dihydrotestosterone in skin homogenates of the treated areas (MAUVAIS-JARVIS et al., 1974).

Some topographic differences have been found in the conversion of testosterone to 5α-dihydrotestosterone in human skin, that is, it is higher in the sexual skin (WILSON and WALKER, 1969; FLAMIGNI et al., 1971; SOMMERVILLE et al., 1971) and in rat skin (DUBE et al., 1975). The conversion of dehydroepiandrosterone and testosterone to 17β-hydroxysteroid metabolites is greater in the axillary than in the facial human skin, while in the latter the conversion to 17-oxosteroids is greater (HAY and HODGINS, 1973).

The utilization of dehydroepiandrosterone but not that of testosterone and the conversion of testosterone to 5α-dihydrotestosterone by the skin are greater in normal men and hirsute women than in normal women (JENKINS and ASH, 1973; THOMAS and OAKE, 1974). The production of 17β-hydroxysteroids, especially 5α-dihydrotestosterone and the 3α- and 3β-androstanediols, which are more active androgens than the 17-oxosteroids, by the skin in vitro is higher in men and hirsute women than in normal women (THOMAS and OAKE, 1974). The 5α-reductase activity in rat skin homogenates is higher in male than in female animals (DUBE et al., 1975). It has been reported that cultured fibroblasts from the skin of human adult females utilize predominantly the 17β-OH pathway of testosterone metabolism, while those cultured from the skin of human adult males utilize the 17C=0 pathway to a much greater extent (SHANIES et al., 1972), but this has not been confirmed by other authors (PINSKY et al., 1972, 1974).

Some alterations in the metabolism of dehydroepiandrosterone have been described in psoriatic patients. Plasma dehydroepiandrosterone levels and urinary excretion of dehydroepiandrosterone, androsterone, and ethiocholanolone are low, red cell glucose-6-phosphate dehydrogenase activity is high, and these values are not modified when dehydroepiandrosterone sulfate is administered to the patients (MENZEL et al., 1971; MORSCHES et al., 1971). In the skin eluates of these patients, if compared to normal controls, there is an increase in 17$\beta$-hydroxysteroids, for example, androstenediol and androstenetriol, and a decrease in 17-oxosteroids, for example, dehydroepiandrosterone, androsterone, and ethiocholanolone (DENNENBAUM et al., 1972).

Several studies in vitro have shown that testosterone metabolism and the conversion of testosterone to dihydrotestosterone are normal, similar to the normal male controls, in the skin of male pseudohermaphrodite patients with testicular feminization syndrome (WILSON and WALKER, 1969; FLAMIGNI et al., 1971; SOMMERVILLE et al., 1971; GILLESPIE, 1971; JENKINS and ASH, 1971; KELCH et al., 1971; ROSENFIELD et al., 1971) and in the preputial glands of male pseudohermaphrodite rats of the Stanley-Grunbach strain (BARDIN et al., 1970). In both cases the cause of the syndrome is a lack of response of the androgen receptors in all target organs to both testosterone and dihydrotestosterone (FRENCH et al., 1966; GWINUP et al., 1966; STRICKLAND and FRENCH, 1969; SHERINS and BARDIN, 1971). Studies performed in kidneys of pseudohermaphrodite mice and in preputial glands of pseudohermaphrodite rats indicate that there is a defect in the cytosol receptor for androgens (BULLOCK and BARDIN, 1972).

Most of the antiandrogens that have been used in therapeutics, such as 17$\alpha$-methyl-19-nortestosterone, chlormadinone acetate, cyproterone acetate, and cyproterone, do not interfere with the metabolism of testosterone in the adrenals, gonads, or the skin. Their main effect consists in interfering with the effect of testosterone and other androgens upon the receptors in the target organs including the skin (WIECHERT et al., 1967; NEUMANN et al., 1970). If cyproterone acetate is given to pregnant rats, it induces male pseudohermaphroditism in the male fetuses (NEUMANN et al., 1966, 1970), that is, it produces the experimental counterpart of the testicular feminization syndrome in man.

## III. Effects of Androgens on Hair Growth in Experimental Animals

### A. Hair Growth in Waves in Rats and Mice

Hair growth in rats and mice is cyclic, each hair follicle having alternate periods of growth and rest. The period of growth lasts approximately 21 days, and it is divided into several phases according to the histologic appearance: anagens I, II, III, IV, V, and VI; then follows a period of regression, or the catagen phase, which lasts 7 days, whereupon the rest period or telogen phase begins. The telogen phase lasts a variable period of time, which increases in each hair cycle (DRY, 1925; BUTCHER, 1934, 1951; CHASE et al., 1951).

Furthermore, the hair follicles develop their activity gradually in waves of hair growth. The waves originate in a limited part of the skin as symmetrical bands where the hair follicles are in the same state of activity. These bands move slowly backward, or sometimes forward and to the sides, and gradually cover the remaining skin. After they pass a region, that region goes into a resting period. Thus, each complete coat of hair is produced by a hair wave that slowly and gradually covers the entire skin surface of the animal in an orderly, symmetrical manner. As the animal grows older, it is covered more slowly by hair waves, and the telogen phase becomes longer (DRY, 1925; BUTCHER, 1934, 1951; DIEKE, 1948; DURWARD and RUDALL, 1949; BORUM, 1953; JOHNSON, 1958a; CHASE and EATON, 1959).

Several morphologic and chemical changes in the skin of rats and mice occur during the waves of hair growth. The principal modifications have been observed in the water, lipid, phosphorus, collagen, and mucopolysaccharide content. The hydroxyproline content is lower in anagen and higher in catagen and telogen. The total glycosaminoglycans increases in early anagen and drops steadily during the rest of the cycle (EBLING and HALE, 1966; MORETTI et al., 1968). It is not possible to speculate on the relation between these chemical changes and the growth of hair, but it is interesting to point out that the intradermal injection of sulfated glycosaminoglycans in the rabbit stimulates the hair follicles and initiates a hair growth cycle (MAYER et al., 1961).

The chemical and morphologic changes in the transition from anagen to telogen in the hair cycle of the rat are similar to those observed in the telogenic loss of hair in the human scalp that characterizes alopecia areata and pattern alopecia (REBORA et al., 1968).

Diverse endocrine factors act upon the hair follicles and sometimes alter growth, texture, or pigmentation of the hair (EMMENS, 1942; RALLI and GRAEF, 1945; HUNDLEY and ING, 1951; RENNELS and CALLAHAN, 1959; DAVIS, 1963). But they generally modify the hair growth waves, accelerating or delaying the rate, or producing irregular or widespread growth (BUTCHER, 1937; DIEKE, 1948; HOUSSAY, 1953b; MOHN, 1958; EBLING and JOHNSON, 1964; HOUSSAY et al., 1965).

## B. Effects of Castration

Gonadectomy of male and female mice stimulates the hair follicles that are in the telogen phase at the time of surgery. The clipped area is covered by a diffuse growth of hair, with the exception of the follicle bands in the anagen phase of the hair waves, which continue their normal growth, which terminates before the diffuse growth (HOUSSAY and HIGGINS, 1949a, b; HOUSSAY, 1953b; HOUSSAY et al., 1965). When the diffuse wave terminates, the hair waves reappear in the subsequent hair generations in these animals, but their rate of advance is faster than before the operation.

The diffuse hair growth is also produced by other endocrine conditions: adrenalectomy in the rat (BUTCHER, 1937; RALLI and GRAEF, 1943; BAKER, 1951) and the mouse (HOUSSAY and HIGGINS, 1951; HOUSSAY, 1953b) and hypophysectomy in the rat (HOUSSAY, 1954; HOUSSAY et al., 1955; MOHN, 1958; EBLING and JOHNSON, 1964) and the mouse (HOUSSAY, 1962; DAVIS, 1963; HOUSSAY et al., 1965). In all these cases surgery results in an immediate stimulation of the resting hair follicles, so that the animal is covered with a new coat of diffuse hair in 21 days. But the waves of hair growth reappear in the subsequent hair generations of these rats and mice.

The diffuse hair growth follows the same pattern and takes the same time to cover the denuded surface of the animal with hair, when it is produced by any of the previously mentioned operations or combinations of them. Thus, in mice the diffuse hair growth is exactly the same if produced by castration in males and females, by adrenalectomy, by hypophysectomy, or by any combination of these operations (HOUSSAY, 1953b, 1962). In rats it is exactly the same if produced by adrenalectomy, by hypophysectomy or by adrenalectomy-hypophysectomy (HOUSSAY et al., 1955).

Castration does not produce diffuse growth in the rat (HOUSSAY, 1953b) with the exception of one strain studied by LETTI (1959). But the hair growth waves are accelerated after gonadectomy, in both male and female animals in most strains of rats (JOHNSON, 1958b).

The diffuse hair growth induced by gonadectomy has been a splendid biological model to study the effect of radiation upon the hair follicles in the different stages of the hair cycles. The graying effect and hair growth delay are more marked when the hair follicles are in telogen, but their sensitivity to rays gradually decreases during the progression of the anagen phase (EPPER et al., 1971).

Spontaneous alopecia is not uncommon in older mice from several strains, so that large denuded areas of skin over the head and back often occur. Gonadectomy induces a prompt growth of grayish hair over these denuded areas in 70% of the alopecic male mice. The duration of this alopecia appears to be a factor in this striking response to the removal of testicles (HOUSSAY and HIGGINS, 1949a; HOUSSAY, 1953b).

## C. Effect of Steroid Hormones

Many gonadal or adrenal steroid hormones have an inhibitory effect on hair growth in rats and mice. This has been demonstrated with estrogens (ZONDEK, 1936; EMMENS, 1942; FORBES, 1942; HOOKER and PFEIFFER, 1943; HOUSSAY and HIGGINS, 1949b; HOUSSAY, 1953a; JOHNSON, 1958b), androgens (KOCHAKIAN, 1940; HOUSSAY and HIGGINS, 1949b; HOUSSAY, 1953b; LIGHT and TORNABEN, 1953a, b; JOHNSON, 1958b; HOUSSAY et al., 1959), and glucocorticoids (BAKER and WHITAKER, 1948; WHITAKER and BAKER, 1948, 1951b; HOUSSAY, 1953b; ANDREASEN, 1958; FUKUYAMA and BAKER, 1958; MOHN, 1958; WHITELEY, 1958; HOUSSAY et al., 1959).

Small physiologic doses of estrogens, testosterone, or glucocorticoids impede the diffuse hair growth caused by gonadectomy in mice (HOUSSAY and HIGGINS, 1949b; HOUSSAY, 1953a, b) or by hypophysectomy in rats (HOUSSAY et al., 1959). Small doses of glucocorticoids inhibit the diffuse hair growth in adrenalectomized rats and mice (WHITAKER and BAKER, 1951a; HOUSSAY, 1953b), whereas large doses of estrogens or androgens are required to inhibit the hair growth produced by adrenalectomy (BAKER and WHITAKER, 1949; HOUSSAY et al., 1959). If the adrenalectomized animals receive subthreshold doses of glucocorticoids, the diffuse hair growth produced by adrenalectomy may be inhibited with small doses of estrogens and androgens (INGLE and BAKER, 1951; HOUSSAY et al., 1959).

Large or pharmacologic doses of estrogens (ZONDEK, 1936; EMMENS, 1942; FORBES, 1942; HOOKER and PFEIFFER, 1943; HOUSSAY, 1953a, b), testosterone (HOUS-

Table 2. Comparison between inhibitory effects of several androgens upon diffuse hair growth produced after castration of $C_3H/Ep$ mice

| | Substitution dose (μg/day) | Inhibitory effect (%) |
|---|---|---|
| Nandrolone phenylpropionate | 2.6 | 100 |
| Nandrolone decanoate | 16.5 | 15.6 |
| Androstanolone | 16.5 | 15.6 |
| Testosterone propionate | 190.0 | 1.4 |

SAY, 1953b; LIGHT and TORNABEN, 1953a, b; HOUSSAY et al., 1959), and glucocorticoids (BAKER and WHITAKER, 1948; WHITAKER and BAKER, 1948, 1951b; HOUSSAY, 1953b) inhibit hair growth in normal, castrated, adrenalectomized, or hypophysectomized rats and mice, and hair waves are slow. Administration of higher doses brings about irregular growth or almost complete cessation of hair growth or even falling out of the hair. Adrenal mineralocorticoids such as desoxycorticosterone also have some inhibitory effect on hair growth (RALLI and GRAEF, 1945; HOUSSAY and HIGGINS, 1949b; WHITAKER, 1949; WHITAKER and BAKER, 1951b).

The sensitivity of the hair follicles to the inhibitory effect of estrogens in castrated mice and rats has been compared. The threshold of sensitivity of the hair follicles to estradiol benzoate is 1000 times lower in mice than in rats, but if the dose of this steroid is calculated per 100 g of animal, the threshold difference is reduced to 125 times (HOUSSAY, 1953a). The hair follicles of mice have also a lower threshold to the inhibitory effect of testosterone than those of rats, but the difference is much smaller (HOUSSAY and HIGGINS, 1949b; HOUSSAY et al., 1959).

The results of comparative study of the inhibitory effects of several anabolic-androgenic steroids upon the diffuse hair growth induced by castration of male $C_3H/Ep$ mice are presented in Table 2. They are expressed as the substitution dose necessary to keep the advance of the hair waves at the same rate as that in the noncastrated controls. The following ranking was obtained: (1) nandrolone phenylpropionate; (2) nandrolone decanoate and androstanolone; and (3) testosterone propionate (HOUSSAY and GAMPER, 1974). Other androgenic steroids, such as dehydroepiandrosterone sulfate, have been shown to have some inhibitory effect on hair growth in mice (BOSSE et al., 1972).

## D. Effects in Other Mammals

Administration of pharmacologic doses of estrogens has a well-known inhibitory effect upon growth of hair in many species of mammals, among them the guinea pig (VON WATTENWYL, 1941; BOSSE, 1972) and the dog (GARDNER and DE VITA, 1940; WILLIAMS et al., 1946). Moreover, certain estrogenic-producing tumors in the dog, such as the Sertoli cell tumor of the testis, are associated with hair loss (COFFIN et al., 1953; PIERREPOINT et al., 1967). Administration of androgens has been shown to produce inhibition of hair growth and loss of hair in guinea pigs (BOSSE et al., 1972; BOSSE, 1972).

In some subhuman primates, alopecia often develops in the scalp and forehead of adult animals. It has been found in some monkeys, as the red nacari (UNO et al., 1969), in apes, such as chimpanzees and orangs, and it has been especially studied in the stump-tailed macaque. It develops in both male and female animals; there is great individual variation in the extent of baldness, and the disease is in many ways similar to human alopecia (MONTAGNA and UNO, 1968).

The data already presented show that the hair follicles in many species of mammals are susceptible to the inhibitory effects of estrogens and androgens, in some cases in physiologic doses, but more often in large pharmacologic doses.

## IV. Metabolism of Androgens by the Rat Skin in Relation to the Hair Cycles

Histochemical studies of dehydroepiandrosterone dehydrogenase and glucose-6-phosphate dehydrogenase have been performed in the skin during the hair cycle of the rat. A considerable amount of the activity of both enzymes was found in fibroblasts and the hypodermis during the last part of anagen. A strong reaction for glucose-6-phosphate dehydrogenase, an NAD-dependent enzyme, was found in the paniculus carnosus in this period. Thus the greater part of the transformation of dehydroepiandrosterone and androstenedione would take part during the last part of anagen (CROVATO et al., 1970).

The metabolism of testosterone in vitro during the hair cycle of the rat has also been studied. The principal metabolites are 5α-dihydrotestosterone and $\Delta^4$-androstenedione, but 5α-androstanedione, 3α-androstanediol, androsterone, epiandrosterone, and 3β-androstanediol are also found. An increase in the 5α-reductase activity has been observed in the telogen phase. The maximal 17β-hydroxysteroid dehydrogenase activity has been found in the middle of anagen (RAMPINI et al., 1971).

The binding of testosterone, 5α-dihydrotestosterone and 17β-estradiol has been studied in the supernatant fraction of homogenates of rat skin. The content of soluble proteins is increased in anagen and decreased in telogen. The maximum binding for testosterone is found in catagen and for 5α-dihydrotestosterone and 17β-estradiol in telogen. Minimum binding for both androgens is seen in mid-anagen, whereas for 17β-estradiol, it is seen in late telogen or early anagen (EPPENBERGER and HSIA, 1972).

## V. Effects of Androgens upon Hair Growth in Man

The hair follicles in man follow the same phases of anagen, catagen, and telogen as those of other mammals. In adult men the development of the hair follicles in a given region is not synchronized as in the rat, and the anagen phase is much longer than the telogen phase (MYERS and HAMILTON, 1951; PECORARO et al., 1968; SAITOH et al., 1970). There are several reviews of the literature that show the important role of the endocrine glands in the regulation of hair growth in man (COOPER, 1930; BAKER, 1951; HOUSSAY, 1954; ROOK, 1965). The several types of human hair may be classified according to their relation to steroid hormones, according to GARN (1951): (1) hair not dependent upon steroid hormones, as eyebrows, eyelashes, and hair of the distal segment of the limbs; (2) hair not dependent upon steroid hormones, but sometimes inhibited by androgenic hormones, as the scalp hair; (3) hair dependent upon adult female amounts of steroid hormones, like pubic and axillary hair; and (4) hair dependent upon adult male amounts of steroid hormones, as beard, mustache, and body hair.

## A. Effects on Scalp Hair

The presence of pattern alopecia is a secondary sexual character in adult men. Three important factors participate in its development—age, genetic factors, and androgenic stimulation (HAMILTON, 1951 b; LUDWIG, 1968).

The genetic factor is very important, but there are controversies about the mode of inheritance (ROOK, 1965; SALOMON, 1968). Pattern alopecia is characteristic of the older males of the Caucasian race, but the frequency is very low in the males of other races such as Chinese, Japanese, and american Indians, and it is the same as that in females of the Caucasian race. In male Caucasians, the frequency increases with age (HAMILTON, 1951 b).

The importance of androgenic stimulation as a causal factor of pattern alopecia has been clearly demonstrated by HAMILTON (1942). Alopecia is absent in eunuchoid and prepubertally castrated males and absent or incomplete following castration between the ages of 14 and 19. It is produced by androgenic treatment of those eunuchoid males who have a genetic predisposition. It appears in women with virilism. Castration of adult males prevents further spread of the alopecia, but it does not cause regrowth of hair in the bald areas (HAMILTON, 1942, 1960). However, no differences in urinary excretion of 17-ketosteroids and testosterone or in plasma testosterone levels have been found between adult males with pattern baldness and those with normal hair, in the same age groups (LUDWIG, 1968). That is to say, at the normal male level of steroid hormones, baldness is produced only in those individuals who have a genetic factor.

Pattern alopecia is uncommon in human females, but it may be found in clinically normal women, provided a genetic factor is present. It is often, but not always, associated with increased urinary testosterone excretion levels (MAGUIRE and KLINGMAN, 1963; LUDWIG, 1964, 1966; SMITH and WELLS, 1964). Women who have endocrine disorders inducing hirsutism or virilization may develop more or less severe patterns of male alopecia, but in these cases it is always associated with high androgen levels. This is also found in women treated with testosterone, provided the genetic predisposition is present (LUDWIG, 1968). That is to say, human females may develop pattern baldness if the level of androgenic steroids is increased, but only when genetic factors are present.

A diffuse scalp hair loss postpartum in many pregnant women has been termed telogen effluvium because the scalp hair is converted from anagen to telogen before being shed. The relation of this phenomenon to the high level of steroid hormones (estrogens and progesterone) in pregnancy and to the sudden decrease postpartum is possible (LYNNFIELD, 1960; KLIGMAN, 1961). There are other types of telogen effluvium not caused by endocrine factors (KLIGMAN, 1961; ECKERT et al., 1967). An experimental telogen effluvium has been produced in guinea pigs by administration of estrogens and androgens (BOSSE et al., 1972; BOSSE, 1972). During the pregnancy of these animals, the hair begins to change from anagen to telogen, and after labor a loss of hair in telogen phase is found (BOSSE, 1972).

Alopecia areata is a type of nonpattern alopecia in which scalp hair and sometimes other types of hair are lost in relatively limited areas. It is called alopecia totalis when all scalp hair is lost and alopecia universalis when all types of hair are lost. The cause is unknown and the relation to endocrine factors is doubtful. It is very interesting that the administration of glucocorticoids induces hair growth in the alopecic areas (DILLAHA and ROTHMAN, 1952).

The information presented here shows that the hair follicles of the human scalp, like those of body hair in dogs, rats, mice, and guinea pigs, may be inhibited by physiologic or pharmacologic doses of steroid hormones. Androgenic inhibition is more important in man, estrogenic inhibition in the dog, but both androgenic and estrogenic factors are effective in inhibiting the growth of hair in rats, mice, and guinea pigs. Moreover the scalp hair of some species of monkeys and apes is also inhibited by steroid hormones, but both androgens and estrogens have this inhibitory effect.

Several types of therapy of pattern alopecia have been used in recent years, in both men and women, with the purpose of decreasing or suppressing the androgen effect upon the hair follicles. Estrogen-progesterone mixtures in the form of oral contraceptive pills were used to suppress gonadal function (ZAUN, 1972). Antiandrogens, especially cyproterone acetate, were employed to block the effects of androgens upon the receptors (ZAUN, 1971; WINKLER, 1972). Antiandrogens were combined with estrogens in reverse-sequential therapy (NEALE et al., 1971; ROSSBERG et al., 1971). But, in general, these therapeutic procedures have not been effective in inducing improvement of alopecia, despite some good results in individual cases.

## B. Effects on Axillary and Pubic Hair

A sexual difference exists between the pubic hair of normal women and men, as the upper limit is horizontal in women and extends like a triangle in the hypogastric region in men. This triangle of hypogastric hair is androgenic-controlled hair like male body hair; the lower pubic hair is like axillary hair, common to both sexes, and has a somewhat different hormonal regulation.

Both axillary and pubic hair are secondary sexual characters of both sexes, as they do not exist in children, and they appear at the time of puberty. They are absent in prepubertal castrated men and women, and their amount decreases markedly after castration in men, and after oophorectomy and menopause in women (HAMILTON, 1951a).

The adrenals also participate in the control of axillary and pubic hair, which are markedly decreased in women with Addison's disease with a normal menstrual cycle (KEPLER et al., 1943; MUSSIO-FOURNIER et al., 1949; SUMMERS, 1949) and after bilateral adrenalectomy in some cancer patients (HUGGINS and SCOTT, 1945).

The administration of testosterone to eunuchs or to castrated men and women produces growth of both axillary and pubic hair (ALBRIEUX and MUSSIO-FOURNIER, 1949; HAMILTON, 1951a). Axillary and pubic hair do not exist in patients with ovarian agenesia, and they grow during the administration of estrogens (ALBRIGHT et al., 1942; SUMMERS, 1949).

In patients with pituitary insufficiency the axillary and pubic hair are very scarce or absent, but they are restored with substitution therapy. Testosterone, estrogens, and glucocorticoids stimulate growth of axillary hair if applied topically to the axilla of these patients.

In some children of both sexes, but more frequently in girls, pubic hair may appear before the age of 8 years, without any sign of virilization. This condition is called premature pubarche (SILVERMAN et al., 1952; ARNAL et al., 1961). Puberty is not advanced and occurs at its normal time. This precocious appearance of pubic hair, often accompanied by the axillary hair, has been attributed to adrenal androgens, and urinary C-19 steroids (ZURBRUGG and GARDNER, 1963), and plasma androgens (RAPPAPORT et al., 1974) may be increased.

Patients with testicular feminization syndrome in its complete form lack axillary and pubic hair (FRENCH et al., 1966; GWINUP et al., 1966). This disease is due to a failure of genetic origin in the androgenic receptors, whereby the androgens are not able to act upon their target tissues, in this case the hair follicles in the axillary and pubic regions (STRICKLAND and FRENCH, 1969).

## C. Effects on Masculine Hair

Androgenic hair is a masculine secondary sexual characteristic, as it is present in normal adult men, but not in children, in normal adult women, and in prepubertal castrated or eunuchoid men. It appears in men at the time of puberty (REYNOLDS, 1951). It consists of the beard and mustache and the body hair. The distribution of the latter is variable, depending on genetic factors; the most hirsute males have hair in the thorax, back, hypogastric and gluteal regions, arms, and legs (GARN, 1951).

Administration of androgens to women (MAGUIRE, 1964) or increase in the endogenous production of androgens in women (KIRSCHNER and BARDIN, 1972) may produce symptoms of virilization, among them hirsutism, loss of scalp hair, hypertrophy of clitoris, muscle hypertrophy, increase in sebum production, deepening of the voice (JAILER, 1953; MULLER, 1969). More common is idiopathic hirsutism, called "primary cutaneous virilism," because the symptoms of virilization are only localized in the skin (SHUSTER, 1972). Besides the hirsutism, there is increased sebum production, increased skin thickness, increased collagen content, and greater sweat response to cholinergic stimulation.

The excessive production of androgens, which induces virilization or only hirsutism, may be due to adrenal hyperfunction (adrenal congenital hyperplasia, Cushing's disease, adrenal tumors), to ovarian androgenic hyperfunction (hilar cell hyperplasia, ovarian tumors), or to an undetermined origin (idiopathic hirsutism) (MULLER, 1969; KIRSCHNER and BARDIN, 1972).

When the virilization or the hirsutism is due to adrenal or ovarian tumors or to adrenal hyperplasia, there is a great increase in androgen production, plasma testosterone levels are high, and the urinary excretion of 17-ketosteroids is increased (KELLER et al., 1958; FINKELSTEIN et al., 1961; LIPSETT and WILSON, 1962; RIVAROLA et al., 1967).

A clinical picture much more frequent in women is virilization localized in the skin with hirsutism, characteristic of idiopathic hirsutism and polycystic ovaries syndrome. The plasma testosterone level is normal or slightly increased, the urinary

17-ketosteroids are normal or somewhat increased, but the production rate of testosterone is always increased (LANTHIER, 1960; LIPSETT and RITER, 1960; MAHESH et al., 1964; BARDIN and LIPSETT, 1967; ETTINGER et al., 1971).

The plasma testosterone level may be normal in hirsute women, because the increased testosterone production rate is compensated by an increased metabolic clearance. In hirsute women, the binding of testosterone with the plasma testosterone-estradiol binding globulin is decreased (CORVOL and MAHOUDEAU, 1971). The hepatic clearance of testosterone does not change; the increased metabolic clearance is due to a high clearance by the extrahepatic tissues, especially the skin (BARDIN and LIPSETT, 1967; SOUTHREN et al., 1968; MAHOUDEAU et al., 1971; KIRSCHNER and BARDIN, 1972).

Testosterone is metabolized by the skin and transformed into more active androgens. The rate of their conversion to androstanediols or to dihydrotestosterone is higher in normal men and hirsute women than in normal women (BERCOVICI and MAUVAIS-JARVIS, 1970; JENKINS and ASH, 1973).

The production of androgens by the adrenal glands is increased by administration of adrenocorticotrophin and decreased or even suppressed by administration of glucocorticoids, especially dexamethasone. The latter drug has been employed diagnostically and therapeutically in hirsute patients. When dexamethasone is administered over a long period, urinary testosterone and 17-ketosteroids and plasma testosterone levels are decreased, the metabolic clearance of testosterone decreases, plasma TeBG does not change, but the hirsutism is improved only in a few cases (JEFFERIES and LEVY, 1959; BARDIN et al., 1968; NICHOLS et al., 1970; WIELAND et al., 1973; ISMAIL et al., 1974).

The production of androgens by the ovary is increased by gonadotrophins and decreased or suppressed by administration of estrogens or estrogen-progesterone combinations. This has also been used in diagnostic studies or therapeutic trials in hirsute patients. When estrogens or estrogen-progesterone mixtures are given, urinary 17-ketosteroids and testosterone excretion and the plasma testosterone levels are decreased, the production rate and metabolic clearance of testosterone also decrease, plasma TeBG increases, but the results in the treatment of hirsutism are poor (KIRSCHNER et al., 1970; WIELAND et al., 1973). Topical administration of estrogens does not show the inhibitory effects on hair growth seen in experimental animals, and it is not useful in the treatment of human female hirsutism (MUSSIO-FOURNIER et al., 1935; WHITAKER et al., 1947).

Antiandrogens, specially cyproterone acetate, have been used in the treatment of hirsutism in women. They do not interfere with the production of androgens, plasma binding, or with the metabolism of testosterone by the tissues (GUILLEMANT et al., 1969), but they block the binding of testosterone and dihydrotestosterone with the receptors, in such a way that they interfere with the peripheral effects upon the skin and other tissues (NEUMANN et al., 1970). The results were good in some patients (ZARATE et al., 1966; KAISER et al., 1967; HAMMERSTEIN and CUPCEANCU, 1969; NELSON and RAKOFF, 1970; VAN WAYJEN and VAN DER ENDE, 1971/72).

Treatment with antiandrogens has been combined with estrogens, and it is now used in a schedule called reversed sequential therapy. During the first half of the menstrual cycle, cyproterone acetate is given together with ethynylestradiol; in the second half, ethynylestradiol is given alone but it is stopped before the menstrual period. The results observed in the treatment of hirsutism are better than those obtained by other therapeutic procedures, but more experience is necessary to know the real value of this method (NEALE et al., 1971; ROSSBERG et al., 1971; ISMAIL et al., 1974).

## VI. Metabolism of Androgens by Isolated Human Hair Follicles

Glucose metabolism has been studied in human scalp hair follicles. Changes in the pentose cycle activity, greater in growing than in resting follicles, constituted the major metabolic difference. Other enzymes, such as glucose-6-phosphate dehydrogenase, hexokinase, fumarase, and lactate dehydrogenase, were also higher in growing than in resting follicles (ADACHI and UNO, 1968).

The metabolism in vitro of dehydroepiandrosterone, androstenedione, and testosterone has been studied in isolated human hair follicles taken from the scalp, beard, axilla, breast, arms, legs, and pubic areas. Dehydroepiandrosterone was converted to androst-5-ene-3$\beta$,17$\beta$-diol, 7$\alpha$-hydroxy dehydroepiandrosterone, and 5$\alpha$-androstanedione (FAZEKAS, 1971; FAZEKAS and LANTHIER, 1971; FAZEKAS and SANDOR, 1972, 1973). Testosterone was converted to androstenedione, 5$\alpha$-dihydrotestosterone, 5$\alpha$-androstanedione, and androsterone (FAZEKAS and LANTHIER, 1971; SANSONE-BAZZANO et al., 1972a; FAZEKAS and SANDOR, 1972; TAKAYASU and ADACHI, 1972b). Androstenedione was mainly metabolized to 5$\alpha$-androstanedione and to a minor extent to testosterone (FAZEKAS and SANDOR, 1972; SCHWEIKERT and WILSON, 1974). The patterns of androgen metabolites formed are similar in all types of hair follicles (SCHWEIKERT and WILSON, 1974).

The metabolism of dehydroepiandrosterone was altered in the scalp hair follicles of balding individuals; less androst-5-ene-3$\beta$,17$\beta$-diol was formed in comparison to nonbalding individuals. Many androgens, including androstenediol, androstenedione, androstanedione, androsterone, and dihydrotestosterone, inhibit the glucose-6-phosphate dehydrogenase activity in human hair follicles, while dehydroepiandrosterone does not (FAZEKAS and SANDOR, 1972). Glucose-6-phosphate dehydrogenase is a key enzyme of the pentose cycle, an important source of NADPH, and it is increased in the growing phase of human hair follicles (ADACHI and UNO, 1968).

Particulate fractions obtained from human scalp hair contain adenyl cyclase activity, which is activated by estrone and inhibited by dihydrotestosterone but not by testosterone. It has been suggested that this inhibitory effect may be the initial step in the development of pattern baldness (ADACHI and KANO, 1970).

## VII. Effects of Androgens upon Sebaceous Glands in Experimental Animals

### A. Effect on Sebaceous Glands of Rats

Many studies have been performed upon the endocrine regulation of the sebaceous glands in rats. Morphologic methods have yielded information about glandular and cell size and number of mitoses (EBLING, 1948, 1957), and physiologic methods, about the rate of sebum secretion (EBLING, 1967; EBLING and SKINNER, 1967; AR-

CHIBALD and SHUSTER, 1969, 1970; NIKKARI and VALAVAARA, 1969) or the changes in chemical composition of the sebum (NIKKARI, 1965; WILDE and EBLING, 1969; NIKKARI and VALAVAARA, 1970a).

Castration in male rats produces atrophy of the sebaceous glands with reduction of the size, volume, and number of mitoses (EBLING, 1957) and decrease in sebum production (THODY and SHUSTER, 1970). On the other hand castration in female rats does not affect sebaceous gland size (EBLING, 1955), but it may cause some increase in sebum production (THODY and SHUSTER, 1971b).

Androgens are the principal factor regulating the size and secretory activity of the sebaceous glands of rats (EBLING, 1974; SHUSTER and THODY, 1974). Administration of testosterone to castrated rats increases glandular size (EBLING, 1948), frequency of mitosis (EBLING, 1955, 1957), sebum production (EBLING and SKINNER, 1967; ARCHIBALD and SHUSTER, 1969), and changes in the chemical composition of sebum (NIKKARI, 1965; WILDE and EBLING, 1969; NIKKARI and VALAVAARA, 1970a, b).

The presence of the hypophysis is important in the effect of testosterone upon the sebaceous glands. Some authors consider that the response to testosterone in hypophysectomized-castrated rats is very small (LASHER et al., 1954, 1955; EBLING, 1957) but improves upon administration of somatotrophin and prolactin, which would have a permissive effect (EBLING and SKINNER, 1967; EBLING et al., 1969b; EBLING, 1974). Other authors think that the response to testosterone is relatively normal and that the decreased sebum production in the testosterone-treated hypophysectomized-castrated rats, if compared to the testosterone-treated castrated rats, is due to the lack of the independent effect of other pituitary hormones upon the sebaceous glands (NIKKARI and VALAVAARA, 1969, 1970a, b; THODY and SHUSTER, 1970, 1971a).

Other androgenic steroids increase sebum production in castrated rats: 5α-dihydrotestosterone is as potent as testosterone, while the 5α-androstanediols, androstenedione, and dehydroepiandrostenerone have a weaker effect (ARCHIBALD and SHUSTER, 1969; EBLING et al., 1971; SHUSTER and THODY, 1974). In hypophysectomized-castrated rats, 5α-androstane-3β,17β-diol, and androstenedione are active in increasing sebum production, while testosterone, dehydroepiandrosterone, 5α-androstanedione, and androsterone are weak (EBLING et al., 1971, 1973). In general, the 17β-hydroxy androgenic steroids are more active than the 17-oxo androgenic steroids in increasing sebum production in the rat (EBLING, 1974).

Administration of estrogens decreases the size of the sebaceous gland, it does not affect the number of mitoses, and it decreases sebum production (HOOKER and PFEIFFER, 1943; EBLING, 1948, 1954; EBLING and SKINNER, 1967). The effect would be to reduce intracellular synthesis of sebum without affecting the mitotic rate (EBLING, 1970). If given together with androgens to castrated animals, the estrogens do not antagonize the androgen effect upon the sebaceous glands (HOOKER and PFEIFFER, 1943; EBLING, 1957).

There is some controversy concerning the results of the administration of progesterone. According to some authors (EBLING, 1961; GROOT et al., 1965; EBLING et al., 1969a), the effect upon the sebaceous glands is slight or nil with respect to gland size, number of mitoses or sebum production, while other authors think that it is a powerful stimulus to gland size and sebum production (HASKIN et al., 1953; SHUSTER and THODY, 1974).

Several antiandrogens have been studied for their effect upon the sebaceous glands of rats, among them 17α-methyl-B nortestosterone (EBLING, 1967), 6α,6β-ethylene-17α-methyl-B nortestosterone (SAUNDERS and EBLING, 1969), Δ'-chlormadinone acetate (EBLING, 1970), and cyproterone acetate (EBLING, 1970, 1973). The antiandrogens given to normal male rats or to castrated rats receiving testosterone decrease glandular size, number of mitoses, and sebum production. They are androgen antagonists (unlike the estrogens), because they interfere with the effect of exogenous testosterone administered to castrated animals. The antiandrogens would interfere with the binding of the androgens with their receptors in the sebaceous glands so that the androgens could not produce their stimulating effects.

## B. Effects on Hamster Flank Organ

The hamster flank organ, also called scent gland or costovertebral organ, consists of a group of sebaceous glands associated with some hair follicles and many pigment cells. It is bilaterally localized on the back. There is a sexual difference, in that it is much bigger in males than in females (MONTAGNA and HAMILTON, 1949; HAMILTON and MONTAGNA, 1950).

Castration of male animals decreases the size of the flank organ to that of normal females (MONTAGNA and HAMILTON, 1949; HAMILTON and MONTAGNA, 1950; BELL, 1974), with a decrease in DNA, RNA, and protein content (GIEGEL et al., 1971) and a decrease in the activity of several enzymes, especially the glycolytic ones and also those of the pentose cycle (TAKAYASU and ADACHI, 1970).

Administration of testosterone propionate to castrated males or normal females produces hypertrophy of the flank organ if given parenterally (MONTAGNA and HAMILTON, 1949; HAMILTON and MONTAGNA, 1950; BELL, 1974) or topically (FROST et al., 1973). Administration of other androgens produces hypertrophy of the gland; 5α-dihydrotestosterone, 3α-androstanediol, and testosterone were very active, androstenedione was moderately active, and no effect was found with androsterone, dehydroepiandrosterone, and progesterone (FROST et al., 1973; LUTSKY et al., 1974). There is no correlation between the growth-promoting and the lipolytic effects upon the flank organs of these steroids; for instance, androsterone and androstenedione stimulate lipid synthesis more than dihydrotestosterone (LUTSKY et al., 1974). Administration of testosterone restores the activity of the glycolytic enzymes in 1 or 2 weeks and controls the structure and function of the gland through the regulation of the synthesis of proteins and nucleic acids (TAKAYASU and ADACHI, 1970).

Several antiandrogenic drugs, such as chlormadinone acetate and cyproterone acetate, have been used topically in male hamsters with equivocal results (BURDICK and HILL, 1970), but when used topically in female hamsters, they were able to interfere with the topical effect of testosterone (FROST and GOMEZ, 1972). New antiandrogens with a different mechanism of action, the 3-oxo-androst-4-en-17β-carboxylic acid and its methyl ester, which inhibits specifically the 5α-reductase when applied topically, inhibit the effect of testosterone but not of 5α-dihydrotestosterone, both also applied topically upon the flank organ of female hamsters (VOIGT and HSIA, 1973a; HSIA and VOIGT, 1974).

## C. Effects on Rat Preputial Glands

The preputial gland of the rat is a secondary sex characteristic, being much bigger in male than in female animals. It atrophies following castration of males, but a greater atrophy is obtained after hypophysectomy. The preputial gland of castrated males grows when the animal receives androgenic steroids, but it is not affected by estrogens. Pituitary growth hormone and testosterone are the principal hormonal factors that control its size. The preputial gland of hypophysectomized female rats has been used to assay the androgenic effects of several steroids (HUGGINS et al., 1955). Administration of antiandrogens interferes with the effect of testosterone upon the preputial glands (JONES and WOODBURY, 1965).

Pseudohermaphrodite rats of the Stanley-Grunbach strain, characterized by xy karyotype, inguinal testicles, and female phenotype, have small preputial glands the size of those of normal females. Administration of testosterone results in growth response in normal females but not in the pseudohermaphrodite rats, which only react to doses 10 times higher (BARDIN et al., 1970; SHERINS and BARDIN, 1971).

Administration of estrogens does not affect the size of the preputial gland, but decreases its secretion. The mechanism would be a decrease in testosterone uptake by the glands and in the 5α reduction to dihydrotestosterone (SANSONE-BAZZANO et al., 1972b).

Administration of testosterone to female, immature male, or castrated male mice causes marked synthesis and accumulation of lipids containing fatty alcohol moieties, such as alkyl glycerols, waxes, and alkyl acetates, in the preputial glands (SANSONE-BAZZANO et al., 1972).

## D. Effects on Sebaceous Glands of Other Mammals

Many rodents have androgen-dependent sebaceous glands. Castration of male animals produces atrophy of these glands in guinea pigs (MARTAN, 1962), mongolian gerbils (GLENN and GRAY, 1965), and prairie deer mice (BLUM, 1971).

On the other hand, topical or parenteral administration of testosterone produces hypertrophy of the sebaceous glands in rabbits (MONTAGNA and KENYON, 1949), mice (LAPIÈRE, 1953), guinea pigs (MARTAN, 1962), and mongolian gerbils (GLENN and GRAY, 1965). It increases the sebaceous gland secretion in mice (LAPIÈRE, 1953).

Administration of estrogens decreases the size of the sebaceous glands and their secretion in mice (LAPIÈRE, 1953). Different estrogenic steroids have been ranked according to their ability to produce atrophy of the sebaceous glands and to decrease sebaceous secretion in mice (BULLOUGH and LAWRENCE, 1960).

Antiandrogens such as like cyproterone acetate interfere with the effect of testosterone upon the sebaceous glands in mice (NEUMANN and ELGER, 1966) and the abdominal gland of the mongolian gerbil (NEUMANN et al., 1970).

Several races of monkeys such as the *Galago crassicaudatus* have androgen-stimulated rather than androgen-dependent sebaceous glands. The structure of these glands is not modified by castration, but administration of testosterone produces fine structural changes in their cells (BELL, 1974).

## VIII. Effects of Androgens in Human Sebaceous Glands

The human sebaceous glands are distributed all over the skin except the palms and the soles. Those of the face are bigger and more numerous. Atrophic in children, they develop to their adult structure at the time of puberty, but their development precedes menarche in girls and sexual maturation in boys. After puberty, a sexual difference in sebum production is established, being greater in adult men than in adult women (STRAUSS and POCHI, 1963a; POCHI and STRAUSS, 1974).

Castration of men produces atrophy of the sebaceous glands and markedly decreases sebum production (POCHI et al., 1962; HAMILTON and MESTLER, 1963; POCHI and STRAUSS, 1963), while no effect (STRAUSS and POCHI, 1963a) or a decrease in sebum secretion (HAMILTON and MESTLER, 1963) has been reported after ovariectomy in women.

Administration of testosterone increases glandular size and sebum production when given to children (RONY and ZAKON, 1943; STRAUSS et al., 1962), to eunuchoid or adult castrated men (HAMILTON, 1941; POCHI et al., 1962), and to normal women (STRAUSS and POCHI, 1963a). An increase in sebum secretion has been reported after administration of testosterone to normal men (JARRETT, 1959; SMITH and BRUNOT, 1961), but very careful studies have not detected any change (STRAUSS et al., 1962). This has been explained in terms of the fact that the sebaceous glands of normal men are already maximally stimulated by testosterone.

Administration of estrogens to normal men decreases markedly glandular size and sebum production (JARRETT, 1955; STRAUSS et al., 1962; STRAUSS and POCHI, 1964; PEIRONE and FENOGLIO, 1965). When given to normal women (STRAUSS and POCHI, 1963b; POCHI and STRAUSS, 1966) and to castrated men (POCHI et al., 1962; STRAUSS and POCHI, 1963a), sebum production is also decreased. Estrogens have been applied topically to normal men on one side of the forehead, but a decrease in sebum secretion was detected on both the treated and nontreated sides (STRAUSS et al., 1962). When estrogens are given together with testosterone, they do not interfere with the effect of the androgen upon the sebaceous glands (STRAUSS et al., 1962). It has been suggested that estrogens have a central effect upon sebaceous secretions, that is, depression of androgen production by the gonads, but a peripheral effect upon the sebaceous glands cannot be discarded (STRAUSS and POCHI, 1964).

The sebaceous glands of normal men, whose sebum production was inhibited by the prolonged administration of ethynyl estradiol, have been proposed as a biological model to study drugs that stimulate sebum secretion, and it has been employed to evaluate progestational and anabolic-androgenic steroids (STRAUSS and POCHI, 1963a).

Despite some reports that progesterone stimulates sebaceous glands (SMITH, 1959), the general consensus is that progesterone in clinical doses does not affect the secretion of sebum (JARRETT, 1959; STRAUSS and KLIGMAN, 1961a). No changes in sebum secretion were detected during administration of other progestational drugs, such as 17α-hydroxyprogesterone caproate (SMITH and BRUNOT, 1961; STRAUSS and KLIGMAN, 1961a), medroxyprogesterone acetate, and norethynodrel (STRAUSS and KLIGMAN, 1961a). But norethindrone acetate, a progestational compound with weak androgenic effects, produces hypertrophy of the glands and increased production of sebum (STRAUSS and KLIGMAN, 1961b).

Several anabolic-androgenic steroids have been assayed for their capacity to stimulate sebum secretion in man. Norethandrolone and methandrostenolone increase sebum secretion, while no effect was detected with fluoxymesterone (STRAUSS and POCHI, 1963a).

The adrenal androgens also participate in the regulation of sebaceous secretion. In patients with Addison's disease or after total bilateral adrenalectomy in Cushing's disease or in cancer patients, there is atrophy of the sebaceous glands and decreased sebum secretion (POCHI et al., 1963). The administration of several adrenal androgens, for example, dehydroepiandrosterone and androstenedione, increases sebum production, but the effect is weaker than that of testosterone (POCHI et al., 1963; STRAUSS and POCHI, 1968; POCHI and STRAUSS, 1969; DRUCKER et al., 1972). Androsterone does not increase sebum secretion (POCHI et al., 1963). Cortisol and cortisone do not stimulate sebum secretion, but they have a permissive effect upon it in adrenalectomized patients (POCHI et al., 1963). Since the pubertal increase in sebum secretion, called sebarche, occurs at 8–10 years of age and precedes pubertal gonadal stimulation, it has been attributed to adrenal androgens (RAMASASTRY et al., 1970).

In hirsute women with idiopathic hirsutism or the Stein-Leventhal syndrome, there is increased sebum secretion (BURTON et al., 1972).

## Androgens and Acne

The relationship between androgens and the production of acne vulgaris in human subjects has been demonstrated by HAMILTON (1941) with the following conclusions: (1) eunuchoid patients do not develop acne; (2) the acne develops when androgens rise in body fluids and tissues; (3) some eunuchoid patients, children, and castrated men or women develop acne during testosterone treatment, but this acne disappears when the treatment is withdrawn; and (4) when androgenic hormones increase (as in puberty, adrenal hirsutism, and ovarian arrhenoblastoma), sebaceous gland secretion increases, which is the background for the increased frequency of acne in these conditions.

Increased sebum secretion (seborrhea) in patients with acne vulgaris is well established (KILE et al., 1950; POCHI and STRAUSS, 1964; CUNLIFFE and SHUSTER, 1969; POWELL and BEVERIDGE, 1970; BURTON and SHUSTER, 1971; GLOOR et al., 1972). When the acne disappears, the seborrhea continues. The androgens are only related to sebum secretion; an increase in sebum secretion facilitates the appearance of acne (POCHI and STRAUSS, 1974).

In order to decrease sebum secretion, acne has been treated with estrogens (TORRE and KLUMPP, 1957) but principally with estrogen-progestin combinations (STRAUSS and POCHI, 1963b; PALITZ et al., 1964; STRAUSS and POCHI, 1964; PEIRONE and FENOGLIO, 1965; TORRE, 1965; POCHI and STRAUSS, 1967; TRONNIER and HOPPE-SEYLER, 1969; BONELLI et al., 1970). The results have been generally good: sebum production was decreased, and acne vulgaris was improved in most patients.

Antiandrogens have been used to treat acne vulgaris because they interfere with the peripheral effects of androgen upon the sebaceous glands. They decrease sebum production in normal men or in hirsute women and interfere with the effects of testosterone administered to castrated subjects. The drugs most frequently employed include: 17$\alpha$-methylnortestosterone (ZARATE et al., 1966; STRAUSS et al., 1969), chlormadinone acetate (WINKLER, 1968a; STRAUSS and POCHI, 1970), and cyproterone and cyproterone acetate (CUNNLIFE et al., 1969; APOSTOLAKIS et al., 1967; WINKLER, 1968a, b, 1971; TAMM et al., 1970; NEUMANN, 1971). The results have been good when the drugs were administered orally or parenterally and variable when topical application was tried. Sebum production was decreased and the acne improved, but the seborrhea reappearead when the treatment was stopped.

Treatment with estrogens and antiandrogens has been combined, because the effects of both types of drugs appear to be additive, in the schedule called reversed sequential therapy (NEALE et al., 1971; ROSSBERG et al., 1971; LEBEAU and VOKAER, 1972). The results appear to be better than those observed when the drugs are given separately; sebum secretion rapidly decreases and the acne lesions improve.

## IX. Metabolism of Androgens in the Sebaceous Glands

Histochemical studies have revealed the presence of hydroxysteroid dehydrogenases in human skin, which are localized in the sebaceous glands. The reports indicate high 3β-, 16β- and 17β-; low 3α- and 11β-; and traces of 6β- and 20β-hydroxysteroid dehydrogenase activities in the sebaceous glands of the face and neck (BAILLIE et al., 1965, 1966). A semiquantitative histochemical method for demonstrating 3β-hydroxysteroid dehydrogenase has been devised (MUIR et al., 1968a), and it has been shown that cyproterone or cyproterone acetate inhibits the reaction (MUIR et al., 1968b). The distribution of 3β-, 16β- and 17β-hydroxysteroid dehydrogenases has been studied in human skin and their presence shown in the sebaceous glands of scalp, perineal region, arms, legs, and abdomen, but principally in face and chest (CALMAN et al., 1970). The hydroxysteroid dehydrogenases have also been demonstrated in the sebaceous glands of rats, guinea pigs, hamsters, and gerbils, in the rat preputial glands, the hamster costovertebral organs, and the gerbil ventral glands (MUIR et al., 1970).

The metabolism of dehydroepiandrosterone, androstenedione, and testosterone has been studied in rat preputial glands (WILSON and GLOYNA, 1970; RICHARDSON and AXELROD, 1971; HODGINS and HAY, 1973), in hamster costovertebral organs (TAKAYASU and ADACHI, 1972a), in the ventral sebaceous gland on the mongolian gerbil (HODGINS, 1973), and the chin gland of the male cuis (HOLT and TAM, 1973). The principal metabolites of testosterone are 5α-dihydrotestosterone, 5α-androstane-3β,17β-diol, 5α-androstane-3α,17α-diol, 5α-androsterone, 5α-androstane-3,17-dione, and androstenedione, and the principal enzymes involved are 3α-, 3β-, and 17β-hydroxysteroid dehydrogenases, 5α-reductase, and sulphotransferase (RICHARDSON and AXELROD, 1971; HODGINS and HAY, 1973; HODGINS, 1973; LUTSKY et al., 1974). The conversion rate of testosterone to 5α-dihydrotestosterone in rat preputial glands is comparable to that in prostate and seminal vesicles and higher than in the skin (WILSON and GLOYNA, 1970; HODGINS and HAY, 1973). The skin of patients with acne vulgaris converted testosterone to 5α-dihydrotestosterone 2 to 20 times faster than that of nonacne subjects (SANSONE-BAZZANO and REISNER, 1971).

In the preputial gland of castrated mice, pentose phosphate shunt activity is tripled after testosterone administration (SANSONE-BAZZANO et al., 1971). The pentose phosphate shunt, especially the glucose-6-phosphate dehydrogenase enzyme, is a major source of NADPH. In the flank organ of hamsters, it has been shown that the limiting factor in the conversion from testosterone to 5α-dihydrotestosterone is the endogenous level of NADPH (TAKAYASU and ADACHI, 1972a). The steroid pathways in rat preputial glands have been discussed and the various steroid interconversions mediated by 5α-reductase and 17β-hydroxysteroid dehydrogenase analysed. It has been suggested that the effect of testosterone upon the sebaceous glands is mediated through an increase in the NADPH intracellular level (SANSONE-BAZZANO and REISNER, 1974).

The receptor proteins for androgens have been studied in hamster flank organs. The cytosol receptor protein (7–8.5) binds dihydrotestosterone, which is subsequently transferred to the nuclear receptor protein (3–4 S). Antiandrogens inhibit the binding of dihydrotestosterone to the receptor proteins. Both cytosol and nuclear receptor proteins decrease rapidly after castration, but reactivation by testosterone in vitro rapidly initiates new protein synthesis (ADACHI and KANO, 1972; ADACHI, 1974).

In human patients with testicular feminization syndrome, the sebaceous glands produce low sebum levels, which do not increase after testosterone administration (GWINUP et al., 1966). In the pseudohermaphrodite rats of the Stanley-Grunbach strain, the preputial glands are small like those of normal females and do not respond to testosterone (BARDIN et al., 1970; SHERINS and BARDIN, 1971). Testosterone metabolism is normal in these preputial glands; the defect resides in the androgen receptor in the cytosol, which does not react with testosterone or dihydrotestosterone (BULLOCK and BARDIN, 1972).

## References


ADACHI, K.: Receptor proteins for androgen in hamster sebaceous glands. J. invest. Derm. **62**, 217—223 (1974).

ADACHI, K., KANO, M.: Adenil cyclase in human hair follicles: its inhibition by dihydrotestosterone. Biochem. biophys. Res. Commun. **41**, 884—890 (1970).

ADACHI, K., KANO, M.: The role of receptor proteins in controlling androgen action in the sebaceous glands of hamsters. Steroids **19**, 567—574 (1972).

ADACHI, K., UNO, H.: Glucose metabolism of growing and resting human hair follicles. Amer. J. Physiol. **215**, 1234—1239 (1968).

ALBRIEUX, A. S., MUSSIO-FOURNIER, J. C.: The local action of testosterone propionate on the development of axillary hair in man. J. clin. Endocr. Metab. **9**, 1434—1436 (1949).

ALBRIGHT, F., SMITH, P. H., FRASER, R.: Syndrome characterized by primary ovarian insufficiency and decreased stature: report of 11 cases with disgression on hormonal control of axillary and pubic hair. Amer. J. med. Sci. **204**, 625—648 (1942).

ALLALOUF, D. A., BER, A.: Effect of percutaneous application of testosterone propionate and of estradiol benzoate on the total acid mucopolysaccharides of rat skin. Endocrinology **69**, 210—216 (1961).

ANDREASEN, E.: Influence of cortisone on hair succession in the albino mouse. Acta path. microbiol. scand. **44**, 392—394 (1958).

APOSTOLAKIS, M., TAMM, J., VOIGT, K. D.: Biochemical and clinical studies with cyproterone. Acta endocr. (Kbh.) **56** (Suppl. 199), 56 (1967).

ARCHIBALD, A., SHUSTER, S.: The bioassay of androgen and antiandrogens using sebum secretion in the rat. Proc. roy. Soc. Med. (Lond.) **62**, 887—888 (1969).

ARCHIBALD, A., SHUSTER, S.: The measurement of the sebum secretion in the rat. Brit. J. Derm. **82**, 146—151 (1970).

ARNAL, M., DRESCH, C., PRADER, A.: Etude de 8 cas de pilosité sexuelle précoce isolée ou « premature pubarche». Helv. paediat. Acta **16**, 186—197 (1961).

BAILLIE, A. H., CALMAN, K. C., MILNE, J. A.: Histochemical distribution of hydroxysteroid dehydrogenases in human skin. Brit. J. Derm. **77**, 610—616 (1965).

BAILLIE, A. H., THOMSON, J., MILNE, J. A.: The distribution of hydroxysteroid dehydrogenase in human sebaceous glands. Brit. J. Derm. **78**, 451—457 (1966).

BAKER, B. L.: The relationship of the adrenal, thyroid and pituitary glands to the growth of hair. Ann. N.Y. Acad. Sci. **53**, 690—707 (1951).

BAKER, B. L., WHITAKER, W. L.: Growth inhibition in the skin following direct application of adrenal cortical preparations. Anat. Rec. **102**, 333—347 (1948).

BAKER, B. L., WHITAKER, W. L.: Relationship of the adrenal cortex to inhibition of the hair growth by estrogen. Amer. J. Physiol. **159**, 118—123 (1949).

BARBERIS, M. L., AMBROGGIO, G.: Sull trattamento delle soluzioni di continuo della cute col 4-clorotestosterone acetato neomicin solfato. Minerva med. (Torino) **61**, 5241—5246 (1970).

BARDIN, C. W., BULLOCK, L., SCHNEIDER, G., ALLISON, J. E., STANLEY, A. J.: Pseudohermaphrodite rat: End organ insensitivity to testosterone. Science **167**, 1136—1137 (1970).

BARDIN, C. W., HEMBREE, W. C., LIPSETT, M. B.: Suppression of testosterone and androstenedione production rates with dexamethasone in women with idiopathic hirsutism and polycystic ovaries. J. clin. Endocr. **28**, 1300—1306 (1968).

BARDIN, C. W., LIPSETT, M. B.: Testosterone and androstenedione blood production rates in normal women and women with idiopathic hirsutism or polycystic ovaries. J. clin. Invest. **46**, 891—902 (1967).

BARTHE, P. L., BULLOCK, L. P., MOWSZOWICZ, I., BARDIN, C. W., ORTH, O. N.: Submaxillary gland epidermal growth factor: A sensitive index of biologic androgen activity. Endocrinology **95**, 1019—1025 (1974).

BAVETTA, L. A., BEKHOR, I., NIMNI, M. E.: Effects of hormone administration on collagen biosynthesis in the rat. Proc. Soc. exp. Biol. (N.Y.) **110**, 294—297 (1962).

BELL, M.: Effect of androgen on the ultrastructure of the sebaceous glands in two species. J. Invest. Dermat. **62**, 202—210 (1974).

BERCOVICI, J. P., MAUVAIS-JARVIS, P.: Physiologie et pathologie du récepteur pilo-sébacé. Presse méd. Paris **78**, 2229—2232 (1970).

BERLINER, D. L., PASQUALINI, J. R., GALLEGOS, A. J.: The formation of water soluble steroids by human skin. J. invest. Derm. **50**, 220—224 (1968).

BLACK, M. M., SHUSTER, S., BOTTOMS, E.: Osteoporosis, skin collagen and androgen. Brit. med. J. **1970 IV**, 773—774.

BLUM, S. L.: Atrophy of the ventral gland of the male deermouse after castration. J. Endocr. **49**, 695—696 (1971).

BONELLI, M., ALESSI, E., TOMASINI, C., PICCININI, S.: Ulteriores observaciones sobre el tratamiento del acné vulgar y el comportamiento de los lípidos de superficie. Sem. méd. (B. Aires) **136**, 1052—1063 (1970).

BORUM, K.: Hair pattern and hair succession in the albino mouse. Acta path. microbiol. scand. **34**, 521—541 (1953).

BOSSE, K.: Kontrazeptiva und Haarwachstum unter tierexperimentellen Aspekten am Modell der Trächtigkeit. Arch. derm. Forsch. **244**, 507—510 (1972).

BOSSE, K., FEHR, R., FRANZ, E., WIONTZEK, H. J.: Die Wirkung von Dehydroepiandrosteron-Sulfat auf das Haarwachstum der Maus und des Meerschweinchens. Arch. derm. Forsch. **243**, 177—187 (1972).

BRÄUER, R.: Ergebnisse der Klinischen Prüfung eines neuen Dermatotherapeutikers. Med. Welt **26**, 1087—1090 (1971).

BULLOCK, L. P., BARDIN, C. W.: Androgen receptors in testicular feminization. J. clin. Endocr. **35**, 935—937 (1972).

BULLOUGH, W. S., HOMAN, J. D. H., LAURENCE, E., OVERBEEK, G. A.: Prevention by nandrolone decanoate of dexamethasone induced mitotic inhibition in mouse ear epidermis in vitro. J. Endocr. **41**, 453—454 (1968).

BULLOUGH, W. S., LAWRENCE, E. B.: Experimental sebaceous gland suppression in the adult male mouse. J. clin. Invest. **35**, 37—42 (1960).

BURDICK, K. H., HILL, R.: The topical effect of the antiandrogen chlormadinone acetate and some of its chemical modifications on the hamster costovertebral organ. Brit. J. Derm. **82** (Suppl. 6), 19—25 (1970).

BURTON, J. L., JOHNSON, C., LIBMAN, L., SHUSTER, S.: Skin virilism in women with hirsutism. J. Endocr. **53**, 349—354 (1972).
BURTON, J. L., SHUSTER, R.: The relationship between seborrhea and acne vulgaris. Brit. J. Derm. **84**, 600—601 (1971).
BUTCHER, E. O.: The hair cycles in the albino rat. Anat. Rec. **61**, 5—19 (1934).
BUTCHER, E. O.: Hair growth in adrenalectomized and adrenalectomized thyroxin-treated rats. Amer. J. Physiol. **120**, 427—434 (1937).
BUTCHER, E. O.: Development of the piliary system and the replacement of hair in mammals. Ann. N.Y. Acad. Sci. **53**, 508—516 (1951).
BYNNY, R. L., ORTH, D. N., COHEN, S., DOYNE, E. S.: Epidermal growth factor. Effects of androgens and androgenic agents. Endocrinology **95**, 776—782 (1974).
CACCIALANZA, P., TOMMASINI, M.: Gli steroidi anabolizzanti sulla cicatrizzazione delle ferite cutanee sperimentali nell' uomo. Minerva med. (Torino) **52**, 372—374 (1961).
CALMAN, K. C., MUIR, A. V., MILNE, J. A., YOUNG, H.: Survey of the distribution of steroid dehydrogenases in sebaceous glands of human skin. Brit. J. Derm. **82**, 567—571 (1970).
CERIMELE, D., GIANETTI, A., BRUNI, L., PAZZAGLIA, A.: Sex hormones and cutaneous connective tissue. G. ital. Derm. **46**, 55—57 (1971).
CHAKRABORTY, J., THOMSON, J., MC SWEEN, M. P., MUIR, A. V., CALMAN, K. C., GRANT, J. K., MILNE, J. A.: The in vitro metabolism of 7-alpha-$^{3}$H-dehydroepiandrosterone by human skin. Brit. J. Derm. **83**, 477—482 (1970).
CHASE, H. B., EATON, G. J.: The growth of hair follicles in waves. Ann. N.Y. Acad. Sci. **83**, 365—368 (1959).
CHASE, H. B., RAUCH, H., SMITH, V. W.: Critical stages of hair development and pigmentation in the mouse. Physiol. Zool. **24**, 1—7 (1951).
COFFIN, D. L., MUNSON, T. D., SCULLY, R. E.: Endocrine diseases of the dog associated with hair loss, Sertoli cell tumor of testis, hypothyroidism, canine Cushing's syndrome. J. Amer. vet. med. Ass. **133**, 402—408 (1953).
COOPER, Z. K.: The relation of the endocrine glands to the growth and distribution of hair—A review of the literature. Arch. Derm. Syph. **21**, 1007—1029 (1930).
CORVOL, P., MAHOUDEAU, J. A.: Le transport plasmatique des androgènes et ses conséquences physio-pathologiques. Presse méd. (Paris) **79**, 1327—1330 (1971).
CROVATO, F., MORETTI, G., GIRARD, A.: Dehydroepiandrosterone and glucose-6-phosphate dehydrogenases in the hair cycle of the rat. Arch. derm. Forsch. **240**, 55—60 (1970).
CUNLIFFE, W. J., SHUSTER, S.: Pathogenesis of acne. Lancet **1**, 685—687 (1969).
CUNLIFFE, W. J., SHUSTER, S., CASSELS-SMITH, A. J.: The effect of topical cyproterone acetate on sebum secretion in patients with acne. Brit. J. Derm. **81**, 200—201 (1969).
DAVIDSON, E., SMALL, W.: Metabolism in vivo of connective tissue mucopolysaccharides. II. Chondroitin sulphate B and hyaluronic acid of skin. Biochim. biophys. Acta (Amst.) **69**, 453—458 (1963).
DAVIS, B. K.: Quantitative morphological studies upon the influence of the endocrine system on the growth of hair by white mice. Acta endocr. (Kbh.) **44** (Suppl. 85), 9—102 (1963).
DAVIS, R. P., RAMPINI, E., HSIA, S. L.: 17$\beta$-hydroxysteroid dehydrogenase of rat skin. Substrate specificity and inhibitors. J. biol. Chem. **247**, 1407—1413 (1972).
DE BRUX, J., DU BOISTESSELIN, R.: Modifications histologiques et histoquimiques du tissu conjonctif sous l'influence de différents actions hormonales. Presse méd. (Paris) **61**, 600—602 (1953).
DENNENBAUM, R., HOFFMANN, G., OERTEL, G. W., MORSCHES, B., HOLZMANN, H.: Steroid and skin. XI. C-19 steroids in skin eluates from psoriatics. Horm. Metab. Res. **4**, 383—385 (1972).
DIEKE, S.: The effect of removing various endocrine glands on the hair cycles of black rats. Endocrinology **42**, 315—319 (1948).
DILLAHA, C. J., ROTHMAN, S.: Therapeutic experiments in alopecia areata with orally administered cortisone. J. Amer. med. Ass. **150**, 546—550 (1952).
DOYLE, J., SZIRMAY, J. A., DE TYSSONK, B. H.: Connective tissue changes in the rooster comb during regression. Acta endocr. (Kbh.) **45**, 457—475 (1964).
DRUCKER, W. D., BLUMBERG, J. M., GANDY, H. M., DAVID, R. R., VERDE, A. L.: Biologic activity of dehydroepiandrosterone sulfate in man. J. clin. Endocr. **35**, 48—54 (1972).
DRY, F. W.: The coat of the mouse, Mus musculus. J. Genet. **16**, 287—324 (1925).

DUBE, J. Y., NGO-THI, N. H., TREMBLAY, R. R.: Testosterone 5α-reductase in rat skin homogenates: measurement in the skin at various anatomical sites in males and females. Endocrinology **96**, 235—237 (1975).

DURWARD, A., RUDALL, K. M.: Studies on hair growth in the rat. J. Anat. (Lond.) **83**, 325—335 (1949).

EBLING, F. J.: Sebaceous glands. I. The effect of sex hormones on the sebaceous glands of the female albino rat. J. Endocr. **5**, 297—302 (1948).

EBLING, F. J.: Changes in the sebaceous glands and epidermis during the oestrous cycle of the albino rat. J. Endocr. **10**, 147—154 (1954).

EBLING, F. J.: Endocrine factors affecting cell replacement and cell loss in the epidermis and sebaceous glands of female albino rats. J. Endocr. **12**, 38—49 (1955).

EBLING, F. J.: The action of testosterone on the sebaceous glands and epidermis in castrated and hypophysectomized male rats. J. Endocr. **15**, 297—306 (1957).

EBLING, F. J.: Failure of progesterone to enlarge sebaceous glands in the female rat. Brit. J. Derm. **73**, 65—68 (1961).

EBLING, F. J.: The action of an anti-androgenic steroid, 17α-methyl-B-nortestosterone on sebum secretion in rats treated with testosterone. J. Endocr. **38**, 181—185 (1967).

EBLING, F. J.: Factors influencing the response of the sebaceous glands of the rat to androgen. Brit. J. Derm. **82** (Suppl. 6), 9—14 (1970).

EBLING, F. J.: The effects of cyproterone acetate and oestradiol upon testosterone stimulated sebaceous activity in the rat. Acta endocr. (Kbh.) **72**, 361—365 (1973).

EBLING, F. J.: Hormonal control and methods of measuring sebaceous gland activity. J. invest. Derm. **62**, 161—171 (1974).

EBLING, F. J., EBLING, E., MC CAFFERY, V., SKINNER, J.: The response of the sebaceous glands of the hypophysectomized-castrated male rat to 5-dihydrotestosterone, androstenedione, dehydroepiandrosterone and androsterone. J. Endocr. **51**, 181—190 (1971).

EBLING, F. J., EBLING, E., MC CAFFERY, V., SKINNER, J.: The responses of the sebaceous glands of the hypophysectomized-castrated male rat to 5-androstanedione and 5α-androstane-3β,17β-diol. J. invest. Derm. **60**, 183—187 (1973).

EBLING, F. J., EBLING, E., SKINNER, J.: The influence of pituitary hormones on the response of the sebaceous glands of the rat to testosterone. J. Endocr. **45**, 245—256 (1969a).

EBLING, F. J., EBLING, E., SKINNER, J.: The influence of pituitary on the response of the sebaceous and preputial glands of the rat to progesterone. J. Endocr. **45**, 257—263 (1969b).

EBLING, F. J., HALE, P. A.: The composition of female rat skin in relation to region, age, hair growth cycle and hormones. J. Endocr. **36**, 177—201 (1966).

EBLING, F. J., JOHNSON, E.: The action of hormones on spontaneous hair growth cycles in the rat. J. Endocr. **29**, 193—201 (1964).

EBLING, F. J., SKINNER, J.: The measurement of sebum production in rats treated with testosterone and estradiol. Brit. J. Derm. **79**, 386—392 (1967).

ECKERT, J., CHURCH, R. E., EBLING, F. J., MUNRO, D. S.: Hair loss in women. Brit. J. Derm. **79**, 543—548 (1967).

EMMENS, C. W.: The endocrine system and hair growth in the rat. J. Endocr. **3**, 64—78 (1942).

ENGELHARDT, D., KARL, H. J.: Metabolism of testosterone by human skin with reference to sex chromosome anomalies. Acta endocr. (Kbh.) **66** (Suppl. 152), 12 (1971).

EPPENBERGER, U., HSIA, S. L.: Binding of steroid hormones by the 105000 × g supernatant fraction from homogenates of rat skin and variations during the hair cycle. J. biol. Chem. **247**, 5463—5469 (1972).

EPPER, C. E., HOUSSAY, A. B., SANCHO, O.: Effects of X-rays in different phases of the hair cycles after castration in mice. Acta physiol. latinoamer. **21**, 303—309 (1971).

ESPOSITO, C.: Contributo clinico-statistico sull'efficacia di una preparazione topica cutanea ad azione anabolizante e antibiotica. Rass. int. Clin. Ter. **50**, 491—497 (1970).

ETTINGER, B., WERDER, K., THENAERS, G. C., FORSHAM, P. H.: Plasma testosterone stimulation-suppression dynamics in hirsute women. Amer. J. Med. **51**, 170—175 (1971).

FAREDIN, I., FAZEKAS, A. G., TOTH, I., KOKAI, K., JULESZ, M.: Transformation in vitro of (4-$^{14}$C)-dehydroepiandrosterone into 7-oxygenated derivatives by normal human male and female skin tissue. J. invest. Derm. **52**, 357—361 (1969).

FAZEKAS, A. G.: Metabolism of dehydroepiandrosterone by isolated human hair follicles. Acta endocr. (Kbh.) **67** (Suppl. 155), 119 (1971).

FAZEKAS, A. G., LANTHIER, A.: Metabolism of androgens by isolated human beard hair follicles. Steroids **18**, 367—379 (1971).
FAZEKAS, A. G., SANDOR, T.: Metabolism of androgens by isolated human hair follicles. J. Steroid Biochem. **3**, 485—491 (1972).
FAZEKAS, A. G., SANDOR, T.: The metabolism of dehydroepiandrosterone by human scalp follicles. J. clin. Endocr. **36**, 582—586 (1973).
FINKELSTEIN, M., FORCHIELLI, E., DORFMAN, R. I.: Estimation of testosterone in human plasma. J. clin. Endocr. **21**, 98—101 (1961).
FLAMIGNI, C., COLLINS, W. P., KOULLAPNIS, E. N., CRAFT, I., DEWHURST, C. J., SOMMERVILLE, I. F.: Androgen metabolism in human skin. J. clin. Endocr. **32**, 737—743 (1971).
FORBES, T. R.: Sex hormones and hair changes in rats. Endocrinology **30**, 465—468 (1942).
FRENCH, F. S., VAN WYK, J. J., BAGETT, B., EASTERLING, W. E., TALBERT, L. M., JOHNSTON, F. R., FORCHIELLI, E., DEY, A. C.: Further evidence of a target organ defect in the syndrome of testicular feminization. J. clin. Endocr. **26**, 493—503 (1966).
FROST, P., GIEGEL, J. L., WEINSTEIN, G. D., GOMEZ, E. C.: Biodynamic studies of hamsters flank organ growth: hormonal influences. J. invest. Derm. **61**, 159—167 (1973).
FROST, P., GOMEZ, E. C.: Inhibitors of sex hormones: development of experimental models. In: MONTAGNA, W., VAN SCOTT, E. J., STOUGHTON, R. B. (Eds.): Advances in biology of skin, pp. 403—420. New York: Appleton-Cenury-Crofts 1972.
FUKUYAMA, K., BAKER, B. L.: The comparative potency of hydrocortisone analogs in suppressing growth of hair in the rat. J. invest. Derm. **31**, 327—330 (1958).
GABRILOVE, J. L., SCHWARTZ, A., CHURG, J.: Effects of hormones on the skin in endocrinologic disorders. J. clin. Endocr. **22**, 688—692 (1962).
GALLEGOS, A. J., BERLINER, D. L.: Transformation and conjugation of dehydroepiandrosterone by human skin. J. clin. Endocr. **27**, 1214—1218 (1967).
GARDNER, W. U., DE VITA, J.: Inhibition of hair growth in dogs receiving estrogens. Yale J. Biol. Med. **13**, 213—215 (1940).
GARN, S. M.: Types and distribution of the hair in man. Ann. N. Y. Acad. Sci. **53**, 498—507 (1951).
GIEGEL, J. L., STOLFI, L. M., WEINSTEIN, G. D., FROST, P.: Androgenic regulation of nucleic acid and protein synthesis in the hamster flank organs and other tissues. Endocrinology **89**, 904—909 (1971).
GILLESPIE, A.: Testosterone metabolism in skin of 46 xy phenotypic females. Gynec. Invest. **2**, 297—299 (1971).
GLENN, E. M., GRAY, J.: Effect of various hormones on the growth and histology of the gerbil (Meriones unguiculatus) abdominal sebaceous gland pad. Endocrinology **76**, 1115—1123 (1965).
GLOOR, M., GRAUMANN, U., KIONKE, M., WIEGAND, I., FRIEDERICH, H. C.: Menge und Zusammensetzung der Hautoberflächenlipide bei Patienten mit Acne vulgaris und gesunden Vergleichspersonen. I. Mitteilung. Arch. derm. Forsch. **242**, 316—322 (1972).
GOMEZ, E. C., HSIA, S. L.: In vitro metabolism of testosterone 4-$^{14}$C and delta-4-androstene-3,17-dione-4-$^{14}$C in human skin. Biochemistry **7**, 24—32 (1968).
GOMEZ, E. C., HSIA, S. L., FROST, P.: Cutaneous transformation of testosterone into 5-androstane-3$\beta$,17$\beta$-diol. J. clin. Endocr. **34**, 417—419 (1972).
GROOT, C. A. DE, LELY, M. A. V. D., KOOIJ, R.: The effect of progesterone on the sebaceous glands of the rat. Brit. J. Derm. **77**, 617—621 (1965).
GUILLEMANT, S., CORVOL, P., CREPY, O., MAUVAIS-JARVIS, P.: Action du cyprotérone sur la liaison plasmatique et le metabolisme de la testosterone. C.R. Acad. Sci. (Paris) **268**, 2595—2598 (1969).
GWINUP, G., WIELAND, R. G., BESCH, P. K., HANWI, G. J.: Studies on the mechanism of production of the testicular feminization syndrome. Amer. J. Med. **41**, 448—452 (1966).
HAMILTON, J. B.: Male hormone substance; a prime factor in acne. J. clin. Endocr. **1**, 570—592 (1941).
HAMILTON, J. B.: Male hormone stimulation is prerequisite and an incitant to common baldness. Amer. J. Anat. **71**, 451—480 (1942).
HAMILTON, J. B.: Patterned loss of hair in man: types and incidence. Ann. N.Y. Acad. Sci. **53**, 708—728 (1951 a).
HAMILTON, J. B.: Quantitative measurement of a secondary sex character, axillary hair growth. Ann. N.Y. Acad. Sci. **53**, 585—599 (1951 b).

Hamilton, J. B.: Effect of castration in adolescent and young adult males upon further changes in the proportions of bare and hairy scalp. J. clin. Endocr. **20**, 1309—1318 (1960).

Hamilton, J. B., Mestler, G. E.: Low values for sebum in eunuchs and oophorectomized women. Proc. Soc. exp. Biol. (N.Y.) **112**, 374—378 (1963).

Hamilton, J. B., Montagna, W.: The sebaceous gland of the hamster. I. Morphological effects of androgens on integumentary structures. Amer. J. Anat. **86**, 191—233 (1950).

Hammerstein, J., Cupceancu, B.: Behandlung des Hirsutismus mit Cyproteronacetat. Dtsch. med. Wschr. **94**, 829—834 (1969).

Haskin, D., Lasher, N., Rothman, S.: Some effects of ACTH, cortisone, progesterone and testosterone on sebaceous glands in the white rat. J. invest. Derm. **20**, 207—211 (1953).

Hay, J. B., Hodgins, M. B.: The androgen metabolism of human skin. J. Endocr. **51**, XXIII—XXIV (1971).

Hay, J. B., Hodgins, M. B.: Metabolism of androgens in vitro by human facial and axillary skin. J. Endocr. **59**, 475—486 (1973).

Heite, H. J., Cardona, F.: Prüfung einer Kombinations-Salbe aus anabolen und katabolen Lokal-Steroiden an der experimentellen Akanthose des Meerschweinchens. Arzneimittel-Forsch. **23**, 1473—1477 (1973).

Heite, H. J., Irro, H.: Ablauf der Wundheilung unter Anwendung anaboler und kataboler Steroide an Kaninchenohr-Modell nach Heite. Arzneimittel-Forsch. **23**, 1341—1346 (1973).

Heite, H. J., Meyer, J.: Experimentelle Prüfung einer Kombinationssalbe aus anabolen und katabolen Localsteroiden anhand des Pyrifer-Erythems der menschlichen Haut. Arzneimittel-Forsch. **23**, 1602—1605 (1973).

Herrick, E. H.: Tensile strength of tissues as influenced by male sex hormone. Anat. Rec. **73**, 145—149 (1945).

Hodgins, M. B.: In vitro metabolism of dehydroepiandrosterone and dehydroepiandrosterone sulphate in breast skin of women. Steroids **18**, 11—23 (1971).

Hodgins, M. B.: Androgen metabolism in the ventral sebaceous gland patch of the Mongolian gerbil. J. Endocr. **59**, 487—493 (1973).

Hodgins, M. B., Hay, H. B.: The metabolism of androgens in rat preputial glands. Steroids **21**, 307—322 (1973).

Holt, W. V., Tam, W. H.: Steroid metabolism by the chin gland of the male cuis, Galea musteloides. J. Reprod. Fertil. **33**, 53—59 (1973).

Holzmann, H., Korting, G. W., Kobelt, D., Vogel, H. G.: Prüfung der mechanischen Eigenschaften von menschlicher Haut in Abhängigkeit von Alter und Geschlecht. Arch. klin. exp. Derm. **239**, 355—367 (1971).

Hooker, C. W., Pfeiffer, C. A.: Effects of sex hormones upon body growth, skin, hair and sebaceous glands in the rat. Endocrinology **32**, 69—76 (1943).

Houssay, A. B.: Umbral de acción de los estrógenos sobre el crecimiento del pelo en ratas y ratones. Rev. Soc. argent. Biol. **29**, 221—227 (1953a).

Houssay, A. B.: The relationship of the gonads and adrenals to the growth of hair in mice and rats. Acta physiol. latinoamer. **3**, 232—246 (1953b).

Houssay, A. B.: Hormonas y crecimiento del pelo. Cienc. Invest. **10**, 291—308 (1954).

Houssay, A. B.: Hormonal regulation of growth waves of the hair follicles. In: Cori, C. E., Foglia, V. G., Leloir, L. F., Ochoa, S. (Eds.): Perspectives in Biology, pp. 86—93. Amsterdam: Elsevier Publ. Co. 1962.

Houssay, A. B., Epper, C. E., Pazo, J. H.: Neurohormonal regulation of the hair cycles in rats and mice. In: Lyne, A. G., Short, B. F. (Eds.): The Biology of the Skin and Hair Growth, pp. 641—654. Sidney: Argus and Robertson Ltd. 1965.

Houssay, A. B., Gamper, C. R.: Comparison between the inhibitory effect of several anabolic steroids upon hair growth in mice. Unpublished data (1974).

Houssay, A. B., Higgins, G. M.: The effect of gonadectomy on the growth of hair in mice. Proc. Mayo Clin. **24**, 269—275 (1949a).

Houssay, A. B., Higgins, G. M.: Hormonal influences upon the rate of growth of hair in gonadectomized mice. Proc. Mayo Clin. **24**, 597—604 (1949b).

Houssay, A. B., Higgins, G. M.: The growth of hair in gonadectomized mice bearing adrenal gland tumors. Proc. Mayo Clin. **26**, 323—326 (1951).

Houssay, A. B., Nallar, R., Saurer, E. I.: Action of testosterone propionate upon the growth of hair in rats. Acta physiol. latinoamer. **9**, 35—49 (1959).

HOUSSAY, A. B., PENHOS, J. C., FOGLIA, V. G.: Hipófisis y crecimiento del pelo en la rata blanca. Rev. Soc. argent. Biol. **31**, 7—12 (1955).

HSIA, S. L., VOIGT, W.: Inhibition of dihydrotestosterone formation: an effective means of blocking androgen action in hamster sebaceous gland. J. invest. Derm. **62**, 224—227 (1974).

HUGGINS, C., PARSONS, F. M., JENSEN, C. V.: Promotion of growth of preputial glands by steroids and the pituitary growth hormone. Endocrinology **57**, 25—32 (1955).

HUGGINS, C. A., SCOTT, W. W.: Bilateral adrenalectomy in prostatic cancer. Clinical features and urinary excretion of 17-ketosteroids and estrogen. Ann. Surg. **122**, 1031—1041 (1945).

HUNDLEY, J. M., ING, R. B.: Effect of adrenalectomy and hypophysectomy on achromotrichia in copper deficient rats. Endocrinology **48**, 482—484 (1951).

INGLE, D. J., BAKER, B. L.: The inhibition of hair growth by estrogens as related to adrenal cortical function in the male rat. Endocrinology **48**, 764—771 (1951).

ISMAIL, A. A., DAVIDSON, D. W., SOUKA, A. R., BARNES, E. W., IRVINE, W. J., KILIMNIK, H., VANDERBECKEN, Y.: The evaluation of the role of androgens in hirsutism and the use of a new antiandrogen "cyproterone acetate" for therapy. J. clin. Endocr. Metab. **39**, 81—95 (1974).

JAILER, J. W.: Virilism. Bull. N.Y. Acad. Med. **29**, 377—394 (1953).

JARRETT, A.: The effect of stilbestrol on the surface sebum and upon acne vulgaris. Brit. J. Derm. **67**, 166—179 (1955).

JARRETT, A.: The effects of progesterone and testosterone on surface sebum and acne vulgaris. Brit. J. Derm. **71**, 102—116 (1959).

JEFFERIES, W. MCK., LEVY, R. P.: Treatment of ovarian dysfunction with small doses of cortisone and hydrocortisone. J. Clin. Endocr. **19**, 1069—1080 (1959).

JENKINS, J. S., ASH, S.: The metabolism of testosterone by skin in normal subjects and in testicular feminization. J. Endocr. **49**, 512—520 (1971).

JENKINS, J. S., ASH, S.: The metabolism of testosterone by human skin in disorders of hair growth. J. Endocr. **59**, 345—351 (1973).

JOHNSON, E.: Quantitative studies of hair growth in the albino rat. I. Normal males and females. J. Endocr. **16**, 337—350 (1958a).

JOHNSON, E.: Quantitative studies of hair growth in the albino rat. II. The effects of gonadal hormones. J. Endocr. **16**, 351—359 (1958b).

JONES, E. L., WOODBURY, L.: The effect of antiandrogens on the response of rat preputial glands to testosterone. J. invest. Derm. **43**, 165 (1965).

JULESZ, M., FAREDIN, I., TOTH, I.: Steroids in human skin. Budapest: Akademiai Kiado 1971.

KAISER, E., SCHMIDT-ELMENDORFF, H., ORSSICH, P.: Der Einfluß eines Antiandrogens auf die Steroidausscheidung bei Frauen mit Hirsutismus. Arch. Gynaek. **204**, 135—136 (1967).

KELCH, R. P., LINDHOLM, U. B., JAFFE, R. B.: Testosterone metabolism in target tissues. 2. Human fetal and adult reproductive tissues, perineal skin and skeletal muscle. J. Clin. Endocr. **32**, 449—456 (1971).

KELLER, M., HAUSER, A., WALSER, A.: Virilizing adrenal adenoma associated with increased urinary excretion of estrogens and 17-ketosteroids. Isolation of steroids from the tumor. J. clin. Endocr. **18**, 1384—1398 (1958).

KEPLER, E. J., PETERS, G. A., MASON, H. L.: Addison's disease associated with pubic and axillary alopecia and normal menses. J. clin. Endocr. **3**, 497—499 (1943).

KILE, R. L., SNYDER, F. H., HAEFELE, J. W.: Nature of skin lipids in acne. Arch. Derm. **61**, 792—799 (1950).

KIM, H. M., HEERMANN, W. L.: In vitro metabolism of dehydroepiandrosterone sulphate in foreskin, abdominal skin and vaginal mucosa. J. clin. Endocr. **28**, 187—191 (1968).

KIRSCHNER, M. A., BARDIN, C. W.: Androgen production and metabolism in normal and virilized women. Metabolism **21**, 667—688 (1972).

KIRSCHNER, M. A., BARDIN, C. W., HEMBREE, W. C., ROSS, G. T.: Effect of estrogen administration on androgen production and plasma luteinizing hormone in hirsute women. J. Clin. Endocr. **30**, 727—732 (1970).

KLIGMAN, A. M.: Pathologic dynamics of human hair loss. I. Telogen Effluvium. Arch. Derm. **83**, 175—198 (1961).

KOCHAKIAN, C. D.: The tolerance of male and female mice respectively to estrogens and androgens. Endocrinology **26**, 54—60 (1940).

KOFOED, J. A., BOZZINI, C. E., TOCCI, A. A.: The concentration and synthesis of glycosaminoglycans in the skin and tracheal cartilage of normal and orchidectomized rats injected with testosterone. Acta endocr. (Kbh.) **63**, 193—198 (1970).

KOFOED, J. A., HOUSSAY, A. B.: The effect of anabolic steroids upon skin glycosaminoglycans in castrated male $C_3H/Ep$ mice. Unpublished data, 1974.

KOWALEWSKI, K.: Effect of an anabolic steroid upon various fractions of tissue hydroxyproline in cortisone treated rats. Acta endocr. (Kbh.) **53**, 73—78 (1966).

LÄUPPI, E., STUDER, A.: Inhibition of experimental, testosterone induced epidermis proliferation in the female rat by a synthetic phenanthrene derivative with antiandrogenic effect. Experientia (Basel) **15**, 264—265 (1959).

LANTHIER, A.: Urinary 17-ketosteroids in the syndrome of polycystic ovaries and hypertrichosis. J. clin. Endocr. **20**, 1587—1600 (1960).

LAPIÈRE, C.: Modifications des glandes sébacées par des hormones sexuelles appliquées localment sur la peau de Souris. C.R. Soc. Biol. (Paris) **147**, 1302—1309 (1953).

LASHER, N., LORINCZ, A. L., ROTHMAN, S.: Hormonal effects on sebaceous glands in the white rat. II. The effect of pituitary-adrenal axis. J. invest. Derm. **22**, 25—29 (1954).

LASHER, N., LORINCZ, A. L., ROTHMAN, S.: Hormonal effects on sebaceous glands in the white rat. III. Evidence for the presence of a pituitary sebaceous gland tropic factor. J. invest. Derm. **24**, 499—505 (1955).

LEBEAU, M., VOKAER, R.: Traitement de l'hirsutisme et de l'acné. Leur thérapeutique séquentielle inversée anti-androgene-oestrogène. J. Gynecol. Obstet. Biol. Reprod. (Paris) **1**, 715—721 (1972).

LETTI, N.: Crescimento do pelo em ratos normais, gonadectomizados, suprarrenalectomizados e tiroidectomizados. Arq. Inst. Anat., Porto Alegre **2**, 34—41 (1959).

LIGHT, A. E., TORNABEN, J. A.: Effects of prolonged percutaneous administration of methyl testosterone and estradiol on growing male rats. J. **51**, 51—63 (1953a).

LIGHT, A. E., TORNABEN, J. A.: Effects of prolonged percutaneous administration of methyl testosterone and estradiol on growing female rats. J. Nutr. **51**, 365—379 (1953b).

LIPSETT, M. B., RITER, B.: Urinary ketosteroids and pregnanetriol in hirsutism. J. clin. Endocr. **20**, 180—186 (1960).

LIPSETT, M. B., WILSON, H.: Adrenocortical cancer: Steroid biosynthesis and metabolism evaluated by urinary metabolites. J. clin. Endocr. **22**, 906—925 (1962).

LODDI, L., MOGGI, L.: L'azione del propionate di testosterone sul processo di guarigione delle ferite chirurgiche sperimentali della cute. Acta chirug. Patav. **9**, 623—627 (1953).

LUDWIG, A. W., BOAS, N. F.: Effect of testosterone on the connective tissue of the comb of the cockerel. Endocrinology **46**, 291—298 (1950).

LUDWIG, E.: Die androgenetische Alopecie bei der Frau. Arch. klin. exp. Derm. **219**, 558—564 (1964).

LUDWIG, E.: Über das endokrine Substrat der diffusen, weiblichen (andro-genetischen) Alopecie. Arch. klin. exp. Derm. **227**, 468—477 (1966).

LUDWIG, E.: The role of sexual hormones in pattern alopecia. In: BACCAREDA-BOY, A., MORETTI, G., FREY, J. R. (Eds.): Biopathology of pattern alopecia, pp. 50—60. Basel-New York: Karger 1968.

LUTSKY, B. N., KOZIOL, P., GERSHAW, P.: Metabolism and effects of androgens in hamster flank organs. Endocrinology **95**, 882—890 (1974).

LYNNFIELD, Y. L.: Effect of pregnancy on the human hair cycle. J. invest. Derm. **35**, 323—327 (1960).

MAGUIRE, H. C.: Facial hair growth over site of testosterone injection in women. Lancet **1964 I**, 864—865.

MAGUIRE, H. C., KLINGMAN, A. M.: Common baldness in women. Geriatrics **18**, 329—334 (1963).

MAHESH, V. B., GLEENBLATT, R. B., AYDAR, C. K., ROY, S., PUEBLA, R. A., ELLEGOOD, J. O.: Urinary steroid excretion patterns in hirsutism. I. Use of adrenal and ovarian suppression tests in the study of hirsutism. J. clin. Endocr. **24**, 1283—1294 (1964).

MAHOUDEAU, J. A., BARDIN, C. W., LIPSETT, M. B.: The metabolic clearance rate and origin of plasma dihydrotestosterone in man and its conversion to the 5α-androstanediols. J. clin. Invest. **59**, 1338—1344 (1971).

MARTAN, J.: Effect of castration and androgen replacement on the supracaudal gland of the male guinea pig. J. Morph. **110**, 285—297 (1962).

MAUVAIS-JARVIS, P., KUTENN, F., BAUDOT, N.: Inhibition of testosterone conversion to dihydrotestosterone in men treated percutaneously by progesterone. J. clin. Endocr. **38**, 142—147 (1974).

MAYER, K., KAPLAN, P., STEIGLEDER, G. K.: Effect of acid mucopolysaccharides on hair growth in the rabbit. Proc. Soc. exp. Biol. (N.Y.) **108**, 59—63 (1961).

MENZEL, P., MORSCHES, B., BREGENZER, M., HOLZMANN, H., OERTEL, G. W.: $C_{19}$-Steroid bei Psoriasis. Acta endocr. (Kbh.) **66** (suppl. 152), 11 (1971).

MOHN, M. P.: The effects of different hormonal states on the growth of hair in rats. In: MONTAGNA, W., ELLIS, R. A. (Eds.): The biology of hair growth, pp. 335—398. New York: Academic Press 1958.

MONTAGNA, W., HAMILTON, J. B.: The sebaceous glands of the hamster. II. Some cytochemical studies in normal and experimental animals. Amer. J. Anat. **84**, 365—395 (1949).

MONTAGNA, W., KENYON, P.: Growth potentials and mitotic division in sebaceous glands of rabbits. Anat. Rec. **103**, 365—379 (1949).

MONTAGNA, W., UNO, H.: The phylogeny of baldness. In: BACCAREDA-BOY, A., MORETTI, G., FREY, Y. R. (Eds.): Biopathology of Pattern Alopecia, pp. 9—24. Basel-New York: Karger 1968.

MORETTI, G.: Lo spessore del cuio capelluto nelle varie etá dell'uomo. G. ital. Derm. **46**, 324—325 (1971).

MORETTI, G., CIPRIANI, C., REBORA, A.: Biochemical aspects of hair growth in the rat. In: BACCAREDA-BOY, A., MORETTI, G., FREY, J. R. (Eds.): Biopathology of Pattern Alopecia, pp. 177—190. Basel-New York: Karger 1968.

MORGAN, C. F.: A study of estrogenic action on the collagen, hexosamine and nitrogen content of skin, uterus and vagina. Endocrinology **73**, 11—19 (1963).

MORSCHES, B., HOLZMANN, H., OERTEL, G. W., BREGENZER, M.: Steroide und Haut. VI. Ausscheidung von $C_{19}$-Steroiden im Harn bei Psoriasis vulgaris. Arch. derm. Forsch. **240**, 204—211 (1971).

MUIR, A. V., CALMAN, K. C., THOMSON, J., MCSWEEN, M. P., MILNE, J. A., CHAKRABORTY, J., GRANT, J. K.: Hydroxysteroid dehydrogenase activity in human skin. I. Histochemical techniques for the demonstration of 3$\beta$-hydroxysteroid dehydrogenase and a method for the semiquantitation of its activity. Brit. J. Derm. **80**, 594—600 (1968a).

MUIR, A. V., CALMAN, K. C., THOMSON, J., MCSWEEN, M. P., MILNE, J. A., CHAKRABORTY, J., GRANT, J. K.: Hydroxysteroid dehydrogenase in human skin. II. Factors influencing hydroxysteroid dehydrogenase activity in sebaceous glands of human skin. Brit. J. Derm. **80**, 652—659 (1968b).

MUIR, A. V., HODGINS, M. B., MILNE, J. A., YOUNG, H.: The presence of hydroxysteroid dehydrogenases in mammalian sebaceous structures. Brit. J. Derm. **83**, 291—295 (1970).

MULAY, S., FINKELBERG, R., PINSKY, L., SOLOMON, S.: Metabolism of 4-$^{14}$C-testosterone by serially subcultured human skin fibroblasts. J. clin. Endocr. **34**, 133—143 (1972).

MULLER, S. A.: Hirsutism. Amer. J. Med. **46**, 803—807 (1969).

MUSSIO-FOURNIER, J. C., BERTOLINI, A., MORATÓ MANARO, J., BUÑO, W.: Depilatory action of pomade containing folliculin in women with hypertrichosis. Bull. Acad. Med. **113**, 352—355 (1935).

MUSSIO-FOURNIER, J. C., POLLACK, E., LUSSICH SIRI, J. J.: Loss of axillary and pubic hair in a patient with Addison's disease and regular menstruation. J. clin. Endocr. **9**, 555—556 (1949).

MYERS, R. J., HAMILTON, J. B.: Regeneration and rate of growth of hairs in man. Ann. N.Y. Acad. Sci. **53**, 562—568 (1951).

NEALE, C., KREBS, D., BETTENDORF, G.: Behandlung der Akne, des Hirsutismus und der Alopezie mit Cyproteron-acetat und Äthinyl-oestradiol. Acta endocr. (Kbh.) **66** (Suppl. 152), 13 (1971).

NELSON, R. M., RAKOFF, A. E.: Hirsutism and acne treated by an androgen antagonist. Obstet. Gynec. **36**, 748—752 (1970).

NEUMANN, F.: Use of cyproterone acetate in animal and clinical trials. Gynec. Invest. **2**, 150—179 (1971).

NEUMANN, F., ELGER, W.: The effect of a new antiandrogenic steroid, 6-chloro-17-hydroxy-1$\alpha$, 2$\alpha$-methylenepregna-4-6-dione-3,20-dione acetate (cyproterone acetate) on the sebaceous glands of mice. J. invest. Derm. **46**, 561—572 (1966).

NEUMANN, F., ELGER, W., KRAMER, M.: Development of a vagina in male rats by inhibiting androgen receptors with an antiandrogen during the critical phase of organogenesis. Endocrinology **78**, 628—632 (1966).

NEUMANN, F., VON BERSWORDT-WALLRABE, R., STEINBECK, H., HAHN, J. D., KRAMER, M.: Aspects of androgen-dependent events as studied by antiandrogens. Recent Progr. Horm. Res. **26**, 337—410 (1970).

NICHOLS, T., NUGENT, C. A., TYLER, F. H.: Glucocorticoid suppression of urinary testosterone excretion in patients with idiopathic hirsutism. J. clin. Endocr. **26**, 79—86 (1970).

NIKKARI, T.: Composition and secretion of the skin surface lipids on the rat: Effects of dietary lipids and hormones. Scand. J. clin. Lab. Invest. **17** (suppl. 85), 1—140 (1965).

NIKKARI, T., VALAVAARA, M.: The production of sebum in young rats: effects of age, sex, hypophysectomy and treatment with somatotrophic hormone and sex hormones. J. Endocr. **43**, 113—118 (1969).

NIKKARI, T., VALAVAARA, M.: The influence of age, sex, hypophysectomy and various hormones on the composition of the skin surface lipids of the rat. Hormones and sebum composition in the rat. Brit. J. Derm. **83**, 459—472 (1970a).

NIKKARI, T., VALAVAARA, M.: Effects of androgens and prolactin on the rate of production and composition of sebum in hypophysectomized female rats. J. Endocr. **48**, 373—378 (1970b).

PALITZ, L. L., MILBERG, I. L., KANTOR, I.: Enovid for acne in the female. Skin **3**, 243—252 (1964).

PEARCE, C. W., FOOT, N. C., JORDAN, G. L., LAW, S. W., WANTZ, G. E.: The effect and interrelation of testosterone, cortisone and protein nutrition on wound healing. Surv. Gynec. Obstet. **3**, 274—278 (1960).

PECORARO, V., ASTORE, I., BARMAN, J. M.: The prenatal and postnatal hair cycles in man. In: BACCAREDA-BOY, A., MORETTI, G., FREY, J. R. (Eds.): Biopathology of Pattern Alopecia, pp. 29—38. Basel-New York: Karger 1968.

PEIRONE, F., FENOGLIO, S.: Effetti dell'Enovid sulla dermatosi acneica femminile. Studio sperimentali col sebo-test. Arch. ital. Dermat. Vener. Sessuol. **33**, 455—466 (1965).

PIERREPOINT, C. G., GALLEY, J. MC. I., GRIFFITHS, K., GRANT, J. K.: Steroid metabolism of a Sertoli cell tumour of the testis of a dog with feminization and alopecia and of normal canine testis. J. Endocr. **38**, 61—70 (1967).

PINSKY, L., FINKELBERG, R., STRAISFELD, C., ZILAHI, B., KAUFMAN, M., HALL, G.: Testosterone metabolism by serially subcultured fibroblasts from genital and non genital skin of individual human donors. Biochem. biophys. Res. Commun. **46**, 364—369 (1972).

PINSKY, L., KAUFMAN, M., STRAISFELD, C., SHANFIELD, B.: Lack of difference in testosterone metabolism between cultured skin fibroblasts of human adult males and females. J. clin. Endocr. **39**, 395—398 (1974).

POCHI, P. E., STRAUSS, J. S.: Sebaceous gland function before and after bilateral orchiectomy. Arch. Derm. **88**, 729—731 (1963).

POCHI, P. E., STRAUSS, J. S.: Sebum production, casual sebum levels, titratable acidity of sebum and urinary fractional 17-ketosteroid excretion in males with acne. J. invest. Derm. **43**, 383—388 (1964).

POCHI, P. E., STRAUSS, J. S.: Effect of cyclic administration of conjugated equine estrogens on sebum production in women. J. invest. Derm. **47**, 582—585 (1966).

POCHI, P. E., STRAUSS, J. S.: Effect of sequential mestranol-chlormadinone on sebum production. Arch. Derm. **95**, 47—49 (1967).

POCHI, P. E., STRAUSS, J. S.: Sebaceous gland response in man to the administration of testosterone, $\Delta^4$-androstenedione and dehydroisoandrosterone. J. invest. Derm. **52**, 32—36 (1969).

POCHI, P. E., STRAUSS, J. S.: Endocrinologic control of the development and activity of the human sebaceous gland. J. invest. Derm. **62**, 191—201 (1974).

POCHI, P. E., STRAUSS, J. S., MESCON, H.: Sebum secretion and urinary fractional 17-ketosteroid and total 17-hydroxycorticoid excretion in male castrates. J. invest. Derm. **39**, 475—483 (1962).

POCHI, P. E., STRAUSS, J. S., MESCON, H.: The role of adrenocortical steroids in the control of human sebaceous gland activity. J. clin. Endocr. **41**, 391—399 (1963).

POWELL, E. W., BEVERIDGE, G. W.: Sebum excretion and sebum composition in adolescent man with and without acne vulgaris. J. invest. Derm. **82**, 243—249 (1970).

RALLI, P., GRAEF, I.: Stimulating effect of adrenalectomy on hair growth and melanin deposition in black rats fed diets adequate and deficient in the filtrate factors of vitamin B. Endocrinology **32**, 1—12 (1943).

RALLI, P., GRAEF, I.: The effects of the synthetic and natural hormone of the adrenal cortex on melanin deposition in adrenalectomized black rats fed diets adequate and deficient in the filtrate factors of vitamin B. Endocrinology **37**, 252—261 (1945).

RAMASASTRY, P., DOWNING, D.T., POCHI, P.E., STRAUSS, J.S.: Chemical composition of human skin surface lipids from birth to puberty. J. invest. Derm. **54**, 139—144 (1970).

RAMPINI, E., VOIGT, W., DAVIS, B.P., MORETTI, G., HSIA, H.L.: Metabolism of testosterone-4-$^{14}$C by rat skin: variations during the hair cycle. Endocrinology **89**, 1506—1514 (1971).

RAPPAPORT, R., FOREST, M.G., BAYARD, F., DUVAL-BEAUPÈRE, G., BLIZARD, R.M., MIGEON, C.J.: Plasma androgens and LH in scoliotic patients with premature pubarche. J. clin. Endocr. **38**, 401—406 (1974).

REBORA, A., CIPRIANI, C., MORETTI, G.: Biochemical aspects of male pattern alopecia. In: BACCAREDA-BOY, A., MORETTI, G., FREY, J.R. (Eds.): Biopathology of Pattern Alopecia, pp. 200—207. Basel-New York: Karger 1968.

RENNELS, E.G., CALLAHAN, W.P.: The hormonal basis for pubertal maturation of hair in the albino rat. Anat. Rec. **135**, 21—31 (1959).

RENZI, A.A., CHART, J.J.: Interaction of methandrostenolone and adrenocortical hormones. Proc. Soc. exp. Biol. (N.Y.) **110**, 259—262 (1962).

REYNOLDS, E.L.: The appearance of adult patterns of body hair in man. Ann. N.Y. Acad. Sci. **53**, 576—584 (1951).

RICHARDSON, G.S., AXELROD, L.R.: Metabolism of labeled testosterone by minces and nuclei of preputial glands of male rats. Endocrinology **88**, 890—894 (1971).

RIVAROLA, M.A., SAEZ, J.M., MIGEON, C.J.: Studies on androgens in patients with congenital adrenal hyperplasia. J. clin. Endocr. **27**, 624—630 (1967).

ROMAN-CALDERÓN, J.: Investigación del efecto de un nuevo anabólico en pacientes quemados. Prensa méd. Mex. **33**, 51—53 (1968).

RONGONE, E.L.: Testosterone metabolism by human male mammary skin. Steroids **7**, 489—504 (1966).

RONY, H.R., ZAKON, S.J.: Effect of androgen on the sebaceous glands of human skin. Arch. Derm. **48**, 601—604 (1943).

ROOK, A.: Endocrine influences on hair growth. Brit. med. J. **1965 I**, 609—614.

ROSENFIELD, R.L., LAWRENCE, A.M., LIAO, S., LANDAU, R.L.: Androgens and androgen responsiveness in the testicular feminization syndrome. Comparison of complete and incomplete forms. J. clin. Endocr. **32**, 625—632 (1971).

ROSSBERG, I.L., LAUR, S., ZIELSKE, F., HAMMERSTEIN, J.: Reversed sequential therapy of hirsutism using cyproterone acetate. I. Further clinical observations. Acta endocr. (Kbh.) **66** (Suppl. 152), 14 (1971).

SAITOH, M., USUKA, M., SAKAMOTO, M.: Human hair cycle. J. invest. Derm. **54**, 65—81 (1970).

SALOMON, T.: Genetic factors in male pattern alopecia. In: BACCAREDA-BOY, A., MORETTI, G., FREY, J.R.: (Eds.): Biopathology of Pattern Alopecia, pp. 39—49. Basel-New York: Karger 1968.

SANSONE, G., DAVIDSON, W., CUMMINGS, B., REISNER, R.M.: Sebaceous gland lipogenesis induced by testosterone; early metabolic events. J. invest. Derm. **57**, 144—148 (1971).

SANSONE, G., REISNER, R.M.: Differential rates of conversion of testosterone to dihydrotestosterone in acne and in normal human skin. A possible pathogenic factor in acne. J. invest. Derm. **56**, 366—372 (1971).

SANSONE-BAZZANO, G., BAZZANO, G., REISNER, R.M., HAMILTON, J.G.: The hormonal induction of alkyl glycerol, wax and alkyl acetate synthesis in the preputial gland of the mouse. Biochim biophys. Acta (Amst.) **260**, 35—40 (1972a).

SANSONE-BAZZANO, G., REISNER, R.M.: Steroid pathways in sebaceous glands. J. invest. Derm. **62**, 211—216 (1974).

SANSONE-BAZZANO, G., REISNER, R.M., BAZZANO, G.: Conversion of testosterone-1,2-$^{3}$H to androstenedione-$^{3}$H in the isolated hair follicle of man. J. clin. Endocr. **34**, 512—515 (1972b).

SANSONE-BAZZANO, G., REISNER, R.M., BAZZANO, G.: A possible mechanism of action of estrogen at the cellular level in a model sebaceous gland. J. invest. Derm. **59**, 299—304 (1972c).

SAUNDERS, H. L., EBLING, F.: The antiandrogenic activity and sebaceous gland inhibitory activity of 6α,6β-ethylene-17α-methyl-B-nortestosterone. J. invest. Derm. **52**, 163—168 (1969).

SCHILLER, S., BENDITT, E. P., DORFMAN, A.: Effect of testosterone and cortisone on the hexosamine content and metachromasia of chick combs. Endocrinology **50**, 504—510 (1952).

SCHELL, H., SCHMID, G. H.: Autoradiographische Untersuchungen über den Einfluß von Cyproteronacetat auf die epidermale Proliferation der männlichen Ratte. Arch. Derm. Forsch. **248**, 89—98 (1973).

SCHWEIKERT, H. V., WILSON, J. D.: Regulation of human hair growth by steroid hormones. II. Androstenedione metabolism in isolated hairs. J. clin. Endocr. **39**, 1012—1019 (1974).

SHANIES, D., HIRSCHHORN, D., NEW, M. I.: Metabolism of testosterone-$^{14}$C by cultured human cells. J. clin. Invest. **51**, 1459—1468 (1972).

SHERINS, R. J., BARDIN, C. W.: Preputial gland growth and protein synthesis in the androgen-insensitive male pseudohermaphrodite rat. Endocrinology **89**, 835—841 (1971).

SHUSTER, S.: Primary cutaneous virilism or idiopathic hirsuties? Brit. med. J. **1972 II**, 285—286.

SHUSTER, S., BLACK, M. M., BOTTOMS, E.: Skin collagen and thickness in women with hirsuties. Brit. med. J. **1970 IV**, 772.

SHUSTER, S., BOTTOMS, E.: Senile degeneration of skin collagen. Clin. Sci. **25**, 487—491 (1963).

SHUSTER, S., THODY, A. J.: The control and measurement of sebum secretion. J. Invest. Derm. **62**, 172—190 (1974).

SILVERMAN, S. H., MIGEON, C., ROSENBERG, E., WILKINS, L.: Precocious growth of sexual hair without other secondary sexual development; "Premature pubarche", a constitutional variation of adolescence. Pediatrics **10**, 426—432 (1952).

SLOPER, J. C., PEGRUM, G. D.: Regeneration of crushed mammalian skeletal muscle and effects of steroids. J. Path. Bact. **93**, 47—63 (1967).

SMITH, J. G.: The aged human sebaceous gland. The effect of hormone administration and a comparison with adolescent gland function. Arch. Derm. **80**, 663—671 (1959).

SMITH, J. G., BRUNOT, F. R.: Hormonal effects on aged human sebaceous glands. Acta Derm. Venereol. (Stockh.) **41**, 61—65 (1961).

SMITH, M. A., WELLS, R. S.: Male-type alopecia, alopecia areata and normal hair in women. Arch. Derm. **89**, 95—98 (1964).

SOMMERVILLE, I. F., COLLINS, W. P., FLAMIGNI, C., KOULLAPIS, N., DEWHURST, C. J.: Androgen metabolism in human skin. Proc. roy. Soc. Med. (Lond.) **64**, 845—847 (1971).

SORENSEN, B. M.: Influence of glucocorticoids and anabolic steroids on the wound healing process. In: ASBOE-HANSEN, G. (Ed.): Hormones and Connective Tissue, pp. 106—123. Baltimore: Williams and Wilkins 1966.

SOUTHREN, A. L., GORDON, G. G., TOCHIMOTO, S.: Further study of factors affecting the metabolic clearance of testosterone in man. J. clin. Endocr. **28**, 1105—1112 (1968).

STEVANOVIC, D. V.: Investigation on the effect on skin healing of a cream containing dexamethasone, chlorhexidine hydrochloride and nandrolone decanoate. Arzneimittel-Forsch. **21**, 1225—1227 (1971).

STRAUSS, J. S., KLIGMAN, A.: The effect of progesterone and progesterone-like compounds on the human sebaceous gland. J. invest. Derm. **36**, 309—318 (1961 a).

STRAUSS, J. S., KLIGMAN, A.: Androgenic effects of a progestational compound, 17α-ethynyl-19-nortestosterone (norlutin), on the human sebaceous gland. J. clin. Endocr. **21**, 215—219 (1961 b).

STRAUSS, J. S., KLIGMAN, A. M., POCHI, P. E.: The effect of androgens and estrogens on human sebaceous glands. J. invest. Derm. **39**, 139—155 (1962).

STRAUSS, J. S., POCHI, P. E.: The human sebaceous gland: its regulation by steroidal hormones and its use as an end organ for assaying androgenicity in vivo. Rec. Progr. Horm. Res. **19**, 385—444 (1963 a).

STRAUSS, J. S., POCHI, P. E.: Effects of enovid on sebum production in females. Arch. Derm. **87**, 366—368 (1963 b).

STRAUSS, J. S., POCHI, P. E.: The effect of cyclic progestin-estrogen therapy on sebum production and acne in women. J. Amer. med. Ass. **190**, 815—819 (1964).

STRAUSS, J. S., POCHI, P. E.: Sebaceous gland response in man to the administration of testosterone, $\Delta^4$-androstenedione and dehydroepiandrosterone. J. invest. Derm. **50**, 358—359 (1968).

STRAUSS, J. S., POCHI, P. E.: Assay of antiandrogens in man by the sebaceous gland response. Brit. J. Derm. **82** (Suppl. 6), 33—42 (1970).

STRAUSS, J. S., POCHI, P. E., SARDA, I. R., WOTIZ, H. H.: Effect of oral and topical 17α-methyl-β-nortestosterone on sebum production and plasma testosterone. J. invest. Derm. **52**, 95—99 (1969).

STRICKLAND, A. L., FRENCH, F. S.: Absence of response to dihydrotestosterone in the syndrome of testicular feminization. J. clin. Endocr. **29**, 1284—1286 (1969).

SUMMERS, V. K.: The role of the adrenal cortex and gonads in the control of sexual hair distribution. Acta med. scand. **136**, 105—111 (1949).

TAKACS, I.: Einfluß von Gravidität, Estrogen und Progesteron auf den Kollagengehalt der Rattenhaut. Acta physiol. Acad. Sci. hung. **39**, 63—69 (1971).

TAKAYASU, S., ADACHI, K.: Hormonal control of metabolism in hamster costovertebral glands. J. invest. Derm. **55**, 13—19 (1970).

TAKAYASU, S., ADACHI, K.: The in vivo and in vitro conversion of testosterone to 17-hydroxy-5-androstan-3-one (dihydrotestosterone) by the sebaceous glands of hamsters. Endocrinology **90**, 73—80 (1972a).

TAKAYASU, S., ADACHI, K.: The conversion of testosterone to 17-hydroxy-5-androstan-3-one (dihydrotestosterone) by human hair follicles. J. clin. Endocr. **34**, 1098—1101 (1972b).

TAMM, J., VOIGT, K. D.: Einige klinisch relevante Ergebnisse der neueren Androgenforschung. Schweiz. med. Wschr. **101**, 1078—1083 (1971).

TAMM, J., VOIGT, K. D., SCHONBROK, M., LUDWIG, E.: The effect of orally administered cyproterone on the sebum production in human subjects. Acta endocr. (Kbh.) **63**, 50—58 (1970).

THODY, A. J., SHUSTER, S.: The effect of hypophysectomy and testosterone on the activity of the sebaceous glands of castrated rats. J. Endocr. **47**, 219—224 (1970).

THODY, A. J., SHUSTER, S.: The effect of hypophysectomy on the response of the sebaceous gland to testosterone propionate. J. Endocr. **49**, 329—333 (1971a).

THODY, A. J., SHUSTER, S.: Increased sebum secretion in adult female rats after neonatal treatment with testosterone propionate. J. Endocr. **49**, 677—681 (1971b).

THOMAS, J. P., OAKE, R. J.: Androgen metabolism in the skin of hirsute women. J. clin. Endocr. **38**, 19—22 (1974).

TORRE, D.: Hormones and the skin. Med. Clin. N. Amer. **49**, 549—564 (1965).

TORRE, D., KLUMPP, M. M.: Cyclic estrogen hormone therapy of acne vulgaris. J. Amer. med. Ass. **164**, 1447—1450 (1957).

TRONNIER, H., HOPPE-SEYLER, G.: Klinischpharmakologische Studie zur Frage von Nebenwirkungen hormoneller Ovulationshemmer im Bereich der Dermatologie. Therapiewoche **19**, 1024—1033 (1969).

UNO, H., ADACHI, K., MONTAGNA, W.: Baldness of the red uacari (Carajao rubicundis); Histological properties and enzyme activities of hair follicles. J. Geront. **24**, 23—27 (1969).

VAN WAYJEN, R. G. A., VAN DEN ENDE, A.: Clinical-pharmacological investigation of cyproterone acetate. Gynec. Invest. **2**, 282—289 (1971/72).

VILJANTO, J., ISOMAKI, H., KULONEN, E.: Effect of an anabolic steroid on the tensile strength of granulation tissue in various nutritional states. Acta endocr. (Kbh.) **41**, 395—399 (1962).

VIVARELLI, I., TOMASSINI, C.: Il metandrostenolone per uso topico nella cura delle ulcerazioni torpide cutanea. Dermatologia (Napoli) **15**, 318 (1964).

VOGEL, H. G.: Zur Wirkung von Hormonen auf physikalische und chemische Eigenschaften des Binde- und Stützgewebes—Teil 1. Arzneimittel-Forsch. **19**, 1495—1503 (1969a).

VOGEL, H. G.: Zur Wirkung von Hormonen auf physikalische und chemische Eigenschaften des Binde- und Stützgewebes—Teil 2. Arzneimittel-Forsch. **19**, 1732—1742 (1969b).

VOGEL, H. G.: Zur Wirkung von Hormonen auf physikalische und chemische Eigenschaften des Binde- und Stützgewebes—Teil 3. Arzneimittel-Forsch. **19**, 1790—1801 (1969c).

VOGEL, H. G.: Zur Wirkung von Hormonen auf physikalische und chemische Eigenschaften des Binde- und Stützgewebes—Teil 4. Arzneimittel-Forsch. **19**, 1981—1996 (1969d).

VOGEL, H. G.: Beeinflussung der mechanischen Eigenschaften der Haut von Ratten durch Hormone. Arzneimittel-Forsch. **20**, 1849—1857 (1970).

VOGEL, H. G.: Stress relaxation in rat skin after treatment with hormones. J. Med. (Basel) **4**, 19—27 (1973).

VOGEL, H. G., KOBELT, D., KORTING, G. W., HOLZMANN, H.: Prüfung der Festigkeitseigenschaften von Rattenhaut in Abhängigkeit von Lebensalter und Geschlecht. Arch. klin. exp. Derm. **239**, 296—306 (1970).

VOIGT, W., FERNANDEZ, E. P., HSIA, S. L.: Transformation of testosterone into 17beta-hydroxy-5-alpha-androstan-3-one by microsomal preparations of human skin. J. biol. Chem. **245**, 5594—5599 (1970).

VOIGT, W., HSIA, S. L.: The antiandrogen action of 4-androsten-3-one-17beta-carboxylic acid and its methyl esther on hamster flank organ. Endocrinology **92**, 1216—1222 (1973a).

VOIGT, W., HSIA, S. L.: Further studies on testosterone 5α-reductase of human skin: Structural features of steroid inhibitors. J. biol. Chem. **248**, 4280—4285 (1973b).

VON WATTENWYL, H.: Die Wirkung subkutan implantierter Oestradioltabletten auf das Haarkleid des Meerschweinchens. Schweiz. med. Wschr. **71**, 1331—1334 (1941).

WHITAKER, W. L.: Inhibition of hair growth in the rat by the percutaneous use of desoxycorticosterone and desoxycorticosterone acetate. Anat. Rec. **103**, 518 (1949).

WHITAKER, W. L., BAKER, B. L.: Inhibition of hair growth by the percutaneous application of certain adrenal cortical preparations. Science **108**, 207—209 (1948).

WHITAKER, W. L., BAKER, B. L.: A direct comparison of the direct action of estrogens and adrenal cortical extracts of growth of hair in the rat. J. invest. Derm. **17**, 69—77 (1951a).

WHITAKER, W. L., BAKER, B. L.: The relative effectiveness of adrenal cortical hormones in the inhibition of hair growth in the rat. Univ. Mich. Med. Bull. **17**, 384—390 (1951b).

WHITAKER, W., STODDARD, F. J., GREKIN, J. N., GOFORTH, L.: Estrone and alpha-estradiol, applied locally, ineffective in reducing facial hypertrichosis. J. invest. Derm. **9**, 49—54 (1947).

WHITELEY, H. J.: The effect of adrenalectomy and adrenocortical hormones on the hair growth cycle in the rabbit and rat. J. Endocr. **17**, 167—176 (1958).

WIECHERT, R., STEINBECK, H., ELGER, W., NEUMANN, F.: Wirkungen und Struktur neuer antiandrogener Steroide. Arzneimittel-Forsch. **17**, 1103—1116 (1967).

WIELAND, R. J., ZORN, E. M., HALLBERG, M. C.: Differential studies on the mechanism of serum androgen and androgen-binding abnormalities in hirsutism. Amer. J. Obstet. Gynec. **117**, 983—986 (1973).

WILDE, P. F., EBLING, F. J.: Preliminary observations of skin surface fat from rats treated with testosterone and estradiol. J. invest. Derm. **52**, 362—365 (1969).

WILLIAMS, W. L., GARDNER, W. V., DE VITA, J.: Local inhibition of hair growth by percutaneous application of estrone. Endocrinology **38**, 368—375 (1946).

WILSON, J. D., GLOYNA, K. E.: The intranuclear metabolism of testosterone in the accessory organs of reproduction. Recent Progr. Horm. Res. **26**, 309—330 (1970).

WILSON, J. D., WALKER, J. D.: The conversion of testosterone to 5α-androstan-17β-ol-3-one (dihydrotestosterone) by skin slices of man. J. clin. Invest. **48**, 371—379 (1969).

WINKLER, K.: La valeur des anti-androgènes en Dermatologie. Ann. Derm. Syph. **95**, 147—153 (1968a).

WINKLER, K.: Die Antiandrogene in der Dermatologie (Gravimetrische Fettbestimmungen während Cyproteron-Anwendung). Arch. klin. exp. Derm. **233**, 296—302 (1968b).

WINKLER, K. S.: Der Einfluß von Hormonen, im besonderen von Antiandrogenen, auf Seborrhoe und Haarwachstum. Z. Haut Geschlechtskr. **46**, 75—78 (1971).

WINKLER, K. S.: Antiandrogene bei der Alopecia androgenetica der Frau. Hautarzt **23**, 518—519 (1972).

WINTER, C. A., HOLLINS, H. T., STEBBINS, R. B.: The effect of androgenic hormones upon the adrenal atrophy produced by cortisone injections and upon the anti-inflammatory action of cortisone. Endocrinology **52**, 123—131 (1953).

WOTIZ, H. H., MESCON, H., DOPPEL, H., LEMON, H. M.: The in vitro metabolism of testosterone by human skin. J. invest. Derm. **26**, 113—119 (1956).

ZARATE, A., MAHESH, V. B., GREENBLATT, R. B.: Effect of an antiandrogen, 17α-methyl-β-nortestosterone, on acne and hirsutism. J. clin. Endocr. **26**, 1344—1398 (1966).

ZAUN, H.: Observations concerning the influence of cyproterone acetate on hair growth. Acta endocr. (Kbh.) **66** (Suppl. 152), 17 (1971).

ZAUN, H.: Über den Einfluß der hormonellen Kontrazetiva auf das Haarwachstum des Menschen. Arch. derm. Forsch. **244**, 510—515 (1972).

ZONDEK, B.: The inhibitory effect of follicular hormones on the anterior lobe of the pituitary gland. Lancet **1936 I**, 10—12.

ZURBRUGG, R. P., GARDNER, L. I.: Urinary $C_{19}$ steroids in two girls with precocious sexual hair. J. clin. Endocr. **23**, 704—708 (1963).

# C. Anabolic-Androgenic Steroids and Experimental Mammary Tumors

R. HILF

## I. Introduction

Androgens have been used with success for the treatment of advanced disease states in patients with breast cancer. In a large collaborative study, conducted by the Cooperative Breast Cancer Group (1964), an overall remission rate of 21% was found. The criteria used for defining a remission were quite strict. Another finding from this study was that efficacy was related to menopausal status of the patient, with the highest remission rates found in women who were at least 5 years postmenopausal. This cooperative group clearly established that remission was correlated with prolongation of survival, since the responders to androgen therapy survived on the average twice as long as the patients who failed to respond. The androgens employed in these studies were either testosterone propionate or fluoxymesterone, both potent androgens that produce obvious masculinization (ṠEGALOFF et al., 1951). Active programs were initiated to devise compounds having equal or better antitumor effects, but with less androgenic "side-effects."[1]

In order to exploit the antitumor effect of steroid structures related to the androstanes, reliable experimental tumor systems are necessary. Such an experimental carcinoma should be reproducible in its growth, but rate of growth should not be an important factor; the latter may seem important from a practical standpoint but such considerations may compromise the validity of the results. For example, the earlier experience of the Cancer Chemotherapy National Service Center, which utilized three experimental mouse tumors (L 1210 leukemia, mammary carcinoma Ca 755, and sarcoma 180), provided results that were similar and could have been obtained if only L 1210 leukemia was used as the test system. Furthermore, mammary adenocarcinoma Ca 755 showed little or no responses to androgens or estrogens, classes of steroids that demonstrate efficacy in the human disease. This, then, becomes the most critical aspect for the experimentalist: Does the animal tumor system reflect the human disease in its response to standard agents?

Based on the earlier finding (KORENCHEVSKY and DENNISON, 1935) that estrone could antagonize the androgen-induced response of the seminal vesicles and coagulating glands, LACASSAGNE (1937) studied the effect of androgen on estrone-induced mammary tumors in mice. In this early report using R 3 mice, which were treated with estrone to enhance development of mammary adenocarcinomas, simultaneous treatment with testosterone acetate (500–1000 μg) delayed the appearance of mam-

[1] The use of this term is improper, but does appear often in the literature—the ability of an androgen to cause hirsutism in women is inherent in the definition of an androgen; it is an undesirable effect, rather than a side-effect.

mary tumors and prolonged survival of animals. He concluded that the androgen antagonized the estrogen-induced tumor formation and suggested the potential utility of testosterone in breast cancer. Shortly thereafter, NATHANSON and ANDERVONT (1939) confirmed and extended these findings in C3H mice, which have a 95% incidence of spontaneous breast cancer. Animals injected with 0.5 mg testosterone propionate 3 times per week for 4 months demonstrated a 30% incidence of mammary tumors compared to 100% incidence of cancer in littermate animals receiving no hormone. However, administration of even larger doses of testosterone propionate (1.0–2.5 mg per day for 4 weeks) had no effect on tumor growth rate when administered to animals that had developed tumors at the time therapy was initiated. In general, androgen treatment has been relatively ineffectual in transplantable tumors of mice, even when administered at levels of 1000 mg/kg/day (GRISWOLD et al., 1963); this may, however, be a reflection of hormone-independence of mammary tumors after they become established.

One of the earliest reports indicating therapeutic efficacy of androgen administration appeared in 1939 (ULRICH), and LOESER (1941) reported on one patient who received testosterone beginning in 1937. In 1944, FELS reported on treatment of three patients with breast cancer, one of whom demonstrated an excellent response. NATHANSON (1946) reviewed the status of early attempts to treat the clinical disease with hormones. It is interesting to note that, with the rather unimpressive effect in mice of androgen on established tumors reported by NATHANSON and ANDERVONT (1939), testosterone propionate was used for treatment of breast cancer in women with advanced (well-established) disease. This re-emphasizes the need for experimental tumor systems that can accurately predict antitumor activity in the clinical disease.

No attempt will be made here to cite every androgen (or structurally related analogue in the androstane family) that has been tried for effects on mammary tumor growth. The reader is referred to Chapter 4, Volume 3 (ROOKS, 1964) and Chapter 7 of Volume 4, Methods in Hormone Research (DORFMAN, 1965), or to the thorough review by GRISWOLD et al. (1963) on the effects of compounds on L1210 leukemia, carcinoma 755, and sarcoma 180, and to the many volumes of data resulting from screening of compounds through the aegis of the Cancer Chemotherapy National Service Center, which appear as supplements to the journal Cancer Research. An additional useful volume details endocrine bioassay data for ~1700 compounds tested for androgenic and myotrophic activity (HILGAR and HUMMEL, 1964). It will be the purpose of this survey to discuss more recent data on the effects of androgens on experimental mammary tumors and to present information on attempts to elucidate the effects of androgens on a biochemical basis.

## II. Effect of Androgens on Mouse Mammary Tumors

Since it is known that mouse mammary tumors are usually not responsive to hormonal perturbations, once the neoplastic transformation has occurred, it is not too surprising that most reports show little or no effects on tumor growth during administration of androgens. For example, STOCK and SUGIURA (1958) reported little or no effect of 9α-fluoro-11β-dihydroxy-17α-methyl-androst-4-ene-3-one (halotestin) at a daily dose of 375 mg/kg on the growth of 20 transplantable tumors in mice, two of which were mammary carcinomas. In the article by DORFMAN (1965), summarizing effects of various steroids on experimental tumors, a number of androgens are listed as causing inhibition of mammary tumors in mice. A number of these reports need to be examined carefully for the criteria utilized in defining significant inhibition of tumor growth. For example, STOCK and SUGIURA (1958) consider a reduction of tumor growth of 50% or less as a $\pm$ or negative response; these criteria are reasonably strict and may be interpreted by some as a response and by others as demonstrating little or no response. A thorough review of data on responses of animal tumors to anticancer agents was summarized by HIRSCHBERG (1963) and represents an invaluable guide to results reported through 1958. Unfortunately, the author is unaware of a more recent compilation of data utilizing this approach.

An interesting report, dealing with translation of animal data to clinical trials, is that by TAKATANI et al. (1967). These investigators examined the effects of a series of androgen analogues containing the diosphenol (enol form of $\alpha$-diketone) structure in the A-ring. Using a transplantable mammary adenocarcinoma of the C3H/Heston strain of mice, they selected one compound, 2,17$\beta$-dihydroxy-4,4-dimethylandrosta-1,5-diene-3-one, as possessing good antitumor activity with little or no androgenic activity, the latter properties assayed by effects on accessory sex organs in the rat. The compound was weakly estrogenic, however, as reflected by its ability to induce cornification of the vaginal mucosa. Although this compound demonstrated approximately 50% inhibition of tumor growth, no clear-cut dose-related response was seen even though as little as 0.5 µg total dose for 14 days was used (this dose gave a 47.5% suppression of tumor growth). Some of the other analogues were equally effective, although at higher doses. They then tested the foregoing diosphenol on 20 women with advanced breast cancer and reported that four of the patients responded by objective remission of their disease (they administered 80–100 mg/day by intramuscular injections 3 times per week). Thus, these investigators demonstrated that results from a mouse mammary tumor system could be translated to clinical response.

A report creating considerable interest was that by MINESITA and YAMAGUCHI (1965) in which they described the establishment of an androgen-dependent transplantable mammary tumor in the DD/Sic strain of mice. This tumor, referred to as the Shionogi carcinoma, arose from a spontaneous autonomous neoplasm, and, after several transplantations into male mice, showed an increase in growth rate, a decreased survival time of the host, and an inability to grow in the female host. Further studies indicated that administration of androgen to female or castrated male tumor-bearing hosts was necessary for the Shionogi carcinoma to grow; castration of male tumor-bearing mouse led to regression of the tumor. These authors also reported on the response of tumor growth to various doses of androgens. Testosterone propionate stimulated tumor growth in castrated male mice, but at higher doses of this androgen, less stimulation of tumor growth was observed. For example, at a total dose of 8.0 mg of testosterone propionate for the 20-day test period, tumor weight was 2.2 g; at 2.0 mg testosterone propionate, tumor weight was 5.65 g; and at 0.5 mg testosterone propionate, tumor weight was 1.78 g. This result is somewhat reminiscent of the effects of estrogen on mammary tumors in rats or on the uterotrophic response to estrogen—a dose-related increase at lower doses of the hormone followed by an inhibition of the response at high, pharmacologic doses. The inhibition of tumor growth seen with testosterone propionate was not seen with weaker androgens (dromostanolone propionate and durabolin) given at the same dose levels; androstenediol, a rather weak androgen, was not capable of stimulating tumor growth at the highest dose employed (8.0 mg total for 20 days) and demonstrated essentially no stimulation of ventral prostate weight. Estrogens (estradiol-17β, hexestrol, and TACE) completely prevented tumor growth at doses as low as 0.1 mg total dose for 15-day studies; administration of estradiol-17β to castrate mice treated simultaneously with anediol (2-hydroxymethylene-17α-methyldihydrotestosterone) antagonized the androgen-stimulated carcinoma growth. Thus, there was now a model that could be exploited for study of antiandrogenic compounds as potential inhibitors of mammary tumor growth.

The description of an androgen-dependent mammary tumor was further documented by studies of MATSUMOTO et al. (1972) and YAMAGUCHI et al. (1974). They compared the uptake and retention of injected tritiated-labeled testosterone in the Shionogi androgen-dependent carcinoma 115 and compared the results with those from the Shionogi carcinoma 42, a nondependent tumor, and with those from muscle and liver, two tissues that are not androgen targets. The amount of 5α-dihydrotestosterone found in subcellular fractions, such as nuclear, microsomal plus mitochondrial, and cytoplasmic, in the Shionogi carcinoma 115 and ventral prostate was 10 to 100 times higher than that recovered from the corresponding fractions in liver and muscle, or in the Shionogi carcinoma 42. These data support the conclusion that Shionogi carcinoma 115 is an androgen target tissue since (1) the carcinoma possesses 5α-reductase activity, which is necessary for the conversion of testosterone to 5α-dihydrotestosterone; (2) the radioactivity is recovered in the form of 5α-dihydrotestosterone, which is thought to be the active intracellular androgen; and (3) the intracellular distribution and retention of labeled 5α-dihydrotestosterone in the Shionogi 115 carcinoma is similar to that in ventral prostate and seminal vesicles.

It is now well accepted that steroid hormones exert their effects in target tissues by initial interaction with a cytoplasmic receptor protein, which in turn translocates as a hormone-receptor complex to the nucleus (KING and MAINWARING, 1974). Such properties are found in hormone target tissues, and, in the case of estrogens, receptor proteins have been examined extensively in tumors of accessory sex organs, such as the mammary gland (JENSEN et al., 1972). It was, therefore, interesting to ascertain whether the Shionogi 115 carcinoma possessed androgen receptor proteins similar to those described for an androgen target tissue, such as the prostate. MAINWARING and MANGAN (1973) reported on their studies with the Shionogi 115 tumor, in which they demonstrated significant amounts of nuclear binding of 5α-dihydrotestosterone and of binding to a cytoplasmic receptor; in contrast, the non androgen-dependent Shionogi carcinoma 42 showed less nuclear binding of labeled steroid and absence of the cytoplasmic receptor. However, nuclear binding by Shionogi 42 cells was quite adequate when mixed with labeled cytoplasm from prostate or Shionogi 115 tissues, implying that it was the relative lack of cytoplasmic receptors in the autonomous Shionogi 42 tumor. These investigators demonstrated the specificity of nuclear binding by examining relative binding of labeled steroids to nuclei from Shionogi 115 tumor. As expected, 5α-dihydrotestosterone showed the highest amount of nuclear binding, followed by testosterone (although about 60% of the steroid recovered behaved like 5α-dihydrotestosterone during thin-layer chromatography) and much lower, estradiol-17β. The nuclear localization of estrogen in this tumor is of interest because estrogen treatment inhibited growth of the carcinoma in vivo (MINESITA and YAMAGUCHI, 1965). Further, estrogen receptors have been reported to be found in normal prostate (JUNGBLUT et al., 1971), and this property may be another indicator of an androgen target tissue, although the finding of an estrogen receptor in a mammary carcinoma may be a reflection of the possession of estrogen responsiveness in mammary neoplasias.

Almost simultaneously, BRUCHOVSKY and MEAKIN (1973) published their studies with the Shionogi carcinoma. For their studies, a tumor line able to grow only in male DD/S mice (androgen-dependent) was used and compared with a line that grew as well in males, castrate males, or females, which has been classified as an autonomous tumor. At 30 min after administration of labeled testosterone, both dependent and autonomous tumors had about the same uptake of hormone; however, the level of radioactivity was about 4 times higher in nuclei from the dependent tumor line than in nuclei from the autonomous line. Cytoplasmic steroid-binding proteins were examined by gel filtration and the amount of steroid bound was found to be 4 times higher in cytosol from dependent tumor cells; there was a corresponding difference in the specific activities of the binding proteins. Unlike the findings of MAINWARING and MANGAN (1973), BRUCHOVSKY and MEAKIN reported that 10–90% of the recovered steroid in cytoplasm and nucleus was testosterone and only about 3% was isolated as 5α-dihydrotestosterone. The similarity of steroids isolated from cytoplasm and nucleus suggested that the difference between dependent and autonomous tumors was not attributable to a defect in steroid metabolic pathways.

In a very recent paper, BRUCHOVSKY et al. (1975) extended their studies of androgen receptors to include 9 variant lines of the Shionogi mammary carcinoma. Two of the tumor lines were classified as androgen-dependent and 7 tumor lines were autonomous; however, both of the dependent lines and 2 of the autonomous lines were found to contain androgen receptors. Thus, the presence of receptors per se does not represent an accurate marker of hormonal dependence in vivo. Uptake and retention of radioactively labeled testosterone into whole tumor was as high, if not higher, in autonomous tumor lines versus the two androgen-dependent lines, although nuclear labeling was higher in those lines (autonomous and dependent) that demonstrated the highest levels of cytoplasmic receptors. Thus, the translocation of labeled testosterone to the nucleus was proposed to be dependent on sufficient levels of cytoplasmic receptor, and this was not an exclusive property of tumors demonstrating hormone dependence in vivo. Such results suggest that other defects, i.e., beyond the translocation step, may be quite important in the regulation of tumor growth.

## III. Androgens and Rat Mammary Tumors

Unlike the relative lack of responsiveness reported for mouse mammary tumors, transplantable breast neoplasms of the rat have been utilized as experimental models that demonstrate growth inhibitory responses to administration of androgens (and other sex steroids as well). Furthermore, the ability to induce mammary carcinomas with chemical carcinogens and the high yield of these tumors, which show hormone dependency, has facilitated the investigation of another aspect of mammary tumorigenesis as influenced by alteration of hormonal milieu. These will be discussed separately below.

## A. Transplanted Mammary Tumors in Rats

### 1. Fibroadenomas

One of the first series of studies exploring the effects of androgens on rat tumor growth was conducted by HUGGINS and his colleagues (1956). Although it had been known much earlier that mammary fibroadenoma of Sprague-Dawley rats were transplantable (HEIMAN and KREHBIEL, 1936; MOHS, 1940) and that morphologic changes occurred in the tumors during pregnancy (EMGE and MURPHY, 1938), it was investigations by HUGGINS et al. (1956; HUGGINS and MAINZER, 1957) that characterized the growth and hormone responsiveness of these tumors at a time when synthetic androgens were available for testing. In the first paper of these studies, HUGGINS et al. (1956) reported on techniques that yielded tumor growth in 95% of the recipients; tumor growth was inhibited by ovariectomy and ovariectomy plus adrenalectomy. Tumor growth was stimulated by estrogen and estrogen plus progesterone in ovariectomized rats; the estrogen effect was stimulatory at low doses and inhibitory at high doses. In hypophysectomized rats, estrogen, progesterone, and prolactin (or growth hormone) stimulated tumor growth. The effects of administration of several steroids in the androstane series on fibroadenoma growth were reported by HUGGINS and MAINZER (1957). They found that growth of this tumor was influenced by structural modification of the steroid; maximal growth inhibition was seen for testosterone, dihydrotestosterone, and 17$\beta$-hydroxy-2$\alpha$-methylandrostan-3-one. Addition of more than one methyl group, such as 2,2′-dimethyldihydrotestosterone, or substitution of a methyl group at C-3, C-4 or C-16, reduced or eliminated tumor-inhibitory potency; however, 17$\alpha$-methyltestosterone was a potent inhibitor of growth of the fibroadenoma. At the C-17 position, a $\beta$ orientation of the hydroxyl group was required, as epitestosterone was inactive; at the C-3 position, an $\alpha$ orientation of the hydroxyl group was preferable for tumor inhibition. They proposed that inhibition of tumor growth occurred as a result of inhibition of the stimulatory effects of estrogen directly on the tumor cell as well as suppression of ovarian activity, presumably via pituitary gonadotrophin inhibition.

This fibroadenoma system was further investigated by GLENN et al. (1959a, b) and modified by employing established growing tumors for assay purposes. They confirmed the finding of HUGGINS et al. (1956) that ovariectomy retarded tumor growth and that ovariectomy plus adrenalectomy was more effective. Testosterone propionate was used as a standard and was found to consistently inhibit tumor growth but never to induce tumor regression. After testing numerous steroids, they concluded that inhibition of tumor growth could be affected by compounds that did not exert significant androgenic or estrogenic activities (GLENN et al., 1959b). They did not rule out, however, that the effects of these agents might be indirect via the pituitary rather than a direct effect on the tumor cell population. For example, 11$\alpha$-hydroxy-17-methyltestosterone and 9,11-epoxy-17-methyltestosterone demonstrated weaker androgenic activity than 17-methyltestosterone, yet all produced a similar tumor-inhibitory potency. The same held for 2-methyl-19-nortestosterone versus 19-nortestosterone. A detailed summary of their work, including some studies with a testosterone propionate-resistant line of the fibroadenoma, was published (GLENN et al., 1960). They concluded that a large number of structural modifications on the C-19 steroid molecule could be made without loss of antitumor activity. For example, from the steroids they examined, the presence of a double bond at C-4 was not necessary nor was a ketone configuration at C-3. However, either a ketone or hydroxyl grouping at C-17 was found to be necessary for antitumor activity. They also demonstrated that some of the steroids that inhibited the growth of the fibroadenoma were ineffective in inhibiting the growth of the testosterone propionate-resistant tumor, although 2-methyl-19-nortestosterone was effective against both the responsive and resistant lines.

To develop a more sensitive assay for screening antitumor activity in this fibroadenoma, ABE et al. (1962) measured the 24-h uptake of [$^{14}$C]-2-glycine into proteins of the fibroadenoma after administration of test compounds for 14 days. In their initial publication, these investigators reported that testosterone propionate was effective in reducing glycine incorporation into fibroadenoma proteins and further, 17$\beta$-hydroxy-2$\alpha$-methyl-5$\alpha$-androstan-3-one (2$\alpha$-methyldihydrotestosterone) was more potent than testosterone propionate in inhibiting glycine uptake. ROOKS et al. (1967) summarized the investigations on the use of this assay and compared the relative potency of a variety of steroids. None of the compounds tested was more active than 2$\alpha$-methyldihydrotestosterone, although 4$\alpha$-methyldihydrotestosterone and 17$\alpha$-methyl-17$\beta$-hydroxy-2-oxa-5$\alpha$-androstan-3-one were equal in potency. A number of other compounds tested demonstrated at least 50% of the potency of the above steroids, some of which included 4-hydroxy-17$\alpha$-methyltestosterone, 7$\alpha$,17$\alpha$-dimethyltestosterone and 2-cyano-17$\beta$-hydroxy-5$\alpha$-androstan-3-one. However, several androstane derivatives were inactive, as were some derivatives of testosterone. $\Delta'$-Testololactone had only borderline activity in this test system, although this compound does demonstrate efficacy in clinical use. This assay system has the obvious advantage of demonstrating activity for 2$\alpha$-methyldihydrotestosterone, a compound that has clinical activity about equal to testosterone propionate; the question still remains as to how well the results translate to antitumor activity in the human disease. Concern has also been expressed regarding the fact that this tumor is a fibroadenoma, rather than a carcinoma.

## 2. Carcinomas

FURTH and CLIFTON (1958), in reviewing the problems in developing techniques for screening compounds that possess antitumor activity, suggested that highly inbred strains of rats be used for the maintenance and study of mammary tumors of varying grades of malignancy and responsiveness. The work of DUNNING (1958), who carefully characterized isologously transplanted mammary carcinomas in several inbred strains of rats, added greatly to the development of models for breast cancer. One such tumor, the R 2426 mammary adenocarcinoma in the August rat, was found to be slightly inhibited by administration of testosterone (2 mg/day) and cortisol (2 mg/day), but not affected by estrone, progesterone, or desoxycorticosterone. Shortly after this report, DUNNING (1960) presented studies on a most interesting mammary tumor, the R 3230, of the Fischer rat. This adenocarcinoma arose spontaneously in an old breeding female and upon initial transplantation studies, was found to grow more rapidly in female recipients than in male hosts. The administration of testosterone (10 mg/day) to male tumor-bearing rats resulted in stimulation of tumor growth, which approximated that seen in the intact female host. Of particular interest was the finding that administration of estrogenic compounds produced noticeable secretory activity in the neoplasm, whereas treatment with androgens caused a reduction in the glandular elements of the tumor.

Both the R3230 and a faster-growing subline, R3230-S, were examined by HILF et al. (1963a) for response to estrogens, androgens, progesterone, and hydrocortisone. These tumors were found to be autonomous, since growth rate was only slightly altered in ovariectomized or castrated hosts; growth of the R3230-S carcinoma was reduced when transplanted to adrenalectomized animals or when adrenalectomy was performed 1 week after tumor implantation. Administration of various doses of testosterone propionate to male tumor-bearing rats stimulated growth of the R3230-S tumor line, as did a high level of progesterone (100 mg/kg/day); estrogen treatment altered tumor growth in a biphasic manner, with low doses stimulating and higher doses inhibiting neoplastic growth. These effects were not seen when hormone administration was delayed until the tumor implant became well established (treatment was initiated 8 days after transplantation). An attempt was made to develop an assay for identifying active antitumor agents utilizing biochemical parameters in these tumors. In this study (HILF et al., 1963b), the R3230-S carcinoma was found to be more responsive than the R3230 tumor, and the response in tumor weight was reflected by an increase in RNA/DNA ratios in the neoplasms. Both estrogen and androgen treatment caused a dose-related increase in tumor RNA/DNA ratios. In contrast, the HMC transplantable tumor, which was essentially unresponsive to hormonal therapy, showed no alteration in RNA/DNA ratios. Several compounds were selected and administered for 13 days to animals bearing R3230-S carcinomas, at which time the animals were sacrificed, tumors were removed, weighed, and analyzed for RNA and DNA (total protein was also determined but showed no differences). The RNA/DNA ratios were significantly elevated in tumors from animals that had received diethylstilbestrol, testosterone propionate, $2\alpha$-methyldihydrotestosterone propionate, $16\alpha$-hydroxytestosterone propionate, 5-fluorouracil and $\Delta'$-testololactone. No effects on tumor RNA/DNA ratios were obtained after treatment with progesterone, A-norprogesterone, A-nortestosterone, and $16\alpha$-fluorotestosterone propionate. Although the increased RNA/DNA ratio resulted from somewhat different effects (i.e., a decrease in DNA, an increase in RNA, or both) by some of the compounds, it was interesting to note that all of the compounds (with the exception of $16\alpha$-hydroxytestosterone propionate) showing an effect on the nucleic acid ratios were shown to be carcinolytic in clinical use. Thus, somewhat like the assay described by ABE et al. (1962), another approach to ascertaining tumor inhibition by effects on a biochemical parameter was described.

In a later publication, another line of the R3230 tumor, called R3230AB, was found to lactate in response to estrogen treatment (HILF et al., 1964). Administration of androgen yielded a dose-related decrease in growth of this autonomous, responsive carcinoma, the inhibition occurring in either intact or castrate hosts. Although the growth inhibition produced by testosterone propionate was not reflected by an alteration in RNA/DNA ratios in the neoplasm, there was a significant decrease in the activities of NADP-malic enzyme and NADP-isocitrate dehydrogenase. In a second experiment, administration of $2\alpha$-methyldihydrotestosterone propionate produced a decrease in the activity of NADP-malic enzyme. This decreased metabolic activity was seen as a flattening of the epithelial cells and an increased density of the stroma, results that confirm the earlier report by DUNNING (1960) on the parent tumor.

Still another line, the R3230AC tumor, was derived from the parent R3230 carcinoma. The R3230AC line had slower growth characteristics and was primarily epithelial by microscopic examination. This carcinoma has been extensively studied on the basis of the early finding by HILF et al. (1965) that administration of estrogens reduced tumor growth concomitantly with a marked lactational response in the tumor. A milk-like fluid that contained casein, lactose, and shorter-chain fatty acids, accumulated in the neoplasm in response to estrogen treatment (HILF, 1967). The effects of testosterone propionate were also examined and, depending on the time of initiation of treatment, the androgen could inhibit, show no effect, or stimulate growth of the tumor. In addition, $\Delta'$-testololactone (a compound lacking androgenic activity but possessing antitumor effects in women with breast cancer) was shown to produce modest inhibition of growth of the R3230AC tumor. It was interesting to note that treatment with $\Delta'$-testololactone appeared to augment the estrogen-induced decrease in tumor growth, but $\Delta'$-testololactone did not augment the androgen-induced inhibition of tumor growth. When testosterone propionate was administered the day after tumor implantation, which resulted in growth inhibition, the activities of glucose-6-phosphate dehydrogenase and NADP-malic enzyme were significantly reduced; when the androgen therapy was administered 30 days after tumor implantation, which stimulated neoplastic growth, the RNA/DNA ratio was elevated and only a slight reduction in glucose-6-phosphate dehydrogenase activity was noted. The decrease in glucose-6-phosphate dehydrogenase and NADP-malic enzyme activities produced by testosterone propionate was not augmented by $\Delta'$-testololactone; however, the estrogen-induced increases in activities of these enzymes were augmented by simultaneous administration of $\Delta'$-testololactone. In general, administration of androgen resulted in a decrease in the activities of the above enzymes in tumors from intact, ovariectomized, orchiectomized, or adrenalectomized animals but not in hypophysectomized rats (HILF et al., 1967). These results were opposite to those seen after treatment with estrogens.

To gain further insight into the androgen-induced inhibition of enzyme activity in the tumor, HILF et al. (1968a) employed actinomycin D and cycloheximide administered concomitantly with testosterone propionate. In the neoplasm, cycloheximide completely prevented the androgen-induced decrease in NADP-malic enzyme activity, and actinomycin D completely prevented the androgen-induced decrease in NADP-isocitrate dehydrogenase activity. However, neither antibiotic affected the androgen-induced decrease in glucose phosphate isomerase activity. From these results, it was concluded that protein synthesis *de novo* was required for some of the androgen-induced alterations in enzyme activities in the R3230AC carcinoma. It is possible that the effects of androgen are indirect via inhibition of pituitary gonadotrophins and effectively produce a decrease in estrogen production. However, the findings that inhibitors of protein synthesis or DNA-directed RNA synthesis demonstrated a capability to interfere with the androgen-induced decrease in enzyme activities cannot be readily explained.

In order to circumvent some of the difficulties and variabilities of response found in primary carcinogen-induced rat tumors (see discussion later), SEGALOFF (1966) developed a transplantable tumor line from a primary, DMBA-induced mammary adenocarcinoma. The inbred Fischer rat was employed, enhancing significantly the possibility for obtaining tumors that would be transplantable. Even so, only one of the primary tumors grew on transplantation, and this neoplasm, 13762, has been used extensively as a model for investigation of antitumor agents. This carcinoma responds to both hypophysectomy and adrenalectomy and grows better in intact females than in ovariectomized hosts. Large doses of androgen stimulated growth of the tumor whereas large doses of estrogen inhibited tumor growth and induced lactation, the latter result similar to that reported for the R3230AC tumor (HILF et al., 1965, 1967). SEGALOFF (1966) also developed different tumor lines by exposing the neoplasm to continuous treatment with estradiol-17$\beta$ and testosterone propionate. Thus, he now offered investigators the opportunity to examine the effects of a variety of hormonal manipulations on responsive and dependent tumors that arose from the same line. The normal male line of 13762 grew most rapidly in all animals whereas the normal female line grew at a slower rate and was inhibited by administration of estrogens and androgens. The androgen-conditioned line displayed significant growth only in animals treated with testosterone propionate.

SEGALOFF (1966) then used the normal female 13762 tumor to carefully and thoroughly examine the effects of administered hormones on tumor growth. The most intriguing result was that one could stimulate or inhibit neoplastic growth on the basis of the dose of steroid used. These results were only apparent because of the very wide range of doses studied, such as 1 µg to 1 g of testosterone propionate per kg per day. SEGALOFF reported that 6$\alpha$-fluorotestosterone propionate, in small doses, increased tumor growth, but at 1 mg/kg/day, inhibited tumor growth equally as well as testosterone propionate; 6$\beta$-fluorotestosterone propionate had no significant effect on tumor growth at any dose tested. Epitestosterone propionate decreased tumor growth as the dose was increased, as did a number of methyltestosterone derivatives given orally. An interesting observation made by SEGALOFF was that $\Delta'$-testololactone, while it did not alter tumor growth when administered alone, did, with diethylstilbestrol, inhibit tumor growth. In this respect, the 13762 tumor resembled the R3230AC tumor. The 13762 is now being used as a primary test system for breast cancer, and it has been shown to be a useful model for investigation of chemotherapeutic agents, particularly with a group of steroidal alkylating agents such as estradiol mustard (BOGDEN et al., 1974). A most exciting finding by BOGDEN and his colleagues was the observation and maintenance of a line of 13762 that routinely metastasizes to other organ sites; such a model has been long sought by experimental oncologists. It is hoped that the data being accumulated by the Endocrine Evaluation Branch of the National Cancer Institute, through its various contractors, will be published shortly; this will add greatly to our knowledge of responsiveness of this 13762 mammary adenocarcinoma.

HILF et al. (1968b) examined a spectrum of enzymes in the 13762 and its sublines as influenced by treatment of the host with androgen or estrogen. They reported that, when growing in their respective hosts, the normal female line had the highest enzyme activities and the normal male line had the lowest enzyme activities. Transplantation of these neoplasms into animals of the opposite sex increased or decreased enzyme activities in the direction of the normal tumor lines of the respective hosts. The androgen-conditioned subline, which grew more rapidly in the presence of exogenous androgen, showed marked elevations in enzyme activities. In general, the alterations in enzyme activities reflected the increased or decreased growth rate of the tumor. Testosterone propionate, which inhibited growth of the normal female tumor line, decreased the activities of glucose-6-phosphate dehydrogenase, NADP-isocitrate dehydrogenase, NADP-malic enzyme, glucose phosphate isomerase, and phosphoglucomutase. Thus, androgens appear to be inhibitory at various metabolic steps, and the responses seen with 13762 indicate that this tumor line holds great promise as an experimental model.

## B. Carcinogen-Induced Mammary Tumors in Rats

The ability of certain chemical carcinogens to induce mammary carcinomas in rats has led to a considerable effort by investigators to exploit fully the potential that such a system offers. Obviously, one can explore the hormonal requirements necessary for tumorigenesis and then examine the hormonal requirements for maintenance of tumor growth. The disadvantage of such a system is the uncertainty of tumor production but even more important, the uncertainty of the properties of each of the multiple tumors that arise in the same animal. As investigators become more sophisticated with the system, many turn their attention to a more careful analysis of the growth patterns of each neoplasm and the response of each lesion to a particular hormonal manipulation. Unfortunately, many of the tumors arising after administration of the carcinogen demonstrate a mixed response, suggesting that the lesions are composed of a spectrum of cell types with somewhat different hormonal and growth properties. Nevertheless, these tumor models have the advantage of arising in situ, rather than the artificial placement of a tumor transplant, and the variety of responses may more accurately reflect the variety seen in the human disease. For the purposes here, it seems logical not to separate the studies performed with either of the two most widely used chemical carcinogens, 3-methylcholanthrene (MCA) or 7,12-dimethylbenz(a)anthracene (DMBA). The pioneering work of SHAY and his colleagues with MCA-induced tumors and of HUGGINS and his collaborators with both MCA- and DMBA-induced carcinomas was largely responsible for drawing the attention of cancer researchers to the value that these systems offer.

Although early studies demonstrated that MCA would elicit mammary cancers in mice (MAISIN and COOLEN, 1936; ORR, 1943; PROHASKA et al., 1939), SHAY et al. (1949) first reported that daily administration of MCA by oral intubation resulted in a high incidence of breast carcinoma in Wistar rats, but that a lower incidence was obtained in male rats. In a series of papers during the succeeding 10 years, SHAY and his colleagues (SHAY et al., 1952, 1957, 1961) refined the techniques to improve the reproducibility of tumor yield. While adrenalectomy had little effect on tumor formation, ovariectomy produced a dramatic decrease in tumorigenesis. Administration of testosterone (via an implanted pellet) at the time MCA treatment was started resulted in a significant decrease in cancer development by increasing the latent period, results similar to those reported by NATHANSON and ANDERVONT (1939) in mice. HUGGINS et al. (1959) developed a rapid method for producing carcinomas in 100% of Sprague-Dawley rats by using a weekly intubation schedule starting at 50 days of age. While small doses of estrogen enhanced the incidence of tumors, large doses hindered tumor development. Most of the established tumors were "hormone-dependent" because they regressed after ovariectomy or hypophysectomy. Administration of dihydrotestosterone (1–2 mg daily) was quite effective in causing tumor regression. However, 10–20% of the tumors continued to grow during hormonal treatment, and these tumors were defined as "hormone-independent." HUGGINS et al. (1959) noted that the cell population of a single tumor was not always uniform, since some cells underwent atrophy while adjacent cells continued to grow. SHAY et al. (1961) confirmed the finding of HUGGINS et al. (1959) regarding modification of dose regimen of MCA to improve the yield of tumors, but did point out that toxicity of the carcinogen was different in Wistar versus Sprague-Dawley rats. SHAY and his co-workers (1961) also pointed out the importance for consideration of not only tumor yield but also the length of time for tumor appearance. Testosterone, when administered (implanted pellet) at the initiation of MCA treatment, not only reduced the number of tumors appearing but also significantly delayed the appearance time. However, when androgen treatment was delayed until 4 months after initiation of carcinogen intubation, no significant alteration in the tumor incidence was noted. The most effective hormonal modification was ovariectomy plus testosterone treatment at 4 months after carcinogen administration. Thus, androgens appear to prevent tumorigenesis when present at the time of carcinogenic insult; MCA does not induce mammary tumors in male rats (DAO and SUNDERLAND, 1959).

About this time, HUGGINS et al. (1961 a, b) reported on the ability of a single dose of DMBA to induce mammary adenocarcinomas in Sprague-Dawley rats. The tumors produced were hormone-dependent; the obvious advantage offered was the ease with which the neoplasms could be generated and subsequently studied. This DMBA-induced tumor model was examined extensively with the goal of adapting it for large-scale screening of hormonal compounds. Two very thorough reports appeared almost simultaneously, one from TELLER et al. (1966) and the other from GRISWOLD et al. (1966), in which the DMBA-induced tumors were carefully evaluated for their potential use in testing antitumor drugs. TELLER et al. (1966) found that tumor growth was rather unpredictable, with about 60% of the carcinomas increasing in size during a 12-week observation period, about 30% remaining relatively static and 10% of the tumors regressing spontaneously; the sample examined contained 253 tumors from untreated or sesame oil-treated rats. Ovariectomy produced regression in 60% of the neoplasms. At a dose of 40 mg/kg/day, the following androgens produced complete remission in 40% or more of the treated rats: 5α-androstan-3α,17β-diol dipropionate, 2α-methyldihydrotestosterone, 2α-methyldihydrotestosterone propionate, 9α-fluoro-11β-hydroxy-17-methyltestosterone, 6β-methyldihydrotestosterone propionate, and the standard employed, testosterone propionate. Δ′-Testololactone was classified as relatively inactive. In the extensive report by GRISWOLD et al. (1966), in which multiple doses of agents were used to examine dose-responsive relationships, they classified the following steroids as demonstrating antitumor activity: 2α-methyldihydrotestosterone, 2α-methyldihydrotestosterone propionate, 17-methyl-19-nortestosterone, 5α-androstan-3α,17β-diol dipropionate, androst-4-ene-3β,17β-diol diacetate, A-nortestosterone propionate, 17β-hydroxy-17α-methyl-5β-androstan-3-one and Δ′-testololactone. The standard used, testosterone propionate, demonstrated antitumor activity as expected. The data by GRISWOLD et al. (1966) indicate some variability in obtaining dose-response curves, such that tumor inhibition was sometimes inconsistent with increasing hormone doses. While both groups of investigators suggest that the DMBA-induced tumor system is a useful model for testing antitumor agents, reservations were expressed on the basis of variability of tumor growth behavior and tumor growth response.

HEISE and GORLICH (1966), studying animals with tumors induced by 3 weekly intubations of DMBA, which yielded more than 50% growing tumors and less than 20% spontaneously regressing neoplasms, found that daily doses of 30 mg/kg of testosterone propionate did not inhibit growth of the carcinomas, and even stimulated tumor growth in some instances. However, at dose levels of 3–6 mg/kg, they reported that tumors regressed in more than 80% of the cases. These results are somewhat difficult to interpret, although one possibility that may be proposed is conversion of the androgen to estrogens, resulting in stimulation of the tumors. In extending their studies, GRISWOLD and GREEN (1970) reported that there was a gradual decrease in the response of DMBA-induced tumors as the time from carcinogen exposure to initiation of treatment was increased. Androgen therapy caused less regression of larger tumors than of smaller tumors and also, androgen treatment initiated immediately after ovariectomy was less effective than ovariectomy alone. It is of interest that $\Delta'$-testololactone did not show this early effect but appeared to augment (or extend) the effects of ovariectomy; this latter effect was also seen with $2\alpha$-methyldihydrotestosterone but not testosterone propionate. Thus, there is general agreement that androgen therapy can induce regression of DMBA-induced tumors, leading one to conclude that this tumor system is a useful model for investigation of agents belonging to the androstane family. It is also interesting that the response of the tumor is not strictly related to androgenic potency, as clearly exemplified by the antitumor activity demonstrated by testosterone propionate, $2\alpha$-methyldihydrotesterone, and $\Delta'$-testololactone.

There are fewer studies of the biochemistry of carcinogen-induced tumors as influenced by administration of androgens than there are investigations of the effects of hormonal deprivation (ovariectomy or hypophysectomy) or of treatment with estrogen. A review of this area appeared recently (HILF, 1973). One of the earliest investigations along these lines was by REES and HUGGINS (1960), who examined several biochemical parameters in MCA-induced tumors regressing as a result of administration of 1 mg/day of dihydrotestosterone for 21 days. They reported that the activities of glucose-6-phosphate dehydrogenase, 6-phosphogluconate dehydrogenase, and lactate dehydrogenase, but not isocitrate dehydrogenase, were decreased in tumors regressing in response to androgen therapy. They concluded that the biochemical responses to androgen therapy were similar to, but lesser in magnitude than, those seen in tumors regressing after ovariectomy. HILF et al. (1969) examined several enzymes in MCA- and DMBA-induced tumors after the tumor-bearing hosts had been treated with 50 mg/kg/day of testosterone propionate for 21 days. In MCA-induced tumors, administration of androgen caused a decrease in glucose-6-phosphate dehydrogenase, NADP-isocitrate dehydrogenase, phosphoglucomutase, and glutamate dehydrogenase activities; these results were similar to those obtained after ovariectomy. Examination of DMBA-induced tumors after treatment with testosterone propionate showed decreased RNA and DNA levels as well as decreased activities of glucose-6-phosphate dehydrogenase and NADP-isocitrate dehydrogenase. Thus, inhibition of tumor growth by androgen treatment resulted in decreases in certain enzymes of carbohydrate metabolism in both carcinogen-induced tumors, results that were similar to the effects of ovariectomy. HEISE and GORLICH (1966) found a decrease in aspartate transcarbamylase activity in DMBA-induced tumors that were regressing as a result of treatment with testosterone propionate, whereas aspartate aminotransferase activity was not decreased; this latter result was confirmed by HILF et al. (1969).

## IV. Mammary Tumors in Vitro

For a long time, there has been an attempt to develop a tumor line in vitro that would demonstrate a parallel behavior to mammary tumors in vivo. For instance, if one could develop and maintain a mammary tumor line that required the presence of a steroid hormone for growth, one might be able to explore in depth those growth-promoting properties without consideration of other hormonal factors. However, caution must be tendered to studies in vitro, particularly with hormone-dependent or hormone-responsive tumors. Such tumors are frequently influenced by multiple hormonal and nonhormonal factors in vivo, and these are usually not readily duplicated in the experiments in vitro. One such simple consideration is the hormone "delivery" system; steroids are transported in the vascular system as both free and protein-bound substances, the latter perhaps acting as a dissociable reservoir or as a regulator inactivating the steroid.

SMITH and KING (1972) reported on their studies of the androgen-dependent Shionogi carcinoma grown in tissue culture. They demonstrated that the tumor cells grew well in the absence of steroids, although growth rate was increased by the addition of testosterone or 5α-dihydrotestosterone (0.01 μg/ml). Addition of 5β-dihydrotestosterone, androst-4-ene-3α,17β-diol, 5α-androstan-3α,17β-diol, 5α-androstan-3β,17β-diol, or androsterone at the same levels (0.01 μg/ml) had no stimulatory effect on tumor cell growth in vitro. In addition, they found that the effects of testosterone were greater when serum (fetal calf serum) concentration was reduced. The effects of testosterone and 5α-dihydrotestosterone on labeling with [$^3$H]-thymidine were observed within 24 h after addition of these hormones, the effect seen being a decrease in the amount of thymidine incorporated per nucleus. A question of importance remains regarding the ability of any hormone(s) to influence growth in an environment that was designed to encourage cell growth.

In a recent study, SUTHERLAND et al. (1974) carefully examined growth of Shionogi carcinoma in male and female mice and then made simple cell suspensions from tumors in male mice and adapted them to grow in tissue culture conditions. They observed that the effects of androgens, either testosterone or dihydrotestosterone, were greatest under conditions that prevented attachment of tumor cells to surfaces, results that were similar to the conclusions reached by SMITH and KING (1972) on the basis of cell density effects on growth. SUTHERLAND et al. (1974) also found a dose-response relationship between colony formation in agar and hormone concentrations over a range of $10^{-4}$ to $10^{-1}$ μg/ml. It should be noted that tumor cells do grow in vitro in the absence of androgens, although growth is much reduced, and SUTHERLAND et al. (1974) concluded from these results that there might be an androgen-independent subpopulation of tumor cells, a conclusion supported by their findings of tumor growth in female mice. Thus, these studies indicate a potentially useful system in vitro for additional studies of hormone-tumor interactions.

Although this literature survey is directed primarily toward rodent tumors, several studies of human tumors in vitro have been performed during the last 5 years. Since the experimental design for studies in vitro, regardless of the cells employed, is at least a simulation of the conditions in vivo, and these conditions are quite similar, human cells studied in vitro can be considered in the same light as rodent cells. Early studies by RIENITS (1959) indicated that testosterone in vitro could depress respiration rates in some human mammary tumors. As tissue culture techniques developed to a sophisticated level, examination of human mammary tumors in vitro became more common (CHAYEN et al., 1970). One such study, published by RILEY et al. (1973), demonstrated the influence of various steroid hormones on the rate of DNA synthesis of cells cultured from 14 breast tumors. At $10^{-5}$ M, testosterone had a variable effect on incorporation of labeled thymidine into DNA; it stimulated, inhibited, or did not effect the rate of DNA synthesis, depending on the individual tissue sample. On overall average, the androgen effect was seen as a slight decrease in DNA synthesis. FLAX et al. (1973), employing methods that utilize tissue slices (organ culture) of human tumors and then ascertaining cell viability by pentose phosphate pathway activity, reported that 14/131 breast cancer tissues were dependent on testosterone. This conclusion of dependence is based on the androgen-induced increase in enzyme activity; the authors suggest that antiandrogen therapy should be successful and claim that this was the case for six of eight patients. BURSTEIN and CAREY (1974) have reported on the effects of various steroid hormones upon the uptake of labeled thymidine into breast carcinoma in vitro and correlated their results with clinical response to therapy. The best correlation seen was with the androgens; three patients, whose tumors were inhibited by testosterone in vitro, responded favorably to androgen therapy. Three other patients received androgen therapy and showed no clinical response; their tumors also showed no response to testosterone in vitro. These results are very encouraging and indicate that responses in vitro may be readily translatable to therapy in vivo.

FINKELSTEIN and his co-workers (1975) examined the effects of testosterone and estradiol-17$\beta$ on the synthesis of DNA, RNA, and protein in organ explants of human breast cancer and nonmalignant diseases. In cultures of fibrocystic disease, testosterone ($6.9 \times 10^{-5}$ M) uniformly inhibited the incorporation of thymidine into DNA, uridine into RNA, and L-amino acids into proteins; testosterone has similar effects on fibroadenomas. In cultures of cancerous tissues, however, testosterone depressed in most cases the incorporation of these labeled precursors into their respective macromolecules, but in 4/17 cancers studied, testosterone stimulated the precursor incorporation. Although they presented the clinical course of 7 patients, any correlation seen at this time was admittedly preliminary. It will be of interest to learn more concerning the attempt by these investigations to correlate clinical course with anticipated response on the basis of the studies in vitro.

## V. Effects of Androgens on Other Experimental Tumor Systems

Since androgen therapy of breast cancer has been utilized rather extensively in women with disseminated disease, the major effort of investigators has been directed to studies on experimental mammary tumors. However, androgens have been investigated for antitumor effects in neoplasms arising from other tissues. A very thorough survey was written by WADE (1967), in which he reviewed the effects of several androgens on a variety of experimental tumors. Antitumor activity has been reported against lymphomas, Ehrlich carcinoma (solid and ascites forms), leukemia and adrenal cortical tumors in mice, against hepatoma, uterine, prostatic, ovarian, and lymphosarcomas in rats, a renal carcinoma in hamsters, uterine fibroids in the guinea pig, and several human tumors grown in chick embryo. The problem that one faces is the difficulty in drawing any general conclusions and the need for more thorough study of the effects of androgens on any well-characterized tumor system.

## VI. Conclusions

From the studies cited here, one can conclude that there are experimental rodent mammary tumors that are inhibited by treatment with androgens or structurally related analogues. It is also apparent that some of these systems, like the results obtained in the clinical setting, are inhibited by compounds that possess weaker androgenicity than the standard, testosterone. These observations reinforce the proposal that antitumor activity may be achieved without the unwanted hormonal properties, masculinization in the female patient. How much further this separation of activities will progress toward finding of new agents cannot be predicted with certainty.

The availability of different tumor models should allow investigation of mechanism of action of androgens. Certainly, the early discovery that testosterone could prevent or delay the appearance of spontaneous mammary tumors in mice would point toward a re-examination of the role of this hormone in neoplastic conversion, either by virus, x-radiation, or chemical carcinogen. Whether this effect is direct, or indirect via other hormonal alterations, needs to be clarified. The finding of an androgen-dependent mouse mammary carcinoma, or the androgen-dependent line of the transplantable 13762 DMBA-induced rat carcinoma, represents two models possessing interesting properties; further studies with these systems should lead to a better understanding of the mechanisms whereby tumor growth is stimulated as well as to the elucidation of the events of tumor regression that occur after withdrawal (or antagonism) of the required hormone. It will be of interest to examine compounds possessing antiandrogenic activity to determine if they can effect tumor regression.

Utilization of transplantable tumors which, although autonomous, will respond to androgen therapy may represent models somewhat akin to the metastatic or recurrent disease in women that responds to treatment with androgens. The biochemical events that are altered by such therapy require more thorough examination, an approach that may shed light on those metabolic events that are critical for maintenance of tumor growth. It is also possible that differences in biochemical characteristics will be identified for dependent tumors and responsive tumors, results that might lead to a predictive index of the best therapeutic modality. Studies such as those performed with estrogen receptors may be extended to include androgen receptors as one approach to definition of hormonal characteristics.

*Acknowledgments:* Studies on the characterization of responses and biochemistry of several mammary tumors investigated in the author's laboratory were supported by Contracts SA-43-

ph-2395 and PH-43-65-1050, Endocrine Evaluation Branch, National Cancer Institute and currently by USPHS Grants CA-12836 and CA-11198. The competent technical assistance of C. BELL, J. J. FREEMAN, H. GOLDENBERG, and I. MICHEL is particularly noted. Appreciation is expressed to Miss MARGERY PALMER for her help with the bibliography and for typing the manuscript.

## References

ABE, O., HERRANEN, A., DORFMAN, R. I.: Influence of 2α-methyl-17β-hydroxy-5α-androstan-3-one on incorporation of glycine-2-$C^{14}$ into proteins of a rat mammary fibroadenoma. A possible bioassay method. Proc. Soc. exp. Biol. (N.Y.) **111**, 706—708 (1962).

BOGDEN, A. E., ESBER, H. J., TAYLOR, D. J., GRAY, J. H.: Comparative study on the effects of surgery, chemotherapy, and immunotherapy, alone and in combination, on metastases of the 13762 mammary adenocarcinoma. Cancer Res. **34**, 1627—1631 (1974).

BRUCHOVSKY, N., MEAKIN, J. W.: The metabolism and binding of testosterone in androgen-dependent and autonomous transplantable mouse mammary tumors. Cancer Res. **33**, 1689—1695 (1973).

BRUCHOVSKY, N., SUTHERLAND, D. J. A., MEAKIN, J. W., MINESITA, T.: Androgen receptors: relationship to growth response and to intracellular androgen transport in nine variant lines of the Shionogi mouse mammary carcinoma. Biochim. biophys. Acta (Amst.) **381**, 61—71 (1975).

BURSTEIN, N. A., CAREY, R. W.: In vitro assay for human breast cancer hormone responsiveness. Oncology **29**, 470—483 (1974).

CHAYEN, J., ALTMANN, F. P., BITENSKY, L., DALY, J. R.: Response of human breast cancer tissue to steroid hormones in vitro. Lancet **1970 I**, 868—870.

Cooperative Breast Cancer Group: Testosterone propionate therapy of breast cancer. J. Amer. med. Assoc. **188**, 1069—1072 (1964).

DAO, T. L., SUNDERLAND, H.: Mammary carcinogenesis by 3-methylcholanthrene. I. Hormonal aspects in tumor induction and growth. J. nat. Cancer Inst. **23**, 567—585 (1959).

DORFMAN, R. I.: Inhibition of tumor growth by steroids. Methods in Hormone Research, Vol. 4, Chap. 7, pp. 165—192. New York: Academic Press 1965.

DUNNING, W. F.: Response of some isologously transplanted rat neoplasms to steroids. Ann. N.Y. Acad. Sci. **76**, 696—704 (1958).

DUNNING, W. F.: Steroid-responsive neoplasms in rats and mice. Biological Activities of Steroids in Relation to Cancer, pp. 225—256. New York: Academic Press 1960.

EMGE, L. A., MURPHY, K. M.: Effect of rapidly repeated pregnancies on transplantable mammary rat adenofibromas. Proc. Soc. exp. Biol. (N.Y.) **37**, 620—621 (1938).

FELS, E.: Treatment of breast cancer with testosterone propionate. A preliminary report. J. clin. Endocr. **4**, 121—125 (1944).

FINKELSTEIN, M., GEIER, A., HORN, H., LEVIJ, I. S., EVER-HADANI, P.: Effect of testosterone and estradiol-17β on synthesis of DNA, RNA and protein in human breast in organ culture. Int. J. Cancer **15**, 78—90 (1975).

FLAX, H., SALIH, H., NEWTON, K. A., HOBBS, J. R.: Are some women's breast cancers androgen dependent? Lancet **1973 I**, 1204—1207.

FURTH, J., CLIFTON, K. H.: Screening techniques and problems of hormone-responsive tumors. Ann. N.Y. Acad. Sci. **76**, 681—695 (1958).

GLENN, E. M., RICHARDSON, S. L., BOWMAN, B. J.: A method of assay of antitumor activity using a rat mammary fibroadenoma. Endocrinology **64**, 379—389 (1959a).

GLENN, E. M., RICHARDSON, S. L., BOWMAN, B. J., LYSTER, S. C.: Steroids and experimental mammary cancer Biological Activities of Steroids in Relation to Cancer, pp. 257—305. New York: Academic Press 1960.

GLENN, E. M., RICHARDSON, S. L., LYSTER, S. C., BOWMAN, B. J.: Inhibition of mammary fibroadenoma of female rats by steroids of the androstane series. Endocrinology **64**, 390—399 (1959b).

GRISWOLD, D. P., GREEN, C. H.: Observations on the hormone sensitivity of 7,12-dimethylbenz-(a)anthracene-induced mammary tumors in the Sprague-Dawley rat. Cancer. Res. **30**, 819—826 (1970).

Griswold, D. P., Laster, W. R., Snow, M. Y., Schabel, F. M., Skipper, H. E.: Experimental evaluation of potential anticancer agents. XII. Quantitative drug response of SA 180, CA 755 and leukemia L 1210 systems to a standard list of active and inactive agents. Cancer Res. **23**, Part 2, 271—519 (1963).

Griswold, D. P., Skipper, H. E., Laster, W. R., Jr., Wilcox, W. S., Schabel, F. M., Jr.: Induced mammary carcinoma in the female rat as a drug evaluation system. Cancer Res. **26**, 2160—2180 (1966).

Heiman, J., Krehbiel, O. F.: The influence of hormones on breast hyperplasia and tumor growths in white rats. Amer. J. Cancer **27**, 450—473 (1936).

Heise, E., Gorlich, M.: Growth and therapy of mammary tumors induced by 7,12-dimethylbenzanthracene in rats. Brit. J. Cancer **20**, 539—545 (1966).

Hilf, R.: Milk-like fluid in a mammary carcinoma: biochemical characterization. Science **155**, 826—827 (1967).

Hilf, R.: Biochemical studies of experimental mammary tumors as related to human breast cancer. Meth. Cancer Res. **7**, 55—114 (1973).

Hilf, R., Freeman, J. J., Johnson, M. M., Stagg, R., Borman, A.: Effects of steroids on three transplantable mammary tumors in the Fischer rat. Cancer Chemother. Rpts. **30**, 1—8 (1963a).

Hilf, R., Freeman, J. J., Michel, I., Borman, A.: Characterization of a transplantable lactating mammary tumor: endocrinological, morphological and biochemical aspects. Cancer Res. **24**, 812—824 (1964).

Hilf, R., Goldenberg, H., Michel, I., Carrington, M. J., Bell, C., Gruenstein, M., Meranze, D. R., Shimkin, M. B.: Biochemical characteristics of mammary glands and mammary tumors of rats induced by 3-methylcholanthrene and 7,12-dimethylbenz(a)anthracene. Cancer Res. **29**, 977—988 (1969).

Hilf, R., Johnson, M. M., Breuer, C., Freeman, J. J., Borman, A.: Comparative biochemistry of three transplantable mammary tumors as influenced by steroid therapy. J. nat. Cancer Inst. **31**, 541—555 (1963b).

Hilf, R., Lerner, L. J., Harris, D. N.: The effect of actidione and actinomycin D on androgen-induced biochemical changes in the R 3230 AC mammary carcinoma and uterus of the rat. Trans. N.Y. Acad. Sci. **30**, 794—803 (1968a).

Hilf, R., Michel, I., Bell, C.: Biochemical and morphological responses of normal and neoplastic mammary tissue to hormonal treatment. Recent Progr. Hormone Res. **23**, 229—295 (1967).

Hilf, R., Michel, I., Bell, C., Freeman, J. J., Borman, A.: Biochemical and morphological properties of a new lactating mammary tumor line in the rat. Cancer Res. **25**, 286—299 (1965).

Hilf, R., Segaloff, A., Lerner, L. J.: Influence of hormonal treatment of Fischer rats on the biochemistry of a transplantable dimethylbenz(a)-anthracene-induced mammary carcinoma and its sublines. Cancer Res. **28**, 1550—1558 (1968b).

Hilgar, A. G., Hummel, D. J.: Androgenic and myogenic endocrine bioassay data. Bethesda, Md.: Cancer Chemotherapy National Service Center, 1964.

Hirschberg, E.: Patterns of response of animal tumors to anticancer agents. Cancer Res. **23**, Part 2, 521—980 (1963).

Huggins, C., Briziarelli, G., Sutton, H., Jr.: Rapid induction of mammary carcinoma in the rat and the influence of hormones on the tumors. J. exp. Med. **109**, 25—42 (1959).

Huggins, C., Grand, L. C., Brillantes, F. P.: Mammary cancer induced by a single feeding of polynuclear hydrocarbons, and its suppression. Nature (Lond.) **189**, 204—207 (1961a).

Huggins, C., Mainzer, K.: Hormonal influences on mammary tumors of the rat. II. Retardation of growth of a transplanted fibroadenoma in intact female rats by steroids in the androstane series. J. exp. Med. **105**, 485—501 (1957).

Huggins, C., Morii, S., Grand, L. C.: Mammary cancer induced by a single dose of polynuclear hydrocarbons: routes of administration. Ann. Surg. **154**, 315—318 (1961b).

Huggins, C., Torralba, Y., Mainzer, K.: Hormonal influences on mammary tumors of the rat. I. Acceleration of growth of transplanted fibroadenoma in ovariectomized and hypophysectomized rats. J. exp. Med. **104**, 525—543 (1956).

Jensen, E. V., Block, G. E., Smith, S., Kyser, K., Desombre, E. R.: Estrogen receptors and hormone dependency. Estrogen Target Tissues and Neoplasia, pp. 23—58. Chicago: University of Chicago Press 1972.

JUNGBLUT, P. W., HUGHES, S. F., GORLICH, L., GOWERS, U., WAGNER, R. K.: Simultaneous occurrence of individual estrogen and androgen receptors in female and male target organs. Hoppe-Seylers Z. Physiol. Chem. **352**, 1603—1610 (1971).

KING, R. J. B., MAINWARING, W. I. P.: Steroid-Cell Interactions. Baltimore: University Park Press 1974.

KORENCHEVSKY, V., DENNISON, M.: Histological changes in the organs of rats injected with oestrone alone or simultaneously with oestrone and testicular hormone. J. Path. Bact. **41**, 323—337 (1935).

LACASSAGNE, A.: Tentatives pour modifier, par la progesterone ou par la testosterone, l'apparition des adenocarcinomes mammaires provoques par l'oestrone chez la souris. C. R. Soc. Biol. (Paris) **126**, 385—387 (1937).

LOESER, A. A.: Mammary carcinoma: response to implantation of male hormone and progesterone. Lancet **2**, 698—700 (1941).

MAINWARING, W. I. P., MANGAN, F. R.: A study of the androgen receptors in a variety of androgen-sensitive tissues. J. Endocr. **59**, 121—139 (1973).

MAISIN, J., COOLEN, M.-L.: Au sujet du pouvoir cancerigene du methylcholanthrene. C. R. Soc. Biol. (Paris) **123**, 159—160 (1936).

MATSUMOTO, K., KOTOH, K., KOSAI, H., MINESITA, T., YAMAGUCHI, K.: Subcellular localization of radioactive steroids following administration of testosterone-$^{3}$H in the androgen dependent mouse tumor, Shionogi carcinoma 115. Steroids **20**, 311—320 (1972).

MINESITA, T., YAMAGUCHI, K.: An androgen-dependent mouse mammary tumor. Cancer Res. **25**, 1168—1175 (1965).

MOHS, F. E.: Effect of estrogen and androgens on growth of mammary fibroma in rats. Proc. Soc. exp. Biol. (N.Y.) **43**, 270—272 (1940).

NATHANSON, I. T.: Endocrine aspects of human cancer. Recent Progr. Hormone Res. **1**, 261—291 (1946).

NATHANSON, I. T., ANDERVONT, H. B.: Effect of testosterone propionate on development and growth of mammary carcinoma in female mice. Proc. Soc. exp. Biol. (N.Y.) **40**, 421—422 (1939).

ORR, J. W.: Mammary carcinoma in mice following the intranasal administration of methylcholanthrene. J. Path. Bact. **55**, 483—488 (1943).

PROHASKA, J. V., BRUNSCHWIG, A., WILSON, H.: Oral administration of methylcholanthrene to mice. Arch. Surg. **38**, 328—333 (1939).

REES, E. D., HUGGINS, W.: Steroid influences on respiration, glycolysis, and levels of pyridine nucleotide-linked dehydrogenases of experimental mammary cancers. Cancer Res. **20**, 963—971 (1960).

RIENITS, K. G.: The effects of estrone and testosterone on respiration of human mammary cancer in vitro. Cancer **12**, 958—961 (1959).

RILEY, P. A., LATTER, T., SUTTON, P. M.: Hormone assays on breast tumor cultures. Lancet **1973II**, 818—821.

ROOKS, W. H.: Anti-mammary tumor activities in rats and mice. Methods in Hormone Research, Volume III, Chapter 4, pp. 139—184. New York: Academic Press 1964.

ROOKS, W. H., BABA, S., ABE, O., HARADA, T., DORFMAN, R. I.: Influence of various compounds, particularly steroids, on a transplantable rat mammary fibroadenoma and a transplantable mouse mammary adenocarcinoma. Current Concepts in Breast Cancer, pp. 63—79. Baltimore: Williams and Wilkins Co. 1967.

SEGALOFF, A.: Hormones and breast cancer. Recent Progr. Hormone Res., **22**, 351—379 (1966).

SEGALOFF, A., GORDON, D., HORWITT, B. N., SCHLOSSER, J. V., MURISON, P. J.: Hormonal therapy in cancer of the breast. I. The effect of testosterone propionate therapy on clinical course and hormonal excretion. Cancer **4**, 319—323 (1951).

SEGALOFF, A., HORWITT, B. N., CARABASI, R. A., MURISON, P. J., SCHLOSSER, J. V.: Hormonal therapy in cancer of the breast. V. The effect of methyltestosterone on clinical course and hormonal excretion. Cancer **6**, 483—487 (1953).

SHAY, H., AEGERTER, E. A., GRUENSTEIN, M., KOMAROV, S. A.: Development of adenocarcinoma of the breast in the Wistar rat following the gastric instillation of methylcholanthrene. J. nat. Cancer Inst. **10**, 255—266 (1949).

SHAY, H., GRUENSTEIN, M., KESSLER, W. B.: Experimental mammary adenocarcinoma of rats: some considerations of methylcholanthrene dosage and hormonal treatment. J. nat. Cancer Inst. **27**, 503—513 (1961).

Shay, H., Harris, C., Gruenstein, M.: Influence of sex hormones on the incidence and form of tumors produced in male or female rats by gastric instillation of methylcholanthrene. J. nat. Cancer Inst. **13**, 307—331 (1952).

Shay, H., Harris, C., Gruenstein, M.: Studies in prophylaxis of methylcholanthrene induced breast cancer in the rat. Proc. II Internat. Symp. Mammary Cancer, pp. 753—764. Univ. Perugia, 1957.

Smith, J. A., King, R. J. B.: Effects of steroids on growth of an androgenindependent mouse mammary carcinoma in cell culture. Exp. Cell. Res. **73**, 351—359 (1972).

Stock, C. C., Sugiura, K.: Screening steroids against a spectrum of tumors. Ann. N.Y. Acad. Sci. **76**, 720—728 (1958).

Sutherland, D. J. A., Robins, E. C., Meakin, J. W.: Effects of androgens on Shionogi carcinoma 115 cells in vitro. J. nat. Cancer Inst. **52**, 37—48 (1974).

Takatani, O., Kumaoka, S., Abe, O., Sakauchi, N., Sato, R., Mori, H.: The effect of steroidal diosphenol agents on advanced breast cancer. Endocr. Jap. **14**, 195—208 (1967).

Teller, M. N., Stock, C. C., Stohr, G., Merker, P. C., Kaufman, R. J., Escher, G. C., Bowie, M.: Biologic characteristics and chemotherapy of 7,12-dimethylbenz(a)-anthracene-induced tumors in rats. Cancer Res. **26**, 245—252 (1966).

Ulrich, P.: Testosterone (hormone mâle) et son rôle possible dans le traitement de certains cancers du sein. Acta Union Internat. contre Cancer **4**, 377 (1939).

Wade, R.: Hormones. Experimental Chemotherapy, Chemotherapy of Neoplastic Diseases, Vol. 5, pp. 133—331. New York: Academic Press 1967.

Yamaguchi, K., Kasai, H., Minesita, T., Kotch, K., Matsumoto, K.: 5α-Reduction and binding of testosterone in androgen-dependent and -independent mouse mammary tumors. Endocrinology **95**, 1424—1430 (1974).

# IV. Mechanism of Action

## A. Regulation of Protein Synthesis by Anabolic-Androgenic Steroids

C. D. KOCHAKIAN

### I. Introduction

The many studies on the changes in weight of the muscles and several organs with proportionate changes in protein content established that the positive nitrogen balance stimulated by testosterone and related compounds was an expression of the synthesis of protein in many tissues other than the accessory sex organs. A more direct approach to the nature of the stimulation of protein synthesis by testosterone was made possible by the availability of radioactively labeled amino acids and the biochemical characterization in vitro of the various steps in protein biosynthesis.

### II. Studies in Vivo

#### A. Mouse

The early studies by NOVAK (1957) were limited to one mouse per determination because of the excessive cost of the radioactively labeled amino acids. The mice, however, were adults of uniform weight and genetic strain. The exact age or body weight of the mice was not reported. Castration was performed 3–6 days before the injection of 0.2 mg/day of testosterone propionate and a simultaneous daily injection of $^{14}C$-labeled (μCi not given) leucine, glycine, histidine, or lysine. The mice were killed at 1, 3, 5, and 7 days, and in another set of experiments at 7, 14, and 21 days, to determine the rate of synthesis of protein. The rate of protein degradation was determined in two different experiments: (1) the daily injection of [$^{14}C$]-lysine was terminated on day 7, and one mouse from each group was killed on subsequent days; (2) normal mice of the same age were injected daily with [$^{14}C$]-leucine (300300 cpm) for 9 days, at which time half of the mice were castrated, and one normal and one castrated mouse were killed at intervals of up to 21 days. All of the viscera and the muscles of the hind legs were made into two respective pools for analysis. In general, the expected increase with time of amino acid incorporation into protein, except for histidine, was accentuated after castration in both the muscles and the pooled viscera. With testosterone propionate, incorporation returned to normal levels in the viscera but was increased further in the muscles. In another set of experiments, neither testosterone propionate nor 17α-methylandrost-5-ene-3β,17β-diol at 0.25 mg/day for up to 15 days significantly altered the rate of incorporation of the amino acids into the tissues of the castrated mice. Testosterone propionate treatment produced a suggestive decrease in the rate of degradation of both muscle and visceral protein. It is unfortunate that the visceral organs were pooled for analysis. An effect in the weight-responsive kidney most likely was masked by the presence of the other tissues.

KOCHAKIAN (1971, unpublished data) castrated mice at 1 month of age and about 2 months later implanted subcutaneously a 15-mg pellet of testosterone propionate for 2 days. Two hours before autopsy, 30 µCi or [2-$^{14}$C]-glycine was injected intraperitoneally. The incorporation of radioactivity into the protein of the kidney was not changed by castration but was increased by approximately 40% in the treated mice.

ICHII and IKEDA (1970) determined the half-life of protein in the homogenate, homogenate acid-soluble fraction, and the mitochondrial and postmitochondrial fractions of the liver, kidney, and spleen of mice castrated at 21–23 g body weight. 2 µCi of [guanidino-$^{14}$C]-L-arginine (45.7 mCi/mmole) were injected intraperitoneally 5–7 days after the operation, and groups of 5 mice were killed 1, 3, 5, and sometimes 7 days later. The organs were pooled separately and fractionated for analysis. The half-lives of the protein of the homogenates of each organ were increased (20% in liver and kidney, 70% spleen), but those of the cellular fractions were unchanged except possibly those of the acid-soluble and postmitochondrial fractions of the spleen. No explanation was provided for the apparent discrepancy between the homogenates and the cellular fractions.

KOTHS et al. (1972) determined the effect of castration and testosterone on the rate of protein synthesis and degradation in the mitochondrial and microsomal proteins of mouse kidney. The mice were castrated at 8 weeks of age and were killed at intervals of up to 32 h and 5 µCi of a mixture of [U-$^{14}$C] amino acids was injected 30 min prior to autopsy. The specific radioactivity of the protein of the mitochondria from the castrated mice rapidly decreased to about 65% of normal by the 8th day and then rapidly increased to normal at about 21 days and continued to increase above normal. The injection of 2.5 mg of testosterone at day 5 reversed the decrease within 14 h. Degradation of mitochondrial and microsomal protein was studied by the injection of [guanidino-$^{14}$C]-arginine at 0, 7, and 14 days after castration. The decay of mitochondrial and microsomal radioactivities was determined at 4 different times between each 7-day interval. No change in the half-life of the protein of either cellular fraction was observed. The decrease in kidney weight following castration, therefore, can be attributed primarily, if not entirely, to the decreased rate of protein synthesis. The protein of the liver fractions also showed no change.

## B. Rat

### 1. [$^{14}$C] Amino Acids

BUTENANDT et al. (1960) applied 10 mg of testosterone in 0.2 ml alcohol to the shaven abdomen of anesthetized (Evipan 15 mg/20 g) juvenile (22–28 days old) male rats and 10 h later injected subcutaneously 1 µCi [$^{14}$C]-L-leucine (1 µCi/mmole). The incorporation of radioactivity in the protein of the rectus femoris muscle did not change in the following 30–270 min. In contrast, a 100% increase in the incorporation of the leucine in the protein of the seminal vesicles was noted within 90 min and a second increase after 360 min of testosterone application.

NIMNI and BAVETTA (1961) castrated rats at 2 months of age and 15 days later implanted subcutaneously a sterile polyvinyl sponge to study the rate of synthesis of a newly formed protein, e.g., collagen. Two weeks later, 2.5 mg/day of testosterone propionate was injected intramuscularly for 3 days. Simultaneously with the last injection, 5 μCi of [1-$^{14}$C]-glycine (4.5 mCi/mmole) was injected. Groups of rats were autopsied, 5, 10, 20, 40, and 60 h later. The incorporation of the amino acids was greatly increased in the protein and acid-soluble fraction of the seminal vesicles and the perineal muscles of the treated rats. No consistent significant difference in the rate of amino acid incorporation was detected in the serum, liver, kidney, heart, diaphragm, and gluteus medius; the skin and polyvinyl sponge showed a small increase. DOEG et al. (1971) failed to detect any change in the rate of incorporation of a mixture of [U-$^{14}$C] amino acids into the purified mitochondrial protein of the liver of rats, but that of the accessory sex tissues was greatly increased. The rats were castrated at 6–7 weeks of age and 2–3 weeks later were injected subcutaneously with 1 mg of testosterone per 100 g body weight and were killed 3, 6, 12, 18, 24, and 72 h later. One h before autopsy, 10 μCi of a mixture of [U-$^{14}$C] amino acids was injected intravenously.

MOSEBACH et al. (1967) injected immature (25 days of age) male rats with 0.1 mg/day testosterone for 5 days followed by an intraperitoneal injection of 2 μCi of $^{14}$C-ring-labeled L-histidine; the rats were starved for 15 h before the histidine injection. The incorporation of radioactivity into liver and kidney protein was not affected but that of the thymus was decreased and that of the seminal vesicles and "levator ani" increased. SEIFULLA et al. (1974), however, observed an increase in the incorporation of [1-$^{14}$C]-glycine into the protein of the liver and a decrease in that of the thymus 4 h after the injection of 5 or 10 mg/kg methandrostenolone to rats. A decrease occurred also in the spleen but not until 48 h postinjection.

The oral administration of one of the newer steroids is reported to accelerate the incorporation of [2-$^{14}$C]-glycine into certain tissues of the rat (YUDAEV and POKROVSKII, 1966). Methandrostenolone (17β-hydroxy-17α-methylandrosta-1,4-diene-3-one) was given orally at 1 mg/100 g body weight to male rats for 7 days. On the 8th day, 1.5 μCi/100 g body weight of [2-$^{14}$C]-glycine was given, and the incorporation rate was accelerated in the tibial muscles of the adult rats and in the tibial muscles, kidneys, and heart of the immature rats.

FERRARI et al. (1968) found no change in the specific radioactivity of the "levator ani" muscle or the iliopsoas muscle of 60-day-old male rats injected with a single i.m. dose of 50 mg of testosterone propionate or 17β-hydroxy-13β,17α-diethylgon-4-en-3-one (norbolethone) and 12 h later with 10 μCi [$^{14}$C]-L-leucine (305 μCi/mmole) followed by autopsy after another 4 h. The steroids doubled the weight of the seminal vesicles. DE RUGGIERI (1963) observed that the rate of incorporation of [$^{14}$C]-aminoisobutyric acid was not changed after the oral administration of 17α-methyltestosterone but was increased 2.3-fold by Dimetazine. The rats were castrated at 156 g body weight. The steroids (2 mg) were administered on the day of castration and 24 and 34 h later. At 48 h after castration, 2 μCi [1-$^{14}$C]-aminoisobutyric acid was injected subcutaneously, and the rats were killed 9 h later.

## 2. [$^{14}$C]-Formate

FARABOLLINI (1962) injected adult (250 g) male rats for 4 days with 5 mg/day of testosterone. Four h after the last injection, 10 μCi of formate (111 μg) was injected intraperitoneally, and 2 h later the liver and kidneys were removed. The radioactivity in the protein of the liver and kidney was 7 and 18% lower in the treated rats. These changes probably were not significant. JAKUBOVIC and CEKAN (1966) also used [$^{14}$C]-formate, but in infantile castrated rats, for their studies. Castration was performed 4 days before 1 or 8 mg of testosterone phenylpropionate was injected. Autopsy was 6, 12, 24, and 72 h and 7, 21, and 35 days later. One hour before autopsy the rats were injected intraperitoneally with 20 μCi [$^{14}$C]-formate. A 20% increase in kidney weight was observed after 7 and 21 days, with a significant increase in the radioactivities of the acid-soluble fraction, nucleic acids, and lipids but not of the protein at the 24-h interval. The liver showed a significant (27%) increase in weight at the 7-day interval and an increase in the radioactivities of the nucleic acids, lipids, and proteins at the 72-h interval. The weight of the soleus muscle was not increased by the steroid, and the radioactivity of only the nucleic acids was increased at the 12-, 24-, and 72-h intervals. The radioactivities in the different fractions of the "levator ani" and accessory sex organs were markedly increased. The increase in radioactivity in all of the organs always preceded the observed increases in weight.

### 3. Anticatabolic Effects

BRESLOFF et al. (1974) studied the ability of nandrolone phenylpropionate to prevent or reverse the decrease in the rate of [U-$^{14}$C]-leucine incorporation into muscle protein observed after hypophysectomy, fasting, or corticosteroid treatment of rats. Young male rats were injected with 1.25 mg of nandrolone phenylpropionate, food was removed 3 days later, and after 20 h 1.5 μCi (50 μg) [U-$^{14}$C]-leucine was injected intravenously 2 h before autopsy. Amino acid incorporation was slightly decreased in the gastrocnemius muscle but was greater than normal in the "levator ani." Immature female rats were hypophysectomized, and 1.0 mg of the steroid was injected after 10 days and another 1.0 mg 2 days later. The amino acid incorporation was determined on the following day. The steroid did not modify the decrease in amino acid incorporation in the proteins of the gastrocnemius muscle. The decrease in protein synthesis produced in the gastrocnemius and soleus muscles of rats by the injection of 50 or 100 μg of dexamethasone sodium phosphate for 2 days was not modified by the injection of 2.5 mg of nandrolone phenylpropionate 3 days preceding the injection of the [U-$^{14}$C]-leucine, but that of the "levator ani" was reversed. Similar results were obtained by the combination of 1.0 or 5 mg of the steroid with 200 μg of the corticosteroid.

## C. Guinea Pig

COSTA et al. (1962) castrated guinea pigs at 490–540 g body weight. Three weeks later the guinea pigs were injected subcutaneously for 3 weeks with 10 mg/day of testosterone propionate. The food was removed on the 21st day of injection, and 12 h later each guinea pig was injected intraperitoneally with 50 μCi of [2-$^{14}$C]-glycine (1.24 μCi/mmole). After another 12 h, all of the organs, 47 of the skeletal muscles, the retractor penis, the seminal vesicles, and the prostates were removed separately and extracted to give the protein and acid-soluble fraction. The specific radioactivities (amino acid incorporation) of the liver, ileum, stomach, urinary bladder, kidney, spleen, and heart were not changed by either castration or the testosterone propionate treatment. Castration produced a decrease in the specific radioactivity of only one muscle, the diaphragm. The injection of testosterone propionate restored the value to normal and increased the specific radioactivities of the protein of the temporal muscle and the retractor penis above normal. None of the rest of the 47 muscles showed significant changes.

In a subsequent, similar study (unpublished, cf. KOCHAKIAN, 1963), the testosterone was implanted subcutaneously into castrated guinea pigs as two 15-mg pellets for 14 days, and 50 μCi of [2-$^{14}$C]-glycine was injected 4 h before autopsy. The specific radioactivity of the protein of the temporal muscle was 17% lower in the castrated and 113% higher in the treated guinea pigs than in the normal adults. In contrast with the muscles, the seminal vesicles and prostates showed a significant decrease in the specific radioactivities of not only the protein but also the nonprotein (acid-soluble) fraction after castration and an increase markedly above normal after testosterone propionate treatment. The radioactivity of the protein in the seminal vesicle fluid also was greater (approx. twofold) in the treated than in the normal guinea pigs (unpublished, cf. KOCHAKIAN, 1963).

As further proof of incorporation of amino acid, the protein of the responsive tissues was hydrolyzed, the amino acids were separated, and the specific radioactivity of each amino acid was determined. Most of the radioactivity was in the glycine and serine. Small amounts were also present in two other metabolites—glutamic and aspartic acids. The specific radioactivities of the individual amino acids changed in direct proportion to the changes noted in the weight of the responsive muscles and the accessory sex organs.

## D. Rabbit

MEISSNER (1964) reported an increased incorporation of [$^{35}$S]-methionine into the thigh muscles of the rabbit (age and sex were not given) after the injection of one of the newer steroids, methenolone. The steroid was administered at 9 and 5 days preceding the intravenous injection of 0.9 mg (80 μCi/kg) of [$^{35}$S]-methionine. No change was detected after 3 h, but a definite increase was apparent after 6 h, which was further increased after 12 h. Methenolone was twice as effective at the 33 mg/kg as at the 6 mg/kg dose. Liver, bone marrow, and thymus did not show any changes from the controls at the different time periods and also at the two doses of the steroid.

The inability of nandrolone phenylpropionate to reverse the decrease in the incorporation of [U-$^{14}$C]-leucine by dexamethasone sodium phosphate in the skeletal muscles of the rat prompted BRESLOFF et al. (1974) to repeat the studies in immature male rabbits to determine whether there was a species difference. The decrease in amino acid incorporation was produced with several doses (50–100 μg/day) of dexamethasone sodium phosphate for 3 and 7 days. A single or daily injection of 5 mg of nandrolone phenylpropionate partially reversed the effect of the corticosteroid in the gastrocnemius and triceps muscles, completely reversed it in the diaphragm, and produced a small, nonsignificant reversal in the heart and soleus muscles.

In view of the large effect on the diaphragm, a study on the normal immature rabbit was conducted (BRESLOFF et al., 1974). The nandrolone phenylpropionate at 5 mg/day increased the incorporation of [U-$^{14}$C]-leucine into the protein of the diaphragm on the 4th day. Thus, the diaphragm of the rabbit, as in the guinea pig (COSTA et al., 1962), appears to be the most sensitive muscle to amino acid incorporation under steroid stimulation. Indeed, BRESLOFF et al. (1974) suggested that the diaphragm was a more appropriate test object than the "levator ani" muscle for the evaluation of the anabolic activity of steroids.

# III. Studies in Vivo/in Vitro

## A. Rat

### 1. Liver and Kidney Slices

BERNELLI-ZAZZERA et al. (1958a, b) castrated rats at 200–250 g body weight; 5 days later a group of the rats was partially hepatectomized, and 2 h later 2 mg of either testosterone propionate or 4-chlorotestosterone was injected and repeated daily. Seventy-two h after the first injection, the rats were killed, and the livers were removed and sliced. The slices were placed in a Warburg flask containing 3.0 ml of phosphate saline. After 10 min of equilibration, 2 μCi (3.4 μmoles) of [1-$^{14}$C]-glycine in 0.4 ml of medium was transferred to the slices. After 1 h of incubation, the flask contents were extracted for protein, and the radioactivity was determined. The treatment with the hormones did not affect the incorporation of radioactivity (amino acid) into the liver protein of the castrated rat but further enhanced the already increased incorporation into that of the partially hepatectomized-castrated rat. Similar results were obtained when [2-$^{14}$C]-phenylalanine was substituted for glycine. The steroids had no influence on the incorporation of either of the amino acids by slices of kidney from the same rats.

PICENI-SERENI et al. (1961) utilized the amino acid incorporating system of BERNELLI-ZAZZERA et al. (1958a, b) to study the effect of 4-chlorotestosterone acetate on the enhanced amino acid incorporation into liver protein of 50-g rats by a low (6%) protein diet. The daily injection of 0.5 mg/100 g body weight had no effect after 10, 20, or 30 days.

CAREDDU et al. (1961) injected 100-g albino rats daily with 0.5 mg/100 g body weight of an aqueous suspension of 4-chlorotestosterone acetate and determined the incorporation of [2-$^{14}$C]-DL-phenylalanine into the protein of the liver and kidneys by the method of BERNELLI-ZAZZERA et al. (1958a, b). The rate of amino acid incorporation was not changed in either tissue after 5 and 15 days. If the rats were made nephrotic by the injection of antikidney serum, the rate of amino acid incorporation was markedly increased. The injection of 4-chlorotestosterone acetate for 5 and 15 days partly reversed the increase in the kidney slices induced by the nephrosis and had a questionable and inconsistent effect in the liver slices.

### 2. Kidney Cell-free System

KOBLET and RIVA (1963) devised a cell-free amino acid-incorporating system that was specific for the nuclei and showed only a trace of activity with the mitochondrial or postmitochondrial fractions of rat liver and kidney. The authors used this system to study the effect of testosterone propionate on the incorporation of [1-$^{14}$C]-glycine into the protein of the isolated nuclei of kidney of 20- to 25-day-old female rats. Injection of the hormone for 2 days at 2.5 mg/day greatly increased the rate of amino acid incorporation. Extension of the injection to 5 days or an increase of the dose to 5 mg/day did not stimulate a further increase.

### 3. Heart Slices

FAZZINI and FONNESU-SEVERI (1965) castrated rats at 100–110 g body weight and 15 days later injected daily 10 mg/kg body weight of either testosterone propionate or 4-chlorotestosterone acetate. The rate of incorporation of [1-$^{14}$C]-DL-leucine into slices of heart muscle of both the castrated and normal rats was not significantly increased after 5 or 35 days of injections. Castration also was ineffective.

### 4. Skeletal Muscles

*a) Strips, Slices*

DE LOECKER (1965) castrated rats at 125–150 g body weight and 8 days later injected 0.1, 1.0, or 10 mg/day of testosterone propionate for 3 days. Strips of the adductor muscles of the hind legs were prepared and incubated in Krebs-Ringer bicarbonate buffer containing 1 μCi of [U-$^{14}$C]-glycine. No significant difference in the incorporation of radioactivity between treated and control rats was noted on a wet-muscle-weight basis. A significant increase was apparent at the highest dose of hormone, if the incorporated radioactivity was calculated on the basis of nitrogen or DNA concentration in the muscle. This difference disappeared when the 15000 g supernatant fraction of a homogenate instead of strips of muscle was used, and the addition of an ATP-generating system to the incubation medium did not alter the results.

*b) Cell-free System, Anabolic Activity*

BREUER and FLORINI (1965) carefully studied conditions for the preparation of ribosomes and transferases from rat skeletal muscle and adapted the Zamecnik-Keller incubation system for the incorporation of leucine from [$^{3}$H]-L-leucyl—t-RNA into protein by the polysomes. Immature rats were used because their muscles had a higher concentration of ribosomes than those of adult rats and a greater rate of amino acid incorporation due to the higher concentration of polysomes. Castration was at 50–60 g body weight, and beginning on the following day the rats received daily injections of different doses of testosterone propionate for 7 days. The thigh muscles of 12 rats were pooled for each study. Single experiments were made for each group except that there were two experiments for the castrated controls and the castrated rats, which received 0.5 mg/day of testosterone propionate, and four experiments with the normal rats.

The incorporation of radioactivity into the ribosomal protein decreased to 50% of normal after castration, returned to normal after 0.1 mg/day of testosterone propionate, increased slightly above normal by 0.5 mg/day, but decreased to less than normal at 2.5 mg/day. The increase in body weight of the rats showed parallel changes. The same doses of the steroid to the intact rats produced a progressive decrease in the incorporation of radioactivity so that at the highest dose the rate of incorporation was slightly lower than that for the castrated rats.

The use of 17α-ethyl-19-nortestosterone gave results similar to those of testosterone propionate except that higher doses were used, and the increase in the castrated rats was less and the decrease in the intact rats was greater.

The rate of incorporation of radioactivity was observed to be related to the concentration of polysomes in the ribosomes. The administration of 10 μg/day of actinomycin D for the last 3 days of 0.25 mg/day testosterone propionate treatment greatly reduced the concentration of polysomes in the ribosomes with a concomitant inability of the steroid to restore the amino acid-incorporating activity of the ribosomes to the normal level. These changes are more likely due to the toxicity of the drug than to an inhibition of RNA synthesis. The rats lost an average of 25 g of body weight after 3 days of actinomycin D treatment.

FLORINI (1970) incubated the soleus muscle from castrated and testosterone propionate-treated rats in Krebs-Ringer bicarbonate buffer containing either [$^3$H]-leucine or [U-$^{14}$C]-L-leucine and combined the muscles from the two rats for homogenization and extraction of the soluble protein. The submission of the protein fraction to disc electrophoresis exhibited an identical pattern of $^3$H and $^{14}$C radioactivity, indicating that the proteins synthesized under testosterone propionate stimulation were qualitatively identical to those already being produced by the castrated rat.

*c) Cell-free System, Anticatabolic Activity*

WHITE (1967) studied the ability of methandrostenolone (17β-hydroxy-17α-methyl-androsta-1,4-dien-3-one) to reverse the reduction in the amino acid-incorporating ability of ribosomes from skeletal (psoas) muscles of rats injected with glucocorticoids. The intraperitoneal injection of a saline suspension of triamcinolone ($\Delta^1$-9α-fluorohydrocortisone-16α,17α-acetonide) at 10 mg/kg for 2 days or 20 mg/kg at 1 day before autopsy decreased by about 50% the ability of the ribosomes from the psoas muscle to incorporate [U-$^{14}$C]-L-leucine into protein. The preinjection of methandrostenolone at 5 mg/kg/day partially reversed the effect of the glucocorticoid after 3 days and completely after 5 days. A comparison of different doses of methandrostenolone indicated that the reversal was dose dependent. The decrease in amino acid incorporation induced by 1 mg/kg for 2 days of triamcinolone was partially reversed by 1.0 mg/kg/day of methandrostenolone, completely by 5, and no further effect was induced by 10 mg/kg/day. An analysis of the free amino acids excluded a change in the pool of individual amino acids as a regulating factor.

In a later report (BULLOCK et al., 1968), the studies were extended to another anabolic steroid, 17β-hydroxy-7α,17α-dimethyl-A-nor-B-homoestrane-3,6-dione. The ribosomes were prepared from: (1) the postnuclear fraction (P-ribosomes), and (2) the microsomal fraction (M-ribosomes) of the psoas muscles. The decrease in amino acid-incorporating activity by the P-ribosomes after glucocorticoid administration apparently was not due to the induction of a difference in the proportion of the various ribosomes. The area under the sucrose gradient profile was the same for the ribosomes from the psoas muscle of the treated and control rats. The profile of the ribosomes of the treated rats, however, did not exhibit the characteristic resolution of the ribosomes. In contrast to the P-ribosomes, triamcinolone (20 mg/kg, 24 h previous to autopsy) did not decrease the amino acid-incorporating activity of the M-ribosomes.

The new anabolic steroid at 0.1 and 1.0 mg/kg/day for 4 days was not able to stimulate protein synthesis in the P-ribosomes of the psoas muscle of normal rats but had a partial effect at the 1.0 mg/kg/day dose on the P-ribosomes from the bulbocavernous muscle. On the other hand, preinjection of the rats for 2 or 4 days with the anabolic steroid partly reversed the decrease in amino acid incorporation in the psoas muscle ribosomes induced by the triamcinolone.

Chronic studies were made in male and female rabbits (age not given). The amino acid-incorporating activity of the P-ribosomes from the rectus femoris muscle was decreased after 4 weeks of cortisone acetate at 15 mg/kg/day. Testosterone at 5.0 mg/kg/day injected simultaneously with the cortisone for the last 2 weeks was not able to reverse the inhibitory effect of the cortisone, but the new steroid at 0.5 and also at 0.05 mg/kg/day reversed the inhibitory effect. The authors concluded that the ability of the newer steroid to counteract the effect of the glucocorticoid was not due to its androgen activity. Further substantiation of this conclusion was provided by similar results obtained with P-ribosomes from rat psoas muscle by another steroid, methandienone (Dianabol) (BULLOCK et al., 1969).

### 5. Perineal Muscle ("Levator Ani") Slices

SHEPPARD et al. (1964) compared the uptake of [U-$^{14}$C]-L-leucine into protein with that of [$^{3}$H]-thymidine by slices of muscle incubated in Krebs-Ringer bicarbonate buffer. The injection of immature rats with 4.0 mg methandrostenolone 25 h prior to autopsy did not influence the incorporation of radioactivity into slices of the "levator ani" muscle or the seminal vesicles and ventral prostate. The incorporation of [$^{3}$H]-thymidine into the hot soluble extract (DNA), however, was greatly enhanced.

ARVILL (1967), however, incubated the intact "levator ani" (including the bulbocavernous) of 21-day-old mice injected 24 h previously with 5 mg of testosterone propionate and obtained a significant increase in the incorporation of [$^{3}$H]-glycine and [U-$^{14}$C]-L-leucine. He also observed that testosterone stimulated the intracellular accumulation of [$^{14}$C]-D-xylose, [$^{14}$C]-aminoisobutyric acid, and [$^{3}$H]-glycine from the incubation medium but only after at least 6 h of the hormone injection.

## B. Mouse

### 1. Kidney, Liver, and Muscle Slices

FRIEDEN et al. (1957) studied the rate of incorporation of [1-$^{14}$C]-glycine into slices of mouse tissues incubated in Krebs-Ringer phosphate solution. Three 1.0 mg injections of testosterone propionate at 48-h intervals into male and female mice (age not given) increased the rate of incorporation of the [1-$^{14}$C]-glycine 40 and 45% respectively, into the protein of the kidney slices but had no effect on that of the diaphragm, liver, or skeletal muscle. The daily injection of testosterone propionate produced a maximum effect at about 48 h after the first injection and thereafter, continued injections led to a gradual decrease to the normal level after 96–120 h. The removal of food 24 h before autopsy resulted in a slight decrease in amino acid incorporation in the kidney slices of the control mice, but the effect of the hormone was not altered.

The fractionation of the slices after incubation revealed amino acid incorporation into the protein of the nuclear, mitochondrial, and postmitochondrial fractions. The greatest increase was in the mitochondria. Alterations in the substrate concentration by varying the quantity of inert glycine indicated that the maximum effect of testosterone propionate was obtained at 1.0 mM glycine or lower.

### 2. Kidney Homogenate

FRIEDEN et al. (1961) extended their studies to a homogenate of kidney of female mice (3–5 months of age) prepared in Krebs-Ringer phosphate buffer containing 0.4 μCi/0.5 mM [1-$^{14}$C]-glycine. The mice were injected subcutaneously with 1 mg/day of testosterone propionate, testosterone, androsterone, or 17α-methylandrost-5-ene-3β,17β-diol. All of the steroids produced a similar maximal increase of about 30% except testosterone propionate (65%), with a restoration to normal by the 6th day. Also the maximum rate of amino acid incorporation occurred after 2 days for testosterone propionate and androsterone and after 4 days for testosterone and 17α-methyltestosterone. The minimum effective dose for testosterone propionate was found to be 0.10 mg/day for 2 days. The stimulatory effect of testosterone propionate was counterbalanced by the simultaneous injection of 1.0, but not 0.5 mg/day of hydrocortisone.

### 3. Kidney Cell-free Systems

KOCHAKIAN (1962) and KOCHAKIAN et al. (1963) utilized the cell-free amino acid-activating system of HOAGLAND et al. (1956) and the protein biosynthesis system of ZAMECNIK and KELLER (1954) for the stepwise exploration of the mechanism of the stimulatory effect of androgens on protein biosynthesis in the mouse kidney. The amino acid-activating enzymes were present in the cytosol, and their specific activity was not changed 90 days after castration of 26-day-old mice or by the subcutaneous implantation of a 15-mg pellet of testosterone propionate for the last 7 days before autopsy. The rate of protein biosynthesis ([1-$^{14}$C]-DL-leucine incorporation), however, was greatly reduced to about 30% of normal by castration and was increased several fold above normal by the testosterone propionate, to a maximum effect at about 14 days, at which point it was maintained for the duration (42 days) of the experiment. The pattern of increase progressed in parallel with that of the increase in kidney weight. Testosterone propionate also increased the incorporation of [1-$^{14}$C]-DL-lysine at the same rate as the [1-$^{14}$C]-DL-leucine, and [1-$^{14}$C]-DL-alanine, [3,4-$^{14}$C]-DL-glutamic acid, and [2-$^{14}$C]-glycine at about $^1/_4$ to $^1/_3$ that of leucine. Testosterone and 17β-hydroxy-5α-androstan-3-one implanted subcutaneously as a single 15-mg pellet were as effective as testosterone propionate in the stimulation of the incorporation of leucine, and also of valine, lysine, and phenylalanine. The inhibitor, chloramphenicol, was without effect on the stimulatory effect of testosterone propionate, but puromycin markedly decreased the rate of protein biosynthesis in not only the treated mice but also the normal and castrated control mice. Fractionation of the postmitochondrial fraction indicated that the cellular site of amino acid incorporation and of androgen stimulation was in the microsomal material.

KOCHAKIAN (1969a) carried out a series of experiments to determine the earliest appearance of the decrease after castration and increase after androgen administration to castrated mice, of amino acid incorporation by the postmitochondrial fraction of the kidney. The earliest detectable significant decrease in amino acid incorporation was evident between 2 and 4 days after castration. Different doses of testosterone propionate and androstane-3α,17β-diol were injected at various time intervals at different sites of administration—subcutaneous and intraperitoneal; as aqueous suspension, oil solution, or subcutaneous implantation of pellets; into young and old mice; and immediately and at various times after castration. The earliest response to the different steroids was elicited between 21 and 24 h, and the minimum effective dose was 25 μg of testosterone propionate.

KOCHAKIAN et al. (1963), in their initial studies, noted that the postmitochondrial fraction rapidly lost its amino acid-incorporating activity even at 4° C. KOCHAKIAN and HAMA (1969) found the loss in activity to be due to at least two labile factors. One was in the microsomal fraction and could be replaced by the addition of the synthetic messenger RNA, polyuridylic acid. The other factor(s) was in the cytosol, that is, the supernatant fraction after centrifugation of the postmitochondrial fraction at 105000 g for 1 h. The loss in activity was prevented by immediate storage at −20° C. Castration or testosterone propionate did not affect these labile factors.

Since polysomes contain the sites of protein synthesis and the cytosol contains at least two factors—transferase I (initiation) and transferase II (elongation)—other than the amino acid-activating and pH 5 enzymes, KOCHAKIAN et al. (1969, 1974) extended their studies to determine the contribution of the several known factors of the amino acid-incorporating system to the decrease after castration and the increase after testosterone treatment. The concentration of the polysomes and also of the monosomes in the postmitochondrial fraction of the mouse kidney was markedly decreased after castration and rapidly increased above normal after testosterone propionate treatment (KOCHAKIAN et al., 1969), thus providing at least a partial explanation for the changes in amino acid-incorporating activity by the postmitochondrial fraction after castration and testosterone treatment. Furthermore, electron microscopy revealed conformational changes. The polysomes from the castrated mice were linear while those from the treated mice were helical, suggesting alterations in the quality of the polysomes. When a fixed quantity of polysomes was added with a fixed quantity of cytosol to the incubation system, the rate of incorporation of [U-$^{14}$C]-L-leucine or [U-$^{14}$C]-L-phenylalanine with or without poly U was decreased to a minimum value after 4–5 days of castration, where it was maintained for the duration (15 days) of the study (KOCHAKIAN et al., 1974). The implantation of a 15-mg pellet of testosterone propionate rapidly increased the incorporation of both of the amino acids to maxima above normal levels within about 4 days. A series of cross mixing experiments of polysomes and cytosol from the kidneys of castrated and treated mice further indicated that the polysomes of the castrated mice were less able to incorporate amino acids into protein and that the deficiency could not be corrected by the addition of poly U to the incorporating system. The role of the cytosol proved to be even greater than that of the polysomes, both after castration and testosterone propionate treatment. The removal of most of the low molecular weight substances, e.g., amino acids, nucleotides, etc., by filtration of the cytosol through Sephadex G-25 enhanced the activity of the cytosol, but to equivalent degrees for the normal, castrated, and treated mice. The addition of 0.05 μmole of each of the 19 amino acids to the reaction mixture (1 ml) further enhanced the activity of the incorporating system but again to equivalent degrees for the three groups of mice. Thus neither the amino acids nor other low molecular weight substances were regulating factors.

Further support for the nonregulating action of the free amino acids was provided by the analysis of the kidneys for free amino acids. Castration of 2-month-old mice produced, after 40 days, a significant decrease in the concentration of only glycine and tryptophan in the perchloric acid extract of the kidneys (KOCHAKIAN, 1974). The subcutaneous implantation of a 15-mg pellet of testosterone propionate for 4 days 1 month after castration restored the values to normal; extension of the hormone treatment to 11 and 25 days did not induce further changes. When the kidneys were homogenized in the sucrose-buffer-salt medium used for protein biosynthesis studies and then extracted with perchloric acid, the concentrations of 13 of the amino acids of the testosterone propionate-treated mice were greater than those of the castrated controls. The difference was even greater when the cytosols were compared. The normal mice did not exhibit this same phenomenon. Apparently testosterone propionate induced an accumulation of a reserve of amino acids for protein synthesis not in a free but in a loosely bound or easily hydrolyzed form, e.g., protein-bound amino acids or aminoacyl-t-RNA. The difference in the concentrations of leucine and phenylalanine, the substrates used in the cell-free protein synthesizing system, in the three groups of mice was not sufficient to affect the rate of amino acid incorporation in vitro.

Fractionation of the cytosol (unpublished) indicated that the specific activities of the transferases and to a lesser degree of the pH 5 fraction were decreased by castration and increased after testosterone propionate administration. The changes in transferase activity were in transferase II. Thus, the androgen regulated peptide bonding.

KOTHS et al. (1972) explored the effect of castration and testosterone propionate administration on the rate of protein synthesis by the mitochondria of the mouse kidney. The mitochondria were carefully purified to remove cytoplasmic polysomes, and aseptic precautions were employed. Castration at 8 weeks of age resulted in a decrease in the rate of amino acid ($^{14}C$ amino acid mixture) incorporation into mitochondrial protein to a minimum of about 60% of normal by 8 days, followed by a "spontaneous" recovery to normal after about 3 weeks. The pattern of response was practically identical with that obtained in the studies in vivo (see above). The injection of 2.5 mg of testosterone 6 days after castration completely restored the ability of the isolated mitochondria to incorporate amino acids into protein after 18 h. A small, 19%, rise above the castrated level was noted after 11 h, which agreed with the earliest, 14-h, in vivo effect. It appeared that the testosterone propionate had an earlier stimulatory effect on the mitochondria than the polysomes. The earliest significant stimulatory effect induced in the postmitochondrial fraction by testosterone was between 21 and 24 h (KOCHAKIAN, 1969a). The injection of 2.5 mg of an aqueous suspension of testosterone 67 h after the castration of 8-week-old mice induced increases in [U-$^{14}C$]-L-leucine incorporation by the postmitochondrial fraction of 8, 15, and 17% after 8, 12, and 18 h, respectively (KOCHAKIAN, 1969a). These results are in agreement with those of KOTHS et al. (1972), but are smaller and are not statistically significant. Furthermore, castration at this time interval produced only an 8% decrease. KOTHS et al. (1972) attributed the earlier effect of the androgen in the mitochondria to treatment at a time after castration when the mouse kidney was most sensitive to androgen stimulation. They injected the androgen 6 days after castration as compared with 3 days by KOCHAKIAN (1969a). The "spontaneous" recovery in protein synthesis by the isolated mitochondria was not observed for the postmitochondrial fraction of the kidneys from 4- and 10-month-old mice (KOCHAKIAN, 1965, 1969a) or for the polysomes from 3-month-old mice (KOCHAKIAN et al., 1974). The rate of incorporation of both [U-$^{14}$C]-leucine and [U-$^{14}$C]-phenylalanine into protein by the postmitochondrial fraction was even possibly decreased somewhat further after 16 days of castration (KOCHAKIAN, 1965, 1969a).

Actinomycin D is known to inhibit RNA synthesis (REICH et al., 1961), and various forms of RNA play important roles in protein biosynthesis. Therefore, KOCHAKIAN et al. (1971) explored the effect of actinomycin D on amino acid incorporation in the postmitochondrial fraction of the mouse kidney. The intraperitoneal injection of 0.5, 0.75, 1.0, 1.25, 2.0, or 4.0 µg/g of the drug 1, 2, or 4 h before autopsy had no significant effect on the rate of incorporation of [U-$^{14}$C]-L-phenylalanine with or without poly U and of [U-$^{14}$C]-L-leucine. Prolongation of the period after injection of 4 µg/g into the toxic phase produced a rapid decrease in the rate of incorporation of the amino acids, so that after 18 h only a trace of activity was apparent in normal, castrated, and testosterone propionate-treated (7-day) mice. A smaller dose (0.85 µg/g body weight) after 24 h abolished the increase in [U-$^{14}$C]-L-phenylalanine incorporation into the postmitochondrial fraction induced after 7 days of subcutaneous implantation of a 15-mg pellet of testosterone propionate. The intraperitoneal injection of 0.85 µg/g body weight of actinomycin D also prevented the early increase, after 24 h, in the incorporation rate of [U-$^{14}$C]-L-phenylalanine produced by the simultaneous injection of 25 µg of testosterone propionate. The actinomycin D also decreased the rate of incorporation in the postmitochondrial fraction of the normal and, to a lesser degree, of the castrated mice with concomitant decreases in the concentration of RNA. It is questionable whether these results are due to a direct inhibition of RNA production or are an indirect effect due to the generalized toxicity that occurs 24 h and even sooner after the injection of actinomycin D.

FRIEDEN et al. (1964) reported that actinomycin D injected intraperitoneally at 200 µg/kg/day beginning 2 days before the injection of testosterone propionate at 1 mg/day for 2 days did not abolish the increased (31% vs. 39%) incorporation of [1-$^{14}$C]-leucine by kidney slices of the treated mice. The increase in incorporation of [1-$^{14}$C]-glycine and [1-$^{14}$C]-arginine also was not changed. In contrast, the increases in $\beta$-glucuronidase and D-amino acid oxidase activity after testosterone propionate injection were abolished by the actinomycin D.

## C. Guinea Pig

### 1. Kidney Cell-free System

The rate of incorporation of [1-$^{14}$C]-DL-leucine in vitro by the postmitochondrial fraction of guinea pig kidney was not influenced by either castration or testosterone administration (KOCHAKIAN, 1964). The guinea pigs were castrated at 470–700 g body weight. The testosterone was implanted subcutaneously as two 15-mg pellets 3 weeks after castration, and the rate of amino acid incorporation was determined after various times from 2–35 days. These results are consistent with the failure of castration or testosterone administration to influence the weight of the kidneys.

### 2. Muscle Cell-free System

KOCHAKIAN et al. (1961) adapted the ATP-$^{32}$PP exchange system of HOAGLAND et al. (1956) to study the activity of the amino acid-activating enzymes of the temporal muscle of the guinea pig. Castration of guinea pigs at 750–900 g body weight resulted in a decrease in weight of the temporal muscles to about 50% of normal within about 3 weeks, where it was maintained for another 6 weeks. The specific activity of the amino acid-activating enzymes was not changed. The implantation of one or two 15-mg pellets of testosterone or 2 pellets of 17β-hydroxy-5α-androstan-3-one 35 days after castration increased the activity of the amino acid-activating enzymes in direct proportion to the increase in weight of the temporal muscle. These effects are similar to those observed in the mouse kidney.

In contrast to the activity of the amino acid-activating enzymes, the rate of protein biosynthesis by cell-free preparations of the temporal muscle of the guinea pig was influenced by castration and testosterone administration (KOCHAKIAN, 1966). The guinea pigs were castrated at 500–660 g body weight and two 15-mg pellets of testosterone were implanted subcutaneously after 64 and 83 days. The animals were killed 41 days later, and the microsomal and pH 5 fractions were separated. Castration decreased (−35%) the rate of incorporation of [U-$^{14}$C]-L-leucine and [U-$^{14}$C]-L-phenylalanine by the combination of these two fractions in the amino acid-incorporating system. Testosterone administration restored the values to or above normal. Similar results were obtained with the postmitochondrial fraction of the temporal muscle after 51–58 days of castration and 2 days of testosterone implanted subcutaneously (KOCHAKIAN, 1969b). The rates of incorporation of [U-$^{14}$C]-L-leucine and [U-$^{14}$C]-L-phenylalanine by the postmitochondrial fraction and the microsomal plus pH 5 fractions of the rectus femoris muscle, however, were not changed by either castration or testosterone administration. The addition of poly U to the phenylalanine system greatly increased the rate of incorporation of phenylalanine, though to a proportionate degree, by the preparations from normal, castrated, or treated guinea pigs. In a later study, MARCAIS and KOCHAKIAN (1971) isolated the polysomes from the microsomal fraction of the temporal muscles and on incubation with saturating amounts of partially purified pH 5 enzymes and transferases observed that the incorporation of [U-$^{14}$C]-L-leucine or [U-$^{14}$C]-L-phenylalanine was decreased after castration and restored to normal by testosterone. The activities of the polysomes from thigh and masseter muscles were not changed.

## IV. Studies in Vitro

### A. Liver Slices

HAUSCHILDT and GROSSMAN (1953) studied the incorporation of [1-$^{14}$C]-glycine into protein by slices from rat and human liver after the addition of crystalline testosterone and other steroids. The incubations were performed in Krebs-Henseleit buffer (2.0 ml) with 100 mg of liver slices, 30–300 μg of steroid, and 0.05 mg [1-$^{14}$C]-glycine (0.5 μCi/mmole). The testosterone had a uniform, depressing effect on glycine incorporation that was not altered by the addition of 2.3 mg of DPN to the incubation medium. Desoxycorticosterone and estradiol-17β also had a depressing effect, while cholesterol and cortisone did not produce significant changes.

## B. Perineal Muscles

### "Levator Ani"

ARVILL (1967) was unable to detect any change in the incorporation of [$^3$H]-glycine into the isolated "levator ani" muscle of immature male rats on addition of testosterone ($35 \times 10^{-9}$ to $35 \times 10^{-3}$ M) to the incubation medium. BURESOVA et al. (1969) modified the incubation procedure and obtained positive effects. The muscle was stretched at optimal resting tension in a chamber with a synthetic tissue culture medium at 20° C and testosterone at a final concentration of $10^{-1}$–$10^{-4}$ mg/ml. After 24 h incubation, the muscle was transferred to flasks containing 30 ml of Krebs-Ringer bicarbonate buffer pH 7.4, 2.5 mg glucose/ml, and [U-$^{14}$C]-L-leucine 0.2 μCi/ml. The incubation flasks were aerated for 1 min with 95% $O_2$ + 5% $CO_2$, sealed, and incubated for 2 h at 37° C. The muscles were homogenized and extracted with TCA, and the radioactivity of the protein was determined. The incorporation of leucine into protein was increased by preincubation with testosterone at $10^{-4}$ mg/ml and further increased (approx. 100%) at $10^{-3}$ mg/ml, and then decreased to the normal level after $10^{-2}$ mg/ml and 50% below normal after $10^{-1}$ mg/ml testosterone. The effect of the testosterone was not observed if the muscle was not stretched during the 24-h preincubation period. The stimulatory effect of testosterone was abolished by the denervation of the muscle of mature or castrated rats (BURESOVA et al., 1972).

## V. Summary

Radioactively labeled protein precursors have been used to study the rate of protein synthesis in various tissues of the mouse, rat, guinea pig, and rabbit after castration and the administration of anabolic-androgenic steroids. Tissues that markedly decreased in weight after castration and increased after administration of anabolic-androgenic steroids exhibited a decrease and an increase, respectively, in the incorporation of the labeled precursors administered in vivo and also in cell-free systems. Several parameters of the biochemical system were affected. The concentration and specific activities of the polysomes, transferase II, and the pH 5 enzyme fraction of the mouse kidney decreased after castration and increased after hormone treatment. The amino acid-activating enzymes of the mouse kidney and guinea pig temporal muscle, on the other hand, changed in direct proportion to the decrease in weight after castration and the increase after hormone treatment. The addition of the synthetic m-RNA, poly U, did not increase the rate of protein biosynthesis by cell-free preparations of castrated mouse kidney or guinea pig temporal muscle to that of the testosterone-treated animals.

## References


ARVILL, A.: Effects of testosterone on the metabolism of the isolated levator ani muscle of the rat. Acta endocr. (Kbh.) **56** (Suppl. 122), 1—14 (1967).

BERNELLI-ZAZZERA, A., BASSI, M., COMOLLI, R., LUCCHELLI, P.: Azione del testosterone e del 4-chlorotestosterone sulle sintesi proteiche "in vitro". Sperimentale **108**, 291—299 (1958a).

BERNELLI-ZAZZERA, A., BASSI, M., COMOLLI, R., LUCCHELLI, P.: Action of testosterone propionate and 4-chlorotestosterone acetate on protein synthesis in vitro. Nature (Lond.) **182**, 663 (1958b).

BRESLOFF, P., FOX, P. K., SIM, A. W., VAN DER VIESS, J.: Effect of nandrolone phenylpropionate on $^{14}$C-leucine incorporation into muscle protein in the rat and rabbit in vivo. Acta endocr. (Kbh.) **76**, 403—416 (1974)

BREUER, C. B., FLORINI, J. R.: Amino acid incorporation into protein by cell-free systems from rat skeletal muscle. IV. Effects of animal age, androgens, and anabolic agents on activity of muscle ribosomes. Biochemistry **4**, 1544—1550 (1965).

BULLOCK, G. R., PETERS, R. F., WHITE, A. M.: Changes in the mitochondrial structure and ribosomal activity in muscle as a consequence of the interaction between a glucocorticoid and some anabolic steroids. Biochem. J. **115**, 47p (1969).

BULLOCK, G., WHITE, A. M., WORTHINGTON, J.: The effects of catabolic and anabolic steroids on amino acid incorporation by skeletal muscle ribosomes. Biochem. J. **108**, 417—425 (1968).

BURESOVA, M., GUTMANN, E., HANZLIKOVÁ, V.: Differential effects of castration and denervation on protein synthesis in the levator ani muscle of the rat. J. Endocr. **54**, 3—14 (1972).

BURESOVA, M., GUTMANN, E., KLICPERA, M.: Effect of testosterone on the incorporation of amino acids into the proteins of the levator ani muscle in vitro. Physiol. bohemoslov. **18**, 137—139 (1969).

BUTENANDT, A., GUNTHER, K., TURBA, F.: Zur primären Stoffwechselwirkung des Testosterons. Z. physiol. Chem. **322** 28—37 (1960).

CAREDDU, P., PICENI-SERENI, L., SERENI, F.: Incorporazione in "vitro" di glicina e di fenilalanina-C$^{14}$ nelle proteine del fegato e del rene di ratto normale e di ratto nefrotico e modificazione indotte dagli steroidi anabolizzanti. Minerva pediat. **13**, 10—12 (1961).

COSTA, G., KOCHAKIAN, C. D., HILL, J.: Effect of testosterone propionate on the incorporation of glycine-2-$^{14}$C in guinea pig tissues. Endocrinology **70**, 175—181 (1962).

DELOECKER, W.: The effects of testosterone on the incorporation of glycine-U-$^{14}$C into the proteins and nucleic acids of skeletal muscle. Arch. int. Pharmacodyn. **153**, 69—78 (1965).

DERUGGIERI, D. P., MATSCHER, R., GONDOLFI, C., CHIARMONTI, D., LUPO, C., PIETRA, E., CAVALLE, R.: Un nuova steroide avente azione proteoanabolica: La dimetazina. Arch. Sci. Biol. **47**, 1—19 (1963).

DOEG, K. A., POLOMSKI, L. L., DOEG, L. H., DUBOS, P. M.: Androgen control of mitochondrial protein biosynthesis in male sex accessory tissue and liver of rats. Endocrinology **89**, 338—346 (1971).

FARABOLLINI, F.: Incorporazione di formiato-$^{14}$C nelle proteine, RNA, e DNA di fegato e rene di ratti trattati con testosterone o cortisone. Rend. Ist. Sci. Univ. Camerino **3**, 160—165 (1962).

FAZZINI, G., FONNESU-SEVERI, C.: Incorporazione "in vitro" di leucine-1-$^{14}$C nelle proteine di cuore di ratti integri e castratti trattati con testosterone e 4-chlorotestosterone. Sperimentale **115**, 104—112 (1965).

FERRARI, W., BERTOLINI, A., MUCCI, P., STERNIERI, E.: Azione di steroidi anabolizzanti sullo sviluppo dell' apparato genitale del coniglio V. Influenza del 13$\beta$-etil-7$\beta$-idrossi-17$\alpha$-etil-4-gonen-3-one (Norbolethone) sull'incorporazione della $^{14}$C-leucine nel levator ani e nelle vescicole seminali del caniglio. Boll. Soc. ital. Biol. sper. **44**, 1309—1310 (1968).

FLORINI, J. R.: Effects of testosterone on qualitative pattern of protein synthesis in skeletal muscle. Biochemistry **9**, 909—912 (1970).

FRIEDEN, E. H., COHEN, E. H., HARPER, A. A.: The effects of steroid hormones upon amino acid incorporation into mouse kidney homogenates. Endocrinology **68**, 862—866 (1961).

FRIEDEN, E. H., HARPER, A. A., CHIN, F., FISHMAN, W. H.: Dissociation of androgen-induced enzyme synthesis and amino acid incorporation in mouse kidney by actinomycin D. Steroids **4**, 777–786 (1964).

FRIEDEN, E. H., LABY, M. R., BATES, F., LAYMAN, N. W.: The effect of testosterone propionate upon incorporation of labeled glycine into mouse kidney slices. Endocrinology **60**, 290—297 (1957).

HAUSCHILDT, J. D., GROSSMAN, C. M.: Effects of testosterone and other steroids on incorporation of glycine-1-$^{14}$C into protein by rat and human liver slices. Endocrinology **53**, 306—310 (1953).

HOAGLAND, M. B., KELLER, E. B., ZAMECNIK, P. C.: Enzymatic carboxyl activation of amino acids. J. Biol. Chem. **218**, 345—358 (1956).

ICHII, S., IKEDA, A.: Effect of adrenalectomy and gonadectomy on protein turnover of liver, kidney and spleen of mouse. Endocr. Jap. **17**, 273—276 (1970).

JAKUBOVIC, A., CEKAN, Z.: The effect of testosterone phenylpropionate on the incorporation of formate-$^{14}$C-Na into various organs of castrated rats. Acta endocr. (Kbh.) **53**, 234—244 (1966).

KOBLET, H., RIVA, G.: Beschleunigung der Aminosäureninkorporation in Zellkerneiweiße durch Androgene. Schweiz. med. Wschr. **93**, 1811—1815 (1963).

KOCHAKIAN, C. D.: Intracellular regulations in the kidney by androgens. Amer. Zool. **2**, 361—366 (1962).

KOCHAKIAN, C. D.: Androgen regulation of nucleic acid and protein biosynthesis in the prostate. Nat. Cancer Inst. Monogr. **12**, 263—273 (1963).

KOCHAKIAN, C. D.: Effect of castration and testosterone on protein biosynthesis in guinea pig tissue preparations. Acta endocr. (Kbh.) **46** (Suppl. 92), 3—16 (1964).

KOCHAKIAN, C. D.: Mechanism of anabolic actions of androgens. In: Mechanisms of Hormone Action, pp. 192—213. New York: Academic Press 1965.

KOCHAKIAN, C. D.: Regulation of muscle growth by androgens. In: The Physiology and Biochemistry of Muscle as a Food, pp. 81—111, Madison, Wisconsin: University of Wisconsin Press 1966.

KOCHAKIAN, C. D.: Early effects of androgens on the biosynthesis of protein and RNA by the postmitochondrial fraction of the mouse kidney. Steroids **14**, 77—90 (1969a).

KOCHAKIAN, C. D.: Effect of androgens on cellular growth and metabolism. In: Adolescent Nutrition and Growth, pp. 207—224. New York: Appleton-Century-Crofts 1969b.

KOCHAKIAN, C. D.: Partial characterization of the changes in mouse kidney RNA after castration and androgen. In: The Sex Steroids, pp. 414—419. New York: Appleton-Century-Crofts 1971.

KOCHAKIAN, C. D.: The free amino acids of the mouse kidney. Effect of castration and androgen. Ala. J. med. Sci. **11**, 333—339 (1974).

KOCHAKIAN, C. D., HAMA, T.: Lability of the protein biosynthesis system of the mouse kidney and its relationship to the regulatory effect of androgens. Acta endocr. (Kbh.) **62**, 328—338 (1969).

KOCHAKIAN, C. D., HILL, J., AONUMA, S.: Regulation of protein biosynthesis in mouse kidney by androgens. Endocrinology **72**, 354—363 (1963).

KOCHAKIAN, C. D., HIRONE, T., KUSUNOKI, H., DUBOVSKY, J., STRICKLAND, B.: Ultrastructural changes in the mouse kidney after castration and androgen administration. Effect of actinomycin D and correlation with parameters of RNA and protein synthesis. Ala. J. med. Sci. **8**, 149—167 (1971).

KOCHAKIAN, C. D., NISHIDA, M., HIRONE, T.: Ribosomes of mouse kidney: regulation by androgens. Amer. J. Physiol. **217**, 383—391 (1969).

KOCHAKIAN, C. D., TANAKA, R., HILL, J.: Regulation of amino acid activating enzymes of guinea pig tissue by androgens. Amer. J. Physiol. **201**, 1068—1072 (1961).

KOCHAKIAN, C. D., TOMANA, M., STRICKLAND, B.: Role of cytosol and polysomes in the stimulation by androgen of protein biosynthesis in the mouse kidney. Mol. Cell. Endocr. **1**, 129—138 (1974).

KOTHS, K. E., GODCHAUX, W., III., DOEG, L. H., DOEG, K. A.: The effect of testosterone on the synthesis of mouse kidney mitochondrial and microsomal proteins in vivo and in vitro. Endocrinology **91**, 125—134 (1972).

MARCAIS, J., KOCHAKIAN, C. D.: Influence of castration and testosterone on guinea pig skeletal muscle ribosomes and protein synthesis. Fed. Proc. **30**(3), 1214 (1971).

MEISSNER, J.: Tierexperimentelle Untersuchung mit $^{32}$P-Phosphate und $^{35}$S-Methionin zum Metabolismus unter Einfluß anaboler Steroide. In: Wirkung und Anwendung anaboler Steroide, pp. 23—26. Berlin: Medicus Verlag 1964.

MOSEBACH, K. O., KUELPMANN, W. R., DIRSCHERL, W., ROMMEL, A.: Einfluß von Testosteron auf den Einbau von [Ring-2-$^{14}$C]-L-Histidine in Organproteine unreifer Ratte nach Nahrungsentzug. Z. physiol. Chem. **348**, 9—16 (1967).

NIMNI, M. E., BAVETTA, L. A.: Effect of testosterone propionate on tissue protein synthesis in the castrated male rat. Proc. Soc. exp. Biol. **106**, 738—740 (1961).

NOVAK, A.: Effects of androgenic hormones on the rate of incorporation of labeled amino acids in male adult mice. Amer. J. Physiol. **191**, 306—312 (1957).

PICENI-SERENI, L., CAREDDU, P., SERENI, F.: Incorporazione in "vitro" di amino acidi marcati nelle proteine del fegato nella distrofia sperimentale del ratto da carenze proteica ed effeti del trattamento con steroidi anabolizzanti. Minerva Pediat. **13**, 7—9 (1961).

REICH, E., FRANKLIN, R. M., SHATKIN, A. J., TATUM, E. L.: Effect of actinomycin D on cellular nucleic acid synthesis and virus production. Science **134**, 556—557 (1961).

SEIFULLA, R. D., CHERMNYKH, N. S., SERFEEV, P. V.: Effect of prednisolone and methandrostenolone on the incorporation of glycine-1-$C^{14}$ in the proteins of liver, thymus, and spleen. Farmakol. Toksikol. **37**, 57—60 (1974).

SHEPPARD, H., TSIEN, W. H., MAYER, P., SEYMOUR, G.: Uptake of thymidine-$H^3$ by accessory sex organs of the immature male rat. Effect of methandrostenolone. Biochem. Biophys. Res. Commun. **15**, 546—550 (1964).

WHITE, A. M.: The anti-catabolic effects of synthetic anabolic androgenic steroids. In: Proc. Int. Congr. Hormonal Steroids, 2nd ed. 1967, pp. 576—585.

YUDAEV, N. A., POKROVSKII, B. V.: Anabolic and androgenic effects of methandrostenolone in male rats. Vôp. med. Khim. **12**, 527—532 (1966).

ZAMECNIK, P. C., KELLER, E. B.: Relation between phosphate energy donors and incorporation of labeled amino acids into proteins. J. biol. Chem. **209**, 337—354 (1954).

# B. Regulation of Nucleic Acid Synthesis by Anabolic-Androgenic Steroids

C. D. KOCHAKIAN

## I. Introduction

The pioneer studies of BRACHET (1947, 1950) and CASPERSON (1950) demonstrated a correlation between RNA and accelerated growth of tissues. Subsequent studies (cf. ZAMECNIK, 1960; WATSON, 1963) demonstrated that protein biosynthesis required at least three different types of RNA: (1) transfer (t-RNA); (2) messenger (m-RNA); and (3) ribosomal (r-RNA). Since androgens stimulated protein synthesis in the various target tissues, it seemed reasonable to expect changes in the concentration of one or more of the pertinent RNAs. Among the many tissues other than the accessory sex organs, the mouse kidney has undergone the most intensive study (KOCHAKIAN, 1962, 1965, 1967a, b, 1969a, b, 1971).

## II. Mouse

### A. Kidney

#### 1. Total Changes

RABINOVITCH (1952) castrated normal adult mice and 30 days later implanted subcutaneously 10 mg of powdered testosterone propionate into the castrated and also into normal mice. Castration produced a small (less than 10%) insignificant decrease in the RNA-P of the separated kidney cortex. The testosterone propionate produced statistically insignificant increases of 20% after 4 days and 15% after 8 days in the castrated mice and an insignificant increase of 13% after 4 days, but a significant, and the only increase, of 36% after 8 days in the normal mice. The average body weights (21–28.5 g) of the mice were unusually small for adult mice and also varied greatly for each group and could explain the inability to detect clear-cut changes.

KOCHAKIAN and HARRISON (1962) and KASSENAAR et al. (1962) simultaneously reported that castration of adult mice resulted in a decrease in the amount of RNA of the kidney and that administration of testosterone propionate resulted in a restoration to or above normal. The decrease after castration reached a minimum level after about 2–3 weeks and was greater than the decrease in kidney weight. Thus, not only the total amount of but also the concentration of RNA was decreased. Testosterone propionate implanted as a subcutaneous pellet (KOCHAKIAN and HARRISON, 1962) or injected at 100 μg/day for 21 days (KASSENAAR et al., 1962) reversed the effect of castration. RNA was restored to normal levels after about 10 days, rose to an above-normal maximum after about 21 days, and then apparently decreased toward normal levels. The last effect was not substantiated by later studies (KOCHAKIAN, 1969a, c). Practically identical results were obtained when a 15-mg pellet of 17α-methyltestosterone or two 15-mg pellets of 17β-hydroxy-5α-androstan-3-one were used instead of the testosterone propionate (KOCHAKIAN and HARRISON, 1962). A single pellet of 17β-hydroxy-5α-androstan-3-one was less effective because of the lower rate of absorption of this steroid from the pellet.

The DNA decreased slightly to a minimum value approximately 2 weeks after castration and returned gradually after a lag period of about 1 week, to the normal level within 3–4 weeks after administration of the different steroids (KOCHAKIAN and HARRISON, 1962).

TRUCKENBRODT (1969) and HEIM et al. (1969), in a series of careful and detailed experiments, studied the changes not only in the RNA and DNA but also in the kidney and body weights of male and female mice with age and after administration of 3-oxo-1-methyl-5α-androst-1-en-17β-yl,enanthate. The normal kidney exhibited two growth phases: (1) an initial period of intensive tissue formation up to 18–25 days of life, during which time tissue growth was mainly a matter of cell multiplication with considerable protein synthesis and DNA duplication; and (2) a second phase, 25–60 days, when tissue growth was slower and was mainly due to enlargement of the cells; protein synthesis was stimulated to a much greater extent than DNA synthesis. The second phase was more pronounced in the male than in the female. These results are consistent with those reported earlier by KOCHAKIAN (1959). The concentrations of DNA and RNA were highest at birth and after about 2 weeks of age decreased slightly; after 32 days of age, the concentration of DNA in cells was higher in the female and that of RNA was higher in the male kidney. Castration of male mice at 18 days of age retarded the rate of increase and decreased the maximum level of RNA to below that of the female kidney. The maximum levels and differences in RNA, DNA, and also protein were reached at 46 days of age. The concentration of RNA per cell increased from the 18th to the 32nd day and then decreased at the 46th day. The changes were parallel but lower in the female and much lower in the castrated mice.

The response of 4-day-old mice and 6-month-old male and female mice to a single injection of 0.5 mg/g body weight of the long-acting new steroid, 3-oxo-l-methyl-androst-l-en-17β-yl,enanthate, was compared at weekly intervals for 8 and 6 weeks, respectively. The RNA of the female kidney increased to a maximum of 107% above that of the control after 14 days and that of the males to 68% of control after 28 days. The DNA increased after 28 days to a maximum of approximately 40% in the female, and 25% in the male kidney. The increases in the 4-day-old mice were smaller and less prolonged. The RNA increased 40% and the DNA 10% after 14 days, decreased to the control levels between the 21st and 28th days, and slightly below after 42 and 56 days. The smaller response in the young mice was attributed to the fact that they were in the normally most active growth phase. The greater percent increase in the female than in the male adult mouse is likely due to the lower normal values in the female. HEIM et al. (1969) compared the responses of 6-week-old with those of the 4-day and 6-month-old female mice to a single injection of 0.5 mg/g of 3-oxo-l-methylandrost-1-en-17β-yl,-heptanoate (enanthate). The maximum increases of RNA, DNA, protein, and cell size were greater and occurred sooner (7 days) in the 6-week-old mice than in either the younger or older mice. HEIM et al. (1969) also studied the effect of the injection of 30 mg/kg of the same steroid every 2 days for 14 days in 6- to 8-month-old female mice. The RNA increased at a greater rate than the kidney weight, the protein at the same rate, and the DNA at a lower rate. CHANG and LEI (1965) reported that 17α-methyl-5α-androstan-17β-ol at 50 mg/kg increased both the weight and RNA in the kidneys of castrated mice.

Further information on the regulation of nucleic acids by androgens was provided by studies on heminephrectomized mice. Both KASSENAAR et al. (1962) and JELINEK et al. (1964) reported that heminephrectomy of intact or castrated adult mice resulted in a similar compensatory hypertrophy with a parallel increase in RNA but no change in the content of DNA. KASSENAAR et al. (1962) observed that the injection of 100 μg/day of testosterone for 21 days, beginning on the day of heminephrectomy of mice castrated 5 weeks previously at 2.5 months of age, resulted in a further increase in kidney weight with a proportionate increase in RNA but no change in DNA content. JELINEK et al. (1964) castrated adult mice and 19–24 days later performed the heminephrectomy; they then immediately injected subcutaneously 3.2 mg of 19-nortestosterone phenylpropionate. Significant compensatory hypertrophy was evident in castrated and normal mice after 48 and 72 h, respectively. Compensatory hypertrophy did not change the DNA content and increased the concentration of RNA in the normal but not in the castrated mice. The steroid induced a similar increase in RNA concentration in both the normal and castrated heminephrectomized mice.

## 2. Subcellular Changes

Since the RNAs in different parts of the cell have different and specific functions in protein biosynthesis (cf. WATSON, 1963), KOCHAKIAN (1969c) determined the RNA concentration of the subcellular kidney components of normal, castrated and testosterone propionate-treated mice. Castration was performed at 26–28 days of age and autopsy, at intervals between 100 and 156 days later. The total RNA in the nuclear, mitochondrial, and cytosol fractions was reduced essentially in proportion to the decrease in kidney weight. The microsomal fraction, however, was decreased in concentration as well as in total amount. In another set of experiments, the mice were not killed until approximately 9 months after castration, and one group of castrated mice was implanted subcutaneously with a 15-mg pellet of testosterone propionate 2 days before autopsy. Again, castration produced a decrease in the concentration of RNA in the total homogenate, which was due to a parallel decrease in the microsomal RNA. The testosterone propionate produced the expected small increase in kidney weight after 2 days but restored the concentrations of RNA in the total homogenate and of microsomal RNA to normal. Therefore, another series of experiments was made in which castrated mice were treated with testosterone propionate and analysed various intervals for 42 days thereafter. As in the initial studies (KOCHAKIAN and HARRISON, 1962), the DNA increased, after a lag period, to normal levels at about 5 weeks. The mitochondrial and soluble RNA increased in direct proportion to the increase in kidney weight and reached normal levels within 3 weeks. The nuclear RNA increased at a slightly lower rate than the kidney weight and attained the normal level within 4 weeks. The microsomal RNA rapidly (within 1 week) attained the normal level and continued to increase to twice the normal level. The soluble and mitochondrial RNA increased in parallel with the kidney weight to above the normal level. Practically identical results were obtained when two 15-mg pellets of 17β-hydroxy-5α-androstan-3-one were implanted in the castrated mice. Furthermore, the implantation of another pellet of testosterone propionate or two more of 17β-hydroxy-5α-androstan-3-one at the 31st day of treatment did not enhance the effect of the initial treatment after 11 days. Thus, the maximal effect was being induced by one pellet of testosterone propionate or two pellets of 17β-hydroxy-5α-androstan-3-one. The time after castration did not modify the changes in the nucleic acids produced by testosterone propionate. Two series of experiments begun 100 days after castration gave results practically identical to those begun 21 days after castration.

The establishment of the changes in RNA concentration after castration and steroid treatment in the microsomal fraction strongly suggested that the ribosomes were involved. KOCHAKIAN et al. (1969) separated the monosomes and polysomes at various intervals between 2 and 15 days after castration of adult mice and observed that these components decreased in parallel with, but at a greater rate and degree than, the weight of the kidney. The maximum decrease was attained about 10 days after castration. The implantation of a pellet of testosterone propionate 21 days after castration of 2.5-month-old mice induced a rapid increase in both the polysomes and the monosomes at a greater rate than the weight of the kidneys. Sucrose gradient analysis indicated an increase in the proportion of monosomes after castration with a decrease after testosterone propionate treatment with a shift to the heavier polysomes. Electron microscopy of the polysomes confirmed the sucrose gradient analysis and provided the additional information that the polysomes from the castrated mice were primarily linear while those from the treated mice were helical in conformation. Electron microscopic study of sections of the kidney (KOCHAKIAN et al., 1971) confirmed the fractionation studies. Castration of adult mice produced a slight decrease in the free polysomes and a concomitant increase in the free monosomes of the proximal convoluted tubules; testosterone propionate reversed these changes. The rough endoplasmic reticulum was not detectably changed after castration or testosterone propionate treatment.

TESSMAN (1968) noted a suggestive, but not definite, increase in the concentration of free monosomes and polysomes in his careful and detailed histologic and electron microscopic study of the kidneys of normal adult (25–30 g) male and female mice after 0.5 mg/day of testosterone propionate for 3, 10, and 16 days.

The sucrose gradient profiles of the phenol-extracted RNA from the postmitochondrial and microsomal fractions (KOCHAKIAN, 1969a, b) and the polysomes and monosomes (KOCHAKIAN et al., 1969) of mouse kidneys were not altered by castration and testosterone propionate treatment.

### 3. Biosynthesis

Radioactively labeled RNA precursors were employed to obtain more detailed information on the regulation of the RNA content of the mouse kidney by androgens. KOCHAKIAN (unpublished) in initial studies utilized [$^{32}$P]-sodium orthophosphate but was not able to detect any change in the rate of incorporation after castration or administration of testosterone propionate for 2–42 days. More specific precursors were then utilized. The first of these was [6-$^{14}$C]-orotic acid, which gave positive but unexpected results (KOCHAKIAN and HILL, 1966). Adult mice were castrated at 93–125 days of age and were injected intraperitoneally with 3 μCi [6-$^{14}$C]-orotic acid (5.2 μCi/mmole) 17–18 h after withdrawal of food and 2 h before autopsy. The concentration of RNA showed the expected decrease, but the specific radioactivity progressively increased over the 30-day postcastration period. The specific radioactivity of the acid-soluble fraction (free nucleotide, nucleosides, and bases) increased more rapidly than that of the RNA. Fractionation of the kidney demonstrated that the increase in specific radioactivity occurred in the RNA of all of the subcellular components, especially in the M-RNA. The acid-soluble fractions of the subcellular components, especially of the postmitochondrial fraction, also were increased. Kinetic studies indicated that the increase in specific radioactivity of the RNA in the nuclear, mitochondrial, microsomal, and cytosol fractions did not become evident until sometime between 30 min and 2 h after injection of the orotic acid. The differences in the acid-soluble fractions, however, were evident in the nuclear, mitochondrial, and cytosol fractions within 15 min but only much later, between 30 min and 2 h, in the microsomal fraction. The difference in the microsomal fraction was small and questionable. The subcutaneous implantation of a 15-mg pellet of testosterone propionate in the castrated mice had no effect after 2 days on the specific radioactivity of either the RNA or the acid-soluble fractions of the total homogenate or the separate subcellular components. After 7 days of testosterone propionate, however, the specific radioactivities of all of the components were restored to normal, where they were maintained on extension of hormone treatment to 21 days. 17β-Hydroxy-5α-androstan-3-one produced results identical to those of testosterone propionate.

Analysis for base composition revealed no differences among the three groups of mice, and all of the radioactivity was in the pyrimidines. The ratio of the specific radioactivities of UMP and CMP was decreased slightly after castration and restored to normal (2.3) after 7 days of testosterone propionate absorption from a subcutaneously implanted pellet, suggesting a change in the rate of conversion of UMP to CMP. The increase in specific radioactivity of the RNA in the nuclear fraction after castration appeared in the CMP. No difference was detected in the UMP. The separation of the compounds in the acid-soluble fraction by ion exchange chromatography demonstrated that all of the radioactivity was contained in the pyrimidines. The specific radioactivities of these compounds were doubled after castration and restored to or near normal by testosterone propionate treatment. The concentrations of the nucleotides, nucleosides, and bases were not changed by castration and testosterone propionate. HEIM (1964) noted that the ATP concentration in the kidneys of 20-g female mice remained constant 6 weeks after the injection of 0.35 mg of the long-acting steroid, methenolonacetate, and concluded that it was an indication of new tissue formation.

Since the specific radioactivities of the RNA and the acid-soluble compounds are dependent not only on the rate of synthesis but also on that of degradation, KOCHAKIAN et al. (1972) initiated studies to compare the rates of degradation of the RNA and of the acid-soluble fraction. The mice were 88 days of age and the duration of castration was 55 days. The food was removed just before the intraperitoneal injection of 3 μCi of [6-$^{14}$C]-orotic acid (5.2 mCi/mmole). The specific radioactivity of the acid-soluble fraction decreased rapidly so that after 24 h only a small fraction

(10–15%) of that at 4 h remained. The specific radioactivity of the acid-soluble fractions in the castrated mice was only slightly greater than that of the normal mice. The specific radioactivity of the RNA surprisingly was the same in both groups of mice and decreased only slightly between 8 and 24 h after the injection of the orotic acid. Since the only procedural difference of this study was that the mice were not fasted for 17–18 h prior to injection of the [6-$^{14}$C]-orotic acid, simultaneous studies on fasted and fed mice were made. No difference was detected between the specific radioactivities of the kidney RNA of normal and castrated mice 2 h after the simultaneous removal of food and the injection of the orotic acid and only a small increase in the castrated mice 4 h later. If the mice were deprived of food 16 h prior to injection of the orotic acid, the specific radioactivities of the RNA and of the acid-soluble fraction of the kidneys of the normal mice were not changed but that of the castrated mice was significantly increased after 2 h and further increased after 4 h, as in the initial studies (KOCHAKIAN and HILL, 1966). Hydrolysis of the acid-soluble fractions indicated no significant difference between the concentrations of the total cytosine and uracil compounds of the castrated and normal mice, but the specific radioactivities of both of these types of compounds were greater in the castrated than the normal mice. It appeared that the food was providing extra orotic acid to the pool, thus diluting the specific radioactivity of the injected precursor. To test this hypothesis, the mice were fasted and injected with increasing amounts of orotic acid but a constant amount of [5-$^{3}$H]-orotic acid. As the amount of unlabeled orotic acid was increased, the difference in the incorporation of radioactivity in the RNA and acid-soluble fraction of the fasted normal and castrated mice gradually disappeared. The difference was greatly reduced at 0.25 μmole of orotic acid and was no longer evident at 0.5–4.0 μmoles. As a further experiment, the rate of disappearance of radioactivity in the RNA and acid-soluble fraction was compared in fed and fasted normal and castrated mice after the injection of [5-$^{3}$H]-orotic acid of high specific activity, 2.5 μCi/0.29 nmole. The specific radioactivity was greater in both the normal and castrated fasted than in the fed mice, and the greater incorporation of radioactivity in the castrated than in the normal mice was now evident in the fed as well as in the fasted mice. Furthermore, the specific radioactivities of the two sets of mice decreased in a parallel logarithmic fashion, so that after 24 h only a fraction of the initial radioactivity remained. The specific radioactivity of the RNA increased in a parallel fashion in each group to a maximum at about 12 h, at which it still remained after 24 h.

As a further delineation of the sites of incorporation of radioactivity after the injection of [6-$^{14}$C]-orotic acid (34 μCi/μmole) 2 h before autopsy, the polysomes and monosomes of the kidneys were isolated and submitted to sucrose gradient sedimentation (KOCHAKIAN et al., 1969). Castration of adult mice for 2 days produced a marked, uniform increase in the radioactivity gradient profiles of the polysomes and monosomes, which were further enhanced at 4 days with no further change after 7 and 16 days. The subcutaneous implantation of a pellet of testosterone propionate 2 weeks after the castration of adult (2.5-month-old) mice, had, after 2 days, no effect on the enhanced incorporation of radioactivity, produced a uniform restoration to normal after 7 days, and a possible slight decrease below normal after 16 days. Thus, no detectable difference in the incorporation of radioactivity was evident among the different-sized polysomes. The sucrose gradient profiles of the phenol-extracted RNA from the polysomes gave the expected 16S and 28S peaks. The radioactivity was greater in the smaller 16S peak than in the larger 28S peak, but both increased uniformly after castration and were restored to normal by testosterone propionate.

The incorporation of radioactivity from 0.125 μCi of [6-$^{14}$C]-orotic acid (specific activity not given) in the RNA of the kidneys of adult (14- to 17-week-old) female mice after the injection of 1 mg testosterone propionate apparently is different than in the castrated male mice. FRIEDEN and KU (1971) removed the food from the mice 2 h before and killed the mice 2 h after the injection of the orotic acid. The testosterone propionate induced an approximately 30% increase in the specific radioactivity of the total kidney RNA within 6 h, which was not changed any further after 12 and 24 h and which was back to normal after 68 h. Hydrolysis of the RNA revealed the radioactivity in the pyrimidines. The specific radioactivity of the UMP was greater than that of the CMP, but the percent increase after testosterone propionate was greater in the CMP than the UMP, not only in the total kidney but also in the nuclear, microsomal-mitochondrial, and cytosol fractions. The injection for 5 days of 1.0 mg/day of DL-ethionine or 0.6 mg/day of cycloheximide did not influence the increase in the incorporation of radioactivity from 0.125 μCi of [6-$^{14}$C]-orotic acid 15–18 h after the injection of 1 mg testosterone propionate. Actinomycin D at 4 μg/day for 4 days also was ineffective. Unfortunately, the acid-soluble fraction was not studied in any of these experiments.

KOCHAKIAN et al. (1972) extended their studies to radioactively labeled uridine to compare the role of the salvage with the anabolic pathway in the regulation of RNA synthesis by castration and testosterone. In the initial experiments [2-$^{14}$C]-uridine was injected simultaneously with [5-$^{3}$H]-orotic acid. No difference was detected for periods of 4–24 h as a result of fasting or after castration and testosterone propionate treatment in contrast to the results obtained with orotic acid (see above). In a subsequent study (DUBOVSKY and KOCHAKIAN, 1973), the failure of castration to affect the incorporation of uridine was confirmed, and similar results were obtained with [5-$^{3}$H]-cytidine (26.6 Ci/mmole) at a tracer (0.002 μmole) or a high (4 μmoles) dose. More than 80% of the cytidine was recovered within 10 min as uracil derivatives in the acid-soluble fraction of both castrated and normal mice. The injection of testosterone (2.5 mg in aqueous suspension) produced a small increase in the incorporation of [5-$^{14}$C]-uridine (20 μCi/2.5 μmoles) in the acid-soluble fraction and RNA of the mouse kidney after 24 h, which returned to the control value after 2 and 4 days. The use of a larger dose of [5-$^{3}$H]-uridine (50 μCi, 5 μmoles) produced a greater incorporation (25%) in a shorter period of time, 12 h, and was further (35%) increased after 24 h. The seminal vesicles of these same animals showed a 300% increase in the RNA and an almost 100% increase in the acid-soluble fraction after 12 h with a further substantial increase after 24 h. The increases in both the kidneys and seminal vesicles probably occur even before 12 h after the injection of testosterone.

The livers from the same mice used in several of the experiments with orotic acid and uridine were examined, but no changes in the incorporation of radioactivity into the RNA or the acid-soluble fraction were noted after castration or testosterone treatment. The incorporation of tracer doses of orotic acid was 10–20 times lower in the liver than in the kidney, but no difference was noted with tracer doses of uridine.

## B. Liver

KOCHAKIAN and HARRISON (1962) determined the RNA and DNA contents of the liver at the same time as that of the kidney of the mice. The liver weight and RNA of the adult mice decreased to the same degree and to a minimum level after 14 days of castration. The subcutaneous implantation of a 15-mg pellet of testosterone, 17α-methyltestosterone, or one or two pellets of 17β-hydroxy-5α-androstan-3-one restored both values to normal levels within 2–3 weeks. The total amount of DNA was not altered by either castration or steroid treatment except for an increase after 28 and 42 days of treatment in the two studies with 17β-hydroxy-5α-androstan-3-one. This apparent difference between testosterone propionate and 17β-hydroxy-5α-androstan-3-one needs further exploration.

TRUCKENBRODT (1969) (see also HEIM et al., 1969) reported that the weight of the liver of immature normal mice increased at a faster rate than that of mice castrated at 18 days of age. The protein and DNA contents increased in proportion to the weight, but the RNA showed a greater increase so that the concentration in the liver of the normal mouse was 12% greater than in that of the castrated mouse at 60 days of age.

## C. Thymus

The RNA and DNA of the thymus of the mice studied by KOCHAKIAN and HARRISON (1962) for changes in the nucleic acids of the kidney and liver increased in proportion to the increase in weight after castration and were similarly reversed after the subcutaneous implantation into the castrated mice of testosterone propionate, 17α-methyltestosterone, or 17β-hydroxy-5α-androstan-3-one. Thus, there was a gain and loss and not a hypertrophy and atrophy of the cells.

## D. Perineal Muscle

### "Levator Ani"

VENABLE (1966a, b) made a careful light and electron microscopic study of the "levator ani" muscle after castration and testosterone administration (see Chap. III). The mice were castrated when full grown (40 days of age). Forty-five days later the "levator ani" muscle was one-third the normal size, and no further decrease was evident at 6 months of age. The injection of 0.5 mg of Depo-Testosterone (19-nortestosterone cyclohexylpropionate) produced an enlargement of the muscle to twice that of the controls (or two-thirds that of normal) within 9 days. The total DNA content was practically identical in the normal, castrated and Depo-Testosterone-treated mice.

# III. Rat

## A. Liver and Kidney

CHEEK et al. (1965), in their studies on growth, found no change in the normal increase of DNA of the liver of rats castrated at 3 weeks of age for the subsequent 5 weeks in comparison with normal animals.

LIMANOWSKI and BRZEZINSKI (1966) studied the changes in RNA in the livers of castrated and testosterone propionate-treated castrated and normal adult (180–200 g) rats by the staining of histologic sections with Brachet's reaction (pyronin) for RNA. Castration decreased the intensity of the reaction, and the injection of testosterone propionate at 1 mg twice per week for 6 weeks beginning at castration greatly increased the reaction, which now was visible in all of the hepatocytes. A similar result was produced in normal rats receiving 2.5 mg testosterone propionate twice per week for 6 weeks.

A number of studies utilizing different radioactively labeled RNA precursors provide further evidence for changes in nucleic acid metabolism in the liver after castration and androgen administration. CANTAROW et al. (1958) injected adult male rats every other day for 12 days with a large dose, 10 mg, of testosterone propionate and 18 h before autopsy with 20 mg of [2-$^{14}$C]-uracil (500 μCi/mmole) in 10 ml of 0.85% NaCl solution intraperitoneally. The specific radioactivity of the RNA of the treated rats was increased 6-fold, with a decrease in the specific radioactivity ratio of UMP: CMP from 2.0 to 0.6.

FARABOLLINI (1962) injected adult (250 g) male rats also with a large dose, 5 mg/day testosterone, for 4 days. Four h after the last injection, 10 μCi (111 μg) [$^{14}$C]-formate was injected intraperitoneally. Two hours later the kidneys and livers were removed and extracted for RNA, DNA, and protein. The RNA content of the tissues was unchanged, but the radioactivity of the RNA of the kidney was decreased (−20%) and that of the liver increased (+373%). The radioactivity of the DNA of the liver exhibited a nonsignificant increase of 13%. CORTI et al. (1970) confirmed and extended the changes in liver r-RNA detected by formate labeling. The rate of incorporation of [$^{3}$H]-formate (100 μCi were injected 24 h before sacrifice) into not only ribosomal but also nuclear RNA was much higher in castrated rats (200–250 g body weight) injected with 1 mg/100 g/day of testosterone propionate for 4 or 8 days than in control, castrated rats. The androgen decreased the nucleotide triphosphate levels, indicating an enhanced utilization for RNA synthesis. No variations in nuclear DNA synthesis were detected.

SHEPPARD et al. (1965) also noted a decrease in the uptake of a DNA precursor by the kidney after steroid treatment. They reported a decrease in the incorporation of [$^{3}$H]-thymidine in the kidney of immature rats 2 days after the injection of 4 mg/day of methandrostenolone with a return to normal after 7 days of treatment. SHEPPARD et al. (1964), however, found no effect of the foregoing steroid treatment on the incorporation of [$^{3}$H]-thymidine by liver slices.

YAMANAKA et al. (1969) castrated adult (200–250 g) male rats and 2 weeks later also injected a large dose, 5 mg, of testosterone propionate in 1.2 ml sesame oil. The rats were killed at various intervals thereafter and were injected intraperitoneally with 2 mCi [$^{32}$-P]-sodium phosphate 60 min before autopsy. The RNA of the kidneys was phenol-extracted and separated into three fractions by Kieselguhr column chromatography. The incorporation of $^{32}$P into the ribosomal RNA fraction nearly doubled 3 h after the injection of testosterone and returned to normal after 6 h. The soluble RNA fraction did not show a significant change. The change in the ribosomal RNA was enhanced when the phenol extractions were carried out at 25° and 65° C in addition to extraction at 4° C, suggesting that DNA-like RNA was present with the ribosomal RNA. This was confirmed by base analysis of the RNA. The third fraction, which was eluted by the addition of SDS (sodium dodecyl sulfate) to the buffer, proved by phenol extraction and analysis of base composition to be similar to the second or ribosomal fraction. In another experiment, the nuclei of the kidney were separated 30 min after the injection of the testosterone propionate. The phenol-extracted RNA, as the RNA of the total kidney, possessed a higher specific radioactivity than the controls in the second or ribosomal fraction. The base composition of this RNA was similar to that of r-RNA, which led the authors to postulate that the androgen induced the synthesis of r-RNA in the nuclei, which was then transported to the cytoplasm.

IWAMOTO et al. (1963) compared the uptake of $^{32}P$ by the RNA in the different subcellular components of the liver. The rats were 100–150 g in body weight. The various fractions of the liver were extracted for RNA by the phenol method, and then the RNAs were separated by DEAE-cellulose chromatography. The $^{32}P$ (50 μCi/100 g) was injected intraperitoneally 5 times at 30 min-intervals to maintain a constant blood level, and the rats were killed 30 min after the last injection. The normal female rats exhibited a higher specific radioactivity in the nuclear RNA than the normal male rats. Consequently, the authors studied the effect of various hormones. The rats were gonadectomized, and the subsequent daily injections for 10 days of 50 μg/100 g of β-estradiol benzoate in ovariectomized rats and 0.2 mg/100 g of 17β-hydroxy-5α-androstan-3-one in normal male rats accelerated the incorporation of radioactivity in the nuclear RNA but not that of the other fractions. In contrast, castration increased, and testosterone propionate after 10 days suppressed, the incorporation of radioactivity in the nuclear RNA. The radioactivity of the RNA in the other fractions was either too low for significant comparison or unchanged. On the other hand, TISHENINA and GOLBER (1968) reported that 1 month after castration of immature or mature rats, the biosynthesis of liver nuclear RNA was decreased and was then increased by the injection into castrated rats every other day for 14 days of testosterone propionate at 4 mg/100 g. The intraperitoneal injection of actinomycin D at 340 μg/100 g body weight 15 min before the injection of the androgen completely eliminated the stimulatory effect of the androgen. Similar results were reported by KROLIKOWSKA-PRASAL (1970). Male white rats were castrated at 250 g body weight. 17α-Methyltestosterone (5 mg) was injected twice weekly in both castrated and normal rats. The RNA and DNA of the liver were determined after 2 weeks and after 1 and 3 months. Two h before autopsy, 100 μCi of $^{32}P$ potassium phosphate per 100 g body weight was injected intraperitoneally. The RNA was extracted from the liver by the perchloric acid procedure. Two weeks after castration, the concentration and specific radioactivity of the liver RNA were greatly decreased; both then gradually (1- and 3-month periods) increased but did not attain the levels in the normal rats. The injection of 17α-methyltestosterone was without effect after 2 weeks; it produced a slight increase after 1 month, but after 3 months the concentration and specific radioactivity of the RNA had approached values near to those of the normal. The delay and slowness of the restoration to normal values in the castrated rats by methyltestosterone could be due to the great loss in efficacy of the steroid by the relatively long interval between injections. The content and specific radioactivity of the DNA were not changed. The injection of 17α-methyltestosterone into normal rats was without effect on either RNA or DNA.

The effect of the androgens on ribosomal RNA receives some support from electron microscopic studies (KAWADA and SHIODA, 1964). An oral dose of 2.0 mg/

day of stanozolol for 35 days in male rats produced a significant increase in the endoplasmic reticulum of the liver.

FUJII and VILLEE (1968) studied the early effects of a single injection of testosterone (1 mg in 0.2 ml propylene glycol) in 3-week-old male rats as indicated by the intraperitoneal injection of 10 μCi/50 g body weight of [$^3$H]-cytidine 30 min before autopsy. The prostate, seminal vesicles, and thymus were studied simultaneously with the liver. The homogenized tissues were separated by centrifugation into nuclei and a cytoplasmic fraction. The specific radioactivities of both fractions of the liver were approximately doubled 2 h after the testosterone injection, were returned to nearly normal levels after 12 h, and were normal after 38 h. The RNA of the thymus of the same rats did not show any significant changes. A comparison of the sucrose gradient centrifuge patterns of the liver and cytoplasmic RNA of rats treated for 2, 12, and 38 h showed no significant difference in sedimentation. The labeling of both the n-RNA and cytoplasmic RNA, however, was increased 2 and 12 h after injection of testosterone. The RNA fraction of the thymus showed no changes.

## B. Muscle

### 1. Skeletal

A few of the skeletal muscles have been studied, usually for comparative purposes with the "levator ani," a muscle of the perineal complex. The skeletal muscles of the rat show only small changes and in direct proportion to the small changes in body weight after castration and androgen treatment (KOCHAKIAN et al., 1956).

CHEEK et al. (1965) in their studies on growth noted that the DNA content of the quadriceps muscles of rats castrated at 3 weeks of age was not changed 5 weeks later. SAUNDERS et al. (1962) found no change in the RNA:DNA ratio of the rectus femoris muscle 5 weeks after castration at 3 weeks of age. The injection of testosterone, 20 mg/kg/day, for 7 days beginning 4 weeks after castration, produced a slight but nonsignificant increase in the RNA:DNA ratio. Subsequent experiments did not verify the small increase. SAUNDERS et al. (1962) also studied the effect of diabetes and insulin alone and in combination with castration and testosterone replacement. Diabetes was induced in 90-day-old intact and castrated rats by the intraperitoneal injection of alloxan. Two weeks later, rats showing blood glucose values of 300 mg/100 ml or greater were injected subcutaneously with insulin at 10 U/100 g for the first 2 days, followed by 5 units/100 g for 5 days. Testosterone was injected at 20 mg/kg/day. The RNA:DNA ratio was decreased by diabetes in the normal and castrated rats, presumbaly by the loss of muscle weight and RNA. The RNA/mg N was not changed. Insulin increased the RNA:DNA ratio above normal with a comparable increase in the RNA/mg N. Testosterone administration, however, was without effect in the diabetic castrated rat and produced a questionable, small enhancement of the effect of insulin.

GALAVAZI and SZIRMAI (1971 b) included the tensor fasciae latae muscle in their study of the "levator ani" muscle because, like the "levator ani," it is one of the whitest of the skeletal muscles. Castration was at 6 weeks of age, and the rats were killed with normals of the same age at 7, 12, 28, 96, and 334 days after castration. Testosterone was injected into castrated rats at 1.25 mg/day for 1, 4, 7, 11, and 15 days preceding autopsy at 138 days of age. Neither castration nor subsequent testosterone treatment resulted in an observable effect on the ribosomal concentration between the fibrils or in the paranuclear cones of the tensor fasciae latae muscle.

## 2. Perineal

*"Levator Ani"*

SAUNDERS et al. (1962) reported that in contrast with the rectus femoris muscle, the "levator ani" muscle showed marked nucleic acid changes after castration and testosterone treatment. Castration did not change the concentration of RNA but decreased the RNA:DNA ratio, presumably because of an increase in concentration of DNA. The injection of testosterone for 7 days increased the concentration of RNA and the RNA:DNA ratio to twice that in the castrated animals. Alloxan diabetes in intact rats produced a small increase in the concentration of RNA but a 50% decrease in the RNA:DNA ratio, presumably accompanied by a loss in muscle and body weight. The administration of insulin increased the concentration of RNA to twice that of the normal and caused a 50% increase in the RNA:DNA ratio. Diabetes in the castrated rats resulted in a 2-fold increase in the concentration of RNA and a small decrease in the RNA:DNA ratio. The injection of testosterone into the diabetic-castrated rat induced a further 2-fold increase in the concentration of RNA and a 3-fold increase in the RNA:DNA ratio. Insulin also increased both the concentration of RNA and the RNA:DNA ratio, but not to quite the same degree as testosterone. The injection of insulin simultaneously with testosterone gave essentially the same results as testosterone alone. The authors do not give the body or "levator ani" weights of any of the groups of rats. These data would be especially helpful in the interpretation of the results in the diabetic rats, which presumably lost body and muscle weight.

SHEPPARD et al. (1964, 1965) studied the effect of castration and methandrostenolone (17$\beta$-hydroxy-17$\alpha$-methylandrosta-1,4-dien-3-one) on several parameters of the "levator ani" muscle in comparison with the seminal vesicles and ventral prostate in immature (50 g body weight) rats. DNA biosynthesis was determined by the intraperitoneal injection of 50 μCi/50 g body weight of [$^3$H]-thymidine and autopsy 2 h later. The steroid-treated rats received the [$^3$H]-thymidine 24 h after the last injection. Castration was without effect during the subsequent 7 days on DNA, RNA, wet weight, and incorporation of thymidine in the levator ani muscle. The injection of methandrostenolone for 7 days at doses of 0.25–2.0 mg/day produced a dose-dependent increase in weight, RNA, DNA, and protein. The RNA increase was progressively greater than that of DNA, to give an approximately 2-fold increase in the RNA:DNA ratio with the 1.0 and 2.0 mg/day doses.

The injection of 4.0 mg/day of the steroid into the intact and castrated immature rats produced a progressive increase in the weight, protein, and RNA of the "levator ani" muscle. The DNA reached a maximum increase in quantity and also in incorporation of [$^3$H]-thymidine on the 2nd day. Thereafter, the quantity of DNA was maintained, and the specific radioactivity decreased to the control level by the 7th day. Autoradiograms and histologic examination suggested that the observed changes in the incorporation of the [$^3$H]-thymidine occurred in parallel with the rate of mitosis. The incorporation of [$^3$H]-thymidine by slices of "levator ani" muscle of intact immature male rats also was increased after 2 days of 4 mg/day of methandrostenolone, and the specific radioactivity was back to normal after 7 days as in the in vivo experiments. ARVILL (1967) observed a 100% increase in the incorporation of [8-$^{14}$C]-adenine, when the "levator ani" muscle of the 21-day-old male rat was removed and incubated 24 h after the injection of 5 mg of testosterone propionate.

The electron microscopic studies of GALAVAZI and SZIRMAI (1971a) supported the results of SHEPPARD et al. (1964) on the changes in DNA of the "levator ani" muscle. The rats were castrated at 42 days of age (125 g body weight) and were injected subcutaneously with 1.25 mg/day of testosterone propionate for 15 days, beginning 46 days after castration. Castration inhibited the normal formation of further nuclei by approximately 50% and testosterone propionate increased the number to 83% of the normal adult level. The increase occurred after a lag period of 4 days and reached a maximum sometime between the 4th and 7th days. These authors concluded that castration produced aplasia as well as atrophy of the fibers of the "levator ani" muscle and that testosterone propionate elicited both hyperplasia and hypertrophy. This was recognized as contrary to the results of VENABLE (1966a, b) in mice (see above). GALVAZI and SZIRMAI (1971a, b) suggested that the difference was due to the degree of maturity at the time of castration. The mice were full grown while the rats were still in the active growth phase and their cells had not attained the full complement of nuclei and DNA. Thus, castration retarded the growth of the myonuclear population, and testosterone induced myonuclear multiplication to make up for the arrears due to castration. These authors also emphasized that DNA determinations in muscle homogenates cannot be used as a measure of myonuclei, since a large proportion of the nuclei in skeletal muscle consists of connective tissue cell nuclei (ENESCO and PUDDY, 1964; VENABLE, 1966a).

HEIM et al. (1969) injected rats intraperitoneally 1.5 years after castration for 14 days every other day with 10 mg/kg of 3-oxo-4-chlorandrost-4-en-17$\beta$-yl-acetate. The DNA and protein increased in proportion to the increase in weight of the "levator ani" muscle, and the RNA increased at a more rapid rate resulting in a higher concentration than in the controls.

Electron microscopic studies were conducted by two different laboratories to obtain information concerning the ultrastructural aspects of the decrease after castration and the increase after androgen administration in the weight of the "levator ani" muscle. PELLEGRINO and POLLERA (1967) in a preliminary report, and GORI et al. (1967, 1969), in later, more detailed reports, observed that castration produced a rapid weight loss of the muscle to one-third the normal size after 2 months of castration with a decrease in the diameter of the fibers to one-third of normal and a decrease in the number and diameter of the myofibrils. The nuclei were deeply indented, and the nucleoli were composed mainly of those in the fibrillar phase; their ribosomes (granules) were notably reduced. The injection of a large dose of 2.5 mg/day of testosterone propionate induced a rapid increase in size of the nuclei during the first 2 days and a second rapid increase, beginning at the 5th day of injections, to normal size at about the 15th day, after which it was maintained. The nucleoli showed a more rapid response. The nucleolar volume was restored to normal within 24 h, and with continued injections, it increased further, reaching a maximum of about twice normal size on the 5th day and a return to normal by the 15th day, which was maintained for the remainder (2 weeks) of the injection period. The fibrillar phase, after 24 h, was already more abundant and appeared as a loose reticulum that contained ribosomes loose or packed together. The ribosomes in the sarcoplasmic spaces were increased to form thick clusters of polyribosomes in rosettes, long strings, or helices. The ribosomes and polysomes continued to increase; after 5 days packed masses of ribosomes occupied every sarcoplasmic space, and figures of rough endoplasmic reticulum were evident. The maximum increase of cytoplasmic ribosomes and polyribosomes occurred later than the peak of nucleolar modifications, which is in accord with the demonstration that the nucleolus is the site of synthesis of ribosomal precursors (PERRY, 1966). As the muscle reached its normal size, the number of ribosomes gradually decreased. The glycogen and mitochondria also decreased after castration. An increase in glycogen deposits was one of the earliest events after testosterone propionate injection, in agreement with previous quantitative studies (LEONARD, 1952; BERGAMINI et al., 1965). The glycogen deposits decreased after the muscle attained normal size, but the number and volume of mitochondria continued to increase to above normal levels.

GALAVAZI and SZIRMAI (1971a, b) reported results essentially similar to those of GORI et al. (1967, 1969). They castrated rats at 6 weeks of age and studied them at intervals for the next 334 days. Testosterone propionate, 1.25 mg/day, was injected at 1, 4, 7, 11, and 15 days preceding autopsy at 138 days of age. Castration for 96 days produced a decrease in the ribosomes in the paranuclear cone of the sarcoplasm but no detectable difference in the intermyofibrillar spaces. An increase in the ribosomes of the paranuclear cones of the sarcoplasm was evident within 24 h after the injection of testosterone propionate. These authors also noted that the concentration of the glycogen granules increased in both the intermyofibrillar spaces and the myofibrils. The increase was complete in 24 h, and the authors concluded, as did GORI et al. (1969), that this change belonged among the earliest effects of testosterone. BERGAMINI (1975) has reported a detailed analysis of gylcogen metabolism in this muscle of the rat under testosterone stimulation.

## IV. Guinea Pig

### Skeletal Muscle

The growth of the muscles of the guinea pig in the absence or presence of testosterone varies widely among the individual muscles. KOCHAKIAN et al. (1964), therefore, compared the concentrations of RNA and DNA in representative muscles of 5- and 10-month-old guinea pigs. The DNA concentrations of the temporal, gastrocnemius, and oblique muscles were essentially the same and that of the masseter was approximately twice that of these muscles. The concentration of RNA, on the other hand, was only about 20% greater in the masseter muscle than in the other muscles. Thus, there was no correlation between responsiveness and nonresponsiveness to testosterone. The difference between the masseter and the other muscles probably is a reflection of the type ("red" vs. "white") of muscle (GUTMANN et al., 1970).

Castration of guinea pigs at 490–540 g body weight had no effect on the RNA and DNA contents of the oblique and gastrocnemius muscles but retarded not only the growth of the temporal and masseter muscles (two of the most responsive muscles) but also the increase in their RNA and DNA contents. A slow increase in weight was observed between the 55th and 230th days after castration, accompanied by a proportionate increase in RNA and an increase of the DNA to approximately the normal level in both muscles. The muscles of the normal animals showed similar (approximately 50%) increases in weight and RNA but no further increase in DNA. The cells of the normal guinea pigs apparently had fully differentiated by 5 months of age, and the further increase in size was due to hypertrophy.

The implantation of two 15-mg pellets of testosterone in 10-month-old guinea pigs castrated at 3 months of age had no effect on the RNA and DNA contents of the gastrocnemius or oblique muscles. Testosterone, however, induced a rapid increase within 5 days in the RNA of both the temporal and masseter muscles and then a slow, gradual (temporal) or no further (masseter) increase for the duration (30 days) of the experiment. The weight of the muscles increased in a linear fashion. Thus, a rapid synthesis of RNA was induced to provide a high concentration of RNA for the synthesis of new protein. MARCAIS and KOCHAKIAN (1971) found in subsequent studies that the polysomes of the temporal muscle were decreased by castration and restored by testosterone administration.

Testosterone administration to guinea pigs 1 or 2 months after castration produced essentially the same effect on the nucleic acids as described above. Furthermore, testosterone propionate and 17β-hydroxy-5α-androstan-3-one gave results essentially the same as testosterone.

## V. Rabbit

### Heart

NOWY and FRINGS (1963) injected young adult male rabbits for 10–17 weeks with 10 mg/day of 4-chlorotestosterone acetate. The heart weight and concentration of DNA were the same as those of the controls, but the RNA concentration was slightly lower in the left ventricle of the treated rabbits. NOWY and FRINGS (1963) conducted a similar study in rabbits with experimentally produced aortic insufficiency. The 4-chlorotestosterone acetate, 10 mg/day, was injected for 3–5 months. The concentration of DNA was unchanged, but that of the RNA showed a statistically significant small decrease after steroid treatment. These studies were extended by NOWY et al. (1965) to testosterone propionate and a new steroid, Emdabol (1α,7α-bis-[acetylthio]-17β-hydroxy-17α-methylandrost-4-en-3-one) at 3, 10, and 30 mg/day for 5 weeks. A small decrease in the concentration of RNA was observed after the 3 mg/day dose of Emdabol and a small decrease in the DNA after 10 and 30 mg/day of testosterone propionate. No significant changes in heart weight were evident in the six treated groups.

## VI. Summary

The RNA and DNA contents of many tissues in the mouse, rat, guinea pig, and rabbit are influenced by castration and by treatment with various anabolic-androgenic steroids. The greatest changes in DNA occurred in the thymus where it changed in direct proportion to the changes in weight after castration and steroid treatment. The mouse kidney DNA decreased only slightly after castration (concentration increased) and was restored to normal by steroid treatment after a lag period.

The rate of DNA development in the hormone-dependent guinea pig muscles was delayed by castration. The DNA of the skeletal muscles of the rat was not affected, but that of the "levator ani" muscle was decreased by castration and increased by hormone treatment. The liver DNAs of the rat and mouse were not significantly affected by either type of hormonal manipulation.

The RNA of the mouse kidney, rat "levator ani" muscle and temporal and masseter muscles of the guinea pig decreased in concentration after castration and increased initially at a more rapid rate than the weight of the tissues on hormone stimulation. The changes were in the microsomal ribosomes of the kidney and muscles. The RNA of the other subcellular elements of the mouse kidney changed in direct proportion to the changes in weight. Induction of an increase in rat liver RNA is reported, but similar changes were not detected in the liver of castrated mice. The rates of biosynthesis of RNA in the various tissues have been explored using radioactive precursors. Studies with radioactively labeled orotic acid indicated a decrease in the endogenous pool of orotic acid after castration, resulting in higher specific activities of the nucleotide precursors and RNA in the castrated mouse kidney with a restoration to normal by testosterone propionate after a lag period of more than 2 and less than 7 days. The incorporation of radioactively labeled uridine, on the other hand, was not affected by castration but was slightly accelerated by the administration of testosterone propionate. The liver of the mice showed no change in the rate of incorporation of either orotic acid or uridine after castration or hormone treatment.

## References


ARVILL, A.: Effects of testosterone on the metabolism of the isolated levator ani muscle of the rat. Acta endocr. (Kbh.) **56** (Suppl. 122), 1—14 (1967).

BERGAMINI, E.: Different mechanisms in testosterone action on glycogen metabolism in rat perineal and skeletal muscles. Endocrinology **96**, 77—84 (1975).

BERGAMINI, E., BOMBARA, G., PELLEGRINO, C.: The effect of testosterone on glycogen content of levator ani muscle of castrated rats. Biochem. J. **97**, 11 P (1965).

BRACHET, J.: Embryologie Chimique, Paris: Masson 1947.

BRACHET, J.: Chemical Embryology, New York: Interscience Publ. 1950.

CANTAROW, A., WILLIAMS, T. L., MELNICK, I., PASCHKIS, K. E.: Influence of acetylaminofluorene, growth hormone, testosterone, and hypothyroidism on incorporation of uracil-2-$^{14}$C in liver RNA in the rat. Cancer Res. **18**, 818—821 (1958).

CASPERSON, T. O.: Cell Growth and Cell Function. New York: W. W. Norton 1950.

CHANG, C. T., LEI, H. P.: Pharmacology of 17$\alpha$-methyl-5$\alpha$-androstan-17$\beta$-ol. Yao Hsueh Pao **12**, 734—739 (1965).

CHEEK, D. B., POWELL, G. K., SCOTT, R. E.: Growth of muscle cells (size and number) and liver DNA in rats and Snell-Smith mice with insufficient pituitary, thyroid or testicular function. Bull. Johns Hopk. Hosp. **117**, 306—321 (1965).

CORTI, A., MONTI, M. G., MORUZZI, M. S.: Action of testosterone on nucleic acids of rat liver subcellular fractions. Ital. J. Biochem. **19**, 277—283 (1970).

DUBOVSKY, J., KOCHAKIAN, C. D.: Early effect of testosterone in the incorporation of uridine into RNA of mouse kidney and seminal vesicles. Endocrinology **92**, 917—920 (1973).

ENESCO, M., PUDDY, D.: Increase in the number of nuclei and weight in skeletal muscle of rats of various ages. Amer. J. Anat. **114**, 235—244 (1964).

FARABOLLINI, F.: Incorporazione di formiato-C$^{14}$nelli proteine, RNA e DNA di fegato e rene di ratti trattati con testosterone o cortisone. Rend. Inst. Sci. Univ. Camerino **3**, 160—165 (1962).

FRIEDEN, E. H., KU, C.-C.: Acute effects of testosterone propionate upon RNA synthesis in mouse kidney. Proc. Soc. exp. Biol. **137**, 1110—1114 (1971).

FUJII, T., VILLEE, C. A.: Effect of testosterone on ribonucleic acid metabolism in the prostate, seminal vesicle, liver and thymus of immature rats. Endocrinology **82**, 463—467 (1968).

GALAVAZI, G., SZIRMAI, J. A.: Cytomorphometry of skeletal muscle: The influence of age and testosterone on the rat M. levator ani. Z. Zellforsch. **121**, 507—530 (1971 a).

GALAVAZI, G., SZIRMAI, J. A.: The influence of age and testosterone on the ribosomal population in the M. levator ani and a thigh muscle of the rat. Z. Zellforsch. **121**, 548—560 (1971 b).

GORI, Z., PELLEGRINO, C., POLLERA, M.: The castration atrophy of the dorsal bulbocavernosus muscle of rat: An electron microscopic study. Exp. molec. Pathol. **6**, 197—198 (1967).

GORI, Z., PELLEGRINO, C., POLLERA, M.: The hypertrophy of levator ani muscle of rat induced by testosterone: An electron microscope study. Exp. molec. Pathol. **10**, 199—218 (1969).

GUTMANN, E., HANZLIKOVÁ, V., LOJDA, Z.: Effect of androgens on histochemical fibre type. Differentiation in the temporal muscle of the guinea pig. Histochemie **24**, 287—291 (1970).

HEIM, F.: Der Einfluß von Methenolonazetat auf den Eiweiß-Glykogen- und ATP-Gehalt mehrerer Organe weißer Mäuse. In: Wirkung und Anwendung anaboler Steroide, pp. 27—30. Berlin: Medicus Verlag 1964.

HEIM, F., SCHWARZLOSE, W., TRUCKENBRODT, H., MITZNEGG, P.: Desoxyribonucleinsäure-, Ribonucleinsäure- und Eiweißgehalt, Zellneubildung und Zellgröße in Abhängigkeit von Alter, Geschlecht und Ernährung und deren Beeinflussung durch anabole und antianabole Stoffe. Z. Gerontol. **2**, 25—47 (1969).

IWAMOTO, T., MORIYAMA, A., TETSUKA, T., MIURA, Y.: Intracellular transfer of nucleic acids. II. Effect of hormones on ribonucleic acid metabolism in rat liver cells. J. Biochem. **53**, 408—415 (1963).

JELINEK, J., VESELÁ, H., VALOVÁ, B.: The effect of nortestosterone phenylpropionate on compensatory hypertrophy of the remaining kidney after unilateral nephrectomy. Acta endocr. (Kbh.) **46**, 352—360 (1964).

KASSENAAR, A., KOUWENHOVEN, A., QUERIDO, A.: On the metabolic action of testosterone and related compounds. V. The effect of testosterone on the nucleic acid composition of seminal vesicles and kidneys of mice. Acta endocr. (Kbh.) **39**, 223—233 (1962).

KAWADA, S., SHIODA, K.: Effect of stanozolol on rat liver. An electron microscopic observation. A. R. Nat. Inst. Nutr. (Tokyo), 23—24 (1964).

KOCHAKIAN, C. D.: Mechanism of androgen actions. Lab. Invest. **8**, 538—556 (1959).

KOCHAKIAN, C. D.: Intracellular regulations in the kidney by androgens. Amer. Zoologist **2**, 361—366 (1962).

KOCHAKIAN, C. D.: Mechanism of anabolic action of androgens. In: Mechanisms of Hormone Action, pp. 192—213. New York: Academic Press 1965.

KOCHAKIAN, C. D.: Androgens, RNA and protein synthesis. In: Conference on Cellular Dynamics. Proc. 3rd and 4th Interdisciplinary Conf. N.Y. Acad. Sci., pp. 186—197, 1967a

KOCHAKIAN, C. D.: Regulation of ribonucleic acid biosynthesis in the mouse kidney by androgens. In: Proc. 2nd Internatl. Cong. Hormonal Steroids 1966. Milan: Excerpta Medica Internat. Congr. Series **132**, 794—802 (1967b).

KOCHAKIAN, C. D.: Effect of androgen on cellular growth and metabolism. In: Adolescent Nutrition and Growth, pp. 207—224. New York: Appleton-Century-Crofts 1969 a.

KOCHAKIAN, C. D.: Androgen regulation of mouse kidney RNA. In: Metabolic Effects of Gonadal Hormones and Contraceptive Steroids, pp. 328—338. New York-London: Plenum Press 1969 b.

KOCHAKIAN, C. D.: Intracellular regulation of nucleic acids of mouse kidney by androgens. Gen. comp. Endocr. **13**, 146—150 (1969 c).

KOCHAKIAN, C. D.: Partial characterization of the changes in mouse kidney RNA after castration and androgen. In: The Sex Steroids, pp. 411—419. New York: Appleton-Century-Crofts 1971.

KOCHAKIAN, C. D., DUBOVSKY, J., BROULIK, P.: Effect of nutritive state and precursor dose on the incorporation of orotic acid and uridine into kidney and liver of normal, castrated and androgen-treated mice. Endocrinology **90**, 531—537 (1972).

KOCHAKIAN, C. D., HARRISON, D. G.: Regulation of nucleic acid synthesis by androgens. Endocrinology **70**, 99—108 (1962).

KOCHAKIAN, C. D., HILL, J.: Effect of androgen on the incorporation of orotic-acid-6-$^{14}$C into the ribonucleic acids and free nucleotides of mouse kidney. Biochemistry **5**, 1696—1701 (1966).

KOCHAKIAN, C. D., HILL, J., HARRISON, D. G.: Regulation of nucleic acids of muscles and accessory sex organs of guinea pigs by androgens. Endocrinology **74**, 635—642 (1964).

KOCHAKIAN, C. D., HIRONE, T., KUSUNOKI, H., DUBOVSKY, J., STRICKLAND, B.: Ultrastructural changes in the mouse kidney after castration and androgen administration. Effect of actinomycin D and correlation with parameters of RNA and protein synthesis. Ala. J. med. Sci. **8**, 149—167 (1971).

KOCHAKIAN, C. D., NISHIDA, M., HIRONE, T.: Ribosomes of mouse kidney: Regulation by androgens. Amer. J. Physiol. **217**, 383—391 (1969).

KOCHAKIAN, C. D., TILLOTSON, C., ENDAHL, G. L.: Castration and the growth of muscles in the rat. Endocrinology **58**, 226—231 (1956).

KROLIKOWSKA-PRASAL, I.: Isotope studies on the influence of testicular androgens on nucleic acids in hepatic tissue in white rats. Polish Endocr. **21**, 147—154 (1970).

LEONARD, S. L.: A glycostatic effect of testosterone on perineal muscle of the rat. Endocrinology **50**, 199—205 (1952).

LIMANOWSKI, A., BRZEZINSKI, R.: Effect of gonadectomy and testosterone administration on certain histochemical reactions in rat liver. Polish. Endocr. **17**, 211—218 (1966).

MARCAIS, J., KOCHAKIAN, C. D.: Influence of castration and testosterone on guinea pig skeletal muscle ribosomes and protein synthesis. Fed. Proc. **30**, 1214 (1971).

NOWY, H., FRINGS, H. D.: Effekte von Anabolica auf den Herzmuskel. 2. Nucleinsäurezusammensetzung des Kaninchenmyocards nach Behandlung mit 4-chlortestosteron-acetat. Arzneimittel-Forsch. **13**, 571—573 (1963).

NOWY, H., FRINGS, H. D., KUDERNATSCH, M.: Effekte von Anabolica auf den Herzmuskel. 6. Untersuchungen über die Wirkung von $1\alpha,7\alpha$ bis-(acetylthio)-$17\beta$-hydroxy-$17\alpha$-methylandrosten(4)-on-(3) und Testosteron auf den Ribonucleinsäure- und Desoxyribonucleinsäuregehalt von Kaninchenherzen. Arzneimittel-Forsch. **15**, 1254—1255 (1965).

PELLEGRINO, C., POLLERA, M.: Modifications of RNA-containing structures of dorsal bulbocavernosus muscle of rat after castration and testosterone administration. Experientia (Basel) **23**, 134—135 (1967).

PERRY, R. P.: On ribosome biogenesis. In: Internat. Symp. The Nucleolus, Its Structure and Function. Natl. Cancer Inst. Monograph No **23**, 527—545, NCI, Bethesda (1966).

RABINOVITCH, M.: "Alkaline" and "acid" phosphatases activities and ribonucleic acid phosphorus content of the kidney cortex of the mouse as influenced by castration and testosterone propionate administration. Folia clin. biol. **18**, 115—119 (1952).

SAUNDERS, H. L., STECIW, B., KLINE, A.: Effect of testosterone and insulin on rat skeletal muscle RNA. Endocrinology **71**, 314—320 (1962).

SHEPPARD, H., TSIEN, W. H., MAYER, P., HOWIE, N.: Metabolism of the accessory sex organs of the immature male rat: changes in nucleic acid composition and uptake of thymidine-$^3$H induced by castration and methandrostenolone. Biochem. Pharmacol. **14**, 41—51 (1965).

SHEPPARD, H., TSIEN, W. H., MAYER, P., SEYMOUR, G.: Uptake of thymidine-H$^3$ by accessory sex organs of the immature male rat: Effect of methandrostenolone. Biochem. Biophys. Res. Commun. **15**, 546—550 (1964).

TESSMANN, D.: Über die Wirkung von Testosteron auf die normale Mäuseniere. Z. Zellforsch. **86**, 61—73 (1968).

TISHENINA, R. S., GOLBER, L. M.: Effect of androgens on the biosynthesis of the nuclear fractions of nucleic acids in rat liver. Probl. Endokr. **14**, 100—104 (1968).

TRUCKENBRODT, H.: Tierexperimentelle Untersuchungen zur Regulation der DNS-, RNS- und Eiweiß-Synthese während des physiologischen Wachstums und deren Beeinflussung durch anabole und antianabole Pharmaka. Arzneimittel-Forsch. **23**, 121—138 (1969).

VENABLE, J. H.: Constant cell population in normal, testosterone deprived and testosterone stimulated levator ani muscles. Amer. J. Anat. **119**, 263—270 (1966a).

VENABLE, J. H.: Morphology of the cells of normal, testosterone deprived and testosterone stimulated levator ani muscles. Amer. J. Anat. **119**, 271—301 (1966b).

WATSON, J. D.: Involvement of RNA in the synthesis of proteins. Science **140**, 17—26 (1963).

YAMANAKA, H., FURUYA, N., SHIMAZAKE, J., SHIDA, K.: Effects of testosterone administration on RNA synthesis in kidneys of castrated rats. Endocr. Jap. **16**, 29—36 (1969).

ZAMECNIK, P. C.: Historical and current aspects of the problems of protein synthesis. Harvey Lectures (1958—1959) **54**, 256—281 (1960).

# C. Regulation of Tissue Enzymes

C. D. KOCHAKIAN

## I. Introduction

The establishment of a definite protein anabolic effect by testosterone and related steroids in nonsexual tissues stimulated a search for the mechanism of action of these hormones. Changes in enzyme activity and production were investigated in target tissues by a large number of investigators. The most striking and frequent changes were observed in the mouse kidney, with lesser effects in the kidneys of other experimental animals (KOCHAKIAN, 1947, 1959, 1962, 1976). Changes in activities of several enzymes after castration and hormone treatment have been reported in other tissues. In most instances the changes are small and questionable.

## II. Kidney

### A. Amino Acid-Related

#### 1. Transaminases and Glutamic Acid Dehydrogenase

*a) Rat*

MARSILII and D'ALLESSANDRO (1955) reported a 20% decrease in the GPT activity of adult male rats 37 days after castration. Weekly injection of 20 mg testosterone cyclopentylpropionate during the period of castration gave irregular results. KOCHAKIAN et al. (1959) found that both the GPT and GOT activities in the kidney of the rat changed in direct proportion to the decrease in weight after castration and the increase after testosterone administration. Duration of castration or testosterone treatment and age of the animals did not influence the results. Glutamic acid dehydrogenase activity was determined on the same animals. The enzyme concentration in the rat kidney of 40-day-old rats was unchanged after castration for 53 days. Subcutaneous implantation of a 15-mg pellet of testosterone propionate for the last 28 days of castration induced an approximately 50% increase in enzyme activity.

PUCHOL and CARBILLIDO (1961) gonadectomized male and female rats at 12 weeks of age and found no significant change in the kidney GOT specific activity 3 months later. Injection of a massive dose, 25 mg/day, of testosterone propionate for 5 days was without effect. The specific activity of GPT, on the other hand, increased 70–95% after castration, and the increase was not altered by administration of testosterone propionate. A similar hormone treatment of normal male and female rats increased the GPT, but did not significantly change the GOT specific activity of the kidneys.

KAWADA et al. (1965a) administered by stomach tube 0.5, 1.0, and 2.0 mg/day of stanozolol for 14 days to castrated rats and observed identical but not significant decreases (−14%) in the specific GPT activity of the kidneys. Subcutaneous injection for 2 days of 1.0 mg of stanozolol produced a small (14%) increase in kidney weight and a statistically nonsignificant decrease (−26%) in the specific activity of GPT. TAKAGI (1963) reported that the single injection of 10 mg of 19-nortestosteronephenylpropionate or testosterone propionate into male rats of 50 g body weight produced a fourfold increase in the specific activity of the kidney GPT between 5 and 14 days later, with a return to normal after another week.

*b) Mouse*

KOCHAKIAN et al. (1959) observed that the activities of both GPT and GOT changed in direct proportion to the decrease in weight of the kidney after castration and the increase upon testosterone administration. The specific activity of GDH in the kidney of mice castrated at 40 days of age was increased slightly 30 days later and gradually restored to normal on implantation of a 15-mg pellet of testosterone propionate at 30 days after castration.

SCHWARZLOSE and HEIM (1973) reported that the GPT specific activity of the kidneys of male mice increased fivefold from 18–42 days of age. Castration at 18 days of age reduced the increase to threefold and prevented the small increases of GOT and GDH activities. Injection of testosterone propionate at 16 mg/kg/day into 42-day-old female mice produced small increases in GPT and GOT activities after 4 days. Injection of 100 mg/kg/day of testosterone increased GDH specific activity about 50% after 2 days with no further increase after 4 days. The simultaneous injection of 100 µg/kg/day of actinomycin D prevented the hormonally induced increases in the enzyme activities.

*c) Guinea Pig*

KOCHAKIAN and ENDAHL (1956) castrated guinea pigs at 510 g body weight and 30 days later implanted subcutaneously one or more 15-mg pellets of various steroids. The GOT activity of the kidney was unchanged after 21 days but that of GPT was increased to a similar degree, about 30%, by 7 different steroids varying in known biological activity from the slightly active epiandrosterone to the very potent testosterone.

### 2. D-Amino Acid Oxidase

*a) Mouse*

CLARK et al. (1943) reported a decrease of −54% in the concentration of D-amino acid oxidase activity of the mouse kidney 130 days after castration at 18 g body weight. Subcutaneous implantation of a pellet of testosterone propionate increased the specific activity of the enzyme slightly above normal in both the castrated and normal mice. VAN BEKKUM and KASSENAAR (1954) obtained a doubling of the D-amino acid oxidase activity of gonadectomized adult male and female mice after the injection of 2.5 mg of testosterone propionate twice per week. The maximum effect was obtained within about 2 weeks, and extension of the hormone treatment to as long as 70 days did not produce a further increase. ENDAHL and KOCHAKIAN (1956) extended their studies to 17 steroids and noted that the increase in enzyme activity was related to dose, duration of treatment, and chemical structure of the steroids. The mice were castrated at 24–26 g body weight and 2 months later, the steroids were implanted subcutaneously as pellets to provide different dosages. Enzyme activity increased progressively in an identical manner to a 3-fold increase between 14 and 20 days with testosterone and 17α-methyltestosterone, and no further increase was obtained upon continuation of hormone treatment for another 2 weeks. A dose response increase in enzyme activity to a maximum of 3 times that of the controls was produced by the following steroids: testosterone propionate; 17β-hydroxy-5α-androstan-3-one; 17β-hydroxy-17α-methyl-5α-androstan-3-one; 5α-androstane-3α,17β-diol; and 17α-methyl-5α-androstane-3α,17β-diol. A large dose of 24 mg/21 days of the weakest steroid, dehydroepiandrosterone, produced a significant (79%) increase. The other steroids gave intermediate maximum values.

FRIEDEN et al. (1964) reported that actinomycin D injected into adult mice at 200 μg/kg/day beginning 1 day before the daily injection of 1 mg testosterone propionate partially prevented the increase in D-amino acid oxidase activity. If the actinomycin D injections were begun after the injection of the testosterone propionate, no inhibition was observed.

*b) Rat*

In contrast to the mice, rats exhibited inconsistent changes in the D-amino acid oxidase activity of the kidney after castration. VAN BEKKUM and KASSENAAR (1954) castrated rats at 300–400 g body weight. Injection of 2.5 mg testosterone propionate every other day for 10 days gave widely variable results. ENDAHL and KOCHAKIAN (1956) conducted 6 separate series of experiments on rats castrated at 25–36 days of age except for one series at 70 days of age. The rats were implanted subcutaneously with 1 or 2 pellets of testosterone, 17α-methyltestosterone, or testosterone propionate at different times after castration and for 6–60 days. Both castration and hormone treatment gave inconsistent changes. The change after castration varied from +50 to −30%. When a decrease after castration was present, it was restored to normal by the steroid treatment.

*c) Guinea Pig*

Castration of young adult guinea pigs for 6 months or adults for 2 months did not alter the D-amino acid oxidase activity (ENDAHL and KOCHAKIAN, 1956). Furthermore, subcutaneous implantations of pellets, containing 9 different steroids, 30 days after castration of guinea pigs at 510 g body weight did not induce statistically significant changes in the D-amino acid oxidase activity of the kidney 21 days later. A few of the more potent steroids, e.g., 17α-methyltestosterone, appeared to induce a questionable increase of about 30%.

### 3. Arginase

*a) Mouse*

KOCHAKIAN (1944) noted no change in the total arginase activity of mouse kidney 60 days after castration at 16.0–19.5 g body weight. Since the kidney weight decreased, the concentration of the enzyme increased as a result of castration. Subcutaneous implantation of 14-mg pellet of 32 different steroids, gave a wide range of responses. Known biologically potent steroids, e.g. testosterone, 17α-methyltestosterone, increased the total arginase activity 1300% after 30 days of treatment while less active compounds were without effect or restored the arginase value to normal; several compounds gave intermediate values. Five of the most active steroids—testosterone; 17α-methyltestosterone; testosterone propionate; 5α-androstane-3α,17β-diol and 17β-hydroxy-5α-androstan-3-one—gave after 10 days nearly the same maximal effect as after 30 days of treatment. Estradiol-17β also was quite effective in stimulating an increase in arginase activity, but progesterone was ineffective.

A comparison of the increases in arginase activity with the amount of steroid absorbed among the potent compounds suggested a dose-response relationship. KOCHAKIAN (1945a), therefore, extended the studies to a wide range of doses of the six most active steroids and four of the less active steroids. The apparent differences in maximal effect among the most active steroids—testosterone; testosterone propionate; 17α-methyltestosterone; 17β-hydroxy-5α-androstan-3-one and 5α-androstane-3α,17β-diol—were no longer evident when compared in terms of equivalent doses. As the doses of these steroids were increased, the kidney weight was initially restored to normal without any change in total enzyme activity. When the kidney weight was stimulated to above normal levels, a progressive increase in arginase activity was induced, which continued long after the kidney weight had reached a plateau. Addition of a 17α-methyl group to the various steroids—17α-methyltestosterone; 17α-methyl-5α-androstane-3α,17β-diol; 17β-hydroxy-17α-methyl-5α-androstan-3-one—hastened the initial rapid increase in enzyme activity. The less active steroids restored the kidney weight to or near normal without affecting the total arginase activity.

Since testosterone and estradiol-17β are antagonistic in several biological actions but show some similarity in inducing increase in kidney arginase activity, KOCHAKIAN et al. (1948) studied the simultaneous administration of these two steroids and obtained a summation of their respective effects. KOCHAKIAN and STETTNER (1948) found that the growth hormone of the anterior pituitary induced a small increase in kidney weight after 10 days without further enhancement after 20 or 34 days and no consistent change in arginase activity. Simultaneous administration of growth hormone and testosterone propionate for 20 and 34 days, however, produced a 50% decrease in the ability of testosterone propionate to increase the kidney arginase activity.

FRIEDEN and FISHEL (1968) reported an increase of approximately 100% in the kidney arginase of 6–12-weeks-old female mice and of 200% in 6-month-old female mice after 5 days of testosterone propionate at 1 mg/day. Simultaneous injection of 200 μg/kg/day of actinomycin D did not affect the increase induced after 5 days by 2.5 mg of testosterone propionate every other day. Cycloheximide at 300 mg/kg/day and ethionine at 50 mg/kg/day, on the other hand, essentially abolished the stimulatory effect of testosterone propionate at 1 mg/kg/day for 5 days.

SWANK et al. (1973) found no difference in the increase of arginase activity in the kidneys of female mice with low and high inducibility of β-glucuronidase activity on the injection of an oil suspension of 10 mg of 17β-hydroxy-5α-androstan-3-one for 2 days and then 5 mg every 3 days for 12 days.

*b) Rat*

KOCHAKIAN and CLARK (1942) reported an average increase of approximately 200% in the kidney arginase activity of rats castrated at 70 g body weight and 5 days later injected with 5 mg/day of testosterone propionate for 16 days and then implanted subcutaneously with a 20-mg pellet of the steroid for 2 weeks. KOCHAKIAN and ROBERTSON (1950) elaborated on this pilot observation. Injection of testosterone propionate at 0.125 mg/day for 21 days doubled the specific activity of the arginase of rats castrated at 270–300 g body weight and adjusted to an isocaloric food intake during the following 6 months. Increase of the dose to 1.0 mg/day produced a further increase to approximately 4 times the control level, and extension of the treatment to 46 days did not produce a further enhancement in enzyme activity. Shortening of the period of injection to 13 days diminished the increase in activity. In another series of experiments young adult rats fed *ad libitum* were castrated and 30 days later injected with 2.5 mg/day of testosterone propionate. After 8 days, the arginase specific activity was 3.5 times that of the control and 4–5 times after 16 days. Subcutaneous implantation of 2 pellets (15 mg each) of testosterone or 17α-methyltestosterone produced an approximately similar 3-fold increase in the specific activity of the arginase after 10 and 30 days. When a single pellet of testosterone propionate was implanted to provide a smaller dose (about one-fourth), the increase was approximately one-half that induced by testosterone and 17α-methyltestosterone.

KOCHAKIAN and ROBERTSON (1950) reported that hypophysectomy of adult (300 g) rats produced a 50% decrease in kidney arginase activity. Subcutaneous implantation of a 14–15-mg pellet of testosterone on the 79th day of hypophysectomy increased the arginase level after 2 weeks to twice the normal value without an effect on kidney weight.

Thyroidectomy also greatly reduced the arginase activity of the kidney of the castrated adult rat (KOCHAKIAN et al., 1954). Injection of 1.0 mg/day of testosterone propionate for 14 days increased the arginase activity to above the control value without a significant effect on kidney weight. Adrenalectomy of male rats (150 g body weight) produced a decrease in arginase activity of approximately 30% within 2 days (KOCHAKIAN and VAIL, 1947). Injection of 1 mg/day of desoxycorticosterone acetate partially prevented the decrease. Injection of 2.5 mg/day of testosterone propionate for 2 days increased the enzyme activity to approximately 2 times normal. Intensive treatment with adrenal corticol extract at hourly intervals for 8 h before autopsy did not influence the effect of testosterone propionate.

Injection of 0.5 mg/day of testosterone propionate for 5 days into alloxan-diabetic castrated adult rats produced the expected increase in arginase activity despite the sclerosis due to the alloxan (KOCHAKIAN et al., 1952). Phlorhizinized normal adult rats exhibited a decrease in kidney arginase activity that was increased to above normal levels by 1.0 mg/day for 5 days of testosterone propionate. Phlorhizin treatment of castrated adult rats on a 70% protein diet increased the kidney arginase activity, which was increased slightly more by injection of 0.5 mg/day for 5 days of testosterone propionate and doubled after injection of 1.0 mg/day for 13 days.

SVEJCAR (1956) compared the effects of age and sex on the response of kidney arginase to injection of 2.5 mg/day of testosterone propionate for 5–8 days. The steroid induced an increase in both male and female rats in the 40–46 and 72–79-day-old groups but only in the males of the 104–120 and 130–147-day-old groups without an increase in kidney weight. Gonadectomy equalized the two sexes, and now testosterone propionate had an equal stimulatory effect on kidney arginase activity.

ZAK et al. (1971) reported a very early effect of testosterone on kidney arginase activity. Male rats of 21–22 days of age were injected intraperitoneally with 1 mg of testosterone suspended in 0.5 ml saline. Arginase activity increased 137% after 1 h and reached a peak of 208% after 3 h, then gradually returned toward normal. Injection of 200 μg actinomycin D 30 min before testosterone prevented the hormonally induced arginase activity. This effect of testosterone is the earliest biochemical change reported in the literature and needs further exploration.

*c) Guinea Pig*

HUMM et al. (1948) castrated guinea pigs at 250 g body weight and 35 days later implanted subcutaneously one or more pellets of either testosterone; 17β-hydroxy-5α-androstan-3-one; 5α-androstane-3α,17β-diol and their 17α-methyl derivatives and also testosterone propionate for 30 days. Kidney arginase activity was reduced about 30% after 65 days of castration, but the reduction was no longer apparent after another 60 days. Testosterone was ineffective at doses of 7.4, 18.9, and 32.2 mg/30 days. The other steroids, especially the 17 methylated compounds, induced increases of 27–79% at the higher dose levels. Injection of a large dose, 12.5 mg/day, of testosterone propionate for 14 days induced a 38% increase in the arginase activity.

NATAF (1953) ovariectomized adult female guinea pigs and after 30 days implanted subcutaneously a 5-mg pellet of testosterone. The arginase activity of the kidney was increased about 30% after 14 days without any change in kidney weight.

The arginase activity of the guinea pig kidney, in contrast to the mouse and rat, is relatively unresponsive to hormonal manipulation.

*d) Hamster*

KOCHAKIAN et al. (1948) castrated hamsters at 45–75 g body weight and observed an increase of about 20% in the arginase activity of the kidneys after 50 days, which was not further enhanced after another 120 days. The arginase activity was restored to normal by injection of 1.0 mg/day of testosterone propionate for 20 or 60 days or by implantation of 3 pellets, 15 mg each, of either testosterone or 17α-methyltestosterone, for 20 days.

### 4. Transcarbamylases

TARNAWSKI and ZAK (1970) injected intraperitoneally 5 mg of testosterone suspended in 0.5 ml saline into 4-month-old rats that had been castrated 4 weeks previously. The aspartate transcarbamylase activity of the kidney was increased after 1 h and significantly doubled after 2 and 3 h, after which it gradually decreased to the normal level at 8 h. Injection of 200 μg actinomycin D 30 min before injection of testosterone prevented the increase in enyzme activity. VAN RYEMENANT et al. (1971), on the other hand, in 12 experiments with male rats weighing 300–400 g did not detect any change in aspartate transcarbamylase activity after 8 days of castration or after the daily injection of 1 mg of testosterone, 10 mg testololactone or a combination of the two steroids in 0.2 ml of saline from the 3rd to 8th day after castration.

ZAK et al. (1971) reported that the injection of a suspension of 1 mg testosterone in 0.5 ml saline into 21–22-day-old male rats induced, as in the case of arginase and aspartate transcarbamylase, an increase in the activity of ornithine transcarbamylase activity within 1 h and reached and maintained a maximum of about 350% from the 2nd to the 5th hour, after which it gradually receded to the normal level at 8 h. Injection of 200 µg actinomycin D 30 min before the testosterone abolished the increases.

### 5. Amidotransferases

SVEJCAR (1956) reported a decrease in the formation of glycocyamine (transamidinase activity) by kidney preparations from adult normal and gonadectomized male and female rats but not normal female rats after injection of testosterone propionate 2 mg/100 g/day for 3–7 days. KRISKO and WALKER (1966) conducted a more detailed study and reported that the transamidinase activity of the kidney of the rat was the same until puberty (about 40 days of age), after which that of the male increased while that of the female decreased. At 60 days of age, therefore, the activity of the male kidney was twice that of the female. A similar sex difference was reported earlier by FITCH (1963) and later by VAN PILSUM and UNGAR (1968). KRISKO and WALKER (1966) found that gonadectomy slightly increased the enzyme activity in the female and decreased that in the male so that both values were similar and intermediate. Injection of 1.0 mg followed 24 h later by 0.5 mg of testosterone propionate induced a 60% increase in the transamidinase activity of the kidneys of adult (173 days old) female rats 72 h after the first injection. A similar experiment with $^1/_{10}$th the dose was ineffective. Injection of 0.5 mg/day for 2 days of either testosterone propionate or 17α-ethylnortestosterone into adult (270 g) female rats produced practically identical increases of 50–60% in the enzyme activity 96 h after the first injection.

UNGAR and VAN PILSUM (1966) found that the transamidinase activity of the kidneys of male rats one month after castration at weaning was no different from that of normal controls. Furthermore, injection of 500 µg/day of testosterone for the last 8 days of the castration period was ineffective as was a similar treatment of weanling male rats. In a later report, VAN PILSUM and UNGAR (1968) extended the injections to 14 days and still did not detect a change in the enzyme activity in the weanling males but noted a questionable, small increase in the female rats. Gonadectomy of adult (200 g) rats also did not change the kidney enzyme activity in the males one month later but increased that of the females. Injection of 500 µg/day of testosterone for 10 days did not change the kidney transamidinase activity of either the male or female gonadectomized adult rats.

### 6. Histamine Methyltransferase

SNYDER and AXELROD (1965) noted that several investigators had reported that female rats excreted 9–10 times as much free histamine as males (LEITCH et al., 1956) and that after the injection of [$^{14}$C]-histamine or [$^{14}$C]-histidine, male rats excreted more [$^{14}$C]-methylhistamine than females (WESTLING and WETTERQUIST, 1962). Since the major pathway for inactivation of histamine is by N-methylation, they studied the activity of histamine methyltransferase in the kidney. The enzyme activity in the kidney of the adult (200 g) male rat was twice that of the female while those of brain, liver, and intestine showed no sex difference. The inability of NETTER et al. (1961) to previously detect this difference was shown to be due to the use of relatively large amounts of tissue for assay.

The injection of male rats with 50 μg estradiol-17$\beta$ or of female rats with 5 mg of testosterone for 1 or 7 days did not change the enzyme activity. Castration decreased the enzyme activity 35% after 17 days, and a single injection of testosterone (5 mg) restored the enzyme value to normal 3 days later. Ovariectomy did not significantly change the enzyme activity 2 weeks later, and a single injection of testosterone increased the enzyme activity 40%.

In contrast to the rat, methyltransferase activity in the mouse was twice as great in the female as the male and in the guinea pig showed no sex difference.

## 7. Histidine Decarboxylase

ROSENGREN and STEINHARDT (1961) observed that the female mouse kidney had a much higher (30-fold) capacity to form histamine than the male. HENNINGSSON and ROSENGREN (1972) investigated the hormonal influence on the sex difference in histamine formation. Injection of 10 μg/day of estradiol-17$\beta$ greatly increased, and injection of 0.5 mg/day of testosterone for 10 days nearly abolished, the urinary excretion of free histamine by 3- and 7-month-old female mice. Injection of 0.5 mg testosterone on 2 successive days decreased the ability of the kidneys of ovariectomized mice to form histamine by 50%. 12 h after the first injection and practically abolished the enzyme activity after 48 h. Analysis of the urine indicated a strong reduction in excretion of histamine that was not due to a conversion to methylhistamine because this substance was slightly reduced instead of increased. The effect of testosterone on histidine decarboxylase activity was confirmed in a subsequent report (GRAHN et al., 1973), and extremely wide variations in the enzyme activity were noted among mice from different litters, suggesting genetic as well as hormonal regulation.

## 8. Ornithine Decarboxylase

GRAHN et al. (1973) investigated the nature of an unknown compound detected by HENNINGSSON and ROSENGREN (1972) in their studies on sex differences in histamine excretion. The unknown substance was identified as putrescine, a metabolite of ornithine. GRAHN et al. (1973) found that injection of 0.5 mg/day of testosterone increased the excretion of putrescine by female mice within 24 h and reached a maximum after about 5 days. Analysis of ornithine decarboxylase activity of the kidneys revealed a twofold increase within 3 days (the earliest determination) and a continued but variable 2- to 10-fold increase for the duration of the injection period. The average increase was 3.7-fold.

## 9. Cathepsin

JUNQUEIRA and ROTHSCHILD (1953) attempted to relate cathepsin activity to the increase in kidney weight after testosterone propionate treatment. Adult mice were castrated, 30 days later 15 mg of testosterone propionate was implanted subcutaneously and 6 days later, 2.5 mg in oil was injected. On the 10th day, the kidney weight was nearly doubled, and the cathepsin activity was slightly but not significantly reduced on a kidney wet weight basis and slightly increased when related to DNA.

### 10. Acetylating Enzyme

The ability to acetylate fed p-aminobenzoic acid is greater in male than in female rats, and the administration of testosterone increases this ability in the female rat (FISHKIN and LATA, 1958). LUUKKAINEN (1961) tried to find an explanation in the enzyme activities of the kidney and liver. Normal and castrated male and female rats were injected one month after castration with 500 μg/day of testosterone propionate for 28 days, then with 3.5 mg testosterone enanthate weekly for 6 weeks, and finally with 500 μg/day of testosterone propionate for 20 days. Castration and testosterone treatment were without significant effect on the ability of the kidney and also the liver mince to acetylate p-aminobenzoic acid. No difference, moreover, was noted between the control and testosterone-treated rats when they were tested in vivo with p-aminobenzoic acid twice a month during treatment.

### 11. Aminopeptidase

DEIMLING et al. (1965) made a histochemical study of the cytological distribution of aminopeptidase activity in the mouse kidney. The proximal tubule exhibited weak activity in both the adult male and female mouse. The activity in the distal portion, however, was 3 times greater in the female than in the male mouse. Castration of the male mouse gave the female pattern, while injection of testosterone reduced the activity in the female to that in the male.

COBET et al. (1967) reported an increase in leucine aminopeptidase activity in the kidneys of male rats (200–250 g) after heminephrectomy and the daily injection of 3 mg/100 g body weight of 19-nortestosterone-17$\beta$-decanoate for 14 days. A summation of the respective effects occurred in the heminephrectomized-hormone treated rats.

## B. Hydrolases

### 1. Phosphatases

#### *a) Mouse*

KOCHAKIAN and FOX (1944) implanted subcutaneously a pellet of testosterone propionate into normal and castrated mice of $17 \pm 2$ g body weight and determined the "alkaline" and "acid" phosphatase activities of the kidneys after 35 and 115 days. The results were similar for the two time intervals. Castration did not change the concentration of the "alkaline" or "acid" phosphatase activities. Testosterone propionate reduced both the total and concentration of "alkaline" phosphatase activity and increased the "acid" phosphatase activity in proportion to the increase in weight of the kidney. KOCHAKIAN (1945b) confirmed the effect of castration, and the use of a more dilute homogenate revealed that the total "alkaline" phosphatase activity was not changed by castration and that the concentration was actually increased because of the decrease in weight of the kidney. The effects of several of the known biologically very active steroids were studied at different doses. Pellets of the following were implanted subcutaneously to provide several dose levels to mice castrated 30 days previously at 16.0–19.5 g body weight: testosterone; testosterone propionate; 17α-methyltestosterone; 17β-hydroxy-5α-androstan-3-one; 5α-androstane-3α,17β-diol; 17α-methyl-5α-androstane-3α,17β-diol. Phosphatase determinations 30 days later exhibited the same dose response for all of these steroids. The total and concentration of "alkaline" phosphatase activity decreased as the kidney increased in weight and reached a constant value in parallel with the plateau of increase in kidney weight. The "acid" phosphatase activity, on the other hand, increased essentially in proportion to the increase in kidney weight. Androst-4-ene-3,17-dione and 5α-androstane-3α,17β-diol in single doses induced comparable effects on the phosphatases. KOCHAKIAN (1948) examined sections from the kidneys of the above mice by the histochemical technique of Gomori. "Alkaline" phosphatase exhibited a progressive decrease from the distal end of the proximal convoluted tubule toward the glomerulus with the various steroids. The degree of enzyme depletion was related to the dose of steroid. Several weak androgens—dehydroisoandrosterone, isoandrosterone, androsterone, 5α-androstane-3,17-dione—were studied by the colorimetric method. The small responses in kidney weight elicited by these steroids were accompanied by increases in the specific activities of both phosphatases. Simultaneous administration of estradiol-17β (KOCHAKIAN et al., 1948) or growth hormone (KOCHAKIAN and STETTNER, 1948) did not alter the effect of testosterone on either phosphatase.

KOCHAKIAN and WILLIAMS (1973) localized the subcellular site of most of the "alkaline" phosphatase activity in the microsomes. Castration increased and testosterone propionate decreased the enzyme activity not only of the microsomes but also of the mitochondria and cytosol. Approximately 75% of the "acid" phosphatase activity was distributed nearly equally between the microsomes and cytosol and changed in direct proportion to the change in kidney weight.

RABINOVITCH (1952) related the changes in phosphatase activities to the number of nuclei. Adult mice were castrated and 30 days later were implanted subcutaneously with 10 mg of powdered testosterone propionate. Normal mice were similarly treated. The cortex of the kidney was separated 4 and 8 days later. The concentration of "acid" phosphatase was unchanged in the castrated and treated mice. The concentration of "alkaline" phosphatase activity, however, was increased after castration and greatly (−50%) decreased in both the normal and castrated treated mice. A plot of the "alkaline" phosphatase activity against the number of nephron tubule cell nuclei per unit volume of kidney cortex gave a linear relationship.

DEIMLING et al. (1965) compared the histochemical distribution of "alkaline" phosphatase in the adult mouse kidney. The female exhibited the same pattern as the castrated (40 days) male and was not changed by ovariectomy. Injection of testosterone at 100 μg for 15 days or 160 μg for 40 days into ovariectomized or normal female mice produced the male type of enzyme distribution. The difference between the male and female mouse is in agreement with the earlier report of DUNN (1948), who noted that after maturity the "alkaline" phosphatase staining appeared in the brush border of an additional segment in the male mouse kidney. The rat, guinea pig, and rabbit did not show this difference.

HUNTER et al. (1964) exploited the effect of testosterone on "alkaline" phosphatase activity to assess the longevity of the action of long-acting androgens. Immature castrated mice were injected with a single dose of 19-nortestosterone decanoate. No change in enzyme activity was detected after injection of 0.25 or 0.5 mg, but 5.0 mg depressed the enzyme activity and sustained the depression for 20 days. These investigators also determined the effect of dose, other steroids, and strains of mice. Injection on alternate days for 14 days of 0.1, 0.2, 0.4, and 0.8 mg of testosterone propionate progressively and markedly decreased the "alkaline" phosphatase activity of the kidney, in agreement with the earlier observation of KOCHAKIAN (1948). Injection of Dianabol at 1 mg in oil or suspension as above did not affect the "alkaline" phosphatase activity, while $\Delta^1$-testosterone undecylenate produced an effect comparable to that of testosterone propionate. The same results were obtained in the Swiss and $C_3H$ strains of mice.

AKIKUSA and YAMAMOTO (1968) reported that the activity of transport ATPase activity in the male mouse kidney was twice that in the female and was decreased to that of the female 2 weeks after castration. A single injection of 2 mg/100 g of testosterone propionate restored the decreased activity to the normal level.

KOCHAKIAN and WILLIAMS (1973) studied the effect of castration and testosterone propionate on the glucose-6-phosphatase and nucleotidase activities simultaneously with that of the phosphatases. The enzymes were localized in the microsomal fraction and changed in direct proportion to the change in weight of the kidney after castration and hormone treatment. DEIMLING et al. (1965) found no clear difference in the cytological distribution of the glucose-6-phosphatase activity between normal and gonadectomized adult male and female mice or after testosterone treatment of female, and estradiol-17$\beta$ treatment of male mice.

### b) Rat

KOCHAKIAN and ROBERTSON (1950) observed a 40% increase in the "alkaline" phosphatase activity of adult castrated rats upon injection of 2.5 mg/day of testosterone propionate for 8 days beginning 30 days after castration. Extension of the injection to 16 days yielded a further small increase. Similar increases were obtained in adult castrated rats on constant food intake, injected for 13 and 21 days with 1.0 mg/day of testosterone propionate. Extension of the injections to 46 days did not give a further increase. A decrease of the dose to 0.125 mg/day for 21 days diminished the response in enzyme activity to +20%. The subcutaneous implantation of two 15-mg pellets of testosterone into adult castrated rats (200–230 g) produced a 25% increase in enzyme activity after 15 days and no further change after 30 days. Similar treatment with 17$\alpha$-methyltestosterone elicited a 20% increase after 10 days but no increase after 30 days. A single pellet of testosterone propionate for 30 days was slightly more effective than the two pellets of testosterone.

Subcutaneous implantation of a single pellet of testosterone in adult rats 79 days after hypophysectomy stimulated a 45% increase in the concentration of "alkaline" phosphatase activity without a change in kidney weight (KOCHAKIAN and ROBERTSON, 1950). KOCHAKIAN et al. (1954) found that injection of 1.0 mg/day of testosterone propionate for 14 days into adult castrated rats thyroidectomized for 25, 40, or 328 days not only restored the decrease in "alkaline" phosphatase activity due to thyroidectomy but increased it above normal. VAIL and KOCHAKIAN (1947) found a

36% increase in the "alkaline" phosphatase activity of the kidneys of adrenalectomized rats (150 g) after the injection of 2.5 mg/day of testosterone propionate for 2 days. The concentration of "acid" phosphatase was not altered.

The relatively small but consistent increases in the concentration of the "alkaline" phosphatase at first seemed to be contrary to the initial observations in the mouse but was consistent with the later studies with the weaker androgens and suggested a species difference in dosage requirement. TISSIERES (1948) castrated rats at 30 days of age and 20 days later injected 10 mg of a crystalline suspension of testosterone at 10-day intervals for 30 days. Histochemical study of the kidneys showed no changes as a result of castration, but the large doses of testosterone resulted in a depletion of the "alkaline" phosphatase similar to that observed in the mouse (KOCHAKIAN, 1948). The "acid" phosphatase was diminished after castration and restored by testosterone treatment.

MUHLENFELD (1969), in a study of the chemodifferentiation of the rat kidney by enzyme histochemistry, observed a sex differentiation in the "alkaline" and "acid" phosphatases between the 29th and 35th postnatal days. WENK (1968) also noted a sex differentiation in "alkaline" phosphatase activity in histochemical studies of rat kidneys. Injection of testosterone propionate at 1 mg/day for 12 days into normal and ovariectomized rats did not change the enzyme intensity or pattern. Ovariectomy, however, converted the enzyme pattern to the male type.

*c) Guinea Pig*

HUMM et al. (1948) castrated guinea pigs at 250 g body weight; 65 days later the concentration of "alkaline" phosphatase activity was decreased (−40%) and after another 60 days the decrease was −29%. Implantation of pellets of several active steroids restored the enzyme activity to or nearly to normal without any definite correlation with dose or specific steroid. Injection of a large dose, 12.5 mg/day, of testosterone propionate for 14 days had no effect on the "alkaline" phosphatase activity. The "acid" phosphatase was not affected either by castration or by any of the steroids.

*d) Hamster*

KOCHAKIAN et al. (1948) found that castration of hamsters at 45–75 g body weight did not decrease the concentration of the kidney "alkaline" phosphatase activity 50 days later but induced a definite decrease after another 40 days. A decrease also was noted in another experiment 140 days after castration at 48 g body weight. Injection of 1 mg/day of testosterone propionate for 20 days increased the enzyme value above normal; a further increase was observed on extension of the injections to 60 days. Subcutaneous implantation of 3 pellets of testosterone or 17α-methyltestosterone at the time of castration (48 g body weight) produced identical increases in the enzyme activity after 20 days. Injection of 2.5 mg/day of testosterone propionate for a long period, 140 days, gave essentially the same increase in "alkaline" phosphatase activity as obtained by implantation of the steroids.

*e) Rabbit*

KAMADA (1965) reported that the injection of 0.5 mg/kg of testosterone propionate for 10 days into normal female rabbits gave a slight increase in the histochemical "alkaline" phosphatase activity and a larger dose, 5 mg/kg, decreased the enzyme activity. Similar treatment beginning the 11th day after castration gave normal histochemical findings after the low dose but a depressive picture after the higher dose.

## 2. Esterases

### a) Mouse

The mouse kidney exhibits a large number (25–30) of electrophoretically distinct naphthyl-acetate esterases. Sharp genetic differences exist between mouse strains (RUDDLE, 1966), and certain isozymes show definite sex differences (RUDDLE and HARRINGTON, 1967).

SHAW and KOEN (1963) found a highly active esterase isozyme in adult male mouse kidney. It was absent in immature and in most adult female mice. The enzyme occurred in the five strains of mice examined but did not show any sex difference in the field mouse. Castration of male mice decreased the intensity of the isozyme after 6 weeks. Injection of 0.5 mg/day testosterone propionate into female mice produced a moderate increase and after 7 days an intensity equal to that in the male. Testosterone caused the appearance of the isozyme also in immature male and female mice. MANDA and OKI (1967) obtained identical results in another strain of mice. KOMEDA et al. (1971) observed no differences in the intensity of the isozymes of the kidney of male and female immature mice. After maturity, however, five of the bands were more intense, and three less intense, in the kidneys of the male than in the female mice. Ovariectomy caused no change, but castration changed the isozyme pattern to that of the female. Injection of testosterone propionate into castrated males or normal females induced the male isozyme pattern. These changes were evident in five different strains of mice with variable modifications, suggesting that sensitivity to testosterone propionate differs in inbred strains of mice.

WEHLER et al. (1973) studied the postnatal development of esterase activity. The enzyme activity progressively increased to 25 days of age at the same rate in male and female mice. The appearance of a definite sex difference was detected at 35 days of age. The increase in esterase activity was hastened by injection of testosterone, but the degree of stimulation was decreasingly effective in the younger mice and was completely ineffective in newborn mice. Furthermore, not only the injection of testosterone at 4 mg/100 g/day for 4 days but also that of testosterone propionate and of 17$\beta$-hydroxy-5$\alpha$-androstan-3-one was more effective in adult than in 25-day-old female mice. A single injection of testosterone was ineffective. Repeated daily injections were necessary to obtain a maximum effect in 6–10 days.

DEIMLING et al. (1965) studied the cytological distribution of the esterase by histochemical methods. A change of distribution was observed only in the proximal convoluted tubules. Castration produced a diminution to that of the female, and the daily injection of 200 µg of testosterone propionate for 10 days increased the enzyme intensity in the maximal tubule of the adult female kidney to the same level as in the male. The enzyme change induced by testosterone appeared to precede the weight increase. KOCHAKIAN and WILLIAMS (1973) studied the esterase distribution in the subcellular components by ultracentrifugation of kidney homogenates. Approximately 70% of the enzyme activity was found in the microsomal fraction. Castration of mice at 2 months of age resulted in an approximately 40% decrease in specific activity of the esterase in both the total homogenate and the microsomal fraction. Subcutaneous implantation of a 15-mg pellet of testosterone propionate 2 weeks before autopsy restored the enzyme activity to the normal level. The small amount of esterase activity in the other subcellular fractions did not detectably respond to castration or hormone treatment.

*b) Rat*

In contrast to their observations in the mouse, SHAW and KOEN (1963) could not detect a sex difference in the isozyme esterase pattern in the adult rat kidney. SCHWARK and ECOBICHON (1967) also observed no sex difference in the isozyme pattern of the kidneys from their strain of rats, and READ and MCKINLEY (1965) observed only a small, nonsignificantly lower total esterase value in the female than in the male rat kidney. MANDA and OKI (1967), however, found two highly active isozymes in the extracts of male rat kidney that were absent in the female. These isozymes disappeared after castration of the male rats and reappeared to the normal intensity after injection of 1 mg/day of testosterone for 10 days. Ovariectomy of the female rat resulted in the appearance of the 2 isozymes, and injection of 10 μg/day of estradiol-17β-benzoate for 10 days caused a disappearance of the isozyme.

*c) Hamster*

LI et al. (1969) found no sex difference in the 13 different esterase isozymes of the kidneys of hamsters less than 3 weeks of age. The adult male, however, showed marked intensity in one of the isozymes, which was absent or barely detectable in the female. Castration of males at 6 weeks or older produced a marked diminution in another isozyme. The subcutaneous implantation of a 20-mg pellet of testosterone propionate 3 weeks after gonadectomy restored both isozymes to normal levels after 3 weeks. An identical response was produced by daily injection of 0.3 mg testosterone, and an increase above normal was produced by 0.5 mg/day of testosterone. Identical responses were produced in ovariectomized hamsters by the same hormone treatment. Subcutaneous implantation of a 20-mg pellet of estradiol-17β or diethylstilbestrol also induced an increase in the 2 isozymes in ovariectomized and to a greater degree in castrated hamsters.

## 3. Ribonucleases

KOCHAKIAN et al. (1964) castrated mice at 25 or 28 days of age and found that the "acid" ribonuclease activity decreased in proportion to the decrease in kidney weight, while the total "alkaline" ribonuclease was unchanged and therefore increased in concentration. Subcutaneous implantation of a 15-mg pellet of testosterone propionate restored the kidney and ribonuclease activities to normal levels after about 14 days. KOCHAKIAN and WILLIAMS (1973) separated the kidney homogenates into the subcellular components by centrifugation. The majority (55%) of the "alkaline" ribonuclease activity was localized in the microsomal fraction. The specific activity was increased almost twofold after 24 days of castration at 2 months of age and was reduced to below normal upon subcutaneous implantation of a 15-mg pellet of testosterone propionate 2 weeks before autopsy. The "acid" ribonuclease activity was localized primarily in the heavy mitochondria (39%), but appreciable activity was also present in the light mitochondria (22%), microsomes (15%), and cytosol (19%). The activity in each fraction changed in direct proportion to the changes in kidney weight.

### 4. β-Glucuronidase

MORROW et al. (1951) found a strain (genetic) difference in the β-glucuronidase activity of the mouse kidney. In all strains, however, the activity in the male was 1.3–2.6 times that of the female. The sex difference was not apparent in immature mice 5–7 weeks of age. Gonadectomy at 13 weeks of age reduced the enzyme activity in the male kidney at 25 weeks of age to that of the female and had no effect in the female. The sex difference was due primarily to the enzyme activity in the kidney cortex. The medulla did not show a sex difference.

FISHMAN (1951) injected 1 mg/day of testosterone propionate and after 5 days observed a greater than 3-fold increase in the specific β-glucuronidase activity of the kidneys of both adult male and female mice. Injection of 1.25 mg of 17α-methylandrost-5-ene-3β,17β-diol every other day into adult male mice stimulated a progressive increase to a maximum of 50 times the control after about 50 days. Castrated male or normal female mice showed the same effect. White Swiss male mice showed only a 10-fold increase on similar hormonal treatment. FISHMAN and FARMELANT (1953) extended these studies for a comparison of 7 different strains of male mice (2 months old) to the injection of 0.5 mg testosterone propionate every other day for 14 days. Quantitative differences in the response by the different strains ranged from 4- to 13-fold. RIOTTON and FISHMAN (1953) found a parallel increase in the urinary excretion of β-glucuronidase activity with that in the kidney after the daily injection of 2.5 mg of testosterone propionate for 16 days in DBA female mice. The earliest effect was noted after 3 days. Histochemical study of the kidney of the treated mice showed a localization of the enzyme activity in the cortex. FISHMAN et al. (1955) compared the effect of 41 steroids related to testosterone and 49 C-21 steroids. Most of the testosterone-related compounds showed some stimulatory activity, and an attempt was made to correlate chemical structure with the response in kidney β-glucuronidase activity. 17α-Methyltestosterone and 17β-hydroxy-5α-androstan-3-one were the most active androgens and were twice as potent as testosterone. Seven of the 49 C-21 steroids showed some stimulatory activity. Progesterone, stilbesterol, and estradiol-17β were ineffective. In a subsequent report, FISHMAN and LIPKIND (1958) refined their assay method for glucuronidase activity and found that esters of testosterone had a much higher stimulatory effect than the unesterified testosterone, confirmed the enhancement by the introduction of a 17α-methyl group, and detected an increase in enzyme activity upon administration of estradiol-3,16β and 16-ketoestrone. HUNTER et al. (1964) confirmed the testosterone propionate stimulating effect on kidney β-glucuronidase in Swiss and $C_3H$ mice. Dianabol in crystalline suspension also was effective but was ineffective in oil solution. The single injection of 0.25 mg of the long-acting steroid, 19-nortestosterone decanoate, in Swiss mice 2 weeks after castration at 8–10 g body weight was ineffective, but 0.5 mg increased the kidney glucuronidase activity after 5 days, and 5 mg produced a great increase that was maintained for the 20 days of the study. SWANK et al. (1973) determined the inducibility of β-glucuronidase in 29 strains of mice. Adult female mice were injected for 2 successive days with an oil suspension of 10 mg 17β-hydroxy-5α-androstan-3-one. The degree of induction of glucuronidase activity in the different strains was divided into a low and a high inducible group. Crossbreeding of mice from the two groups produced progeny with intermediate responses to hormone stimulation. The induction of an increase in the low and high groups by different steroids occurred in parallel fashion. Thus, although the maximum effect was lower in the low than in the high group, the relative effectiveness of the various steroids in the 2 groups was the same. Testosterone propionate, and 17β-hydroxy-5α-androstan-3-one and 5α-androstane-3α,17β-diol were the most potent. Testosterone was much less effective, and 5α-androstane-3,17-dione and androst-4-ene-3,17-dione were barely effective only in the high group.

PETTENGILL and FISHMAN (1962) compared the labeling of protein with β-glucuronidase activity in the mouse kidney by the injection of [1-$^{14}$C]-glycine. If the mice were stimulated maximally by testosterone propionate injection before injection of [1-$^{14}$C]-glycine, the specific radioactivity of the β-glucuronidase was greater than that of the controls. When the [1-$^{14}$C]-glycine was injected prior to endogenous testosterone stimulation by the injection of gonadotropin, the specific radioactivity of the β-glucuronidase decreased more rapidly than in the controls, presumably due to dilution with newly synthesized enzyme. Support for the de novo synthesis of β-glucuronidase was presented by FRIEDEN et al. (1964); they observed that actinomycin D prevented the stimulatory effect of testosterone propionate. Further support was presented by FRIEDEN and FISHEL (1968) in studies with 6- to 10-week-old female mice. Injection of 200 μg/kg/day of actinomycin D inhibited the stimulatory effect of 2.5 mg of testosterone propionate every day for 5 days. Ethionine at 50 mg/kg/day also was inhibitory. Injection of the inhibitors 2 days prior to the beginning of androgen injections was more effective than beginning 2 days after androgen injection. Cycloheximide at 300 mg/kg/day did not influence the enzyme stimulatory effect of the androgen.

KATO et al. (1970) reported that the synthesis of β-glucuronidase under androgen stimulation occurred in the microsomes and that the enzyme subsequently was transported to the lysosomes. Injection of 2.5 I.U. gonadotropin produced an increase in the β-glucuronidase activity only in the microsomal fraction of the kidney during the first 36 h. Then, the enzyme activity increased in both the microsomal and lysosomal fractions at the same rate for the next 44 h. D-[1-$^{14}$C]-glucosamine or L-[U-$^{14}$C]-leucine was injected intraperitoneally 69 h after the injection of 2.5 I.U. gonadotropin, and the kidney was fractionated into its subcellular components 3 h later. The partial purification of the enzyme in the respective fractions demonstrated the incorporation of these labels in β-glucuronidase, with 80% appearing in the microsomal and 20% in the lysosomal enzyme. As a further elucidation of the transportation of the β-glucuronidase from the microsomes to the lysosomes, KATO et al. (1972) conducted time studies on the appearance of the enzyme in submicrosomal fractions after stimulation by gonadotropin. Male mice 2 months old were injected with 2.5 I.U. of human chorionic gonadotropin. The enzyme activity in the subcellular fractions was determined at intervals for 168 h. The results suggested that the newly synthesized enzyme was formed at the site of the membrane-bound ribosome, transferred across the membrane into cisternae of the rough endoplasmic reticulum and then to the lysosomes via the smooth endoplasmic reticulum. Several properties of the enzyme activity from the microsomes and lysosomes were compared and found to be the same. Disc gel electrophoresis, however, exhibited two bands for the microsomal but only one band for lysosomal enzyme activity.

SWANK and PAIGEN (1973) observed that mouse kidney and also liver and lung exhibited 5 isozymes on polyacrylamide gel electrophoresis, which they designated as L, $M_1$, $M_2$, $M_3$, and $M_4$. L was found primarily in the lysosomes and was a tetramer of 260000 molecular weight. $M_1$ to $M_4$ were present exclusively in the microsomes, and the molecular weights ranged from 310000 to 470000. The increase in molecular weight was due to the sequential addition of an accessory protein chain. Induction of an increase in $\beta$-glucuronidase activity in adult female mice by the injection of a suspension of 10 mg 17$\beta$-hydroxy-5$\alpha$-androstan-3-one was accompanied by the appearance of another isozyme (X) in the microsomes with a molecular weight of 260000. All of the isozymes began to increase 30 and 32 h after the injection of the steroid, but as the induction proceeded, there were changes in the relative proportions of the different isozymes.

The specific subcellular localization of the isozymes prompted SWANK and PAIGEN (1973) to study the kidneys of the YBR strain of mice, which lack the $M_1$-$M_4$ isozymes due to a genetic mutation (SWANK et al., 1973). The 20-fold hormonal induction of the $\beta$-glucuronidase activity was in the L and X forms, and none of the M forms appeared. Thus, there is a genetic block in the conversion of the X-form to the M isozymes. SWANK and PAIGEN, on the basis of these data, could not accept KATO et al. (1970, 1972) suggestions that the L isozyme is derived from the M isozymes. Furthermore, they found that the L and M forms increased simultaneously on induction by 17$\beta$-hydroxy-5$\alpha$-androstan-3-one and that the studies on isotopic labeling of the $\beta$-glucuronidase isolated from the subcellular fractions were not valid because the enzymes were not purified to constant specific radioactivities.

DOFUKU et al. (1971) utilized the recently discovered (LYON and HAWKES, 1970) strain of mice with an X-linked testicular feminization (Tfm) mutation to study the genetic regulation of enzyme induction. Male and female mice were castrated at 7–9 weeks of age, and 2 weeks later each received an intraperitoneal injection of 10 mg 17$\beta$-hydroxy-5$\alpha$-androstan-3-one suspended in 0.1 ml cottonseed oil. No induction of glucuronidase activity or increase in kidney weight was observed in either the male or female gonadectomized mice. On the other hand, similar treatment of normal mice induced a progressive, large increase in the $\beta$-glucuronidase activity to about the 10th day, after which the enzyme level returned toward normal. Castration also was without effect on the enzyme activity of the kidneys from the mice with the mutation and decreased it in the normal mice. TETTENBORN et al. (1971) injected 6–10-week-old heterozygous male and female mice with a massive dose, 10 mg, of 17$\beta$-hydroxy-5$\alpha$-androstan-3-one, and 24 h later 10 mg of depotestosterone cypionate, each suspended in 0.1 ml cottonseed oil. The degree of induction of $\beta$-glucuronidase activity 5–10 days later varied between that of the noninducible and the inducible types in accordance with the genetic pattern.

## 5. Glycosidases

CONCHIE and FINDLAY (1959) studied the effect of castration and testosterone administration on the activities of three other glycosidases in comparison with $\beta$-glucuronidase. Castration of male mice after 4 weeks decreased the kidney specific activity of α-mannosidase, β-galactosidase, and β-N-acetylglucosaminidase to approximately one-half that of the normal males and to the same as that of the female. Testosterone, 4 mg/kg every 3 days for 4 weeks, restored the α-mannosidase activity in the kidneys of the castrated mice but did not have a significant effect on the activity of the other two glycosidases. Testosterone, also, was without effect on ovariectomized females. OHNO et al. (1971 a) were unable to detect any change in the concentration of β-glucosidase in the kidneys of castrated mice during the 15 days following a single intraperitoneal injection of 10 mg of 17β-hydroxy-5α-androstan-3-one. The enzyme activity increased in direct proportion to the increase in kidney weight.

## C. Oxidative and Glycolytic Enzymes

ENDAHL and KOCHAKIAN (1959) found that the activities of several of the enzymes of the citric acid cycle—malic, succinic, and isocitric dehydrogenases and fumarase—changed in direct proportion to the decrease in kidney weight after castration and the increase after various doses and durations of testosterone in mice, rats, and guinea pigs. TAGLIAPIETRA and GAZZARINI (1956) found no difference in the succinic dehydrogenase activities of adult male and female rat kidney.

DEIMLING et al. (1965) reported a slightly greater succinic dehydrogenase activity by histochemical staining in the distal portion but not the proximal portion of the proximal convoluted tubule of the male than of the female mouse kidney. Castration decreased the activity to the female level, and testosterone increased that of the female to the male level.

COBET et al. (1967) did not detect any influence of heminephrectomy or the injection of 3 mg/100 g of 19-nortestosterone-17β-decanoate on the activities of lactate or malate dehydrogenases of the kidneys of male rats of 200–250 g body weight. Aldolase activity, on the other hand, did not increase in proportion with the induced increase in kidney weight after heminephrectomy, and the injection of the steroid resulted in a further decrease in the concentration of the aldolase activity.

KOCHAKIAN and WILLIAMS (1973) found the majority of succinic dehydrogenase activity in the mitochondria and all of the glucose-6-phosphate dehydrogenase activity in the cytosol of the mouse kidney. Both enzymes decreased after castration and increased after testosterone propionate administration in direct proportion to the changes in kidney weight.

## D. β-Hydroxybutyrate Dehydrogenase

WENK (1968) observed a sex differentiation in the cellular distribution of β-hydroxybutyrate after 46–70 days of age in the rat kidney. Ovariectomy rendered the histochemical enzyme pattern to the male type. The injection of testosterone propionate at 1 mg/day for 12 days resulted in a male type enzyme pattern. Similar treatment of normal female rats, however, was ineffective. MUEHLENFELD (1969) confirmed the sex differentiation in the histochemical intensity and localization of the β-hydroxybutyrate dehydrogenase activity of the rat kidney after 2 months of age. The subcortical zone was more intense in the male than in the female due to more enzyme in the proximal and distal tubules.

## E. Alcohol Dehydrogenase

OHNO et al. (1970a) reported that (1) alcohol dehydrogenase (ADH) in the liver and kidney of the mouse was specified by the same autosomally inherited gene locus, (2) liver ADH was made continuously but kidney ADH required an inducer, (3) testosterone was the most potent inducer of kidney ADH, (4) kidney ADH was higher in the male than in the female and castration depressed kidney ADH activity to the female level without affecting liver activity, (5) when fully induced by injected testosterone, kidney ADH activity rose to a level equal to but never above that of the liver, and (6) aldehyde dehydrogenase activity of both liver and kidney did not appear to be affected by an inducer. OHNO et al. (1970b) made a more detailed study of kidney ADH induction by testosterone and other steroids. The kidney ADH concentration was twice as great in the male than in the female adult mouse. Gonadectomy produced an immediate decrease in 1–2 days to approximately one-half the normal level, where it was maintained for the duration (16 days) of the experiment. Adrenalectomy of the castrated mice completely abolished the remaining ADH activity. A decrease in room temperature from 22 to 10–15° C also greatly reduced the ADH activity of the kidney, but the effect of castration was still apparent. The ADH activity of the liver was not affected by gonadectomy or the lower room temperature. When a single dose of 10 mg of testosterone propionate suspended in oil was injected intraperitoneally into castrated males 20 or more days after castration, a rapid steep rise in ADH activity was induced until the 7th day, after which it rapidly decreased to the control level on the 10th day. The same effect was induced with smaller doses, 2 and 5 mg, of testosterone propionate and also in ovariectomized female mice. Liver ADH activity was not changed. Intraperitoneal injection of 0.1 mg of actinomycin D 30 h after the injection of testosterone completely halted a further rise in ADH activity after 18 h. This dose of actinomycin was very lethal and killed the mice at 20 h. A smaller dose, 0.01 mg, of actinomycin D had no apparent effect on the mice and produced an inconsistent inhibition in the rise of ADH activity. Injection of actinomycin D 78 h after injection of testosterone propionate merely decreased the rate of increase of ADH activity. Liver ADH was not affected by actinomycin D.

Since ovariectomy also decreased kidney ADH activity, OHNO et al. (1970b) injected gonadectomized male and female mice with estradiol-17$\beta$. A massive dose, 10 mg, did not produce a clear-cut increase in kidney ADH activity in either the male or female gonadectomized mice. A smaller dose, 1 mg, induced a small but definite increase that was completely halted by the injection of 0.01 or 0.1 mg of actinomycin D 24 h after the estradiol injection.

As a further clarification of the regulation of ADH activity, OHNO et al. (1970b) studied the development of ADH activity with age. Both the male and female mice were devoid of kidney ADH activity during the first 14 days of life, a small identical increase occurred during the next 6 days, and at puberty a definite sex differentiation became evident due to an increase in the male kidney enzyme. Injection of testosterone at the 8th day of life induced a definite increase of ADH activity in both male and female mice, indicating that the lack of ADH activity in the infantile mouse kidney was due to the absence of an inducer. The induction of ADH activity apparently was specific for the kidney. No ADH activity could be detected in the prostate or uterus at various periods after injection of testosterone in gonadectomized mice.

OHNO and LYON (1970) observed that the induction of ADH activity was independent of an increase in kidney weight. Injection of 10 mg of cyproterone acetate 24 h before injection of depotestosterone cypionate completely suppressed kidney hypertrophy but totally failed to prevent kidney ADH induction. As a further exploration of the genetic nature of the induction of ADH activity, OHNO and LYON (1970) mated female mice heterozygous for Tfm with normal males. When the progeny were 6 weeks old, they were injected with 20 mg of depo-testosterone cypionate, and ADH activity was determined 5 days later. The normal progeny showed no increase in ADH activity or kidney weight, and the heterozygous mice showed variable intermediate responses even in the individual kidneys. Injection of 10 mg 17$\beta$-hydroxy-5$\alpha$-androstan-3-one into Tfm hemizygotes also did not induce an increase in ADH activity. OHNO and LYON (1970) and OHNO et al. (1971 b) concluded that the Tfm mutation is a noninducible mutation of the repressor locus analogous to the dismutation of the lac-repressor locus of *E. coli* (JACOB and MONOD, 1963).

OHNO et al. (1971 a) compared the ability of several androgens to induce an increase in both ADH and $\beta$-glucuronidase activity in the mouse kidney and concluded that 5$\alpha$-androstane-3$\alpha$. 17$\beta$-diol, an intracellular metabolite of 17$\beta$-hydroxy-5$\alpha$-androstan-3-one, was the true inducer.

SWANK et al. (1973) found no difference in the ability of 17$\beta$-hydroxy-5$\alpha$-androstan-3-one to induce ADH and arginase activity in the kidneys of a low and high $\beta$-glucuronidase inducible strain of mice, indicating that the strain difference was specific for $\beta$-glucuronidase.

# III. Liver

## A. Amino Acid-Related

### 1. Transaminases and Glutamic Acid Dehydrogenase

Glutamic-aspartic (GOT) and glutamic pyruvic (GPT) transaminase activities of the liver of the guinea pig (KOCHAKIAN and ENDAHL, 1956) and the mouse (KOCHAKIAN et al., 1959) were not altered by castration or the administration of several C-19 steroids for different periods of time. A small decrease in the GOT activity of the rat liver, however, was detected 123 days after castration at 86 g body weight. Subcutaneous implantation of 2 pellets of testosterone (15 mg each) did not restore the activity to normal. Furthermore, injection of 0.5 or 1.0 mg/day of testosterone propionate for 14 days did not influence the decrease in GOT activity produced by thyroidectomy of the rat. In a subsequent series of 3 separate studies (KOCHAKIAN et al., 1959), the small decrease in GOT of rat liver after castration was confirmed and an even greater decrease was noted in the GPT activity. The rats were castrated at 48, 40, and 90 days of age. Subcutaneous implantation of a pellet of testosterone, 17$\alpha$-methyltestosterone, or testosterone propionate at the time of castration or 28 or 34 days later, restored the enzyme activities to normal after 28 days. Similar results were reported by APOSTOLAKLIS et al. (1963).

RINDI and PERRI (1955 a, b) did not detect any change in GOT activity of the liver 20–30 days after castration of rats at 100–120 g body weight but found an increase in the GPT activity. A comparable increase, however, was obtained upon injection of oil. Injection of 1 mg/day testosterone propionate for 8–10 days beginning 3 days after castration reversed the effect of castration on GPT activity and lowered it below normal in the normal rat.

PUCHOL and CARBALLIDO (1961) gonadectomized male and female rats at 140–195 g body weight and found no difference in the GOT or GPT activities from those of normal controls. Injection of a massive dose, 25 mg/day, of testosterone propionate for 5 days greatly depressed both the GOT and GPT activities in the gonadectomized and normal male and female rats. Similar treatment with testosterone acetate produced a comparable depression in the normal but not in the gonadectomized rats.

SECCHI et al. (1960) and SECCHI (1962) found no changes in the liver glutamic acid dehydrogenase (GDH), GOT, and GPT activities of adult male rats injected with 1 mg/day 4-hydroxy-17α-methyltestosterone for 15 days while on a low protein diet or during repletion from the low protein diet. RINDI et al. (1955) found that the injection of 10 mg and 5 mg testosterone propionate at 48 and 24 h respectively, after partial hepatectomy of 140–200 g rats did not influence the regeneration of the liver or the GOT activity. GHIZARI and BUTCULESCU (1972) also found that 4-chlorotestosterone at 0.7 mg/kg/day did not influence the activities of GOT, GPT, or catalase of the liver of partially hepatectomized rats (100–150 g) during the subsequent 14 days. BENGMARK and OLSSON (1963) failed to detect any changes in the GPT activity of partially hepatectomized male and female rats after the injection of 8 mg of the long-acting steroid, androdurin (testosterone-3-p-hexyloxiphenyl-propionate), or in the regeneration of the weight and other components of the liver (BENGMARK et al., 1966, 1967).

## 2. D-Amino Acid Oxidase

MARSILLI and BRANZI (1953) castrated male rats and 14–15 days later surgically removed a piece of liver, injected 2 mg of testosterone propionate, and 2 h later removed another piece of liver. The D-amino acid oxidase activity was determined on liver slices by $O_2$ consumed and $NH_3$ and urea formed. Castration did not change any of the parameters studied. Testosterone propionate had no effect on $O_2$ consumption but produced a decrease in $NH_3$ and urea production. Addition of D-alanine to the incubation medium resulted in a marked increase in $NH_3$ production after the hormone treatment. Addition of DL-alanine gave a much higher production of $NH_3$ without any detectable effect after hormone treatment. Apparently, the L-alanine either masked or inhibited the effect of the testosterone propionate on the metabolism of D-alanine.

VAN BEKKUM and KASSENAAR (1954) injected 2.5 mg of testosterone propionate every other day into 300–400 g male rats castrated 3 weeks previously and observed an approximately 50% increase in the D-amino acid oxidase activity ($NH_3$ production) of the liver. ENDAHL and KOCHAKIAN (1956) found an inconsistent and occasional small decrease in the D-amino acid oxidase activity ($O_2$ uptake and $NH_3$ and pyruvate formation) of the rat liver after castration, with a restoration to normal by the administration of either testosterone propionate or 17α-methyltestosterone. The variations in D-amino acid oxidase activity and the apparent inconsistencies in the effect of castration and the hormones among the several experiments could not be associated with age at and duration of castration, diet, or season of the year.

The D-amino acid oxidase activity of the guinea pig liver was not influenced by castration at 2 or 6 months of age. Administration of 9 different C-19 steroids as subcutaneously implanted pellets to the castrated guinea pigs for 21 days did not influence the enzyme activity.

D-amino acid oxidase activity was not present in the liver of the mouse.

### 3. Catalase

ADAMS (1952) observed that the catalase activity of the liver of adult male mice was about 50% greater than that of the female. Castration of the male reduced the enzyme activity within 48 h to or slightly below that of the level in the female. Injection of 125 µg/day of testosterone propionate beginning on the 6th day was without effect after 2 days but after another 2 days restored the enzyme activity to the normal male level. Cessation of injections was without effect during the first 2 days, but by the 4th day the enzyme activity was back to the castration level. Injection of various doses of testosterone propionate into normal or adrenalectomized female mice produced a sharp increase in enzyme activity up to doses of 40 µg/day without any further increase with doses up to 640 µg/day. ADAMS (1960) determined the relative distribution of the catalase activity between the sediment and supernatant fraction after centrifugation of the liver homogenate at 10000 g. The ratio of sediment/supernatant was 3.3 in the male and 4.9 in the female. Injection of either 17α-methyltestosterone or testosterone into the female mice converted the ratio to that of the male type. Hormone treatment of male mice did not influence the distribution ratio. ADAMS concluded that the hormones influenced the migration of the catalase from the "granules" (sediment) to the cytoplasm (supernatant fraction).

GHIZARI and BUTCULESCU (1972) reported that 4-chlorotestosterone injected at 0.7 mg/100 g into partially hepatectomized male rats (100–150 g) increased the catalase activity during the first 3 days and was restored to normal after 14 days.

### 4. Tyrosine Oxidase

SECCHI (1962) observed that 1 mg/day of 4-hydroxy-17α-methyltestosterone injected for 15 days into rats on a protein-deficient diet and castrated 10 days previously did not influence the tyrosine oxidase activity of the liver. If the rats were placed on a repletion diet, the enzyme activity was increased about 30% by the steroid.

### 5. Tryptophan Pyrrolase

APOSTOLAKLIS et al. (1963) observed a large increase in tryptophan pyrrolase activity of the liver of rats (106–140 g) 30 days after castration and an inhibition after injection of 1 or 2 mg/day of testosterone propionate for 21–30 days.

### 6. Phenylalanine Hydroxylase

KAWADA et al. (1965a) administered stanozolol by stomach tube to female (200 g) and male (300 g) rats for 14 days at 0.5, 1.0, and 2.0 mg/day and reported a small but statistically nonsignificant increase in phenylalanine hydroxylase activity. The depression in phenylalanine hydroxylase activity produced by the addition of 5% L-phenylalanine to the diet for 2 weeks was restored during the following 2 weeks by the simultaneous administration of 1 mg/day stanozolol by stomach tube (KAWADA et al., 1965b).

## 7. Monoamine Oxidase

LEMAY and BERLINGUET (1967) confirmed earlier reports of a slightly lower monoamine oxidase activity in the liver of the female than in the male rat. Gonadectomy at 45 days of age and injection for 10 days of 2.5 mg/kg twice a day of testosterone propionate beginning 21 days after gonadectomy did not influence the enzyme activity in the liver of the female rats but produced a small (20%) decrease in that of the male rats. Parallel results were observed in the treatment of 70-day-old normal male and female rats.

## 8. Arginase

Because the effect of testosterone and related compounds induced a shift from protein catabolism to anabolism with a decrease in urea production, a change in liver arginase activity was anticipated. Consequently this was one of the first enzymes submitted to investigation. The extensive studies of KOCHAKIAN and associates (see studies on kidney enzymes, p. 249) revealed a lack of influence of castration or of the administration of a large number of anabolic steroids for various periods of time, doses, and endocrine conditions on the arginase activity of the liver of the mouse, rat, guinea pig, and hamster. In contrast to these observations, DE ANGELIS et al. (1952) reported that the intraperitoneal injection of 1 mg/day of testosterone propionate into 22–23-day-old female rats increased after 52 days the liver arginase activity by 83%. Simultaneous administration of growth hormone slightly and questionably inhibited the effect of testosterone propionate, while 20 µg/day of thyroxine prevented the effect of testosterone propionate. These hormones administered alone were ineffective. The results with testosterone propionate are similar but greater than those in the initial preliminary report of KOCHAKIAN and CLARK (1942) with a high dose of steroid. The rats were castrated at 70 g body weight and were injected 5 days later with 5 mg/day of testosterone propionate for 15 days when a 20-mg pellet of testosterone propionate was implanted subcutaneously. Arginase activity was determined after another 2 weeks.

ZAK et al. (1971) sought an early effect and injected 1 mg of a suspension of testosterone intraperitoneally into immature (21–22-day-old) male rats and reported a small (20%) questionable increase after 5 h. MARSILLI (1951) also sought an early effect. He castrated rats at 110–130 g and 14–15 days later removed a piece of liver under anesthesia and immediately injected subcutaneously 10 mg of testosterone propionate. After 2 or 3 h the rats were exsanguinated, and another piece of liver was obtained. Arginase activity was much higher in the fed than in the fasted rats. Castration did not change the arginase activity but the hormone treatment decreased the arginase activity 35–40% in the castrated rats and increased it about 30% in normal rats whether they were fasted or fed.

### 9. Arginine Synthetase

RIGGS and WALKER (1963) studied the system that catalyzes the conversion of citrulline to arginine, since this system limits the rate of urea formation in liver. Female rats of 174–267 g were ovariectomized 1 week previous to the injection of 1 mg/day of testosterone propionate for 4–10 days. The concentration and the total activity of the arginine synthetase system were decreased 33 and 23% respectively.

### 10. Aspartate- and Ornithine-Transcarbamylases

Carbamyl phosphate plays a key role in both protein (urea formation) and pyrimidine (RNA synthesis) metabolism but has received attention only recently. TARNAWSKI and ZAK (1970) castrated male rats at 3 months of age and 4 weeks later injected intraperitoneally 5 mg of a microsuspension of testosterone. Aspartate-carbamyl transferase activity (RNA synthesis) was increased about 50% after 2 h, maintained at the elevated level for 2 h, and returned to normal by 6 h. Intraperitoneal injection of 200 μg actinomycin D 30 min before injection of testosterone lowered the effect of testosterone. VAN RYEMENANT et al. (1971) castrated rats at 300–400 g and observed an insignificant decrease in the enzyme activity 8 days later. Injection from the 3rd to the 8th day after castration with a crystalline suspension of 1 mg/day testosterone produced a 20% increase in enzyme activity, and 10 mg/day of testolo-lactone, an inactive steroid, produced an even greater response (+80%). Simultaneous injection of the two steroids decreased the enzyme activity to about 60% of normal.

ZAK et al. (1971) found a questionable 23% increase in the ornithine-transcarbamylase activity of the liver of immature male rats (21–22 days old) 3 h after intraperitoneal injection of 1 mg testosterone in saline suspension.

### 11. Creatine-Synthesizing Enzymes

VIVIANI et al. (1955) reported that injection of 25 mg of testosterone propionate or the cyclopentylpropionate 12 weeks after castration of rats increased methionine-guanidine acetic transmethylase activity in the liver above normal within 48 h and maintained the increase for 15 days. The cyclopentylpropionate had a delayed effect. The increase was apparent after 7 days. KANNO (1960) found no changes in the activity of this enzyme of the liver of rats after castration. VIVIANI et al. (1955) reported a 30% decrease 12 weeks after castration at 40 days of age.

### 12. Histidase

FEIGELSEN (1971) observed that castration of rats at 23 days of age did not influence the rapid increase in liver histidase activity during the first 60 days of age when the enzyme activity of the normal rat reached a plateau but that of the castrated rat continued to increase to a 20% higher plateau beginning at about 65 days of age. Injection of the castrated rat with 25 mg/kg/day of testosterone propionate resulted in a rise in enzyme activity exactly as in the normal rat liver.

### 13. Decarboxylases

GRAHN et al. (1973) found that injection of 0.5 mg/day of testosterone for 3–7 days did not change the activities of the histidine or ornithine decarboxylases of the liver of 2- to 4-month-old female mice.

SLOANE-STANLEY (1949) reported that adult male rats contained twice the cysteic acid decarboxylase activity as adult female rats. Castration of the male rats did not produce any change in the enzyme activity 30 days later. Ovariectomy, however, increased the enzyme activity to that of the males. Injection of 2 μg estrone per 100 g body weight every second or third day for 55 days beginning 3 weeks after ovariectomy partially reversed the increase in enzyme activity. CHATAGNER and BERGERET (1956, 1957) reported parallel results when cysteine sulfinic acid was used as substrate in place of cysteic acid. HOPE (1955) found that the two substrates are decarboxylated by the same enzyme.

## B. Hydrolases

### 1. Phosphatases

KOCHAKIAN and associates, in their extensive studies on tissue enzymes (see kidney enzymes, p. 255), were unable to detect significant changes in the activity of "alkaline" and "acid" phosphatases and also of ATPase after castration or treatment with a large number of steroids in the livers of mice (KOCHAKIAN and FOX, 1944; KOCHAKIAN and STETTNER, 1948), guinea pigs (HUMM et al., 1948), and hamsters (KOCHAKIAN et al., 1948).

A small increase in "alkaline" phosphatase was detected in the liver of adult castrated rats on constant food intake (KOCHAKIAN and ROBERTSON, 1950) after the injection of 0.125 mg/day of testosterone propionate for 21 days. An increase of the dose to 1.0 mg/day or extension of the period of treatment to 46 days did not produce a further increase. Testosterone did not alter the "alkaline" phosphatase activity of the adult hypophysectomized or thyroidectomized rat (KOCHAKIAN et al., 1954). D'ATRI et al. (1973) noted by histochemical examination an increase in the liver "alkaline" phosphatase of male adult (200 g) rats injected intraperitoneally for 7 days with 20 mg/kg/day of the new steroid, formyldienolone.

### 2. Glucose-6-Phosphatase

SHULL and BAUTISA (1962) found that the glucose-6-phosphatase activity of the liver of adult male rats was 44% higher than that in females. Injection of a very excessive dose, 10 mg/day, of testosterone propionate for 14 days increased the activity in the female rats to that of the male rats. A massive dose, 50 mg/day, for 3 days was without effect.

### 3. Esterases

SCHWARK and ECOBICHON (1967) reported a sex difference in the electrophoretic (isozyme) pattern of the liver esterases in rats. KOMEDA et al. (1971) extended these studies to mice. The difference was not apparent in immature mice and did not appear upon castration at 45 days of age. Injection of 0.5 mg/day of testosterone propionate for 7 days beginning at 60 days of age changed the electrophoretic pattern of both the castrated and female mice to the male type. ALLEN and MOORE (1966) reported comparable changes in specific isoenzyme bands of the electrophoretic pattern of mouse blood serum esterases.

## C. Folate Enzymes

MARCHETTI et al. (1972), to obtain information on the mechanism of stimulation of nucleic acid synthesis, studied the effect of castration and testosterone on the metabolism of folic acid and the activity of the folic acid coenzymes in liver, as well as the seminal vesicles and prostate of adult male rats. The rats were castrated at 15 weeks of age, 4 weeks later, 1.0 mg testosterone propionate was injected every other day for 5 days. Castration produced a decrease in the folic acid coenzymes: $H_4$ folate dehydrogenase, 10–HCO–$H_4$ folate synthetase, serine hydroxymethyltransferase, and 5,10–$CH_2$–$H_4$ folate dehydrogenase. The alterations in the activity of these coenzymes resulted in a lower utilization of intraperitoneally injected folic acid (200 µg/100 g) 36 h after the last injection of the hormone. Testosterone propionate injections restored both the coenzyme levels and the utilization of folic acid to normal.

## D. N-Methylnicotinamide Oxidase

HUFF and CHAYKIN (1967) found a genetic and sex difference in the activity of N-methylnicotinamide oxidase activity in the liver of mice. Seven to ten times more activity was observed in adult C57Bl/6 mice than in DBA/2 mice of the same sex, and the enzyme was 2–3 times more active in adult males than in females of the same strain. The sex difference was apparent in seven other strains of mice but was not present in immature mice. The enzyme activity was the same in 15-day-old male and female mice and increased to their respective adult levels at about 43 days of age. Castration of male mice at 31 days of age gave an enzyme value similar to that of females of the same age. Injection of 0.1 mg/day of testosterone propionate for 2 weeks increased the enzyme activity in the female and restored that of the castrated mice toward the normal male level. By a more detailed study of the effects of castration and testosterone propionate administration, HUFF and CHAYKIN (1968) further established that testosterone was a controlling factor in the expression of $N^1$-methylnicotinamide oxidase activity in the liver of mice.

## E. Oxidative and Glycolytic Enzymes

GAZZARINI and ZANOTTI (1955) reported that castration of male rats produced a decrease of 31% in the succinic dehydrogenase activity of the liver. In a subsequent report, TAGLIAPIETRA and GAZZARINI (1957, 1958) found that subcutaneous implantation of crystalline testosterone or progesterone into castrated female rats reduced the liver succinic dehydrogenase activity 17–20%. FUJII et al. (1963) noted a greater succinic dehydrogenase activity in the liver mitochondria of female than male rats after 40 days of age. Injection of a saline suspension (5 mg/ml) of testosterone propionate from the 10th to 40th days of life depressed the enzyme activity in the females to that of the males and produced a questionable, small increase in the males.

RINDANI and VALANJU (1957) reported that castration of 28–31-day-old rats decreased after 50 days the succinic dehydrogenase activity and the $O_2$ uptake of the liver. MARINARI et al. (1965) also reported a decrease in succinic dehydrogenase activity and a concomitant decrease in the number of mitochondria and in the P/O ratio of the liver after castration; administration of testosterone propionate produced an increase to above normal levels. ENDAHL and KOCHAKIAN (1959) were unable to detect a significant change in the activities not only of succinic but also of malic and isocitric dehydrogenases in rats, mice, and guinea pigs, after various periods of castration and testosterone administration. The fumarase activity of mouse liver, however, was promptly decreased after castration of adult mice, and the rate of increase with age was retarded after castration of immature mice. ROSENCRANTZ et al. (1962) castrated rats at 24 days of age and found no change in succinic dehydrogenase activity of the liver 7 days later. Injection of 0.343 mg/day of a crystalline suspension of testosterone for 7 days after castration decreased the enzyme activity 27%.

In male rabbits of 1 kg (HANNA et al., 1958), injection of 5 mg/day of testosterone propionate for 10 days induced an increase in the liver succinic dehydrogenase activity. A larger dose, 10 mg/day, for 7 days, had the reverse effect.

HUGGINS and YAO (1959) castrated rats at 45–59 days of age and 3 weeks later found no change in the activities of lactic (LAD), 6-phosphogluconic (6-PGD), and glucose-6-phosphate (G-6-PD) dehydrogenases. Injection of 0.1 mg/day of dihydrotestosterone for 3 weeks greatly decreased the G-6PD activity and modestly decreased that of 6-PGD in the female rat liver. The same influence was exerted by 5α-androstane-3α,17β-diol and androst-4-ene-3α,17β-diol. Estradiol-17β increased the activity of these enzymes, and the simultaneous injection of dihydrotestosterone produced a stoichiometric inhibition in the increase produced by the estrogen.

SECCHI et al. (1960) studied the effect of 4-hydroxy-17α-methyltestosterone at 1 mg/day for 15 days in male rats on a low protein diet or during repletion from the low protein diet. The activities of none of the several glycolytic or Krebs cycle enzymes were affected by the hormone treatment.

KOWALEWSKI (1963a) obtained an approximately 20% increase in liver malic dehydrogenase and DPN-cytochrome-C reductase activities after oral administration of 5 mg/day of methandrostenolone for 14 days to adult (250–275 g) male rats. In a subsequent study KOWALEWSKI (1963b) was unable to detect any change in succinic dehydrogenase or cytochrome oxidase activities of the liver mitochondria of the rats after oral administration of 5 mg/day for 14 days of either Dianabol or Nilevar. APOSTOLAKLIS et al. (1963) found no significant changes in the activities of liver lactic and malic dehydrogenases 30 days after castration of rats at 106–140 g body weight. Injection of 1 or 2 mg/day of testosterone propionate into the castrated and normal rats was without effect on the activities of these two enzymes. The activity of 6-phosphoglycerodehydrogenase was increased by the 2 mg/day but not by the 1 mg/day injections of testosterone propionate into the castrated rats.

## F. Fatty Acid Dehydrogenase

FUJII et al. (1963) found a great increase in the fatty acid dehydrogenase activity (Formazan method) in the liver of rats from the 10th to the 40th days of life without any sex differentiation up to the 100th day of life. Injection of a saline suspension of testosterone propionate (5 mg/ml) every 48 h did not influence the enzyme activity in the liver of either the male or female rats.

# IV. Muscle

## A. Amino Acid-Related

KOCHAKIAN and ENDAHL (1956) castrated guinea pigs at 450–580 g body weight; approximately 30 days later, groups of guinea pigs were implanted subcutaneously with 20 different steroids. The glutamic-oxalacetic (GOT) and glutamic-pyruvic (GPT) transaminase activities of the temporal muscle increased in direct proportion to the increase in weight stimulated by the different hormones. The hearts of these same animals showed no detectable differences in weight and enzyme activities. But the hearts of the rats and mice, which decreased in weight after castration and increased after administration of potent anabolic-androgenic steroids, responded in a manner analogous to that of the temporal muscle (KOCHAKIAN et al., 1959). The transaminase activities decreased after castration and increased after administration of testosterone, testosterone propionate, or 17α-methyltestosterone. The time after castration or duration of treatment did not influence the results. Testosterone propionate at 0.5 or 1.0 mg/day for 14 days, however, did not restore the decrease in weight or the transaminase and ATPase activities of the heart produced after 25 or 40 days of thyroidectomy of rats castrated at 82–119 g body weight.

APOSTOLAKLIS et al. (1963) noted that, in contrast to the guinea pig skeletal muscles, the specific activities of the GOT and GPT of the "levator ani" of rats castrated at 106–140 g were increased after 31 days. Injection of testosterone propionate at 1 or 2 mg/day for the last 10 days reversed the effect of castration. These enzymes of the biceps femoris of the same rats were not changed. The results with the "levator ani" confirmed the earlier reports of RINDI and PERRI (1955a, b) who, in addition, noted that testosterone propionate injected at 1 mg/day for 7 days decreased the activities of these enzymes in the normal rat.

## B. Hydrolases

DIECKOFF (1966) studied the role of anabolic-androgenic steroids in the differentiation of several tissues of the rabbit by determining the activities of several enzymes of the fetal heart after injection of 4-chlorotestosterone acetate for the first 20–28 days of pregnancy. The specific activities of both the glutamic-pyruvic (GPT) and glutamic-oxalacetic (GOT) transaminase of the heart were greatly reduced. The author extended these studies to include injection of 4-chlorotestosterone acetate and decadurabolin into newborn rabbits for the first 10, 20, and 30 days after birth and observed a decrease in the specific activities of GPT and GOT of the heart.

D'ARTI et al. (1973) castrated rats at approximately 200 g body weight. Daily intraperitoneal injection for 7 days of formyldienolone increased the histochemical alkaline phosphatase activity of the myocardium but not that of the gastrocnemius muscle. The change was not related to the serum level of the enzyme activity.

## C. Oxidative and Glycolytic Enzymes

### 1. Rat

The energy-producing enzymes were studied by several investigators. EISENBERG et al. (1949) noted that the $QO_2$ of the "levator ani" and diaphragm of rats castrated at 30 days of age and injected for the following 30 days with 1 mg/day of testosterone propionate was no different than that of the normal or castrated control rats of the same age. In support of these observations, LEONARD (1950) determined the activity of succinic dehydrogenase by the manometric method and also found no change in the "levator ani" 30 days after castration of young (63–103 days of age) or adult (5–6 months of age) rats. Injection of 1.0 mg/day of testosterone propionate for 10 days into 50–55 g male rats also did not change the enzyme ($QO_2$) activity. The enzyme activity of the dorsal thigh muscles also were not changed 30 days after castration (2–4 months old). The succinic dehydrogenase activity of the thigh muscles was much higher (threefold) than that of the "levator ani" muscle. LORING et al. (1961) found no difference in succinic dehydrogenase activity ($QO_2$) between rats castrated at 140 g body weight and those injected intraperitoneally with 5 mg testosterone at weekly intervals for "three or more" weeks. They did not differentiate among the muscles and pooled the highly responsive perineal complex with the nonresponsive temporal and masseter muscles for the preparation of the homogenate that was used as the enzyme preparation. Assay for malic acid dehydrogenase and citric acid dehydrogenase revealed enhancement of these enzymes by testosterone, which was confirmed in a second series of experiments. Succinic dehydrogenase activity was now also increased by the hormone, as was lactic acid dehydrogenase. Furthermore, testosterone enhanced the incorporation of $^{32}P$ into ATP and the formation of keto acids in all of the enzyme systems except the succinic acid dehydrogenase system.

APOSTOLAKLIS et al. (1963) studied several of the aerobic and anaerobic enzymes in the "levator ani" muscle and also the biceps femoris of rats castrated at 106–140 g body weight. After 31 days of castration, the specific activities of malic acid dehydrogenase (MDH), glyceraldehydephosphate dehydrogenase (GAPDH), diphosphofructose aldolase (ALD) were increased and reversed by the injection of 1 mg/day testosterone propionate. A twofold increase in the dose of the steroid was without further effect. The specific activities of lactic acid dehydrogenase (LDH), 6-phospho-glyceraldehyde dehydrogenase, and creatine phosphokinase (CPK) were unchanged. The $\alpha$-glycerophosphate dehydrogenase (GDH) was greatly reduced and that of tryptophan pyrrolase (TP) disappeared after castration. Treatment with testosterone propionate restored the GDH activity to normal and increased that of TP in normal rats to 2–4 times normal. The specific activity of only GAPDH of the biceps femoris was changed (increased) by castration and reversed by treatment with testosterone propionate. The injection of 2 mg/day but not of 1 mg/day increased the specific activity of TP 2.5–3.5 times. The specific activities of LDH, MDH, GDH, and ALD were questionably (10–20%) increased after castration and returned to or near normal by the injection of the hormone.

ECKSTEIN et al. (1960) found no change in the activity of succinic and DPN-linked isocitric dehydrogenases 30 days after castration of rats at 200 g body weight or the injection of 1 mg/day of testosterone propionate during the 30-day period.

### 2. Rabbit

DIECKOFF (1966) found a great decrease in the specific activities of LDH, MDH, and ALD of the heart of the rabbit fetus after daily injection of the mothers for the first 20 or 28 days of pregnancy with 1 mg/kg of 4-chlorotestosterone acetate. The decreases were comparable with those in GOT and GPT. Similar decreases were observed in the heart of newborn rabbits after the injection of not only 4-chlorotestosterone acetate but also deca-Durabolin for 10, 20, or 30 days.

### 3. Guinea Pig

GUTMANN et al. (1970) reported that the temporal muscle of adult (2.5 months old) guinea pig exhibited a sexual differentiation in the histochemical enzyme patterns. The smaller "red" female muscle showed high activity of the mitochondrial enzymes (succinic dehydrogenase and glycerophosphodehydrogenase) while the larger "white" male muscle exhibited higher activity of the glycolytic enzymes (phosphorylase and lactic acid dehydrogenase). Injection of the female guinea pig with 40 mg/kg/day of testosterone propionate from 1–2.5 months of age not only markedly enlarged the temporal muscle but also converted it to the white type with an enzyme pattern similar to that in the male. These investigators (BASS et al., 1971) extended their studies with a more detailed comparison of the glycolytic and aerobic enzyme patterns. The enzyme activities now were determined on extracts of the muscles. The concentration of myofibrillar ATPase activity of the temporal muscle was the same in adult (2.5 months old) male and female guinea pigs and also in female guinea pigs injected with 40 mg/kg/day of testosterone propionate from 14 days to 1.5 months of age. The energy-supplying enzymes showed no sexual difference at 14 days of age, but as the guinea pigs matured a differentiation in the enzyme activities appeared. In the male muscle an increase occurred in the glycolytic enzymes (triose phosphate dehydrogenase, lactate dehydrogenase, and glycerol-3-phosphate dehydrogenase), but there were no significant changes in the specific activities of the aerobic enzymes (malate dehydrogenase, citrate synthetase) and hexokinase. Injection of testosterone propionate at 40 mg/kg/day from 14 days to 1.5 months of age converted the enzyme pattern in the muscle of the female guinea pig to the male type. ENDAHL and KOCHAKIAN (1959) had reported earlier that the concentrations of the activities of the aerobic enzymes (malate and succinate dehydrogenases) of the guinea pig temporal muscle and also of the rat gastrocnemius muscle were not altered by castration or by testosterone treatment.

# V. Gastrointestinal Tract

## A. Oxidases, Dehydrogenases, Phosphatases

### 1. Rat

KSIAZKIEWICZ-SZAPIRO (1970) investigated the effect of castration and testosterone propionate on various enzymes of the small intestine of the rat, especially those concerned with absorption, simultaneously with histologic studies (cf. Chap. III, p. 140). The rats were castrated at 190–210 g body weight and 2 days later were injected with a large dose, 40 mg/kg/day, for 42 days; the enzymes in the proximal segment of the intestine were studied by histochemical methods. Cytochrome oxidase and succinate dehydrogenase activities were diminished by castration and further diminished by testosterone propionate. GIORDANO et al. (1962) had previously reported an increase in the latter enzyme after hormone treatment.

The activities of monoamine oxidase and $\alpha$-glycerophosphate dehydrogenase were not affected by either castration or the hormone. 3$\beta$- and 17$\beta$-Hydroxysteroid dehydrogenase activities, however, were decreased after castration and increased to or above normal by testosterone propionate. This effect is similar to that observed for the 17$\beta$-hydroxysteroid dehydrogenase activity in the guinea pig kidney (KOCHAKIAN and ENDAHL, 1960; KOCHAKIAN and STEVENSON, 1973). The hydrolytic enzymes—"alkaline" and "acid" phosphatases, thiamine pyrophosphatase, and nonspecific esterases—were decreased after castration and increased above normal after testosterone propionate. An increase in "alkaline" phosphatase after hormone treatment is in agreement with the earlier observation of GIORDANO et al. (1962). TISSIERES (1948), however, found no change in "alkaline" phosphatase activity of the small intestine by the histochemical method in 200-g male rats after castration at 30 days of age or after injection of 10 mg/kg of testosterone propionate at 10-day intervals for 30 days.

KSIAZKIEWICZ-SZAPIRO attempted to correlate the enzyme changes, especially that of the phosphatase, with the absorption of fat. Therefore, on the last day of the experiment, five rats from each group were tube-fed a mixture of oil and egg yolk $2^1/_2$ h before autopsy. Fat absorption was not influenced by castration but was greatly accelerated by testosterone propionate. An abundance of fatty bodies was found in the cytoplasm of the intestinal epithelium and also in the considerably dilated lymphatic vessels of the stroma of the villi. This effect of the hormone was interpreted as being mediated through the enhanced activities of the hydrolytic enzymes.

### 2. Mouse

KOCHAKIAN and FOX (1944) did not observe any change in the concentration of either "alkaline" or "acid" phosphatase activity in the small intestine of the mouse 35 or 115 days after castration at 17 g body weight. Subcutaneous implantation of a pellet of testosterone propionate at the time of castration also was ineffective.

## B. Glycolytic Enzymes

STIFEL and associates in a series of reports have demonstrated the dependence of glycolytic and folate-metabolizing enzymes of the jejunum on the sex hormones. STIFEL et al. (1970a) found that castration of adult male rats significantly depressed the activities of phosphofructokinase, pyruvate kinase, and fructose-1,6-diphosphate aldolase. The oral administration of 100 µg/100 g testosterone restored to or near normal within 24 h the activities of pyruvate kinase and phosphofructokinase but had no effect on the aldolase. Oral administration of testosterone and its glucosiduronide also increased the activity of pyruvate kinase in the normal adult male rat. STIFEL et al. (1969) had previously demonstrated that oral administration of testosterone at 100 µg/100 g body weight for 2 days increased the activities of phosphofructokinase and pyruvate kinase, did not change the activities of fructose-1,6-diphosphate aldolase and hexokinase, and decreased that of fructose diphosphatase. Intramuscular injection of testosterone (STIFEL et al., 1970b), however, had no effect on the activities of these enzymes in both male and female adult rats.

## C. Folate-Metabolizing Enzymes

STIFEL et al. (1970b) extended their studies to include the folate-metabolizing enzymes. Twenty-four hours after oral administration of testosterone at 5 µg/100 g body weight to adult male rats, the activities of serine hydroxymethyltransferase, glutamate formiminotransferase, and methylenetetrahydrofolatedehydrogenase were significantly increased. An increase of the dose to 25 µg/100 g produced a further increase to 2–3 times the control values. A dose of 100 µg/100 g did not induce a further increase in enzyme activities. A similar response was induced with testosterone glucosiduronide. Intramuscular injection of testosterone for 2 days at 100 µg/day was without effect in normal males and decreased the activities of these three enzymes in the normal females. Castration decreased the activities of serine hydroxymethyltransferase and methylenetetrahydrosynthetase. Oral administration (intubation) of testosterone at 100 µg/100 g for 3 days restored the values to normal levels. Results of the folate-related enzymes were similar to those of the jejunal glycolytic enzymes. STIFEL et al. (1971) found that 25 µg/100 g actinomycin D or 100 mg/100 g ethionine injected simultaneously with the administration of testosterone prevented the hormone-induced increase in the activities of pyruvate kinase, serinehydroxymethyltransferase, and methylenetetrahydrofolate dehydrogenase. The activity of fructose-1,6-diphosphate aldolase was not changed by either testosterone or the inhibitors. LUFKIN et al. (1972) have reported parallel changes in human intestinal enzymes.

## D. Amylase

MISRA and KUMAR (1970) prepared a Thiry-Vella loop in 10 dogs, and 4 days later castration was performed. The Thiry-Vella loop was irrigated with isotonic saline each morning at 8 a.m. while in the postabsorptive state, and $^1/_2$ h later the juice was collected for 1 h and analyzed. The concentration of amylase activity was significantly decreased 3 days after castration; it then rose gradually to a level slightly but significantly above the precastration level 10 days later. At this time, the dogs were injected with a single dose of a mixture of 5 mg testosterone propionate, 10 mg testosterone phenylpropionate, and 10 mg testosterone isocaproate. The concentration of amylase activity gradually decreased during the subsequent 6 days and then returned to the preinjection level in the following 3–4 days. The rate of secretion of the juice and bicarbonate content, on the other hand, decreased to a sustained low value. The injection of testosterone propionate stimulated a rapid rise in these two factors to a maximum between the 6th and 8th days with a subsequent gradual drop to the preinjection level. The temporal effect of the hormone on these two parameters thus was the direct opposite of that on the amylase activity, indicating that the change in amylase activity was a dilution effect. These changes provide at least a partial possible explanation for the improved intestinal utilization of dietary protein by depancreatized-castrated dogs after treatment with testosterone propionate (KOCHAKIAN and COSTA, 1959).

## E. Pepsin

GRIFFIN et al. (1961) studied the effect of testosterone cyclopentyl propionate on gastric secretion by the preparation of male and female dogs with the Heidenhain pouch. A large daily dose, 100 mg/day, of testosterone cyclopentylpropionate was injected for 20 days. Small but not significant decreases in volume and free acid and no changes in pepsin activity were observed. Similar treatment of dogs receiving 30 mg/day of histamine in beeswax accelerated the onset of duodenal ulcer formation and death in both the male and female dogs. Similar results were obtained upon injection of a large dose, 5 mg/day, of stilbestrol. The authors hypothesized that the enhancement of ulcer formation by testosterone might be due to a decrease in the production of mucopolysaccharides on the basis of KOWALEWSKI's (1959) observation in rats.

## F. Arginase

KOCHAKIAN (1944) determined the arginase activity in the small intestine of the mouse at the same time as in the kidney. The enzyme activity was not affected by castration or treatment of the mice with a wide variety of androgens.

# VI. Other Tissues

Limited studies on other tissues have been reported and are included in the pertinent chapters.

# VII. Summary

A large number of enzymes in several of the target tissues have been studied in an attempt to delineate the metabolic effects of testosterone and related compounds. Since the several nonsexual target tissues of the various animals vary widely in their ability to grow under stimulation by the anabolic-androgenic steroids, each tissue for each animal species requires separate consideration. Furthermore, dose and duration of treatment, age, sex, and nutritional status are contributing and modifying factors. In most instances the studies were limited with respect to hormonal dose and duration of treatment. Consequently, assessments of the results were limited.

The two most responsive muscles are the temporal muscle in the guinea pig and the "levator ani" muscle in the rat and mouse. The activities of GOT, GPT, and ATPase changed in direct proportion to the changes in weight of the temporal muscle. The activities of the glycolytic enzymes were greater in the male than in the female after maturity, and the injection of testosterone propionate increased the activities of these enzymes in the female to that of the male. The activity of the oxidative enzymes changed in direct proportion to changes in weight of the muscle. The activities of the glycolytic and oxidative enzymes of the "levator ani" muscle responded in variable fashion to castration and testosterone propionate. The reported changes were not extensive and the studies were limited.

The enzymes of the liver have been studied extensively. In a few instances clear-cut changes have been observed after castration and steroid administration. In most instances, however, no or small and inconsistent changes have been reported. If the hormone has an effect on these enzyme systems of the liver, it is low grade and does not require a major change in the enzyme activity to effect its objective. The activities of several enzymes appear to be regulated by the presence of the hormones. In most instances these enzymes showed sex differentiation and occasionally genetic control. Reports on many enzymes were conflicting and questionable.

The activities of several of the glycolytic and folate-metabolizing enzymes of the small intestine of the rat appear to be partially regulated by testosterone as is the amylase activity of the dog intestine. Several other enzymes in the rat and mouse showed no dependence on testosterone, and others showed controversial responses.

The most striking changes in enzyme activities were encountered in the mouse kidney and to a lesser extent in the rat kidney with small questionable changes in the guinea pig and hamster. The changes, thus, were species dependent. The various enzymes can be placed in three different categories in regard to their response to castration and treatment with testosterone and related compounds: (1) total activity not changed, (2) concentration not changed, (3) concentration changed. Furthermore, many of the enzymes exhibited sex differentiation.

The increase after hormone administration: (1) was related to kidney growth, for example, D-amino acid oxidase, (2) occurred more rapidly than the increase in kidney weight, e.g., $\beta$-glucuronidase, (3) occurred after kidney weight was stimulated to its maximum increase, for example, arginase, and (4) occurred without any increase in kidney weight, e.g., arginase and "alkaline" phosphatase in hypophysectomized rats.

The changes in arginase activity in the mouse kidney have been postulated by KOCHAKIAN (1965) as a mechanism for the regulation of growth of the kidney by diverting aspartate and carbamyl phosphate from pyrimidine synthesis to the urea cycle. The reports of an early increase in aspartate and ornithine transcarbamylase activities by testosterone (TARNAWSKI and ZAK, 1970; ZAK et al., 1971) provided further support for this hypothesis. The very early, 1–2 h detection of these changes, however, needs confirmation, especially since VAN RYMENANT et al. (1971) could not detect any changes after castration of adult rats upon injection of testosterone for 5 days.

Histidine and ornithine decarboxylase and histamine methyltransferase activities exhibited definite sex differentiation in the mouse kidney with dependence on testosterone.

## References

ADAMS, D. H.: Hormonal factors influencing liver catalase activity in mice. Testicular and adrenal factors. Biochem. J. **50**, 486—493 (1952).

ADAMS, D. H.: The mechanisms of action of androgenic hormones on mouse liver catalase. Biochem. J. **74**, 141—148 (1960).

AKIKUSA, Y., YAMAMOTO, K.: Effect of testosterone on renal adenosinetriphosphatase of the mouse. Nippon Seirigaku Zasshi **30**, 873—874 (1968).

ALLEN, R. C., MOORE, D. J.: Sex-associated quantitative differences in the plasma esterases of inbred mice. Endocrinology **78**, 655—658 (1966).

APOSTOLAKLIS, M., MATZELT, D., VOIGT, D.: Der Einfluß von Testosteronpropionat auf glykolytische und transaminierende Enzymaktivitäten in Leber, Musculus biceps femoris und Musculus levator ani der Ratte. Biochem. Z. **337**, 414—424 (1963).

BASS, A., GUTMANN, E., HANZLIKOVA, V., SYROVY, I.: Sexual differentiation of enzyme pattern and its conversion by testosterone in the temporal muscle of the guinea pig. Physiol. bohemoslov. **20**, 423—431 (1971).

VAN BEKKUM, D. W., KASSENAAR, A. A. H.: The metabolic action of testosterone and related compounds. III. Effect on D-amino acid oxidase activity. Acta endocr. (Kbh.) **15**, 9—16 (1954).

BENGMARK, S., EKMAN, B., OLSSON, R., REHNSTROM, B.: Further studies on the effect of testosterone on liver healing after partial hepatectomy in male rats. Scand. J. Gastroent. **1**, 106—110 (1966).

BENGMARK, S., OLSSON, R.: Experimental study of liver healing after partial hepatectomy with special regard to the change in content of liver glutamic pyruvic transaminase. Acta hepato-splenol. (Stuttg.) **10**, 282—293 (1963).

BENGMARK, S., OLSSON, R., REHNSTROM, B.: The effect of testosterone in partially hepatectomized female rats. Scan. J. Gastroent. **2**, 90—94 (1967).

CHATAGNER, F., BERGERET, B.: Decarboxylation de l'acide cysteinesulfinique par la foie et le cerveau du rat mâle, du rat femelle ovariectomise. Bull. soc. chim. Biol. **38**, 1159—1163 (1956).

CHATAGNER, F., BERGERET, B.: Influence de l'oestradiol sur la decarboxylation de l'acide cysteine-sulfinique par le foie de rat adult ovariectomise. C.R. Soc. Biol. (Paris) **244**, 2322—2323 (1957).

CLARK, L. C., JR., KOCHAKIAN, C. D., FOX, R. P.: The effect of castration and testosterone propionate on D-amino acid oxidase activity in the mouse. Science **98**, 89—90 (1943).

COBET, G., DIECKHOFF, J., SIEPELT, H.: Über den Einfluß von 19-nortestosterone-17β-decanoat und einseitiger Nephrektomie auf einige Fermente der verbliebenen Rattennieren. Z. ges. exp. Med. **142**, 180—190 (1967).

CONCHIE, J., FINDLAY, J.: Influence of gonadectomy, sex hormones and other factors, on the activity of certain glycosidases in the rat and mouse. J. Endocr. **18**, 132—146 (1959).

D'ATRI, G., GALIMBERTI, E., GOMARASCA, P.: Aumento di Attivata Enzimatiche Nei Tessuti Indotte da Formildienolone. III. Fosfatasi alcalina. Boll. chim. farm. **112**, 620—631 (1973).

DE ANGELIS, W., VIGNOLO, W., HOLM, N., BARSANTINI, J. C.: Accion de varias hormonas sobre la actividad arginasica del higade. Arch. Sol. Biol. Montevideo **19**, 27—31 (1952).

DEIMLING, O. V., BAUMANN, G., NOLTENIUS, H.: Hormonabhängige Enzymverteilung in Geweben. V. Wirkung von Kastration und Sexual-Hormon auf fünf Enzyme der Mäuseniere. Histochemie **5**, 1—10 (1965).

DIECKOFF, J.: Beeinflussung des Fermentstoffwechsels durch anabole Steroide während der Gravidität im Experiment. Abh. dtsch. Akad. Wissensch. Berlin **3**, 545—549 (1966).

DOFUKU, R., TETTENBORN, U., OHNO, S.: Testosterone-"Regulon" in the mouse kidney. Nature (Lond.) New Biol. **232**, 5—7 (1971).

DUNN, T. B.: Sex difference in the alkaline phosphatase distribution in the kidney of the mouse. (Abs.) Amer. J. Path. **24**, 719—720 (1948).

ECKSTEIN, B., KAHAN, D., BORUT, A., ESHKOL, Z., SOBEL, H.: Effect of steroids on dehydrogenase system of male rat tissues. Bull. Res. Counc. Israel **8 E**, 141—146 (1960).

EISENBERG, E., GORDAN, G. S., ELLIOTT, H. W.: Testosterone and tissue respiration of the castrate male rat with a possible test for myotrophic activity. Endocrinology **45**, 113—119 (1949).

ENDAHL, B. R., KOCHAKIAN, C. D.: Role of castration and androgens in D-amino acid oxidase activity of tissues. Amer. J. Physiol. **185**, 250—256 (1956).

ENDAHL, G. L., KOCHAKIAN, C. D.: Influence of androgens on enzymes of the Krebs cycle. Endocrinology **64**, 833—835 (1959).

FEIGELSON, M.: Androgenic regulation of rat liver histidase. Endocrinology **89**, 625—628 (1971).

FISHKIN, A. F., LATA, G. F.: Some hormonal influences on the acetylation of sulfanilamide in vivo. Endocrinology **63**, 162—168 (1958).

FISHMAN, W. H.: β-Glucuronidase and the action of steroid hormones. Ann. N.Y. Acad. Sci. **54**, 548—557 (1951).

FISHMAN, W. H., ARTENSTEIN, M., GREEN, S.: The renal β-glucuronidase response to androgens. Endocrinology **57**, 646—657 (1955).

FISHMAN, W. H., FARMELANT, M. H.: Effects of androgens and estrogens on β-glucuronidase in inbred mice. Endocrinology **52**, 536—545 (1953).

FISHMAN, W. H., LIPKIND, J. B.: Comparative ability of some steroids and their esters to enhance the renal β-glucuronidase activity of mice. J. biol. Chem. **232**, 729—736 (1958).

FITCH, C. D.: Effect of diet and sex on kidney transamidinase activity in the rat. Proc. Soc. exp. Biol. (N.Y.) **112**, 636—639 (1963).

FRIEDEN, E. H., FISHEL, S. S.: Differential sensitivity of parameters of androgen action to metabolic inhibitors: Arginase and β-glucuronidase. Biochem. Biophys. Res. Commun. **31**, 515—521 (1968).

FRIEDEN, E. H., HARPER, A. A., CHIN, F., FISHMAN, W. H.: Dissociation of androgen-induced enzyme synthesis and amino acid incorporation in mouse kidney by actinomycin D. Steroids **4**, 777—786 (1964).

FUJII, T., KAMEI, T., NEGAMI, S., FUJII, E.: Sex difference of metabolism in rat liver after testosterone propionate administration. Endocr. Jap. **10**, 254—259 (1963).

GAZZARINI, A., ZANOTTI, C.: Variazioni del Tasso della Succinodeihydrogenasi Epatica di Ratte in Seguito a Castrazione. Boll. Soc. ital. Biol. sper. **31**, 235—237 (1955).

GHIZARI, E., BUTCULESCU, I.: Recherches sur l'activité des enzymes dans le foie en voie de regeneration. Action de la 4-chlortestosterone sur le catalase et les transaminases. Rev. Roum. Physiol. **9**, 139—143 (1972).

GIORDANO, C., FOPPIANI, E., ROMANO, R.: Le influenze ormonali sull'apparatto digerents. Arch. Maragliano **18**, 51—86 (1962).

GRAHN, B., HENNINGSSON, S. S. G., KAHLSON, G., ROSENGREN, E.: Alterations in the activities of ornithine and histidine decarboxylases provoked by testosterone in mice. Brit. J. Pharmacol. **48**, 113—120 (1973).

GRIFFIN, W. O., JR., NICOLOFF, D. M., STONE, N. H., WANGENSTEEN, O. K.: Role of the sex hormones in peptic ulcer. Proc. Soc. exp. Biol. (N.Y.) **106**, 101—104 (1961).

GUTMANN, E., HANZLIKOVA, V., LOJDA, A.: Effect of androgen on histochemical fiber type. Differentiation of the temporal muscle of the guinea pig. Histochemistry **24**, 287—291 (1970).

HANNA, T., MOLATY, H. A., KASHEF, H. A.: The histochemical effects of testosterone on the liver of rabbits. Alexandria med. J. **4**, 50—60 (1958).

HENNINGSSON, S. S. G., ROSENGREN, E.: Alterations of histamine metabolism after injections of sex hormones in mice. Brit. J. Pharmacol. **44**, 517—526 (1972).

HOPE, D. B.: Pyridoxal phosphate as the coenzyme of the mammalian decarboxylase for L-cysteinesulfinic and L-cysteic acids. Biochem. J. **59**, 497—500 (1955).

HUFF, S. D., CHAYKIN, S.: Genetic and androgen control of N'-methylnicotinamide oxidase activity in mice. J. biol. Chem. **242**, 1265—1270 (1967).

HUFF, S. D., CHAYKIN, S.: Kinetics of testosterone action, in vivo, on liver N'-methylnicotinamide oxidase activity in mice. Endocrinology **83**, 1259—1267 (1968).

HUGGINS, C., YAO, F.-O.: Influence of hormones on liver. I. Effects of steroids and thyroxine on pyridine nucleotide-linked dehydrogenases. J. exp. Med. **110**, 899—919 (1959).

HUMM, J. H., KOCHAKIAN, C. D., BARTLETT, M. N.: Effect of castration and steroids on arginase and phosphatases of organs of the guinea pig. Amer. J. Physiol. **155**, 251—254 (1948).

HUNTER, F., ODDIS, L., LEATHAM, J. H.: Kidney enzymes and anabolic steroids. Proc. Penn. Acad. Sci. **38**, 36—39 (1964).

JACOB, F., MONOD, J.: Genetic repression, allosteric inhibition, and cellular differentiation. In: Cyto-Differentiation and Macromolecular Synthesis, pp. 30–64. London: Academic Press 1963.

JUNQUEIRA, L. C., ROTHSCHILD, H. A.: Studies on the relation of cathepsins to protein synthesis. The effect of hormones. Acta physiol. latinoamer. **3**, 247—255 (1953).

KAMADA, H.: Influence of sex hormones on the function and histochemistry of the kidney of the female rabbit. Igaku Kenkyu **35**, 314—374 (1965).

KANNO, Y.: Activities of creatine-synthesizing enzymes in skeletal muscle and liver, and effects on them of thyroidectomy, hypophysectomy, adrenalectomy, and castration in rats. Nippon Yakurigaku Zasshi **56**, 1172—1179 (1960).

KATO, K., HIROATA, I., FISHMAN, W. H., TSUKAMATO, H.: Intracellular transport of mouse kidney β-glucuronidase induced by gonadotrophin. Biochem. J. **127**, 425—435 (1972).

KATO, K., IDE, H., SHIRAHAMA, T., FISHMAN, W. H.: Incorporation of [$^{14}$C] glucosamine and [$^{14}$C]-leucine into mouse kidney β-glucuronidase induced by gonadotrophin. Biochem. J. **117**, 161—167 (1970).

KAWADA, S., IMADA, H., TANIGUCHI, Y., IKEGAMI, Y.: Effect of stanozolol on activity of liver phenylalanine hydroxylase and kidney GPT from rats. A.R. Nat. Inst. Nutr. (Tokyo), pp. 13—16 (1965a).

KAWADA, S., IMADA, H., TANIGUCHI, Y., IKEGAMI, Y.: Effect of stanozolol on the rats depressed in activity of liver phenylalanine hydroxylase. A.R. nat. Inst. Nutr. (Tokyo), pp. 17—19 (1965b).
KOCHAKIAN, C. D.: The effect of castration and various steroids on the arginase activity of the tissues of the mouse. J. biol. Chem. **155**, 579—589 (1944).
KOCHAKIAN, C. D.: The effect of dose and nutritive state on kidney arginase after steroid stimulation. J. biol. Chem. **161**, 115—125 (1945a).
KOCHAKIAN, C. D.: The effect of various steroid hormones on the "alkaline" and "acid" phosphatases of the kidney of the mouse. Amer. J. Physiol. **145**, 118—122 (1945b).
KOCHAKIAN, C. D.: The role of hydrolytic enzymes in some of the metabolic activities of steroid hormones. Rec. Progr. Horm. Res. **1**, 177—216 (1947).
KOCHAKIAN, C. D.: Histochemical study of "alkaline" phosphatase of the kidney of the castrated mouse after stimulation with various androgens. Amer. J. Physiol. **152**, 257—262 (1948).
KOCHAKIAN, C. D.: Mechanisms of androgen actions. Lab. Invest. **8**, 538—555 (1959).
KOCHAKIAN, C. D.: Intracellular regulations in the kidney by androgens. Amer. Zoologist **2**, 361—366 (1962).
KOCHAKIAN, C. D.: Regulation of kidney growth by androgens. In: Advances in Steroid Biochemistry and Pharmacology, Vol. VI, in press. New York: Academic Press 1976
KOCHAKIAN, C. D., BARTLETT, M. N., GONGORA, J.: Effect of castration and androgens on body and organ weights and the arginase and phosphatases of kidney and liver of the male Syrian hamster. Amer. J. Physiol. **153**, 210—214 (1948).
KOCHAKIAN, C. D., CLARK, L. C., JR.: The effect of testosterone propionate on the arginase content of the liver, kidney and intestine. J. biol. Chem. **143**, 795—796 (1942).
KOCHAKIAN, C. D., COSTA, G.: The effect of testosterone propionate on the protein and carbohydrate metabolism in the depancreatized castrated dog. Endocrinology **65**(2), 298—309 (1959).
KOCHAKIAN, C. D., ELSAS, F., HARRISON, D. G.: Androgen regulation of mouse kidney ribonucleases. Acta endocr. (Kbh.) **46**, 179—184 (1964).
KOCHAKIAN, C. D., ENDAHL, B. R.: Influence of androgens on transaminase activities of different tissues. Amer. J. Physiol. **186**, 460—462 (1956).
KOCHAKIAN, C.D., ENDAHL, B.R.: Testosterone regulation of 17β-hydroxysteroid dehydrogenases. Proc. Soc. exp. Biol. (N.Y.) **104**, 720—722 (1960).
KOCHAKIAN, C. D., ENDAHL, B. R., ENDAHL, G. L.: Influence of androgens on the transaminases and glutamic dehydrogenase of tissues. Amer. J. Physiol. **197**, 129—134 (1959).
KOCHAKIAN, C. D., FOX, R. P.: The effect of castration and testosterone propionate on the "alkaline" and "acid" phosphatases of the kidney, liver, and intestine of the mouse. J. biol. Chem. **153**, 669—674 (1944).
KOCHAKIAN, C. D., GARBER, E. E., BARTLETT, M. N.: Effect of estrogen alone and in combination with testosterone on the body and organ weights and the arginase and phosphatases of the organs of the mouse. Amer. J. Physiol. **155**, 265—271 (1948).
KOCHAKIAN, C. D., REED, B. A., EISCHEID, A. M.: Changes in tissue weights and enzymes produced by thyroidectomy and testosterone. Amer. J. Physiol. **177**, 413—417 (1954).
KOCHAKIAN, C. D., ROBERTSON, E.: Influence of androgens and hypophysectomy on arginase and phosphatases of the kidneys and liver of the rat. Arch. Biochem. **29**, 114—123 (1950).
KOCHAKIAN, C. D., STETTNER, C. E.: Effect of testosterone propionate and growth hormone on the arginase and phosphatases of the organs of the mouse. Amer. J. Physiol. **155**, 262—264 (1948).
KOCHAKIAN, C. D., STEVENSON, D., MAYUMI, T.: Dependence of the guinea pig kidney 17β-hydroxy-$C_{19}$-steroid dehydrogenase isoenzyme profile on androgen. Biochem. Biophys. Res. Commun. **54**, 519—523 (1973).
KOCHAKIAN, C. D., VAIL, V. N.: The effect of adrenalectomy, adrenal cortical hormones, and testosterone propionate plus adrenal cortical extract on the arginase activity of the liver and kidney of the rat. J. biol. Chem. **169**, 1—6 (1947).
KOCHAKIAN, C. D., WILLIAMS, B. R.: The subcellular localization of the changes in ribonuclease and phosphatase activities of the mouse kidney produced by castration and testosterone. Steroids **21**, 95—110 (1973).
KOCHAKIAN, C. D., WRIGHT, P. M., ROBERTSON, E.: Testosterone propionate and arginase activity in diabetic rats. Arch. Biochem. Biophys. **36**, 221—230 (1952).
KOMEDA, K., MANDA, M., NISHIDA, S.: Genetic and physiological control of esterases in experimental small animals. IX. Influence of gonadal hormones on the kidney esterase isozymes of mice. Tohoku J. Agri. Res. **22**, 96—101 (1971).

KOWALEWSKI, K.: Effect of certain androgenic steroids and cortisone on gastric ulcerogenesis in fasting rats. Proc. Soc. exp. Biol. (N.Y.) **101**, 147—150 (1959).

KOWALEWSKI, K.: Activity of malic dehydrogenase and DPN-cytochrome-C-reductase of liver mitochondria in rats treated with cortisone and with an anabolic androgen. Proc. Soc. exp. Biol. (N.Y.) **113**, 310—312 (1963a).

KOWALEWSKI, K.: Activity of some respiratory enzymes in liver mitochondria of rats treated with cortisone and anabolic androgens. Arch. int. Pharmacodyn. **142**, 9—16 (1963b).

KRISKO, I. M., WALKER, J. B.: Influence of sex hormones on amidinotransferase levels. Metabolic control of creatine biosynthesis. Acta endocr. (Kbh.) **53**, 655—662 (1966).

KSIAZKIEWICZ-SZAPIRO, M. D.: Experimental investigations of the rat small intestine following castration and testosterone administration. Folia Histochem. Cytochem. **8**, 369—388 (1970).

LEITCH, J. L., DEBLEY, V. G., HALEY, T. J.: Endogenous histamine excretion in the rat as influenced by X-ray irradiation and compound 48/80. Amer. J. Physiol. **187**, 307—311 (1956).

LEMAY, A., BERLINGUET, L.: Influence de la castration et des hormones sexuelles sur l'activité de la monoamine oxydase chez le rat. Rev. Can. Biol. **26**, 295—303 (1967).

LEONARD, S. L.: Succinic dehydrogenase levels in striated muscle in relation to male hormone. Endocrinology **47**, 260—264 (1950).

LI, J. J., KIRKMAN, H., HUNTER, R. L.: Sex difference and gonadal hormone influence on Syrian hamster kidney esterase isozymes. J. Histochem. Cytochem. **17**, 386—393 (1969).

LORING, J., SPENCER, J., VILLEE, C.: Some effects of androgens on intermediary metabolism in muscle. Endocrinology **68**, 501—506 (1961).

LUFKIN, E. G., STIFEL, F. B., HERMAN, R. H., ROSENSWEIG, N. S.: Effect of testosterone on jejunal pyruvate kinase activities in normal and hypogonadal males. J. clin. Endocr. **34**, 586—591 (1972).

LUUKKAINEN, T.: The effect of testosterone, progesterone and estrogen on the acetylating activity in rat liver and kidney preparations. Acta physiol. scand. **51**, 254—262 (1961).

LYON, M. F., HAWKES, S. G.: An X-linked gene for testicular feminization in the mouse. Nature (Lond.) **227**, 1217—1219 (1970).

MANDA, M., OKI, Y.: Genetic and physiological control of esterases in experimental small animals. IV. Sex hormone-dependent esterases in mouse and rat kidneys. Tohoku J. Agric. Res. **17**, 211—214 (1967).

MARCHETTI, M., BOVINA, C., TOLOMELLI, B., ROVINETTI, C.: Meccanismo d'azione degli ormoni steroidei: testosterone, acidi nucleici e coenzimi folici. Acta Vitaminol. Enzymol. **26**, 27—42 (1972).

MARINARI, U. M., NOVELLI, A., NANTE, L.: Effects of castration and testosterone on the number, morphology and functional activity of mitochondria. Giorn. Biochim. **14**, 389—398 (1965).

MARSILII, G.: Influenza del propionato di testosterone sull'arginasi epatica di ratti castrati e normali. Sperimentale **101**, 151—154 (1951).

MARSILII, G., BRANZI, G.: Influenza esercitata dal propionato di testosterone sull'attivita aminoacidossidasica epatica di ratti castrati. Sperimentale **103**, 15—23 (1953).

MARSILII, G., D'ALLESSANDRO, A.: Effetto dell castrazione su alcuni sistemi enzymatici I. Transaminasi e glutamminasi cutanea e renale. Boll. Soc. ital. Biol. sper. **31**, 906—907 (1955).

MISRA, N. K., KUMAR, S.: Effect of orchidectomy on the intestinal secretion of dogs. Endocrinology **87**, 1361—1363 (1970).

MORROW, A. G., CARROLL, D. M., GREENSPAN, E. M.: A sex difference in the kidney glucuronidase activity of inbred mice. J. nat. Cancer Inst. **11**, 663—669 (1951).

MUHLENFELD, W. E.: Über die Entwicklung und Chemodifferenzierung der Rattenniere unter besonderer Berücksichtigung der Geschlechtsunterschiede. Histochemie **18**, 97—131 (1969).

NATAF, B.: Modifications de l'activite arginasique du rein de Cobaye au cours de la gestation et sous l'action de certaines hormones sexuelles (oestradiol, progesterone, testosterone). C.R. Soc. Biol. (Paris) **147**, 1701—1704 (1953).

NETTER, K. J., COHN, V. H., JR., SHORE, P. A.: Sex difference in histamine metabolism in the rat. Amer. J. Physiol. **201**, 224—226 (1961).

OHNO, S., DOFUKU, R., TETTENBORN, U.: More about X-linked testicular feminization of the mouse as a noninducible ($i^s$) mutation of a regulatory locus: $5\alpha$-androstan-$3\alpha$,$17\beta$-diol as the true inducer of kidney alcohol dehydrogenase and $\beta$-glucuronidase. Clin. Genetics **2**, 128—140 (1971a).

OHNO, S., LYON, M. F.: X-linked testicular feminization in the mouse as a non-inducible regulatory mutation of the Jacob-Monod type. Clin. Genetics **1**, 121—127 (1970).

OHNO, S., STENIUS, C., CHRISTIAN, L.: Sex difference in alcohol metabolism: Androgenic steroid as an inducer of kidney dehydrogenase. Clin. Genetics **1**, 35—44 (1970a).

OHNO, S., STENIUS, C., CHRISTIAN, L., HARRIS, C., IVEY, C.: More about testosterone induction of kidney alcohol dehydrogenase activity in the mouse. Biochem. Genetics **4**, 565—577 (1970b).

OHNO, S., TETTENBORN, U., DOFUKU, R.: Molecular biology of sex differentiation. Hereditas **69**, 107—124 (1971b).

PETTENGILL, O. S., FISHMAN, W. H.: Influence of testosterone on glycine incorporation into mouse kidney β-glucuronidase. Exp. Cell Res. **28**, 248—253 (1962).

VAN PILSUM, J. F., UNGAR, F.: Effect of castration and steroid sex hormones on rat kidney transamidinase. Arch. Biochem. **124**, 372—379 (1968).

PUCHOL, J. R., CARBILLIDO, A.: Action de quelques androgens sur l'activite traminasique des tissus du rat. Med. Exp. (Madrid) **4**, 45—50 (1961).

RABINOVITCH, M.: "Alkaline" and "Acid" phosphatases activities and ribonucleic acid phosphorus content of the kidney cortex of the mouse as influenced by castration and testosterone propionate administration. Folia Clin. Biol. **18**, 115—120 (1952).

READ, S. I., MCKINLEY, W. P.: The effect of parathion on rat liver and kidney carboxylesterases. Can. J. Biochem. **43**, 1625—1632 (1965).

RIGGS, T. R., WALKER, L. M.: Diminution of arginine synthetase activity in livers of rats treated with testosterone propionate. Endocrinology **73**, 830—831 (1963).

RINDANI, T. H., VALANJU, N. N.: Effect of castration on succinic dehydrogenase activity in male rat liver. Current Sci. **26**, 88—89 (1957).

RINDI, G., ALLEGRI, A., RIZZOLINI, G. F.: Il compartamento dell'attivita transaminasica nel fegato rigenerante dopo parziale epatectomia. Arch. Fisiol. **55**, 152—164 (1955).

RINDI, G., PERRI, V.: Influenza della castrazione e del trattamento con testosterone dull'attivata di alcune transaminasi dei tessuto di ratto. Arch. sci. biol. **39**, 343—351 (1955a).

RINDI, G., PERRI, V.: Influence of castration and testosterone treatments on the glutamic-oxalacetic transaminase activity of rat tissue. Ricerca sci. Suppl. **25**, 307—310 (1955b).

RIOTTON, G., FISHMAN, W. H.: β-Glucuronidase studies in inbred mice: Androgenic hormones and kidney and urinary β-glucuronidase activity. Endocrinology **52**, 692—697 (1953).

ROSENGREN, E., STEINHARDT, C.: Elevated histidine decarboxylase activity in the kidney of the pregnant mouse. Experientia (Basel) **17**, 544 (1961).

ROSENKRANTZ, H., CASTLE, V., MCLAUGHLIN, E. R.: Influence of ablation, hormones and age on the enzyme reducing systems in the tissues of the male rat. Endocrinology **71**, 307—313 (1962).

RUDDLE, F. H.: Kidney esterases of the mouse (Mus musculus). J. Histochem. Cytochem. **14**, 25—32 (1966).

RUDDLE, F. H., HARRINGTON, L.: Tissue specific esterase isozymes of the mouse (Mus musculus). J. exp. Zool. **166**, 51—64 (1967).

VAN RYMENANT, M. E., KEYMOLEN, P., PORCHERET, J., DESCHUTTER, A., KRAM, R.: Action of testosterone and testololactone on isocitric dehydrogenase, aspartate transcarbamylase and gulonate NADP oxidoreductase in the rat in vivo. Acta endocr. (Kbh.) **66**, 498—502 (1971).

SCHWARK, W. S., ECOBICHON, D. J.: Characterization of rat liver and kidney esterases. Can. J. Biochem. **45**, 451—455 (1967).

SCHWARZLOSE, W., HEIM, F.: Effect of testosterone on glutamate-pyruvate-transaminase, glutamate-oxaloacetate-transaminase and glutamate dehydrogenase in mice kidneys. Life Sci. II. **12**, 107—115 (1973).

SECCHI, G. C.: Experimentelle Untersuchungen über die Beziehungen zwischen anabolen Hormonen und Enzymtätigkeit des Eiweißstoffwechsels in der Leberzelle. Klin. Wschr. **40**, 107—109 (1962).

SECCHI, G. C., ZAMPIERI, A., BELLOTTI, C., GHIDONI, A.: Ricerche sperimentali sue rapporti fra ormoni anabolizzanti ed attivita enzimatiche della cellula epatica. Arch. Stud. Fisiopat. Clin. Ric. **24**, 515—524 (1960).

SHAW, C. R., KOEN, A. L.: Hormone-induced esterase in mouse kidney. Science **140**, 70—71 (1963).

SHULL, K. H., BAUTISA, E.: Glucose-6-phosphatase activities in normal and testosterone treated rats. Endocrinology **70**, 842—845 (1962).

SLOANE-STANLEY, G. H.: Amino acid decarboxylases of rat liver. Biochem. J. **45**, 556—559 (1949).

SNYDER, S., AXELROD, J.: Sex differences and hormonal control of histamine methyltransferase activity. Biochim. biophys. Acta (Amst.) **111**, 416—421 (1965).

STIFEL, F. B., HERMAN, R. H., ROSENSWEIG, N. S.: Dietary regulation of glycolytic enzymes. VII. Differential hormonal effects in male and female rat jejunum. Biochim. biophys. Acta (Amst.) **184**, 495—502 (1969).

STIFEL, F. B., HERMAN, R. H., ROSENSWEIG, N. S.: Dietary regulation of glycolytic enzymes. IX. The effect of oral, intramuscular and conjugated steroids on jejunal glycolytic enzyme activities in normal and castrated male and female rats. Biochim. biophys. Acta (Amst.) **222**, 65—70 (1970a).

STIFEL, F. B., HERMAN, R. H., ROSENSWEIG, N. S.: Dietary regulation of glycolytic enzymes. X. The effect of oral intramuscular and conjugated sex steroids on jejunal folate-metabolizing enzyme activities in normal and castrated male and female rats. Biochim. biophys. Acta (Amst.) **222**, 71—78 (1970b).

STIFEL, F. B., HERMAN, R. H., ROSENSWEIG, N. S.: Dietary regulation of glycolytic enzymes. XI. Effect of inhibitors of protein synthesis on the adaptation of certain jejunal glycolytic and folate-metabolizing enzymes to diet and sex steroids. Biochim. biophys. Acta (Amst.) **237**, 484—489 (1971).

SVEJCAR, G.: Vergleich der Wirkung des Testosteronpropionats und des Methylandrostendiols auf den Proteinstoffwechsel. Praxis **45**, 374—377 (1956).

SWANK, R. T., PAIGEN, K.: Biochemical and genetic evidence for a macromolecular β-glucuronidase complex in microsomal membranes. J. molec. Biol. **77**, 371—389 (1973).

SWANK, R. T., PAIGEN, K., GANSCHOW, R. E.: Genetic control of glucuronidase induction in mice. J. molec. Biol. **81**, 225—243 (1973).

TAGLIAPIETRA, B., GAZZARINI, A.: Influenza del sesso sul tasso della succinodeidrase epatica e renale di ratte. Arch. Obstet. Ginecol. **61**, 1—3 (1956).

TAGLIAPIETRA, B., GAZZARINI, A.: Influenza del testosterone e del progesterone sul tasso della succinodeidrase epatica di ratte castrate. Quad. Clin. Obstet. Ginecol. **12**, 551—556 (1957).

TAGLIAPIETRA, B., GAZZARINI, A.: Succinodeidrase ed ormoni sessuali. Quad. Clin. Ostet. Ginecol. **13**, 539—547 (1958).

TAKAGI, Y.: Effect of anabolic steroid in rat tissues. Tokyo Ika Daikai Zasshi **21**, 53—60 (1963).

TARNAWSKI, R., ZAK, T.: The influence of testosterone on aspartate carbamyltransferase activity in the rat liver and kidney. Polish Endocr. **21**, 342—345 (1970).

TETTENBORN, U., DOFUKU, R., OHNO, S.: Non-inducible phenotype exhibited by a proportion of female mice heterozygous for the X-linked testicular feminization mutation. Nature (Lond.) New Biol. **234**, 37—40 (1971).

TISSIERES, A.: L'influence de la castration, du testosterone et de l'oestradiol sur les phosphatases du rein chez le rat. Acta Anat. **5**, 236—242 (1948).

UNGAR, F., VAN PILSUM, J. F.: Hormonal regulation of rat kidney transamidinase. Effect of growth hormone in the hypophysectomized rat. Endocrinology **78**, 1238—1247 (1966).

VAIL, V. N., KOCHAKIAN, C. D.: The effect of adrenalectomy, adrenal cortical hormone, and testosterone propionate plus adrenal cortical extract on the "alkaline" and "acid" phosphatases of the liver and kidneys of the rat. Amer. J. Physiol. **150**, 580—587 (1947).

VIVIANI, R., INFANTE, R., SECHI, A. M.: Influenza di ormoni sessuali maschili sulla metionin-guanidinacetico transmetilasi sel fegato di ratto. Boll. Soc. ital. Biol. sper. **31**, 161—164 (1955).

WEHLER, E., GROSSARTH, E., DEIMLING, O. VON: Esterase. VI. Der Einfluß von Testosteron auf die unspezifische Carboxylesterase der Mäuseniere während der postnatalen Entwicklung. Histochemie **33**, 47—52 (1973).

WENK, H.: Hormonabhängige Enzymverteilung in der Niere. IV. Wirkung von Kastration und Sexualhormonen auf die Aktivität der β-Hydroxy-Buttersäure-Dehydrogenase in der Niere bei weiblichen Albinoratten. Histochemie **12**, 120—129 (1968).

WESTLING, H., WETTERQVIST, H.: Further observations on the difference in the metabolism of histamine in male and female rats. Brit. J. Pharmacol. **19**, 64—73 (1962).

ZAK, T., JOZKIEWICZ, S., TARNAWSKI, R., ADAMSKA, A.: Hormonal regulation of activity of some of the enzymes of the ornithine cycle. Polish Endocr. **22**, 40—44 (1971).

# D. The Metabolism in Vitro of Anabolic-Androgenic Steroids by Mammalian Tissues

C. D. KOCHAKIAN and N. ARIMASA

## I. Introduction

Very early in the development of our knowledge of the biochemistry of the internal secretion of the testis it was recognized that the product of the testis was metabolized during its circulation through the body. The active principle isolated from the urine—androsterone—was chemically and biologically different from the active principle of the testis—testosterone. Furthermore, the injection of potent urinary extracts and later pure testosterone and related compounds resulted in the isolation of only a fraction of the biological activity from the urine due primarily to the conversion to less active metabolites, e.g., 5α-androsterone and 5β-androsterone. This conversion initially suggested a detoxifying process. Since the liver is assumed to be the main organ for the maintenance of homeostasis, the ability of this organ to metabolize these steroids was studied. Subsequent studies, however, have indicated that not only do many other tissues metabolize these steroids but also that the products formed are different and the spectrum of products varies with animal species suggesting specific metabolic roles for the various metabolites in accordance with tissue and animal species.

## II. Sites of Metabolism

### A. Liver

#### 1. **Rabbit** (Table 1)

The first demonstration of the conversion of a steroid hormone to other steroids in vitro was reported by CLARK and KOCHAKIAN (1947). Testosterone suspended in rabbit serum to facilitate solubility was incubated with rabbit liver slices in Ringer-phosphate buffer. The incubation mixture was extracted, purified, and separated by fractional crystallization and adsorption column chromatography to yield androstene-3,17-dione as the main metabolite, smaller quantities of epitestosterone, and several other compounds in amounts too small to identify by the then available methods. One of the unknown compounds was of particular interest because it suggested hydroxylation. The elementary analysis gave the formula $C_{19}H_{28-30}O_3$. An exhaustive search was made for androsterone and its isomers and also estrogens without success. CLARK et al. (1947) repeated these experiments with androst-4-ene-3,17-dione as the substrate and were able to isolate in good yield testosterone as the main metabolite and epitestosterone as the second product. A number of other steroids again were isolated but could not be completely characterized. KOCHAKIAN

Table 1. The metabolism of $C_{19}$ steroids in vitro by adult male rabbit liver preparations

| Substrate | Tissue prepn.[a] | Coenz.[b] | pH | Metabolites | References |
|---|---|---|---|---|---|
| 17β-Hydroxyandrost-4-en-3-one | S | — | 7.4 | Androst-4-ene-3,17-dione<br>17α-Hydroxyandrost-4-en-3-one | CLARK and KOCHAKIAN (1947) |
| | H | — | 7.4 | Androst-4-ene-3,17-dione<br>17α-Hydroxyandrost-4-en-3-one | KOCHAKIAN et al. (1952) |
| 17α-Hydroxyandrost-4-en-3-one | S,H | NA | 7.4 | Androst-4-ene-3,17-dione | KOCHAKIAN and NALL (1953) |
| Androst-4-ene-3,17-dione | S | — | 7.4 | 17β-Hydroxyandrost-4-en-3-one<br>17α-Hydroxyandrost-4-en-3-one | CLARK et al. (1947) |
| Androst-4-ene-3,17-dione | H | — | 7.4 | 17β-Hydroxyandrost-4-en-3-one<br>17α-Hydroxyandrost-4-en-3-one<br>5α-Androstane-3,17-dione<br>3α-Hydroxy-5α-androstan-17-one | KOCHAKIAN and STIDWORTHY (1954) |
| 3β-Hydroxyandrost-5-en-17-one (Sodium hemisuccinate) | S | — | 7.4 | Androst-5-ene-3β,17β-diol<br>Androst-5-ene-3β,16β,17β-triol | SCHNEIDER and MASON (1948a) |
| 3α-Hydroxy-5α-androstan-17-one (Sodium hemisuccinate) | S | — | 7.4 | 5α-Androstane-3α,17β-diol<br>5α-Androstane-3,17-dione<br>3β-Hydroxy-5α-androstan-17-one | SCHNEIDER and MASON (1948b) |
| 3α-Hydroxy-5β-androstan-17-one (Sodium hemisuccinate) | S | — | 7.4 | 5β-Androstane-3α,17β-diol<br>5β-Androstane-3α,17α-diol<br>5β-Androstane-3,17-dione | SCHNEIDER and MASON (1948b) |
| 5α-Androstane-3α,17β-diol (Sodium disuccinate) | S,H | — | 7.4 | 3α-Hydroxy-5α-androstan-17-one<br>3β-Hydroxy-5α-androstan-17-one<br>5α-Androstane-3,17-dione | KOCHAKIAN and APOSHIAN (1952) |
| 3α-Hydroxy-5α-androstan-17-one (Sodium sulfate) | S | — | 7.4 | 5α-Androstane-3α,17α-diol,3-monosulfate | CREPY et al. (1963) |
| 3α-Hydroxy-5α-androstan-17-one | P | — | — | 5α-Androstane-3,17-dione<br>3β-Hydroxy-5α-androstan-17-one | BENARD et al. (1959) |
| 3β-Hydroxy-5α-androstan-17-one | P | — | — | 5α-Androstane-3,17-dione<br>3α-Hydroxy-5α-androstan-17-one | BENARD et al. (1961) |
| 5α-Androstane-3,17-dione | P | — | — | 3α-Hydroxy-5α-androstan-17-one<br>3β-Hydroxy-5α-androstan-17-one | |
| 5α-Androstane-3,17-dione | H | NA,NAD | 7.4 | 3α-Hydroxy-5α-androstan-17-one<br>3β-Hydroxy-5α-androstan-17-one | RUBIN (1957) |

[a] S = slices; H = homogenate; P = perfusion.
[b] NA = nicotinamide.

et al. (1952) found these same conversions to occur if a homogenate was used instead of the slices, and KOCHAKIAN and NALL (1953) demonstrated the conversion of epitestosterone to androst-4-ene-3,17-dione and testosterone by either rabbit liver slices or homogenate.

The introduction of paper chromatography made it possible not only to utilize smaller quantities of the steroids in the incubations but also to more readily separate and identify the metabolites. KOCHAKIAN and STIDWORTHY (1954) now were able to identify not only testosterone and epitestosterone as metabolites of androst-4-ene-3,17-dione after incubation with rabbit liver homogenate but also smaller quantities of 5α-androstane-3,17-dione and 3α-hydroxy-5α-androstan-17-one.

SCHNEIDER and MASON (1948a, b) incubated the sodium hemisuccinates of 3β-hydroxyandrost-5-en-17-one and 3α-hydroxy-5α-androstan-17-one with rabbit liver slices. They were able to isolate and characterizce androst-5-ene-3β,17β-diol and androst-5-ene-3β,16β,17β-triol as metabolites of 3β-hydroxy-androst-5-en-17-one and 5α-androstane-3α,17β-diol, 5α-androstane-3,17-dione, and 3β-hydroxy-5α-androstan-17-one as metabolites of 3α-hydroxy-5α-androstan-17-one. In a third set of experiments the hemisuccinate of 3α-hydroxy-5β-androstan-17-one yielded 5β-androstane-3α,17β-diol; 5β-androstane-3α,17α-diol and 5β-androstane-3,17-dione indicating that steroisomerism of the ring at the 5 carbon did not influence the oxidation of the 3 hydroxyl or reduction of the 17-ketone. KOCHAKIAN and APOSHIAN (1952) found that 5α-androstane-3α,17β-diol was only slightly metabolized by rabbit liver slices or homogenates but the use of the sodium disuccinate as the substrate resulted in a greater though still small conversion to 3α-hydroxy-5α-androstan-17-one, 3β-hydroxy-5α-androstan-17-one, and 5α-androstane-3,17-dione. CREPY et al. (1963) reported that rabbit liver slices reduced 17-oxo-5α-androstan-3α-yl-sulfate to 5α-androstane-3α,17α-diol-3-sulfate.

BENARD et al. (1959) noted a rapid disappearance of 3α-hydroxy-5α-androstan-17-one on perfusion through rabbit liver with the appearance of 5α-androstane-3,17-dione, 3β-hydroxy-5α-androstan-17-one, and two more polar unidentified steroids. The reciprocal conversions occurred (BENARD et al., 1961) when 3β-hydroxy-5α-androstan-17-one or 5α-androstane-3, 17-dione were perfused through the liver indicating an interconversion of these three steroids by enzymes of the rabbit liver. RUBIN (1957) noted a reduction of the 3-oxo group of 5α-androstane-3,17-dione to the 3α- and 3β-hydroxyls on incubation with liver homogenate of male and female rabbits.

## 2. Guinea Pig (Table 2)

KOCHAKIAN and associates extended their studies to the adult male guinea pig because this animal exhibits specific characteristic dependence on anabolic-androgenic steroids for skeletal muscle development and no or slight dependence for the kidney and liver. The incubation of androstane-3α,17β-diol with homogenate of liver at pH 7.4 yielded approximately 1% ketonic material from which androsterone was isolated (KOCHAKIAN and APOSHIAN, 1952). The incubation of androst-4-ene-3,17-dione at pH 7.4–8.0 yielded more metabolites which were detected by the newly available technic of paper chromatography (KOCHAKIAN and STIDWORTHY, 1954). Testosterone was the main metabolite and androsterone, epiandrosterone, and an-

Table 2. The metabolism of $C_{19}$-steroids in vitro by adult male guinea pig liver preparations

| Substrates | Tissue prepn.[a] | Coenz.[b] | pH | Metabolites | References |
|---|---|---|---|---|---|
| Androstane-3α,17β-diol | H | — | 7.4 | 3α-Hydroxy-5α-androstan-17-one<br>3β-Hydroxy-5α-androstan-17-one | Kochakian and Aposhian (1952) |
| Androst-4-ene-3,17-dione | H | NA,NAD | 7.4 | 17β-Hydroxyandrost-4-en-3-one<br>3α-Hydroxy-5α-androstan-17-one<br>3β-Hydroxy-5α-androstan-17-one<br>5α-Androstane-3,17-dione | Kochakian and Stidworthy (1954) |
| 5α-Androstane-3,17-dione | H | NADH | 6.7 | 17β-Hydroxy-5α-androstan-3-one<br>3α-Hydroxy-5α-androstan-17-one<br>3β-Hydroxy-5α-androstan-17-one | Hamm et al. (1956) |
| 5α-Androstane-3,17-dione[f] | H | NA,NAD | 7.4 | 3α-Hydroxy-5α-androstan-17-one<br>3β-Hydroxy-5α-androstan-17-one | Rubin (1957) |
| 17β-Hydroxyandrost-4-en-3-one | H | NA,NAD | 9.0 | Androst-4-ene-3,17-dione | Kochakian et al. (1957a,b) |
| 17β-Hydroxy-5β-androstan-3-one | H | NA,NAD | 9.0 | 5β-Androstane-3,17-dione<br>3α-Hydroxy-5β-androstan-17-one | Kochakian et al. (1957b)<br>Kochakian (1959) |
| 17β-Hydroxy-5α-androstan-3-one | H | NA,NAD | 9.0 | 5α-Androstane-3,17-dione | Kochakian (1959) |
| 17α-Hydroxyandrost-4-en-3-one | H | NA,NAD | 9.0 | Androst-4-ene-3,17-dione | Kochakian (1959) |
| Androst-5-ene-3β,17β-diol | H | NA,NAD | 9.0 | Androst-4-ene-3,17-dione<br>3β-Hydroxyandrost-5-en-17-one | Kochakian (1959) |
| Androst-5-ene-3β,17α-diol | H | NA,NAD | 9.0 | Androst-4-ene-3,17-dione<br>17α-Hydroxyandrost-4-en-3-one<br>3β-Hydroxyandrost-5-en-17-one | Kochakian (1959) |
| 5α-Androstane-3α,17β-diol | H | NA,NAD | 9.0 | 5α-Androstane-3,17-dione<br>3α-Hydroxy-5α-androstan-17-one<br>3β-Hydroxy-5α-androstan-17-one | Kochakian (1959) |
| 5α-Androstane-3β,17β-diol | H | NA,NAD | 9.0 | 5α-androstane-3,17-dione<br>3β-Hydroxy-5α-androstan-17-one | Kochakian (1959) |

Table 2 (continued)

| | | | | | |
|---|---|---|---|---|---|
| 3α-Hydroxy-5α-androstan-17-one | H | NA,NAD | 9.0 | 5α-Androstane-3,17-dione<br>3β-Hydroxy-5α-androstan-17-one | KOCHAKIAN et al. (1957b)<br>KOCHAKIAN (1959) |
| 3β-Hydroxy-5α-androstan-17-one | H | NA,NAD | 9.0 | 5α-Androstane-3,17-dione | |
| 3β-Hydroxyandrost-5-en-17-one | H | NA,NAD | 9.0 | Androst-4-ene-3,17-dione | |
| 5α-Androstane-3,17-dione | H | NADH | 7.0 | 3β-Hydroxy-5α-androstan-17-one<br>3α-Hydroxy-5α-androstan-17-one<br>17β-Hydroxy-5α-androstan-3-one | KOCHAKIAN (1959) |
| 5β-Androstane-3,17-dione | H | NADPH | 7.0 | 3β-Hydroxy-5β-androstan-17-one<br>3α-Hydroxy-5β-androstan-17-one | KOCHAKIAN (1959) |
| 17β-Hydroxyandrost-4-en-3-one | H | NAD; NADP | 9.0 | Androst-4-ene-3,17-dione | KOCHAKIAN et al. (1961) |
| Androst-4-ene-3,17-dione | H | NADH; NADPH | 6.8 | 17β-Hydroxyandrost-4-en-3-one | |
| 17β-Hydroxyandrost-4-en-3-one | Ms | NAD | 9.6 | Androst-4-ene-3,17-dione | ENDAHL et al. (1960) |
| | Cyt | NADP | 10.0 | Androst-4-ene-3,17-dione | |
| 17β-Hydroxyandrost-4-en-3-one[c] | S[e] | NAD(H),<br>NADP(H) | 7.25 | Androst-4-ene-3,17-dione<br>17α-Hydroxyandrost-4-en-3-one | SZAMATOWICZ et al. (1971) |
| Androst-4-ene-3,17-dione[d] | S[e] | NAD(H),<br>NADP(H) | 7.25 | 17β-Hydroxyandrost-4-en-3-one<br>17α-Hydroxyandrost-4-en-3-one | SZAMATOWICZ et al. (1971) |

[a] H = homogenate, S = slices, Ms = microsomes, Cyt = cytosol.
[b] NA = nicotinamide.
[c] [4-$^{14}$C]-labeled.
[d] $^{3}$H-labeled.
[e] Male and female guinea pigs (200–300 g body weight) were injeted for 6 days with human chorionic gonadotropin. The incubation mixture contained the four coenzymes.
[f] Male and female.

drostanedione were obtained in smaller quantities. The metabolites were characterized by m.p., derivatives, and infrared analysis. A similar incubation of 5α-androstane-3,17-dione with the addition of nicotinamide and NADH yielded epiandrosterone (3β-hydroxy-5α-androstan-17-one) as the main metabolite (15–20% conversion) at an optimum pH of approximately 6.7 and NADH 1.0 mg/ml (HAMM et al., 1956). Androsterone and 17β-hydroxy-5α-androstan-3-one were found in much smaller yields which were not affected by changes in pH. RUBIN (1957) in a single experiment also noted a much greater conversion of 5α-androstane-3,17-dione to the 3β- than the 3α-hydroxy-17-oxosteroid on incubation with liver homogenate of both male and female guinea pigs.

KOCHAKIAN et al. (1957a) determined the optimum conditions for the conversion of testosterone to androstenedione in 100 ml of sodium pyrophosphate buffer. The optima were: pH 9.5, incubation time 90 min, NAD 50 mg, liver 8 g, testosterone >200 mg. Under these conditions 50% of the testosterone was converted to androst-4-ene-3,17-dione. The incubation of 17β-hydroxy-5α-androstan-3-one yielded 12% 5α-androstane-3,17-dione; and 17β-hydroxy-5β-androstan-3-one yielded 21–26% 5β-androstane-3,17-dione and 12–23% 3α-hydroxy-5β-androstan-17-one. The high capacity of the homogenate to metabolize testosterone was particularly noteworthy. Testosterone was metabolized at a nearly linear rate up to the maximum quantity, 200 mg, used. Thus, the guinea pig liver contained a very high 17β-hydroxy-$C_{19}$-steroid dehydrogenase activity. KOCHAKIAN et al. (1957b) compared the oxidation of a number of $C_{19}$-steroids. The conversion of testosterone and epitestosterone to androst-4-ene-3,17-dione was 62 and 5% respectively. Saturation of the A ring resulted in a marked decrease in the dehydrogenation of the 17β-hydroxyl; 17β-hydroxy-5α-androstan-3-one was dehydrogenated to 5α-androstane-3,17-dione in only 10% yield. The 5β ring configuration also influenced the 17β-OH dehydrogenase activity. 17β-Hydroxy-5β-androstan-3-one was converted to 5β-androstane-3,17-dione in 21% yield with an apparent subsequent reduction to 3α-hydroxy-5β-androstan-17-one in 12% yield to give a total conversion of 33%. The 3,17-dihydroxysteroids were metabolized to a smaller degree than the 17-hydroxy-3-oxosteroids possibly because of their lower solubility in the incubation medium. Androst-5-ene-3β,17α-diol showed a lower (1.6%) conversion to androst-4-ene-3,17-dione than the 17β analog (5.5%); both compounds yielded comparable quantities of 3β-hydroxyandrost-5-ene-17-one (2.2 and 2.9% respectively); the 17α-steroid also provided 2.3% of 17α-hydroxyandrost-4-en-3-one. The 3α- and 3β,17β-5α-androstanediols yielded as in the earlier studies small quantities of 5α-androstane-3,17-dione (2.1 and 1.2%) and 3β-hydroxy-5α-androstan-17-one (0.3 and 1.8%) but only the 3α-diol was converted (0.4%) to 3α-hydroxy-5α-androstan-17-one. The guinea pig liver also dehydrogenated the 3 hydroxy group of the 17-oxosteroids to the corresponding diketosteroids. In addition, when 3α-hydroxy-5α-androstan-17-one was the substrate, 3β-hydroxy-5α-androstan-17-one was formed presumably from the main metabolite, androstane-3,17-dione, but the 3α-hydroxysteroid was not formed when the 3β-hydroxysteroid was the substrate.

KOCHAKIAN et al. (1961) found that NADP was as effective as NAD in the enhancement of the oxidation of the 17β-hydroxy group of testosterone and NADPH was twice as effective as NADH in the reduction of androst-4-ene-3,17-dione to 17β-hydroxyandrost-4-en-3-one. Moreover, the reductase activity was much less than the oxidase activity. Also, the addition of testosterone to the reductase system (0.1 M sodium phosphate, pH 6.8) inhibited the reduction of androst-4-ene-3,17-dione to 17$\beta$-hydroxyandrost-4-en-3-one while the addition of androstenedione to the oxidase system (0.166 M sodium pyrophosphate, pH 9.0) had no effect on the conversion of 17$\beta$-hydroxyandrost-4-en-3-one to androst-4-ene-3,17-dione. Further evidence for the presence of 3$\alpha$- and 3$\beta$-oxidoreductase activities in guinea pig liver homogenate was presented by the incubation with 5$\alpha$- and 5$\beta$-androstane-3,17-dione in phosphate buffer, pH 7.0, fortified with NADH (KOCHAKIAN, 1959). The major metabolites were the respective 3$\beta$-hydroxy derivatives with a smaller (50%) amount of the 3$\alpha$-hydroxy compounds. The 5$\alpha$-androstane-3,17-dione also was converted to a trace of 17$\beta$-hydroxy-5$\alpha$-androstan-3-one. RUBIN (1957) observed the formation of the 3$\alpha$- and 3$\beta$-derivatives of 5$\alpha$-androstane-3,17-dione by both male and female guinea pig liver homogenate with NAD as coenzyme.

SZAMATOWICZ et al. (1971) confirmed the interconversion of testosterone, androstenedione, and epitestosterone. They injected young (200–300 g) male and female guinea pigs daily for 6 days with human chorionic gonadotropin and incubated liver minces at pH 7.5 with NAD, NADH, NADP, and NADPH and [4–$^{14}$C]-testosterone or [4–$^{14}$C]-androst-4-ene-3,17-dione.

### 3. Rat

*a) Oxidoreductases* (Table 3)

SAMUELS et al. in a series of studies incubated rat liver preparations with testosterone and determined the disappearance of the $\alpha,\beta$-unsaturated-3-oxo group by the Zimmerman colorimetric method. NAD was found to be an essential factor in the formation of 17-ketosteroids (SWEAT and SAMUELS, 1948). OFNER (1955) incubated male rat liver mince fortified with nicotinamide and NAD to observe a conversion of [$^{14}$C]-labeled or unlabeled testosterone to androstenedione. HUBENER (1954) found that the cytosol of the male rat liver fortified with NAD interconverted testosterone and androstenedione and also reduced the 17-oxo group of androst-4-ene-3,11,17-trione to the 17$\beta$-hydroxy group.

WENZEL et al. (1967) and WENZEL and POLLOW (1967) proposed that estradiol is the preferred donor of hydrogen for the reduction of the oxo group of androst-4-ene-3,17-dione. The incubation of androst-4-ene-3,17-dione in the presence of [17$\alpha$-$^{3}$H]-estradiol-17$\beta$ with slices of female rat liver in Ringer-phosphate buffer yielded [3,17-$^{3}$H]-5$\alpha$-androstane-3$\alpha$,17$\beta$-diol. Similar incubation with slices of male liver yielded [17-$^{3}$H]-testosterone, [4,5-$^{3}$H]-5$\alpha$-androstane-3$\beta$,17$\beta$-diol and [4,5-$^{3}$H]-5$\beta$-androstane-3$\alpha$,17$\beta$-diol. When androstenedione was incubated with a NADH-generating system ($^{3}$H-labeled glucose or glutamate) and liver slices, only 0.01% of the radioactivity was found in the metabolites. The transfer of tritium from [17$\alpha$-$^{3}$H] to 5$\alpha$-androst-2-en-17-one and 5$\alpha$-androst-3-en-17-one for reduction of the 17-oxo group by both male and female rat liver slices was more effective than to androst-4-ene-3,17-dione, and [2-$^{3}$H]-lactate or [2-$^{3}$H]-isocitrate were 1000-fold less active than estradiol as reducing agents (WENZEL and MUETZEL, 1970).

WENZEL and WOLF (1969) found that the $^{3}$H transfer from estradiol to androstenedione was effected by both the microsomes and cytosol of the female rat liver and

Table 3. The metabolism of $C_{19}$-steroids in vitro by rat liver preparations

| Sex | Substrates | Tissue prep.[a] | Coenz. | pH | Metabolites | References |
|---|---|---|---|---|---|---|
| M | 17$\beta$-Hydroxyandrost-4-en-3-one | Cyt. | NAD | 7.3 | Androst-4-ene-3,17-dione | HUBENER (1954) |
| | Androst-4-ene-3,17-dione | | | | 17$\beta$-Hydroxyandrost-4-en-3-one | HUBENER (1954) |
| | Androst-4-ene-3,11,17-trione | | | | 17$\beta$-Hydroxyandrost-4-en-3,11-dione | HUBENER (1954) |
| M | 17$\beta$-Hydroxyandrost-4-en-3-one | M | NA,NAD | 7.8 | Androst-4-ene-3,17-dione | OFNER (1955) |
| | 17$\beta$-Hydroxyandrost-4-en-3-one[b] | | | 7.4 | Androst-4-ene-3,17-dione | OFNER (1955) |
| M | Androst-4-ene-3$\beta$,17$\beta$-diol | AP | NAD | 7.4 | Androst-4-ene-3,17-dione<br>17$\beta$-Hydroxyandrost-4-en-3-one<br>17$\alpha$-Hydroxyandrost-4-en-3-one | UNGAR et al. (1957) |
| | Androst-4-ene-3$\alpha$,17$\beta$-diol | | | | Androst-4-ene-3,17-dione<br>17$\beta$-Hydroxyandrost-4-en-3-one<br>17$\alpha$-Hydroxyandrost-4-en-3-one | UNGAR et al. (1957) |
| F | 17$\beta$-Hydroxyandrost-4-en-3-one | H | NA,NAD | 7.4 | 17$\beta$-Hydroxy-5$\alpha$-androstan-3-one | RUBIN and DORFMAN (1956) |
| M,F | 5$\alpha$-Androstane-3,17-dione | H | NA,NAD | 7.4 | 3$\alpha$-Hydroxy-5$\alpha$-androstan-17-one<br>3$\beta$-Hydroxy-5$\alpha$-androstan-17-one | RUBIN et al. (1957) |
| M | 3$\alpha$-Hydroxy-5$\alpha$-androstan-17-one | H | NA,NAD | 7.4 | 3$\beta$-Hydroxy-5$\alpha$-androstan-17-one<br>5$\alpha$-Androstane-3,17-dione | RUBIN et al. (1957) |
| M,F | Androst-4-ene-3,17-dione | Ms | NADPH | ? | 17$\beta$-Hydroxyandrost-4-en-3-one<br>7$\alpha$-Hydroxyandrost-4-ene-3,17-dione<br>6$\beta$-Hydroxyandrost-4-ene-3,17-dione<br>5$\alpha$-Androstane-3,17-dione | ALBAUM and STAIB (1965) |
| M | Androst-4-ene-3,17-dione | Ms | NADPH | 7.5 | 3$\beta$,17$\beta$-Dihydroxy-5$\alpha$-androstan-16-one | GUSTAFSSON and LISBOA (1970) |
| M | 5$\alpha$-Androstane-3,17-dione | Ms | NADPH | 7.5 | 3$\beta$,17$\beta$-Dihydroxy-5$\alpha$-androstan-16-one | GUSTAFSSON and LISBOA (1970) |
| M | 3$\alpha$/3$\beta$-Hydroxy-5$\alpha$-androstan-17-one | Ms | NADPH | 7.5 | 3$\beta$,17$\beta$-Dihydroxy-5$\alpha$-androstan-16-one | GUSTAFSSON and LISBOA (1970) |
| | 5$\alpha$-Androstane-3,17-dione | Ms | NADPH | 7.5 | 5$\alpha$-Androstane-2$\beta$,3$\alpha$,17$\beta$-triol<br>5$\alpha$-Androstane-3$\beta$,16$\alpha$,17$\beta$-triol | GUSTAFSSON and LISBOA (1970) |

Table 3 (continued)

| | | | | | | |
|---|---|---|---|---|---|---|
| | 3α-Hydroxy-5α-androstan-17-one | | | | 5α-Androstane-2β,3α,17β-triol<br>5α-Androstane-3α,17β-diol | GUSTAFSSON and LISBOA (1970) |
| M | 17β-Hydroxyandrost-4-en-3-one | Ms | NADPH | 7.0 | 6β,17β-Dihydroxyandrost-4-en-3-one<br>5α-Androstane-3α,6β,17β-triol | ABRAHAM and STAUDINGER (1966) |
| M | 17β-Hydroxyandrost-4-en-3-one[b] | Ms | NADPH | 7.5 | 2β,17β-Dihydroxyandrost-4-en-3-one<br>6β,17β-Dihydroxyandrost-4-en-3-one<br>7α,17β-Dihydroxyandrost-4-en-3-one<br>16α,17β-Dihydroxyandrost-4-en-3-one<br>6β,16α,17β-Trihydroxyandrost-4-en-3-one | LISBOA et al. (1968) |
| M | 17β-Hydroxyandrost-4-en-3-one[b] | Ms | NADPH | 7.5 | 2β,17β-Dihydroxy-5α-androstan-3-one<br>6β,17β-Dihydroxy-5α-androstan-3-one<br>7α,17β-Dihydroxy-5α-androstan-3-one<br>16α,17β-Dihydroxy-5α-androstan-3-one<br>16α,17β-Dihydroxyandrost-4-en-3-one | GUSTAFSSON et al. (1968) |
| M | 17β-Hydroxy-5α-androstan-3-one[b] | Ms | NADH | 7.5 | 2β,17β-Dihydroxy-5α-androstan-3-one<br>7α,17β-Dihydroxy-5α-androstan-3-one<br>16α,17β-Dihydroxy-5α-androstan-3-one | GUSTAFSSON et al. (1968) |
| M | 2β,17β-Dihydroxyandrost-4-en-3-one[b] | Ms | NADH | 7.5 | 2β,17β-Dihydroxy-5α-androstan-3-one | GUSTAFSSON et al. (1968) |
| | 7α,17β-Dihydroxyandrost-4-en-3-one[b] | | | | 7α,17β-Dihydroxy-5α-androstan-3-one | |
| | 16α,17β-Dihydroxyandrost-4-en-3-one[b] | | | | 16α,17β-Dihydroxy-5α-androstan-3-one | GUSTAFSSON et al. (1968) |
| M | 6β,17β-Dihydroxyandrost-4-en-3-one | Ms | NADPH | 7.0 | 5α-Androstane-3α,6β,17β-triol | ABRAHAM and STAUDINGER (1966) |
| M | Androst-4-ene-3,17-dione[b] | Ms | NADH | 7.5 | 7α-Hydroxyandrost-4-ene-3,17-dione<br>6β-Hydroxyandrost-4-ene-3,17-dione<br>16α-Hydroxyandrost-4-ene-3,17-dione<br>17β-Hydroxyandrost-4-en-3-one<br>3β-Hydroxy-5α-androstan-17-one<br>3β-Hydroxyandrost-4-en-17-one<br>5α-Androstane-3,17-dione | EINARSSON et al. (1973) |

Table 3 (continued)

| Sex | Substrates | Tissue prep.[a] | Coenz. | pH | Metabolites | References |
|---|---|---|---|---|---|---|
| M | 5α-Androstane-3α,17β-diol[b] | Ms | NADH | 7.5 | 5α-Androstane-3α,7α,17β-triol<br>5α-Androstane-2β,3α,17β-triol<br>5α-Androstane-2α,3α,17β-triol<br>5α-Androstane-3β,17β,18-triol<br>5α-Androstane-3α,17β,18-triol<br>3α-Hydroxy-5α-androstan-17-one<br>17β-Hydroxy-5α-androstan-3-one<br>5α-Androstan-3,17-dione | EINARSSON et al. (1973) |
| M | Androst-4-ene-3,17-dione w/estradiol-17β[c] | S | — | 7.4 | 17β-Hydroxyandrost-4-en-3-one<br>5β-Androstane-3α,17β-diol<br>5α-Androstane-3β,17β-diol | WENZEL et al. (1967)<br>WENZEL and POLLOW (1967) |
| F | Androst-4-ene-3,17-dione w/estradiol-17β[c] | Ms, Cyt. | NAD, NADP | 7.4 | 17β-Hydroxyandrost-4-en-3-one<br>5α-Androstane3α,17β-diol | WENZEL and WOLF (1969) |
| M | Androst-4-ene-3,17-dione w/estradiol-17β[c] | S | — | 7.4 | 5α-Androstane-3α,17β-diol | WENZEL and POLLOW (1967) |
| M, F | 17β-Hydroxy-1-methyl-5α-androst-1-en-3-one[c] w/estrone | S | — | 7.4 | 1-Methyl-5α-androst-1-ene-3,17-dione | WENZEL (1971) |
| F | 17β-Hydroxyandrost-4-en-3-one w/estrone | S | — | 7.4 | Androst-4-ene-3,17-dione | WENZEL (1971) |

Table 3 (continued)

| | | | | | | |
|---|---|---|---|---|---|---|
| M, F | 5α-Androst-2-en-17-one w/estradiol-17β[c] | S | — | 7.4 | 5α-Androst-2-en-17(?)-ol | Wenzel and Muetzel (1970) |
| | 5α-Androst-3-en-17-one w/estradiol-17β[c] | S | — | 7.4 | Androst-3-en-17(?)-ol | |
| M, F | 17β-Hydroxyandrost-4-en-3-one | P | | | 6β-Hydroxyandrost-4-ene-3,17-dione | Axelrod and Miller (1954) |
| | Androst-4-ene-3,17-dione | | | | 6β-Hydroxyandrost-4-ene-3,17-dione<br>17β-Hydroxy-5α-androstan-3-one | |
| M | 3β-Hydroxyandrost-5-en-17-one | S | NADPH | 7.4 | 3β,16α-Dihydroxyandrost-5-en-17-one | Colas (1962) |
| F | 17β-Hydroxyandrost-4-en-3-one[b] | H | NA, NAD(P)H | 6.7 | 17β-Hydroxy-5α-androstan-3-one<br>Androst-4-ene-3,17-dione<br>5α-Androstane-3,17-dione<br>3α-Hydroxy-5α-androstan-17-one<br>3β-Hydroxy-5α-androstan-17-one | Arimasa and Kochakian (1972) |
| F | Androst-4-ene-3,17-dione[b] | H | NA, NAD(P)H | 6.7 | 17β-Hydroxyandrost-4-en-3-one<br>17β-Hydroxy-5α-androstan-3-one<br>5α-Androstane-3α,17β-diol<br>5α-Androstane-3β,17β-diol<br>3α-Hydroxy-5α-androstan-17-one<br>3β-Hydroxy-5α-androstan-17-one<br>5α-Androstane-3,17-dione | Arimasa and Kochakian (1972) |

[a] Cyt = cytosol, M = mince, AP = acetone powder, Ms = microsomes, S = slices, H = homogenate, P = perfusion with heparinized blood.
[b] [4-$^{14}$C]-labeled.
[c] 17α$^{3}$H-labeled.

NAD was twice as effective as NADP as the coenzyme intermediate. They postulated that the $^3$H transfer was mediated by a 17β-steroid dehydrogenase and demonstrated a high degree of dependence between estrogens and $C_{19}$-steroids in oxidation-reduction since other $^3$H-labeled substances which produced NADP $^3$H were much less effective. In a further study WENZEL (1971) demonstrated the reverse reaction. The 17α-$^3$H of 17β-hydroxy-1-methyl-5α-androst-1-en-3-one and testosterone was transferred to estrone to form [17α-$^3$H]-estradiol-17β. Male rat liver slices were more efficient than female rat liver slices. The liver slices were more active than slices from testis, prostate, and seminal vesicles.

UNGAR et al. (1957) incubated androst-4-ene-3α,17β-diol and androst-4-ene-3β,17β-diol with a 0.1 M phosphate buffer (pH 7.4) extract of acetone powder of rat liver in Krebs-Ringer solution containing 0.002 M NAD and separated the metabolites by paper chromatography. Both steroids yielded the same products in a time sequence of testosterone, androst-4-ene-3,17-dione, and epitestosterone. Trace amounts of ring-A reduced oxosteroids were indicated.

RUBIN (1957) observed the conversion of 5α-androstane-3,17-dione to 3α- and 3β-hydroxy-5α-androstan-17-one by homogenates of adult male and female rat liver. The 3α-hydroxy-5α-androstan-3-one in turn was converted to the dione and the 3β-hydroxysteroid.

*b) Hydroxylases*

AXELROD and MILLER (1954) perfused testosterone and androst-4-ene-3,17-dione in heparinized blood separately through isolated rat liver and identified 6β-hydroxy-androst-4-ene-3,17-dione in both instances as one of the metabolites indicating the presence of 6β-hydroxylase as well as 17β-hydroxy dehydrogenase activity. COLAS (1962) observed 16α-hydroxylation of 3β-hydroxyandrost-5-en-17-one by adult male rat liver slices incubated in Krebs-Ringer phosphate at pH 7.4. The addition of NADPH to the incubation medium increased the yield of metabolite. Castration for 4–6 weeks slightly reduced the hydroxylation and female rat liver slices showed only a trace of activity. ALBAUM and STAIB (1965) further observed that the microsomes from male or female rat liver fortified with NADPH converted androstenedione not only to testosterone but also to 7α-hydroxy- and 6β-hydroxy-androst-4-ene-3,17-dione and 5α-androstane-3,17-dione. ABRAHAM and STAUDINGER (1966) isolated by thin layer chromatography and identified by IR spectrophotometry 6β-hydroxytestosterone and also the 5α reduced hydroxylated compound, 5α-androstane-3α,-6β,17β-triol, after incubation of testosterone with liver microsomes from adult male rats at pH 7.4 and an excess of NADPH. The triol was also obtained after incubation of 6β-hydroxytestosterone suggesting that this compound served as an intermediate in the transformation of testosterone to the triol. GUSTAFSSON and LISBOA (1970) utilized thin-layer chromatography and gas chromatography-mass spectrometry to isolate and detect a large number of metabolites after the incubation of androst-4-ene-3,17-dione, 5α-androstane-3,17-dione, androsterone or epiandrosterone with rat liver microsomes and a NADPH generating system at pH 7.5. The four steroids gave a common metabolite, 3β, 17β-dihydroxy-5α-androstan-16-one, and several hydroxylated 17-oxosteroids. Several other 17β-hydroxy- and 17-oxosteroids occurred in trace quantities and were identified only by gas chromatography-mass spectrometry.

The authors did not detect (at least did not mention) any 17α- or 17β-hydroxy metabolites, e.g., androst-4-ene-3,17-dione to 17β-hydroxyandrost-4-ene-3-one. Presumably, these compounds are not formed by the microsomes of rat liver contrary to the report of ALBAUM and STAIB (1965) [but see WENZEL et al. (see above)].

LISBOA et al. (1968) made a more detailed search and identification of the hydroxylated metabolites of testosterone after incubation of liver microsomes from adult male rats. They identified four hydroxylated derivatives of testosterone and tentatively characterized 6β,16α,17β-trihydroxyandrost-4-en-3-one. In addition a number of saturated highly polar metabolites were observed and were characterized in the following report. GUSTAFSSON et al. (1968) extended the hydroxylation by microsomes to 17β-hydroxy-5α-androstan-3-one because earlier studies by CHAMBERLAIN et al. (1965) indicated that the reduced metabolites of testosterone were further metabolized (hydroxylated) on continuation of the incubation beyond about 4 min. The microsomes were prepared from adult male liver and the incubation system was fortified with a NADPH generating system. Four hydroxylated (2β-, 6β-, 7α-, or 16α-) derivatives of 17β-hydroxy-5α-androstan-3-one were identified as metabolites of testosterone or the corresponding hydroxylated derivatives of testosterone. With the exception of the 6β-hydroxylated compound, these same metabolites were formed on incubation of 17β-hydroxy-5α-androstan-3-one. These results demonstrated that the hydroxylated products of testosterone can be further metabolized to form the hydroxylated derivatives of 17β-hydroxy-5α-androstan-3-one. These studies were repeated in germ-free male adult rats (EINARSSON et al., 1973) with [4-$^{14}$C]-androst-4-ene-3,17-dione and [4-$^{14}$C]-androstane-3α,17β-diol as the substrate. The microsomes from the germ-free rats metabolized androstenedione to approximately twice as much of the respective hydroxylated metabolites as that from the conventional rats, possibly due to their higher concentration of cytochrome P-450. On the other hand, the 5α-reduction of androstenedione was less in the germ-free rats and the difference in the metabolism of the androstanediol was not statistically significant.

*c) Reductases*

RUBIN and DORFMAN (1956) reported 5α-reductase activity in female rat liver homogenate. 17β-Hydroxy-5α-androstan-3-one was isolated and characterized by paper chromatography as presumably the only oxometabolite after the incubation of testosterone at pH 7.4 in the presence of NAD. BREUER et al. (1963) incubated androst-4-ene-3,17-dione with the microsomes of rat liver at pH 7.2 and in the presence of NADH to obtain a complete conversion to 5α-androstane-3,17-dione and 3α-hydroxy-5α-androstan-17-one. Under identical conditions androst-4-ene-3α (or 3β),17β-diol gave no reduced metabolites. ARIMASA and KOCHAKIAN (1972) conducted a more detailed study with [4-$^{14}$C]-testosterone and [4-$^{14}$C]-androst-4-ene-3,17-dione and female rat liver under optimum conditions for reductase activity and obtained not only the above but also several other metabolites. The major metabolites were 5α steroids, but oxidoreductase activities also were evident at the 17 position even though the pH and coenzymes were not favorable for such reactions. The formation of metabolites was small when no coenzyme was present in the incubation solution. The addition of NADH, NADPH, or a NADPH generating system greatly enhanced the transformations. NADPH or the generating system was more effective than NADH. The 17β-hydroxyl of testosterone was dehydrogenated in the NADH but not the NADPH containing system to a small quantity of androstenedione and in both systems to give several 17-oxo-5α-steroids. The major metabolite of testosterone was 5α-androstane-3α,17β-diol. When [4-$^{14}$C]-androst-4-ene-3,17-dione was the substrate, 3α-hydroxy-5α-androstan-17-one was the main metabolite but significant quantities of 5α-androstane-3α,17β-diol and 17β-hydroxy-5α-androstan-3-one and small quantities of 5α-androstane-3β,17β-diol in addition to the same 17-oxo-5α-steroids from testosterone were produced. The quantities of the various metabolites varied not only with the particular coenzyme but also with variations in pH, time of incubation, and substrate and homogenate concentrations. Most of the transformations occurred in the microsomal fraction but certain quantitative changes were effected by the cytosol and mitochondria. Purified nuclei converted testosterone to only trace quantities of the metabolites when NADH was added as the coenzyme and appreciable amounts of only 17β-hydroxy-5α-androstan-3-one when NADPH was the hydrogen donor. The conversion by the nuclei, however, was a fraction (15%) of that by the microsomes indicating either slight contamination of the nuclei or separate subcellular 5α-reductases.

OTA et al. (1974) reported that the cytoplasm of male rat liver converted testosterone predominantly to 5β reduced metabolites while only trace amounts were formed by female rat liver. The injection of the female rats with testosterone propionate not only increased the production of the 5β-reduced metabolites from both testosterone and androstenedione but also enhanced the conversion of androstenedione to testosterone. The injection of puromycin or actinomycin D prevented the enhancing effect of testosterone propionate. BREUER et al. (1963) incubated androst-4-ene-3β,17β-diol with the liver cytosol of young, 80–100 g, rats in the presence of both NADH and NADPH at pH 6.5 and also characterized by paper chromatography testosterone as the main metabolite but 5β-androstane-3α,17β-diol as a second metabolite. The incubation of the androst-4-ene-3α,-17β-diol also yielded testosterone but no other metabolite was apparent, presumably because of the very small quantity (50 μg) of the substrate.

### 4. **Human** (Table 4)

STYLIANOU et al. (1961) obtained a sample of liver at surgery on a 45-year-old woman for adenocarcinoma of the colon and incubated the homogenate with [4-$^{14}$C]-testosterone in a buffer system at pH 7.0 containing nicotinamide and a NADPH generating system. They identified androst-4-ene-3,17-dione and androsta-4,16-dien-3-one as dehydrogenated products and three reduced steroids. LISBOA et al. (1965) (see also BREUER et al., 1968) reported a series of studies in which they compared normal with cirrhotic liver slices. The incubations were conducted in Krebs-phosphate buffer at pH 7.4 without the addition of cofactors or coenzymes. The metabolites were identified by paper chromatography. The initial experiments with testosterone were confirmed by the use of [4-$^{14}$C]-testosterone as substrate. The oxidation of the 17β-hydroxyl was found in five metabolites, 16α-hydroxylation in two compounds, and 5α and 5β-saturation in three and two steroids, respectively. An intensive search was made for 17-oxo-steroids, e.g., 5α and 5β-androsterone, without success. More testosterone was converted to androstenedione by cirrhotic than normal liver but twice as much of the hydroxylated steroids were formed by the

Table 4. The metabolism of $C_{19}$-steroids in vitro by human liver preparations

| Species | Sex | Substrate | Tissue prep.[a] | Coenz. | pH | Metabolites | References |
|---|---|---|---|---|---|---|---|
| 1. Human | F | 17β-Hydroxyandrost-4-en-3-one[b] | H | NA, NADPH | 7.0 | Androst-4-ene-3,17-dione<br>Androst-4,16-dien-3-one<br>5β-Androstane-3α,17β-diol<br>5α-Androstane-3α,17β-diol<br>5α-Androstane-3β,17β-diol | STYLIANOU et al. (1961) |
| 2. Human (Normal and Cir.) | M, F | 17β-Hydroxyandrost-4-en-3-one | S | — | 7.4 | Androst-4-ene-3,17-dione<br>5α-Androstane-3,17-dione<br>5β-Androstane-3,17-dione<br>6β-Hydroxyandrost-4-ene-3,17-dione<br>6β,17β-Dihydroxyandrost-4-en-3-one<br>16α-Hydroxyandrost-4-en-3,17-dione<br>16α,17β-Dihydroxyandrost-4-en-3-one<br>17β-Hydroxy-5α-androstan-3-one<br>17β-Hydroxy-5β-androstan-3-one<br>5α-Androstane-3α,17β-diol<br>5α-Androstane-3β,17β-diol<br>5β-Androstane-3α,17β-diol | LISBOA et al. (1965)<br>BREUER et al. (1968) |
| 3. Human (Normal, Cir., Fatty) | M, F | 17β-Hydroxyandrost-4-en-3-one[b] | H, S | NAD(H), NADP(H) | 7.4 | Androst-4-ene-3,17-dione<br>3α-Hydroxy-5β-androstan-17-one<br>5β-Androstane-3α,17β-diol<br>5α-Androstane-3,17-dione<br>17β-Hydroxy-5α-androstan-3-one<br>3α-Hydroxy-5α-androstan-17-one<br>5α-Androstane-3α,17β-diol<br>5α-Androstane-3β,17β-diol | ENGELHARDT et al. (1971, 1974) |
| 4. Human (Normal, Cir., Fatty) | | Androst-4-ene-3,17-dione[b] | H | NAD(H), NADP(H) | 7.4 | 17β-Hydroxyandrost-4-en-3-one<br>(5β and 5α steroids as above) | ENGELHARDT et al. (1971) |
| 5. Human (17—18 wks) | M | Androst-4-ene-3,17-dione | Ms, Cyt | NA, NADPH | 7.5 | 17β-Hydroxyandrost-4-en-3-one<br>17α-Hydroxyandrost-4-en-3-one | PLASSE and LISBOA (1973) |

[a] H = homogenate, S = slices, Ms = microsomes, Cyt = cytosol.
[b] [4-$^{14}$C]/labeled, Cir. = cirrhotic.

normal than the cirrhotic liver. The yield of ketols was small with a suggestion of more by the normal than the cirrhotic liver.

ENGELHARDT et al. (1971, 1974) incubated liver obtained by needle biopsy from three normal men and five women, two men with cirrhotic liver and four men with fatty liver with [4-$^{14}$C]-testosterone and [1,2-$^{3}$H]-androstene-3,17-dione in Krebs-Ringer bicarbonate buffer (pH 7.4) containing NAD, NADH, NADP, and a NADPH generating system. The metabolites were identified by gas chromatography after separation of the extract of the incubation by alumina chromatography. The conversion of testosterone to androstenedione was the same in normal male and female and male cirrhotic liver but lower in male fatty liver. The reverse reaction was 3 times higher in normal women than men, was increased to that of the female liver in male cirrhotic liver, and was intermediate in male fatty liver. A number of other metabolites were identified in the studies with normal male and female liver. Testosterone and androstenedione were interconverted and both were converted to 3α-hydroxy-5β-androstan-17-one, 5β-androstane-3α,17β-diol, and to a series of 5α steroids. Thus, the results in general confirm those of LISBOA et al. (1965) with two major differences. The common urinary 17-ketosteroids were identified but the hydroxylated steroids were not detected. These differences between the results of the two laboratories are probably at least partly due to the addition of the four coenzymes to the incubation system by ENGELHARDT et al.

PLASSE and LISBOA (1973), in their search for the site of formation of 17α-steroids during pregnancy, incubated the microsomes and cytosol of human male fetal liver with androst-4-ene-3,17-dione at pH 7.5 and a NADPH generating system. They separated the testosterone-epitestosterone fraction from the many other metabolites by preparative thin-layer chromatography and then separated and characterized the two steroids by gas chromatography-mass spectometry. No attempt was made to quantitate the results but a 20:1 ratio was observed for formation by the cytosol of the 17β- and 17α-hydroxysteroids. The suggestion was made that the 17α-hydroxy-androst-4-en-3-one metabolite was the precursor of the 17α-hydroxylated estrogens during pregnancy.

## 5. **Dog** (Table 5)

AXELROD et al. (1956) perfused testosterone in heparinized blood through an isolated male dog liver and separated 25 metabolites from the perfusion fluid by paper chromatography. Androst-4-ene-3,17-dione and epitestosterone and four hydroxylated metabolites—2β-, 6α- and 16α-hydroxytestosterone, and 6β-hydroxyandrostenedione—were identified. BURCHARDT et al. (1968) perfused male dog liver with testosterone and obtained both androstenedione and epitestosterone and the reverse reaction when epitestosterone was used as the substrate. KLEMPIEN et al. (1961) perfused the isolated livers of four male and two female dogs with 3β-hydroxy-androst-5-en-17-one and separated and identified androst-4-ene-3,17-dione, 17β-hydroxyandrost-4-en-3-one, and androst-5-ene-3β,17β(?)-diol by paper chromatography. Two other metabolites were isolated but could not be identified. A search for phosphate and glucuronide conjugates was unsuccessful. They also did not detect any epitestosterone which might have been formed from the metabolite, androstenedione.

Table 5. The metabolism of $C_{19}$-steroids by perfusion of isolated dog livers

| Sex | Substrates | Metabolites | References |
|---|---|---|---|
| M | 17β-Hydroxyandrost-4-en-3-one | Androst-4-ene-3,17-dione<br>17α-Hydroxyandrost-4-en-3-one<br>2β-, 6α- and 16β-Hydroxytesterone<br>6β-Hydroxyandrost-4-ene-3,17-dione | AXELROD et al. (1956) |
| M | 17β-Hydroxyandrost-4-en-3-one | Androst-4-ene-3,17-dione<br>17α-Hydroxyandrost-4-en-3-one | BURCHARDT et al. (1968) |
| M | 17α-Hydroxyandrost-4-en-3-one | Androst-4-ene-3,17-dione<br>17β-Hydroxyandrost-4-en-3-one | |
| M, F | 3β-Hydroxyandrost-5-en-17-one | Androst-4-ene-3,17-dione<br>17β-Hydroxyandrost-4-en-3-one<br>Androst-5-ene-3β,17β(?)-diol | KLEMPIEN et al. (1961) |

## 6. **Mouse** (Table 6)

The incubation of mouse liver homogenate with [4-$^{14}$C]-androst-4-ene-3,17-dione at pH 6.7 gave testosterone and unidentified highly polar compounds as the major metabolites. 5α-Androstane-3α,17β-diol and its 3β-isomer were formed in trace amounts (ARIMASA and KOCHAKIAN, 1973). The addition of NADH to the incubation system greatly increased the polar compounds and reduced the formation of testosterone. The addition of NADPH resulted in almost complete transformation of the androstenedione to the polar compounds and only trace amounts of testosterone and the diols were detected. The newly formed polar compounds probably were hydroxylated derivatives[1]. This suggestion gains support from the use of testosterone as the substrate. Practically no polar metabolites were formed on incubation with homogenate alone; addition of NADH gave a small increase while NADPH stimulated almost complete conversion to the polar compounds with only a trace of androstenedione. On the other hand, the addition of NADH approximately doubled the small amount of androstenedione produced by the homogenate alone.

JAGARINEC et al. (1967) incubated [4-$^{14}$C]-testosterone with male and female (5–6 weeks old) mouse liver microsomes and excess NADPH at pH 7.4 for 10 min and obtained polar metabolites as the main product, a small amount of androst-4-ene-3,17-dione, and traces of 5α-steroids. Four hydroxylated compounds were isolated from the polar compounds indicating preferential hydroxylation by the mouse liver.[1]

[1] FORD et al. (1975) incubated a mixture of 48 nmoles of [4-$^{14}$C]-testosterone, 266 nmoles of [7β-$^{2}$H]-testosterone and 250 nmoles unlabeled testosterone with microsomes from 7 g female mouse liver and 5.6 μmoles of NADPH in 15 ml phosphate buffer, pH 7.3. The following metabolites were identified by mass spectra and gas chromatography: 1ζ-, 2β-, 6α-, 6β-, 7α-, 15α-, 15β-, 16α-, and 16β-hydroxy testosterone; 6α-, 6β- and 7α-hydroxyandrost-4-ene-3,17-dione and androst-4-ene-3,17-dione. Also a 6- or 7-oxotestosterone. [FORD, H.C., WHEELER, R., ENGEL, L.L.: Hydroxylation of testosterone at carbons 1, 2, 6, 7, 15 and 16 by the hepatic microsomal fraction from adult female C57BL/6J mice. Eur. J. Biochem. **57**, 9—14 (1975)].

Table 6. The metabolism of $C_{19}$-steroids in vitro by liver preparations from several animal species

| Species | Sex | Substrates | Tissue prepn.[a] | Coenz. | pH | Metabolites | References |
|---|---|---|---|---|---|---|---|
| Steer | — | 17β-Hydroxyandrost-4-en-3-one | Aq | NAD | 8.3 | Androst-4-ene-3,17-dione | Sweat et al. (1950) |
| Chick | — | Androst-4-ene-3β,17β-diol | AP | NAD | 7.4 | Androst-4-ene-3,17-dione<br>17β-Hydroxyandrost-4-en-3-one<br>17α-Hydroxyandrost-4-en-3-one | Ungar et al. (1957) |
| Chick | — | Androst-4-ene-3α,17β-diol | AP | NAD | 7.4 | Androst-4-ene-3,17-dione<br>17β-Hydroxyandrost-4-en-3-one<br>17α-Hydroxyandrost-4-en-3-one | Ungar et al. (1957) |
| Hamster | M | 5α-Androstane-3,17-dione | H | NA,NAD | 7.4 | 3α-Hydroxy-5α-androstan-17-one | Rubin (1957) |
| Hamster | F | 5α-Androstane-3,17-dione | H | NA,NAD | 7.4 | 3α-Hydroxy-5α-androstan-17-one<br>3β-Hydroxy-5α-androstan-17-one | Rubin (1957) |
| Turtle | M, F | 5α-Androstane-3,17-dione | H | NA,NAD | 7.4 | 3α-Hydroxy-5α-androstan-17-one<br>3β-Hydroxy-5α-androstan-17-one | Rubin (1957) |
| Mouse | M, F | 5α-Androstane-3,17-dione | H | NA,NAD | 7.4 | 3α-Hydroxy-5α-androstan-17-one<br>3β-Hydroxy-5α-androstan-17-one | Rubin (1957) |
| Mouse | M | Androst-4-ene-3,17-dione | H | NADH,<br>NADPH, | 6.7 | 17β-Hydroxyandrost-4-en-3-one<br>5α-Androstane-3α,17β-diol<br>5α-Androstane-3β,17β-diol | Arimasa and Kochakian (1974) |
| Mouse | M | 17β-Hydroxyandrost-4-en-3-one[b] | H | NADH,<br>NADPH, | 6.7 | Androst-4-ene-3,17-dione<br>5α-Androstane-3α,17β-diol<br>5α-Androstane-3β,17β-diol | Arimasa and Kochakian (1974) |
| Mouse | M, F | 17β-Hydroxyandrost-4-ene-3-one | MS | NADPH | 7.4 | Androst-4-ene-3,17-dione<br>16α,17β-Dihydroxyandrost-4-en-3-one<br>7α,17β-Dihydroxyandrost-4-en-3-one<br>6β,17β-Dihydroxyandrost-4-en-3-one<br>6β-Hydroxyandrost-4-ene-3,17-dione | Jagarinec et al. (1967) |

[a] Aq = aqueous extract, AP = acetone powder, H = homogenate.
[b] [4-$^{14}$C]/labeled.

### 7. **Chick** (Table 6)

UNGAR et al. (1957) incubated androst-4-ene-3α,17β-diol and androst-4-ene-3β,17β-diol with a 0.1 M phosphate buffer, pH 7.4, extract of a chick liver acetone powder in Krebs-Ringer solution containing 0.002 M NAD. The results were identical with a similar study with rat liver (see above) and were confirmed by repeating the experiments with $^{14}C$-labeled steroids.

## B. Kidney

### 1. Rabbit

KOCHAKIAN et al. (1952) demonstrated a conversion of testosterone to androst-4-ene-3,17-dione and epitestosterone on incubation with rabbit kidney homogenate at pH 7.4. The epitestosterone because of its low formation (1%) was not detected when 100 mg of testosterone was used as substrate but was found after the incubation of 1000 mg of testosterone. The metabolites were isolated by chromatography on alumina and characterized by color reactions, m.p. and mixed m.p. and infrared analysis. In a subsequent study (KOCHAKIAN and NALL, 1953), epitestosterone was converted to a small amount of androst-4-ene-3,17-dione but no testosterone was isolated. The incubation of the kidney homogenate with 100 mg of androst-4-ene-3,17-dione and the employment of the newly developed technic of paper chromatography permitted the identification of both testosterone and epitestosterone and in addition 5α-androstane-3,17-dione and epiandrosterone as metabolites (KOCHAKIAN and STIDWORTHY, 1954).

### 2. **Guinea Pig** (Table 7)

The homogenate of adult male guinea pig kidney produced a slight interconversion of testosterone and androstenedione on incubation in Krebs-Ringer bicarbonate buffer (pH 7.4) containing nicotinamide (KOCHAKIAN and STIDWORTHY, 1954; KOCHAKIAN et al., 1957a). The addition of NAD had a questionable small enhancing effect but when the buffer was changed to sodium pyrophosphate at pH 9.0, the NAD had a pronounced enhancing effect; a 50% conversion of testosterone to androstenedione was obtained (KOCHAKIAN et al., 1957a). The maximum conversion was obtained after 90 min incubation and the optimum pH was 9.5. The enzyme in 12 g of homogenate was saturated at 100 mg of testosterone. The substitution of NADP for NAD proved to be equally effective for the conversion of testosterone to androstenedione as was NADPH for NADH in the incubation of androstene-3,17-dione at pH 6.8 to yield testosterone (KOCHAKIAN et al., 1961). The reductase activity was less than the oxidative activity. KOCHAKIAN et al. (1957c) determined the optimum conditions for the conversion of testosterone to androstenedione. Under these conditions, the conversion of testosterone to androst-4-ene-3,17-dione was 50%; 17β-hydroxy-5α-androstan-3-one to 5α-androstane-3,17-dione was 10%; and 17β-hydroxy-5β-androstane-3-one to 5β-androstane-3,17-dione was 18%; in addition 4% of 3α-hydroxy-5β-androstan-17-one was isolated. These comparisons were confirmed and extended to a number of other 17-hydroxy-$C_{19}$-steroids and 17-oxosteroids (KOCHAKIAN et al., 1957b; KOCHAKIAN, 1959). Androst-5-ene-3β,17β-diol gave androst-4-ene-3,17-dione (2%) and 3β-hydroxyandrost-5-en-17-one (7%); 5α-an-

Table 7. The metabolism of $C_{19}$-steroids in vitro by adult male guinea pig kidney homogenate

| Substrates | Coenz. | pH | Metabolites | References |
|---|---|---|---|---|
| 17β-Hydroxyandrost-4-en-3-one | NA,NAD; NADP | 9.5 | Androst-4-ene-3,17-dione | KOCHAKIAN et al. (1961) |
| Androst-4-ene-3,17-dione | NA,NAD | 7.4 | 17β-Hydroxyandrost-4-en-3-one | KOCHAKIAN and STIDWORTHY (1954) |
| Androst-4-ene-3,17-dione | NADH;NADPH | 6.8 | 17β-Hydroxyandrost-4-en-3-one | KOCHAKIAN et al. (1961) |
| 5α-Androstane-3α,17β-diol | — | 7.4 | 3α-Hydroxy-5α-androstan-17-one | KOCHAKIAN and APOSHIAN (1952) |
| 5α-Androstane-3α,17β-diol | NA,NAD | 9.0 | 5α-Androstane-3,17-dione<br>3α-Hydroxy-5α-androstan-17-one | KOCHAKIAN et. al. (1957c),<br>KOCHAKIAN (1959) |
| 5α-Androstane-3β,17β-diol | NA,NAD | 9.0 | 5α-Androstane-3,17-dione<br>3β-Hydroxy-5α-androstan-17-one | KOCHAKIAN et al. (1957c)<br>KOCHAKIAN (1959) |
| Androst-5-ene-3β,17β-diol | NA,NAD | 9.0 | Androst-4-ene-3,17-dione<br>3β-Hydroxyandrost-5-en-17-one | KOCHAKIAN et al. (1957c)<br>KOCHAKIAN (1959) |
| 3α-Hydroxy-5α-androstan-17-one | NA,NAD | 9.0 | 5α-Androstane-3,17-dione | KOCHAKIAN et al. (1957c)<br>KOCHAKIAN (1959) |
| 3β-Hydroxy-5α-androstan-17-one | NA,NAD | 9.0 | 5α-Androstane-3,17-dione | KOCHAKIAN et al. (1957c)<br>KOCHAKIAN (1959) |
| 3β-Hydroxyandrost-5-en-17-one | NA,NAD | 9.0 | Androst-4-ene-3,17-dione | KOCHAKIAN et al. (1957c)<br>KOCHAKIAN (1959) |
| 17β-Hydroxyandrost-4-en-3-one | NA,NAD | 9.0 | Androst-4-ene-3,17-dione | KOCHAKIAN et al. (1957c)<br>KOCHAKIAN (1959) |
| 17β-Hydroxy-5β-androstan-3-one | NA,NAD | 9.0 | 5β-Androstane-3,17-dione<br>3α-Hydroxy-5β-androstan-17-one | KOCHAKIAN et al. (1957c)<br>KOCHAKIAN (1959) |
| 17β-Hydroxy-5α-androstan-3-one | NA,NAD | 9.0 | 5α-Androstane-3,17-dione | KOCHAKIAN et al. (1957c)<br>KOCHAKIAN (1959) |
| 5α-Androstane-3,17-dione | NADH | 7.0 | 3β-Hydroxy-5α-androstan-17-one<br>3α-Hydroxy-5α-androstan-17-one<br>17β-Hydroxy-5α-androstan-3-one | KOCHAKIAN (1959) |
| 5β-Androstane-3,17-dione | NADH | 7.0 | 3β-Hydroxy-5β-androstan-17-one<br>3α-Hydroxy-5β-androstan-17-one | KOCHAKIAN (1959) |

drostane-3α,17β-diol gave 5α-androstane-3,17-dione (1%) and 3α-hydroxy-5α-androstan-17-one (1%); and 5α-androstane-3β,17β-diol gave (3β-hydroxy-5α-androstan-17-one (3%) and 5α-androstane-3,17-dione (3%). The small yield of metabolites from the diols was anticipated from the earlier study with 5α-androstane-3α,17β-diol (KOCHAKIAN and APOSHIAN, 1952). The failure to detect 17α-hydroxy metabolites in the above studies probably was due to the smaller quantities of substrate, 3 vs 100 to 1000 mg and the very low formation of such steroids under these conditions (KOCHAKIAN et al., 1952; KOCHAKIAN and STIDWORTHY, 1954). The 3-hydroxy group of the 17-oxosteroids was dehydrogenated by the kidney homogenate to give the corresponding 3,17-dioxo steroids. The degree of conversion was 3α-hydroxy > 3β-hydroxy > $\Delta^5$-3β-hydroxy-$C_{19}$-steroids. It is noteworthy that the metabolism of all of the steroids by not only the kidney but also the liver homogenates was much greater than that by the homogenates of the prostate or testes.

The presence of 3α- and 3β-oxidoreductase activities were demonstrated by the reduction of the 3-oxo group of both 5α- and 5β-androstane-3,17-dione to the respective 3α- and 3β-hydroxy-steroids on incubation in phosphate buffer, pH 7.0, fortified with NADH (KOCHAKIAN, 1959). A trace amount of 17β-hydroxy-5α-androstan-3-one was formed from 5α-androstane-3,17-dione.

WEST and SAMUELS (1951) reported the conversion of the 17-hydroxy group of testosterone to the 17-oxo group (colorimetric) by kidney mince at pH 7.4 and that the enzyme system was NAD-dependent.

### 3. **Rat** (Table 8)

BREUER and DAHM (1964) reported that the cytosol of rat kidney contained an enzyme which interconverted testosterone and androstenedione. The oxidoreductase had optimum pHs of 8.7 and 6.6, utilized either NAD(H) or NADP(H), and was also active on phenolic steroids. BREUER et al. (1963) in an earlier report observed that the precipitate obtained between 40–60% $(NH_4)_2SO_4$ saturation of the cytosol of the kidneys of young male rats converted androst-4-ene-3,17-dione in the presence of NADH and NADPH at pH 6.6 not only to testosterone (15%) but also 3β-hydroxy-androst-4-en-17-one (2%).

SUZUKI and TAMAOKI (1973) observed that the rat kidney homogenate fortified with NADPH in Tris buffer, pH 7.4, not only exhibited 5α- and 3α-reduction of testosterone but also 17β-hydroxy-dehydrogenation (androstenedione). WENZEL and MEUTZEL (1970) reported that rat kidney slices preferentially transferred the 17α-$^3$H of [17α-$^3$H]estradiol-17β to the 17-oxo group of either 5α-androst-2-en-17-one or 5α-androst-3-en-17-one.

MEIGS and RYAN (1966) found that a partially purified 16α-hydroxy-$C_{18}$-steroid dehydrogenase of adult female rat kidney (KING, 1960) was also able to reduce 16-ketotestosterone and 16-ketoandrost-5-ene-3β,17β-diol to their respective 16-hydroxy compounds with NADH as coenzyme.

### 4. **Dog** (Table 8)

Kidney minces converted testosterone to a 17-oxosteroid (colorimetric) with enhancement by the addition of NAD to the incubation system (WEST and SAMUELS,

Table 8. The metabolism of $C_{19}$-steroids in vitro by mammalian kidney preparations

| Species | Sex | Substrate | Prepn.[a] | Coenz. | pH | Metabolites | References |
|---|---|---|---|---|---|---|---|
| Rat | M,F | 17β-Hydroxyandrost-4-ene-3-one | Cyt. | NAD; NADP₁ | 8.7 | Androst-4-ene-3,17-dione | BREUER and DAHM (1964) |
| Rat | | Androst-4-ene-3,17-dione | Cyt. | NADH, NADPH | 6.6 | 17β-Hydroxyandrost-4-en-3-one<br>3β-Hydroxyandrost-4-en-17-one | BREUER et al. (1963) |
| Rat | M,F | 17β-Hydroxyandrost-4-en-3-one | Cyt. | NADH | 7.4 | 17β-Hydroxy-5α-androstan-3-one<br>5α-Androstane-3α,17β-diol<br>Androst-4-ene-3,17-dione | SUZUKI and TAMAOKI (1973) |
| Rat | F | 17β-Hydroxyandrost-4-ene-3,16-dione | Cyt. | NADH; NADPH | 6.1 | 16β,17β-Dihydroxyandrost-4-en-3-one | MEIGS and RYAN (1966) |
| Rat | | 3β,17β-Dihydroxyandrost-5-en-16-one | Cyt. | NADH; NADPH | | Androst-5-ene-3β,16β,17β-triol | MEIGS and RYAN (1966) |
| Rat | M,F | 5α-Androst-2-en-17-one | S | — | 7.4 | 5α-Androst-2-en-17β-ol | WENZEL and MUETZEL (1970) |
| Rat | M,F | 5α-Androst-3-en-17-one | S | — | 7.4 | 5α-androst-3-en-17β-ol | WENZEL and MUETZEL (1970) |
| Dog | — | 17β-Hydroxyandrost-4-en-3-one | P | — | — | Androst-4-ene-3,17-dione | HUDSON and YANDELL (1956) |
| Monkey | — | 17β-Hydroxyandrost-4-en-3-one[c] | S | NADPH | 7.0 | Androst-4-ene-3,17-dione | FREIRE et al. (1974) |
| Mouse | M (N,C,T)[b] | Androst-4-ene-3,17-dione[c] | H | NADH, NADPH | 6,7 | 17α-Hydroxyandrost-4-en-3-one<br>5α-Androstane-3α,17β-diol<br>17β-Hydroxyandrost-4-en-3-one | ARIMASA and KOCHAKIAN (1973) |
| Mouse | M | 17β-Hydroxyandrost-4-en-3-one[c] | H | NADH, NADPH | 6.7 | 17α-Hydroxyandrost-4-en-3-one<br>5α-Androstane-3α,17β-diol<br>Androst-4-ene-3,17-dione | ARIMASA and KOCHAKIAN (1973) |
| Mouse | M | Androst-4-ene-3,17-dione[c] | Cyt. | NADH, NADPH | 6.7 | 17α-Hydroxyandrost-4-en-3-one<br>17β-Hydroxyandrost-4-en-3-one<br>5α-Androstane-3α,17β-diol | ARIMASA and KOCHAKIAN (1973) |
| Mouse | C | Androst-4-ene-3,17-dione[d] | Cyt. | — | 7.4 | 17β-Hydroxyandrost-4-en-3-one | BULLOCK and BARDIN (1974) |
| | C | 17β-Hydroxy-5α-androstan-3-one[d] | Cyt. | — | 7.4 | 5α-Androstane-3α,17β-diol<br>5α-Androstane-3β,17β-diol | BULLOCK and BARDIN (1974) |

[a] Cyt = cytosol, P = perfusion, S = slices, H = homogenate.
[b] N = normal, C = castrated, T = testosterone propionate-treated.
[c] [4-$^{14}$C]-labeled.
[d] $^{3}$H-labeled.

1951). The perfusion of the isolated dog kidney with testosterone yielded androst-4-ene-3,17-dione as well as unmetabolized testosterone which were characterized by paper and silica gel chromatography (HUDSON and YANDEL, 1956).

**5. Monkey** (Table 8)

Monkey kidney plays an active role in the metabolism of testosterone (FREIRE et al., 1974). The incubation of [$^{14}$C]-testosterone with kidney slices and NADPH at pH 7.0 gave up to 8.3% of the recovered radioactivity as androst-4-ene-3,17-dione as identified by radio-gas chromatography. The chromatographic fractions corresponding to $\Delta^4$/5α-androstane-3α,17β-diol, and $\Delta^4$/5α-androstane-3β,17β-diol contained 1.0 and 1.6% of the radioactivity respectively; the rest of the radioactivity (4.5%) was recovered in the conjugated fraction.

**6. Mouse** (Table 8)

The mouse kidney is one of the most responsive tissues to the presence or withdrawal of testosterone and metabolized testosterone and androst-4-ene-3,17-dione to a different spectrum of metabolites (ARIMASA and KOCHAKIAN, 1973). Epitestosterone was the major metabolite on incubation of androst-4-ene-3,17-dione with kidney homogenate at pH 6.7. Approximately 70% of the substrate was converted to epitestosterone when saturating quantities of homogenate were used. The addition of NADPH but not NADH at the lower concentrations of homogenate enhanced the transformation. Testosterone was detected in very small yield and was not increased by the addition of the coenzymes or greater amounts of homogenate but was increased with increase in substrate concentration. 5α-Androstane-3α,17β-diol was isolated in small yield which was increased progressively but at a low rate with the use of additional homogenate and its production was not noticeably enhanced by either of the coenzymes. Several 5α-reduced steroids were detected in trace amounts. The conversion of testosterone to epitestosterone, androst-4-ene-3,17-dione, and 5α-androstane-3α,17β-diol was very small and was enhanced only 2- to 3-fold by the coenzymes. Castration of the mice further reduced the formation of the metabolites in the coenzyme fortified as well as the nonfortified systems. The decrease in epitestosterone formation probably was due to the unfavorable conditions (pH 6.7, reduced coenzymes) for the formation of the intermediate androstenedione.

The metabolism of androst-4-ene-3,17-dione was localized in the cytosol except more testosterone was formed in the microsomal fraction in the NADH system. Of further interest was the fact that the production of testosterone was almost twice that of epitestosterone in the cytosol-NADPH system but epitestosterone production in the cytosol-NADH system was much greater than that of testosterone and was 3 times that with the homogenate. BULLOCK and BARDIN (1974), and MOWSZOWICZ and BARDIN (1974) also found practically no metabolism of [$^3$H]-testosterone on incubation with the cytosol (no coenzyme) or slices of castrated or normal mouse kidney but found most of the androst-4-ene-3,17-dione converted to testosterone (6.6%); and 17β-hydroxy-5α-androstan-3-one to androstanediols (80%). The failure to detect epitestosterone after androstenedione incubation could be due to the paper chromatography technic. The testosterone probably was primarily epitestosterone (cf. ARIMASA and KOCHAKIAN, 1973).

### 7. Human

ENGELHARDT et al. (1974) could not detect an "appreciable" metabolism of testosterone by kidney slices incubated by the same system that exhibited high activity by liver slices. These results are in marked contrast with kidneys of experimental animals (see above).[2]

## C. Blood (Table 9)

### 1. Initial Observations in Several Species

LINDNER (1959) found epitestosterone in the spermatic vein blood of calves but not in the testis. The failure to find this steroid in the testis suggested that it might be formed in the blood. Incubation of androst-4-ene-3,17-dione with sterile whole blood from ox or sheep rapidly produced epitestosterone and a small quantity of testosterone. The metabolism did not occur when plasma or washed erythrocytes were used. When testosterone, however, was incubated with whole ox blood (LINDNER, 1961), no epitestosterone or any other metabolites were detected by paper chromatography. The conversion of androst-4-ene-3,17-dione to epitestosterone, moreover, was confirmed for normal male and female and castrated sheep and male and female oxen. On the other hand, the blood of the female pig, stallion, and man transformed androst-4-ene-3,17-dione to small quantities of testosterone but no epitestosterone. The reducing activity was dependent on the presence of the cellular fraction. Sheep plasma alone or with either a NADH or NADPH generating system was ineffective. Erythrocytes washed with saline lost their reductive activity and the addition of glucose to the washed erythrocytes restored the enzyme activity (LINDNER, 1965). Furthermore, the washed cells without glucose tranformed epitestosterone to androst-4-ene-3,17-dione and the addition of glucose to the incubation medium strongly inhibited the oxidative reaction. The ovine blood also converted 11-oxygenated-17-oxosteroids to corresponding 17α-hydroxy metabolites. VAN DER MOLEN and GROEN (1968) confirmed the observations of LINDNER with sheep and human blood both with respect to the interconversions and the localization and characteristics of the enzyme systems.

### 2. Rat

NICOL et al. (1968) found that whole blood of adult male rats effectively reduced 3β-hydroxyandrost-5-en-17-one to androst-5-ene-3β,17β-diol and localized the enzyme activity in the red blood cells. These studies were extended (NICOL et al., 1969) to a number of other steroids and the enzyme activity was demonstrated to be in the cytoplasm of the RBC. The metabolites were characterized by thin-layer chromatog-

2 MARBERGER et al. (1976) incubated 300 mg of normal human kidney slices with 200 μmoles [$^3$H]-testosterone and 1 and 2 μg of gestonorone caproate. Androst-4-ene-3,17-dione and epitestosterone were the main metabolites with lesser ($^1/_{16}$) amounts of the reduced steroids, 5α-androstane-3α,17β-diol and 17β-hydroxy-5α-androstan-3-one. Slices of renal carcinoma were about $^1/_3$ as effective as the normal tissue. [MARBERGER, M., JR., ALTWEIN, J. E., ORESTANO, F.: Influence of the Progestogen, Gestonorone Caproate on testosterone turnover in renal cell carcinoma. In vitro study. Invest. Urol. 13, 302—306 (1976)].

raphy. The erythrocytes were washed twice with a glucose-phosphate buffer (pH 7.4) solution, the membranes were ruptured by freezing twice at −30° C, and the soluble fraction was used as the enzyme source. The cytoplasm fortified with NADH was less effective than the whole blood in the transformations except in the case of androsterone, when it was much more effective. On the other hand, NADPH enhanced the metabolism of all of the steroids to at least twice that of the whole blood and in addition formed a relatively large quantity of 5α-androstane-3α,17β-diol whereas with NADH none was formed and only a trace of the 17β-hydroxy-5α-androstan-3-one. The transformations of the 5α-steroids was much less than that of the unsaturated compounds and the 5β-steroid (3α-hydroxy-5β-androstan-17-one) was not metabolized by either the whole blood or the coenzyme fortified systems. Dehydroepiandrosterone sulfate also was not metabolized by any of the three systems. The reverse (oxidative) reactions were observed when NADP instead of NADPH was added to the incubation system. Similar oxidoreductase activities were observed with the estrogens (estrone, estradiol-17β). HEYNS and DE MOOR (1974) confirmed the presence of not only 3β- but also 17β-hydroxysteroid dehydrogenase activity in rat blood by the incubation of several $^3$H-labeled steroids with whole blood and partially purified hemolysates. The reduction of the 3-oxo group was limited to 5α-steroids. Estrone and estradiol-17β also were interconverted by the partially purified preparations.

### 3. Rabbit

The whole blood of rabbits exhibited a strong ability to convert androstenedione to testosterone (VAN DER MOLEN and GROEN, 1968), but practically no conversion of testosterone to androstenedione. When the RBC were washed twice with 0.9% NaCl, the oxidative activity was equal to the reductase activity. The addition of NAD(H) and NADP(H) and glucose to the incubation enhanced the reductase activity but greatly reduced the oxidative activity. Identical results were obtained with dog blood. MAHOUDEAU et al. (1973) found that the interconversion of testosterone and androstenedione by whole blood of adult rabbits in vitro was the same as by perfusion in vivo and was time-dependent. The conversion of androstenedione to testosterone was very much greater than the reverse reaction as determined by the ratio of the radioactivity ($^3$H) of the recovered substrate and metabolite.

### 4. Human

BLAQUIER et al. (1967) confirmed the initial observations of LINDNER (1959, 1961), that human RBC converted androstenedione to testosterone, and extended the observation to a number of other $^3$H-labeled $C_{19}$-steroids. Pools of blood were obtained from male and female donors, aliquots were incubated with the added cofactors NAD, NADPH, ATP, and nicotinamide. The metabolites were separated and characterized by paper chromatography. The interconversions of the several steroids not only indicated the presence of several enzymes but also demonstrated the ability of the blood to transform less biologically active to more active steroids. NICOL et al. (1971) observed that human erythrocytes converted 3β-hydroxyandrost-5-en-17-one to androst-5-ene-3β,17β-diol but were only about one-tenth as effective as rat RBC.

Table 9. The metabolism of $C_{19}$-steroids in vitro by blood

| Species | Sex | Substrate | Tissue prep.[a] | Coenz. | pH | Metabolites | References |
|---|---|---|---|---|---|---|---|
| Ox, sheep | M,F | Androst-4-ene-3,17-dione | B | | | 17α-Hydroxyandrost-4-en-3-one<br>17β-Hydroxyandrost-4-en-3-one | Lindner (1959, 1961) |
| Sheep | M,F | Androst-4-ene-3,17-dione[b] | B | | | 17α-Hydroxyandrost-4-en-3-one<br>17β-Hydroxyandrost-4-en-3-one | Lindner (1965) |
| Sheep | M,F | 17α-Hydroxyandrost-4-en-3-one[b] | W-RBC | | | Androst-4-ene-3,17-dione | Lindner (1965) |
| Sheep | M,F | Androst-4-ene-3,17-dione | W-RBC + glucose | | | 17α-Hydroxyandrost-4-en-3-one | Lindner (1965) |
| Pig | F | Androst-4-ene-3,17-dione | B | | | 17β-Hydroxyandrost-4-en-3-one | Lindner (1961) |
| Horse | M | Androst-4-ene-3,17-dione | B | | | 17β-Hydroxyandrost-4-en-3-one | Lindner (1961) |
| Human | M | Androst-4-ene-3,17-dione | B | | | 17β-Hydroxyandrost-4-en-3-one | Lindner (1961) |
| Human | M,F | 3β-Hydroxyandrost-5-en-17-one[c] | B | NAD, NADPH | | Androst-4-ene-3,17-dione<br>Androst-5-ene-3β,17β-diol<br>17β-Hydroxyandrost-4-en-3-one | Blaquier et al. (1967) |
| Human | M,F | Androst-5-ene-3β,17β-diol[c] | B | NAD, NADPH | | 3β-Hydroxyandrost-5-en-17-one<br>17β-Hydroxyandrost-4-en-3-one<br>Androst-4-ene-3β,17β-diol | Blaquier et al. (1967) |
| Human | M,F | 17β-Hydroxyandrost-4-en-3-one[c] | B | NAD, NADPH | | Androst-4-ene-3,17-dione<br>Androst-4-ene-3β,17β-diol | Blaquier et al. (1967) |
| Human | M,F | Androst-4-ene-3,17-dione[c] | B | NAD, NADPH | | 17β-Hydroxyandrost-4-en-3-one<br>Androst-4-ene-3,17-dione | Blaquier et al. (1967) |
| Human | M,F | Androst-4-ene-3β,17β-diol[c] | B | NAD, NADPH | | 17β-Hydroxyandrost-4-ene-3-one<br>Androst-4-ene-3,17-dione | Blaquier et al. (1967) |
| Human | M | 3β-Hydroxyandrost-5-en-17-one[c] | RBC | — | — | Androst-5-ene-3β,17β-diol | Nicol et al. (1971) |
| Human | | Androst-4-ene-3,17-dione[b] | L | NADPH | 7.4 | 17β-Hydroxyandrost-4-en-3-one<br>5α-Androstane-3,17-dione<br>3α-Hydroxy-5α-androstan-17-one | Raith and Karl (1974) |
| Human | M | 17β-Hydroxyandrost-4-en-3-one[b] | L | NADPH | 7.4 | Androst-4-ene-3,17-dione<br>(17β-Hydroxy-5α-androstan-3-one)[d] | Raith and Karl (1974) |
| Rat | M | 3β-Hydroxyandrost-5-en-17-one | RBC-Cyt | NAD, NADPH | 7.4 | Androst-5-ene-3β,17β-diol | Nicol et al. (1968, 1969) |
| | | Androst-4-ene-3,17-dione | RBC-Cyt | NAD, NADPH | 7.4 | 17β-Hydroxyandrost-4-en-3-one | Nicol et al. (1969) |

Table 9 (continued)

| | | | | | | | |
|---|---|---|---|---|---|---|---|
| | | (Estrone) | RBC-Cyt | NAD, NADPH | 7.4 | (Estradiol-17β) | NICOL et al. (1969) |
| | | 5α-Androstane-3,17-dione | RBC-Cyt | NAD, NADPH | 7.4 | 5α-Androstane-3α,17β-diol<br>17β-Hydroxy-5α-androstan-3-one | NICOL et al. (1969) |
| Rat | M | 3α-Hydroxy-5α-androstan-17-one | RBC-Cyt | NADH, NADPH | 7.4 | 5α-Androstane-3α,17β-diol | NICOL et al. (1969) |
| Rat | M | 3β-Hydroxyandrost-5-en-17-one[c] | RBC | — | — | Androst-5-ene-3β,17β-diol | NICOL et al. (1971) |
| Rat | M,F | (Estradiol-17β[e]) | h-RBC | NADP | 7.1 | (Estrone) | HEYNS and DEMOOR (1974) |
| Rat | M,F | 5α-Androstane-3α,17β-diol[c] | h-RBC | NADP | 7.1 | 3β-Hydroxy-5α-androstan-3-one | HEYNS and DEMOOR (1974) |
| Rat | M,F | 17β-Hydroxyandrost-4-en-3-one[c] | h-RBC | NADP | 7.1 | Androst-4-ene-3,17-dione | HEYNS and DEMOOR (1974) |
| Rat | M,F | 17β-Hydroxy-5α-androstan-3-one[c] | h-RBC | NADP | 7.1 | 5α-Androstane-3,17-dione | HEYNS and DEMOOR (1974) |
| Rat | M,F | 3β-Hydroxy-5α-androstan-17-one[c] | h-RBC | NADP | 7.1 | 5α-Androstane-3,17-dione | HEYNS and DEMOORE (1974) |
| Rat | M,F | 17β-Hydroxy-5α-androstan-3-one[c] | h-RBC | NADPH | 7.1 | 5α-Androstane-3α,17β-diol | HEYNS and DEMOOR (1974) |
| | | (Estrone[e]) | h-RBC | NADPH | 7.1 | (Estradiol-17β) | HEYNS and DEMOOR (1974) |
| | | 3β-Hydroxyandrost-5-en-17-one[c] | h-RBC | NADPH | 7.1 | Androst-5-ene-3β,17β-diol | HEYNS and DEMOOR (1974) |
| | | 3α-Hydroxy-5α-androstan-17-one[c] | h-RBC | NADPH | 7.1 | 5α-Androstane-3α,17β-diol | HEYNS and DEMOOR (1974) |
| Dog | — | Androst-4-ene-3,17-dione[b] | B, RBC | — | — | 17β-Hydroxyandrost-4-en-3-one | VAN DER MOELLEN and GROEN (1968) |
| | | 17β-Hydroxyandrost-4-en-3-one[b] | RBC | — | — | Androst-4-ene-3,17-dione | VAN DER MOELLEN and GROEN (1968) |
| Rabbit | — | Androst-4-ene-3,17-dione[c] | B, RBC | — | — | 17β-Hydroxyandrost-4-en-3-one | VAN DER MOELLEN and GROEN (1968) |
| Rabbit | | 17β-Hydroxyandrost-4-en-3-one[c] | B | | | Androst-4-ene-3,17-dione | MAHOUDEAU et al. (1973) |

[a] B = whole blood, W-RBC = washed red blood, Cyt = cytosol, h-RBC = hemolyzed red blood cells, L = leukocytes.
[b] [4-$^{14}$C]-labeled.
[c] [7α-or 1,2-$^{3}$H]-labeled.
[d] Only by leukemic cells.
[e] [6,7-$^{3}$H]-labeled.

When the substrate was bound to the serum protein prior to incubation, the reduction by both human and rat erythrocytes was decreased to about one-third.

Human leukocytes also are able to metabolize testosterone and androstenedione. RAITH and KARL (1974) incubated [4-$^{14}$C]-androst-4-ene-3,17-dione in Krebs-Ringer bicarbonate buffer containing a NADPH generating system. Androstenedione was converted to testosterone, 5α-androstane-3,17-dione, and 3α-hydroxy-5α-androstan-17-one by both normal and leukemic cells. Testosterone was converted to androstenedione by normal and leukemic cells and also to 17β-hydroxy-5α-androstan-3-one only by leukemic cells.

## D. Skeletal Muscle (Table 10)

Testosterone was only slightly metabolized by homogenates of the gastrocnemius muscle from normal, castrated, or testosterone-treated guinea pigs (KINSON and KOCHAKIAN, 1964). The use of [4-$^{14}$C]-testosterone and separation of the extract of the incubation mixture by paper chromatography, however, demonstrated the formation of [4-$^{14}$C]-androstenedione which was enhanced by the addition of NAD or NADP. The temporal muscle, a highly testosterone-responsive muscle, exhibited essentially the same 17β-hydroxy dehydrogenase activity as the gastrocnemius muscle which is only slightly dependent on testosterone for growth. The rat and dog gastrocnemius muscle homogenates were as active as those of the guinea pig. Furthermore, NAD had a much greater effect than NADP in the dog gastrocnemius muscle but the reverse in the rat and guinea pig gastrocnemius muscle. The overall conversions were small and no other metabolites could be detected on a radioactive scan of the paper chromatograms. SHIMAZAKI et al. (1965) were able to identify both androst-4-ene-3,17-dione and 5α-androstane-3,17-dione on incubation of [4-$^{14}$C]-17β-hydroxyandrost-4-en-3-one with a mince of dog gluteal muscle in Krebs-Ringer phosphate containing nicotinamide, glucose, and sodium fumarate.

The incubation of androst-4-ene-3,17-dione (100 mg) with 12 g of rabbit skeletal muscle at pH 9.0 yielded a trace of testosterone and a 0.1% conversion to 5α-androstane-3,17-dione (KOCHAKIAN and STIDWORTHY, 1954). THOMAS and DORFMAN (1964a) incubated strips of the hind leg muscles of adult rabbits with [4-$^{14}$C]-androst-4-ene-3,17-dione and identified by paper chromatography not only 5α-androstane-3,17-dione and testosterone but also 3β-hydroxy-5α-androstan-17-one as metabolites. The 5α-reductase activity was localized in the microsomal fraction and was NADPH-dependent (THOMAS, 1968), and the dehydrogenase activity was in the cytosol (THOMAS and DORFMAN, 1964b). Incubation of the cytosol with [4-$^{14}$C]-androst-4-ene-3,17-dione at pH 7.5 resulted in conversion to testosterone and a new steroid, 3β-hydroxyandrost-4-en-17-one. The addition of NAD, NADH, or NADP to the incubation did not change the degree of formation of either metabolite but NADPH formation by a generating system increased both metabolites by 3- to 4-fold. The use of [4-$^{14}$C]-testosterone as a substrate in the NADPH generating system gave androst-4-ene-3β,17β-diol as the metabolite and [4-$^{14}$C]-3β-hydroxyandrost-5-en-17-one gave androst-5-ene-3β,17β-diol.

YASUE et al. (1974) maintained strips of quadriceps muscle from immature (22–30 days) male rats in culture medium for 24 h; the culture medium was removed and Eagle's minimum medium containing 4 mM glutamine, 0.002 ml ethyl alcohol, and

Table 10. The metabolism of $C_{19}$-steroids in vitro by muscle preparations

| Muscle | Substrates | Tissue prep.[a] | Coenz. | pH | Metabolites | References |
|---|---|---|---|---|---|---|
| 1. Gastrocnemius (Dog, rat, guinea pig) | 17β-Hydroxyandrost-4-en-3-one | H | NAD; NADP | 9.0 | Androst-4-ene-3,17-dione | Kinson and Kochakian (1964) |
| 2. Temporal (guinea pig) | 17β-Hydroxyandrost-4-en-3-one | H | NAD; NADP | 9.0 | Androst-4-ene-3,17-dione | Kinson and Kochakian (1964) |
| 3. Gluteal (dog) | 17β-Hydroxyandrost-4-ene-3-one[b] | M | NA | 7.4 | Androst-4-ene-3,17-dione<br>5α-Androstane-3,17-dione | Shimazaki et al. (1965) |
| 4. Hind Leg (rabbit) | Androst-4-ene-3,17-dione | H | — | 9.0 | 5α-Androstane-3,17-dione<br>17β-Hydroxyandrost-4-en-3-one | Kochakian and Stidworthy (1954) |
| 5. Hind Leg (rabbit) | Androst-4-ene-3,17-dione[b] | St | — | 7.4 | 5α-Androstane-3,17-dione<br>17β-Hydroxyandrost-4-en-3-one | Thomas and Dorfman (1964a) |
| 6. Hind Leg (rabbit) | | Cyt. | NADPH | 7.5 | 17β-Hydroxyandrost-4-en-3-one<br>3β-Hydroxyandrost-4-en-3-one | Thomas and Dorfman (1964b) |
| 7. Hind Leg (rabbit) | 3β-Hydroxyandrost-5-en-17-one[b] | Cyt | NADPH | 7.5 | Androst-5-ene-3β,17β-diol | Thomas and Dorfman (1964b) |
| 8. Hind Leg (rabbit) | 17β-Hydroxyandrost-4-en-3-one[b] | Cyt | NADH | 7.5 | Androst-4-ene-3β,17β-diol | Thomas and Dorfman (1964b) |

Table 10 (continued)

| Muscle | Substrates | Tissue prep.[a] | Coenz. | pH | Metabolites | References |
|---|---|---|---|---|---|---|
| 9. Hind Leg (rat) 22—30 days | 17β-Hydroxyandrost-4-en-3-one[b] | Culture | | | 5α-Androstane-3α,17β-diol<br>17β-Hydroxy-5α-androstan-3-one<br>5α-Androstane-3β,17β-diol<br>3α-Hydroxy-5α-androstan-17-one<br>5α-Androstane-3,17-dione<br>Androst-4-ene-3,17-dione | YASUE et al. (1974) |
| 10. Heart (ventricle) (rabbit) | Androst-4-ene-3,17-dione | H | — | 7.4 | 5α-Androstane-3,17-dione | KOCHAKIAN and STIDWORTHY (1954) |
| 11. Heart (guinea pig) | 17β-Hydroxyandrost-4-en-3-one | H | NA, NAD | 9.0 | Androst-4-ene-3,17-dione | KOCHAKIAN et al. (1957c) |
| 12. Heart (guinea pig) | 17β-Hydroxy-5β-androstan-3-one | H | NA, NAD | 9.0 | 5β-Androstane-3,17-dione<br>3α-Hydroxy-5β-androstan-17-one | KOCHAKIAN et al. (1957c) |
| 13. Heart (guinea pig) | Androst-5-ene-3β,17β-diol | N | NA, NAD | 9.0 | 3β-Hydroxyandrost-5-en-17-one | KOCHAKIAN et al. (1957c) |
| 14. Heart (guinea pig) | 5α-Androstane-3α,17β-diol | H | NA, NAD | 9.0 | 3α-Hydroxy-5α-androstan-17-one | KOCHAKIAN et al. (1957c) |
| 15. Urinary bladder | 17β-Hydroxyandrost-4-en-3-one | P | | | Androst-4-ene-3,17-dione<br>5α-Androstane-3,17-dione | MORFIN et al. (1970) |

[a] M = minces, H = homogenate, St = strips, Cyt = cytosol, P = perfusion.
[b] [$4^{14}C$] = labeled.

[$^3$H]-testosterone (1 nmole/μCi) were added and incubation continued for 2–8 h. The metabolites were separated by paper chromatography and elution chromatography on Amberlite IRC-50. 5α-Androstane-3α,17β-diol was the main metabolite at a conversion of 0.4–1.0%. Several other 5α-metabolites and androst-4-ene-3,17-dione at less than 0.4% conversion also were detected.

In contrast to the above studies, adult male rat skeletal and also "levator ani" muscles seem to possess a trace or no 5α-reductase activity (GLOYNA and WILSON, 1969; BECKER et al., 1972; KRIEG et al., 1974; MASSA and MARTINI, 1974). On the other hand, slices of skeletal muscle of rat and rabbit (newborn or embryo) incubated in Krebs-Ringer-bicarbonate-glucose converted testosterone to 17β-hydroxy-5α-androstan-3-one (WILSONand LASNITZKI, 1971). The rat muscles were more effective than the rabbit muscles. The incubation of fragments of skeletal or "levator ani" muscle with [4-$^{14}$C]-17β-hydroxy-5α-androstan-3-one in Krebs-Ringer-bicarbonate buffer yielded appreciable amounts of androstane-3α,17β-diol indicating the presence of 3α-oxido-reductase activity (MASSA and MARTINI, 1974).

Human striated muscle in contrast to human liver exhibited practically no ability to metabolize testosterone in vitro (ENGELHARDT, 1974). BECKER et al. (1972) also noted that no 5α-steroids were present in muscle 30 min after i.v. injection of [1,2-$^3$H]-testosterone into men but 5α-androstane-3α,17β-diol and 3α-hydroxy-5α-androstan-17-one were readily detected after injection of [1,2-$^3$H]-17β-hydroxy-5α-androstan-3-one.

## E. Heart (Table 10)

KOCHAKIAN and STIDWORTHY (1954) detected a trace of testosterone and androsterone and a small amount (0.1%) of 5α-androstane-3,17-dione after incubation of 100 mg of androst-4-ene-3,17-dione with a homogenate of 12 g of rabbit heart (ventricle) at pH 7.4. In a later study under more optimal conditions, KOCHAKIAN et al. (1957c) obtained a 6% conversion of testosterone to androst-4-ene-3,17-dione and smaller conversions of several other 17β-hydroxysteroids to their corresponding 17-oxosteroids by homogenate of male guinea pig heart ventricle.

WILSON and LASNITZKI (1971) reported the conversion of [1,2-$^3$H]-testosterone to trace amounts of 17β-hydroxy-5α-androstan-3-one on incubation in Krebs-Ringer-phosphate-glucose solution by slices of heart from rabbit newborn and embryo. A somewhat greater conversion was obtained with rat heart.

## F. Urinary Bladder (Table 10)

MORFIN et al. (1970) infused [4-$^{14}$C]-testosterone in 0.9% NaCl through the bladder and prostate of dogs for 2–10 min and analyzed the organs for metabolites. The major metabolite in the bladder was androst-4-ene-3,17-dione, and 5α-androstane-3,17-dione was found in smaller amounts. In contrast, the prostate yielded 17β-hydroxy-5α-androstan-3-one and 5α-androstane-3α,17β-diol as the major metabolites. Metabolism by the blood was ruled out as a contributing factor.

Table 11. The metabolism of $C_{19}$-steroids in vitro by intestine, thymus, and bone

| Tissue | Substrate | Tissue prep.[a] | Coenz. | pH | Metabolites | References |
|---|---|---|---|---|---|---|
| 1. Intestine (dog) | 17β-Hydroxyandrost-4-en-3-one | H | NADP, G-6-P | 7.4 | Androst-4-ene-3,17-dione<br>5α-Androstane-3,17-dione<br>3(?)-Hydroxy-5α-androstan-17-one<br>5α-Androstane-3β,17β-diol | HARRI et al. (1970) |
| Intestine (dog) | 17β-Hydroxyandrost-4-en-3-one | Ms | NADP, G-6-P | 7.4 | Androst-4-ene-3,17-dione<br>5α-Androstane-3,17-dione<br>17β-Hydroxy-5α-androstan-3-one<br>5α-Androstane-3α,17β-diol | HARRI et al. (1970) |
| Intestine (dog) | 17β-Hydroxyandrost-4-en-3-one | Cyt | NADP, G-6-P | 7.4 | Androst-4-ene-3,17-dione<br>Androst-4-ene-3β,17β-diol<br>5β-Androstane-3α,17β-diol | HARRI et al. (1970) |
| 2. Gastric mucosa (human) | 17β-Hydroxyandrost-4-en-3-one[b] | S | — | 7.4 | Androst-4-ene-3,17-dione | ENGELHARDT et al. (1974) |
| 3. Thymus (human, M, F) | Androst-4-ene-3,17-dione[b] | H | NA, NAD | 7.0 | 17β-Hydroxyandrost-4-en-3-one | JUNGMAN et al. (1967) |
| 4. Femur and tibia (chick) | 17β-Hydroxyandrost-4-en-3-one[b] | H | NADPH | 7.4 | 5α-Androstane-3α,17β-diol<br>17β-Hydroxy-5α-androstan-3-one | MUROTA and TAMAOKI (1967) |
| 5. Femur[d] (rat) | 17β-Hydroxyandrost-4-en-3-one[c] | M | — | 7.4 | Androst-4-ene-3,17-dione<br>5α-Androstane-3α,17β-diol<br>17β-Hydroxy-5α-androstan-3-one | JAFFE (1969) |
| 6. Mandibler (rat)[1] | 17β-Hydroxyandrost-4-en-3-one[c] | H | NADPH | 7.4 | Androst-4-ene-3,17-dione<br>3α-Hydroxy-5α-androstan-17-one<br>5α-Androstane-3,17-dione<br>5β-Androstane-3,17-dione<br>5α-Androstane-3α,17β-diol<br>5β-Androstane-3α,17β-diol<br>17β-Hydroxy-5α-androstan-3-one | VITTEK et al. (1974) |

[a] H = homogenates, Ms = microsomes, Cyt = cytosol, S = slices.
[b] [$4^{14}C$] = labeled.
[c] [7α-$^{3}H$) = labeled.
[d] growing rat.

## G. Intestine (Table 11)

The intestinal mucosa of the dog actively metabolized testosterone. Incubation of [4-$^{14}$C]-testosterone with homogenates of the mucosa at pH 7.4 resulted in 17β-hydroxy dehydrogenation, 5α-reduction, and 3β-reduction plus the formation of polar and nonpolar metabolites (HARRI et al., 1970). The same metabolites were produced by the microsomes except that 17β-hydroxy-5α-androstan-3-one was found but no 3α,(orβ)-hydroxy-5α-androstan-17-one. The cytosol produced only a slight amount of androst-4-ene-3,17-dione; 5β-androstane-3β,17β-diol and androst-4-ene-3β,17β-diol were the two main identifiable metabolites. Both subcellular fractions produced unidentified highly polar and to a lesser extent very nonpolar metabolites. The ribosomes revealed no steroid-metabolizing activity and the mitochondria exhibited activity similar to but much lower than the microsomal fraction and was attributed to microsomal contamination.

The presence of 3β- and 17β-hydroxysteroid dehydrogenase activity was demonstrated by histochemical methods in the mucosa of the small intestine of the rat (KSIAZKIEWICZ-SZAPIRO, 1970). Castration decreased and testosterone increased the activity of the enzymes to or above normal. The enzyme activities were localized in the cytoplasm with the greatest intensity in the apical part of the villus.

Slices of human gastric mucosa (as well as of lung and fat) actively (27–70%) transformed testosterone to androstenedione (ENGELHARDT et al., 1974). Other metabolites were not detected.

## H. Thymus (Table 11)

Incubation of [4-$^{14}$C]-androst-4-ene-3,17-dione with homogenates of thymus from two boys, 3 and 8$^1/_2$ years old, and a 13-year-old girl resulted in the formation of 3–12% of testosterone and several unidentified metabolites (40–54%) (JUNGMANN et al., 1967). The homogenate also was able to interconvert estrone and estradiol-17β.

## I. Bone (Table 11)

Testosterone is metabolized not only to a number of other $C_{19}$ steroids but also to three estrogens and unidentifiable metabolites by the crushed and homogenized mandible of adult castrated rats (VITTEK et al., 1974). The inclusion of a NADPH generating system in the incubation medium (pH 7.4) enhanced the conversions. 5α-Androstane-3α,17β-diol represented more than 50% of all metabolites. Dehydrogenation of the 17β-hydroxyl was low not only for the $C_{19}$ steroids but also the estrogens. Androst-4-ene-3,17-dione was the only 17-oxosteroid isolated in measurable quantity. Estradiol-17β and estriol were the principal measurable estrogens; estrone was present but not measurable. On the other hand, a mince of the distal femoral epiphyses of the growing rat transformed testosterone to androstenedione as the major metabolite and smaller but equal quantities of 5α-androstane-3α,17β-diol and 17β-hydroxy-5α-androstan-3-one (JAFFE, 1969). The incubation of a homogenate of the epiphyses of the tibia and femur of chick embryos (13 days) with [4-$^{14}$C]-testosterone in a NADPH generating system yielded only 17β-hydroxy-5α-androstan-3-one and 5α-androstane-3α,17β-diol as metabolites (MUROTA and TAMAOKI, 1967).

## J. Neural Tissues (Table 12)

### 1. Rat

The incubation of brain homogenates of adult male and female rats with [4-$^{14}$C]-testosterone at pH 7.4 in Krebs-Ringer buffer containing a NADPH generating system gave an unidentified less polar compound as the major metabolite and a lesser quantity of more polar metabolites (SHOLITON et al., 1966). The less polar metabolite was formed by only the microsomal fraction while the more polar metabolites were formed by the cytosol as well as the microsomal fractions. JAFFE (1969) was able to characterize by paper chromatography the metabolites formed by minces of not only the cerebral cortex but also the pituitary and hypothalamus. The tissue minces from 59-day-old male rats were incubated with [7α-$^{3}$H]-testosterone in Krebs-Ringer bicarbonate buffer, pH 7.4, without added cofactors. The major metabolite was 17β-hydroxy-5α-androstan-3-one and androst-4-ene-3,17-dione occurred in somewhat smaller quantities. The pituitary was more active than either the hypothalamus or cerebral cortex. The same transformations were detected with minces of human fetal pituitary and hypothalamus.

MASSA et al. (1972) compared the ability of slices of the various parts of the brain of the adult male rat with the prostate and seminal vesicles in the metabolism of testosterone. The major metabolite for all the tissues was 17β-hydroxy-5α-androstan-3-one. The greatest transformation was produced by the prostate which was 3 times as active as the anterior pituitary. The cerebral cortex and amygdala showed a trace of activity of the anterior pituitary. 5α-Androstane-3α,17β-diol, androst-4-ene-3,17-dione, and 5α-androstane-3,17-dione were also identified as metabolites produced by the different tissues. Castration enhanced the transformation of testosterone to 17β-hydroxy-5α-androstan-3-one by the anterior pituitary and to a lesser degree and consistency by the hypothalamus and not by the cerebral cortex or amygdala. The injection of 2 mg/day testosterone propionate prevented the increase in transformation but again the results with the hypothalamus were small, delayed, and variable. Indeed, an actual increase was noted after 2 days of treatment. The androgen did not influence the transformations by the cerebral cortex.

Since castration increases and testosterone restores the level of FSH and LH, and numerous steroid hormones are present in blood, MASSA et al. (1972) added different pituitary and steroid hormones to the incubation medium. FSH, but not LH or ACTH, produced a very significant increase in the 5α-reductase activity of the anterior pituitary. When progesterone was added to the incubation mixture, it actively competed for the 5α-reductase enzyme and was able to completely inhibit the transformation of testosterone to 17β-hydroxy-5α-androstan-3-one by the anterior pituitary and also the hypothalamus. 11-Desoxycorticosterone, 11-deoxycortisol, and corticosterone were inhibitory but to a lesser degree. The 5α-reductase activity was age-dependent. The transformation of testosterone to 17β-hydroxy-5α-androstan-3-one by the anterior pituitary and also the hypothalamus, cerebral cortex, and amygdala progressively decreased from 7–21 days of age and then no further or a slight decrease was apparent up to 60 days of age. This same relationship was not observed for the prostate.

Minces of the hypothalamus and cerebral cortex of 7-day-old female rats converted [1,2-$^{3}$H]-androstene-3,17-dione to predominantly testosterone and smaller

Table 12. The metabolism of $C_{19}$-steroids in vitro by preparations of neural tissue

| Species[b] | Substrates | Tissue prep.[a] | Coenz. | pH | Metabolites | References |
|---|---|---|---|---|---|---|
| 1. Rat (M,F) (Pit. Hythl., Cer.) | 17β-Hydroxyandrost-4-en-3-one[c] | M | — | 7.4 | 17β-Hydroxy-5α-androstan-3-one<br>Androst-4-ene-3,17-dione | JAFFE (1969) |
| 2. Rat (M) (Pit. Hythl., Cer. Amyg.) | 17β-Hydroxyandrost-4-en-3-one[b] | S | — | 7.4 | 17β-Hydroxy-5α-androstan-3-one<br>5α-Androstane-3α,17β-diol<br>Androst-4-ene-3,17-dione<br>5α-Androstane-3,17-dione | MASSA et al. (1972) |
| 3. Rat (♂) (Pineal) | 17β-Hydroxyandrost-4-en-3-one | Organ | — | 7.4 | 17β-Hydroxy-5α-androstan-3-one<br>5α-Androstane-3α,17β-diol<br>5α-Androstane-3,17-dione<br>Androst-4-ene-3,17-dione | CARDINALI et al. (1974) |
| 4. Rat (M) (Pit. Hythl.) | Androst-4-ene-3,17-dione[c] | M | — | 7.4 | 17β-Hydroxyandrost-4-en-3-one<br>17β-Hydroxy-5α-androstan-3-one<br>3α-Hydroxy-5α-androstan-17-one<br>5α-Androstane-3α,17β-diol | PEREZ et al. (1975) |
| 5. Rat (♀, 7d) (Hythl., Cer.) | Androst-4-ene-3,17-dione[c] | M | NADPH | 7.4 | 17β-Hydroxyandrost-4-en-3-one<br>5α-Androstane-3α,17β-diol<br>17β-Hydroxy-5α-androstan-3-one<br>Estrone<br>Estradiol-17β | WEISZ and GIBBS (1974) |
| 6. Dog (Pit., Hythyl., limbic), | Androst-4-ene-3,17-dione[b] | M | NAD, NADP, G-6-P | 7.4 | 17β-Hydroxyandrost-4-en-3-one<br>5α-Androstane-3,17-dione<br>17β-Hydroxy-5α-androstan-3-one | PEREZ-PALACIOS et al. (1970) |
| | 17β-Hydroxyandrost-4-en-3-one[c] | M | NAD, NADP, G-6-P | 7.4 | 17β-Hydroxy-5α-androstan-3-one<br>Androst-4-ene-3,17-dione<br>5α-Androstane-3,17-dione | PEREZ-PALACIOS et al. (1970) |

Table 12 (continued)

| Species[b] | Substrates | Tissue prep.[a] | Coenz. | pH | Metabolites | References |
|---|---|---|---|---|---|---|
| 7. Human fetus | 3β-Hydroxyandrost-5-en-17-one[c] | M | glucose | 7.3 | Androst-5-ene-3β,16β,17β-triol<br>Androst-5-ene-3β,17β-diol<br>3β,17β-Dihydroxyandrost-5-en-16-one<br>16β,17β-Dihydroxyandrost-5-en-17-one<br>Estradiol-17β<br>Estrone | KNAPSTEIN et al. (1968) |
| 8. Human fetus (Pit., Hythl.) | 17β-Hydroxyandrost-4-en-3-one[c] | M | — | 7.4 | 17β-Hydroxy-5α-androstan-3-one<br>Androst-4-ene-3,17-dione | JAFFE (1969) |
| 9. Guinea Pig (♂) (Cer., Hythl., Pit.) | 17β-Hydroxy-5α-androstan-3-one[c] | M | — | ? | 5α-Androstane-3,17-dione<br>5α-Androstane-3α,17β-diol<br>5α-Androstane-3β,17β-diol | SHOLL et al. (1975) |
| | 17β-Hydroxyandrost-4-en-3-one[c] | M | — | ? | Androst-4-ene-3,17-dione<br>5(α,β)-Androstane-3,17-dione<br>17β-Hydroxy-5α-androstan-3-one | SHOLL et al. (1975) |

[a] M = mince, S = slices, H = homogenate.
[b] Pit. = pituitary, Hythl. = hypothalamus, Cer. = cerebrum, Amyg. = Amygdala.
[c] [7α-$^{3}$H]-labeled.
[d] [4-$^{14}$C]-labeled.

but similar quantities of 5α-androstane-3α,17β-diol and 17β-hydroxy-5α-androstan-3-one on incubation in Hanks medium containing NADP and glucose-6-phosphate. In addition, estrone equal to the latter 2 metabolites and a much smaller quantity of estradiol-17β were formed by the hypothalamus but only negligible amounts by the cerebrum. Testosterone was converted to the estrogens by several parts of the brain, but $C_{19}$ metabolites were not detected (WEISZ and GIBBS, 1974). A previous report (NAFTOLIN et al., 1972) had indicated the transformation of [7α-$^{3}$H]-androst-4-ene-3,17-dione to small amounts of estrone (0.014% males, 0.0072% females) on incubation with the anterior hypothalamus of adult rats.

PEREZ et al. (1975) incubated [2,2-$^{3}$H]-androst-4-ene-3,17-dione with minces of pituitary, hypothalamus, and hippocampus from mature male rats. 5α-Androstane-3,17-dione was the major metabolite and smaller quantities of testosterone, androsterone, and 17β-hydroxy-5α-androstan-3-one were identified. The pituitary showed the highest conversions followed in decreasing order by the hippocampus and hypothalamus. Castration 3 weeks prior to the experiments resulted in a significant decrease in 17β-hydroxyl formation by the pituitary with a concomitant increase in the 5α-steroids. No significant changes were observed after castration in the hypothalamus and the hippocampus.

Since the pineal gland is involved in the regulation of gonadal function, CARDINALI et al. (1974) studied the uptake and metabolism of [$^{3}$-H]-testosterone by rat whole pineal gland in vitro by incubation in Krebs-Ringer bicarbonate buffer, pH 7.4. 17β-Hydroxy-5α-androstan-3-one and 5α-androstane-3α,17β-diol were the main metabolites with small amounts of androst-4-ene-3,17-dione, and 5α-androstane-3,17-dione. Incubation with cytosol and nuclei yielded only the two main metabolites.

## 2. **Monkey** (Table 12)

The female rhesus monkey brain is able to metabolize dehydroepiandrosterone (3β-hydroxyandrost-5-en-17-one) to not only several $C_{19}$ steroids but also to estrogens (KNAPSTEIN et al., 1968). The brains of two females of 5–6 years of age were perfused in vivo with [4-$^{14}$C]-3β-hydroxyandrost-5-en-17-one and the $^{3}$H-labeled sulfate for 1 h or the [$^{3}$H]-steroid for 13 min. The brain and plasma from the jugular vein were extracted and the metabolites separated by paper chromatography. Approximately 16% of the steroids were metabolized to form androst-5-ene-3β,17β-diol as the main metabolite and smaller amounts of 3β,16-dihydroxyandrost-5-en-17-one, 3β,17β-dihydroxyandrost-5-en-16-one, and androst-5-ene-3β,16β,17β-triol. Trace amounts of estradiol-17β and even smaller amounts of estrone were found in the brain cortex. Similar trace conversions of androstenedione to estrone and testosterone to estradiol were obtained by incubation of the [$^{14}$C]- and [$^{3}$H]-labeled $C_{19}$ steroids with homogenates of the hypothalamus and limbic system of an immature male and a mature male and female rhesus monkey in Krebs-Ringer phosphate buffer containing NAD, ATP, and a NADPH generating system (FLORES et al., 1973). The hypothalamus was more active than the limbic system. Apparently no search was made for $C_{19}$ steroid metabolites.

### 3. **Human** (Table 12)

The same metabolites as for the monkey were obtained on the incubation of [4-$^{14}$C]-3β-hydroxyandrost-5-en-17-one with the brain mince of a 12-week-old human fetus (KNAPSTEIN et al., 1968). [7α-$^{3}$H]-Testosterone was converted to androst-4-ene-3,17-dione and 17β-hydroxy-5α-androstan-3-one by minces of human fetal hypothalamus and pituitary on incubation in Krebs-Ringer bicarbonate buffer (JAFFE et al., 1969).

### 4. **Dog** (Table 12)

The brain tissues of the dog exhibit the ability to effect transformations similar to the tissues of the rat. PEREZ-PALACIOS et al. (1970) incubated minces of pituitary, hypothalamus, and hippocampus (limbic area) with [7α-$^{3}$H]-testosterone or [4-$^{14}$C]-androst-4-ene-3,17-dione in Krebs-Ringer-bicarbonate-glucose, pH 7.4, fortified with NAD and NADPH generating systems. The metabolites formed from androstenedione were: testosterone > 5α-androstane-3,17-dione > 17β-hydroxy-5α-androstan-3-one. The greatest conversion, as in the rat, was by the pituitary. When testosterone was the substrate, androstenedione and the same two other metabolites were obtained. The prevalence of formation of testosterone by the dog in contrast to the rat tissues which produced predominantly 17β-hydroxy-5α-androstan-3-one might suggest a species difference, but incubation conditions were different and must be carefully considered prior to drawing any quantitative or even qualitative differences.

### 5. **Guinea Pig** (Table 12)

The neural tissues of the adult castrated male guinea pig also exhibited differential uptake and metabolism of $C_{19}$ steroids (SHOLL et al., 1975). Minces of the tissues were incubated with either [1,2-$^{3}$H]-testosterone or [1,2-$^{3}$H]-17β-hydroxy-5α-androstan-3-one, and the metabolites were isolated by thin-layer chromatography. The incubation of 17β-hydroxy-5α-androstan-3-one yielded androstane-3,17-dione as the major metabolite and its formation was greater in the cerebral cortex than the pituitary or the hypothalamus. The reverse conversion was noted for the diols. The greatest conversion of testosterone was to androst-4-ene-3,17-dione; the highest occurred on incubation with the cerebral cortex.

## K. **Submaxillary Gland** (Table 13)

WEINER et al. (1970) perfused [4-$^{14}$C]-testosterone through the submaxillary gland of anesthetized 5-year-old male dogs for 10 min and 10 min later extracted the glands and separated the metabolites by thin-layer chromatography. Androst-4-ene-3,17-dione represented nearly 80% of the metabolites. The other 20% contained partially identified 5α- and/or 5β-reduced steroids.

The rat submaxillary gland also has an active 17β-hydroxysteroid oxidoreductase system. BALDI and CHARREAU (1972) incubated minces of submaxillary gland from 3-month-old rats with [4-$^{14}$C]-testosterone in Krebs-Ringer-bicarbonate buffer and obtained androst-4-ene-3,17-dione as the main (80%) metabolite and

Table 13. The metabolism of $C^{19}$-steroids in vitro by the submaxillary gland

| Species | Sex | Substrate | Tissue prep.[a] | Coenz. | pH | Metabolites | References |
|---|---|---|---|---|---|---|---|
| Dog | M | 17β-Hydroxyandrost-4-en-3-one[b] | P | — | — | Androst-4-ene-3,17-dione<br>5α-Androstane-3,17-dione<br>5α/5β-Androsterone/<br>17β-Hydroxy-5α-androstan-3-one<br>5α-Androstane-3α(β),17β-diol | WIENER et al. (1970) |
| Rat | M,F | 17β-Hydroxyandrost-4-en-3-one[b] | M | — | 7.4 | Androst-4-ene-3,17-dione<br>3α-Hydroxy-5α-androstan-17-one<br>5α-Androstane-3,17-dione | BALDI and CHARREAU (1972) |
| Rat | M | 17β-Hydroxyandrost-4-en-3-one[b] | M | — | 7.4 | 5α-Androstane-3α,17β-diol<br>17β-Hydroxy-5α-androstan-3-one | MASSA and MARTINI (1974) |
| Mouse | M | Androst-4-ene-3,17-dione[c] | H | NADPH | 7.4 | 3α-Hydroxy-5α-androstan-17-one<br>17β-Hydroxy-5α-androstan-3-one<br>17β-Hydroxyandrost-4-en-3-one | COFFEY (1973) |
| Mouse | M | 17β-Hydroxyandrost-4-en-3-one[c] | H | NADPH | 7.4 | 5α-Androstane-3α,17β-diol<br>Androst-4-ene-3,17-dione<br>17β-Hydroxy-5α-androstan-3-one<br>3α-Hydroxy-5α-androstan-17-one | COFFEY (1973) |
| Human | F | Androst-4-ene-3,17-dione[b] | S | NAD(H), NADP(H) | 7.4 | 17β-Hydroxyandrost-4-en-3-one | EL ATTAR (1974a) |

[a] P = perfusion, M = mince, H = homogenate, S = slices.
[b] [$^{14}C$]-labeled.
[c] [$^{3}H$]-labeled.

androsterone and 5α-androstane-3,17-dione in smaller amounts. The formation of the two lesser metabolites were not affected by the sex of the rats or by gonadectomy. Androstenedione production, however, was greater by the submaxillary glands of the female than the male rats and castration increased the transformation to that of the females but ovariectomy was without effect. The addition of NAD and a NADPH generating system to a homogenate of the gland did not change the difference between male and female rats. The 17β-hydroxysteroid dehydrogenase activity gradually decreased with age. It was highest in the fetus, had already decreased in the newborn, and was greatly decreased after 1 year. MASSA and MARTINI (1974) incubated 15-mg specimens of submaxillary glands from adult male rats with [4-$^{14}$C]-testosterone in Krebs-Ringer-bicarbonate solutions and obtained an appreciable amount of 5α-reduced metabolites (approximately one-fourth that produced by the prostate). Androstane-3α,17β-diol represented 36% of the 5α-reduced metabolites indicating 3α-oxidoreductase as well as 5α-reductase activity.

EL ATTAR (1974a) incubated slices of a submandibular gland tumor and of the normal tissue from a 63-year-old woman with [4-$^{14}$C]-androst-4-ene-3,17-dione in Krebs-Ringer-phosphate buffer pH 7.4 fortified with NAD, NADH, and a NADPH generating system. Approximately 4% of the substrate was converted to testosterone by the normal tissue and 12% by the tumor tissue. A similar conversion of estrone to estradiol-17β was obtained.

The development and function of the mouse submaxillary gland is very sensitive to the absence or presence of androgens and also shows a different pattern of metabolism of the steroid hormones. COFFEY (1973) incubated [1,2-$^{3}$H]-androst-4-ene-3,17-dione with homogenate of the submaxillary glands from 6-month-old male mice in phosphate buffer containing a NADPH generating system. The substrate was rapidly metabolized with a 54% conversion in 15 min and 95% in 180 min. The major metabolite was androsterone, 37 and 58% respectively. Much smaller amounts of testosterone (3–4%) and 17β-hydroxy-5α-androstan-3-one (1–2%) were detected. Only a trace amount (0.8%) of 5α-androstane-3α,17β-diol was present at 15 min but 7% after 180 min. Omission of the NADPH generating system resulted in the formation of only trace amounts of the various metabolites even after 180 min of incubation. Testosterone was not metabolized as actively as androstenedione. The major metabolite was 5α-androstane-3α,17β-diol and only traces of androst-4-ene-3,17-dione,17β-hydroxy-5α-androstan-3-one and androsterone were detected. Omission of the NADPH generating system greatly decreased the production of the 5α-androstane-3α,17β-diol and again only traces of the three other metabolites were produced.

The submaxillary gland of the boar is of interest because it contains high concentrations of testosterone and even higher (4 times) concentration of 17β-hydroxy-5α-androstan-3-one, while these steroids were not detected in the female pig (BOOTH, 1972).

## L. Oral Mucosa (Table 14)

The gingiva of the rat and man are reported to metabolize testosterone. VITTEK et al. (1974) incubated [7α-$^{3}$H]-testosterone with minces of gingiva from adult (300 g) male rats at pH 7.4 in a NADPH generating system. Most of the metabolites were 5α- and

Table 14. The metabolism of $C_{19}$-steroids in vitro by the oral mucosa

| Species | Sex | Substrate | Tissue prep.[a] | Coenz. | pH | Metabolites | References |
|---|---|---|---|---|---|---|---|
| Rat | M | 17β-Hydroxyandrost-4-en-3-one[b] | M | NADPH | 7.4 | 5α-Androstane-3α,17β-diol<br>5α-Androstane-3β,17β-diol<br>5β-Androstane-3α,17β-diol<br>17β-Hydroxy-5α-androstan-3-one<br>17β-Hydroxy-5β-androstan-3-one<br>3α-Hydroxy-5α-androstan-17-one<br>5α-Androstane-3,17-dione<br>5β-Androstane-3,17-dione<br>Androst-4-ene-3,17-dione | Vittek et al. (1974) |
| Human | M,F | Androst-4-ene-3,17-dione[b] | S | NAD(H), NADP(H) | 7.4 | 17β-Hydroxyandrost-4-en-3-one<br>3α-Hydroxy-5α-androstan-17-one | El Attar (1974b) |

[a] M = minces, S = slice.
[b] [$^3$H]-labeled.

5β-reduced steroids, while androst-4-ene-3,17-dione and unidentified metabolites were found in lesser amounts. No difference was found after castration for 2 weeks. The injection of 2 mg/100 g medroxyprogesterone acetate every other day for 2 weeks caused a 2-fold increase in the 5α-reductase activity.

EL ATTAR (1974b) compared the ability of the normal with the inflamed gingiva in humans to metabolize [1,2-$^3$H]-androst-4-ene-3,17-dione. The tissues from males and females were sliced on a cryostat and incubated with [1,2-$^3$H]-androst-4-ene-3,17-dione in Krebs-Ringer-phosphate solution at pH 7.4 containing NAD(H), NADP(H), and antibiotics. The inflamed gingiva produced twice as much testosterone as the normal gingiva and androsterone, the second metabolite, was detected only in 4 out of 12 incubations with inflamed gingiva, and in each of 12 incubations with normal gingiva.

## III. Subcellular Distribution of $C_{19}$-Steroid Metabolizing Enzyme Activities in Tissues

In several of the studies subcellular fractions of the tissues were used to study the metabolism of the various steroids. More comprehensive studies have been made by several investigators.

### A. 17β-Hydroxy-$C_{19}$-Steroid Oxidoreductases

ENDAHL et al. (1960) separated the NAD from the NADP-linked 17β-hydroxy-$C_{19}$-steroid dehydrogenase activity of the guinea pig liver by centrifugation. The former was located in the microsomes and the latter in the cytosol. They established the optimum requirements of the respective activities and formulated a quantitative assay system by the spectrophotometric determination of the quantity of formed reduced coenzyme. AOSHIMA and KOCHAKIAN (1963) utilized this assay system to determine the NAD- and NADP-linked activities and their subcellular distribution in the liver and kidney of adult male guinea pig, rabbit, mouse, rat, hamster, and dog. The enzyme activity varied markedly among the species and tissues. The NAD-17β-hydroxy-$C_{19}$-steroid dehydrogenase activity of the guinea pig and rat liver was primarily in the microsomal fraction but that of the other animals was distributed in varying proportions between the microsomes and cytosol. The mitochondria did not exhibit detectable activity and the nuclei were not purified. The NADP-linked activity was localized in the cytosol of all of the animals. In contrast with the liver[3], both coenzyme-related dehydrogenase activities were found in the cytosol of the kidney of all of the animals. The pH optimum was about 9.6 for both enzymes in both tissues. The microsomal NAD dehydrogenase was precipitated at pH 5.2, by $Mg^{++}$ and by incubation with RNase. The enzymes of the cytosol were precipitated at pH 4.2. Triton-X-100, Duponol C, and desoxycholate readily solubilized the microsomal enzyme. Excess amounts of these reagents, however, inhibited enzyme activity.

[3] The enzymes of dog prostate exhibited still another pattern of enzyme distribution. Both the NAD- and the NADP-linked 17β-hydroxy-$C_{19}$-steroid dehydrogenase activities were localized in the microsomal fraction (HUSSEIN and KOCHAKIAN, 1968a, b).

GHRAF et al. (1974) investigated the presence of sexually differentiated and undifferentiated 17$\beta$-hydroxysteroid dehydrogenase activities in the liver and kidney as well as in the adrenals and gonads of adult rats. The tissues were separated into cytosol, microsomal, and mitochondrial fractions. Several [4-$^{14}$-C]-labeled steroids (estradiol-17$\beta$, testosterone, 17$\beta$-hydroxy-5$\alpha$-androstan-3-one, 17$\beta$-hydroxy-5$\beta$-androstan-3-one, and androst-5-ene-3$\beta$,17$\beta$-diol) were compared as substrates. The incubations were in sodium pyrophosphate buffer, pH 9.5, with either NAD or NADP as coenzyme. The 17-oxo metabolites were isolated by paper chromatography and assayed by determination of the radioactivity. 17$\beta$-Hydroxy-5$\beta$-androstan-3-one was also reduced at the 3 position. The 17$\beta$-hydroxy-$C_{19}$-steroid dehydrogenase activity of the liver was primarily in the microsomal fractions and was 3–4 times more active with NAD than NADP as the coenzyme. 17$\beta$-Hydroxy-5$\alpha$-androstan-3-one was the optimal substrate and no sex difference was noted. The enzyme activity in the cytosol was less than in the microsomes, used NAD and NADP equally well, and the activity was 5-fold greater in the male than the female liver. The enzyme activity of the kidney was greater in the cytosol than the microsomal fraction, showed a slight sex differentiation favoring the female, and used both coenzymes equally well.

## B. 3-Hydroxy-$C_{19}$-Steroid Oxidoreductases

AOSHIMA et al. (1964) extended their spectrophotometric type studies to the 3$\alpha$-hydroxy- and the $\Delta^5$-3$\beta$-hydroxy-$C_{19}$-steroid dehydrogenases. The respective substrates were androsterone and dehydroepiandrosterone. NAD- and NADP-linked enzymes were present in the liver and kidney of all of the animals but the activities were much lower than the 17$\beta$-hydroxy-$C_{19}$-steroid dehydrogenase activities. The NAD-3$\alpha$-dehydrogenase activity of the dog and mouse liver was localized in the microsomes but that of the other animals was distributed almost equally between the microsomes and cytosol. On the other hand, the NAD-$\Delta^5$-3$\beta$- and NADP-3$\alpha$-dehydrogenase activities were localized in the cytosol of all of the animals except the NAD-$\Delta^5$-3$\beta$ activity of the mouse was in the microsomes and the NADP-3$\alpha$-activity of the guinea pig was distributed equally between the two subcellular fractions. NADP-$\Delta^5$-3$\beta$-activity of the rabbit liver was detected only in the cytosol. The presence and distribution of the $\Delta^5$-3$\beta$-dehydrogenase activity of the liver of the various animals was further substantiated by isolation and quantitation of the metabolite, androstenedione. A stoichiometric relation was found between the amounts of metabolite and reduced coenzyme.

The kidney exhibited a very variable distribution of the 3$\alpha$- and $\Delta^5$-3$\beta$-$C_{19}$-steroid dehydrogenase activities (AOSHIMA et al., 1964). The NAD dehydrogenases of the mouse, rat, and hamster kidney required both the microsomes and the cytosol to exhibit activity. A pH optimum of about 9.6 was shown by the kidney and liver except that the rat liver postnuclear fraction gave a pH optimum at 8.5 for both NAD-linked enzymes with a formation of more reduced coenzymes than metabolites. The reaction, however, was stoichiometric if the incubation medium was at pH 9.6. The interfering enzyme was in the mitochondria; the microsomes and cytosol gave optimum enzyme activity at pH 9.6. The microsomal enzymes were solubilized optimally at 0.15% Triton X-100 or 0.2% desoxycholate. All of the enzymes rapidly lost activity on storage at 4° C in contrast to the excellent stability of the 17β-hydroxydehydrogenase. The loss in activity was partly prevented by EDTA at an optimum concentration of 0.5–1.0 mM. The microsomal enzymes were precipitated by $Mg^{++}$ and also on incubation with RNase. The failure to detect enzyme activity in certain instances may be due to the assay method. The use of isotopically labeled substrates and isolation of metabolites is more sensitive and could reveal low enzyme activities.

REPKE and SAMUELS (1964) also found by spectrophotometric studies that 3α-hydroxysteroid dehydrogenase activity was present in the cytosol and microsomal fraction of the liver of adult male rats with a pH optimum of 9.6 for NADP and 8.6 for NAD. The enzyme activity was greater with NADP than NAD. The cytosol also contained NAD-dependent 3β-hydroxysteroid dehydrogenase activity with a pH optimum at 10.2. HOFF and SCHRIEFERS (1973) in a comprehensive study compared the 3α- and 3β-hydroxysteroid oxidoreductase activity of the liver of male and female rats (110–125 days of age) and the subcellular distribution of these enzyme activities. [4-$^{14}$C]-labeled 5α-androstane-3,17-dione, 5β-androstane-3,17-dione, or 17β-hydroxy-5α-androstan-3-one were incubated with the tissue fractions in Tris buffer, pH 7.4, containing either NADH or NADPH as coenzymes. The metabolites were separated by paper chromatography and assayed by the determination of the radioactivity. The 3α-hydroxysteroid dehydrogenase activity was greater than the 3β-activity in the liver of the male and female rats and both enzymes were distributed between the microsomes and cytosol. The total activity and relative distribution was affected not only by the sex of the animals but also the substrate and the coenzyme indicating sexual differentiation in some instances. The cytosol contained two 3α- and two 3β-hydroxysteroid dehydrogenase activities; one of each was sexually differentiated (3α-activity, female > male; 3β-activity, male > female). The sexual differences were more pronounced in the microsomal fraction. Three 3α- and two 3β-hydroxysteroid specific activities were indicated and only one 3α was not sexually differentiated. In all instances the enzyme activities were higher in the male than the female rats.

LAX and SCHRIEFERS (1974) extended their comparative studies to the $\Delta^4$-3β-hydroxysteroid dehydrogenase activity of rat liver. They established an optimal pH 9.6 (pyrophosphate buffer) with 18 mM NAD or NADP, androst-4-ene-3β,17β-diol as substrate and specific limits of tissue and an incubation period of less than 20 min. The formed $\Delta^4$-3-oxosteroids were determined by the color developed on reaction with isonicotinic acid hydrazide. No reduction of steroid ring A occurred under the prescribed conditions. Enzyme activity was found in both the microsomal fraction and the cytosol; the activity was higher with NAD than NADP. The microsomal activity was higher in the male than the female rats and the difference was greater with NADP (5.8:1) than NAD (1.9:1). No sex difference was detected in the cytosol. The authors postulated the existence of two NAD-dependent enzyme activities and possibly a separate NADP-dependent microsomal enzyme on the basis of the large male: female activity ratios.

The nuclei in the above studies were not assayed for 3-oxidoreductase activities. SIEBERT et al. (1966) incubated nuclei from rat liver at pH 8.6 with [$^{14}$C]-3β-hydroxyandrost-4-en-17-one and obtained androst-4-ene-3,17-dione as the metabolite. NAD but not NADP was the required coenzyme. NAD-17β-hydroxy-$C_{18}$-steroid dehydrogenase activity also was present. The purity of the nuclei was not established.

ARIMASA and KOCHAKIAN (1972) fractionated the liver of adult female rats into the subcellular components and determined the completeness of the separations by assaying the activity of characteristic enzymes. The incubations were carried out with [4-$^{14}$C]-testosterone at pH 6.7 optimum for reductase activity. Only a questionable trace of reductase activity was found in the nuclei and mitochondria. The majority of the enzyme activity was distributed between the microsomes and cytosol. NADPH was much more effective than NADH and 3α-activity was greater than 3β-activity. In the NADH and NADPH systems the 3α-activity (5α-androstane-3α,17β-diol and androsterone formation) and the 3β-NADPH system were much greater in the microsomes than the cytosol while the 3β-NADH-activity (5α-androstane-3β,17β-diol) gave the reverse.

ARIMASA and KOCHAKIAN (1973) conducted identical studies to the above with mouse kidney and obtained 3α-reductase activity in only the cytosol on incubation of the various subcellular components with [4-$^{14}$C]-androst-4-ene-3,17-dione. NADH was more effective than NADPH. The formation of 5α-androstane-3α,17β-diol represented only about 2% of the total metabolites. No 3β-reductase activity was detected.

## C. 5-Reductases ($\Delta^4$-Hydrogenases)

### 1. Liver

Reductases have been reported to be present in the nuclei, microsomes, and cytosol of rat liver. The enzymes appear to be not only different in the separate subcellular components but also have suggestive heterogeneity. The original studies were made with the adrenal cortical steroids and subsequent studies have indicated that the same or closely related enzymes act on the $C_{19}$-steroids. FORCHIELLI and DORFMAN (1956) reported that incubation of the cytosol (75000 *g* supernatant fraction) of rat liver homogenates prepared and incubated in Krebs-Ringer-bicarbonate (pH 7.35) with 11-deoxycortisol yielded 3β,17α,21-trihydroxy-5β-pregnan-20-one and 3α-hydroxy-5β-androstan-17-one while the microsomal fraction produced 3α,17α,21-trihydroxy-5α-pregnan-3-one and 5α-androstane-3,17-dione. TOMKINS (1957) reported the presence of 5β-reductase activity in the postmitochondrial fraction of male rat liver which acted on a variety of steroids. In a later report, MCGUIRE and TOMKINS (1959b) presented evidence for multiple and specific 5α-reductase activities in the rat liver microsomes. The reductase activities for not only the $C_{21}$- but also the $C_{19}$-steroids were greatly enhanced by the addition of NADPH to the incubation medium (MCGUIRE and TOMKINS, 1959a, 1960). The 5α-reductase had an absolute requirement for NADPH. The slight enhancing effect obtained in a NADH generating system was attributed to the generation of NADPH from endogenous NADP (MCGUIRE et al., 1960). The enzymes had a high substrate specificity and catalyzed an apparently irreversible reaction. The reduction of a given steroid was competitively inhibited by a less substituted steroid, e.g., cortisone by androst-4-ene-3,17-dione.

ROY (1971) fractionated the liver from adult female rats and elected to determine the quantity of total 5α-steroid metabolites as an assay of 5α-reductase activity. The degree of purity of the subcellular components was not established. The mitochondria and microsomes each contained 48% of the 5α-reductase activity and the nuclei only 3%. In contrast, the prostate nuclei contained 41%, the mitochondria 23%, and the microsomes 35%. No activity was found in the cytosol of either tissue. In confirmation of MCGUIRE et al. (1960), 5α-reductase activity demonstrated an absolute requirement for NADPH; no 5α-reductase activity was detected when NADH was added as the coenzyme. NADP, a product of the reaction, inhibited the reaction; about 60% inhibition occurred when NADP and NADPH were both present at the same (0.1 mM) concentration in the incubation mixture. The conversion of 17β-hydroxy-5α-androstan-3-one to testosterone was not detected. ARIMASA and KOCHAKIAN (1972) also fractionated the liver from adult female rats and incubated the total homogenate, the mixture of the subcellular components and the individual subcellular components separately at pH 6.7 with [4-$^{14}$C]-testosterone and NADPH as coenzyme. The metabolites were isolated by thin-layer and paper chromatography. Approximately 70% of the 5α-metabolites was formed by the microsomal fraction, 14% by the mitochondria, 6% by the cytosol, and 5% by the nuclei. The marker enzyme determination indicated a slight contamination of the mitochondria nuclei, and cytosol with microsomal enzyme activities and could explain the 5α-reductase activity in these subcellular components. When NADH was used as the coenzyme, the formation of 17β-hydroxy-5α-androstan-3-one, androsterone, epiandrosterone, and 5α-androstane-3,17-dione was much less than with NADPH, but that of 5α-androstane-3α,17β-diol and 5α-androstane-3β,17β-diol was only about 50% lower. Only a trace of activity was evident when the liver homogenate was incubated with testosterone and no coenzyme.

GUSTAFSSON and POUSETTE (1974) studied the properties of the 5α-reductase activity of 8-week-old male and female rats. Electron microscopy and the determination of cytochrome P 450 indicated the lack of or only a trace of microsomal contamination in the purified nuclei. The incubations were performed in a NADPH generating system with [1,2-$^{3}$H]-androstene-3,17-dione as substrate. When the incubation period was for 10 min, only 5α-androstane-3,17-dione was produced. An increase of the concentration of the nuclei and extension of the incubation period yielded a series of 5α-reduced steroids with a disappearance of the substrate and 5α-androstane-3,17-dione. The metabolites were identified as 3α-hydroxy-5α-androstan-17-one, 5α-androstane-3β,17β-diol, 3α,7α-dihydroxy-5α-androstan-17-one, 7α,17β-dihydroxy-5α-androstan-3-one, and 5α-androstane-3α,7α,17β-triol. The appearance of the 7α-hydroxylated metabolites indicated the presence of trace amounts of microsomal 5α-reductase in the purified nuclei. In contrast with the female, the nuclei from male rat liver converted androstenedione after 10 min incubation not only to 5α-androstane-3,17-dione but also a number of other metabolites—3β-hydroxy-5α-androstan-17-one, 3β-hydroxyandrost-4-en-17-one, and 17β-hydroxyandrost-4-en-3-one. The 5α-reductase activity was much lower in the nuclei of the male than the female rat liver. The investigators estimated that the contamination with microsomal 5α-reductase (JARASCH and FRANKE, 1974) could not account for more than 13% of the measured 5α-reductase activity. Several kinetic properties of the 5α-reductase activity were studied and compared with those obtained by various investigators for nuclei from prostate, liver, and kidney.

## 2. Kidney

VERHOEVEN and DE MOOR (1971-1972, 1971) prepared nuclei from adult rat kidney and checked their purity by phase-contrast and electron microscopy. The nuclei were incubated with [1,2-$^3$H]-testosterone and a NADPH-generating system. Female nuclei formed only 17β-hydroxy-5α-androstan-3-one while the male nuclei produced in addition 5α-androstane-3α,17β-diol. The formation of 17β-hydroxy-5α-androstan-3-one was linear with time, required NADPH but could not utilize NADH. The addition of several other steroids to the incubation mixture produced various degrees of inhibition. The major part of the 5α-reductase activity was distributed between the mitochondria and the microsomes. The nuclei of the male rat contained 15–18% of the total 5α-reductase activity and was contaminated with 11% of the microsomal enzyme, glucose-6-phosphatase.

In contrast to the rat, mouse kidney (ARIMASA and KOCHAKIAN, 1973) exhibited no 5α-reductase activity in the purified nuclei and trace activity in the microsomes on incubation with [4-$^{14}$C]-androst-4-ene-3,17-dione at pH 6.7. 5α-Androstane-3α,17β-diol was formed in measurable amounts and several other 5α-steroids in trace amounts by the cytosol. NADH was more effective than NADPH as coenzyme but the activity was low in both instances, about 2% of the total metabolism.

## IV. Regulation of the Activities of the $C_{19}$-Steroid Metabolizing Enzymes in Peripheral Tissues

### A. Guinea Pig

KOCHAKIAN and ENDAHL (1960) castrated guinea pigs at 68 days of age. After 65 days, homogenates of the liver and kidney were assayed under optimal conditions by the spectrophotometric method (ENDAHL et al., 1960). Both the NAD- and NADP-linked 17β-hydroxy-$C_{19}$-steroid dehydrogenase activities of the liver were unchanged as a result of castration or daily injections of testosterone propionate. The dehydrogenase activities of the kidney, however, were reduced by castration and restored to normal by the hormone treatment. Increases in the dose of testosterone propionate from 5–10–25 mg/day did not produce any further increase in the activities of the enzymes. The injection of testosterone propionate 36 days after castration for 13–16 days also restored the enzyme activities to normal. The decrease and restoration of enzyme activity proved to be due to the disappearance and reappearance of one of the 17β-hydroxysteroid dehydrogenase isozymes (KOCHAKIAN et al., 1973). The disappearance of the specific isozyme was apparent 36 days but not 3 and 6 days after castration. The testosterone was implanted subcutaneously as two cylindrical pellets of 15 mg each 21 days after castration and the analyses were performed 14 days later.

### B. Rat

Extensive and detailed studies have been conducted on the regulation of the metabolism of the $C_{19}$ steroids by the level of steroid and other hormones.

#### 1. Sex Differences

RUBIN (1957) found that male and female rat liver homogenates reduced the 3-oxo group of 5α-androstane-3,17-dione to the same degree but the proportion of the two metabolites, 3α-hydroxy-5α-androstan-17-one and 3β-hydroxy-5α-androstan-17-

one, differed greatly. The males produced a greater amount of the 3β-steroid (3β:3α=2.3–3.5) and the females a greater amount of the 3α-steroid (3β:3α=0.12–0.16). The difference was apparent in several strains of rats but in varying degrees. Castration produced a decrease in the 3β:3α ratio within 2 weeks and to as low as that of the females after 9 weeks. The injection of 0.5 mg/day testosterone propionate for 14 days restored the ratio to that of normal males. Methyltestosterone, 17-ethyl-19-nortestosterone, and 17β-hydroxy-5α-androstan-3-one also restored the ratio toward normal. The injection of a large dose, 2.5 mg/day, of testosterone propionate greatly increased the ratio to 13.3. Estradiol-17β at 0.1 mg/day for 14 days was without effect on the decreased ratio in the castrated rats. The 3β:3α ratio of the female rats was not altered by ovariectomy or by injection of estradiol-17β and was increased to 2.0 only after injection of a large dose, 2 mg/day, of testosterone propionate for a very long period, 84 days. When androst-4-ene-3,17-dione was the substrate, the same metabolites were obtained with the same sex difference in the 3β:3α ratio. Testosterone also gave the same metabolites on incubation with male liver but gave 3α-hydroxy-5α-androstan-17-one and 17β-hydroxy-5α-androstan-3-one with female liver. 5β-Androstane-3,17-dione in contrast with its 5α isomer yielded only 3α-hydroxy-5β-androstan-17-one with both male and female liver.

Preliminary experiments with 5α-androstane-3,17-dione as substrate were made with liver homogenates of male and female mice, rabbits, hamsters, and guinea pigs (RUBIN, 1957). These studies also showed a tendency for the 3β:3α ratio to be greater in the males than the females except in the guinea pig which exhibited ratios of 4.0 and 9.0 for males and females, respectively. The liver of the American turtle also showed a larger 3β:3α ratio in the female than the male. In order to establish the interconversion of the metabolites and the substrate, 3α-hydroxy-5α-androstan-17-one, and 3β-hydroxy-5α-androstan-17-one were incubated with rat liver homogenates. The 3β-steroid was not metabolized to either 5α-androstane-3,17-dione or 3α-hydroxy-5α-androstan-17-one by liver homogenates of male or female, castrated or castrated rats treated with testosterone propionate. On the other hand, the 3α-hydroxy-5α-androstan-17-one was converted primarily to the 3β-isomer and smaller amounts of 5α-androstane-3,17-dione by male rat liver. The conversion to the 3β-steroid was reduced to about one-third after 9 weeks of castration and restored to normal by daily injections of testosterone propionate during the last 2 weeks of castration. The conversion to 5α-androstane-3,17-dione was not significantly changed. The female rat liver produced a small amount of the diketone but practically no 3β-steroid.

RUBIN and STRECKER (1961) and RUBIN et al. (1963) conducted kinetic enzymatic studies on homogenates and subcellular fractions in an attempt to elucidate the sex difference in 3β-hydroxysteroid dehydrogenase activity.

YATES et al. (1958) in their studies on the role of physiologic factors in the reduction of the $\Delta^4$-oxo group of adrenal cortical steroids noted that testosterone also was reduced more rapidly (5 times) by the homogenate of female than male rat liver. Practically identical results were reported by LEYBOLD and STAUDINGER (1959) and in addition an unidentified more polar (hydroxylated) compound was detected after incubation of not only testosterone but also 5α-androstane-3α,17β-diol. BULLOCK et al. (1971) confirmed the difference between male and female rats and noted that the injection of either testosterone or 17β-hydroxy-5α-androstan-3-one reduced the $\Delta^4$-reductase activity toward that of the male but was without effect on the enhanced enzyme activity of the pseudohermaphroditic rat. Castration increased the enzyme activity to the female level and androgen administration reduced it to the male level.

SCHRIEFERS et al. (1971) extended the studies on the influence of sex and also age on the metabolism of the $C_{19}$ steroids to a simultaneous comparison of the activities of four different enzymes. The 5α-reductase activity was determined by the disappearance of absorption at 241 nm when testosterone was incubated with liver microsomes in a NADPH generating system. Hydroxylase activity was determined by the incubation of [4-$^{14}$C]-testosterone in the above system and separation of the metabolites by paper chromatography with analysis for radioactivity. The 3α- and 3β-hydroxysteroid dehydrogenase activities were determined by incubation of the whole homogenate with [4-$^{14}$C]-androstane-3,17-dione in a NADPH generating system, separation of androsterone and epiandrosterone by paper chromatography, and quantitative determination of radioactivity. All of the enzyme activities showed a sex difference beginning at about 30 days of age, reached their respective plateaus at about 60 days of age where they were then maintained except that the 5α-reductase activity in the male liver showed a gradual decrease after about 90 days of age. The 5α-reductase and 3α-hydroxysteroid dehydrogenase activities were greater in the female while the hydroxylases and 3β-hydroxysteroid dehydrogenases (as well as 20-ketoreductase) activities were greater in the male liver. The enzymes showed variable differences in activities and different courses of development in two strains of rats. HOFF et al. (1973) separated the microsomes and cytosol and delineated the development from 1–20 days of age in the respective enzyme activities.

Castration at 21 days of age, the injection of 1 mg estradiol benzoate at 1 day of age, and the application of 50 mg Testoviron-Depot to the neonatally estrogenized rat at 90 days of age modified the activities of the above enzymes and also of 17β-hydroxy-$C_{19}$-steroid dehydrogenase (SCHRIEFFERS et al., 1973). The neonatally produced androgen deficiency by the injection of the estrogen induced a failure of the sexual differentiation of all of the enzymes except the 3α-hydroxysteroid dehydrogenase of the cytosol. The administration of the testosterone preparation restored the normal sexual differentiation. The androgen had no influence on the distribution of enzyme activities between the cytosol and microsomes and its influence was not confined to either cell compartment. The extent of androgen dependency varied among the different enzymes. Castration was without influence on the cytosol 3α-hydroxysteroid and 17β-hydroxysteroid dehydrogenase activities but inhibited the development of the microsomal 3β-hydroxysteroid dehydrogenase activity.

GHRAF et al. (1974) conducted further studies on the effect of castration and androgens on the level of several steroid metabolizing enzymes in the liver of male rats. The rats were castrated at 22 days of age and the injections were for 15 days beginning at 74 days of age. Testosterone, 17β-hydroxy-5α-androstan-3-one, and 19-nortestosterone were compared at 0.3 mg/100 g/day. In addition, cyproterone acetate, 3 mg/100 g/day, was injected simultaneously with the first two steroids. The enzyme activities were determined on the microsomal fractions. The activities of 5α-reductase (substrate testosterone) and total hydroxylase activity (substrate testosterone) were determined in the presence of a NADPH generating system. The 3α- and 3β-hydroxysteroid dehydrogenase activity was determined with 17β-hydroxy-5α-androstan-3-one as substrate and NAD as coenzyme and $\Delta^4$-3$\beta$-hydroxysteroid dehydrogenase activity was determined with androst-4-ene-3$\beta$,17$\beta$-diol as substrate and NAD and NADP as coenzymes. Castration decreased the 3$\alpha$- and 3$\beta$-hydroxysteroid dehydrogenase activities to the female level and the $\Delta^4$-3$\beta$-hydroxysteroid dehydrogenase activities nearly to the female level. The 5$\alpha$-reductase activity was lower in the male than the female and was slightly increased by castration. The administration of testosterone, nortestosterone, and 17$\beta$-hydroxy-5$\alpha$-androstan-3-one reversed the effects of castration. The effect of nortestosterone was the same as that of testosterone although its effect on the seminal vesicles weight was only 30% of testosterone. Cyproterone acetate did not influence the effect of either testosterone or 17$\beta$-hydroxy-5$\alpha$-androstan-3-one on the liver enzyme activities in spite of the marked (60–70%) inhibition on the weight of the seminal vesicles. The total hydroxylase activity was relatively small but twice as great in the female and the 7$\alpha$-hydroxylase activity was small and the same in both sexes. Castration decreased the activities of the hydroxylases. The administration of testosterone restored the total hydroxylase activity to normal and 17$\beta$-hydroxy-5$\alpha$-androstan-3-one to almost twice normal. Cyproterone acetate produced a further increase and testosterone and 17$\beta$-hydroxy-5$\alpha$-androstan-3-one now stimulated identical increases. 16$\alpha$-Hydroxylase was the only enzyme in which the enhancing effects of the two steroids was completely blocked by cyproterone acetate. 7$\alpha$-Hydroxylase activity showed no dependency on androgens but its activity was almost doubled when cyproterone acetate was administered with the steroids. Nortestosterone was without effect on any of the hydroxylase activities.

GUSTAFSSON and collaborators in a series of studies have further elaborated on the endocrine regulation of steroid metabolizing enzymes in rat liver. EINARSSON et al. (1973) investigated the influence of pre- and postpubertal castration and androgen treatment on the metabolism of steroids by the liver microsomal fraction of male and female rats in an incubation medium containing NADPH or a NADPH generating system and under quantitative enzymatic conditions. The metabolites were separated by thin-layer chromatography and quantitated by the determination of the radioactivity. The enzyme activities were grouped into three categories: (a) the 2$\beta$-hydroxylase (5$\alpha$-androstane-3$\alpha$,17$\beta$-diol as substrate), the 6$\beta$-hydroxylase (androst-4-ene-3,17-dione as substrate), and the 18-hydroxylase (5$\alpha$-androstane-3$\alpha$,17$\beta$-diol substrate) were more active in male than female rats. This sexual difference was abolished by neonatal and postpubertal castration and was stimulated by testosterone propionate treatment. (b) 2$\alpha$-hydroxylase (5$\alpha$-androstane-3$\alpha$,17$\beta$-diol as substrate), 16$\alpha$-hydroxylase (androst-4-ene-3,17-dione as substrate), $\Delta^{16}$ C19 steroid epoxidase (androsta-4,16-dien-3-one as substrate), and 17$\beta$-hydroxysteroid oxidoreductase (androst-4-ene-3,17-dione as substrate) were generally much more active in male than in female rats. Neonatal castration completely and postpubertal castration only partially abolished the difference between males and females[4]. Testosterone propionate treatment increased the activities of these enzymes. (c) 7$\alpha$-Hydroxylase (androst-4-ene-3,17-dione as substrate) was generally slightly more active in female than male rats. Castration or testosterone propionate treatment did not significantly change

[4] THALER-DAO and BREUER (1974) found no change in the 17$\beta$-hydroxy-$C_{18}$-steroid dehydrogenase activity of male rat liver after castration. The enzyme activity of female rat liver, moreover, was less than that of the male, and after ovariectomy increased toward that of the male suggesting that the level of this enzyme activity in the liver was regulated by the estrogens.

the enzyme activity. In order to investigate further the hormonal regulation of the liver microsomal hydroxylases in male and female rats, BERG and GUSTAFSSON (1973) selected 5α-androstane-3α,17β-diol as the substrate because it was acted upon by the above three systems. The microsomes from male rats hydroxylated the substrate at the 2α, 2β, 7α, 7β, and 18 positions while that from the females acted only at the 2α and 7α positions. Castration at 56 days of age decreased the activities of all of the enzymes except the 7α-hydroxylase in the male, and ovariectomy was ineffective on the two enzymes in the female. The decreased enzyme activities were restored by the injection of 100 μg/day for 14 days of testosterone propionate, 17β-hydroxy-5α-androstan-3-one, or 5α-androstane-3α,17β-diol, but these steroids were ineffective in ovariectomized or postpubertal female rats. The 7α-hydroxylase was more active in the female than the male rats. The injection of estradiol benzoate at 100 μg/day for 14 days produced the female pattern of enzyme activity in the castrated rats.

GUSTAFSSON and STENBERG (1974a) explored the neonatal influence from the testes on the sexual differences in androgen responsiveness by comparing male rats castrated at birth and at 14 days of age. Testosterone propionate injections at 100 μg/day were begun at 56 days of age and continued to 69 or 79 days. The 2β- and 7β-hydroxylation of [4-$^{14}$C]-androstane-3α,17β-diol was evident after testosterone propionate treatment only in the rats castrated at 14 days of age. Also 18-hydroxylation of 5α-androstane-3α,17β-diol and 6β-hydroxylation of androst-4-ene-3,17-dione was increased twice as much by androgen treatment of the rats castrated at 14 days as those castrated at birth. On the other hand, the cytosol 5β-reductase activity was increased 3.9 times by testosterone propionate treatment of the neonatally castrated rats and was slightly (20%) decreased in the 14 days castrated rats. GUSTAFSSON and STENBERG (1974) attempted to further define the neonatal imprinting of the enzyme activities by a study of the metabolism of androst-4-ene-3,17-dione and 5α-androstane-3α,17β-diol in the microsomal fraction, and androst-4-ene-3,17-dione in the liver cytosol of male and female rats castrated at various intervals after birth and neonatally castrated rats given a single subcutaneous injection of 0.145 or 1.45 μmoles of testosterone propionate on the day after castration. The authors classified the changes into three categories: (a) A relatively small dose (0.145 μmole) of testosterone propionate imprinted 5α-reductase activity (reduction of androst-4-ene-3,17-dione) and the process was completed by 7 days of age. (b) A relatively high (1.45 μmoles) dose of testosterone propionate was required to imprint 2α-hydroxylase action on 5α-androstane-3α,17β-diol, 3β-hydroxysteroid reductase action on 5α-androstane-3,17-dione and 17α-hydroxysteroid and 5β-reductase actions on androst-4-ene-3,17-dione. The imprinting was complete or almost complete by 7 days of age. (c) The imprint of 16α-hydroxylase action on androst-4-ene-3,17-dione and 7α-hydroxylase action on 5α-androstane-3α,17β-diol required 14–30 days after birth.

GUSTAFSSON (1974) studied the effect of age on the influence of testosterone propionate, 2 mg/kg/day, for 5 days beginning at castration, on the various hydroxylations of 5α-androstane-3α,17β-diol by the liver microsomal fraction of male and female rats. The male and female rats were castrated at birth and at intervals up to 55 days of age. The 7α-hydroxylase activity was not affected by the androgen treatments. The 2α-, 2β-, and 18-hydroxylase activities of both male and female rats showed relatively no response to the steroid treatment up to 19 days of age. From 27–33 days of age the male and female rats responded equally to the androgen injections. From about 40 days of age the female rats showed a progressive decrease in response to the androgen while the male rats retained their responsiveness.

HEINRICHS et al. (1966) observed a marked increase in the 16α-hydroxylation of 3β-hydroxyandrost-5-en-17-one (COLAS, 1962) of male rat liver between 6 and 10 weeks of age and no change in that of female rat liver. Several other steroids also were hydroxylated but the degree of hydroxylation varied with chemical structure. The liver microsomes of adult rhesus monkeys did not exhibit a sex difference.

TABEI and HEINRICHS in a series of subsequent reports have studied the factors regulating the oxidation and reduction of the 16 and 17 position of 3β-hydroxyandrost-5-en-17-one by rat liver microsomes. The $^{14}C$-labeled steroid was incubated with the microsomes in Krebs-Ringer-phosphate buffer, pH 7.4, containing NADPH. The metabolites were separated by thin-layer chromatography and gas liquid chromatography. In addition to the expected products—androst-5-ene-3β,17β-diol, 3β,16α-dihydroxyandrost-5-en-17-one, and androst-5-ene-3β,16α,17β-triol—, 3β,17β-dihydroxyandrost-5-en-16-one was identified. The three 16 oxygenated compounds were formed in higher yield by the microsomes from the male than the female liver of 70-day-old rats while androst-5-ene-3β,17β-diol was produced at the same rate by both sexes (TABEI and HEINRICHS, 1972). The formation of the 16 oxygenated compounds appeared at puberty in the males (but not the females) and increased thereafter (TABEI and HEINRICHS, 1973). The rate of biosynthesis of 3β,17β-dihydroxyandrost-5-en-16-one was greater than that of the 3β,16α-dihydroxyandrost-5-en-17-one and the 3β,16α,17β-triol. In contrast, the rate of formation of androst-5-ene-3β,17β-diol was not affected by age as well as sex. TABEI and HEINRICHS (1974) demonstrated that 7α-hydroxylation of 3β-hydroxyandrost-5-en-17-one by rat liver microsomes also progressively increased with age in males and decreased in females. Gonadectomy within the first 10 days of life reduced the 7α-,7β-, and 16α-hydroxylase and the 7- and 16α-hydroxysteroid dehydrogenase activities in males but not females. The 17β-hydroxysteroid dehydrogenase activity had identical activities in both sexes and was not influenced by gonadectomy. The effect of neonatal castration was reversed by the administration of testosterone propionate (TABEI and HEINRICHS, 1975). The male rats were castrated on the first day of life and the females at 7 days of age. The injection of 50 μg TP on the first day of life and then increasing doses from 50–400 μg weekly from 18–70 days to maintain a normal growth curve enhanced the rate of formation of the 16-oxygenated steroids in the male and the female rats. A larger dose of TP, 500 μg every other day from 18–40 days and then 1000 μg to day 70 completely restored the rate of 16 oxygenation to the adult male level. The single neonatal injection or the injection from 18–70 days separately were without effect. In contrast, enhanced 7-oxygenation was apparent after injections begun at 18 days alone and also at the lowest doses of testosterone propionate if an injection was also made on the first day of life. Neonatal treatment alone was ineffective. The 16α-hydroxylation was stimulated to a greater degree by testosterone propionate in the prepubertal females than males but was reversed after 70 days of age. It was of interest that restoration of these enzymes was accomplished with a smaller dose of testosterone propionate than required for restoration of the accessory sex organs.

The regulation of 5-reductase activity in liver and other tissues has received special attention by several investigators because of the suggestion that the enzyme in the prostate is essential for the conversion of testosterone to 17β-hydroxy-5α-androstan-3-one to effect stimulation of growth in the prostate. ALTMAN et al. (1972) assayed the $\Delta^4$-reductase activity of the liver homogenate of male (100–200 g) rats by two methods (a) noting the disappearance of absorbance at 240 nm on incubation with testosterone at pH 5.6 and NADPH as the coenzyme, and (b) using [4-$^{14}$C]-testosterone as the substrate and isolating the 5α-radiometabolites by thin-layer chromatography and quantitating by determination of the radioactivity. A single dose of 25 or 50 mg/100 g of medroxyprogesterone acetate (17-acetoxy-6α-methyl-pregn-4-ene-3,20-dione) or at least 2 weeks of 5 mg every other day produced a 2-fold increase in the $\Delta^4$-reductase activity as determined by both types of assay. 5α-Androstane-3α,17β-diol was the main metabolite and represented the major increase after treatment. 17β-Hydroxy-5α-androstan-3-one plus 3α-hydroxy-5α-androstan-17-one occurred in smaller amounts and were only slightly increased by medroxy-progesterone acetate treatment. The injection of 17α-hydroxyprogesterone and also progesterone had a stimulatory effect but progesterone required a high dose and prolonged treatment. Castration produced a small increase in the reductase activity but did not influence the stimulatory effect of the medroxyprogesterone acetate. Adrenalectomy and cortisol and phenobarbital administration were without effect on the basal or stimulated activity. The injection of medroxyprogesterone acetate also increased the $\Delta^4$-reduction of androst-4-ene-3,17-dione, cortisol, progesterone, and even medroxyprogesterone acetate itself. The stimulatory effect of medroxypro-gesterone acetate was localized in the microsomal fraction. The increase in $\Delta^4$-reductase activity on testosterone was decreased in competitive fashion by the addition of medroxyprogesterone acetate to the incubation mixture and by the injection of actinomycin D (0.2 mg/100 g) 10 min prior to injection of the medroxypro-gesterone acetate.

PATTERSON et al. (1974) utilized the disappearance in absorbance at 240 nm when testosterone was incubated with liver preparations in the presence of a NADPH generating system at pH 6.4 to investigate hormonal regulation of $\Delta^4$-reductase activity in rat liver. As expected, a much higher enzyme activity was found in the female than in the male rat liver homogenate. The liver of 15-day pregnant rats exhibited the same activity as that of the normal female. Practically all of the enzyme activity was in the microsomal fraction. The cytosol contained a small amount of activity and the nuclei and mitochondria were not examined. Estradiol valerate at 0.1 mg/day for 14 days increased the $\Delta^4$-reductase activity in the male rat liver. Estriol and chlorotrianisene at the same dose were ineffective but an increase of the dose of chlorotrianisene to 0.2 μg/day doubled the enzyme activity and a further increase was obtained with a further increase in dose. Castration of adult rats for 16 days quadrupled the microsomal reductase activity in agreement with the earlier observation of KATO et al. (1970). Hypophysectomy nearly abolished and adrenalectomy was without effect on the enzyme activity. Testosterone enanthate at 2.25 mg/day for 4 days decreased the microsomal $\Delta^4$-reductase activity in the castrated and adrenalectomized rats but was without effect in the hypophysectomized rats. Estradiol valerate at 0.1 mg/day for 4 days had no effect on the enhanced enzyme activity of the castrated rat or the greatly reduced activity of the hypophysectomized rats but increased that of the adrenalectomized rats. The simultaneous injection of both hormones gave an intermediate effect in the castrated rats, cancelled each other in the adrenalectomized rats, and was ineffective in the hypophysectomized rats. Prolactin was ineffective in normal or hypophysectomized rats and did not influence the effects of either steroid hormone. Both actinomycin D (1.0 μg/g) and cycloheximide (1.5 μg/g) inhibited the increase in $\Delta^4$-reductase activity induced by estradiol valerate.

The presence of 5α-reductase activity in liver nuclei (Gustafsson and Pousette, 1974a), as well as the microsomes and the higher activity in the female than the male rat, suggested a different type of regulation for the liver enzyme than for the prostatic nuclear enzyme. In order to elucidate this possibility, Gustafsson and Pousette (1974b) studied the influence of age, sex, castration, and treatment with androgens and estrogens. The nuclei were checked for purity by electron microscopy. Incubation of the nuclei was conducted in a NADPH generating system providing a linear conversion of [1,2-$^3$H]-androst-4-ene-3,17-dione. Incubation with kidney and prostatic nuclei also contained dithiothreitol at a concentration to give optimum 5α-reductase activity. As described earlier (Gustafsson and Pousette, 1974a) only 5α-androstane-3,17-dione was formed by female rat liver nuclei but male nuclei also formed small amounts of 3β-hydroxy-5α-androstan-17-one and testosterone. The adult (56 days) female rat liver, however, exhibited about 30 times more 5α-reductase activity than the male liver nuclei. No difference in enzyme activity was noted between the 28-day-old and 56-day-old males but the 28-day-old females showed only 12% of the activity of the 56-day-old female rats. Castration of male rats at birth or 56 days of age induced a 10-fold increase in the nuclear 5α-reductase activity. The injection of 400 μg/day testosterone propionate or 100 μg/day estradiol benzoate for 7 days beginning the day after castration at 56 days of age was without effect on the elevated 5α-reductase activity. Ovariectomy of adult female rats did not change the 5α-reductase activity of the nuclei. Kidney nuclei from both male and female rats transformed [1,2-$^3$H]-androst-4-ene-3,17-dione to 5α-androstane-3,17-dione and also testosterone. No sex difference or effect of castration was detected. The injection of testosterone propionate also was ineffective but estradiol benzoate induced a 5-fold increase in the castrated rats. The 17β-hydroxysteroid reductase activity also was unaffected except estradiol benzoate decreased the enzyme activity 40%. In contrast to the liver and kidney nuclear 5α-reductase activity, the prostatic enzyme activity in confirmation of earlier reports was found to be androgen-dependent. Furthermore, the response of the nuclear 5α-reductase activity was different from that of liver microsomal activity to hormone manipulation and the kidney 5α-reductase activity showed still another pattern of response suggesting that the respective 5α-reductase systems are different.

## 2. Thyroxine

McGuire and Tomkins (1959b) observed an increase in the reduction of the 4–5 double bond of androst-4-ene-3,17-dione and testosterone as well as that of several adrenal cortical steroids by the liver homogenate of immature (45–55 g) male rats after the injection of a growth depressing dose of thyroxine for 3 days. The increase was explained by the stimulation of an increase in NADPH. The reductase activity was determined by the disappearance of absorption at 240 nm of the $\Delta^4$-3-oxo-group. The increase occurred in both the 5β-reductase activity of the soluble fraction and the 5α-reductase activity of the microsomal fraction. Isolation of the metabolites formed on incubation of the microsomal fraction after 16 days of thyroxine treatment were in agreement with the spectrophotometric studies. KATO et al. (1970) confirmed the results of MCGUIRE and TOMKINS in the liver microsomal fraction in slightly older rats (8–9 weeks) after the injection of 1 mg/kg thyroxine for 10 days but obtained the reverse effect, a decrease, in the $\Delta^4$-reductase activity of the liver microsomes from female rats. Furthermore, thyroxine treatment was without effect on the reductase (5β) activity of the cytosol of either the male or female rats and also no sex difference in enzyme activity was detected. Castration after 23 days markedly increased the reductase activity in the liver microsomes, and testosterone, 1 mg/kg/day, during the last 10 days decreased the enzyme activity to the normal level. Thyroxine treatment did not induce a further increase in the elevated reductase level of the castrated rats or have an effect on the enzyme activity of the testosterone-treated castrated rats. The reductase activity of the cytosol was not affected by any of these hormonal manipulations. Mice 7 weeks old, in contrast to the rats, showed no clear sex difference in liver microsomal reductase activity. Thyroxine, 1 mg/kg/day, for 10 days decreased the $\Delta^4$-reductase activity in not only the female but also the male mice.

### 3. Pituitary

DENEF (1974) investigated the role of the pituitary in the sexual differentiation of $\Delta^4$-oxosteroid reduction in the rat liver. The liver slices were incubated with [1,2-$^3$H]-testosterone in Krebs-Ringer phosphate buffer, pH 7.4, under nitrogen. The metabolites were separated by thin layer chromatography. The 17β-hydroxy metabolites—17β-hydroxy-5α-androstan-3-one, 5α-androstane-3α,17β-diol, 5α-androstane-3β,17β-diol, and 5β-androstane-3α,17β-diol—represented 80% of the metabolites. The changes in radioactivity in the 17-oxosteroids—3α-hydroxy-5α-androstan-17-one, 3β-hydroxy-5α-androstan-17-one, and 3α-hydroxy-5β-androstan-17-one—were in parallel with those of the 17-hydroxy metabolites. The rats were gonadectomized at 25 days of age and liver reductase activity was determined 35–65 days later. In agreement with earlier observations, males converted more testosterone to the 3α,5β- and 3β,5α-isomers and formed less 17β-hydroxy-5α-androstan-3-one and 3α,5α-isomers than females. Hypophysectomy abolished the sex difference. In both sexes the formation of the 3α,5β- and 3β,5α-isomers became the main pathway and the formation of 3α,5α-steroids and 17β-hydroxy-5α-androstan-3-one was drastically reduced. These changes represented a shift to a male-like type of metabolism. When [1,2-$^3$H]-17β-hydroxy-5α-androstan-3-one was used as the substrate, essentially the same effects were observed. The implantation of a male or female adult pituitary under the kidney capsule of gonadectomized-hypophysectomized rats produced in both instances the female type of metabolism of testosterone or 17β-hydroxy-5α-androstan-3-one after 45 days suggesting that hypothalamic control of the pituitary is required to imprint sex differentiation.

LAX et al. (1974) found that the pituitary played an important role in the regulation of sex differences in the activities of several rat liver $C_{19}$-steroid metabolizing enzymes. On the basis of the previous studies, all of the enzyme activities were determined on the microsomal fraction except the 17β-hydroxysteroid dehydrogenase activity was determined in the cytosol. Twenty-five days after hypophysectomy at 50 days of age, the 17β-hydroxysteroid dehydrogenase activity in both male and female rats was typical of the normal female and the activities of the 20-hydroxysteroid and the 16α-hydroxylase were intermediate to those of the normal male and female. On the other hand, the 3β- and $\Delta^4$-3β-hydroxysteroid dehydrogenase exhibited activities in both sexes typical of the normal male. The 3α-hydroxysteroid dehydrogenase was increased in both sexes above the normal male values. The subcutaneous administration of 75 mg testoviron depot (20 mg testosterone propionate and 55 mg testosterone enanthate) at 60 days of age to the male and female hypophysectomized rats, did not produce any significant changes in the enzyme activities 15 days later. When the rats were gonadectomized at 25 days of age and hypophysectomized 25 days later, the 3α- and 3β- and $\Delta^4$-3β-hydroxysteroid dehydrogenase activity lost after 15 days their residual sex difference due to a decrease in enzyme activities in the male rats.

GUSTAFSSON and STENBERG (1975) carried out experiments with LH and FSH administration to male and female normal and castrated rats to delineate the role of the anterior pituitary in the regulation of the hepatic metabolism of the $C_{19}$ steroid hormones. The rats were gonadectomized at 56 days of age and injected for 14 days with (a) either FSH or LH, (b) 100 μg/day testosterone propionate, or (c) simultaneously with treatments (a) and (b). The administration of LH gave significant decreases in 17-hydroxysteroid reduction and 7α-hydroxylation of androstane-3,17-dione in both male and female rats and in 6β-hydroxylation of androstene-3,17-dione in female rats. FSH, on the other hand, specifically stimulated (masculinized) the following sex-dependent hydroxylations: 16α-hydroxylation of androstane-3,17-dione was induced in female rats and 2α-, 2β-, and 18-hydroxylation of 5α-androstane-3α,17β-diol was increased 2-fold in male rats. The administration of testosterone propionate did not affect the FSH response of the female rats but the sex-dependent 2β- and 16α-hydroxylases were induced.

## 4. Diabetes

GOLD and GARREN (1964) compared the metabolism of [4-$^{14}$C]-androst-4-ene-3,17-dione by the liver microsomes and cytosol of normal and diabetic female rats on incubation for 5 min at pH 7.5 with a NADPH generating system. The microsomes of the diabetic rats produced less of the 5α-reduced steroids—3α-hydroxy-5α-androstan-17-one, 5α-androstane-3,17-dione, and 5α-androstane-3α,17β-diol—, more testosterone and about the same amount of 7α-hydroxyandrost-4-ene-3,17-dione and 17β-hydroxy-5α-androstan-3-one. The major metabolite formed by the cytosol was 3α-hydroxy-5β-androstan-17-one and its formation was not influenced by the diabetes. Smaller amounts of 3α-hydroxy-5α-androstan-17-one were formed by the normal cytosol and even less by the cytosol from the diabetic rats. Testosterone was present in small amounts.

## 5. Drugs

It is of particular interest that certain drugs enhance the activity of the liver NADP microsomal enzymes concerned with their metabolism. The same enzymes are also stimulated by the administration of testosterone (BOOTH and GILLETTE, 1962) and the drugs in turn stimulate the liver microsomal enzymes which hydroxylate $C_{19}$ steroids. CONNEY and KLUTCH (1963) injected young (40–45 g) male rats i.p. with sodium phenobarbital (37 mg/kg) or chlorcyclizine hydrochloride (12 mg/kg) twice daily for 4 days and obtained a several-fold increase in the formation of polar metabolites on incubation of the microsomes with either [4-$^{14}$C]-testosterone or [4-$^{14}$C]-androst-4-ene-3,17-dione. On the other hand, the injection of 3-methylcholanthrene only slightly stimulated the formation of polar metabolites and inhibited the formation of nonpolar metabolites. The drugs had no effect on the metabolism of the steroids by the nuclei, mitochondria, or cytosol. NADPH or a regenerating system was required. Maximum effect was obtained when both NADPH and NADH were present in the incubation mixture. NADH alone had only a slight stimulatory effect. The hydroxylase systems were identified by isolation and characterization of the metabolites. Testosterone was metabolized to 7α- 6β-, 16α-, and 2β-hydroxytestosterone and androst-4-ene-3,17-dione to 7α- and 6β-hydroxyandrost-4-ene-3,17-dione. BERG and GUSTAFSSON (1973) obtained increased hydroxylation of [4-$^{14}$C]-5α-androstane-3α,17β-diol at the 2α-, 2β-, 7α-, 7β-, and 18 positions by the liver microsomes of male and female rats gonadectomized at 56 days of age and injected for 5 days with 20 mg phenobarbital or 10 mg 16α-cyanopregnenolone. Injection of 5 mg methylcholanthrene for 5 days specifically increased 7α-hydroxylation.

WELCH et al. (1967) extended their studies to several insecticides. The addition to the incubation system of organic phosphorothionate insecticides or halogenated hydrocarbon insecticides markedly inhibited the hydroxylation of testosterone by the liver microsomes from adult male rats. Isolation of the metabolites indicated the inhibition of metabolism of testosterone to 16α- and 6β-hydroxytestosterone but not to 7α-hydroxytestosterone. The degree of inhibition varied with the type of insecticide. The injection of chlorthion for 10 days into immature rats, 45–50 g, inhibited the microsomal formation of 16α-hydroxytestosterone and to a lesser extent 6β- or 7α-hydroxytestosterone. Chlordane or DDT, on the other hand, markedly stimulated the transformation of testosterone to 16α-hydroxytestosterone with a lesser effect on the formation of 6β- and 7α-hydroxytestosterone.

SHOLITON et al. (1967) studied the influence of diphenylhydantoin in vitro on the metabolism of [4-$^{14}$C]-testosterone by liver slices from adult rats. The drug was added to the incubation solution of Ringer-bicarbonate (pH 7.4) containing a NADPH generating system, liver slices, and [4-$^{14}$C]-testosterone. The presence of the drug enhanced the formation of androsterone and 5α-androstane-3α,17β-diol 4- to 5-fold by female liver slices while 3α-hydroxy-5β-androstan-17-one was increased by the male liver slices. More polar metabolites were not identified but were produced in greater amounts by the female liver slices, and the lower values produced by the male liver slices were increased to that of the females.

CARPENTER and HOWARD (1974) studied the interrelationship of vitamin E and several steroid hormones in the induction of increased hydroxylation by rat liver microsomes. They reported that α-tocopherol replaced the requirement of the steroid hormones in the vitamin E-steroid-depleted (castrated) rat.

These striking changes in rat liver microsomal hydroxylation do not necessarily mean that the same results occur in other animal species (see BRODIE, 1956).

# V. Purification of $C_{19}$-Steroid Enzymes

## A. 17β-Hydroxy-$C_{19}$-Steroid Oxidoreductase

### 1. Liver

*a) Rabbit*

Although the initial demonstration of the metabolism of the $C_{19}$ steroids in vitro was performed on male rabbit liver, no attempts at isolation and characterization of the several indicated enzymes have been made. Recently, however, a partial purification of a 17β-hydroxysteroid NAD(P) oxidoreductase from female rabbit liver cytosol has been reported. THALER-DAO et al. (1972) separated the subcellular components of the livers of adult female rabbits by ultracentrifugation. Both the cytosol and the microsomal fraction contained 17β-hydroxysteroid oxidoreductase activity with NAD or NADP as coenzymes. The cytosol activity was 3 times that of the microsomal fraction and the purification was continued on the cytosol. The enzyme activities were assayed by the spectrophotometric method with three different substrates (testosterone, estradiol-17β and Des-A-estradiol-17β) and with the two coenzymes, NAD and NADP. Most of the enzyme activity was present in the precipitate formed between 40–80% $(NH_4)_2SO_4$ saturation. DEAE-cellulose chromatography separated the $(NH_4)_2SO_4$ fraction into three separate enzyme peaks containing 50, 30, and 20% of the oxidoreductase activity. A second chromatography on DEAE-cellulose separated the second peak into two fractions similar to the 1st and 3rd peaks. Peak 1 was weakly anionic, contained other dehydrogenases, and had very weak activity with estradiol-17β. Peak 3 was strongly anionic and contained low NAD-related activity. Further purification of these two fractions was accomplished by Sephadex G 100 column chromatography. The first peak was separated into two peaks with oxidoreductase activity. The first subfraction showed a high nonspecific reduction of NAD, no estradiol-17β NADP-oxidoreductase activity, and contained alcohol dehydrogenase activity. The molecular weight was 65000–100000. The second subfraction was devoid of alcohol dehydrogenase activity and contained both estradiol-17β and testosterone oxidoreductase activity. The Km for testosterone in the presence of NADP was $5 \times 10^{-4}$ M. The molecular weight was $30000 \pm 5000$. Analytical polyacrylamide gel electrophoresis separated six proteins, three of which corresponded with testosterone NADP-oxidoreductase activity. SDS gel electrophoresis gave two proteins of about 25000 and 30000 respectively.

Sephadex G-100 column chromatography of the third peak from DEAE-cellulose chromatography effected a 20–30-fold purification. Polyacrylamide gel electrophoresis gave five to six protein bands: one contained the major enzyme activity but three others showed minor activity; no activity was observed in the blank (substrate omitted). SDS polyacrylamide gel electrophoresis showed only one protein band with a molecular weight of 35000. UV absorption and fluorescence spectra indicated a tryptophan-containing protein. Km was $3 \pm 0.5 \times 10^{-6}$ M at pH 9.5 with NADP as cofactor. The optimum pH was 9.4 and increase of temperature between 15 and 38° C gave a linear increase of enzyme velocities. The highest enzyme activity was at 38–40° C. Thiol reagents inhibited enzyme activity which was reversed by the addition of β-mercaptoethanol. The enzyme demonstrated broad specificity with both NAD and NADP as coenzymes. Testosterone elicited the greatest activity but 3α-hydroxy-5$\beta$-androstan-17-one was almost as effective (84%) and had a similar Km; $3 \pm 0.5 \times 10^{-6}$ and $2.5 \pm 0.5 \times 10^{-6}$ M respectively. The other steroids tested gave smaller responses (estradiol-17$\beta$, 18%; epiandrosterone, <20%; dehydroepiandrosterone <16%; and androsterone, 45%. No 17$\alpha$-oxidoreductase activity was indicated; estradiol-17$\alpha$ was not dehydrogenated but served as a competitive inhibitor for both 17$\beta$- and 3$\alpha$-activity. 17$\alpha$-Hydroxyandrost-4-en-3-one was not studied. HASNAIN and WILLIAMSON (1974), however, demonstrated that an estradiol-17$\alpha$ dehydrogenase was present in female rabbit liver cytosol and was separated from the 17$\beta$ dehydrogenase activity during the DEAE-cellulose chromatography. The ability to use $C_{19}$ steroids as substrates was not studied.

*b) Steer*

Ground steer liver was mixed 1:1 (w/v) with water and after 24 h at 2–4° was filtered through cheesecloth and clarified by centrifugation (SWEAT et al., 1950). The supernatant fraction converted testosterone to androstenedione on incubation at pH 7.8. The addition of NAD enhanced the enzyme activity. Half-saturation with $(NH_4)_2SO_4$ did not precipitate the enzyme but full saturation did. The pH optimum of this partially purified preparation was 8.3 and Km was $3.3 \times 10^{-5}$ M.

*c) Guinea Pig*

The soluble NADP-17$\beta$-hydroxy-$C_{19}$-steroid dehydrogenase of the male guinea pig liver (ENDAHL et al., 1960) has been partially purified and studied by several laboratories. ENDAHL and KOCHAKIAN (1962) effected a partial purification by two separate procedures. A 5-fold purification was accomplished by precipitation of 65% of the enzyme activity between 50–60% $(NH_4)_2SO_4$ saturation of the cytosol and a further purification by chromatography on DEAE-cellulose. The active material exhibited only a negative migrating band on paper electrophoresis. A greater recovery and similar purification was attained by curtain paper electrophoresis. The major enzyme activity exhibited a single protein peak with a corresponding single enzyme activity peak on reelectrophoresis of the major protein of the enzyme activity but gave three distinct peaks on Tiselius electrophoresis. The partially purified enzyme was free of several other NADP-linked enyzmes and showed no activity with epitestosterone, androsterone, epiandrosterone, dehydroepiandrosterone, and 5$\beta$-androsterone. The 17$\beta$-hydroxy-$C_{19}$-steroids, 17$\beta$-hydroxy-5$\alpha$-androstan-3-one and 17$\beta$-hydroxy-5$\beta$-androstan-3-one, were effective substrates but with lower Km values (Testosterone $5.6 \times 10^{-5}$, 17$\beta$-hydroxy-5$\beta$-androstan-3-one $2.07 \times 10^{-5}$ and 17$\beta$-hydroxy-5$\beta$-androstan-3-one $1.08 \times 10^{-5}$ M). The enzyme did not exhibit any requirement for the several metallic ions studied. The previous studies (ENDAHL et al., 1960) indicated an inhibiton by thiol reagents and also cyanide. The NAD enzyme was more sensitive to thiol reagents than the NADP-linked enzyme and EDTA did not affect the enzyme activity.

VILLEE and SPENCER (1960) confirmed the presence of a NADP-specific 17$\beta$-hydroxy-$C_{19}$-steroid dehydrogenase enzyme in the cytosol and a NAD-specific enzyme in the microsomal fraction of guinea pig liver. Presumably the livers from male and female guinea pigs were pooled for enzyme separation. No differentiation was presented. The NADP-linked enzyme was precipitated from the cytosol by 60–75% $(NH_4)_2SO_4$ saturation. Chromatography on calcium phosphate gel and electrophoresis on starch block gave a 12-fold purification. Another 17$\beta$-hydroxy-steroid dehydrogenase which reacted more rapidly with estradiol-17$\beta$ than testosterone and utilized either NAD or NADP as coenzyme was precipitated below 40% saturation with $(NH_4)_2SO_4$. The reductase activity of the NADP-$C_{19}$-steroid enzyme was favored at pH 6.0, the oxidase activity at pH 9.0 and the reactions were equal at pH 7.4 in the presence of either NADP or NADPH. The Km values were $2.7 \times 10^{-4}$ M for testosterone and $1.1 \times 10^{-5}$ M for NADP.

JOSHI et al. (1963) effected a 200–300-fold purification of the NADP-specific steroid dehydrogenase activity by successively treating the 24000 *g* supernatant fraction with $(NH_4)_2SO_4$ precipitation, DEAE cellulose chromatography, and hydroxylapatite chromatography. The enzyme activity was precipitated between 50 and 75% $(NH_4)_2SO_4$ saturation. The partially purified enzyme was free of other NAD and NADP dehydrogenases. 3$\beta$-Hydroxy-5$\alpha$-androstan-17-one, however, was dehydrogenated in the presence of NAD but not NADP; other 3-hydroxysteroids of both the 5$\alpha$- and 5$\beta$-series did not serve as substrates. The pH optimum of 10.3 for testosterone dehydrogenation was in agreement with the value reported by ENDAHL et al. (1960). The purified enzyme utilized several 17$\beta$-hydroxy steroids as substrate but the enzyme activity was affected by the chemical structure of the steroid. Estradiol-17$\beta$ also was dehydrogenated but at only 20% of the rate of testosterone.

On partial purification of NADP-specific 17$\beta$-hydroxy-$C_{19}$-steroid dehydrogenase from the cytosol of guinea pig liver by $(NH_4)_2SO_4$ precipitation followed by DEAE cellulose chromatography, two enzyme active fractions were obtained by KAGEURA and TOKI (1971). One of the fractions also contained 3-hydroxyhexobarbital dehydrogenase activity. NAD and NADP served as coenzymes for either fraction with sodium phosphate-NaOH (pH 10.7) but only NADP with Tris buffer at pH 9.0. A further purification of the enzymes was accomplished by submitting the liver cytosol to $(NH_4)_2SO_4$ precipitation (50–75% saturation) followed by two separations by Sephadex-G-100 filtration and DEAE-cellulose DE-32 and hydroxylapatite column chromatography. The enzyme with only steroid dehydrogenase activity was separated by DEAE-cellulose chromatography. The enzyme with dual substrate activity showed a constant 1.60 greater NAD-3-hydroxyhexobarbital than NAD-testosterone dehydrogenase activity and a ratio of 3.0 when NADP was the coenzyme. The NADP-linked activity was greater than the NAD-linked activity with both substrates. Phosphate ions stimulated the NAD-linked enzyme activity. The activity of the enzyme with both substrates was similarly inhibited by an increase in temperature and p-chloromercuribenzoate. The optimum pH was different. The 3-hydroxyhexobarbital activity was optimum at pH 8.9 and that of testosterone at 10.7. The addition of testosterone at the pH optimum for dehydrogenation of 3-hydroxyhexobarbital resulted in competitive inhibition even though testosterone was only slightly metabolized at pH 8.9. The authors concluded that the two substrates were dehydrogenated by the same enzyme. KAGEURA and TOKI (1974) purified 3-hydroxy-hexobarbital dehydrogenase as a single homogeneous protein from the soluble fraction of guinea pig liver and found that it still demonstrated 17$\beta$-hydroxy-$C_{19}$-steroid dehydrogenase activity[5]. The fraction with only the 17$\beta$-hy-

[5] In contrast to the guinea pig liver TAKENOSHITA and TOKI (1974) found that the purified 3-hydroxyhexobarbital dehydrogenase of rabbit liver cytosol did not have 17$\beta$-hydroxy-$C_{19}$-steroid dehydrogenase activity.

droxy-$C_{19}$-steroid dehydrogenase activity also was purified by TOKI and KAGEURA (1974) by $(NH_4)_2SO_4$ precipitation, followed by gel filtration on Sephadex-G-100 and G-75 and DEAE-cellulose-DE-32 and hydroxylapatite column chromatography. The purified enzyme gave a single band on polyacrylamide gel electrophoresis, and utilized NADP as the preferred coenzyme. The molecular weight was estimated at 39000. The Km value was $2.7 \times 10^{-5}$ for testosterone and $4.7 \times 10^{-5}$ for NADP. Vmax for testosterone was 0.75 μmole/min/mg/protein. The pH optima were 10.6 with NADPH and 6.1 with NADPH. The activity was strongly inhibited by p-chloromercuribenzoate, $Ag^+$ and $Hg^+$.

VILLEE and SPENCER (1960) solubilized the enzyme in the 24, 500 *g* sediment by treatment with 1% digitonin or by sonic disruption and precipitated the enzyme activity between 30 and 50% saturation with $(NH_4)_2SO_4$. The partially purified enzyme reacted somewhat more readily with testosterone than estradiol-17β. The Km for testosterone was $3.3 \times 10^{-5}$ M and $2.8 \times 10^{-5}$ M for NAD. BLOMQUIST et al. (1974) determined some further properties of the microsomal enzyme. The relative activities with testosterone and estradiol-17β as substrate were not changed after washing the microsomes successively with 0.14 M NaCl, 1.0 M NaCl, and 0.1 M $Na_2CO_3$-0.1 M $NaHCO_3$. On separation by centrifugation in 1.23 M sucrose, 90% of the enzyme activity was in the "smooth" microsomes. The enzyme activity was inhibited by 6.3 mM o-phenanthroline and 0.16 mM 1.8 ANS but not by 2, 2′bipyridine, isobutyramide, or pyrazole suggesting inhibition by hydrophobic interaction at the active site. Vmax was the same for testosterone and estradiol-17β at 1.6 mM indicating a common active center.

### *d) Horse*

Crystalline alcohol dehydrogenase prepared from horse liver exhibited 3β-hydroxy- and 17β-hydroxy-$C_{19}$-steroid dehydrogenase activity. PIETRUSZKO et al. (1966) detected most of the steroid dehydrogenase activity in one of the multiple zones separated on starch electrophoresis of crystalline alcohol dehydrogenase from two commercial sources. Furthermore, the proportion of steroid to alcohol dehydrogenase activities was very different for the two preparations. THEORELL et al. (1966) conducted a systematic reinvestigation of their isolation method and found that one (the basic) of the fractions was responsible for the steroid dehydrogenase activity. It was recrystallized and found to be electrophoretically homogeneous. Amino acid analysis gave a nearly identical composition as that for the alcohol dehydrogenase except for possibly threonine. A higher amide concentration was found which could explain the difference in electrophoretic mobility and substrate specificity. The enzyme presumably utilizes either a 17β-hydroxy- or a 3α-hydroxy steroid.

## 2. Kidney

### *a) Rat*

The 17β-hydroxysteroid oxidoreductase of rat kidney has been partially purified by BREUER and DAHM (1964) from the 20000 *g* 60 min supernatant fraction of 6–8-week-old male and female rats. The major portion of the enzyme activity was found in the precipitate after 40–80% $(NH_4)_2SO_4$ saturation. The specific activity was increased 12-fold. The enzyme utilized either phenolic or neutral steroids and either NAD or

NADP as coenzymes. The optimum pH for conversion of testosterone to androstenedione was 8.7 and for the reverse reaction 6.6. The Km's were $18.5 \times 10^{-5}$ M for testosterone and $40 \times 10^{-5}$ M for NAD. The equilibrium constant was $9.1 \times 10^{-9}$ M. The activation energy was 13.2 kcal/mole and inactivation energy −24.0 kcal/mole; the $Q_{10}$ values varied between 1.6 and 1.5. SH blocking agents inhibited the activity but zinc ions in low concentrations ($10^{-5}$ M) activated the enzyme.

MEIGS and RYAN (1966) effected a partial purification of the 16α-hydroxysteroid oxidoreductase of mature female rat kidney by submitting the cytosol to Sephadex-G-25 M filtration and then by passage through a column of carboxymethyl cellulose to give up to an 18-fold purification. The enzyme activity was very labile and was only partially protected by 20% glycerol. The preparation was free of 17β-hydroxy- and was nearly completely separated from 16β-hydroxysteroid oxidoreductase. The partially purified enzyme utilized NADPH as well as NADH as coenzyme, exhibited a pH optimum of 6.1 to 6.2, and was inhibited by low concentrations of heavy metal ions, sulfhydryl reagents, and metal chelators. A number of $C_{18}$ and $C_{19}$ steroids were able to serve as substrate.

*b) Guinea Pig*

The 17β-hydroxy-$C_{19}$-steroid dehydrogenase activity of the adult male guinea pig is localized in the cytosol and utilizes either NAD or NADP (ENDAHL et al., 1960). Purification of this enzyme has been accomplished. LIU and KOCHAKIAN (1972a) obtained a nearly complete purification by streptomycin sulfate precipitation to remove RNA, separation of the enzyme activity at 40–80% $(NH_4)_2SO_4$ saturation followed by Sephadex G-100 filtration and DEAE-cellulose chromatography. The enzyme activity was contained in three major and two minor bands on gel electrophoresis with corresponding protein bands and traces of nonenzyme proteins. The ratio of enzyme activity with NADP and NAD was unchanged during the various purification steps and favored NADP at 8:1. The different components of the isozyme had similar molecular weights but different electrical charges (LIU and KOCHAKIAN, 1972b). The molecular weight was determined by ultracentrifugation (35100), Sephadex filtration (31600), and acrylamide gel electrophoresis (31200). Testosterone was the most effective substrate, estradiol-17β elicited a trace of activity, 17β-hydroxy-5α-androstan-3-one was one-half and 17β-hydroxy-5β-androstan-3-one was one-seventh as effective as testosterone. Thiol reagents inactivated the enzyme; the simultaneous addition of dithiothreitol prevented the inhibition but could not restore the activity after inactivation. The optimum pH was about 9.9 with NADP and 9.4–9.7 with NAD. The purified enzyme was unstable but addition of 7 mM mercaptoethanol and 0.25 M sucrose or 20% glycerol prevented the loss in activity. The enzyme activity was labile to heat but was stable at −20° C. STEVENSON and KOCHAKIAN (1974) removed the trace of protein impurities by submitting the nearly purified enzyme to chromatography on CM-Sephadex-C50 and also separated one of the major isoenzymes. The purified isoenzymes gave a single band on SDS-gel electrophoresis and a molecular weight of 31000 identical with that obtained with the partially purified enzyme. The enzyme exhibited a differential response to treatment with urea. Treatment of the partially purified enzyme with 4 M urea inactivated the enzyme but all of the activity was restored by dialysis at 4° C. Treatment with 6 M urea permitted restoration of only 25% of the enzyme activity and 8 M urea permanently inactivated the enzyme.

### 3. Blood

#### a) Human

MULDER et al. (1972) localized the 17$\beta$-hydroxy-$C_{19}$-steroid dehydrogenase activity of the erythrocytes in the cytoplasm. The membrane showed no activity. The erythrocytes from freshly drawn heparinized or out-dated blood-bank blood (male and female?) was washed 3 times with 0.9% NaCl and hemolyzed by the addition of 2.5 vol of water. The soluble fraction was separated by centrifugation and the enzyme precipitated by the addition of an equal amount of saturated $(NH_4)_2SO_4$ solution. The precipitate was dissolved in a solution with a final concentration of $(NH_4)_2SO_4$ of 20% saturation. Insoluble material was removed by centrifugation and the solution was made to 50% concentration of $(NH_4)_2SO_4$. The procedure was repeated. The enzyme solution then was submitted to chromatography on Sephadex G-150 and the enzyme active fractions were pooled. Attempts at further purification on a DEAE Sephadex-A-50 column resulted in loss of most of the enzyme activity. Most of the glucose-6-phosphate dehydrogenase activity was separated by Sephadex-G-150 chromatography but small amounts tailed into the 17$\beta$-hydroxy-$C_{19}$-steroid dehydrogenase eluates. The partially purified enzyme gave a molecular weight of 64000 on Sephadex-G-150 chromatography. The enzyme activity was specific for NADP(H). The pH optimum was broad and dependent on the buffer. The optimum pH in Tris-HCl (0.04 M) was 8.5 and about 10.0 in 0.04 M glycine buffer. The enzyme had a broad specificity. It not only interconverted testosterone and androstenedione but a number of other $C_{19}$-steroids and also estradiol-17$\beta$ and estrone, indicating a similarity to the 17$\beta$-hydroxy-$C_{18}$-steroid dehydrogenase partially purified by JACOBSOHN and HOCHBERG (1968) from human erythrocytes.

#### b) Rat

HEYNS and DEMOOR (1974) effected a partial purification of steroid dehydrogenase activity from the hemolysate of the erythrocytes of adult (male and female?) rats and observed that 3$\beta$-hydroxy-and 17$\beta$-hydroxy- (testosterone and estradiol-17$\beta$) dehydrogenase activity were not separated. The enzyme had a molecular weight of 70000 and was precipitated by $(NH_4)_2SO_4$ at 25–33 g/100 ml. A 7200-fold purification was obtained by a combination of affinity chromatography with KCl elution, gel filtration on Sephadex-G-100, and affinity chromatography with NADP elution. Polyacrylamide gel electrophoresis gave a weak single protein band which exhibited both 3$\beta$- and 17$\beta$-hydroxysteroid dehydrogenase activity. The authors postulated a single enzyme. The enzyme was not active on 5$\beta$-androstane compounds. RONGONE et al. (1957) reported earlier, however, that commercial bovine blood albumin contained a 5$\beta$-enzyme which not only converted androst-4-ene-3,17-dione to 5$\beta$-androstane-3,17-dione but also cortisone and progesterone to their respective 5$\beta$-compounds.

## B. 3$\alpha$-Hydroxy-$C_{19}$-Steroid Oxidoreductase

### Rat Liver

TOMKINS (1956) partially purified 3$\alpha$-hydroxysteroid dehydrogenase activity from the liver of rats (male and female?) by centrifuging the homogenate in 0.1 M $NaH_2PO_4$, pH 7, at 10000 $g$ and adding $(NH_4)_2SO_4$ to the supernatant fraction (microsomes and cytosol) to 55% saturation. Subsequent treatment of the active fraction with $Ca_3(PO_4)_2$ gel and 20–30% ethanol effected an overall purification of about 75-fold. The enzyme utilized either NAD(H) or NADP(H) as coenzyme and was most effective with 5β-androstane-3,17-dione or dihydrocortisone which were 5 times as effective as 5α-androstane-3,17-dione. The microsomal fraction also contained a 3α-hydroxysteroid dehydrogenase with properties similar to the enzyme in the soluble fraction (HURLOCK and TALALAY, 1959; AOSHIMA et al., 1964).

KOIDE (1963) effected a purification of approximately 100-fold in the 3α-hydroxysteroid dehydrogenase activity in the soluble fraction of male rat liver by precipitation at 35–50 and 50–70% $(NH_4)_2SO_4$ saturation, adsorption on Cγ-alumina, and hydroxylapatite-DEAE-cellulose chromatography. The NAD- and NADP-linked activities were concentrated to the same degree. The enzyme activity was very labile especially after the $(NH_4)_2SO_4$ purification. The activities were stabilized by the addition of glycerol at 30% to the media (KOIDE, 1965a). The two $(NH_4)_2SO_4$ fractions contained 3β-hydroxysteroid oxidoreductase activity with 3β-hydroxyandrost-5-en-17-one of less than 0.1% and with 3β-hydroxy-5α-androstan-17-one less than 2% that with 3β-hydroxy-5β-androstan-17-one as substrate. NADP gave less than 1% the activity of NAD as coenzyme. KOIDE (1965b) obtained a further 2- to 3-fold purification of the NAD(P) 3α-hydroxysteroid oxidoreductase by submitting the active fractions from Sephadex-G-100 filtration to DEAE-cellulose chromatography. Analysis on disc gel electrophoresis gave different mobilities for the 3α- and 3β-hydroxysteroid dehydrogenases on enzyme staining (DOMAN and KOIDE, 1966). Both the NAD- and NADP-linked 3α-hydroxysteroid oxidoreductases were in the same band. Protein staining of the 3α-hydroxy enzyme gave a major band, which corresponded to the enzyme band, and three minor protein bands. Electrophoretic analysis of a mixture of the 3α- and 3β-oxidoreductases suggested that there were two distinct 3α-oxidoreductases; a NAD-specific and a dual pyridine nucleotide-linked enzyme.

PIETRUSZKO and BARON (1965) also reported a 100-fold purification of the NAD(P) 3α-hydroxysteroid oxidoreductase activity from rat (male and female?) liver by a procedure very similar to that of KOIDE. The cytosol was submitted to $(NH_4)_2SO_4$ fractionation; the 50–80% $(NH_4)_2SO_4$ precipitate was submitted successively to DEAE-cellulose chromatography, second fractionation with $(NH_4)_2SO_4$, and chromatography on Sephadex-G-25, CM-cellulose, and calcium phosphate *chromatography*. The pH optimum for NAD-linked dehydrogenase activity (androsterone as substrate) was 8.0 and 9.6 for NADP-linked activity. The reduction of androstane-3,17-dione with NADH or NADPH was at the broad pH range of 6.5–8.5. Some differences in the response of the NAD- and NADP-linked activities to inhibitors and salts were observed.

## C. 5β-Reductase ($\Delta^4$-5β-Hydrogenase)

### Pig

VAN DOORN et al. (1973) partially purified the 5β-reductase activity in the cytosol of the female pig liver by $(NH_4)_2SO_4$ precipitation at 40–70% saturation and Sephadex-G-100 chromatography. The enzyme preparation was very labile even after the $(NH_4)_2SO_4$ step. The addition of EDTA or several SH compounds did not stabilize the enzyme activity. Sulfhydryl reagents inhibited the enzyme activity. The partially purified preparation exhibited a number of other proteins on disc gel electrophoresis. The protein with testosterone $\Delta^4$-$\beta$-reductase activity eluted from a column of Sephadex G-100 had a molecular weight corresponding to 25000–30000. A pH optimum of approximately 6.4 was observed in Tris-Maleate buffer. The enzyme was active with a wide variety of $C_{19}$ and $C_{21}$ steroids.

## VI. Summary

The androgenic-anabolic steroids are metabolized by practically every peripheral tissue in the body. The qualitative and quantitative conversions are different for the various tissues and animals and are further modified by age, sex, endocrine, and physiologic status of the animal and certain drugs. The metabolism of the steroids is effected by specific enzymes which are localized in different parts of the cell. The subcellular localization of these enzymes is different for many of the tissues. Several of the enzymes have been partially purified and one has been purified. The wide variety of metabolites formed by the various tissues indicates the presence of numerous specific enzymes, e.g., 17$\beta$-, 17$\alpha$-, 3$\alpha$-, 3$\beta$-; and $\Delta^4$- and $\Delta$-$^4$ and $\Delta^5$-3$\beta$-oxidoreductases, 5$\alpha$ and 5$\beta$-reductases, and hydroxylases for apparently every available carbon on the ring system of the steroid molecule. The oxidoreductases and 5-reductases are specific for either NAD(H) or NADP(H) or may utilize both enzymes with different effectiveness. The hydroxylases utilize NADPH and are located primarily in the microsomal fraction. Some of the tissues also form sulfate or glucuronide conjugates of the steroids which are also susceptible to metabolism by the various steroid enzymes. The variation in the formation of the metabolites by the different tissues and by different animal species probably is for the regulation and induction of specific metabolic events in the specific tissue.

## References

ABRAHAM, R., STAUDINGER, HJ.: Isolierung und Identifizierung eines neuen Testosteronmetaboliten 5$\alpha$-Androstantriol-(3$\alpha$,6$\beta$,17$\beta$). Z. physiol. Chem. **346**, 198—207 (1966).

ALBAUM, G., STAIB, W.: Über den Metabolismus von $\Delta^4$-Androsten-3,17-dion bei Inkubation mit Rattenlebermikrosomen. Biochem. Z. **342**, 120—128 (1965).

ALTMAN, K., GORDON, G. G., SOUTHREN, A. L., VITTEK, J., WILKER, S.: Induction of hepatic testosterone A-ring reductase by medroxy-progesterone acetate. Endocrinology **90**, 1252—1260 (1972).

AOSHIMA, Y., KOCHAKIAN, C. D.: Activity, intracellular distribution and some properties of 17$\beta$-hydroxy-$C_{19}$-steroid dehydrogenases in liver and kidney. Endocrinology **72**, 106—114 (1963).

AOSHIMA, Y., KOCHAKIAN, C. D., JADRIJEVIC, D.: TPN- and DPN-specific 3$\alpha$-hydroxy- and $\Delta^5$-3$\beta$-hydroxysteroid dehydrogenases of liver and kidney. Endocrinology **74**, 521—531 (1964).

ARIMASA, N., KOCHAKIAN, C. D.: Subcellular localization of the $C_{19}$-steroid reductase activities of female rat liver. Steroids **19**, 325—355 (1972).

ARIMASA, N., KOCHAKIAN, C. D.: Epitestosterone and 5$\alpha$-androstane-3$\alpha$,17$\beta$-diol: the characteristic metabolites of androst-4-ene-3,17-dione produced by mouse kidney in vitro. Endocrinology **92**, 72—82 (1973).

AXELROD, L. R., MILLER, L. L.: The bioxidation of steroids at carbon 6 by isolated perfused rat livers. Arch. Biochem. **49**, 248—249 (1954).

Axelrod, L. R., Miller, L. L., Herling, F.: The metabolism of testosterone in the isolated perfused dog liver. J. biol. Chem. **219**, 455—461 (1956).

Baldi, A., Charreau, E. H.: 17$\beta$-Hydroxysteroid dehydrogenase activity in rat submaxillary glands. Its relation with sex and age. Endocrinology **90**, 1643—1646 (1972).

Becker, H., Kaufmann, J., Klosterhalfen, H., Voigt, K. D.: In vivo uptake and metabolism of $^3$H-testosterone and $^3$H-5$\alpha$-dihydrotestosterone by human benign prostatic hypertrophy. Acta endocr. (Kbh.) **71**, 589—599 (1972).

Bénard, H., Cruz-Horn, A., David, H.: Sur le mécanisme de la transformation de l'androsterone et de l'épiandrostérone au cours de la perfusion du foie isolé de Lapin, en aérobiose. C.R. Soc. Biol. **155**, 235—237 (1961).

Bénard, H., Cruz-Horn, A., David, H., Seeman, A.: Transformation de l'androstérone en autres composés 17-cétostéroids au cours de la perfusion du foie isolé du Lapin, en aérobiose. C.R. Soc. Biol. **153**, 982—984 (1959).

Berg, A., Gustafsson, J.-Å.: Regulation of hydroxylation of 5$\alpha$-androstane-3$\alpha$,17$\beta$-diol in liver microsomes from male and female rats. J. biol. Chem. **248**, 6559—6567 (1973).

Blaquier, J., Forchielli, E., Dorfman, R. I.: In vitro metabolism of androgens in whole human blood. Acta endocr. (Kbh.) **55**, 697—704 (1967).

Blomquist, C. H., Nelson, R. M., Hakanson, E. Y.: Structural and kinetic properties of 17$\beta$-hydroxysteroid dehydrogenase. J. Steroid Biochem. **5**, 319 (1974). (abstract).

Booth, W. D.: The occurrence of testosterone and 5$\alpha$-dihydrotestosterone in the submaxillary salivary gland of the boar. J. Endocr. **55**, 119—125 (1972).

Booth, J., Gillette, J. R.: The effect of anabolic steroids on drug metabolism by microsomal enzymes in rat liver. J. Pharm. exp. Ther. **137**, 374—379 (1962).

Breuer, H., Breuer, J., Lisboa, B. P.: Intermediary metabolism of testosterone in human liver under normal and pathological conditions. In: Testosterone. Proc. workshop conf. April 20–22, 1967 at Tremsbuttel, pp. 76—79. Stuttgart: Georg Thieme 1968.

Breuer, H., Dahm, K.: Anreicherung und Charakterisierung einer 17$\beta$-hydroxysteroid: NAD(P)-oxydoreductase der Rattenniere. Biochim. biophys. Acta (Amst.) **85**, 29—37 (1964).

Breuer, H., Dahm, K., Norymberski, J. K.: Enzymic formation of 3$\beta$-hydroxyandrost-4-en-17-one. J. Endocr. **27**, 357—358 (1963).

Brodie, B. B.: Pathways of drug metabolism. J. pharm. Pharmacol. **8**, 1—17 (1956).

Bullock, L. P., Bardin, C. W.: Androgen receptors in mouse kidney: A study of male, female and androgen-insensitive (tfm-y) mice. Endocrinology **94**, 746—756 (1974).

Bullock, L. P., Bardin, C. W., Gram, T. E., Schroeder, D. H., Gillette, J. R.: Hepatic ethylmorphine demethylase and $\Delta^4$-steroid reductase in the androgen-insensitive pseudohermaphroditic rat. Endocrinology **88**, 1521—1523 (1971).

Burchardt, P., Tamm, J., Voigt, K. D.: Metabolism of testosterone and epitestosterone in perfused isolated canine liver. Symp. Dtsch. ges. Endokr. **13**, 122—125 (1968).

Cardinali, D. P., Nagle, C. A., Rosner, J. M.: Metabolic fate of androgens in the pineal organ: Uptake, binding to cytoplasmic proteins and conversion of testosterone into 5$\alpha$-reduced metabolites. Endocrinology **95**, 179—187 (1974).

Carpenter, M. P., Howard, C. N., Jr.: Vitamin E, steroids, and liver microsomal hydroxylations. Amer. J. clin. Nutr. **27**, 966—979 (1974).

Chamberlain, J., Jagarinec, N., Ofner, P.: Catabolism of $C_{19}$-steroids by subcellular fractions of mammalian and avian tissues. I. Hydroxylation of ring A-saturated substrates by rat-liver microsomes. Steroids, Suppl. II, 1—12 (1965).

Clark, L. C., Jr., Kochakian, C. D.: The in vitro metabolism of testosterone to $\Delta^4$-androstenedione-3,17, cis-testosterone and other steroids by rabbit liver slices. J. biol. Chem. **170**, 23—33 (1947).

Clark, L. C., Jr., Kochakian, C. D., Lobotsky, J.: The in vitro metabolism of $\Delta^4$-androstenedione-3,17 to testosterone, cis-testosterone and several unidentified steroids. J. biol. Chem. **171**, 493—500 (1947).

Coffey, J. C.: Steroid metabolism by mouse submaxillary glands. I. In vitro metabolism of testosterone and 4-androstene-3,17-dione. Steroids **22**, 247—257 (1973).

Colás, A.: The 16$\alpha$-hydroxylation of dehydroepiandrosterone (3$\beta$-hydroxyandrost-5-en-17-one) by rat-liver slices. Biochem. J. **82**, 390—394 (1962).

CONNEY, A. H., KLUTCH, A.: Increased activity of androgen hydroxylases in liver microsomes of rats pretreated with phenobarbital and other drugs. J. biol. Chem. **238**, 1611—1617 (1963).
CREPY, O., LACHESE, B., JAYLE, M. F.: Transformation du sulfate d'androstane-3α-yl-17α-ol en presence de coupes de foie de lapin. Description de la technique utilisée. C.R. Acad. Sci. (Paris) **257**, 2163—2165 (1963).
DENEF, C.: Effect of hypophysectomy and pituitary implants at puberty on the sexual differentiation of testosterone metabolism in rat liver. Endocrinology **94**, 1577—1582 (1974).
DOMAN, E., KOIDE, S. S.: Analysis of 3α- and 3β-hydroxysteroid oxidoreductases of rat liver by disc electrophoresis. Biochim. biophys. Acta (Amst.) **128**, 209—211 (1966).
EINARSSON, K., GUSTAFSSON, J.-Å., GUSTAFSSON, B. E.: Differences between germ-free and conventional rats in liver microsomal metabolism of steroids. J. biol. Chem. **248**, 3523—3630 (1973).
EINARSSON, K., GUSTAFSSON, J.-Å., STENBERG, Å.: Neonatal imprinting of liver microsomal hydroxylation and reduction of steroids. J. biol. Chem. **248**, 4987—4997 (1973).
ELATTAR, T. M. A.: In vitro metabolism of estrone 2, 4, 6, 7-$^3$H and 4-androstene-3,17-dione-1,2-$^3$H in submandibular gland and submandibular gland cancer tumor. Steroids **24**, 519—526 (1974a).
EL ATTAR, T. M. A.: The in vitro conversion of male sex steroid, [1,2-$^3$H]-androstenedione in normal and inflamed human gingiva. Arch. oral. Biol. **19**, 1185—1190 (1974b).
ENDAHL, G. L., KOCHAKIAN, C. D.: Partial purification and further characterization of the triphosphopyridine nucleotide specific $C_{19}$-17β-hydroxysteroid dehydrogenase of guinea pig liver. Biochim. biophys. Acta (Amst.) **62**, 245—250 (1962).
ENDAHL, G. L., KOCHAKIAN, C. D., HAMM, D. I.: Separation of a triphosphopyridine nucleotide-specific from a diphosphopyridine nucleotide-specific 17β-hydroxy (testosterone) dehydrogenase of guinea pig liver. J. biol. Chem. **235**, 2792—2796 (1960).
ENGELHARDT, D., EISENBURG, J., UNTERBURGER, P., KARL, H. J.: Untersuchungen über den Stoffwechsel von Testosterone und $\Delta^4$-Androstendion in der Leber des Menschen. Klin. Wschr. **49**, 439—440 (1971).
ENGELHARDT, D., UNTERBURGER, P., KARL, H. J.: Studies on testosterone metabolism in liver and extrahepatic tissues of man. J. Steroid Biochem. **5**, 318—319 (1974). (abstract).
FLORES, F., NAFTOLIN, F., RYAN, K. J.: Aromatization of androstenedione and testosterone by rhesus monkey hypothalamus and limbic system. Neuroendocrinology **11**, 177—182 (1973).
FORCHIELLI, E., DORFMAN, R. I.: Separation of $\Delta^4$-5α-and $\Delta^4$-5β-hydrogenases from rat liver homogenates. J. biol. Chem. **223**, 443—448 (1956).
FREIRE, O., BRECKWOLDT, M., LISBOA, B. P.: In vitro study on the metabolism of oestrone and testosterone by the rhesus monkey kidney. Acta endocr. (Stockh.) Suppl. **184**, 155 (1974).
GHRAF, R., VETTER, U., SCHRIEFERS, H.: Die Organspezifität der sexuellen Differenzierung der 17β-hydroxysteroid-dehydrogenase-Aktivitäten der Ratte. Z. physiol. Chem. **355**, 543—550 (1974).
GHRAF, R., LAX, E. R., HOFF, H.-G., SCHRIEFERS, H.: Regulation of the enzymes involved in the metabolism of steroid hormones in rat liver: The effect of 19-nortestosterone and the influence of cyproterone acetate on the action of testosterone and 5α-dihydrotestosterone. Acta endocr. (Kbh.) **77**, 287—297 (1974).
GLOYNA, R. E., WILSON, J. D.: A comparative study of the conversion of testosterone to 17β-hydroxy-5α-androstan-3-one (dihydrotestosterone) by prostate and epididymis. J. clin. Endocr. **29**, 970—977 (1969).
GOLD, N. I., GARREN, L. D.: Effect of experimental diabetes on steroid metabolism. II. Alterations in androst-4-ene-3,17-dione and cortisol metabolism. J. biol. Chem. **239**, 2796—2803 (1964).
GUSTAFSSON, J.-Å.: Androgen responsiveness of the liver of the developing rat. Biochem. J. **144**, 225—229 (1974).
GUSTAFSSON, J.-Å., LISBOA, B. P.: Studies on the metabolism of $C_{19}$-steroids in rat liver. Biosynthesis of saturated 17-oxo-$C_{19}O_3$ steroids in rat liver microsomes. Europ. J. Biochem. **16**, 475—480 (1970).
GUSTAFSSON, J.-Å.: Different mechanisms of regulation of nuclear reduced nicotinamide-adenine dinucleotide phosphate-dependent 3-oxo steroid 5α-reductase activity in rat liver, kidney and prostate. Biochem. J. **142**, 273—277 (1974a).
GUSTAFSSON, J.-A., POUSETTE, A.: Properties of nuclear 5α-reductase in rat liver. Biochemistry **13**, 875—881 (1974b).

GUSTAFSSON, J.-Å., STENBERG, Å.: Irreversible androgenic programming at birth of microsomal and soluble rat liver enzymes active on 4-androstene-3,17-dione and 5α-androstane-3α,17β-diol. J. biol. Chem. **249**, 711—718 (1974b).

GUSTAFSSON, J.-Å., STENBERG, Å.: Neonatal programming of androgen responsiveness of liver of adult rats. J. biol. Chem. **249**, 719—723 (1974a).

GUSTAFSSON, J.-Å., STENBERG, Å.: Partial masculinization of rat liver enzyme activities following treatment with FSH. Endocrinology **96**, 501—504 (1975).

GUSTAFSSON, J.-Å., LISBOA, P. B., SJÖVALL, J.: Studies on the metabolism of $C_{19}$ steroids in rat liver. 2. Biosynthesis of hydroxylated derivatives of 17β-hydroxy-5α-androstane-3-one in rat liver microsomes. Europ. J. Biochem. **5**, 437—443 (1968).

HAMM, D. I., KOCHAKIAN, C. D., CARROLL, B. R.: Reduction of androstane-3,17-dione by liver homogenates. Proc. Soc. exp. Biol. (N.Y.) **93**, 493—496 (1956).

HARRI, M. P., NIENSTEDT, W., HARTIALA, K.: Testosterone metabolism by canine intestine. Suomen Kemistilehti **43**, 395—399 (1970).

HASNAIN, S., WILLIAMSON, D. G.: The separation and partial purification of the soluble 17α- and 17β-hydroxysteroid dehydrogenases of rabbit liver. Can. J. Biochem. **52**, 120—125 (1974).

HEINRICHS, W. L., FEDER, H. H., COLAS, A.: The steroid 16α-hydroxylase system in mammalian liver. Steroids **7** (1), 91—98 (1966).

HEYNS, W., DE MOOR, P.: A 3(17)β-hydroxysteroid dehydrogenase in rat erythrocytes. Conversion of 5α-dihydrotestosterone into 5α-androstane-3β,17β-diol and purification of the enzyme by affinity chromatography. Biochim. biophys. Acta (Amst.) **358**, 1—13 (1974).

HOFF, H.-G., SCHRIEFERS, H.: Sexuell differenzierte und sexuell undifferenzierte 3α- und 3β-hydroxysteroid-dehydroganse-Aktivitäten und ihre intrazellulare Lokalisation in der Rattenleber. Z. physiol. Chem. **354**, 507—513 (1973).

HOFF, H.-G., GHRAF, R., RAIBLE, M., SCHRIEFERS, H.: Ontogenses van Hydroxysteroid-Dehydrogenase-Aktivitäten in der Rattenleber. Z. physiol. Chem. **354**, 306—311 (1973).

HUBENER, H. J.: Enzymatische Umwandlung von Steroiden. IV. Umwandlungen und Abbau von $C_{21}$- und $C_{19}$-Steroiden in der Leber. Z. physiol. Chem. **298**, 283—290 (1954).

HUDSON, P. B., YANDEL, F., JR., LOMBARDO, M. E.: Studies on the in vitro perfusion of steroids through the dog kidney. J. biol. Chem. **220**, 699—711 (1956).

HURLOCK, B., TALALAY, P.: Microsomal 3α- and 11β-hydroxysteroid dehydrogenases. Arch. Biochem. **80**, 468—470 (1959).

HUSSEIN, K. A., KOCHAKIAN, C. D.: DPN- and TPN-17β-hydroxy-$C_{19}$-steroid dehydrogenases: Intracellular localization in dog prostate. Acta endocr. (Kbh.) **59**, 459—474 (1968a).

HUSSEIN, K. A., KOCHAKIAN, C. D.: 17β-Hydroxy-$C_{19}$-steroid dehydrogenase activity of dog prostate, solubilization and partial purification and characterization. Steroids **12**, 589—605 (1968b).

JACOBSOHN, G., HOCHBERG, R. B.: 17β-Hydroxysteroid dehydrogenase from human red blood cells. J. biol. Chem. **243**, 2985—2994 (1968).

JAFFE, R. B.: Testosterone metabolism in target tissues. Steroids **14**, 483—498 (1969).

JAGARINEC, N., CHAMBERLAIN, J., OFNER, P.: Metabolism of [4-$^{14}$C] testosterone by mouse liver microsomes. Biochim. biophys. Acta (Amst.) **144**, 479—481 (1967).

JARASCH, E.-D., FRANKE, W. W.: Is cytochrome oxidase a constituent of nuclear membranes? J. biol. Chem. **249**, 7245—7254 (1974).

JOSHI, S., DUNCAN, E. L., ENGEL, L. L.: Soluble guinea pig liver TPN dependent 17β-hydroxysteroid (testosterone) dehydrogenase: Partial purification and substrate specificity. Steroids **1**, 508—527 (1963).

JUNGMANN, R. A., KOT, E., SCHWEPPE, J. S.: In vitro 17β-hydroxysteroid dehydrogenase activity in human thymus. Steroids **10**, 397—409 (1967).

KAGEURA, E., TOKI, S.: New aspect of guinea pig liver 17β-hydroxysteroid (testosterone) dehydrogenase. Life Sci. **10**, 469—474 (1971).

KAGEURA, E., TOKI, S.: Guinea pig liver 3-hydroxyhexobarbital dehydrogenase as a 17β-hydroxysteroid dehydrogenase. Biochim. biophys. Acta (Amst.) **341**, 172—177 (1974).

KATO, R., ONODA, K.-I., OMORI, Y.: Mechanism of thyroxine-induced increase in steroid $\Delta^4$-reductase activity in male rats. Endocr. jap. **17**, 215—219 (1970).

KING, R. J. B.: The demonstration of a 16-hydroxysteroid dehydrogenase in rat kidney. Biochem. J. **76**, 7 P (1960).

KINSON, G. A., KOCHAKIAN, C. D.: DPN- and TPN-17$\beta$-hydroxy-$C_{19}$-steroid dehydrogenase activities in skeletal muscles. Endocrinology **74**, 233—235 (1964).

KLEMPIEN, E. J., VOIGT, K. D., TAMM, J.: Der Umsatz von Dehydroisoandrosteron in der Hundeleber. Acta endocr. (Kbh.) **36**, 498—510 (1961).

KNAPSTEIN, P., DAVID, A., WU, C.-H., ARCHER, D. F., FLICKINGER, G. L., TOUCHSTONE, J. C.: Metabolism of free and sulfoconjugated DHEA in brain tissue in vivo and in vitro. Steroids **11**, 885—896 (1968).

KOCHAKIAN, C. D.: Metabolism of androgens by tissue enzymes. In: Proc. Fourth International Congress of Biochemistry, Biochemistry of Steroids **4**, 196—207 (1959). Pergamon Press. London-New York.

KOCHAKIAN, C. D., APOSHIAN, H. V.: The in vitro metabolism of 3$\alpha$,17$\beta$-androstanediol by liver and kidney. Arch. Biochem. **37**, 442—448 (1952).

KOCHAKIAN, C. D., ENDAHL, B. R.: Testosterone regulation of 17$\beta$-hydroxysteroid dehydrogenases. Proc. Soc. exp. Biol. (N.Y.) **104**, 720—722 (1960).

KOCHAKIAN, C. D., NALL, D. M.: The metabolism of epitestosterone by rabbit tissues in vitro. J. biol. Chem. **204**, 91—94 (1953).

KOCHAKIAN, C. D., STIDWORTHY, G.: Metabolism of $\Delta^4$-androstene-3,17-dione by tissue homogenates. J. biol. Chem. **210**, 933—939 (1954).

KOCHAKIAN, C. D., CARROLL, B. R., UHRI, B.: Comparisons of the oxidation of $C_{19}$-hydroxysteroids by guinea pig liver homogenates. J. biol. Chem. **224**, 811—818 (1957b).

KOCHAKIAN, C. D., CARROLL, B. R., UHRI, B.: Metabolism of $C_{19}$ steroids by homogenates of kidney, heart, testes and prostate. Amer. J. Physiol. **189**, 83—85 (1957c).

KOCHAKIAN, C. D., GONGORA, J., PARENTE, N.: Metabolism of testosterone by homogenates of rabbit liver and kidney. J. biol. Chem. **196**, 243—246 (1952).

KOCHAKIAN, C. D., RAUT, V. S., KINSON, G.: Use of coenzymes I and II in interconversion of testosterone and androstenedione. Amer. J. Physiol. **200**, 348—350 (1961).

KOCHAKIAN, C. D., RAUT, V., NALL, D. M.: Metabolism of testosterone by guinea pig liver and kidney homogenates. Amer. J. Physiol. **189**, 78—82 (1957a).

KOCHAKIAN, C. D., STEVENSON, D., MAYUMI, T.: Dependence of the guinea pig kidney 17$\beta$-hydroxy-$C_{19}$-steroid dehydrogenase isoenzyme profile on androgen. Biochem. biophys. Res. Commun. **54**, 519—523 (1973).

KOIDE, S. S.: Purification of 3$\alpha$-hydroxysteroid dehydrogenase obtained from the soluble fraction of rat liver. Arch. Biochem. **101**, 278—285 (1963).

KOIDE, S. S.: Separation of 3$\alpha$- and 3$\beta$-hydroxysteroid oxidoreductases from the soluble fraction of rat liver by gel filtration. Steroids **6**, 123—128 (1965a).

KOIDE, S. S.: Further study on the purification of 3$\alpha$-hydroxysteroid dehydrogenase of rat liver. Biochim. biophys. Acta (Amst.) **110**, 189—194 (1965b).

KRIEG, M., SZALAY, R., VOIGT, K. D.: Binding and metabolism of testosterone and of 5$\alpha$-dihydrotestosterone in bulbocavernosus levator ani (BCLA) of male rats: in vivo and in vitro studies. J. Steroid Biochem. **5**, 453—459 (1974).

KSIAZKIEWICZ-SZAPIRO, M. D.: Experimental investigations of the rat small intestine following castration and testosterone administration. Folia Histochem. Cytochem. **8**, 369—388 (1970).

LAX, E. R., SCHRIEFERS, H.: $\Delta^4$-3$\beta$-Hydroxysteroid dehydrogenase activity in rat liver. Intracellular distribution and sex dependency. Europ. J. Biochem. **42**, 561—566 (1974).

LAX, E. R., HOFF, H.-G., GRAF, R., SCHRODER, E., SCHRIEFERS, H.: The role of the hypophysis in the regulation of sex differences in the activities of enzymes involved in hepatic steroid hormone metabolism. Z. physiol. Chem. **355**, 1325—1331 (1974).

LEYBOLD, K., STAUDINGER, HJ.: Geschlechtsunterschiede im Steroidstoffwechsel von Rattenlebermikrosomen. Biochem. Z. **331**, 389—398 (1959).

LINDNER, H. R.: Androgens in the bovine testis and spermatic vein blood. Nature (Lond.) **183**, 1605—1606 (1959).

LINDNER, H. R.: Androgens and related compounds in the spermatic vein blood of domestic animals. II. Species linked differences in the metabolism of androstenedione in blood. J. Endocr. **23**, 161—166 (1961).

LINDNER, H. R.: The 17$\alpha$-hydroxy-$C_{19}$-steroid dehydrogenase activity of ovine blood. Steroids Suppl. II. 133—148 (1965).

LISBOA, B. P., DROSSE, I., BREUER, H.: Stoffwechsel von Testosteron in Leberschnitten des Menschen. Z. physiol. Chem. **342**, 123—131 (1965).

LISBOA, B. P., GUSTAFSSON, J.-Å., SJÖVALL, J.: Studies on the metabolism of $C_{19}$-steroids in rat liver. I. Hydroxylation of testosterone in rat liver microsomes. Europ. J. Biochem. **4**, 496—505 (1968).

LIU, D. K., KOCHAKIAN, C. D.: Partial purification and some properties of guinea pig kidney 17$\beta$-hydroxy-$C_{19}$-steroid dehydrogenase. Steroids **19**, 701—719 (1972a).

LIU, D. K., KOCHAKIAN, C. D.: Heterogeneity of guinea pig kidney 17$\beta$-hydroxy-$C_{19}$-steroid dehydrogenase activity observed by disc gel electrophoresis. Steroids **19**, 721—729 (1972b).

MAHOUDEAU, J. A., CORVOL, P., BRICAIRE, H.: Rabbit testosterone-binding-globulin. II. Effect on Androgen metabolism in-vivo. Endocrinology **92**, 1120—1125 (1973).

MASSA, R., MARTINI, L.: Testosterone metabolism: A necessary step for activity? J. Steroid Biochem. **5**, 941—947 (1974).

MASSA, R., STUPNICKA, E., KNIEWALD, Z., MARTINI, L.: The transformation of testosterone into dihydrotestosterone by the brain and the anterior pituitary. J. Steroid Biochem. **3**, 385—399 (1972).

MCGUIRE, J. S., TOMKINS, G. M.: The multiplicity and specificity of $\Delta^4$-3-ketosteroid hydrogenases (5$\alpha$). Arch. Biochem. Biophys. **82**, 476—477 (1959a).

MCGUIRE, J. S., TOMKINS, G. M.: The effects of thyroxin administration on the enzymic reduction of $\Delta^4$-3-ketosteroids. J. biol. Chem. **234**, 791—794 (1959b).

MCGUIRE, J. S., TOMKINS, G. M.: The heterogeneity of $\Delta^4$-3-ketosteroid reductases (5$\alpha$). J. biol. Chem. **235**, 1634—1638 (1960).

MCGUIRE, J. S., HOLLIS, V. W., JR., TOMKINS, G. M.: Some characteristics of the microsomal steroid reductases (5$\alpha$) of rat liver. J. biol. Chem. **235**, 3112—3117 (1960).

MEIGS, R. A., RYAN, K. J.: 16$\alpha$-Hydroxysteroid dehydrogenase of rat kidney. Purification, assay, and properties. J. biol. Chem. **241**, 4011—4015 (1966).

MORFIN, R. F., ALIAPOULIOS, M. A., CHAMBERLAIN, J., OFNER, P.: Metabolism of testosterone-4-$^{14}$C by the canine prostate and urinary bladder in vivo. Endocrinology **87**, 394—405 (1970).

MOWSZOWICZ, I., BARDIN, C. W.: In vitro androgen metabolism in mouse kidney: high 3-keto-reductase (3$\alpha$-hydroxysteroid dehydrogenase) activity relative to 5$\alpha$-reductase. Steroids **23**, 793—807 (1974).

MULDER, E., LAMERS-STAHLHOFEN, G. J. M., VAN DER MOLEN, H. J.: Isolation and characterization of 17$\beta$-hydroxysteroid dehydrogenase from human erythrocytes. Biochem. J. **127**, 649—659 (1972).

MUROTA, S., TAMAOKI, B.: Metabolism of progesterone and testosterone by chick embryo cartilage in vitro. Biochim. biophys. Acta (Amst.) **137**, 347—355 (1967).

NAFTOLIN, F., RYAN, K. J., PETRO, Z.: Aromatization of androstenedione by the anterior hypothalamus of adult male and female rats. Endocrinology **90**, 295—298 (1972).

NICOL, M. M., SAVOURÉ, M., LERAY, G.: Conversion érythrocytaire de la déhydroépiandrostérone chez l'Homme et le Rat; influence des protéines sériques. C.R. soc. Biol. **165**, 1771—1774 (1971).

NICOL, M. M., SAVOURÉ, N., RICO, S.: Sur le specificité de la 17$\beta$-hydroxysteroide-deshydrogenase erythrocytaire du Rat Wistar. C.R. Acad. Sci. [D] (Paris) **268**, 1552—1555 (1969).

NICOL, M. M., SAVOURÉ, M., RICO, S., DESPLANQUES, D.: Conversion de la transdéhydroépiandrostérone en $\Delta^5$-androstènediol par l'erythrocyte du rat Wistar. C.R. Acad. Sci. (Paris) [D] **266**, 1663—1664 (1968).

OFNER, P.: A preliminary investigation of the products of the metabolism of testosterone by rat liver in vitro. Biochem. J. **61**, 287—297 (1955).

OTA, M., SATO, N., OBARA, K.: Induction of androgen-metabolizing enzymes by testosterone in female rat liver. J. Steroid Biochem. **5**, 319 (1974).

PATTERSON, D. C., CLARK, A. F., BIRD, C. E.: Testosterone $\Delta^4$-reductase activity in rat liver: Hormonal control in vivo. J. Endocr. **63**, 181—189 (1974).

PEREZ, A. E., ORTIZ, A., CABEZA, M., BEYER, C., PEREZ-PALACIOS, G.: In vitro metabolism of $^3$H-androstenedione by the male rat pituitary, hypothalamus, and hippocampus. Steroids **25**, 53—62 (1975).

PEREZ-PALACIOS, G., CASTANEDA, E., GOMEZ-PEREZ, F., PEREZ, A. E., GUAL, C.: In vitro metabolism of androgens in dog hypothalamus, pituitary, and limbic system. Biol. Reprod. **3**, 205—213 (1970).

PIETRUSZKO, R., BARON, D. N.: Dehydrogenation of androsterone by purified 3α-hydroxy steroid-dependent nicotinamide-adenine dinucleotide (phosphate)-transhydrogenating enzyme of rat liver. Biochem. J. **96**, 557—566 (1965).

PIETRUSZKO, R., CLARK, A., GRAVES, J. M., RINGOLD, H. J.: The steroid activity and multiplicity of crystalline horse liver alcohol dehydrogenase. Biochem. biophys. Res. Commun. **23**, 526—534 (1966).

PLASSE, J.-C., LISBOA, B. P.: Studies on the metabolism of steroids in the foetus. Biosynthesis of epitestosterone in the humanfoetal liver. Europ. J. Biochem. **39**, 443—447 (1973).

RAITH, L., KARL, H. J.: Metabolism of testosterone and androstenedione in human leucocytes. J. Steroid Biochem. **5**, 319—320 (1974).

REPKE, K., SAMUELS, L. T.: Evidence favoring the nonspecificity of 3-hydroxysteroid dehydrogenases in relation to steroid conformation. Biochemistry **3**, 685—689 (1964).

RONGONE, E. L., STRENGTH, D. R., BOCKLAGE, B. C., DOISY, E. A.: The reduction of the 4, 5 double bond of steroids by bovine blood proteins. J. biol. Chem. **225**, 959—967 (1957).

ROY, A. B.: The steroid 5α-reductase activity of rat liver and prostate. Biochimie **53**, 1031—1040 (1971).

RUBIN, B. L.: Sex differences in orientation of reduction products of 3-keto-$C_{19}$ steroids by rat liver homogenates. J. biol. Chem. **227**, 917—927 (1957).

RUBIN, B. L., DORFMAN, R. I.: In vitro conversion of testosterone to 17β-hydroxyandrostan-3-one. Proc. Soc. exp. Biol. (N.Y.) **91**, 585—586 (1956).

RUBIN, B. L., STRECKER, H. J.: Further studies on the sex difference in 3β-hydroxysteroid dehydrogenase activity of rat livers. Endocrinology **69**, 257—267 (1961).

RUBIN, B. L., STRECKER, H. J., KOFF, E. B.: Further observations on sex-influenced activity of 3β-hydroxysteroid dehydrogenase of rat liver. Endocrinology **72**, 764—770 (1963).

SCHNEIDER, J. J., MASON, H. L.: Studies on intermediary steroid metabolism. I. Isolation of $\Delta^5$-androstene-3(β),17(α)-diol and $\Delta^5$-androstene-3(β),16(β),17(α)-triol following the incubation of dehydroisoandrosterone with surviving rabbit liver slices. J. biol. Chem. **172**, 771—782 (1948a).

SCHNEIDER, J. J., MASON, H. L.: Studies on intermediary steroid metabolism. II. Compounds isolated following the incubation of androsterone and etiocholan-3(α)-ol-17-one with surviving rabbit liver slices. J. biol. Chem. **175**, 231—240 (1948b).

SCHRIEFERS, H., HOFF, H.-G., GHRAF, R.: Androgenabhängigkeit der Ontogenese des Aktivitätsmusters von Enzymen des Steroidhormonstoffwechsels in der Rattenleber. Z. physiol. Chem. **354**, 501—506 (1973).

SCHRIEFERS, H., GHRAF, R., HOFF, H.-G., OCKENFELS, H.: Einfluß von Alter und Geschlecht auf die Entwicklung und Differenzierung der Aktivitätsmuster von Enzymen des Steroidhormon-Stoffwechsels in der Leber von Ratten zweier verschiedener Tierstämme. Z. physiol. Chem. **352**, 1363—1371 (1971).

SHIMAZAKI, J., KURIHARA, H., IRO, Y., SHIDA, K.: Metabolism of testosterone in muscular tissue. Gumna J. med. Sci. **14**, 100—106 (1965).

SHOLITON, L. J., MARNELL, R. T., WERK, E. E.: Metabolism of testosterone-4-$^{14}$C by rat brain homogenates and subcellular fractions. Steroids **8**, 265—275 (1966).

SHOLITON, L. J., WERK, E. E., MACGEE, J.: The effect of diphenylhydantoin in vitro on the metabolism of testosterone by rat liver slices. Acta endocr. (Kbh.) **56**, 490—498 (1967).

SHOLL, S. A., ROBINSON, J. A., GOY, R. W.: Neural uptake and metabolism of testosterone and dihydrotestosterone in the guinea pig. Steroids **25**, 203—215 (1975).

SIEBERT, G., DAHM, H., BREUER, H.: Aktivitäten von Steroidenzymen in Zellkernen der Rattenleber. Naturwissenschaften **23**, 615 (1966).

STEVENSON, D., KOCHAKIAN, C. D.: Purification of male guinea pig kidney 17β-hydroxy-$C_{19}$-steroid dehydrogenase. Endocrinology **96**, 766—770 (1974).

STYLIANOU, M., FORCHIELLI, E., TRUMMILLO, M., DORFMAN, R. I.: Metabolism in vitro of 4-$^{14}$C-testosterone by a human liver homogenate. J. biol. Chem. **236**, 692—694 (1961).

SUZUKI, K., TAMAOKI, B.: Testosterone metabolism and 5α-dihydrotestosterone-binding macromolecule in rat kidney. Steroids Lipids Res. **4**, 266—276 (1973).

SWEAT, M.L., SAMUELS, L.T.: Diphosphopyridine nucleotide as an essential factor in the metabolism of testosterone by the liver. J. biol. Chem. **173**, 433—434 (1948).

SWEAT, M.L., SAMUELS, L.T., LUMRY, R.: Preparation and characterization of the enzyme which converts testosterone to androstenedione. J. biol. Chem. **185**, 75—84 (1950).

SZAMATOWICZ, M., DROSDOWSKY, M., JAYLE, M.F.: The role of testosterone and androstenedione as precursors of epitestosterone in guinea pigs (in vivo and in vitro studies). Acta endocr. **67**, 187—196 (1971).

TABEI, T., HEINRICHS, W.L.: Enzymatic oxidation and reduction of $C_{19}$-$\Delta^5$,3$\beta$-hydroxysteroids by hepatic microsomes. I. Biosynthesis of 3$\beta$,17$\beta$-dihydroxyandrost-5-en-16-one and sex differences in adult rats. Endocrinology **91**, 969—976 (1972).

TABEI, T., HEINRICHS, W.L.: Enzymatic oxidation and reduction of $C_{19}$-$\Delta^5$-3$\beta$-hydroxysteroids by hepatic microsomes. II. Effect of age in rats on 16,17-oxidoreduction of 3$\beta$-hydroxyandrost-5-en-17-one (DHA). Endocrinology **92**, 1161—1164 (1973).

TABEI, T., HEINRICHS, W.L.: Enzymatic oxidation and reduction of $C_{19}$-$\Delta^5$-3$\beta$-hydroxysteroids by hepatic microsomes. III. Critical period for the neonatal differentiation of certain mixed function oxidases. Endocrinology **94**, 97—103 (1974).

TABEI, T., HEINRICHS, W.L.: Enzymatic oxidation and reduction of $C_{19}$-$\Delta^5$-3$\beta$-hydroxysteroids by hepatic microsomes. V. Testosterone as a neonatal determinant in rats of the 7- and 16$\alpha$-hydroxylation and reduction of 3$\beta$-hydroxyandrost-5-en-17-one (DHA). Endocrinology **97**, 418—424 (1975).

TAKENOSHITA, R., TOKI, S.: Rabbit liver 3-hydroxyhexobarbital dehydrogenase. J. biol. Chem. **249**, 5428—5429 (1974).

THALER-DAO, H., BREUER, H.: Gonadal factors regulating the activity of the 17$\beta$-hydroxysteroid oxidoreductase in rat liver. Acta endocr. (Kbh.) **77**, 727—736 (1974).

THALER-DAO, H., DESCOMPS, B., SAINTOT, M., DE PAULET, A.C.: The 17$\beta$-hydroxysteroid NAD(P) oxidoreductases from female rabbit liver cytosol: separation and characterization. Biochimie **54**, 83—91 (1972).

THEORELL, H., TANIGUCHI, S., AKESON, A., SKURSKY, L.: Crystallization of a separate steroid-active liver alcohol dehydrogenase. Biochem. biophys. Res. Commun. **24**, 603—610 (1966).

THOMAS, P.Z.: Metabolism of androst-4-ene-3,17-dione-4-$^{14}C$ by rabbit skeletal muscle in vitro. The presence of a 5$\alpha$-steroid reductase in the particular fraction. J. biol. Chem. **243**, 6110—6114 (1968).

THOMAS, P.Z., DORFMAN, R.I.: Metabolism in vitro of androst-4-ene-3,17-dione-4-$^{14}C$ by rabbit skeletal muscle strips. Isolation of testosterone-$^{14}C$, 5$\alpha$-androstane-3,17-dione-$^{14}C$, and 3$\beta$-hydroxy-5$\alpha$-androstan-17-one-$^{14}C$. J. biol. Chem. **239**, 762—765 (1964a).

THOMAS, P.Z., DORFMAN, R.I.: Metabolism of androst-4-ene-3,17-dione-4-$^{14}C$ by rabbit skeletal muscle supernatant fraction. Isolation of 3$\beta$-hydroxyandrost-4-en-17-one-$^{14}C$ and testosterone-$^{14}C$. J. biol. Chem. **239**, 766—772 (1964b).

TOKI, S., KAGEURA, E.: Purification and properties of newly discovered guinea pig liver 17$\beta$-hydroxysteroid dehydrogenase. Proc. Symp. Drug. Metab. Action. 5th 1973 pp. 175—185 (1974).

TOMKINS, G.M.: A mammalian 3$\alpha$-hydroxysteroid dehydrogenase. J. biol. Chem. **218**, 437—447 (1956).

TOMKINS, G.M.: The enzymatic reduction of $\Delta^4$-3-ketosteroids. J. biol. Chem. **225**, 13—24 (1957).

UNGAR, F., GUT, M., DORFMAN, R.I.: Metabolism of $\Delta^4$-androstane-3$\beta$,17$\beta$-diol and $\Delta^4$-androstene-3$\alpha$,17$\beta$-diol in vitro. J. biol. Chem. **224**, 191—200 (1957).

VAN DER MOLEN, H.J., GROEN, D.: Interconversion of progesterone and 20$\alpha$-dihydroprogesterone and of androstenedione and testosterone in vitro by blood and erythrocytes. Acta endocr. (Kbh.) **58**, 419—444 (1968).

VAN DOORN, E.J., NDUAGUBA, J.C., CLARK, A.F.: Studies on partially purified pig liver steroid $\Delta^4$-5$\beta$-reductase activity. Can. J. Biochem. **51**, 1661—1668 (1973).

VERHOEVEN, G., DE MOOR, P.. Intranuclear metabolism of testosterone in kidneys of male and female rats. Endocrinology **89**, 842—846 (1971).

VERHOEVEN, G., DE MOOR, P.: Nucleus-associated 5$\alpha$-reductase activity in the rat. Gynec. Invest. **2**, 290—296 (1971/1972).

VILLEE, C.A., SPENCER, J.M.: Some properties of the pyridine nucleotide-specific 17$\beta$-hydroxy steroid dehydrogenases of guinea pig livers. J. biol. Chem. **235**, 3615—3619 (1960).

VITTEK, J., ALTMAN, K., GORDON, G. G., SOUTHREN, A. L.: The metabolism of 7α-$^3$H-testosterone by rat mandibular bone. Endocrinology **94**, 325—329 (1974).
VITTEK, J., ALTMAN, K., GORDON, G. G., SOUTHREN, A. L.: The metabolism of 7α-$^3$H-testosterone by rat oral mucosa: Effect of medroxyprogesterone acetate on testosterone A-ring reductase activity. Endocrinology **95**, 1473—1477 (1974).
WEINER, A. L., OFNER, P., SWEENEY, E. A.: Metabolism of testosterone-4-$^{14}$C by the canine submaxillary gland in vivo. Endocrinology **87**, 406—409 (1970).
WEISZ, J., GIBBS, C.: Conversion of testosterone and androstenedione to estrogens in vitro by the brain of female rats. Endocrinology **94**, 616—620 (1974).
WELCH, R. M., LEVIN, W., CONNEY, A. H.: Insecticide inhibition and stimulation of steroid hydroxylases in rat liver. J. Pharmacol. exp. Ther. **155**, 167—173 (1967).
WENZEL, M.: Oxydation des Anabolikums 1-Methylandrost-1-en-17β-ol-3-on mit Wasserstofftransfer auf Östrone. Acta endocr. (Kbh.) **67**, 517—530 (1971).
WENZEL, M., MÜTZEL, W.: Einfluß hormonell unwirksamer Steroide auf die Östradiol-Oxidation. I. Untersuchung über den Wasserstoff-Transfer von Östradiol-17β auf 5α-Androstan-17-on im Rattengewebe. Z. physiol. Chem. **351**, 1221—1231 (1970).
WENZEL, M., POLLOW, K.: Die Reduktion von Androgenen durch [17α-T]-Östradiol. Z. physiol. Chem. **348**, 1667—1676 (1967).
WENZEL, M., POLLOW, K., ZIEBARTH, T.: Oestradiol, ein bevorzugter Wasserstoffdonator bei der Reduktion von Androstendion im Lebergewebe. Z. Naturforsch. [B] **22**, 894—896 (1967).
WENZEL, M., WOLF, S.: Charakterisierung des Enzymsystems für den Wasserstofftransfer von Oestradiol-17β auf Androgene in Rattenleber-Mikrosomen und -Cytoplasma. Z. physiol. Chem. **350**, 1203—1212 (1969).
WEST, C. D., SAMUELS, L. T.: The metabolism of testosterone and related steroids by kidney tissue. J. biol. Chem. **190**, 827—835 (1951).
WILSON, J. D., LASNITZKI, I.: Dihydrotestosterone formation in fetal tissues of the rabbit and rat. Endocrinology **89**, 659—668 (1971).
YASUE, S., YAMADA, M., MIZUTANI, S., MATSUMOTO, K.: 5α-Reduction of testosterone in vitro by seminifersous tubules, prostate and muscle from immature rats. Endocr. jap. **21**, 161—166 (1974).
YATES, F. E., HERBST, A. L., URQUHART, J.: Sex difference in rate of Ring A reduction of $\Delta^4$-3-keto-steroids in vitro by rat liver. Endocrinology **63**, 887—902 (1958).

# V. Dissociation of the Androgenic and Other Hormonal Activities from the Protein Anabolic Effects of Steroids

G. O. Potts, A. Arnold, and A. L. Beyler

## I. Introduction

Much research effort has been expended over the last two decades in laboratories throughout the world in search of a molecule that possesses all the anabolic properties of the male hormone, testosterone, but does not elicit responses generally characterized as virilizing. These efforts have been justified on the premise that a "pure" anabolic agent would be expected to find many more clinical applications and be more generally accepted by the medical specialists than testosterone and its derivatives, particularly for longterm uses associated with debilitating disease states, slow rate of healing of hard or soft tissue, and, in general, for conditions where conservation of the body's quantity of protein or replenishment of depleted protein states were sought. The clinical preference for an anabolic over an androgenic agent in the above-listed uses would derive from the view that the androgenic or masculinizing actions of the latter would be unwanted or considered as side-effects particularly for children and women.

Although major strides in the desired direction have been made, at this writing the authors are not aware of any absolute separation of these activities. Rather the dissociation that has been achieved in most instances has been relative to that of testosterone or methyltestosterone. The degree of separation of activities of some steroids has been sufficient based upon the results of laboratory animal experiments to justify clinical testing in humans and subsequent introduction to the drug marketplace.

It is the intent of the authors to present in this chapter a considerable, if not exhaustive, body of data derived from animal experimentation that provide a basis for assessment of the protein anabolic actions of a host of androstane and androstene derived steroids. Emphasis in most cases will be placed on those data that characterize the extent to which the anabolic effect is dissociated from androgenic, as well as other hormonal properties. It is this aspect of endeavor which has taken on "global" interest. As has been mentioned by others (Krüskemper, 1968; Salvioni, 1962) difficulties arise in drawing meaningful assessments from a compilation of data from laboratory studies throughout the world due to differences in methodology as well as differences in treatment of the data from one laboratory to the next. This situation cannot be corrected until assay procedures have been standardized on an international basis for anabolic androgenic agents. Given the situation as it is, the authors feel that overall impressions can be gained of structure-activity relationships among the myriad of structures studied as long as the problem of lack of uniformity in procedural matters is recognized and acceptance of relative or comparative data is confined to the conditions under which the assessments were made.

This applies especially as it relates to the large body of data from published work comparing the myotrophic effect of various steroids based on levator ani weight responses with androgenic effects based on weight responses of the sex accessory organs of castrated male rats as described by HERSHBERGER et al. (1953) or EISENBERG and GORDAN (1950) or according to variously modified procedures.

Relatively little attention has been focused on nitrogen balance studies in experimental animals as a means of assaying the protein anabolic effect of steroids. In our laboratory we have sought to rectify this situation by submitting to nitrogen balance studies many of the steroids which exhibited significant myotrophic activity as determined by us or as reported by others. The authors feel that the nitrogen excretion data provide a very important dimension to the assessment of anabolic activity in that it is more likely to represent an overall protein anabolic effect rather than an effect that may be specific to a particular target tissue such as the levator ani muscle. Additional parameters which provide a basis for assessing the anabolic properties of a steroid and which we have used to a lesser extent are effects on general body growth, improvement in feed utilization, as well as reversal of the catabolic state induced by intensive glucocorticoid treatments. Data from experiments using these parameters will be included in this report as supplemental to the main body of data.

## II. Experimental Methods for Assessing Anabolic and Androgenic Activities

Of the various methods of evaluation of anabolic or anticatabolic agents available to the laboratory investigator, those which deal with protein metabolism, general body growth, and efficiency of food utilization seem to offer data more applicable to considerations for clinical utility than might those which rely upon specific tissue responses. The choice, however, in many instances is dictated by the experimental objectives. For example, the myotrophic/androgenic assay may have somewhat limited usefulness for projection to clinical measurements in that the levator ani muscle response may not represent an overall response of skeletal muscle and may be sex hormone-linked in terms of sensitivity to androgenic substances (NIMNI and GEIGER, 1957). Yet this method of assay is very useful for initial evaluation of anabolic agents and for the assessment of the degree of dissociation of the anabolic from androgenic responses. As such it has a very definite place in the battery of experimental methods for evaluation of anabolic activity.

The nitrogen balance studies using castrated male rats as first described by KOCHAKIAN (1950) offer the type of data which speaks more directly to protein metabolism and to parameters to be followed in possible human metabolic balance studies. We have extended the meaningfulness of this assay by using androgenic data determined separately in castrated male rats not previously equilibrated to nitrogen and body weight. This permits calculation of the index of dissociation of the two activities. In this instance the usefulness of one test method has been supplemented with results obtained from an allied procedure. Other laboratories have found it helpful to resort to similar combinations of assays. STUCKI and his associates (1960), and LYSTER and DUNCAN (1963) have employed oophorectomized monkeys for assessment of the nitrogen retention activity of steroids, and castrated rats for evaluation of androgenic activity. The fact that the use of these methods has led to the development of clinically useful anabolic steroids with relatively reduced androgenicity attests to their validity and merit.

The methods which we have found useful in the overall assessment of the anabolic and other hormonal properties of steroids are detailed in the following section.

## A. The Evaluation of Anabolic Properties of Steroids Based on Nitrogen Balance Studies

### 1. Assay for Nitrogen Retention

The nitrogen-retaining activity of test agents was determined according to the following procedure (ARNOLD et al., 1963a):

Castrated male rats of the Sprague-Dawley strain which weighed approximately 200 g were placed in individual cages. The initial food allotment of 14–15 g per day was decreased gradually until body weights of the rats reached a plateau. Generally, the time required to reach weight and nitrogen equilibrium was 3–4 months, at which time the food allotments were 10.5–11 g per day and urinary nitrogen excretion ranged from 85–115 mg/rat/day. The percentage composition of the diet was: cerelose, 33.67; dextrin, 33.67; yeast (Fleischmann's 2019), 9.2; hydrogenated vegetable oil, 7.4; lactalbumin, 6.0; methylcellulose, 5.0; Jones and Foster salt mixture, 3.7; choline dihydrogen citrate, 0.9; cod liver oil, 0.25; wheat germ oil, 0.16; and liver extract (Wilson's fraction 0), 0.05. Bulk was supplied by adding 5 parts alphacel per 100 parts diet. This diet supplied 1.4% nitrogen, equivalent to 9.35% protein. Control urinary nitrogen values were determined over a 3-day collection period. The urine and daily cage washings were filtered through glass wool into collecting bottles, and samples were diluted to 500 ml for macro Kjeldahl nitrogen determinations using 25 ml aliquots. Premedication urinary nitrogen excretion values and body weights provided the basis for the distribution of rats into uniform groups. The test compounds were incorporated into the dextrin component of the diet and administered over a 5-day period following the control period. Urine collections were made over the last 3 days of the medication period to permit 2 days for onset of response. The nitrogen-retaining activity of the test agents was determined by taking the difference between individual urinary nitrogen excretion values calculated for the 72-h control period and the respective values obtained during the 72-h test period. Quantification of the nitrogen-retaining activity of anabolic steroids was based on multiple-dose-level assays, usually of $3 \times 3$ design. In many instances, steroids of special interest were assayed more than once to obtain replicated estimates of activity. For evaluation of oral activity, methyltestosterone served as the reference standard. Testosterone propionate, on the other hand, was the reference steroid for parenteral studies.

### 2. Calculations of Relative Anabolic-Androgenic Potencies

The oral anabolic and androgenic activities of the test steroids relative to methyltestosterone were estimated by the US Pharmacopeia procedure (1960) when applicable. When the comparisons were based on unequal numbers of rats in the combined groups or when the amounts of steroids given were not at equal log intervals, the longer and applicable procedure of BLISS (1951) was used. Relative androgenic data were observed to follow log dose:log response characteristic as noted by OVERBEEK et al. (1962).

This method of assay is time-consuming and subject to variability when not rigidly controlled. If one chooses to abide by the special requirements, very satisfactory and reproducible results can be obtained. Use of a liquid diet as reported by TOMARELLI and BERNHART (1964) is a convenient modification of this experimental approach. Laboratory animals other than the rat have also been used, but less extensively. KOCHAKIAN and MURLIN (1935) were the first to report on the anabolic actions of androgens in the dog, and STUCKI et al. (1960) have employed the oophorectomized monkey as the assay animal.

The great advantage of this method is that it gives the investigator a result which represents the overall protein anabolic activity of an agent rather than one which may be restricted to a weight increase of a particular muscle such as the levator ani, which will be discussed subsequently. Possibly of equal importance, this assay may provide results which are readily translatable into follow-up clinical studies of nitrogen balance in humans. It is for these reasons that we have chosen to place great emphasis on nitrogen retention studies in laboratory animals.

## B. The Myotrophic/Androgenic Assay as a Method for Evaluating the Dissociation of Anabolic from Androgenic Activities

### Assay for Myotrophic and Androgenic Activities

A modification of the HERSHBERGER et al. assay (1953) was used for determining myotrophic and androgenic activities. The increase in weight of the levator ani muscle of young castrated male rats served as the index of myotrophic activity. Androgenicity was assessed on the basis of increase in weight of the ventral prostate gland. Subcutaneous injections were made daily, except Sunday, beginning 1 week after castration and continuing for a total of 9 injection days. Oral studies were conducted in the same manner except that total daily dose of the test agent was divided into two equal portions administered by stomach b.i.d. The index of dissociation of myotrophic from androgenic activities was determined as a ratio of the ventral prostate response to the levator ani response. The data were plotted on a log dose-response basis for calculation of relative activities.

The myotrophic/androgenic assay first described by EISENBERG and GORDAN (1950) and modified by HERSHBERGER et al. (1953) is a sensitive, reproducible method for assessing the degree of dissociation of an anabolic effect from an androgenic effect. Possibly this method, more than any other, has hastened the advent of clinically useful anabolic steroids. Most laboratories today engaged in research on anabolic agents use this procedure as a basic test because of its relative simplicity and usefulness as a means of quantifying both parameters in the same animal. As mentioned above, we place great emphasis on nitrogen balance studies for definitive assessment of the protein anabolic property of a test agent. However, the myotrophic/androgenic assay is used as a preliminary test and serves the following three purposes: First, it provides a means for searching for active compounds with favorable activity profiles. Second, the data obtained permit selection of candidate compounds for the more arduous nitrogen balance studies. Third, the androgenic data are used with the nitrogen retention and levator ani data to assess each compound on the basis of its index of dissociation of protein anabolic and myotrophic activities from androgenic activities.

## C. The Effect of Anabolic Steroids on Rate of Growth and Efficiency of Feed Utilization

The ability of testosterone to stimulate the growth of rats has been reported by several investigators (CLAUSEN and FREUDENBERGER, 1939; RUBINSTEIN and SOLOMON, 1940; SELYE, 1941; KOCHAKIAN et al., 1950; KOCHAKIAN and WEBSTER, 1958). This effect generally has been associated with the protein-anabolic property of the male hormone coupled with an observed increase in feed intake of rats fed ad lib. Based on these data, one would expect the anabolic steroids to stimulate the growth of female rats comparable in extent to that observed with potent androgens. Our laboratory has reported (TAINTER et al., 1964; BEYLER et al., 1968) on the growth-promoting properties of various steroids. The procedure used is detailed below.

### Method

Adult female rats of the Sprague-Dawley strain weighing approximately 200 g were placed in individual cages equipped with specially designed feeding jars. A daily record was kept of the amount of coarsely ground laboratory chow consumed during the 2-week experimental period. Body weights were recorded on the 1st, 7th and 14th days. The distribution of rats into experimental and control groups was such as to keep the average body weight and range as uniform as possible, usually approximating $200 \pm 5$ g for each group. The test agents were administered subcutaneously in 10% ethanol-cottonseed oil and given by stomach tube. The control groups received the volume of vehicle used to administer the test agents. The body weight and feed consumption data were used to calculate (1) the growth rate in g gained/rat/day, (2) the feed conversion factor in g of feed consumed per day per each g of weight gained per rat per day, and (3) the efficiency of feed utilization expressed as body weight gained per rat per day as a percent of the feed consumed per rat per day.

## D. Fasting Body Weight Loss and Increments in Urinary Excretion of Nitrogen, Phosphorus, and Potassium as Indices of Catabolic Action of Glucocorticoids

Relatively little attention has been focused on use of the protein catabolic effects of glucocorticoids in laboratory animals as a means of assaying the influence of anabolic steroids on protein metabolism. Some contradiction exists in the literature concerning the ability of androgens to reverse the protein catabolic actions of glucocorticoids. SELYE (1955, 1958) reported that androgens do not influence the cortisol-induced catabolism in intact or adrenalectomized rats. This finding is supported by others (WINTER et al., 1953; HALPERN et al., 1956). Nevertheless, several authors (cf. KRÜSKEMPER, 1968) have reported on the reversal by anabolic androgens of the glucocorticoid-induced weight loss. This discrepancy remains unexplained. The authors' major interest in the effect of steroid hormones on growth and anabolism led to studies of the acute catabolic responses of castrated male rats to glucocorticoids preliminary to experiments designed to assess the influence of anabolic steroids on corticoid-induced catabolism (BURNHAM and FANUELE, 1962; TAINTER et al., 1963a, b; BEYLER et al., 1968).

### Method

Male rats of the Sprague-Dawley strain weighing 90–100 g were castrated. One week postoperatively, the rats were placed in individual cages and given daily weighed portions of coarsely ground laboratory chow in specially designed feeders. After a 3-week equilibration period the rats were placed on an overnight fast and transferred to individual metabolism cages for a 24-h experimental period during which free access to water but no food was permitted. Animals were distributed into experimental groups according to uniform body weight. The catabolic agent to be tested was administered subcutaneously in 0.2 ml of 10% ethanol-cottonseed oil as five evenly spaced injections over the first 8 h of the experimental period. Body weights were recorded at the beginning and at the end of the experimental period. The 24-h urine specimens were analyzed for nitrogen by the micro-Kjeldahl procedure, for phosphorus by the method of TAUSSKY and SHORR (1953), and for potassium by flame photometric analysis. Increments over control values for body weight loss, urinary nitrogen, phosphorus and potassium excretion were used to assess the catabolic action of the test agents.

## E. Reversal of Catabolic Action of Glucocorticoids by Anabolic Steroids

### 1. Acute Experiments

BEYLER et al. (1960) and TAINTER et al. (1963a, b) have reported on the ability of certain anabolic steroids to reverse in castrated rats the catabolic effects of cortisone acetate measured over a 24-h experimental fasting period. The fasting period was employed to assure that any changes in the urinary nitrogen values were referable to tissue changes and not to dietary or gastrointestinal factors. The essential procedural details were as follows: Male rats were castrated at a body weight of 90–100 g and placed on 15 g of ground rat chow each day for 14 days. On the 14th day the rats were fasted overnight. At 8:30 a.m. the following day all rats were placed in metabolism cages designed to collect urine separate from feces. During the ensuing 24 h the animals remained on fast but were given free access to drinking water. Cortisone acetate was injected subcutaneously at 2-h intervals beginning at 9:00 a.m. at a dose of 12 mg/kg. Five injections were given for a total dose of cortisone acetate of 60 mg/kg. In most instances, the anabolic steroid was given orally concurrent with the cortisone acetate with the total dose divided into three equal portions and administered at 2-h intervals beginning at 10:00 a.m. Urine collection was started at 9:00 a.m. of the experimental period and continued for 24 h. Body weight was recorded for each rat at the beginning of the experimental period and again 24 h later. The nitrogen content of the urine was determined by the micro-Kjeldahl procedure. The anticatabolic effectiveness of the anabolic steroid was measured by its ability to lessen the body weight loss and to reduce the urinary excretion of nitrogen during the 24-h experimental period. On several occasions the urinary nitrogen data were supplemented with determinations of the 24-h urinary excretion of phosphorous and potassium.

### 2. Subacute Experiments

Young adult castrated male Sprague-Dawley strain rats, weighing 265–285 g, were maintained on 14 g of a 9% protein diet (ARNOLD et al., 1959) daily for several weeks to stabilize their weights and to bring them into moderate nitrogen equilibrium. At this time they were placed on test.

Prednisone was given daily over a 5-day period. The amounts given were 0.5, 1.0, 2.0, and 4.0 mg/kg. Urine collections were conducted the final 72 h of the test period, 2 days being allowed for adjustment to the effects of the glucocorticoid.

The nitrogen excretion of the rats given the 0.5 mg/kg dose of prednisone, $97.0 \pm 3.6$ mg N per rat, was not different from that of the control rats, $98.7 \pm 4.4$ mg N per rat. While the group given 1 mg prednisone per kg daily excreted more N, $109.3 \pm 3.6$ mg N per rat, than the control rats, the increase was not significant. The excretion at the 2 mg prednisone per kg dose, $132.0 \pm 8.9$ mg N, and at the 4 mg prednisone per kg dose, $158.4 \pm 12.4$ mg N, were significantly more than the control excretion level.

Accordingly, the latter two dosage levels were utilized for the anticatabolic evaluations. For these evaluations, rats maintained continuously on the 14 g 9% protein feed were divided into groups of 10 rats each and were treated with prednisone alone or in combination with the anabolic agent. The anticatabolic activity of the anabolic agent was assessed on its ability to reverse the prednisone-induced increase in urinary nitrogen excretion.

## F. The Anticatabolic Activity of Anabolic Steroids Based on the Suppression of Glucocorticoid Induction of Liver Tryptophan Pyrrolase

HARDING and POTTS (1965) have reported on the ability of various anabolic steroids to suppress the glucocorticoid induction in rats of the liver enzyme, tryptophan pyrrolase, as a measure of anticatabolic action. Their studies were based on the observations of MORAN and SOURKES (1964) and KNOX and AUERBACH (1955) as well as others that increased tryptophan pyrrolase activity in the liver, and elevation of kynurenine excretion occurs in rats following various catabolic conditions and after glucocorticoid treatment. The method used employed male rats weighing between 100 and 150 g. Cortisone acetate was administered as a single subcutaneous injection at a dose of 5 mg/kg. The anabolic steroid was administered orally at 4 h and 1 h prior to the injection of cortisone acetate. Livers were removed 5 h after injection of cortisone, and tryptophan pyrrolase activity was determined by the method of KNOX (1955). This experimental approach to the evaluation of the anticatabolic actions of anabolic steroids has significance in that a visceral organ, the liver, is the selected site of action as distinct from the whole body or a particular muscle.

## G. Assay of Other Hormonal Activities

In instances where the desired hormonal property being sought in a particular molecule is strictly a protein-sparing or anabolic action, all other hormonal actions can be considered as side- or undesired effects. To assess the extent to which any one steroid possesses or lacks these other actions we have generally employed assays for estrogenic, progestational, and pituitary inhibitory activities. The methods employed for assaying these activities are briefly described here. The reader interested in greater detail is referred to the appropriate literature citation.

Estrogenicity was assessed on the basis of (1) stimulation of growth of the uterus of immature female rats as described by LAUSON et al. (1939), and (2) induced cornification of vaginal smears of castrated adult female rats using a modified ALLEN-DOISY method (1923). Progestational activity was assessed by histological grading of the endometrial response of the uterus of immature, estrogen-primed female rabbits to oral or intramuscular injections of the test steroid as described by CLAUBERG (1930).

Possible inhibitory effects of the test steroid on pituitary function were assessed by daily subcutaneous injection of the steroid into intact mature male or female rats over a period of 2 weeks. Within 24 h after the last injection, the animals were autopsied and the changes in the body, gonadal, sex accessory organs, adrenal, thymus, and pituitary weights were recorded. These, together with evidence of modification of the vaginal estrous cycle of the female rats, were used to judge the extent of pituitary inhibitory effects of the test steroids.

A detailed accounting of the results of these secondary studies is not given in this presentation. However in instances where significant activities have been observed, a comment as to their nature will be included as a qualification to the other assessments.

# III. Dissociation of Anabolic and Androgenic Activities of Selected Protein Anabolic Steroids

Following is a discussion and summary listing (Table 1) of selected protein anabolic steroids of interest which have been evaluated experimentally for their relative anabolic and androgenic activities using methods detailed in Section B. Other hormonal activities are included where they have been provided.

## A. Testosterone-Derived Protein Anabolic Steroids

### 1. Testosterone and 17α-Methyltestosterone

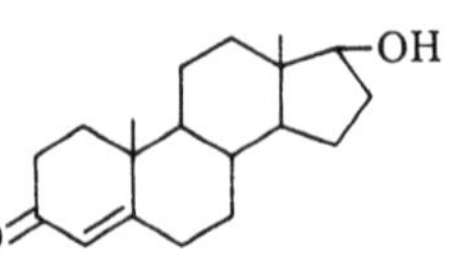


17β-Hydroxyandrost-4-en-3-one

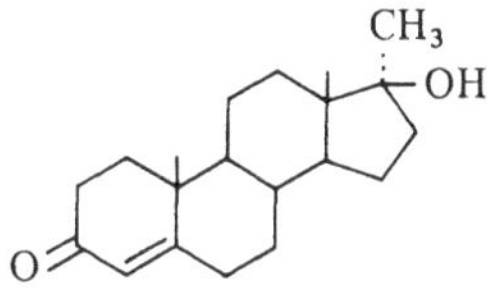


17β-Hydroxy-17-methylandrost-4-en-3-one

Quantitative comparisons of anabolic activities of these reference steroids were based early on the growth of the levator ani muscle of the castrated young male rat. Relative to methyltestosterone orally, JUNKMANN and SUCHOWSKY (1962) found testosterone to be more androgenic than it was myotrophic for a dissociation of 0.6. Parenterally, testosterone, relative to methyltestosterone, was more myotrophic than androgenic for a dissociation of 2.0. DORFMAN and KINCL (1963) reported the myotrophic activity of testosterone to be 0.36 relative to orally administered methyltestosterone. The androgenic activity for testosterone relative to methyltestosterone was 0.28 for a dissociation index of 1.3. The dissociation for testosterone, relative to methyltestosterone, based on the myotrophic weight increase was not favorable, (0.36:0.50) 0.72. Parenterally, DORFMAN and KINCL (1963) noted the relative dissociation of the myotrophic:androgenic activities of testosterone and methyltestosterone not to be meaningfully different. The dissociation for methyltestosterone, relative to testosterone myotrophic:ventral prostate was (1.14:1.03) 1.1. Based on the myotrophic:seminal vesicle weight increase it was (1.14:0.94) 1.2.

In addition to a comparison of the myotrophic activities based on the increase in weight of the levator ani muscle, steroids of interest may be evaluated on the basis of their stimulation of nitrogen conservation in the castrate male rat. Thus, KOCHAKIAN (1950), utilizing implanted pellets, estimated the following relative nitrogen-conserving effectiveness to be: testosterone propionate > testosterone > methyltestosterone. Parenterally, POTTS et al. (1960a) found methyltestosterone to be 0.17 as effective as testosterone propionate in conserving nitrogen in the castrate rat. Myotrophically, methyltestosterone was 0.63 times the reference testosterone propionate. Orally, testosterone was found to have a dissociation index relative to methyltestosterone of 0.67 based on a multiple level comparison for nitrogen-conserving and androgenic activities, 0.38 and 0.57, respectively (ARNOLD et al., 1963b).

Thus, testosterone and methyltestosterone differ only modestly in activity orally or parenterally, especially in comparison with other protein anabolic steroids of this discussion as given in the following paragraphs and as summarized in Table 1. No great case appears justifiable for meaningful differences in biological effects for selecting one from the other. Duration of activity may favor the methylated steroid when administered orally, however.

Parenterally (Authors, unpublished), testosterone at 25 mg/kg increased the growth of reproductively mature female rats 100% and produced a 40% improvement in feed conversion. This same dose caused a 76% decrease in the number of vaginal estrous days, which is indicative of pituitary inhibition. Testosterone produces a progestin-like stimulation of the endometrium in the Clauberg assay (KLEIN and PARKES, 1937).

Methyltestosterone at 6 mg/kg increased the growth rate 35% in the same test system and improved the feed conversion 32% when administered orally. A 48%

Table 1. Summary comparison of anabolic and androgenic activities of selected protein anabolic steroids

| Protein anabolic steroid | Reference | Route | Anabolic activity by | | Androgenic | Dissociation index by | |
|---|---|---|---|---|---|---|---|
| | | | Nitr. ret'n | Myotrophic | | Nitr. ret'n | Myotrophic |
| *I. Testosterone- and methyltestosterone-derived protein anabolic steroids* | | | | | | | |
| Testosterone | Methyltestos. | p.o. | | | | | 0.6[1] |
| | Methyltestos. | par. | | | | | 2.0[1] |
| | Methyltestos. | p.o. | | 0.36 | 0.28–0.50 | | 0.7–1.3[2] |
| | Methyltestos. | p.o. | 0.38 | | 0.57 | 0.67[4] | |
| | Testos. prop. | par. | 0.17[3] | 0.63[3] | | | |
| Methyltestos. | Testos. | p.o. | | 1.14 | 0.94–1.03 | | 1.1–1.2[2] |
| Chlorotestos. | Testos. prop. | par. | | 0.68 | 0.14 | | 5[4] |
| acetate | Testos. prop. | par. | | 1.26 | 0.483 | | 3[4] |
| | Testos. | par. | | 0.10 | 0.03 | | 3[7] |
| Fluoxymesterone | Methyltestos. | p.o. | | 20 | 9.5 | | 2[8] |
| | Methyltestos. | p.o. | | 28.2 | 4.5 | | 6.3[9] |
| | Methyltestos. | p.o. | | | | | 0.9[9] |
| | Methyltestos. | p.o. | 3.8 | | 1.5 | 2.7[10] | |
| | Methyltestos. | p.o. | | 0.72 | 0.46 | | 1.5[2] |
| | Methyltestos. | p.o. | | 1.75 | 0.93 | | 1.9[2] |
| | Methyltestos. | p.o. | 5 | | 9 | 0.55[4] | |
| | Testos. | par. | | | | | 1.6[9] |
| | Testos. | par. | | 17.45 | 1.18–7.51 | −7.57 | 2.3–15[2] |
| Oxymesterone | Methyltestos. | par. | | 3.3 | 0.48 | | 6.9[9,12,13] |
| | Methyltestos. | par. | | 3.3 | 0.5 | | 6.6[14] |
| | Methyltestos. | p.o. | | 1.34 | 0.42–0.61 | | 2.2–3.2[2] |
| | Methyltestos. | p.o. | | | | | 4[1] |
| | Methyltestos. | p.o. | 1.8 | | 0.36 | 5.0[10] | |
| | Testos. | par. | | | | | 4.3[1] |
| Chloromethyltestosterone | Methyltestos. | par. | | 0.85 | 0.1 | | 8.5[9] |
| | Methyltestos. | p.o. | | 0.5 | 0.1–0.15 | | 3–5[2] |
| | Methyltestos. | oral | 0.65 | 0.32 | 0.12 | 5.4[15] | 2.7[15] |
| | Testos. | par. | | 0.46 | 0.26–0.27 | | 1.7–1.8[2] |
| Thiomesterone | Methyltestos. | p.o. | | 4.65 | 0.61 | | 7.5[14] |
| Methandrostenolone, | Testos. | par. | | | | | 0.5[1] |
| Methandienone | Testos. | par. | | 2.1 | 0.6 | | 3.5[9,12] |
| | Testos. | par. | | 2.25[16] | | | |
| | Testos. | par. | | 0.89 | 0.45 | | 1.97[17] |
| | Methyltestos. | p.o. | | | | | 1.4[1] |
| | Methyltestos. | p.o. | 1.2 | | 0.35 | 3.4[18] | |
| | Methyltestos. | par. | | 1.9 | 0.64 | | 3[20] |
| | Testos. prop. | par. | | 0.1 | 0.1 | | 1.0[19] |
| | Testos. prop. | par. | | 6.0 | 0.8 | | 7.5[21] |
| Methandienone (cyclopentyl-1'-enyl ether | Methyltestos. | p.o. | | 0.6 | 0.2 | | 3[22] |
| Bolasterone | Methyltestos. | par. | 5.75 (monkeys) | | 3 | 1.9[20–23] | |
| | Methyltestos. | p.o. | 4.2 | | 1.3 | 3.2[10] | |
| | Methyltestos. | par. | | 13.4 | 3.0 | | 4.5[20] |
| | Testos. | par. | | 11.4 | 2.24–6.6 | | 2–5[24] |

Table 1 (continued)

| Protein anabolic steroid | Reference | Route | Anabolic activity by | | Androgenic | Dissociation index by | |
|---|---|---|---|---|---|---|---|
| | | | Nitr. ret'n | Myotrophic | | Nitr. ret'n | Myotrophic |
| *II. Nortestosterone-derived protein anabolic steroids* | | | | | | | |
| Nortestos. | Methyltestos. | p.o. | 0.73 | | 0.074 | 9.9[15] | |
| | Testos. | par. | | | | | 3.3[25] |
| | Testos. | par. | | | | | 10[24] |
| | Testos. | p.o. | | | | | 5[24] |
| | Testos. prop. | par. | | | | | 16[26] |
| | Testos. prop. | par. | | | | | 10[27] |
| Nortestos. decanoate | Testos. prop. | par. | | 3.29–4.92 | 0.41–0.31 | | 12.1–10.6[55] |
| Nortestos. | Testos. prop. | par. | | 2.2 | 0.14 | | 16[19] |
| β-phenyl-propionate | Testos. prop. | par. | | 2.05 | 0.18 | | 11.1[55] |
| Hydroxy-nortestos. cyclopentyl propionate | Testos. prop. | par. | | 1.2–1.6 | 0.25–0.4 | | 3–6[6,13] |
| Chloro-nortestos. acetate | Testos. prop. | par. | | 1.12 | 0.25 | | 4.5[6] |
| | Testos. prop. | par. | | 0.57 | 0.04 | | 14.2[9] |
| Methyl-nortestos. | Testos. prop. | par. | | 1.0 | 0.06 | | 16[26] |
| | Testos. | par. | | 4.5 | 0.37 | | 12[27] |
| | Testos. | p.o. | | 14.2 | 6–1.92 | | 2–7[24] |
| | Testos. | p.o. | | 0.60 | 0.25 | | 2.4[2] |
| | Methyltestos. | p.o. | | 3.25 | 1.25 | | 2.6[28] |
| | Methyltestos. | p.o. | 4.6 | | 1.1 | 4.2[15] | |
| | Methyltestos. | p.o. | | 5.8 | 1.1 | | 5.3[15] |
| Methyl-nortestos. phenyl-proprionate | Testos. | par. | | 1.5 | 0.15 | | 10[9] |
| Norethan-drolone | Testos. prop. | par. | | 1.0 | 0.06 | | 16[26,29] |
| | Testos. prop. | par. | | 0.8 | 0.12 | | 6.6[19,33] |
| | Testos. prop. | par. | | 0.77 | 0.38 | | 2[6] |
| | Testos. prop. | par. | | | | | 3–4[30] |
| | Testos. | par. | | | | | 1.3[2] |
| | Methyltestos. | p.o. | | 2 | 0.25–0.5 | | 4–8[26,29] |
| | Methyltestos. | p.o. | | 0.75 | 0.22 | | 3.4[9] |
| | Methyltestos. | p.o. | 2[31] | | | | |
| | Methyltestos. | p.o. | 3.9 | | 0.19 | 20[18] | |
| | Methyltestos. | p.o. | | | | | 6.7[2] |
| | Methyltestos. | p.o. | | 1.028 | 0.335 | | 3.1[17,32] |
| Norbolethone | Methyltestos. | p.o. | 16.3[36] | | | | |
| | Testos. prop. | par. | | 3.4 | 0.15 | | 20[33–35] |
| | Testos. prop. | par. | | 3.4 | 0.17 | | 20[19] |
| *III. Androstane-derived protein anabolic steroids* | | | | | | | |
| Andro-stanolone | Testos. prop. | par. | | 2.5 | 1.53 | | 1.6[9] |
| | Testos. | par. | | | | | 0.3[1] |
| | Methyltestos. | p.o. | | | | | 0.8[1] |
| | Methyltestos. | p.o. | 0.26 | | 0.53 | 0.13[37] | |

Table 1 (continued)

| Protein anabolic steroid | Reference | Route | Anabolic activity by | | Andro-genic | Dissociation index by | |
|---|---|---|---|---|---|---|---|
| | | | Nitr. ret'n | Myo-trophic | | Nitr. ret'n | Myo-trophic |
| 2α-Methyl-androstanolone | Testos. | par. | | 0.62–1.30 | 0.25–0.4 | | 2.5–3.0[38-40] |
| 2α-Methylandrostanolone 2′-tetrahydropyranyl ether | Methyltestos. | p.o. | | 3.95 | 0.85 | | 4.65[41] |
| Mestanolone | Methyltestos. | p.o. | 0.8 | | 1.0 | 0.8[10] | |
| Oxymetholone | Methyltestos. | par. | | 1.46 | 0.31 | | 4.7[9] |
| | Methyltestos. | p.o. | | | | | 2.5[1] |
| | Methyltestos. | p.o. | | 3.2 | 0.45 | | 7.1[2] |
| | Methyltestos. | p.o. | 1.75 | | 0.2 | 8.75[18] | |
| | Methyltestos. | par. | | 4.4 | 1.0 | | 4.4[20] |
| | Testos. | par. | | | | | 4.3[1] |
| | Testos. | par. | | 2.3 | 0.64–1.0 | | 2.3–3.6[2] |
| | Testos. | par. | | | | | 2[42] |
| *IV. Heterocyclic protein anabolic steroids* | | | | | | | |
| Stanozolol | Methyltestos. | p.o. | | 2 | 0.33 | | 6[3] |
| | Methyltestos. | p.o. | | 2.2 | 0.33 | | 7[43] |
| | Methyltestos. | p.o. | 10 | | 0.33 | 30[18] | |
| | Methyltestos. | p.o. | | 3.2 | 0.3 | | 10.6[12] |
| | Methyltestos. | par. | | 3.7 | 0.52 | | 7.1[9] |
| | Methyltestos. | p.o. | | | | | 2.5[1] |
| | Methyltestos. | p.o. | | | | 0.4[11] | |
| | Testos. | par. | | | | | 4.3[1] |
| | Testos. | par. | | 7.5 | 2.5 | | 3[20] |
| | Testos. prop. | par. | | 0.13 | 0.03 | | 4[33] |
| Androisoxazole | Testos. | par. | | | | | 2.1[1] |
| | Methyltestos. | p.o. | | | | | 1.7[1] |
| | Methyltestos. | p.o. | 1.55 | | 0.22 | 7[4] | |
| 17-Methyl-5α-androstano-[2,3-d]-isoxazol-17β-ol | Methyltestos. | p.o. | 9.7 | 2.0 | 0.24 | 40[18] | 8.3[18] |
| 2′,17-Dimethyl-androst-2-eno-[3,2-d]-thiazol-17β-ol | Methyltestos. | p.o. | | 2.0 | 0.4 | | 5[44] |
| | Testos. | par. | | 0.3 | 0.1 | | 3[44] |
| Methyl-diazinol | Methyltestos. | p.o. | | 3.0 | 0.2 | | 15[45] |
| Dimethazine | Testos. | par. | | 0.27 | <0.06–0.12 | | > 2[2] |
| | Methyltestos. | p.o. | 3.3 | 2.1 | 0.15 | 14[46] | 22[46] |
| Oxandrolone | Methyltestos. | p.o. | 6.3[47] | | | | |
| | Methyltestos. | par. | | 3.22 | 0.24 | | 13[48] |

Table 1 (continued)

| Protein anabolic steroid | Reference | Route | Anabolic activity by | | Androgenic | Dissociation index by | |
|---|---|---|---|---|---|---|---|
| | | | Nitr. ret'n | Myotrophic | | Nitr. ret'n | Myotrophic |
| *V. Miscellaneous structure protein anabolic steroids* | | | | | | | |
| Methylandrostendiol | Methyltestos. | p.o. | | 0.70 | 0.78 | | 0.9[9] |
| | Methyltestos. | p.o. | | 0.8 | 0.68 | | 1.8[32] |
| | Methyltestos. | p.o. | 0.70 | 0.46 | | 1.1[15] | 0.7[15] |
| | Testos. prop. | par. | 0.067[3] | 0.025[3] | | | |
| Methenolone | Methyltestos. | p.o. | | | | | 13[1] |
| | Testos. | par. | | | | | 16[1] |
| | Testos. | par. | | 0.88 | 0.44–0.57 | | 2[51] |
| | Testos. prop. | par. | | 0.30 | 0.01–0.02 | | 15–30[50] |
| Methenolone acetate | Methyltestos. | p.o. | | 0.86 | 0.12 | | 7[46] |
| Ethylestrenol | Testos. | par. | | | | | 2.6[1] |
| | Methyltestos. | p.o. | | | | | 4[1] |
| | Methyltestos. | p.o. | | 4.207 | 0.222 | | 19[17,32,52] |
| | Methyltestos. | p.o. | 1.7 | | 0.21 | 8.1[4] | |
| | Methyltestos. | p.o. | | 2.0 | 0.4 | | 5[2] |
| | Methyltestos. | par. | | 0.4 | 0.17 | | 2.3[53] |
| Ethyldienolone | Methyltestos. | par. | | 1.0 | 0.1 | | 10[53] |
| | Methyltestos. | p.o. | | 1.5 | 0.3 | | 5[53] |
| 17-Methyl-3-methylene-androst-1-en-17β-ol | Methyltestos. | p.o. | | 2.66 | 0.12 | | 22[49] |

[1] JUNKMANN and SUCHOWSKY, 1962; [2] DORFMAN and KINCL, 1963; [3] POTTS et al., 1960a; [4] ARNOLD et al., 1963b; [5] SALA et al., 1956; [6] SALA and BALDRATTI, 1957; [7] CAVALLERO, 1967; [8] LYSTER et al., 1956; [9] CAMERINO and SALA, 1960; [10] ARNOLD et al., 1963c; [11] STUCKI et al., 1962; [12] SALA, 1960; [13] SALA, 1964; [14] KRAFT and BRÜCHNER, 1964; [15] ARNOLD and POTTS, 1966; 16 DESAULLES, 1959; [17] OVERBEEK, 1961; [18] ARNOLD et al., 1963a; [19] EDGREN and SMITH, 1965; [20] CAMPBELL and BABCOCK, 1965; [21] NUTTING et al., 1966; [22] ERCOLI et al., 1962; [23] CAMPBELL and BABCOCK, 1959; [24] SEGALOFF, 1963; [25] HERSHBERGER et al., 1953; [26] SAUNDERS and DRILL, 1956; [27] DESAULLES and KRÄHENBÜHL, 1962; [28] OVERBEEK and DE VISSER, 1956; [29] SAUNDERS and DRILL, 1957; [30] PERRINE, 1961; [31] LIDDLE and BURKE, 1960; [32] OVERBEEK et al., 1962; [33] EDGREN, 1963; [34] BUZBY et al., 1966; [35] STRIKE et al., 1967; [36] TOMARELLI and BERNHART, 1964; [37] ARNOLD and POTTS, 1963d; [38] RINGOLD, 1961; [39] KINCL and DORFMAN, 1962; [40] KINCL, 1965; [41] CROSS et al., 1964; [42] ČEKAN and PELC, 1966; [43] BEYLER et al., 1961; [44] ZDERIC et al., 1963; [45] CHURCH et al., 1965; [46] Authors (unpublished); [47] FOX et al., 1962; [48] LENNON and SAUNDERS, 1964; [49] IRMSCHER et al., 1964; [50] NEUMANN and WIECHERT, 1965; [51] TAIT and BURSTEIN, 1964; [52] OVERBEEK et al., 1961; [53] EDGREN, 1966; [54] OVERBEEK and DE VISSER, 1961.

decrease in the number of vaginal estrous days was noted (i.e., presumptive evidence of pituitary inhibition). Progestin-like (KLEIN and PARKES, 1937) activity has also been reported for methyltestosterone.

Parenterally, testosterone propionate at 2.5 mg/kg (Authors, unpublished) increased the growth of reproductively mature female rats 81% and improved the feed conversion 34%. This same dose of testosterone propionate caused a 74% decrease in the number of vaginal estrous days, which is indicative of pituitary inhibition.

**2. Chlorotestosterone Acetate**

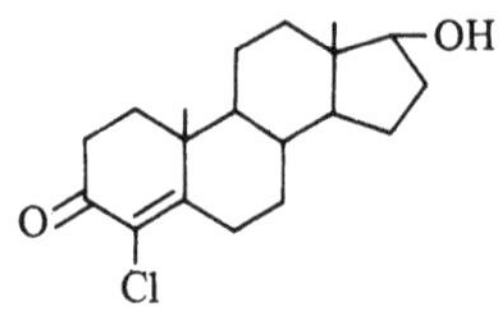


4-Chloro-17β-hydroxyandrost-4-en-3-one (acetate)

SALA et al. (1956) found chlorotestosterone acetate to have a myotrophic:androgenic dissociation of (0.68:0.14) 5 relative to testosterone propionate. It supported growth gains in adult female rats (SALA et al., 1956). SALA and BALDRATTI (1957) gave it a myotrophic:androgenic dissociation of (1.26:0.483) 3 relative to testosterone propionate. Relative to testosterone, CAVALLERO (1967) gave it a myotrophic:androgenic dissociation of (0.10:0.03) 3.

**3. Fluoxymesterone**

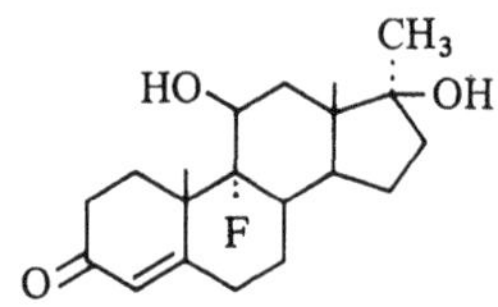


9α-Fluoro-11β,17β-dihydroxy-17-methylandrost-4-en-3-one

Fluoxymesterone (HERR et al., 1956) has shown anabolic activity considerably greater than that of methyltestosterone and has been studied extensively. LYSTER et al. (1956) found fluoxymesterone to be 20-fold as anabolic (levator ani) as methyltestosterone, while being 9.5 times as androgenic. CAMERINO and SALA (1960) noted fluoxymesterone to be 28.2 times as myotrophic as methyltestosterone on oral administration, but only 4.5 times as androgenic for the promising dissociation index of 6.3. JUNKMANN and SUCHOWSKY (1962) reported the dissociation index to be 0.9 relative to methyltestosterone orally and 1.6 relative to testosterone on parenteral administration. Evaluated for anabolic activity orally on the basis of nitrogen retention in rats, ARNOLD et al. (1963c) found fluoxymesterone to be 3.8 times as anabolic as methyltestosterone, 1.5 times as androgenic and to have a dissociation index of 2.7. Parenterally, DORFMAN and KINCL (1963) noted fluoxymesterone to be 17.45 times as myotrophic as testosterone and 1.18–7.57 as androgenic for dissociation index of 2.3 to 15. Relative to methyltestosterone orally, DORFMAN and KINCL (1965) found the myotrophic:androgenic dissociation for fluoxymesterone to be minimally favorable (0.72–1.75:0.46–0.93) 1.5–1.9. The dissociation, based on nitrogen retention in ovariectomized monkeys (STUCKI et al., 1962) was 5 times that of methyltestosterone compared with its androgenicity, 9 times the reference in rats, for a dissociation of 0.55. This was less favorable than that determined solely in rat studies.

Parenterally (Authors, unpublished), fluoxymesterone increased the growth rate of reproductively mature female rats 43% and produced a 28% improvement in feed conversion at 2.0 mg/kg. The same dose caused a 73% decrease in the number of vaginal estrous days, which is indicative of pituitary inhibition.

The enhanced androgenic activity of fluoxymesterone relative to that of the reference steroids has yielded a clinically useful androgen. Its modestly enhanced anabolic activity has not been adequate to provide useful clinical application primarily as an anabolic agent.

## 4. Oxymesterone

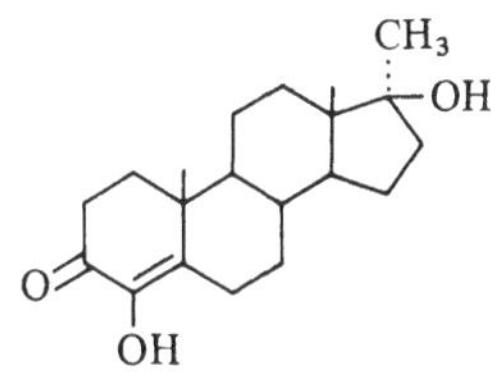

4,17β-Dihydroxy-17-methylandrost-4-en-3-one

CAMERINO and SALA (1960), and SALA (1960, 1964) reported oxymesterone to have a dissociation index of 6.9, relative to methyltestosterone, based on a myotrophic activity of 3.3 times the reference with an androgenic activity of 0.48. KRAFT and BRÜCKNER (1964) noted a myotrophic:androgenic dissociation of (3.3:0.5) 6.6 relative to methyltestosterone. DORFMAN and KINCL (1963) found it to have a dissociation of 2.2–3.2 based on myotrophic activity of 1.34, relative to methyltestosterone, and androgenic activity of 0.61–0.42. Orally, JUNKMANN and SUCHOWSKY (1962) reported oxymesterone to have a dissociation index of 4 relative to methyltestosterone. Parenterally, they found it to have an index of 4.3 relative to testosterone. ARNOLD et al. (1963c) reported oxymesterone to be 1.8-fold methyltestosterone in conserving nitrogen in castrate rats and 0.36-fold as androgenic for a dissociation of 5.0.

Parenterally, oxymesterone at 5 mg/kg (Authors, unpublished) increased the growth of reproductively mature female rats 96% and produced an 89% improvement in feed conversion. This same dose caused a 95% decrease in the number of vaginal estrous days, which is indicative of pituitary inhibition.

## 5. Chloromethyltestosterone

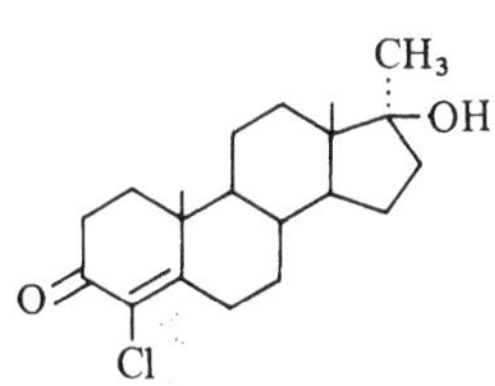

4-Chloro-17β-hydroxy-17-methylandrost-4-en-3-one

CAMERINO and SALA (1960) reported chloromethyltestosterone to have a dissociation index of 8.5 parenterally relative to methyltestosterone propionate, based on a myotrophic activity of 0.85 and an androgenicity of 0.1. Evaluated orally, DORFMAN and KINCL (1963) reported its myotrophic activity to be half that of methyltestosterone with androgenic activity of 0.10–0.15 that of the reference for a dissociation of 3 to 5. Parenterally, the latter investigators found its dissociation, relative to testosterone, to be 1.7–1.8 based on myotrophic activity of 0.46, and an androgenicity of 0.26–0.27. ARNOLD and POTTS (1966) judged its dissociation orally to range from 2.7–5.4 based on myotrophic activity of 0.32, nitrogen-conserving activity of 0.65, and androgenic activity of 0.12, relative to methyltestosterone orally.

### 6. Thiomesterone

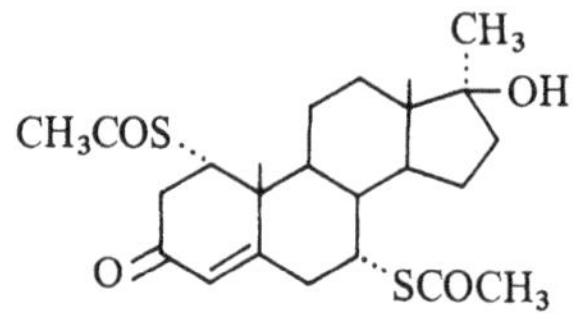


1α,17α-Bis(acetylthio)-17β-hydroxy-17-methylandrost-4-en-3-one

KRAFT and BRÜCKNER (1964), and KRÜSKEMPER (1966) found thiomesterone to be 4.56 times as myotrophic orally as methyltestosterone and 0.61 times as androgenic for the distinctly favorable dissociation of 7.5. This degree of dissociation is among the most favorable noted experimentally.

### 7. Methandrostenolone, Methandienone

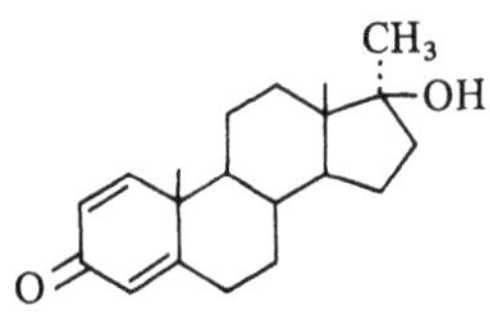


17β-Hydroxy-17-methylandrosta-1,4-dien-3-one

JUNKMANN and SUCHOWSKY (1962) reported the myotrophic:androgenic dissociation of methandrostenolone, relative to methyltestosterone, to be 0.5. SALA (1960), and CAMERINO and SALA (1960) reported the myotrophic:androgenic dissociation to be (2.1:0.6) 3.5. DESAULLES (1959) found methandrostenolone to be 2.25-fold the reference myotrophically. OVERBEEK et al. (1961) noted the myotrophic:androgenic dissociation to be (0.89:0.45) 1.97. ARNOLD et al. (1963a) reported its nitrogen-conserving:androgenic dissociation to be (1.2:0.35) 3.4 relative to methyltestosterone orally. STUCKI et al. (1965) found methandrostenolone to be 1.2 times methyltestosterone orally in supporting weight gains in ovariectomized female monkeys. EDGREN and SMITH (1965) did not find methandrostenolone to differ in dissociation from testosterone propionate parenterally in that is was only a tenth as androgenic but was only a tenth as myotrophic. CAMPBELL and BABCOCK (1965) found methandrostenolone to have a myotrophic:androgenic dissociation of (1.9:0.64) 3 relative to methyltestosterone. NUTTING et al. (1966) found methandrostenolone to have a myotrophic:androgenic dissociation of (6.0:0.8) 7.5 relative to testosterone propionate.

The related cyclopentyl-1′-enyl ether in place of the 17α-methyl group (Quinobolin) has been reported to have a similar myotrophic:androgenic dissociation,

(0.6:0.2) 3 relative to methyltestosterone orally (ERCOLI et al., 1962) with a potentially reduced effect on the liver for the reason that it lacks the 17α-alkyl chemical group.

Orally, methandrostenolone at 10 mg/kg (Authors, unpublished) increased the growth of reproductively mature female rats 32% and improved the feed conversion 24%. This same dose caused a 65% decrease in the number of vaginal estrous days indicative of pituitary inhibition.

**8. Chloro-1-dehydromethyltestosterone**

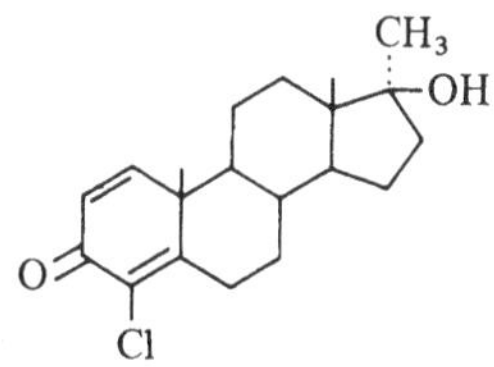


4-Chloro-17β-hydroxy-17-methylandrosta-1,4-dien-3-one

DÖRNER and SCHUBERT (1962) found 4-chloro-$\Delta^1$-methyltestosterone to have a wider myotrophic:androgenic dissociation of myotrophic and androgenic activities than the parent $\Delta^1$-methyltestosterone compound.

**9. Bolasterone, Dimethyltestosterone**

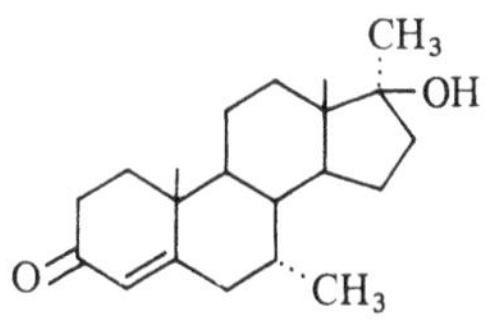


17β-Hydroxy-7α,17α-dimethylandrost-4-en-3-one

STUCKI et al. (1962, 1965) reported dimethyltestosterone, synthesized by CAMPBELL and BABCOCK (1959, 1965), to be 5.75 times as effective as methyltestosterone in conserving nitrogen and potassium, and in promoting body weight gains in ovariectomized female monkeys. It was 3-fold as androgenic as the reference, for a dissociation of 1.9. ARNOLD et al. (1963c) found dimethyltestosterone orally to be 4.2 times as effective in conserving nitrogen in castrate male rats and 1.3 as androgenic as methyltestosterone for a dissociation of 3.2. SEGALOFF (1963) found dimethyltestosterone to have a myotrophic:androgenic dissociation of (11.4:2.24–6.6) 2–5 relative to testosterone. CAMPBELL and BABCOCK (1965) found dimethyltestosterone to exhibit a myotrophic:androgenic dissociation of (13.4:3.0) 4.5 relative to methyltestosterone.

Thiomesterone, with an anabolic:androgenic dissociation of 7.5, appears to have the most interesting clinical promise of the testosterone- and methyltestosterone-derived protein anabolic steroids.

## B. 19-Nortestosterone-Derived Protein Anabolic Steroids

### 1. 19-Nortestosterone, Nandrolone

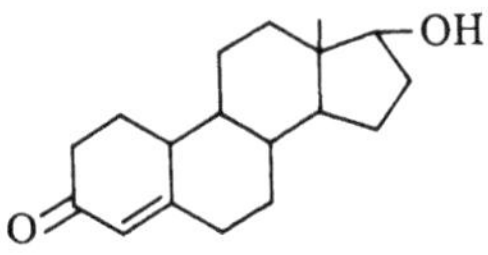

17β-Hydroxyestr-4-en-3-one

Hershberger et al. (1953) contributed the important observation, confirmed by Barnes et al. (1954), that nortestosterone and nortestosterone analogs retained the myotrophic activity of testosterone, but possessed markedly diminished hormonal effects. In their hands the dissociation of myotrophic from androgenic activities was 3.3. Saunders and Drill (1956) found nortestosterone to be 16 times as myotrophic as androgenic relative to testosterone propionate. Desaulles and Krähenbühl (1962) found its dissociation relative to testosterone propionate to be 10. Esterifying nortestosterone to form the β-phenyl propionate yielded an agent with a longer duration of activity and a myotrophic:androgen dissociation, relative to testosterone propionate, of (2.2:0.14) 16 (Edgren and Smith, 1965). In the castrate male rats, nortestosterone exhibited a nitrogen conserving:androgenic index of (0.73:0.074) 9.9, relative to methyltestosterone orally (Arnold and Potts, 1964). Segaloff (1963) found dissociations of 5 (parenterally) to 10 (orally) for nortestosterone relative to testosterone with the 19-nor analog being both more myotrophic than the reference and less androgenic. The shorter duration of biological activity of the 19-nortestosterone compared with its 17α-alkyl homologs has limited its application.

While the duration of activity is too short for 19-nortestosterone to be useful orally, parenteral administration of selected esters have received widespread clinical trial. Additionally, as with esters of testosterone, esters of 19-nortestosterone may be selected to release the active agent over a period of time so that parenteral dosage may be selected to permit weekly to monthly administration. For example, Overbeek and De Visser (1961) showed that the phenylpropionate and decanoate esters of 19-nortestosterone manifest 11–12 times the myotrophic:androgenic dissociation of the corresponding testosterone esters when each is given once weekly s.c. to rats. Thus, parenteral administration favors slower release of the active agent so that its duration of action is prolonged compared with its administration orally.

### 2. Hydroxy-19-nortestosterone cyclopentylpropionate, Oxabolone

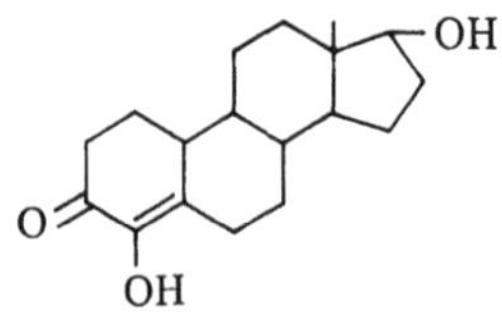

4,17β-Dihydroxyestr-4-en-3-one cyclopentylpropionate

This hydroxy-19-nortestosterone ester was reported to be more myotrophic than testosterone propionate to yield a myotrophic:androgenic dissociation of (1.2–

1.6:0.25–0.4) 3.0–6.0 (SALA and BALDRATTI, 1963; SALA, 1964). Other hormonal effects were lessened for this 19-nor steroid (SALA, 1964). While the dissociation of the unesterified nor-steroid remains favorable (4.5, SALA, 1964), its duration of activity is less.

### 3. Chloro-19-nortestosterone Acetate

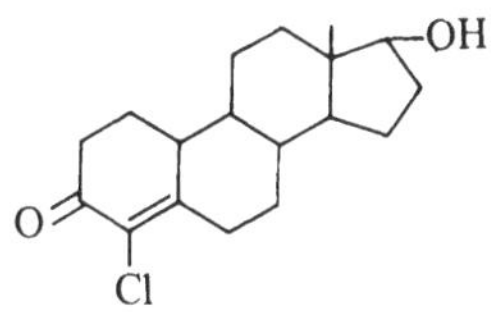


4-Chloro-17β-hydroxyestr-4-en-3-one acetate

SALA and BALDRATTI (1957) found 4-chloro-19-nortestosterone acetate to compare favorably with testosterone propionate with a myotrophic:androgenic dissociation of (1.12:0.25) 4.5 relative to the latter. CAMERINO and SALA (1960) found a dissociation of 14.2 for 4-chloro-19-nortestosterone acetate relative to testosterone propionate parenterally on the basis of a myotrophic index of 0.57 and an androgenic index of only 0.04.
SALA et al. (1956) found the cyclopentylpropionate ester to have wider separation of myotrophic:androgenic effects than the acetate relative to testosterone propionate.

### 4. Methyl-19-nortestosterone

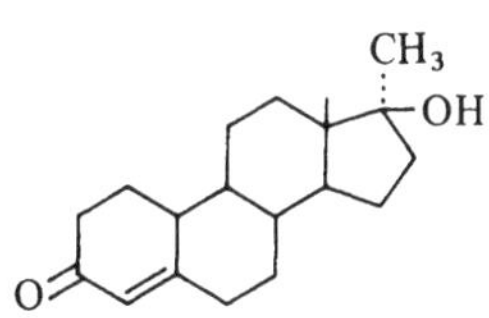


17β-Hydroxy-17-methylestr-4-en-3-one

The 17α-methyl homolog has received widespread evaluation also. SAUNDERS and DRILL (1956) found it to have a myotrophic:androgenic dissociation of (1.0:0.06) 16 relative to testosterone propionate. Relative to methyltestosterone, OVERBEEK and DE VISSER (1956) found its myotrophic:androgenic dissociation to be only (3.25:1.25) 2.6. DESAULLES and KRÄHENBÜHL (1962) noted its dissociation to be (4.5:0.37) 12. SEGALOFF (1963) found it to differ little from testosterone, but to have a myotrophic:androgenic dissociation orally, relative to testosterone, of (14.2:6–1.92) 2–7. The dissociation of its phenylpropionate ester was found to be 10 (CAMERINO and SALA, 1960) on the basis of myotrophic activity of 1.5 and androgenic activity of only 0.15. Relative to testosterone orally, its anabolic activity was 0.60, its androgenic activity 0.25 for a dissociation of 2.4 (DORFMAN and KINCL, 1963). On the basis of nitrogen conservation in rats ARNOLD and POTTS (1966) found methyl-nortestosterone to be 4.6 times as anabolic as methyltestosterone orally with an androgenic ratio of 1.1 for a dissociation of 4.2. The latter investigators found its myotrophic:androgenic dissociation to be (5.8:1.1) 5.3.

Parenterally, this steroid at 10 mg/kg (Authors, unpublished) increased the growth of reproductively mature female rats 149% and improved the feed conversion 65%. This same dose caused an 86% decrease in the number of vaginal estrous days indicative of pituitary inhibition.

### 5. Norethandrolone, 17α-Ethyl-9-nortestosterone

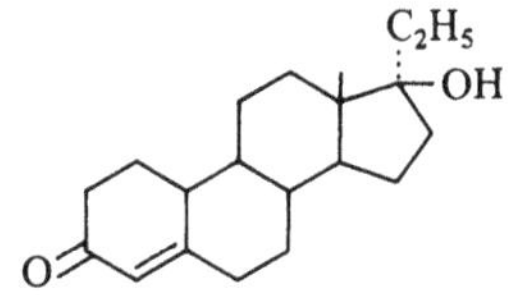


17-Hydroxy-19-nor-17α-pregn-4-en-3-one

Norethandrolone was among the earliest of the anabolic agents to receive widespread clinical trial following its study by HERSHBERGER et al. (1953). SAUNDERS and DRILL (1956, 1957, 1958) reported it to be as anabolic as the reference testosterone propionate but only $^1/_{16}$th as androgenic. Orally, norethandrolone was twice as myotrophic while being only one-half to one-fourth as androgenic as methyltestosterone. SALA and BALDRATTI (1957) found norethandrolone to have a myotrophic:androgenic dissociation of 2 based on myotrophic:androgenic activities of 0.77:0.38 relative to testosterone propionate. PERRINE (1961) estimated its myotrophic:androgenic dissociation to be 3–4 relative to testosterone propionate parenterally. CAMERINO and SALA (1960) found norethandrolone to have an anabolic activity of 0.75, relative to methyltestosterone orally with an androgenic activity of 0.22 for a dissociation of 3.4. In man, LIDDLE and BURKE (1960) reported norethandrolone to be twice as anabolic as methyltestosterone when evaluated on the basis of nitrogen retention studies. In rats, ARNOLD et al. (1963a) found norethandrolone to have an anabolic activity based on nitrogen retention of 3.9 relative to methyltestosterone, and an androgenic index of 0.19 for the markedly favorable dissociation of 20. JUNKMANN and SUCHOWSKY (1962) estimated the dissociation, relative to testosterone parenterally, to be 1.3. Relative to methyltestosterone orally, its dissociation was given as 6.7. The latter investigators judged norethandrolone to be about 4-fold as progestational as it was myotrophic. It also was judged to have significant antiluteinizing effects. OVERBEEK et al. (1961, 1962) reported a myotrophic:androgenic dissociation of (1.028:0.335) 3.1 relative to methyltestosterone. EDGREN (1963), and EDGREN and SMITH (1965) found norethandrolone to have a dissociation of 7 relative to testosterone propionate parenterally with a myotrophic:androgenic index of (0.8:0.12) 6.6 relative to testosterone propionate parenterally.

### 6. Ethyldienolone

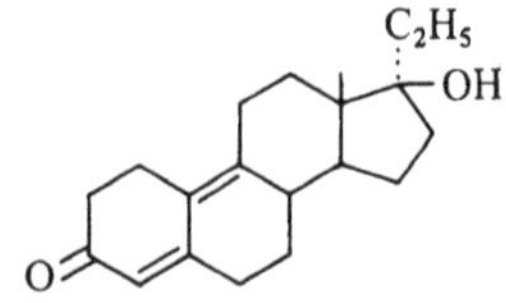


17-Hydroxy-19-nor-17α-pregna-4,9-dien-3-one

Ethyldienolone has been found to have the favorable myotrophic:androgenic dissociation index of (1.0:0.1) 10, relative to methyltestosterone parenterally and (1.5:0.3) 5 orally. EDGREN et al. (1966) did not judge its dissociation to be better than that of ethylnortestosterone, however.

### 7. Norbolethone

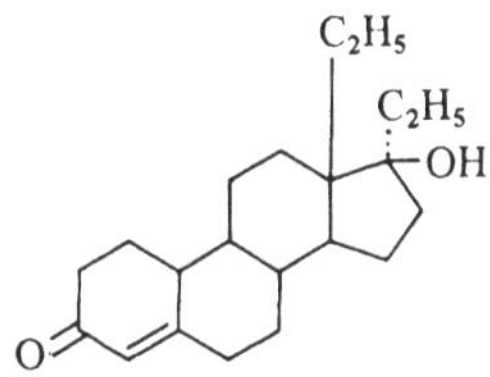


(*dl*-)17*β*-Hydroxy-13*β*,17*α*-diethylgon-4-en-3-one
(*dl*-)13-Ethyl-17-hydroxy-18,19-dinor-17*α*-pregn-4-en-3-one

Norbolethone manifested a myotrophic activity of 3.4 relative to testosterone propionate parenterally and an androgenic activity of only 0.15 for a dissociation of 20 (EDGREN, 1963; BUZBY et al., 1966; STRIKE et al., 1967). Relative to methyltestosterone orally, TOMARELLI and BERNHART (1964) found norbolethone to be 16.3 times as effective in conserving nitrogen in equilibrated rats as was methyltestosterone. EDGREN and SMITH (1965) estimated the myotrophic:androgenic dissociation to be (3.4:0.17) 20 relative to testosterone propionate parenterally.

The 19-nortestosterone-derived protein anabolic steroids have shown promising anabolic:androgenic dissociation and, where side-effects have not limited their prolonged application, have been widely accepted as clinically useful therapeutic agents.

## C. Androstane-Derived Protein Anabolic Steroids

### 1. Androstanolone, Stanolone, Androlone, Dihydrotestosterone

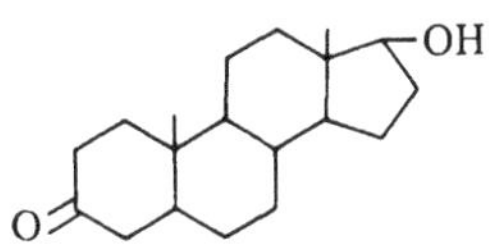


17*β*-Hydroxy-5*α*-androstan-3-one

CAMERINO and SALA (1960) reported androstanolone to have a dissociation index of 1.6 based on a myotrophic activity of 2.5 and an androgenic activity of 1.53 relative to testosterone propionate. JUNKMANN and SUCHOWSKY (1962) found that, relative to methyltestosterone orally, androstanolone had a dissociation index of 0.8. Parenterally, relative to testosterone, its dissociation index was given 0.3. Its dissociation was found less favorable, 0.26, based on nitrogen conservation in rats, 0.13 relative to methyltestosterone orally, with an androgenic activity of 0.50 (ARNOLD et al., 1963d). Dorfman and Dorfman (1963) judged it to have an unfavorable dissociation relative to testosterone.

Parenterally, androstanolone at 10 mg/kg (Authors, unpublished) increased the growth of reproductively mature female rats 68% and improved the feed conversion 32%. This same dose produced an 82% decrease in the number of vaginal estrous days which is indicative of pituitary inhibition.

## 2. Drostanolone, 2α-Methyldihydrotestosterone, 2α-Methylandrostanolone

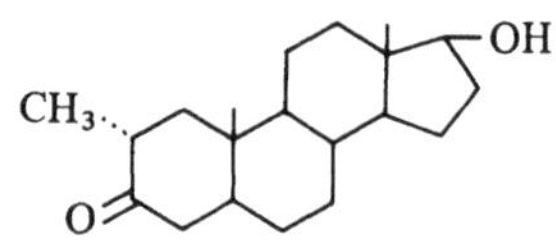


17β-Hydroxy-2α-methyl-5α-androstan-3-one

2α-Methylandrostanolone has been reported (Ringold, 1961; Kincl and Dorfman, 1964; Kincl, 1965) to have moderate myotrophic:androgenic dissociation (0.62–1.3:0.25–0.4) 2.5–3.0 relative to testosterone. Its 2′-tetrahydropyranyl ether (Cross et al., 1964) showed improved dissociation. The myotrophic:androgenic index was found to be (3.95:0.85) 4.65 relative to methyltestosterone orally. Its dissociation appears not to be as favorable as that of some related 5α-androstane-derived steroids.

## 3. Mestanolone, Methylandrostanolone

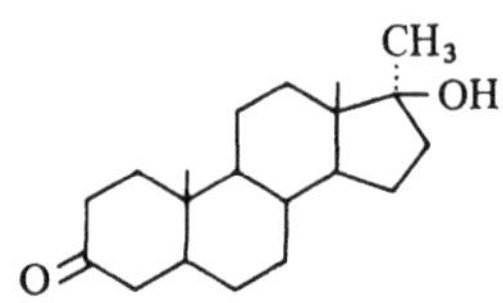


17β-Hydroxy-17-methyl-5α-androstan-3-one

Dorfman and Kincl (1963), Nutting et al. (1966), and Neumann and Wiechert (1965) did not find methylandrostanolone to exhibit favorable myotrophic:androgenic dissociation relative to reference anabolic agents. Similarly, based on its ability to promote nitrogen retention, methylandrostanolone was 0.8 times as anabolic as methyltestosterone while being equal to the reference androgenically for a dissociation of 0.8 (Arnold, et al., 1963c).

## 4. Oxymetholone

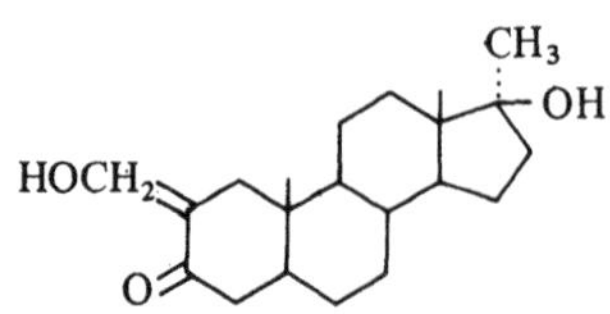


17β-Hydroxy-17-methyl-2-(hydroxymethylene)-5α-androstan-3-one

Oxymetholone is one of the more widely evaluated protein anabolic steroids with a promising dissociation (RINGOLD et al., 1959; ZAFFARONI, 1960). CAMERINO and SALA (1960) estimated the dissociation index orally to be 4.7 based on a myotrophic activity of 1.46 and androgenic activity of 0.31 relative to methyltestosterone. JUNKMANN and SUCHOWSKY (1962) estimated the dissociation index to be 2.5 orally relative to methyltestosterone and 4.3 parenterally relative to testosterone. These investigators judged oxymetholone to have significant hypophysis inhibitory activity. DORFMAN and KINCL (1963) estimated the myotrophic activity, relative to methyltestosterone orally, to be 3.2. Its androgenic activity was 0.45 for a dissociation of 7.1. Relative to testosterone parenterally, its dissociation was 2.3–3.6 based on a myotrophic activity of 2.3 and androgenic activities of 1.0–0.64. Nitrogen retention studies in castrate rats favored oxymetholone to the extent that its protein retention activity, relative to methyltestosterone orally, was 1.75 with an androgenic activity of 0.2 for the favorable dissociation of 8.75 (ARNOLD et al., 1963a). STUCKI et al. (1965) found oxymetholone to be 1.1 times methyltestosterone orally in supporting weight gains in ovariectomized female monkeys. In rats, the myotrophic:androgenic dissociation was (4.4:1.0) 4.4 relative to methyltestosterone according to CAMPBELL and BABCOCK (1965). EDGREN (1963, 1965) noted it to be androgenic at doses that did not support anabolic effects as estimated by body weight gains in female rats. Parenterally, ČEKAN and PELC (1966) noted a dissociation of only about 2 relative to testosterone.

Oxymetholone appears to be the most promising of the androstane-derived protein anabolic steroids on the basis of its anabolic:androgenic dissociation.

## D. Heterocyclic Protein Anabolic Steroids

### 1. Stanozolol

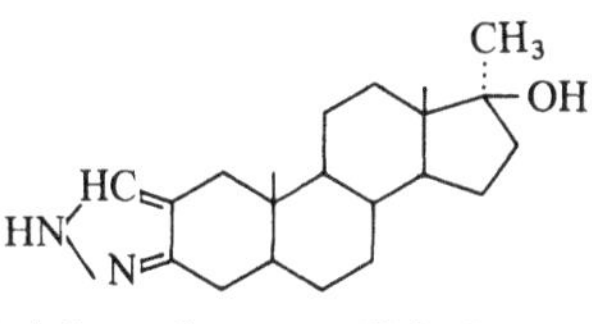


17-Methyl-5α-androstano-[3,2-c]pyrazol-17β-ol

Stanozolol (CLINTON et al., 1959), like oxymetholone, has been evaluated widely. POTTS et al. (1960b) reported the myotrophic activity, relative to methyltestosterone orally, to be 2, its androgenic activity to be 0.33 for a dissociation of 6. BEYLER et al. (1961) found the myotrophic:androgenic dissociation to be (2.2:0.33) 7, relative to methyltestosterone. Anabolic activity evaluated on the basis of nitrogen retention studies in castrate rats was indicated to be 10, relative to methyltestosterone orally (ARNOLD et al., 1963a; POTTS et al., 1962). On the basis of an androgenic activity of 0.33 the dissociation was projected to be 30. SALA (1960) found the myotrophic activity to be 3.2, relative to methyltestosterone orally, the androgenic activity to be 0.3 for a dissociation of 10.6. CAMERINO and SALA (1960) reported the myotrophic activity to be 3.7, the androgenic activity to be 0.52 for a dissociation of 7.1. JUNKMANN and SUCHOWSKY (1962) found the dissociation to be 2.5 orally relative to methyltestosterone and 4.3 parenterally relative to testosterone. The nitrogen and potassium retention activities and body weight gains of ovariectomized female monkeys (STUCKI et al., 1965) given stanozolol orally did not differ from those given the reference methyltestosterone, however. The androgenic activity of stanozolol was estimated to be greater than that of methyltestosterone for a dissociation of 0.4, clearly less than the estimates given above. This suggests that the predictive values with ovariectomized female monkeys appear to conflict seriously with rat data and with favorable clinical findings. CAMPBELL and BABCOCK (1965) reported the rat myotrophic:androgenic dissociation to be (7.5:2.5) 3.0. EDGREN (1963) found the myotrophic:androgenic dissociation to be (0.13:0.03) 4, relative to testosterone propionate parenterally.

Orally, stanozolol at 6 mg/kg (Authors, unpublished) increased the growth of reproductively mature female rats 43% and improved the feed conversion 45%. This same dose produced a 50% decrease in the number of vaginal estrous days indicative of pituitary inhibition.

## 2. Androisoxazole

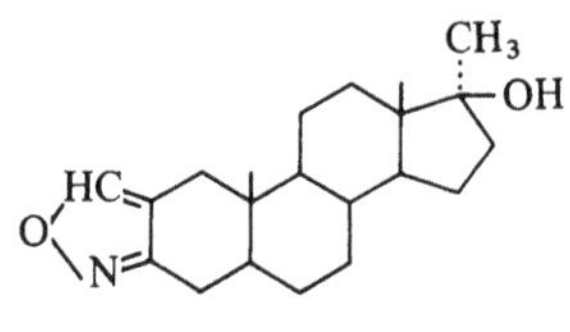


17-Methyl-5α-androstano-[3,2-c]isoxazol-17β-ol

Relative to methyltestosterone orally, JUNKMANN and SUCHOWSKY (1962) reported androisoxazole to have a dissociation of 1.7. Parenterally, relative to testosterone, its dissociation was 2.1. The nitrogen-conserving:androgenic dissociation of androisoxazole was noted to be considerably more promising, (1.55:0.22) 7 as reported by ARNOLD et al. (1963b).

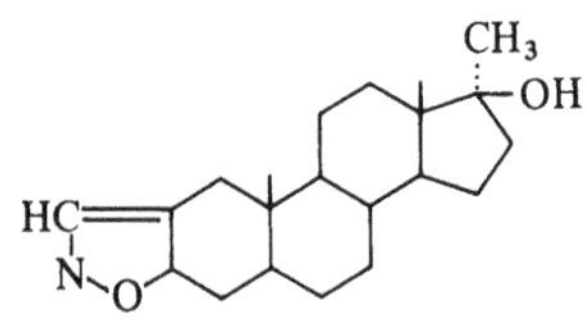


17-Methyl-5α-androstano[2,3-d]-isoxazol-17β-ol

The 2,3-d isoxazole isomer of androisoxazole (MANSON et al., 1963) was reported by ARNOLD et al. (1963a) to have a myotrophic:androgenic dissociation of (2.0:0.24) 8.3, and a nitrogen-conserving:androgenic dissociation of (9.7:0.24) 40.

Orally, the 2,3-d isoxazole (Authors, unpublished) increased the growth of reproductively mature female rats 47% and improved the feed conversion 26% at 5 mg/kg. This same dose produced a 61% decrease in the number of vaginal estrous days indicative of pituitary inhibition.

### 3. 2′,17-Dimethylandrostan-2-eno[3,2-d]thiazol-17β-ol

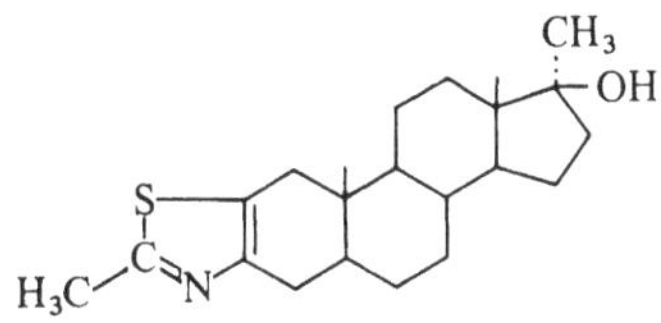


ZDERIC et al. (1963) found the related thiazole, dimethyl-androstano[3,2-d]thiazolol, to have the favorable myotrophic:androgenic dissociation (2.0:0.4) 5 relative to methyltestosterone orally. Relative to testosterone parenterally its dissociation was somewhat less (0.3:0.1) 3.

### 4. Methyldiazinol

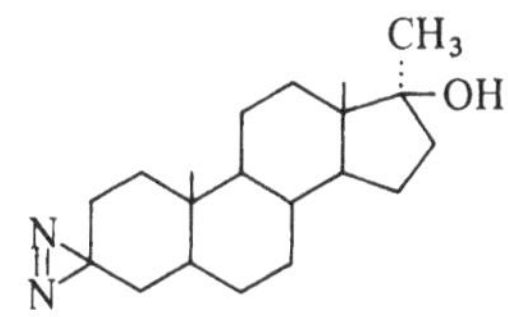


3-Azi-17-methyl-5α-androstan-17β-ol

The 3-azi derivative of methandrostanolone was noted by CHURCH et al. (1965) to have a wide myotrophic:androgenic dissociation (3:0.2) 15 relative to methyltestosterone orally, among the highest noted.

### 5. Dimethazine

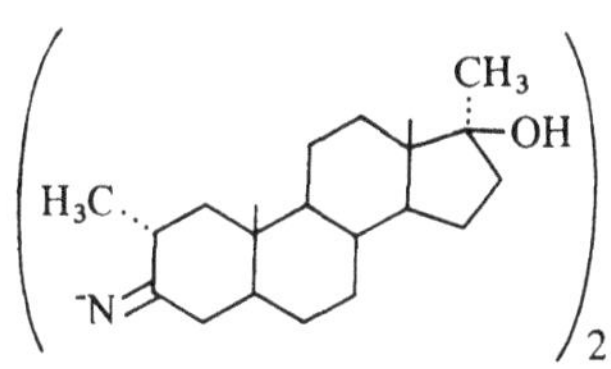


17β-Hydroxy-2α,17-dimethyl-5α-androstan-3-one-3,3′-azine

DORFMAN and KINCL (1963) found dimethazine to have a minimal myotrophic:androgenic dissociation relative to testosterone parenterally with an index of (0.27: <0.06–0.12) over 2. Its dissociation compared with methyltestosterone orally was of a similar order (DE RUGGIERI et al., 1963). Its myotrophic:androgenic ratio was found to be (2.1:0.15) 14 and anabolic by nitrogen retention:androgenic dissociation was (3.3:0.15) 22 (Authors, unpublished), very favorable ratios.

### 6. Oxandrolone

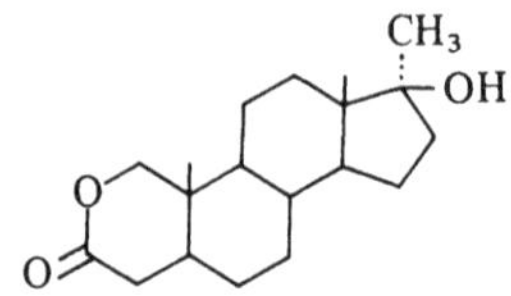

17β-Hydroxy-17-methyl-2-oxa-5α-androst-3-one

Fox et al. (1962) found the hetero oxygen-substituted steroid, oxandrolone, to have a nitrogen retention-conserving effect of 6.3, relative to methyltestosterone, in man. Experimentally, its myotrophic:androgenic dissociation was found to be (3.22:0.24) 13 relative to methyltestosterone, a ratio that compares favorably with the more promising protein anabolic agents (LENNON and SAUNDERS, 1964).

The heterocyclic protein anabolic steroids represent a number of structures that manifest great clinical promise on the basis of their uniformly favorable anabolic:androgenic dissociation shown experimentally.

## E. Miscellaneous Structures

### 1. Methylandrostenediol

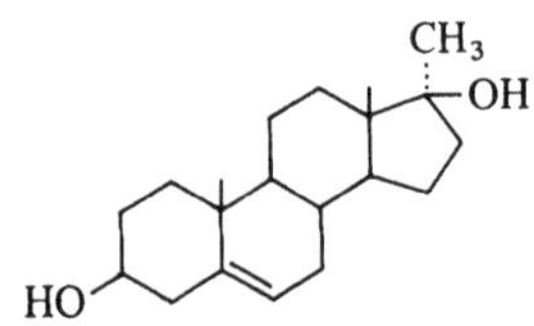

17-Methylandrost-5-ene-3β,17β-diol

Methylandrostenediol was one of the earliest steroids indicated to have potentially useful dissociation of anabolic from androgenic activities (PARTRIDGE et al., 1953; SALA and BALDRATTI, 1957). Relative to methyltestosterone orally, the myotrophic: androgenic dissociation of methylandrostenediol was reported to be (0.70:0.78) 0.9 (CAMERINO and SALA, 1960). OVERBEEK et al. (1962) found the myotrophic:androgenic dissociation, relative to methyltestosterone, to be (0.8:0.68) 1.8. Given parenterally, POTTS et al. (1960) reported methylandrostenediol, relative to testosterone propionate, to be 0.067 as effective in conserving nitrogen and 0.025 as effective myotrophically. Orally, ARNOLD and POTTS (1966) found methylandrostenediol to be 0.46 as myotrophic and 0.70 as effective in conserving nitrogen as methyltestosterone for dissociation indices of 0.7 and 1.1, respectively. KINCL (1965) did not consider its dissociation promising.

### 2. Methenolone

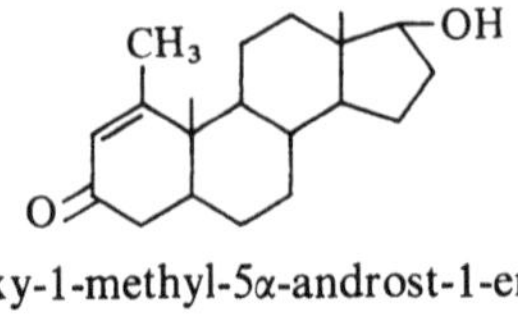

17β-Hydroxy-1-methyl-5α-androst-1-en-3-one

JUNKMANN and SUCHOWSKY (1962), and SUCHOWSKY and JUNKMANN (1965) found favorable myotrophic:androgenic dissociation for the enanthate ester of methenolone. Relative to testosterone parenterally, its dissociation was 16. Relative to methyltestosterone orally its dissociation was 13. The acetate ester gave a dissociation of 8.5 relative to testosterone propionate (SUCHOWSKY and JUNKMANN, 1962). NEUMANN and WIECHERT (1965) reported its dissociation relative to testosterone propionate to be (0.30:0.01–0.02) 15–30. Relative to testosterone TAIT and BURSTEIN (1964) found its myotrophic:androgenic dissociation to be (0.88:0.44–0.57) 2. The myotrophic:androgenic dissociation of its acetate ester was found to be (0.86:0.12) 7, but it did not appear to support nitrogen retention in castrated male rats (Authors, unpublished). KRÜSKEMPER and BREUER (1962) noted its favorable effect on nitrogen retention in the human. The lack of an alkyl group at position 17 may reduce the possibility of an adverse effect on liver function in man.

### 3. Ethylestrenol

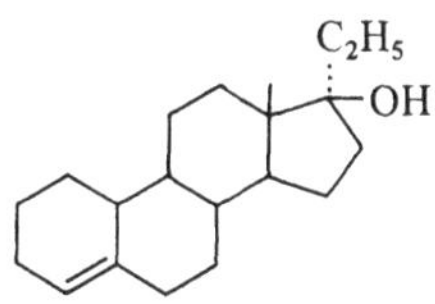


19-Nor-17α-pregn-4-en-17-ol

JUNKMANN and SUCHOWSKY (1962) noted ethylestrenol to have a myotrophic: androgenic dissociation of 4 relative to methyltestosterone orally. Its dissociation relative to testosterone parenterally was 2.6. These investigators noted it to be progestational. OVERBEEK (1961), and OVERBEEK et al. (1961, 1962) found the myotrophic: androgenic dissociation, relative to methyltestosterone, to be (4.207:0.222) 19. On the basis of nitrogen retention: androgenic effects, ARNOLD et al. (1963b) found a favorable dissociation index of (1.7:0.21) 8.1. KINCL (1965) judged the myotrophic: androgenic dissociation, relative to methyltestosterone orally, to be (2.0:0.4) 5. EDGREN et al. (1966) found it slightly more favorable than ethylnortestosterone with a dissociation of (0.4:0.17) 2.3 relative to the latter. VAN WAYJEN et al. (1961) judged the agent to have promising clinical potential.

### 4. 17-Methyl-3-methylene-1-androsten-17β-ol

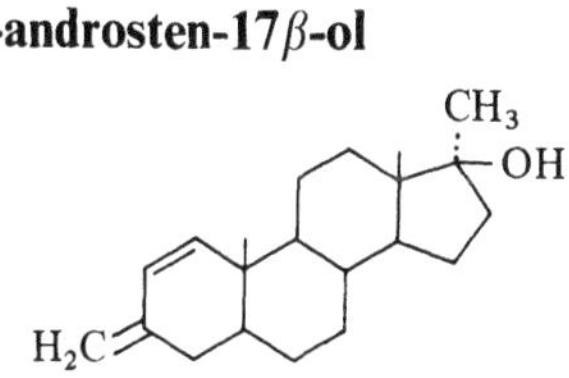


IRMSCHER et al. (1964) noted an especially favorable oral myotrophic:androgenic dissociation for 3-methylene-androst-1-enolone, relative to methyltestosterone, (2.66:0.12) 22.

Other than methylandrostenediol, whose early promise was not confirmed clinically, this group of miscellaneous structures merits consideration for clinical trial in the basis of the experimentally evaluated anabolic:androgenic dissociation indices.

# IV. Reversal of the Acute Catabolic Actions of Glucocorticoids by Protein Anabolic Steroids

## A. The Acute Catabolic Actions of Adrenal Glucocorticoids

The comprehensive experiments of LONG et al. (1940) and ENGEL (1951) established that one of the prominent physiologic effects of glucocorticoids is the mobilization of endogenous protein with its eventual hydrolysis and deamination to ketoacids utilized in gluconeogenesis. This action of adrenal cortical hormones is reflected in fasted laboratory animals by elevated urinary nitrogen output accompanied by substantial decrements in body weight. It seemed reasonable to expect that anabolic steroids, which improve nitrogen retention and increase body weight gains under standard nutritional status, should reverse the catabolic responses to glucocorticoids induced in experimental animals during fasting or limited dietary intake periods.

In preliminary experiments, decrements from control values of body weight and increased urinary nitrogen excretion of fasted rats over the 24-h period were used by BEYLER et al. (1960) to assess the catabolic action of cortisone acetate. The extent of these responses brought about by graded doses of cortisone acetate is presented in Table 2.

The data show that both the body weight loss and urinary nitrogen excretion values increased as the total dose of the glucocorticoid was increased from 6.25 mg/kg to 60 mg/kg. Conversion of the absolute urinary nitrogen values into theoretical lean body weight calculated on the basis that 31.25 mg urinary N derives on the average from one g of lean tissue catabolized during fasting (REIFENSTEIN et al., 1945) indicates that 30–40% of the actual body weight loss induced by each dose can be accounted for on the basis of the amount of nitrogen excreted. It was assumed that

Table 2. Influence of cortisone acetate (Cac) on body weight loss and urinary nitrogen excretion of fasted castrated male rats

| Total dose (Cac) (mg/kg) | No. of rats | Body weight loss g/rat/24 h | Urinary nitrogen excretion mg/100 g B.W./24 h | Theoretical lean body weight loss[a] g/rat/24 h |
|---|---|---|---|---|
| s.c. | | mean ± s.e. | mean ± s.e. | |
| 0 | 6 | 11.8 ± 0.9 | 70.8 ± 4.7 | 4.2 |
| 6.25 | 6 | 13.9 ± 1.2 | 80.2 ± 3.3 | 4.7 |
| 12.5 | 6 | 14.8 ± 1.2 | 91.2 ± 5.7 | 5.4 |
| 25.0 | 6 | 17.2 ± 0.9 | 114.7 ± 4.8 | 6.8 |
| 60.0 | 6 | 25.2 ± 1.3 | 121.8 ± 4.7 | 7.2 |

[a] Calculated on the basis that 31.25 mg of nitrogen are excreted for each g of lean tissue catabolized during fasting.

Table 3. Relative potencies of glucocorticoids based on urinary nitrogen excretion and liver glycogen deposition

| Glucocorticoid | Relative potencies | |
|---|---|---|
| | Nitrogen excretion | Liver glycogen |
| Cortisone acetate | 1.0 | 1.0 |
| Corticosterone | 0.9 | — |
| Hydrocortisone acetate | 1.7 | 2.0 |
| Prednisone | 4.0 | 4.0 |
| 9α-Fluorohydrocortisone acetate | 19.0 | 17.0 |

the remaining 60–70% of the actual weight loss is related, in major part, to fat catabolism and water diuresis.

Using the nitrogen excretion data for quantification of the catabolic activity of various glucocorticoids, the following relative potencies were calculated from log dose-response characterstics: cortisone, 1.0; corticosterone, 0.9; hydrocortisone acetate, 1.7; prednisone, 4.0 and 9α-fluorohydrocortisone acetate, 19.0. Accepting the biochemical evidence (ENGEL, 1951; GOODLAD and MUNRO, 1959) that the urinary nitrogen under the influence of the glucocorticoids is derived from amino acids from proteolysis and that the keto-acids formed as a result of deamination become precursors for the gluconeogenic action of the glucocorticoids, one would anticipate that the glycogenic action of the glucocorticoid should correspond to their relative catabolic action. In Table 3 are presented the relative catabolic activities of several glucocorticoids based on nitrogen excretion increments compared with relative glycogenic activity determined in adrenalectomized rats following the procedure of OLSEN et al. (1944). An excellent correspondence between the two activities was obtained supporting the implied close interdependence of these two metabolic effects of glucocorticoids.

In a series of experiments conducted under identical conditions in the authors' laboratories, the number of parameters for assessment of the catabolic as well as other actions of glucocorticoids was extended to include, in addition to fasting body weight loss and urinary nitrogen excretion, also urine volume, urinary potassium, sodium, phosphorus and calcium excretion along with weight changes in the adrenal and thymus glands.

The catabolic profile of cortisone acetate is shown graphically in Figures 1–3 together with the results obtained with progesterone. Progesterone was included in these studies since it had been observed by LANDAU et al. (1957, 1961, 1961a) to be catabolic in human subjects at moderately high dose levels. The data presented here on progesterone have significance beyond the fact that they serve to point out the relative specificity of the actions of the glucocorticoid. The results show that cortisone acetate produced dose related increases in body weight loss and urinary excretion of nitrogen, phosphorus, and potassium, as well as sodium, water, and calcium indicative of increased tissue breakdown.

Although the changes brought about by progesterone were generally in the same direction as those induced by cortisone, the effects were not nearly as marked in degree.

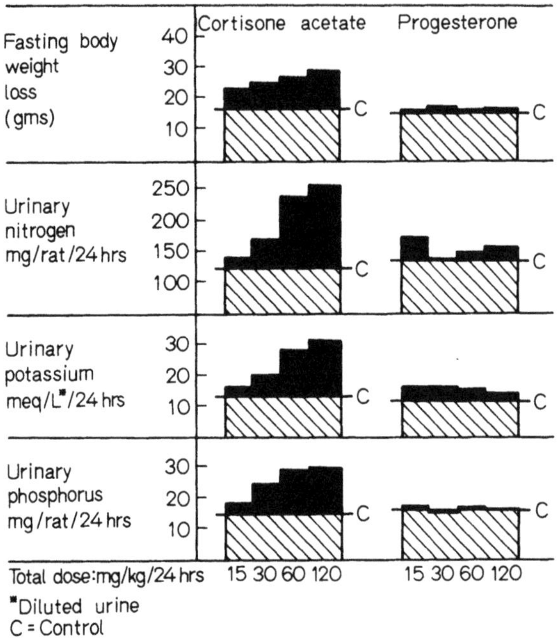


Fig. 1. Comparative catabolic effects (fasting body weight loss, urinary nitrogen, urinary potassium, urinary phosphorus) of progesterone and cortisone acetate in castrated male rats

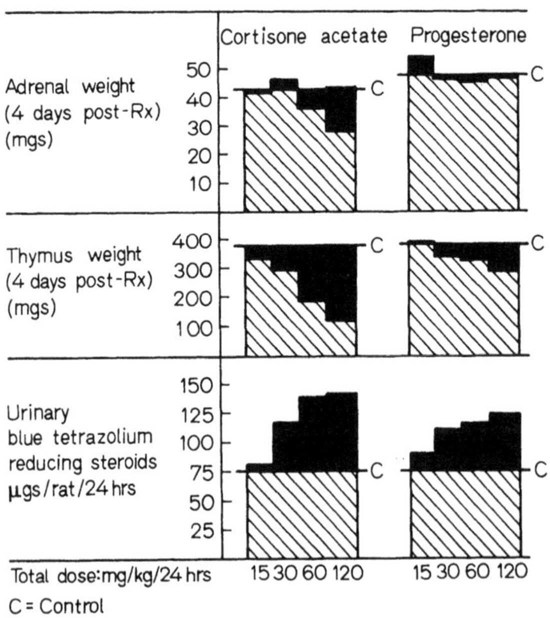


Fig. 2. Comparative catabolic effects (adrenal weight, thymus weight, urinary blue tetrazolium-reducing steroids) of progesterone and cortisone acetate in castrated male rats

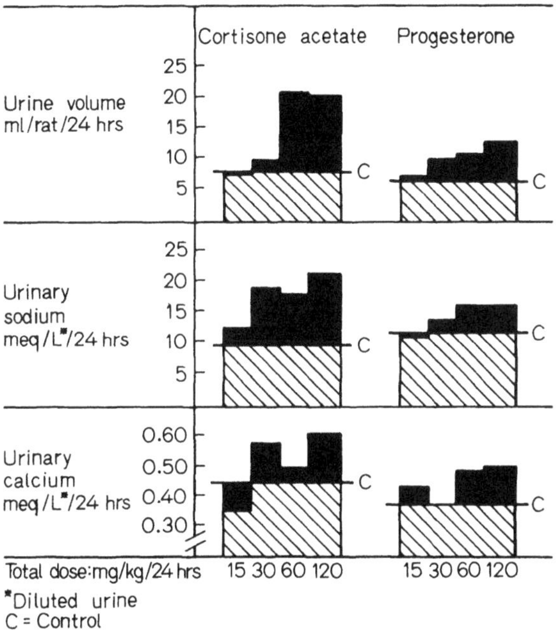


Fig. 3. Comparative catabolic effects (urine volume, urinary sodium, urinary calcium) of progesterone and cortisone acetate in castrated male rats

Table 4. Comparison of metabolic effects of cortisone acetate and progesterone in castrated and adrenalectomized-castrated male rats

| Parameter | Cortisone acetate: 60 mg/kg | | Progesterone: 15—120 mg/kg | |
|---|---|---|---|---|
| | Castrated % change | Adrex castrated % change | Castrated Max. % change | Adrex castrated Max. % change |
| Fasting B.W. loss | + 62[a] | + 43[a] | + 7 | −24 |
| Urinary nitrogen | + 93[a] | +121[a] | +24[a] | +16 |
| Urinary phosphorus | + 99[a] | +305[a] | + 7 | +11 |
| Urinary potassium | +107[a] | +342[a] | +39[a] | +90[a] |
| Urinary sodium | +110[a] | + 87[a] | +41[a] | −15 |
| Urine volume | +145[a] | +120[a] | +81[a] | +15 |
| Thymic weight | − 51[a] | − 72[a] | −25[a] | + 8 |

[a] Statistically significant at $P = .05$ or less.

The data in Table 4 present the changes induced by cortisone and progesterone as percentages of the control values. When expressed in this fashion, the magnitude of the responses is more readily apparent and the differences between cortisone and progesterone are more obvious. Progesterone loses most of its catabolic effect in adrenalectomized-castrated rats, whereas cortisone in regard to its effect on urinary excretion of nitrogen, phosphorus, and potassium in the adrenalectomized-castrated rat is markedly catabolic. This suggests that the catabolic action of progesterone under these experimental conditions may depend upon the presence of the adrenal and thus may be an indirect effect.

## B. Reversal of the Catabolic Actions of Glucocorticoids by Protein Anabolic Steroids

After our laboratories had established the characteristics of the acute catabolic actions of various glucocorticoids in castrated male rats, extensive studies were initiated to determine to what extent the orally active anabolic steroids might reverse or modify these actions. BEYLER et al. (1960) used cortisone acetate as the glucocorticoid and methyltestosterone as the orally active reference anabolic androgen. Reversal of body weight loss and urinary nitrogen excretion was selected as endpoints for assessing the anticatabolic action of the anabolic steroid. The data in Table 5 show that oral doses of methyltestosterone as large as 120 mg/kg were ineffective in reversing the cortisone-induced increase in body weight loss and urinary nitrogen excretion. It was not until a dose of 480 mg/kg was given that significant anticatabolic effects were observed. Stanozolol, on the other hand, at 15 and 60 mg/kg reduced the cortisone-induced body weight loss an average of 13% and 30%, respectively, and decreased the nitrogen excretion 21% and 52% (Table 6). In separate studies by ARNOLD et al. (1963c) this heterocyclic steroid was observed to be 10 times more active than methyltestosterone in providing nitrogen retention in nitrogen-equilibrated, castrated male rats and 2 times more myotrophic and one-third to one-quarter as androgenic as methyltestosterone. It is of interest to note that the dose ratio for achieving comparable effects on nitrogen excretion in these studies (8:1) was quite similar to that obtained in the nitrogen retention assays, i.e. **10**:1. However, the doses were much larger for both substances in the acute anticatabolic studies. Apparently, reversal of the acute glucocorticoid-induced nitrogen excretion under these conditions is a much less sensitive parameter than is that employed in the assay of the nitrogen-retaining effect of steroids using equilibrated castrated male rats. It is

Table 5. Reversal by methyltestosterone (Mt) of catabolic effects of cortisone acetate (Cac) administered to fasting castrated male rats

| Total dose mg/kg | | No. of Rats | Body weight loss | | Urinary nitrogen excretion | |
|---|---|---|---|---|---|---|
| Cac | Mt | | g/rat/24 h | % Reversal | mg/100 g B.W./24 h | % Reversal |
| s.c. | i.g. | | | | Mean ± s.e. | |
| 0 | 0 | 5 | 15.2 (4.7)[a] | — | 81 ± 6.0 | — |
| 60 | 0 | 6 | 19.0 (7.5) | — | 129 ± 8.8 | — |
| 60 | 15 | 5 | 19.0 (7.0) | 0 | 120 ± 6.3 | 19 |
| 60 | 60 | 5 | 20.0 (7.6) | 0 | 131 ± 3.5 | 0 |
| 0 | 0 | 6 | 11.5 (3.7) | — | 77 ± 6.8 | — |
| 60 | 0 | 6 | 17.7 (6.2) | — | 127 ± 8.6 | — |
| 60 | 120 | 6 | 17.3 (7.1) | 6 | 146 ± 14.0 | 0 |
| 0 | 0 | 8 | 17.8 (5.0)[a] | — | 65.9 ± 3.8 | — |
| 60 | 0 | 8 | 26.4 (7.8) | — | 103.0 ± 6.4 | — |
| 60 | 240 | 8 | 27.6 (7.3) | 0 | 96.9 ± 5.4 | 16 |
| 60 | 480 | 8 | 22.6 (6.1) | 44 | 81.6 ± 4.8 | 58 |

[a] Values on parentheses are theoretical lean body weight losses calculated from urinary nitrogen values

Table 6. Reversal by stanozolol (Stz) of catabolic effects of cortisone acetate (Cac) administered to fasting castrated male rats

| Total dose mg/kg | | No. of Rats | Body weight loss | | Urinary nitrogen excretion | |
|---|---|---|---|---|---|---|
| Cac | Stz | | g/rat/24 h | % Reversal | mg/100 g B.W./24 h | % Reversal |
| s.c. | i.g. | | | | mean ± s.e. | |
| 0 | 0 | 31 | 16.7 (4.8)[a] | — | 80.3 ± 4.7 | — |
| 60 | 0 | 34 | 22.4 (7.3) | — | 122.2 ± 6.6 | — |
| 60 | 15 | 25 | 21.1 (6.8) | 13 | 114.0 ± 5.9 | 21 |
| 60 | 60 | 32 | 20.9 (5.9) | 30 | 100.7 ± 3.9 | 52 |

[a] Values in parentheses are theoretical lean body weight losses calculated from urinary nitrogen values.

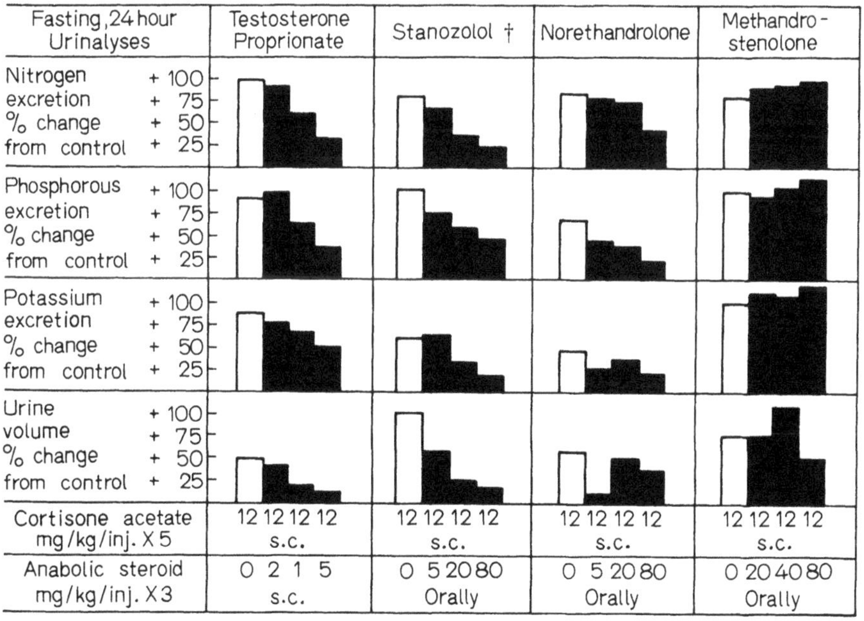


Fig. 4. Influence of anabolic steroids on cortisone-induced diuresis of nitrogen, phosphorus, potassium and water in fasted, castrated male rats

likely that some of the discrepancies referred to above in regard to the claimed ineffectiveness of methyltestosterone to reverse the catabolic effects of glucocorticoid may be due to the relative insensitivity of this experimental approach. From the data presented here it would appear that the dose of the anabolic steroid has to be increased 5 to 10-fold over the dose required to effect nitrogen retention or muscle growth before significant reversal of the glucocorticoid catabolic effects is obtained. The short-term nature of the anticatabolic methods as reported in this section may contribute to this insensitivity.

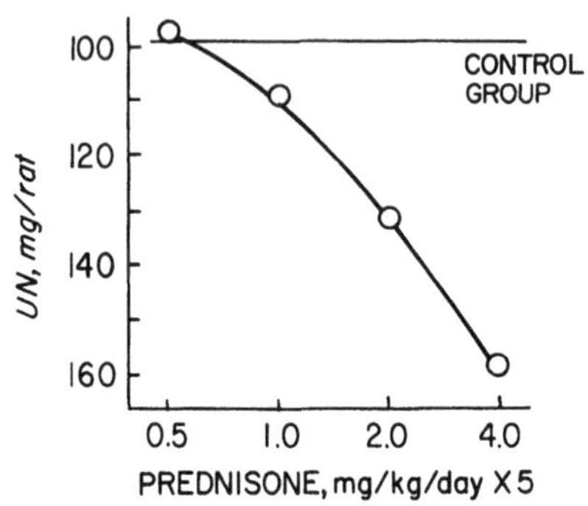


Fig. 5. Effect of increasing doses of prednisone on urinary nitrogen excretion of weight- and nitrogen-equilibrated castrated male rats

The ability of anabolic steroids to decrease the cortisone-induced urinary loss of nitrogen, phosphorus, and potassium has also been used to evaluate their anticatabolic action (TAINTER et al., 1963; BEYLER et al., 1968). The results of representative experiments are shown in Figure 4. Stanozolol and norethandrolone given orally reduced the catabolic responses to cortisone in a pattern similar to that brought about by testosterone propionate given by injection. Methandrostenolone, on the other hand, within the same dose range, appeared to add slightly to the catabolic action of cortisone as measured in these acute studies. It may be of interest to note that methandrostenolone has been reported to enhance the anti-inflammatory properties of glucocorticoids in humans (CLARK and MILLS, 1962) and to be active as an anti-inflammatory agent in animals (RENZI et al., 1961). The anticatabolic action of the orally active steroids tested generally paralleled their relative anabolic potency.

As mentioned above, the sensitivity of the methods employed to assess the ability of anabolic agents to reverse the acute catabolic actions of glucocorticoids appeared to be considerably less than that of the more chronic assay methods. This was reflected in the relatively high oral dose levels of the anabolic steroids required to achieve significant reversal of the cortisone-induced nitrogen loss compared with the dose levels necessary to evoke improvement in the retention of nitrogen of nitrogen-equilibrated castrated rats, as reported by ARNOLD et al. (1963a).

Taking the differences in experimental design into account, ARNOLD has reported (1966) on the anticatabolic effect of stanozolol on nitrogen-equilibrated, castrated male rats given daily doses of prednisone over a 5-day experimental period. Figure 5 shows that under these conditions prednisone at 0.5, 1.0, 2.0, and 4.0 mg/kg/day produced a dose-related increase in the excretion of urinary nitrogen. The nitrogen excretion brought about by the 2 and 4 mg/kg dose levels of prednisone were significantly different from control levels and were chosen for the evaluations of the anticatbolic effects of stanozolol as described in the methods section above.

In Figure 6 the reversal by stanozolol of the catabolic effects of the 2 mg/kg prednisone dose is shown. The nitrogen excretion of the rats decreased with increasing doses of stanozolol. Nitrogen losses were approximately halved by 10 or 30 mg stanozolol per kg. Stanozolol at 90 mg per kg more than completely reversed the catabolic effects of the 2 mg prednisone dose.

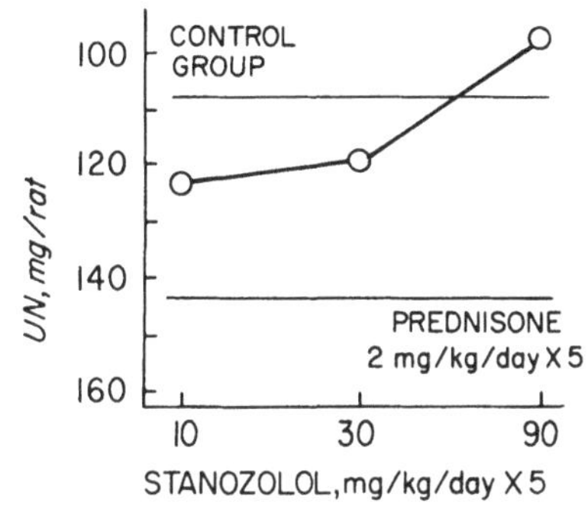


Fig. 6. Effect of increasing doses of stanozolol on weight- and nitrogen-equilibrated castrated male rats given 2 mg prednisone per kg over a 5-day period

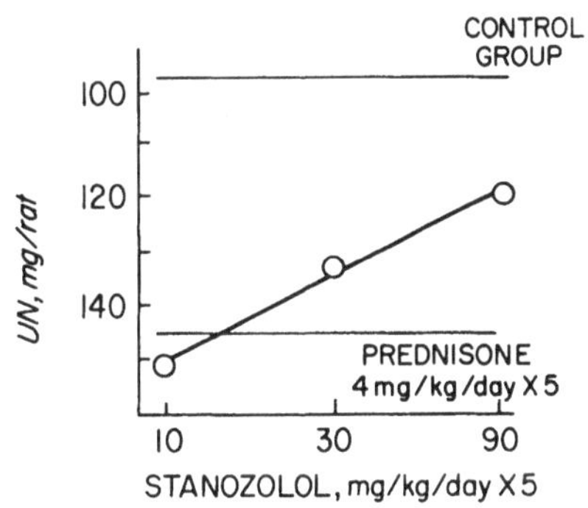


Fig. 7. Effect of increasing doses of stanozolol on weight- and nitrogen-equilibrated castrated male rats given 4 mg prednisone per kg over a 5-day period

The reversal of the catabolic effects of prednisone at 4 mg/kg is shown in Fig. 7. The nitrogen excretion of the rats here also decreased with increasing doses of stanozolol. It should be noted that even 90 mg stanozolol per kg was not able to counteract completely the catabolic effects of the 4 mg prednisone per kg dose.

These data underscore, in terms of tissue wastage as exemplified by urinary nitrogen excretion, the marked catabolic effects of the glucocorticoids. The reversal of these catabolic effects by stanozolol are in accord with the earlier acute studies of BEYLER et al. (1960) who found that at elevated anabolic steroid dosages the acute catabolic effects of parenterally administered cortisone acetate could be ameliorated. It is concluded that under these conditions of test the extent to which the anabolic agent is able to counter the catabolic influence of glucocorticoids is, in part at least, determined by the severity of the catabolic state induced. The sensitivity of the assay is affected accordingly.

## C. The Anticatabolic Activity of Anabolic Steroids Based on the Suppression of Cortisone Acetate Induction of Liver Tryptophan Pyrrolase

Tryptophan pyrrolase has been shown to control the catabolism of the amino acid, tryptophan. This enzyme catalyzes the formation of kynurenine, a major metabolite

Table 7. The effects of anabolic steroids on induction of liver tryptophan pyrrolase by cortisone acetate in intact male rats

| Cortisone acetate mg/kg × 1 s.c. | Anabolic steroid mg/kg × 2 i.g. | Tryptophan pyrrolase activity μmoles kynurenine/g protein/h ± S.E. |
|---|---|---|
| | Methyltestosterone | |
| — | — (5)[a] | 18.2 ± 2.1 |
| 5 | — (5) | 54.2 ± 3.0 |
| 5 | 240 (5) | 50.9 ± 5.5 |
| — | 240 (5) | 42.0 ± 11.8 |
| | Methandrostenolone | |
| — | — (5) | 16.4 ± 1.4 |
| 5 | — (5) | 58.5 ± 5.2 |
| 5 | 200 (5) | 33.6 ± 8.6 |
| — | 200 (5) | 12.6 ± 1.4 |
| | Norethandrolone | |
| — | — (6) | 23.6 ± 2.8 |
| 5 | 120 (6) | 52.3 ± 3.2 |
| 5 | 120 (6) | 34.2 ± 5.1 |
| — | 120 (6) | 28.6 ± 2.4 |
| | Stanozolol | |
| — | – (6) | 26.1 ± 1.1 |
| 5 | — (6) | 98.3 ± 3.5 |
| 5 | 30 (6) | 47.7 ± 8.4 |
| — | 30 (6) | 27.6 ± 1.5 |

[a] Number in parenthesis indicates number of rats per group.
——— Significantly different from cortisone acetate–treated group $p < .02$.
═══ significantly different from cortisone acetate–treated group $p < .0002$.

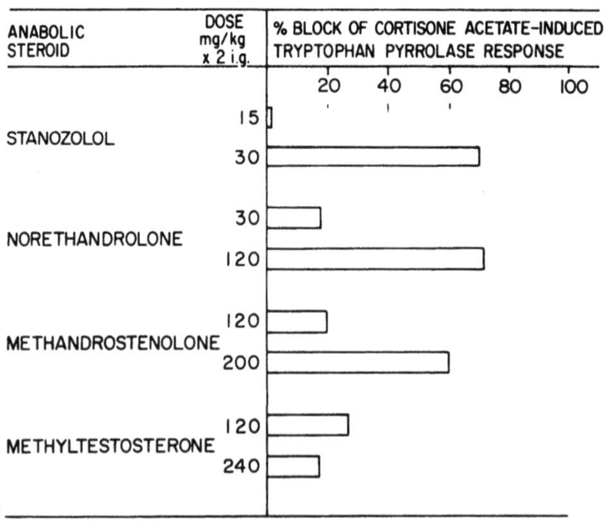


Fig. 8. Effects of various anabolic agents on induction of liver tryptophan pyrrolase by cortisone acetate (5 mg/kg) in intact rats

Table 8. The effects of anabolic steroids on induction of liver tryptophan pyrrolase by cortisone acetate in adrenalectomized rats

| Cortisone acetate mg/kg × 1 s.c. | Anabolic steroid mg/kg × 2 i.g. | Tryptophan pyrrolase activity µmoles kynurenine/g protein/h ± S.E. |
|---|---|---|
| | Methyltestosterone | |
| — | — | 8.2 ± 0.5 |
| 5 | — | 58.4 ± 4.1 |
| 5 | 240 | 14.4 ± 2.1 |
| — | 240 | 7.6 ± 0.8 |
| | Methandrostenolone | |
| — | — | 13.4 ± 1.8 |
| 5 | — | 68.6 ± 3.2 |
| 5 | 200 | 22.7 ± 3.1 |
| — | 200 | 6.7 ± 1.9 |
| | Norethandrolone | |
| — | — | 9.2 ± 0.4 |
| 5 | — | 67.4 ± 7.1 |
| 5 | 120 | 13.5 ± 0.9 |
| — | 120 | 13.5 ± 0.9 |
| | Stanozolol | |
| — | — | 15.8 ± 1.2 |
| 5 | — | 104.9 ± 12.3 |
| 5 | 30 | 44.6 ± 6.5 |
| — | 30 | 13.8 ± 0.7 |

——— Significantly different from cortisone acetate–treated group $p<.001$.

of tryptophan. Increased tryptophan pyrrolase activity in the liver and an elevation of kynurenine excretion have been reported by MORAN and SOURKES (1964) to occur in various catabolic conditions in the rat.

It has been reported by KNOX and AUERBACH (1955) that treatment with glucocorticoids will result in marked increases in tryptophan pyrrolase activity in rat liver. The effects of the glucocorticoids on the activity of this enzyme is consistent with the known fact that these steroids cause an increased urinary nitrogen excretion and body weight loss.

Anabolic steroids have been observed to reduce considerably the catabolic actions of the glucocorticoids as measured by a reversal of urinary nitrogen and body weight losses (see above). In the above-cited studies stanozolol and norethandrolone were about equally effective, while methandrostenolone was weakly active. It became of interest to HARDING and POTTS (1965) to determine if the catabolic action of cortisone acetate, as represented by increased tryptophan pyrrolase activity, could be altered by anabolic steroids.

Four anabolic steroids were studied at several dose levels: methyltestosterone, methandrostenolone, norethandrolone, and stanozolol.

Administration of cortisone acetate alone caused 2- to 4-fold increases in enzyme activity (Table 7). The induction of tryptophan pyrrolase by cortisone acetate was suppressed 40–70% by anabolic agents administered orally. The minimal dose (mg/kg × 2) of each agent which significantly suppressed enzyme induction was as follows: stanozolol, 30; norethandrolone, 120; methandrostenolone, 200. Methyltestosterone produced no suppression at 240 mg/kg × 2. Methyltestosterone alone resulted in elevation of tryptophan pyrrolase activity presumably by increasing corticosterone output, since enzyme suppression was obtained in adrenalectomized rats (Table 8). Quantitatively (cf. Fig. 8) there appears to be wide differences among the steroids tested with the pyrazol steroid exhibiting the highest activity.

The above-described comparative anticatabolic activity of anabolic agents is in agreement with data based on reversal of nitrogen excretion in cortisone-treated rats as cited above.

These data support the concept that anabolic agents should be useful in offsetting the catabolic effects observed clinically following chronic administration of glucocorticoids. Additionally, these data provide evidence that the anticatabolic effects of anabolic agents are elicited in visceral as well as somatic tissues.

## V. Comments and Conclusions

It is well established that androgenic substances are anabolic in varying degrees. From a practical consideration the selection of the anabolic agent for clinical use must take into account the degree of associated androgenic activity. This is particularly true when the agent is considered for administration to women and children. The purpose of providing a comparative evaluation of this class of compounds in terms of both androgenic and anabolic activities is to provide a basis for the selection of the appropriate agent in each clinical situation.

The assessment of anabolic and androgenic activities in various laboratories is based on a variety of experimental tests conducted under different conditions. The estimation of androgenic activity is relatively straightforward with the response of the ventral prostate and seminal vesicles of the castrated immature male rat being the most commonly employed endpoint. This procedure provides an estimate of androgenic potency relative to the appropriate reference standards. The predictability of the degree of clinical manifestation of the androgenic activity of a particular compound has been clearly established on the basis of the results of these tests. The chick and capon comb assay methods are generally less useful in projecting a meaningful estimate of the degree of androgenic activity which may be observed in the clinical evaluation of a new steroid although they provide additional criteria for comparative purposes (SEGALOFF and GABBARD, 1963). The same probably holds true for the lacrimal gland assay described by CAVALLERO (1967).

The assessment of anabolic activity has been based on one or more of the following procedures: increase in the weight of the levator ani muscle, increase in the total body weight, and increase in nitrogen retention. Of these, the growth of the levator ani muscle has become the method most widely used, whereas nitrogen balance studies have been reserved for more conclusive evaluation of anabolic activity. A major advantage of the myotrophic assay is the fact that androgenic activity can be evaluated concomitantly with anabolic activity. There are certain conditions which do modify the nature of the response, as with most assays, and should be chosen in accord with the objectives of the study. The time lapse between castration and initiation of medication is an important factor. If the rat is placed on test the day of castration, one must take into consideration the ability of the test agent to maintain the weights of the ventral prostate, seminal vesicles, and levator ani muscle as well as to stimulate these tissues during the course of the assay period. Some compounds simply maintain the weights of these organs rather than stimulate them. A holding period of 1–2 weeks after castration has become the method most generally employed. Comparisons of test and reference agents may be modified by the duration of test in that the longer the duration the greater is the confidence in the conclusion as to the order of androgenicity.

In addition to the view that an estimate of the nitrogen-retention properties of anabolic agents is the best index of anabolic activity (CAMERINO and SALA, 1960; KINCL, 1965; BEYLER et al., 1968; VIDA, 1969; CAMERINO and SCIAKY, 1975), it may be of interest to note that, in our experience, a number of steroids synthesized as variations of the androstane molecule yielded myotrophic-androgenic dissociations which were favorable but not strikingly different from one another within the series. However, their anabolic activities based on nitrogen balance studies did differ significantly. It would appear that nitrogen balance assays are more discriminating.

From an overall clinical utilitarian viewpoint, the anabolic actions of steroids are important possibly more as they pertain to a capacity to reverse an existing catabolic state than to their ability to promote growth. In this framework of reference the term anticatabolic has particular clinical significance. KRÜSKEMPER (1968) discusses the anticatabolic action of steroids at some length and the reader is referred to this reference for a comprehensive review of the subject. In this connection, we wish to mention here that in the studies reported in this chapter, the term anticatabolic has been used to express an action which is clearly opposite to the catabolic state induced under specified experimental conditions. These conditions have been controlled by castrating the laboratory animal, by restricting the amount of feed intake and by administering glucocorticoids. Obviously, the loss of body weight and the excessive excretion of nitrogen, phosphorous, and potassium that accompany the administration of glucocorticoids during fasting or limited dietary intake constitute a catabolic state with a net loss of body constituents. The extent to which dietary factors, stressful conditions, and the secretions of the adrenal cortex contribute to the economy of the body's metabolism has been a topic of considerable interest to many investigators. Among many of the articles of importance in this area are: ENGEL (1951); MUNRO (1951); KOCHAKIAN and ROBERTSON (1951); SILBER and PORTER (1953); GOODLAD and MUNRO (1959); ALBANESE et al. (1961); IZZO and GLASSER (1961); LINET (1970). From these investigations it is obvious that nutritional and hormonal factors play a major role in maintaining or adjusting the metabolic balance to the needs of the body. The extent to which these factors are able to meet the needs under a wide variety of conditions determine the health of the individual. That anabolic steroids are able to offset the catabolic influence of glucocorticoids as shown by the data in this chapter should provide the clinician with the rationale and means to assist the debilitated or "catabolic" patient in regaining a metabolic balance commensurate with good health. It is suggested that this clinical application of anabolic steroids is not as widely appreciated as it deserves to be.

The observation that numerous steroids have been shown to have a dissociation of anabolic and androgenic activities invites discussion. KASUYA and WOLFE (1973) have provided evidence for three androgenic receptor sites in the ventral prostate.

These findings are based on the relative displacement of dihydrotestosterone from each of the receptor sites by analogous steroids. In practice, the binding affinity of a given steroid for any or all of these receptor sites is reflected in the biological response, i.e., androgenicity. Of equal interest is the report by Steinetz et al. (1971) that the levator ani muscle has two sets of receptors: one for anabolic and one for androgenic activity. Stimulation of either type of receptor leads to muscle growth. From this there appears to be presumptive evidence to explain the observations that various steroids stimulate the sex accessory organs and levator ani muscle disproportionately.

With the evidence that there are multiple receptors in the sex accessory as well as muscle tissue for androgenic steroids, it is tempting to speculate that the dissociation of anabolic from androgenic responses could be mediated through a mechanism of differential affinities. A particular chemical modification of the androstane nucleus may be visualized as affecting the affinity of the molecule for a receptor for muscle growth more than for an androgen receptor in the ventral prostate or seminal vesicles and vice versa. Much has been written in recent years and several theories and hypothesis have emerged relative to the interaction of androgenic hormones with receptors and the structural and configurational requirements for attachment. Vida (1969) has discussed the various theories and has contributed his own to those of others. Knowledge in this area is understandably in a state of flux and the idea of differential affinities with multiple receptor sites for androgens may be a meaningful interpretation of the mass of data available supporting evidence for dissociation of the anabolic from the androgenic activities of steroids.

If one accepts the idea of dissociation of activities based on differential affinity for receptor sites, then one should expect to find instances of dissociation in either direction. An instance of reverse separation, i.e., greater androgenicity than anabolic is represented by unpublished data given below on 17α-methylandrostane-17β-ol-3-one azine: This compound brought about a 4- to 5-fold increase in the ventral prostate weight without stimulating the growth of the levator ani muscle significantly.

| | Dose mg/kg/day × 9 s.c. | Ventral prostate weight (mg) mean ± s.e. | Levator ani muscle weight (mg) mean ± s.e. |
|---|---|---|---|
| Control | — | 7.4 ± 0.2 | 41.3 ± 3.9 |
| Azine | 0.35 | 30.4 ± 2.7 | 43.4 ± 2.7 |
| Azine | 0.70 | 35.7 ± 1.3 | 39.9 ± 2.9 |

Thus, it is apparent that the separation of activities can occur in both directions. The fact that some steroids are relatively more androgenic than anabolic and others more anabolic than androgenic provides the framework for their clinical utility.

As a conclusion to this chapter, we have listed some representative steroids which on the basis of the laboratory animal assays exhibit high anabolic activity with low androgenicity on the one hand and those, on the other hand, which have high anabolic as well as high androgenic activity.

| Steroid | Dissociation index[a] |
|---|---|
| 1. Representative steroids having high anabolic with low androgenic activities: | |
| thiomesterone | 4.56/0.61 (7.5) |
| oxabolone | 1.2–1.6/0.25–0.4 (3–6) |
| norbolethone | 3.4/0.15 (20) |
| oxymetholone | 1.75/0.2 (8.75) |
| stanozolol | 10/0.33 (30) |
| androisoxazole | 1.55/0.22 (7.0) |
| dimethazine | 2.1–3.3/0.5 (14–22) |
| oxandrolone | 3.22/0.24 (13) |
| ethylestrenol | 1.7/0.21 (8.1) |
| 2. Steroids having both high anabolic and androgenic activities: | |
| fluoxymesterone | 20/9.5 (2.1) |
| bolasterone | 5.75/3.0 (1.9) |

[a] Dissociation index is defined as the ratio of relative anabolic to androgenic activity and is given in the parentheses.

Obviously, if androgenic activity is desirable or of no real concern as in hypogonadism, testosterone or methyltestosterone and their close chemical analogs and the steroids listed in category 2 above may be used. The steroids listed under 1 above and others that qualify, should be considered for uses in instances where anabolic activity is the preferred action and masculinization is to be avoided as much as possible as when treatment of women and children is being considered.

## References


ALBANESE, A. A., LORENZE, E. J., ORTO, L. A.: Nutritional and metabolic effects of some newer cortex-steroids. N.Y. State J. Med. **61**, 3998—4002 (1961).

ALLEN, E., DOISY, E. A.: Ovarian hormone preliminary report on its localization, extraction, partial purification, and action in test animals. J. Am. med. Ass. **81**, 819—821 (1923).

ARNOLD, A.: Reversal by stanozolol of glucocorticoid induced nitrogen losses in rats. Proceedings 7th Int'l Congr. Nutrition **5**, 846—852 (1966).

ARNOLD, A., BEYLER, A. L., POTTS, G. O.: Andro-stanzole, a new orally active anabolic agent. Proc. Soc. exp. Biol. (N.Y.) **102**, 184—187 (1959).

ARNOLD, A., POTTS, G. O.: Oral anabolic and androgenic effects of 19-nortestosterone and 17α-methyl-19-nortestosterone. Fed. Proc. Fed. Amer. Soc. exp. Biol. **23**, 412 (1964).

ARNOLD, A., POTTS, G. O.: Oral anabolic:androgenic evaluations of four anabolic steroids. Acta endocr. (Kbh.) **52**, 489—496 (1966).

ARNOLD, A., POTTS, G. O., BEYLER, A. L.: Evaluation of the protein anabolic properties of certain orally active anabolic agents based on nitrogen balance studies in rats. Endocrinology **72**, 408—417 (1963a).

ARNOLD, A., POTTS, G. O., BEYLER, A. L.: Relative oral anabolic to androgenic activity ratios of androisoxazole, ethylestrenol, methylandrostenol-isoxazole and testosterone. Acta endocr. (Kbh.) **44**, 490—498 (1963b).

ARNOLD, A., POTTS, G. O., BEYLER, A. L.: Oral anabolic:androgenic activities of 7α,17α-dimethyltestosterone, oxymesterone, mestanolone and fluoxymesterone. J. Endocr. **28**, 87—92 (1963c).

Arnold, A., Potts, G. O., Beyler, A. L.: Relative oral anabolic: androgenic activities of testosterone, dihydrotestosterone, 4-hydroxymethyltestosterone, 7α,17α-dimethyltestosterone and 9α-fluoro-11β-hydroxymethyltestosterone. 6th Intl. Congress of Nutrition, Edinburgh, Aug. 9—15, 161 (1963d).
Barnes, L. E., Stafford, R. O., Guild, M. E., Thole, L. C., Olson, K. J.: A comparison of myotrophic and androgenic activities of testosterone propionate with 19-nortestosterone and its esters. Endocrinology **55**, 77—84 (1954).
Beyler, A. L., Arnold, A., Potts, G. O.: Methods for evaluating anabolic and catabolic agents in laboratory animals. J. Amer. med. Women Ass. **23**, 708—721 (1968).
Beyler, A. L., Burnham, D., Potts, G. O.: Reversal by androstanozole of catabolic action of cortisone acetate. 1st International Congress of Endocrinology, Copenhagen, Advanced Abstracts of Short Communications (Fuchs, F. ed.) Periodica, Copenhagen. 829 (1960).
Beyler, A. L., Potts, G. O., Arnold, A.: Influence of molecular unsaturation on hormonal activity pattern of certain heterocyclic steroids. Endocrinology **68**, 987—995 (1961).
Bliss, C. I.: Statistical methods in vitamin research. In: György, P. (Ed.): Vitamin Methods, pp. 445—610. New York: Academic Press 1951.
Burnham, D. F., Fanuele, F.: The acute catabolic response of castrated male rats to various glucocorticoids. Fed. Proc. Fed. Amer. Soc. exp. Biol. **21**, 191 (1962).
Buzby, G. C., Jr., Walk, C. R., Smith, H.: Totally synthetic steroid hormones. X. Some (±)-13β-ethyl-7α-methylgonane derivatives. J. med. Chem. **9**, 782—784 (1966).
Camerino, B., Sala, G.: Anabolic steroids. Prog. Drug Res. **2**, 71—134 (1960).
Camerino, B., Sciaky, R.: Structure and effects of anabolic steroids. Pharmac. Therap. B Vol. 233—275 (1975).
Campbell, J. A., Babcock, J. C.: The synthesis of some 7α- and 17α-methyl steroid hormones. J. Amer. chem. Soc. **81**, 4069—4074 (1959).
Campbell, J. A., Babcock, J. C.: 7α-Methyltestosterones, A new class of potent anabolic agents. In: Martini, L., Pecile, A. (Eds.): Hormonal Steroids, Vol. 2, pp. 59—67. New York: Academic Press 1965.
Cavallero, C. P.: Relative effectiveness of various steroids in an androgen assay using the exorbital lacrimal gland of the castrated rat. Acta endocr. (Kbh.) **55**, 119—130 (1967).
Čekan, Z., Pelc, B.: Effect of 1-alkyl substitution on the biological action in a series of androstanes. Steroids **8**, 209—218 (1966).
Church, R. F. R., Kende, A. S., Weiss, M. J.: Diazirines. I. Some observations on the scope of the ammonia-hydroxylamine-0-sulfonic acid diaziridines and diazirines. J. Amer. chem. Soc. **87**, 2665—2671 (1965).
Clark, G. M., Mills, D.: Corticosteroid therapy of rheumatoid arthritis supplemented with methandrostenolone. Arthritis Rheumatism **5**, 156—162 (1962).
Clauberg, C.: Zur Physiologie und Pathologie der Sexualhormone im Besonderen des Hormones des Corpus luteum. I. Mitteilung. Der biologische Test für das Luteohormon (das spezifische Hormon des Corpus luteum) am infantilen Kaninchen. Zbl. Gynaekol. **54**, 2757—2770 (1930).
Clausen, F. W., Freudenberger, C. B.: A comparison of the effects of male and female sex hormones on immature female rats. Endocrinology **25**, 585—592 (1939).
Clinton, R. O., Manson, A. J., Stonner, F. W., Beyler, A. L., Potts, G. O., Arnold, A.: Steroidal [3,2-c] pyrazoles. J. Amer. chem. Soc. **81**, 1513 (1959).
Cross, A. D., Harrison, I. T., Crabbé, P., Kincl, F. A., Dorfman, R. I.: Steroids. CCLXVII. Potent orally-active anabolic steroids. Steroids **4**, 229—235 (1964).
Desaulles, P. A., Krähenbühl, C.: Comparison du spectre d'activité de certains gestagénes de synthése. Acta endocr. (Kbh.) **40**, 217—231 (1962).
Desaulles, P. A., Krähenbühl, C., Schuler, W., Bein, H. J.: Etude expérimentale du Dianabol, un novel anabolisant. 17α-méthyl-17β-hydroxyandrosta-1,4-diene-3-one. Schweiz. med. Wschr. **89**, 1313—1318 (1959).
Dörner, G., Schubert, A.: The myotrophic and androgenic activity of new 4-chlorotestosterone derivatives. Intl. Congress on Hormonal Steroids, Milan, May 14—19, 210 (1962).
Dorfman, R. I., Dorfman, A. S.: The assay of subcutaneously injected androgens in the castrated rat. Acta endocr. (Kbh.) **42**, 245—253 (1963).

DORFMAN, R. I., KINCL, F. A.: Relative potency of various steroids in an anabolic-androgenic assay using the castrated rat. Endocrinology **72**, 259—266 (1963).

EDGREN, R. A.: A comparative study of the anabolic and androgenic effects of various steroids. Acta endocr. (Kbh.) **44**, Suppl. 87, 3—21 (1963).

EDGREN, R. A., PETERSON, D. L., JONES, R. C., NAGRA, C. L., SMITH, H., HUGHES, G. A.: Biological effects of synthetic gonanes. Recent Prog. Horm. Res. **22**, 305—341 (1966).

EDGREN, R. A., SMITH, H.: Wy 3475, A new potent orally active anabolic agent. In: MARTINI, L., PECILE, A. (Eds.): Hormonal Steroids, Vol. 2, pp. 161—171. New York: Academic Press 1965.

EISENBERG, E., GORDAN, G. S.: The levator ani muscle of the rat as an index of myotrophic activity of steroidal hormones. J. Pharmacol. exp. Ther. **99**, 38—44 (1950).

ENGEL, F. L.: A consideration of the roles of the adrenal cortex and stress in the regulation of protein metabolism. Recent Progr. Hormone Res. **6**, 277—313 (1951).

ERCOLI, A., GARDI, R., VITALE, R.: Steroid-17$\beta$-yl acetals and enol ethers. New classes of orally and parenterally active hormonal derivatives. Chem. Ind. (Lond.) **1962**, 1284—1285.

FOX, M., MINOT, A. S., LIDDLE, G. W.: Oxandrolone: A potent anabolic steroid of novel chemical configuration. J. clin. Endocr. **22**, 921—924 (1962).

GOODLAD, G. A. J., MUNRO, H. N.: Diet and action of cortisone on protein metabolism. Biochem. J. **73**, 343—348 (1959).

HALPERN, B., SEAKAPOULOS, P., BRICOT, M.: Inhibition of growth of young rats by cortisone and aggravation of this action by testosterone. C.R. Soc. Biol. (Paris) **150**, 1311—1316 (1956).

HARDING, H. R., POTTS, G. O.: The anticatabolic activity of anabolic steroids based on the suppression of cortisone acetate (EAc) induction of liver tryptophan pyrrolase. VI$^{th}$ Pan American Congress of Endocrinology, Mexico D.F. (1965).

HERR, M. E., HOGG, J. A., LEVIN, R. H.: Synthesis of potent oral anabolic-androgenic steroids. J. Amer. chem. Soc. **78**, 500—501 (1956).

HERSHBERGER, J. G., SHIPLEY, E. G., MEYER, R. K.: Myotrophic activity of 19-nortestosterone and other steroids determined by modified levator ani muscle method. Proc. Soc. exp. Biol. (N.Y.) **83**, 175—180 (1953).

IRMSCHER, K., KRAFT, H. G., BRÜCKNER, K.: Synthesis and anabolic activity of 3-methylene- and 3-methylsteroids of the androstane and 19-norandrostane series. J. med. Chem. **7**, 345—347 (1964).

IZZO, J. L., GLASSER, S. R.: Comparative effects of glucagon, hydrocortisone and epinephrine on the protein metabolism of the fasting rat. Endocrinology **68**, 189—198 (1961).

JUNKMANN, K., SUCHOWSKY, G.: Untersuchung anabol wirksamer Steroide. Arzneimittel-Forsch. **12**, 214—218 (1962).

KASUYA, K., WOLFF, M. E.: Uptake of androgen analogs by prostate tissue. J. med. Chem. **16**, 832—836 (1973).

KINCL, F. A.: Anabolic steroids. Meth. Hormone Res. **4**, 21—76 (1965).

KINCL, F. A., DORFMAN, R. I.: Anabolic-androgenic potency of various steroids in a castrated rat assay. Steroids **3**, 109—122 (1964).

KLEIN, M., PARKES, A. S.: The progestin-like action of testosterone and certain related compounds. Proc. roy. Soc. Biol. (Lond.) **121**, 574—579 (1937).

KNOX, W. E.: L-Tryptophan peroxidase from liver. In: COLOWICK, S. P., KAPLAN, N. O. (Eds.): Methods in Enzymology, Vol. 2, pp. 242—246. New York: Academic Press 1955.

KNOX, W. E., AUERBACH, V. H.: The hormonal control of tryptophan peroxidase in the rat. J. biol. Chem. **214**, 307—313 (1955).

KOCHAKIAN, C. D.: Comparison of protein anabolic property of various androgens in the castrated rat. Amer. J. Physiol. **160**, 53—61 (1950).

KOCHAKIAN, C. D., MOE, J. G., DOLPHIN, J.: Protein anabolic effect of testosterone propionate in adrenalectomized and normal rats. Amer. J. Physiol. **162**, 581—589 (1950).

KOCHAKIAN, C. D., MURLIN, J. R.: Effect of male hormone on protein and energy metabolism of castrate dogs. J. Nutr. **10**, 437—459 (1935).

KOCHAKIAN, C. D., ROBERTSON, E.: Adrenal steroids and body composition. J. biol. Chem. **190**, 495—503 (1951).

KOCHAKIAN, C. D., WEBSTER, J. A.: Effect of testosterone propionate on the appetite, body weight and composition of the normal rat. Endocrinology **63**, 737—742 (1958).

KRAFT, H. G., BRÜCKNER, K.: Anabole und androgene Wirksamkeit von thio-substituierten Steroiden der Androstan-Reihe. Arzneimittel-Forsch. **14**, 328—330 (1964).

KRÜSKEMPER, H. L.: Über den Einfluß von 1α,7α-Bis(acetylthio)-17α-methyltestosteron und 7α,17α-Dimethyltestosteron auf Parameter der Leberfunktion (Serum Enzym, Bromsulphalein-Retention, Gerinnungsfaktoren). Arzneimittel-Forsch. **16**, 608—612 (1966).

KRÜSKEMPER, H. L.: Anabolic steroids, pp. 1—236. New York: Academic Press 1968.

KRÜSKEMPER, H. L., BREUER, H.: Studies on the anabolic effect and metabolism of 1-methyl-1-androsten-17β-ol-3-one. Intl. Congress on Hormonal Steroids, Milan, May 14—19, 209 (1962).

LANDAU, R. L., LUGIBIHL, K.: The effect of dietary protein on the catabolic influence of progesterone. J. clin. Endocr. **21**, 1345—1363 (1961).

LANDAU, R. L., LUGIBIHL, K.: The catabolic and natriuretic effects of progesterone in man. Recent Progr. Hormone Res. **17**, 249—292 (1961 a).

LANDAU, R. L., LUGIBIHL, K., BERGANSTAL, D. M., DIMICK, D. F.: The metabolic effects of progesterone in man: dose response relationships. J. Lab. clin. Med. **50**, 613—620 (1957).

LAUSON, H. D., HELLER, C. B., GOLDEN, J. B., SEVERINGHAUS, E. L.: The immature rat uterus in the assay of estrogenic substances, and a comparison of estradiol, estrone and estriol. Endocrinology **24**, 35—44 (1939).

LENNON, H. D., SAUNDERS, F. J.: Anabolic activity of 2-oxa-17α-methyldihydrotestosterone (Oxandrolone) in castrated rats. Steroids **4**, 689—697 (1964).

LIDDLE, G. W., BURKE, JR., H. S.: Anabolic steroids in clinical medicine. Helv. med. Acta **27**, 504—513 (1960).

LINET, O.: Interaction between androgenic-anabolic steroids and glucocorticoids. Progr. Drug Res. **14**, 140—195 (1970).

LONG, C. N. H., KATZIN, B., FRY, E. G.: The adrenal cortex and carbohydrate metabolism. Endocrinology **26**, 309—344 (1940).

LYSTER, S. C., DUNCAN, C. W.: Anabolic, androgenic and myotrophic activities of derivatives of 7α-methyl-19-nortestosterone. Acta endocr. (Kbh.) **43**, 399—411 (1963).

LYSTER, S. C., LUND, G. H., STAFFORD, R. O.: Androgenic and myotrophic properties of orally administered 9-fluoro-11-oxy-methyltestosterones. Endocrinology **58**, 781—785 (1956).

MANSON, A. J., STONNER, F. W., NEUMANN, H. C., CHRISTIANSEN, R. G., CLARKE, R. L., ACKERMAN, J. H., PAGE, D. F., DEAN, J. W., PHILLIPS, D. K., POTTS, G. O., ARNOLD, A., BEYLER, A. L., CLINTON, R. O.: Steroidal heterocycles. VII. Androstano-[2,3-d]isoxazoles and related compounds. J. med. Chem. **6**, 1—9 (1963).

MORAN, J. F., SOURKES, T. L.: Induction of tryptophan pyrrolase by α-methyltryptophan and its metabolic significance in vivo. J. biol. Chem. **238**, 3006—3008 (1964).

MUNRO, H. N.: Carbohydrate and fat as factors in protein utilization and metabolism. Physiol. Rev. **31**, 449—488 (1951).

NEUMANN, F., WIECHERT, R.: Über anabole und androgene Wirkungen von in $C_1$ und $C_1C_2$ substituierten Steroiden. Arzneimittel-Forsch. **15**, 1168—1184 (1965).

NIMNI, M. E., GEIGER, E.: Non-suitability of levator ani method as an index of anabolic effect of steroids. Proc. Soc. exp. Biol. (N.Y.) **94**, 606—610 (1957).

NUTTING, E. F., KLIMSTRA, P. D., COUNSELL, R. E.: Anabolic-androgenic activity of A-ring modified androstane derivatives. Part I. A comparison of parenteral activity. Part II. A comparison of oral activity. Acta endocr. (Kbh.) **53**, 627—635; 635—643 (1966).

OLSEN, R. E., JACOBS, F. A., RICHERT, D., THAYER, S. A., KOPP, L. J., WADE, N. J.: The comparative bio-assay of several extracts of the adrenal cortex in tests employing four separate physiologic responses. Endocrinology **35**, 430—455 (1944).

OVERBEEK, G. A., DELVER, A., DE VISSER, J.: Anabolic properties of ethylestrenol. Acta endocr. (Kbh.) **41**, 133—139 (1962).

OVERBEEK, G. A., DE VISSER, J. A.: A new substance with progestational activity. II. Pharmacological properties. Acta endocr. (Kbh.) **22**, 318—329 (1956).

OVERBEEK, G. A., DE VISSER, J. A.: A comparison of the myotrophic and androgenic activities of the phenylpropionates and decanoates of testosterone and nandrolone. Acta endocr. (Kbh.) **38**, 285—292 (1961).

OVERBEEK, G. A., DE VISSER, J. A., DELVER, A.: Pharmacological comparisons of anabolic steroids (ethylestrenol, nandrolone esters). Acta endocr. (Kbh.) **39**, Suppl. 63, 7—17 (1961).

PARTRIDGE, J. W., BOLING, L., DE WIND, L., MARGEN, S., KINSELL, L. W.: Metabolic and clinical effects of methylandrostenediol in human subjects. J. clin. Endocr. **13**, 189—202 (1953).

PERELMAN, M., FARKAS, E., FORNEFELD, E. J., KARAY, R. J., RAPALA, R. T.: A new class of active steroids. The 19-nor-$\Delta^{4,9}$-3 ketosteroids. J. Amer. chem. Soc. **82**, 2402—2403 (1960).

PERRINE, J. W.: The anabolic, androgenic and antigonadotrophic properties of five synthetic 19-nortestosterone analogs. Acta endocr. (Kbh.) **37**, 376—384 (1961).

POTTS, G. O., ARNOLD, A., BEYLER, A. L.: Comparative myotrophic and nitrogen retaining effects of several steroids. Endocrinology **67**, 849—854 (1960a).

POTTS, G. O., ARNOLD, A., BEYLER, A. L.: Comparative nitrogen retaining and androgenic activities of certain orally active steroids. Intl. Congress on Hormonal Steroids, Milan, May 14—19, 211 (1962).

POTTS, G. O., BEYLER, A. L., BURNHAM, D. F.: Myotrophic and androgenic activities of androstanazole. A new heterocyclic steroid. Proc. Soc. exp. Biol. (N.Y.) **103**, 383—384 (1960b).

REIFENSTEIN, E. C., JR., ALBRIGHT, F., WELLS, S. L.: The accumulation, interpretation, and presentation of data pertaining to metabolic balances, notably those of calcium, phosphorus, and nitrogen, J. clin. Endocr. **5**, 367—395 (1945).

RENZI, A. A., STEIN, M., REILLY, P., DUBE, V.: The anti-inflammatory activity in rats of $\Delta^1$-17α-methyltestosterone and corticosteroids. Fed. Proc. Fed. Am. Soc. exp. Biol. **20**, 178 (1961).

RINGOLD, H. J.: Electronic alteration of the steroid molecule. In: VILLEE, C. A., ENGLE, L. L.: (Eds.): Mechanisms of Action of Steroid Hormones, pp. 200—232. New York: Pergamon Press 1961.

RINGOLD, H. J. BATRES, E. D., HALPERN, O., NECOECHEA, E.: Steroids. CV. 2-Methyl and 2-hydroxymethylene-androstane derivatives. J. Amer. chem. Soc. **81**, 427—432 (1959).

RUBINSTEIN, H. S., SOLOMON, M. L.: Growth-stimulating effect of testosterone propionate. Proc. Soc. exp. Biol. (N.Y.) **44**, 442—446 (1940).

RUGGIERI, P. DE, MATSCHER, R., GANDOLFI, C., CHIARAMONTI, D., LUPO, C., PIETRA, E., CAVILLI, R.: Un nuovo steroide avente azione proteoanabolica: Archo Sci. biol. (Bologna) **47**, 1—19 (1963).

SALA, G.: Clinical evaluation of the anabolic effect of 4-hydroxy-17α-methyltestosterone. Endocrinology **58**, 567—572 (1956).

SALA, G.: Clinical evaluation of the anabolic effect of 4-hydroxy-17α-methyltestosterone. Helv. med. Acta **27**, 519—522 (1960).

SALA, G.: Biological properties of 4-hydroxy-3-keto-$\Delta^4$-steroids. In: MARTINI, L., PECILE, A. (Eds.): Hormonal Steroids, Vol. 1, pp. 67—75. New York: Academic Press 1964.

SALA, G., BALDRATTI, G.: Myotrophic (anabolic) activity of 4-substituted testosterone analogs. Proc. Soc. exp. Biol. (N.Y.) **95**, 22—24 (1957).

SALA, G., BALDRATTI, G.: A long-acting anabolic steroid: 4-hydroxy-19-nortestosterone 17-cyclopentylpropionate. Endocrinology **72**, 494—495 (1963).

SALA, G., BALDRATTI, G., RONCHI, R., CLINI, V., BERTUZZOLIO, C.: Attivita anabolica di una serie di nuovi derivati del testosterone. Sperimentale **106**, 490—510 (1956).

SALVIONI, E.: Evaluation of the anabolic activity of steroids. Lancet **1962 II**, 453.

SAUNDERS, F. J., DRILL, V. A.: The myotrophic and androgenic effects of 17-ethyl-19-nortestosterone and related compounds. Endocrinology **58**, 567—572 (1956).

SAUNDERS, F. J., DRILL, V. A.: Comparative androgenic and anabolic effects of several steroids. Proc. Soc. exp. Biol. (N.Y.) **94**, 646—649 (1957).

SAUNDERS, F. J., DRILL, V. A.: Some biological activities of 17-ethynyl and 17-alkyl derivatives of 17-hydroxyestrenones. Ann. N.Y. Acad. Sci. **71**, 516—531 (1958).

SEGALOFF, A.: The enhanced local androgenic activity of 19-nor steroids and stabilization of their structure by 7α- and 17α-methyl substituents to highly potent androgens by any route of administration. Steroids **1**, 299—315 (1963).

SEGALOFF, A., GABBARD, R. B.: 14-Dehydrogenation as a means of increasing androgenicity when measured by local applications on the chick comb. Steroids **1**, 77—89 (1963).

SELYE, H.: Effect of dosage on the morphogenetic actions of testosterone. Proc. Soc. exp. Biol. (N.Y.) **46**, 142—146 (1941).

SELYE, H.: Selective alteration of certain ACTH and glucocorticoid effects by androstene derivatives. Metabolism **4**, 403—415 (1955).

Selye, H., Mishra, R. K.: On the ability of methyltestosterone to counteract catabolism in diverse conditions of stress. Arch. int. Pharmacodyn. **117**, 444—451 (1958).
Silber, R. H., Porter, C. C.: Nitrogen balance, liver protein repletion and body composition of cortisone treated rats. Endocrinology **52**, 518—525 (1953).
Steinetz, B. G., Giannina, T., Butler, M., Popick, F.: Dissociation of anabolic and androgenic properties of steroids by anti-anabolic and anti-androgenic agents in rats. Endocrinology **89**, 894—896 (1971).
Strike, D. P., Herbst, D., Smith, H.: Totally synthetic hormones. XVI. The conversion of estr-4-en-17β-ol to testosterone and the total synthesis of some 18-methylandrostane and 18-methyl pregnane derivatives. J. med. Chem. **10**, 446—451 (1967).
Stucki, J. C., Duncan, G. W., Lyster, S. C.: Anabolic and androgenic activities of 7α,17α-dimethyltestosterone (U-19, 763), a new anabolic agent. Int. Congress on Hormonal Steroids, Milan, May 14—19, 65—66 (1962).
Stucki, J. C., Duncan, G. W., Lyster, D. S. C.: Anabolic and androgenic activities of 7α,17α-dimethyltestosterone (U-19763), A new anabolic steroid. In: Martini, L., Pecile, A. (Eds.): Hormonal Steroids, Vol. 2, pp. 119—132. New York: Academic Press 1965.
Stucki, J. C., Forbes, A. D., Northam, J. I., Clark, J. J.: An assay for anabolic steroids employing metabolic balance in the monkey: The anabolic activity of fluoxymesterone and its 11-keto analogue. Endocrinology **66**, 585—598 (1960).
Suchowsky, G. K., Junkmann, K.: Anabole Steroide und ihre Nebenwirkungen. Acta endocr. (Kbh.) **39**, 68—78 (1962).
Suchowsky, G. K., Junkmann, K.: Recent findings in anabolic steroids. In: Martini, L., Pecile, A. (Eds.): Hormonal Steroids, Vol. 2, pp. 155—159. New York: Academic Press 1965.
Tainter, M. L., Arnold, A., Beyler, A. L., Burnham, D. F., Potts, G. O., Roth, C. H.: Synthetic anabolic steroids in geriatrics. Medical Times **91**, 747—759 (1963a).
Tainter, M. L., Arnold, A., Beyler, A. L., Coulston, F., Potts, G. O., Roth, C. H.: A summary of the comparative therapeutic actions of stanozolol, an oral anabolic steroid. Clin. Med. **71**, 2089—2102 (1964).
Tainter, M. L., Arnold, A., Beyler, A. L., Drobeck, H. P., Potts, G. O., Roth, C. H.: Anabolic therapy in the management of cancer: a summary. J. Amer. Geriat. Soc. **11**, 1148—1151 (1963b).
Tait, J. F., Burstein, S.: In vivo studies of steroid dynamics. In: Pincus, G., Thimann, K. V., Astwood, E. B. (Eds.): The Hormones, Vol. 5, pp. 441—557. London: Academic Press 1964.
Taussky, H. H., Shorr, E.: A microcolorimetric method for the determination of inorganic phosphorus. J. biol. Chem. **202**, 675—685 (1953).
Tomarelli, R. M., Bernhart, F. W.: Oral anabolic activity of (±)-13β,17α-diethyl-17β-hydroxygon-4-en-3-one (norbolethone) in a nitrogen retention assay. Steroids **4**, 451—456 (1964).
U.S. Pharmacopeia. Design and analysis of biological assays. 16th Revision, pp. 873—885. Easton, Pa.: Mack Printing Co. 1960.
Van Wayjen, R. G. A., Buyze, G.: Clinical-pharmacological evaluation of certain anabolic steroids. Acta endocr. (Kbh.) **39**, Suppl. 63, 18—36 (1961).
Vida, J. A.: Structure-activity theories. In "Androgens and anabolic agents." Chemistry and pharmacology, 21 H. New York: Academic Press 1969.
Winter, C. A., Hollins, H. L., Stebbins, R. B.: The effect of androgenic hormones upon the adrenal atrophy produced by cortisone injections and upon the anti-inflammatory action of cortisone. Endocrinology **52**, 123—134 (1953).
Zaffaroni, A.: The effect of alkyl- and electronegative-group substitution on steroidal hormone activity. Acta endocr. (Kbh.) Suppl. **50**, 139—145 (1960).
Zderic, J. A., Carpio, H., Ruiz, A., Limen, D. C., Kincl, F., Ringold, H. J.: Steroids. CLXXXIII. New anabolic agents from steroids. J. med. Chem. **1963**, 195—198.

# VI. Animal Husbandry

T. W. PERRY

## I. Sources of Androgens

There are three generally recognized endocrine sources of androgens, namely, testes, ovaries, and adrenals. Normally, neither the adrenals nor the ovaries are considered to be more than token sources of androgens, but this capability varies according to such factors as adrenal tumors and other conditions. By far the largest source of androgens within the body is the testes.

Although multiple androgen-like substances have been identified, probably the two most common types are testosterone and androsterone.

Many assume that the interstitial cells of Leydig represent the principal source of the male sex hormones, although some have postulated that the germinal epithelium is at least a contributing source.

## II. General Role of Androgens in Animals

Perhaps the role of androgens in animal husbandry has not been as well defined as it has been in man—certainly not as well as in the laboratory animal. Undoubtedly, there must exist marked species differences in target responses to androgens.

In contrast to laboratory animals, cattle, swine, and sheep do not lend themselves as well to meticulous laboratory study because of the greatly increased space and management problems involved. The effects of androgens in farm animals are perhaps much more gross and thus more obvious.

Fortunately, gross observations of the effects of androgens in farm animals are readily available because most male animals destined for production of meat are castrated. This custom aids in the comparison of castrate and intact males. However, one must not conclude that such differences are due totally to the lack or presence of androgens, since removal of the testes may deprive the host animal of factors other than androgens.

## III. Cattle

A phenomenon observed almost exclusively in cattle demonstrates the effect of androgens during pregnancy. In this situation twin calves—one a male and the other apparently a female—yield a normal male and a sterile female. In such females, commonly known as freemartins, the internal organs of reproduction usually are predominantly male in character and the external organs are usually of the female type (LILLIE, 1917). The freemartin in cattle has been recognized for at least two centuries, but perhaps its peculiarity to cattle has not been conclusively explained. About 87% of apparently different-sexed bovine twins produce sterile females, whereas about 13% of the males are sterile.

LILLIE (1917) concluded that the freemartin must be the result of one of two situations, that is, (1) that it and its partner are monozygotic twins (identical), or (2) that they are fraternal. Under the first assumption, since the sex of monozygotic twins is identical, the freemartin necessarily would be interpreted as a modified male. The second assumption would necessitate the interpretation that the freemartin is a female whose reproductive system had been modified.

Later research established that a freemartin is a genetic female and that the intersexual condition is caused by a hormone that reaches the female embryo from the male twin embryo through vascular anastomoses between the partially fused placentae (MOORE et al., 1957).

MARTIN et al. (1966) reported a 5-year summary of the comparative performances of bulls (intact males) and steers (castrate males) from the same herd of cows and sired by the same bulls. A total of 157 male calves were weaned at an average age of 231 days in 5 calf crops. They were randomly allotted to bull or steer groups. Calves assigned to the steer group were castrated at weaning time. All animals were slaughtered near 454 kg live weight. Carcasses were graded and weights of chuck, loin, rib, round, and kidney fat were recorded at slaughter. The *longissimus dorsi* muscle and external fat covering were traced on the twelfth rib location (beef carcasses are broken between the twelfth and thirteenth rib). The 9-10-11 rib section was separated physically into fat, lean, and bone. Lean from the 9-10-11 rib section was ground and sampled for moisture, protein, and ether-extractable content.

Bulls gained significantly ($P<0.01$) faster than steers throughout the adjusted 300-day postweaning feeding period. However, the greatest advantage, in terms of gain, of bulls over steers occurred in the first half of the period. Actually, steers gained slightly faster than bulls during the last half of the feeding period, but the advantage gained by the bulls during the first half of the period was too great for the steers to approach.

When weights were adjusted to the same age, bulls weighed 35 kg more than steers (473 vs. 437 kg) at 300 days; when adjusted to 454 kg live weight, bulls reached that weight 30 days more quickly.

Bull carcasses had a higher proportion of lean or primal cuts than steer carcasses, but this was due entirely to increased chuck and round weights. Steer carcasses were fatter than bull carcasses as evidenced by heavier kidney fat weight, greater marbling score, deeper outside fat cover, smaller loin eye area, higher proportion of fat in 9-10-11 rib, lower proportion of lean 9-10-11 rib, more fat in 9-10-11 rib lean, and less moisture and protein in 9-10-11 rib lean.

Bull carcasses graded U.S. Good (USDA), approximately one-half grade lower than steer carcasses, had a "modest" marbling score and approximately 1.4 cm outside fat cover, in contrast to steer carcasses graded U.S. Choice, with a "moderate" marbling score and 2.1 cm outside fat cover. A cookery test performed in the home economics department did not show differences between bull and steer steaks on the basis of tenderness and palatability characteristics.

KLOSTERMAN et al. (1954a, b) compared the weight gain of "early" (1 month of age) and "late" (weaning—6–7 months of age) castrated beef calves with that of bulls in the feedlot. Their data showed that age of castration had no effect on subsequent gain in the feedlot, but that bulls gained at a significantly faster rate than steers. Bulls were consistently more efficient than steers in converting feed to gain. There was no significant difference in the amount of bone expressed as percent of carcass weight, between bulls and steers. Bulls were significantly ($P<0.01$) different than steers in the following respects—bulls had heavier hides, a lower carcass grade (generally indicating less interstitial fat deposition), heavier forequarters, lighter hindquarters, lighter flank, and lighter kidney fat. There were no significant differences in any of the slaughter or carcass determinations between steers castrated shortly after birth or those castrated at weaning age. This would indicate that shortly after castration, beef animals lose androgenic effects rather rapidly.

PILKINGTON et al. (1959) compared the performances of bull, steer, and diethylstilbestrol (DES)-implanted bull calves. Steer calves had been castrated at 3 months of age, bulls implanted with DES were implanted with 12 mg at 3½ months of age and re-implanted with 24 mg at 6½ months. All calves and their dams were in the same pasture throughout the test, where the calves were creep-fed a grain mix. The calves were marketed as fat calves at approximately 8 months of age. Bull calves gained 9 kg more to weaning (159 vs. 150 kg) than steer calves, whereas DES-implanted bulls gained 4.5 kg more than steer calves. Thus, implanted DES tended to depress gains. At time of slaughter, testicles were obtained from bull carcasses. Implantation with DES caused a 50% reduction of testicle size and of weight (361 vs. 194 g).

Steers had a significantly greater amount of external fat covering and a firmer, finer textured lean in the ribeye. Steers were graded one-third grade higher than implanted bulls, and implanted bulls were graded higher than untreated bulls. Untreated bulls had largest rib-eye area, indicative of greater true growth. They also had the greatest carcass lean and lowest carcass fat.

KLOSTERMAN et al. (1955) reported somewhat contrasting results from implanting bulls with much higher doses of DES. In one experiment, an initial dose of 84 mg, plus another 132-mg implant at 98 days, was studied. During the first 98-day-period, the average daily gain of untreated bulls was 1.07 kg, which was 150 g more than steers. DES-treated bulls gained 118 g per day more than nontreated bulls. During the second 98-day-period, steers gained 0.91 kg per day, bulls, 1.06 kg, and DES-implanted bulls, 1.21 kg. Carcass data showed that bulls were graded significantly lower ($P<0.01$) than steer carcasses. MARTIN (1974) used much lower levels of implanted DES with bulls (single initial levels of 36 or 72 mg) and reported a short-term response of 91 g per day over the nonimplanted control bulls over the first 84 days.

NICHOLS et al. (1964) compared the production of slaughter animals from bulls and steers of the Holstein-Friesian breed. In addition, weight at time of slaughter was compared between 364 kg and 454 kg. Bulls reached slaughter weight significantly ($P<0.01$) faster than steers, requiring 34 days less to reach 364 kg and 25 days less to reach 454 kg. Bulls had a 0.9% lower ($P<0.05$) dressing percentage than steers, largely due to a heavier ($P<0.05$) hide of the bulls and fatter ($P<0.01$) steer carcasses. Steer carcasses were scored subjectively higher than those of bulls, indicating finer, closer-grained muscles. Bulls had greater ($P<0.01$) loin eye areas than steers, indicating that bulls developed muscle more rapidly than steers.

Man has not been able to simulate the effect of the intact testis of cattle through the use of injectable or oral hormones. One or more of several explanations might be in order. For example, the testis may supply growth-promoting effects other than that attributed to the androgenic materials contained in the testis. Thus, adding back only testosterone to a castrated male would still leave much to be desired. There is also the possibility that oral or injectable testosterone is either metabolized before reaching its targets or is prevented by other means from reaching its target. There are research data available that demonstrate the superiority of growth and the changes in carcass composition that are shown for the bull (intact male cow) over the steer (castrate male). ANDREWS et al. (1950) implanted subcutaneously 180 mg testosterone in yearling Hereford steers and reported there was no measurable effect on rate of gain, efficiency of feed conversion, or upon routine carcass measurements.

BURRIS et al. (1952) reported a slight response (45 g per day increase) in the gain of steers from weekly injections of 1 mg testosterone per kg body weight. Steers showed some improvement in efficiency of feed conversion as the result of testosterone injections. The injections were made in the form of an aqueous suspension of micropellets. Heifers showed 4 times the response to testosterone injections that was shown by steers. DAVIS et al. (1969) reported no difference in effect between oral administration of DES and subcutaneous implantation of DES plus 120 mg testosterone. DES at 10 mg/day orally, gave a 7.75% increase in daily gain and a 6.20% increase in efficiency of feed conversion. GLIMP and CUNDIFF (1971) implanted yearling and 2-year-old heifers with 120 mg testosterone plus 24 mg DES without any effect on performance. In contrast, the feeding of 0.40 mg melengestrol acetate (MGA) resulted in improved performance.

Many researchers have demonstrated that steers or bulls will gain faster than heifers (RAY et al., 1969). However, the complexity of differences between a male and a female is so great that it would be most difficult to indicate that just the androgenic differential could explain the difference in performance. Furthermore, a steer—which is a castrate—would be expected to produce very little androgen in comparison to a bull.

## IV. Swine

Research data on the comparative performances of castrate and intact male swine are not as readily available as are those for cattle, because there is little interest in raising intact male swine for meat purposes because of the undesirable odors and flavors associated with such meat products. This effect occurs at a very early age in swine, so early slaughter cannot circumvent it, as is possible in cattle.

The effect of castration on growth rate in swine is less marked than in cattle or sheep. TURTON (1962) reviewed several papers on the effect of castration on the performance of swine and concluded there was not sufficient beneficial evidence to justify a break from the traditional practice of castrating male swine for meat production purposes. PALSSON (1955), in a review, pointed out that contrary to sheep, cattle, and horses, female swine mature later than males. Furthermore, carcass measurements show that at similar live weights, females are leaner and have thicker bellies than castrate males. The females also have heavier skeletons and less fat than castrated males.

WINTERS et al. (1942) compared growth rates of intact (boars) and castrate (barrows) male swine, utilizing 86 boar/barrow pairs from 8–24 weeks, or through 100 kg live weight. Poland China boars were slightly heavier than barrows at all ages, but the differences were nonsignificant after 16 weeks of age. Minnesota No. 1 boars were heavier from 8–16 weeks, and the barrows were heavier thereafter. CHARETTE (1961) compared 16 intact females (gilts), 16 boars, and 16 barrows (the barrows were castrated at birth, 6, 12, or 16 weeks of age). Boars consumed less feed per day, had less loin fat and yielded a higher percentage of shoulder than gilts or barrows. The carcasses of boars and gilts were longer and had less fat covering over the shoulder and back, a larger area of loin, and a higher iodine number than those castrated later. Average daily gains were almost identical for all treatments, ranging between 600 and 614 g per day.

BRATZLER et al. (1954) showed that castration of males did not affect growth rate of swine. However, carcass data paralleled that of CHARETTE (1961) in that boars had a higher percentage of lean in the loin, less back fat, and longer body than barrows.

As stated above, the discrimination against boar carcasses is based largely on the fact that they possess an oder or taint. BRATZLER et al. (1954) observed that the pork from boars slaughtered at 100 kg was inferior to that from barrows of a comparable slaughter weight and had an undesirable oder on cooking. In contrast, CHARETTE (1961) found no taint in the meat from boars slaughtered at 100 kg.

CLAUS et al. (1971) isolated a steroid in pigs that was responsible for the odor typical of boars. The steroid (5a-androst-16-ene-3-one) was also identified as a sex attractant for the female hog. Both the odor steroid and testosterone showed a sharp increase with age, both in testicular tissue and in peripheral blood plasma of boars. The values for the odor steroid increased with age from 6–22.3 mg/ml plasma in males, 0.8–2.0 in females, and 1.3–2.7 in the castrated male hog. There were remarkably high levels—not dependent upon age—in the fatty and parotid tissue of boars, but none in the same tissues of the female or the castrate male, indicating that these tissues serve as storage areas for a relatively large amount of the odor steroid.

BRUNER et al. (1958) tabulated the gain performance of 385 littermate gilts and barrows at the Ohio evaluation station. Marked differences in performance were obtained for several parameters: daily gain, 740 vs. 804 g; back fat thickness, 4.0 vs. 4.21 cm; loin eye area 26 vs. 22 $cm^2$; primal cuts as percent of live weight, 49% vs. 48%, for gilts and barrows, respectively. ALLEN and BRAY (1964) reported that barrow carcasses contained 3% more ether-extractable material, a 0.47 higher marbling score in longissimus dorsi muscle, and 2.3% more separable blade and seam fat than gilt carcasses. In fatty acid composition comparisons, ALLEN et al. (1967) found marked increases in C-18 of heavier gilts and barrows as compared to a constant level in the muscle of boars, either at 50 or 94 kg. Furthermore, there were lower ($P<0.01$) C-18-1 values for heavier barrows and gilts than for boars, which resulted in a markedly higher mean ratio of C-18-1 to C-18 for heavier boars. Much of the increase in saturation of intramuscular lipid occurring after 55 kg live weight was attributed to the ratio of C-18-1 to C-18, which is dependent upon the presence of male sex hormones.

KOCH et al. (1968) reported back fat from spayed gilts had less ($P<0.01$) linoleic acid than that from boars or barrows; that from boars had more ($P<0.01$) linoleic acid than the other sexes and that from barrows contained more ($P<0.01$) stearic acid than back fat from gilts.

LIDVALL et al. (1964) followed growth and efficiency-of-feed-conversion patterns of 2 boars, 2 barrows, and 2 gilts from each of 4 litters from 22–104 kg. They reported no difference in daily gain due to sex. However, boars required less ($P<0.05$) feed per unit gain than either females or barrows, and barrows and gilts had a higher dressing percent (indicative of a higher percent of fat in the carcass). Boars had less ($P<0.05$) back fat.

MCMEEKAN (1940) completely dissected the carcasses and separated out bone, muscle, and fat from castrated male (barrows) and intact female (gilts) swine, at 91 kg live weight. His results showed that gilts had heavier skeletons (from 4–14% more bone), more muscle (0–19%), and less fat (4–23%).

Even though the obvious effects on growth rate and body composition exerted by the testis of the pig have been shown in research studies, it has been extremely difficult to cause more than minor alterations in swine through externally administered androgens. Several research groups have studied the route of orally administered androgens. BEESON et al. (1955) fed methyltestosterone at 20 mg per pig per day to weanling pigs through to a market weight of 100 kg, without affecting rate of gain or efficiency of feed conversion. However, some changes in carcass composition were observed. The four lean cuts equalled 62.4% of the total carcass from pigs fed methyltestosterone and 58.8% from the carcasses of control pigs. The three fat cuts then, represented 37.6 and 41.2%, respectively. Chemical analyses of the meat from one-half the carcass of each pig showed 5% less fat and 5% more lean in carcasses from pigs fed methyltestosterone.

BIDNER et al. (1972) studied the effect of incorporating 2.2 mg each of diethylstilbestrol (DES) and methyltestosterone per kg diet. Less ($P<0.05$) back fat and larger ($P<0.06$) longissimus dorsi muscle area characterized the carcasses from pigs fed the combined hormone treatment. However, since the two hormone treatments were not fed separately, it was not possible to segregate any potential individual hormonal effect. Electrophoretic separation of serum proteins indicated that the combined hormonal treatment tended to result in an increased percentage of albumin and decreased globulins. It also resulted in decreased follicle size in gilts, when the treatment was initiated at 23 kg live weight.

PERRY et al. (1956) fed up to 62 mg methyltestosterone per weanling pig per day in a free choice protein supplement, from 23 kg to a slaughter weight of 95 kg. Pigs consuming 27 mg methyltestosterone per day, or more, grew significantly slower and had significantly less back fat. At the same station, THRASHER et al. (1959) introduced a testosterone analog (9α-fluoro-11β-hydroxy-17α-methyltestosterone) but reported it had no effect on gain or carcass composition. However, feeding 11 mg methyltestosterone and 0.7, 2.0, 2.6, and 4.0 mg of 11β-hydroxy-17α-methyltestosterone per kg diet did produce effects suggesting increased carcass leanness. Feeding of methyltestosterone resulted in significantly higher iodine numbers for extracted fat from the edible portion of the carcass.

JOHNSTON et al. (1957) reported five experiments in which 4.1 or 5.8 mg methyltestosterone per kg of diet was fed to pigs. Little effect due to treatment was observed. However, the data indicated decreases in rate of gain, in daily feed consumption, in efficiency of feed conversion, and in thickness of back fat due to methyltestosterone. Treatment also appeared to increase the ratio of lean to fat in the carcass. NOLAND and BURRIS (1956) compared the performances of boars, gilts, barrows, and castrated females, all fed 0, 0.015, 0.15, or 1.5 mg methyltestosterone/kg body weight/day. There was no apparent effect on gain, but a level of 1.5 mg/kg body weight/day resulted in decreased sperm production in boars.

WHITAKER et al. (1959) studied the effect of a combination of 10 or 20 mg methylandrostenediol and 10 mg thyroprotein per day for growing and finishing pigs. Rate of gain was not affected, but the treatment resulted in significantly higher lean content in the carcass. Also, treated pigs exhibited masculine behavior.

Implantion and injection techniques have been employed to administer androgens to pigs. SLEETH et al. (1953) injected weanling pigs with either 0.5 or 1 mg of testosterone/kg body weight and/or 1 mg estradiol benzoate, once per week for 6 weeks and semiweekly thereafter. Neither testosterone nor estradiol benzoate affected gain or carcass composition, but testosterone-injected barrows and gilts ranted and rode each other. BRATZLER et al. (1954) implanted testosterone pellets and WOEHLING et al. (1951) implanted weanling pigs with 15 mg testosterone propionate plus 12 mg DES initially and after 12 weeks. Site of implantation was the hind flank. Both papers reported no effect from the implantation techniques they employed.

## V. Sheep

HAMMOND (1955) described the general characteristic differences between male and female sheep. Females attain a more advanced stage of development earlier in life than males. However, the males reach a greater ultimate size and a higher degree of development at maturity than the females. Carcasses of ewes and castrate males (wethers) were compared at 9 weeks of age and at a carcass weight of 13.5 kg (PALSSON and VERGES, 1952). Ewes had 4.7% higher dressing percentage, and all joints of the carcass, except head and feet, were proportionately better developed than in wether carcasses. At 41 weeks, after the animals had been growing at maximum rate, the situation is reversed, with the wethers not only being 48% heavier than the ewes, but also reaching a proportionately higher degree of development of the later maturing parts and tissues of the body.

HAMMOND (1955) also compared skeletal development of the three sexes (ram, wether, ewe) at 5 months of age. Relative to the cannon bones, all bones except those of the head and the cervical vertebrae were better developed in the wether than in the ewe; all skeletal parts in the ram, except the pelvis, were better developed than in the wether. The ewes matured earlier than wethers and the wethers, earlier than the rams. The meat-to-bone ratio at 5 months was highest in the ewes (7:22), intermediate in the wethers (6:71), and lowest in the rams (6:40). At this age, active growth is still taking place in the early maturing skeleton of the ram at the expense of the later maturing tissues, muscle, and fat, while bone growth in the ewe has slowed down so that nutrients are used for later maturing tissues.

Table 1. Average cutting yields of lambs slaughtered at 20, 36 and 52 weeks of age[a]

| | Rams (%) | Wethers (%) |
|---|---|---|
| Leg, untrimmed | 33.4 | 33.5 |
| Leg, trimmed | 28.7 | 28.9 |
| Shoulder | 20.4 | 19.6 |
| Loin | 9.4 | 9.9 |
| Neck | 4.0 | 3.3 |
| Ribs | 15.3 | 15.8 |
| Breast | 15.2 | 14.8 |
| Kidney fat | .6 | .5 |

[a] HUNT et al. (1938).

HUNT et al. (1938) conducted a 3-year study to compare the performance of ram and wether lambs. A total of 75 rams and 70 wethers was involved. Average daily gains on pasture were 86 and 79 g per day whereas in drylot, rams and wethers gained 159 and 163 g per day. There were no significant differences in chilled carcass weights between rams and wethers (18 vs. 17.9 kg). Wethers had a higher dressing percent than rams except when the two types were slaughtered at 5 and 15 weeks (age of slaughter was 5, 10, 15, 20, 28, 36, 44, and 52 weeks). From 20–52 weeks of age, inclusive, wethers out-dressed rams by approximately 2.25 percentage points. In 8 out of 17 groups the chilling shrink of the carcass was greater for the rams than for the wethers, whereas in 9 groups it was not.

BRADFORD and SPURLOCK (1964), in studies on the effect of castration on growth and body composition of lambs, reported that rams weighed 5% more than wethers at weaning and 15 and 23% more as yearlings on high and low planes of nutrition. Carcasses from yearling rams weighed 4 and 8% more than those from wethers on the low and high planes of nutrition. Proportion of carcass weight in the forequarter was somewhat higher for rams, but the intact males had less kidney fat and a percentage of carcass weight as loin and trimmed leg comparable to that of the wethers. MCCLAUGHERTY et al. (1959) castrated alternate male lambs at 1–2 weeks after birth and subsequently slaughtered them at 39 kg, or more. Ram lambs gained 254 g per day and dressed 48.8%; ewe lambs gained 240 g per day and dressed 50.8%; wethers gained 245 g per day and dressed 50%. Carcasses from ewe and wether lambs were graded U.S. Prime—whereas those from ram lambs were graded U.S. Choice +, indicating a slightly lower percentage of ether-extractable material in the ram carcasses. Earlier-born ram lambs gained more rapidly whereas later-born ram lambs gained more slowly than ewe and wether lambs. KEMP et al. (1970) compared the performances of 30 ram and 30 wether lambs weaned at 18 kg and fed out to predetermined slaughter weights of 36, 45, and 54 kg. Wethers had the higher dressing percentage (50.0 vs. 48.4%) and more ($P<0.01$) carcass weight per day of age (168 vs. 160 g). Wethers had fatter carcasses—increasingly so with increasing age.

Perhaps lambs are more responsive to androgenic implants than other species of animals. ANDREWS et al. (1949) reported, that daily gains increased ($P<0.05$) with implants of 10 or 20 mg testosterone in finishing lambs (control, 159 g; 10 mg testosterone, 195 g; 20 mg testosterone, 186 g, daily gain). Feed utilization was improved accordingly. Testosterone-treated lambs had the most desirable finish at time of slaughter. O'MARY et al. (1952) showed no benefit from a 15-mg pellet implant of testosterone, but weekly injections of 20 mg in alcohol solution resulted in both increase in daily gain (218 vs. 195 g/day) and growth of the seminal vesicles. No response to a 15-mg implant of testosterone propionate reported by POPE et al. (1950) agrees with the no-response reported by O'MARY, above.

RAY et al. (1966) implanted a combination of 25 mg testosterone and 2.5 mg estradiol benzoate in ewe lambs. Their data showed implanted lambs made significantly faster gains, requiring less feed per kg gain. Carcass grade of the lambs was not affected by treatment. Implanted lambs had smaller kidney knobs (fat) at time of slaughter and less external fat than nonimplanted lambs.

DUTT et al. (1968) studied the effect of unilateral castration in ram lambs. Forty ram lambs, 12–48 days of age, were involved. Rams in one group were bilaterally castrated, those in a second and third group were unilaterally castrated on the right or left side, and those in a fourth group served as controls. All were castrated at approximately 36 kg at about 6 months of age. Compensatory growth occurred in the remaining testis when one was removed, since the single testis weighed on the average 71% as much as the combined weight of both testes in intact rams. There were no significant differences between weights of right or left testis, seminal vesicles, or bulbourethral glands within groups. The single testis of the right and left unilateral castrates was heavier ($P < 0.05$) than the right or left testis of the control (right, 145.2 vs. 195.6 g; left, 140.2 vs. 211.1 g). Unilateral castration had no effect on growth rate since average age at slaughter was nearly the same for unilateral castrates and for intact rams. Bilateral castration depressed ($P < 0.05$) growth rate. Unilateral castration did not affect seminal vesicle weight but significantly reduced bulbourethral gland weights (0.83, 0.57, and 0.70 g, control, right unilateral castrate, and left unilateral castrate, respectively), indicating that in the young ram the bulbourethral glands are more sensitive than the seminal vesicles to circulating levels of androgens.

Fatty acid composition of the fat of sheep is affected by androgens. TICHENOR et al. (1970) found that both peripheric and subcutaneous fat of rams contained less stearic acid and more linoleic acid than comparable fat of castrate males. Ram fat had a higher proportion of unsaturated C-18 acids. CROUSE et al. (1972) studied the effect of castration or implanted testosterone on adipose fat composition in sheep. The subcutaneous tissue of ram lambs had higher levels of C-14 and C-18-3, with a decrease in C-18, as compared to wethers. The wether lambs had greater back fat deposition. Implanting wethers with either 50 or 100 mg testosterone had no effect on the fatty acid deposition pattern. In a periodic biopsy technique, CRAMER and MARCHELLO (1964) studied seasonal and sex patterns in Columbia ram-ewe twin lambs. Fat was biopsied at 9 weeks of age and every succeeding 9 weeks until the lambs were 18 months of age. Environment had the greatest effect with the highest iodine number occurring during winter months. Fat from the ewes had a higher iodine number and melting point than that from rams. Ewe fat contained a higher percentage of fatty acids containing 16 or more carbon atoms, regardless of degree of unsaturation, and a lower percent of all fatty acids with less than 16 carbon atoms.

Level of androgen production and circulation is affected by ambient temperature (GOMES et al., 1971). The Ohio researchers elevated the temperature of yearling rams to 32° C in the daytime, cooling down to 28° C at night, for 2 weeks. Control lambs were maintained in an open shed under Ohio spring weather conditions. Those subjected to the elevated temperature had significantly smaller testes (12.4 vs. 9.3 g dry weight) and less ($P<0.05$) testosterone (1.09 vs. 0.35 µg per g dry weight). Spermatic venous blood testosterone level decreased ($P<0.05$) from 8.2–1.9 µg per 100 ml plasma.

## VI. Poultry

The effect of androgens on poultry has been studied perhaps more extensively than in the four-footed farm animals. One of the more easily discernible effects of androgens are their effects—or lack of effect—on plumage. Many people have felt that molting is a direct effect of hormones, or a lack of hormones. This is true in part, but many factors affect molt. Within 6 months after normal molt has been completed, fowl will molt again in response to thyroxine, progesterone, and gonadotropins. Also, after this time they will molt in response to desoxycorticosterone acetate, Enheptin, and prolactin, as well as to starvation (FARNER and KING, 1972).

Domestic chickens and ducks may enter a continuous state of molt after castration (VOITKEVICH, 1966), suggesting that molt is inhibited by the presence of high levels of sex hormones. Administration of exogenous androgens and estrogens to intact birds can have varying effects, based on kind of bird. Administration of exogenous androgens and estrogens to pigeons slows or arrests the normal course of molt, and such inhibitory effect is evidenced in the domestic fowl, which molts after laying has ceased and the ovary has diminished in activity.

Secondary sexual characteristics that differentiate male and female birds include comb size, plumage color, structure of feathers, voice, and temperament (STURKIE, 1965). Androgen produces comb growth in males and females. The hackle and saddle feathers of the rooster are long and tapering, whereas those of the female are shorter and more blunt. The difference is due to the estrogen effect and not to that of androgen, since castrate males and castrate females grow feathers similar to those of the intact cock.

The effect of androgen upon comb growth is spectacular, indeed. BRENEMAN (1939) injected chicks with testosterone propionate and greatly increased comb size, but testis size was decreased.

Castration of the male (caponization) results in delayed rate of growth and greatly increased levels of fat deposition.

## References

ALLEN, E., BRAY, R. W.: Physical and chemical lipid determinations in barrow and gilt carcasses. J. Anim. Sci. **23**, 656—660 (1964).

ALLEN, E., BRAY, R. W., CASSENS, R. G.: Changes in fatty acid composition of porcine muscle lipid associated with sex and weight. J. Food Sci. **32**, 26—29 (1967).

ANDREWS, F. N., BEESON, W. M., HARPER, C.: The effect of stilbestrol and testosterone on the growth and fattening of lambs. J. Anim. Sci. **8**, 578—582 (1949).

ANDREWS, F. N., BEESON, W. M., JOHNSON, F. D.: The effects of hormones on the growth and fattening of yearling steers. (Abs.) J. Anim. Sci. **9**, 677 (1950).

BEESON, W. M., ANDREWS, F. N., PERRY, T. W., STOB, M.: The effect of orally administered stilbestrol and testosterone on growth and carcass composition of swine. J. Anim. Sci. **14**, 475—481 (1955).

BIDNER, T. D., MERKEL, R. A., MILLER, E. R.: Effect of a combination of diethylstilbestrol and methyltestosterone on performance, carcass traits, serum and muscle characteristics of pigs. J. Anim. Sci. **35**, 525—533 (1972).

BRADFORD, G. E., SPURLOCK, G. M.: Effects of castrating lambs on growth and body composition. Anim. Prod. **6**, 291—299 (1964).

BRATZLER, L. J., SOUK, R. P., JR., REINEKE, E. P., PAUL, P.: The effect of testosterone and castration on growth and carcass characteristis of swine. J. Anim. Sci. **13**, 171—176 (1954).

BRENEMAN, W. R.: Effect of androgens on the chick. In: Proc. VII World's Poul. Congr. P 1939, pp. 91—95.

BRUNER, W. H., CAHILL, V. R., ROBINSON, W. L., WILSON, R. L.: Performance of barrow and gilt littermates at the Ohio Swine Evaluation Station. J. Anim. Sci. **17**, 875—878 (1958).

BURRIS, M. J., BOGART, R., OLIVER, A. W., MACKEY, A. O.: The effects of testosterone propionate on rate of gain, feed efficiency and carcass quality of beef heifers and steers. (Abs.) J. Anim. Sci. **11**, 758 (1952).

CHARETTE, L. A.: The effects of sex and age of male at castration on growth and carcass quality of Yorkshire swine. Can. J. Anim. Sci. **41**, 30—39 (1961).

CLAUS, R., HOFFMAN, B., KARG, H.: Determination of $5_a$-Androst-16-en-3-one, a boar taint steroid in pigs, with reference to relationships to testosterone. J. Anim. Sci. **33**, 1293—1297 (1971).

CRAMER, D. A., MARCHELLO, J. A.: Seasonal and sex patterns in fat composition of growing lambs. J. Anim. Sci. **23**, 1002—1010 (1964).

CROUSE, J. D., KEMP, J. D., FOX, J. D., ELY, D. G., MOODY, W. G.: Effect of castration, testosterone and slaughter weight on fatty acid content of ovine adipose tissue. J. Anim. Sci. **34**, 384—392 (1972).

DAVIS, J. K., BORDER, J. R., SYMONDS, E. R.: Testosterone-DES implants plus oral DES for heifers. (Abs.) J. Anim. Sci. **29**, 156 (1969).

DUTT, R. H., DAME, P. R., RENFRO, R. E.: Effect of unilateral castration on body weight and accessory gland development in ram lambs. Ky. Anim. Sci. Res. Progr. Rep. **176**, 37—38 (1968).

FARNER, D. S., KING, J. R.: Mechanisms and control of molt. In: Avian Biology, Vol. II, pp. 119. New York: Academic Press 1972.

GLIMP, H. A., CUNDIFF, L. V.: Effects of oral melengestrol acetate and a testosterone-diethylstilbestrol implant, breed and age on growth and carcass traits of beef heifers. J. Anim. Sci. **32**, 957—961 (1971).

GOMES, W. R., BULTER, W. R., JOHNSON, A. D.: Effect of elevated ambient temperature on testis and blood levels and in vitro synthesis of testosterone in the ram. J. Anim. Sci. **33**, 804—807 (1971).

HAMMOND, J.: Progress in the Physiology of Farm Animals. London: Butterworth 1955.

HUNT, W. E., MEADE, D., CARMICHAEL, B. E.: Effect of castration of lambs on their development and quality of meat. Bull. Md. Agric. Exp. Sta. No. **417**, 259—278 (1938).

JOHNSTON, E. F., ZELLER, J. H., HINER, R. L.: Some effects on swine of orally administered methyl testosterone. (Abs.) J. Anim. Sci. **16**, 1024 (1957).

KEMP, J. D., CROUSE, J. D., DEWEESE, W., MOODY, W. G.: Effect of slaughter weight and castration on carcass characteristics of lambs. J. Anim. Sci. **30**, 348—354 (1970).

KLOSTERMAN, E. W., CAHILL, V. R., KEMKLE, L. E., MOXON, A. L.: The subcutaneous implantation of stilbestrol in fattening bulls and steers. J. Anim. Sci. **14**, 1050—1058 (1955).

KLOSTERMAN, E. W., KUNKLE, L. E., GERLAUGH, P., CAHILL, V. R.: The effect of age at castration upon rate and economy of gain of beef calves. J. Anim. Sci. **13**, 817—825 (1954a).

KLOSTERMAN, E. W., KUNKLE, L. E., GERLAUGH, P., CAHILL, V. R.: Bull calves make more rapid gains than steers in the feedlot. Ohio Farm Home Res. **39**, 70—77 (1954b).

KOCH, D. C., PARR, A. F., MERKEL, R. A.: Fatty acid composition of the inner and outer layer of porcine backfat as affected by energy level, sex and sire. J. Food Sci. **33**, 176—180 (1968).

LIDVALL, E. R., BURGESS, C. R., RAMSEY, C. R., COLE, J. W.: Performance and carcass characteristics of boars, barrows and gilts. (Abs.) J. Anim. Sci. **23**, 851 (1964).

LILLIE, F. R.: The free martin: A study of the action of sex hormones in the foetal life of cattle. J. exp. Zool. **23**, 371—452 (1917).

MARTIN, T. G.: Bulls for beef production. Purdue Univ. Agric. exp. Sta. Cattle Day Rept., 3—6 (1974).

MARTIN, T. G., DOUGLASS, L. W., GARWOOD, V. A.: Growth and carcass characteristics of Angus bulls and steers. Purdue Univ. Agric. Exp. Sta. Res. Prog. Rept. 253 (1966).
McCLAUGHERTY, F. A., CARTER, R. C., GAINES, J. A.: Effect of docking and castration on growth and quality of spring lambs. (Abs.) J. Anim. Sci. **18**, 1159 (1959).
McMEEKAN, C. P.: Growth and development in the pig with special reference to carcass quality characters. J. Agric. Sci. **30**, 276—386 (1940).
MOORE, K. L., GRAHAM, M. A., BARR, M. L.: The sex chromatin of the bovine freemartin. J. exp. Zool. **135**, 101—121 (1957).
NICHOLS, J. R., ZIEGLER, J. H., WHITE, J. M., KESLER, E. M., WATKINS, J. L.: Production and carcass characteristics of Holstein-Friesian bulls and steers slaughtered at 800 or 1000 pounds. J. Dairy Sci. **47**, 179—185 (1964).
NOLAND, P. R., BURRIS, M. J.: The effect of oral administration of methyl testosterone on swine growth and development. J. Anim. Sci. **15**, 1014—1019 (1956).
O'MARY, C. C., POPE, A. L., WILSON, G. D., BRAY, R. W., CASIDA, L. E.: The effects of diethylstilbestrol, testosterone and progesterone on growth and fattening and certain carcass characteristics of western lambs. J. Anim. Sci. **11**, 656—673 (1952).
PALSSON, H.: Conformation and body composition. In: HAMMOND, J. (Ed.): Progress in the Physiology of Farm Animals, Vol. II, pp. 430—542. London: Butterworth Sci. Publ. 1955.
PALSSON, H., VERGES, J. B.: Effects of the plane of nutrition on growth and development of carcass quality in lambs. J. Agric. Sci. **42**, 93—149 (1952).
PERRY, T. W., BEESON, W. M., MOHLER, M. T., ANDREWS, F. N., STOB, M.: The effect of various levels of orally administered methyl testosterone on growth and carcass composition of swine. J. Anim. Sci. **15**, 1008—1013 (1956).
PILKINGTON, D. H., WALTERS, L. E., POPE, L. S., ODELL, G. V., STEPHENS, D. F.: Carcass studies with steers, bulls and stilbestrol implanted bulls sold as slaughter calves. Okla. Agric. Exp. Sta. Misc. Pub. **55**, 62—64 (1959).
POPE, A. A., O'MARY, C. C., BOTTERMAN, W. E., BRAY, R. W., CASIDA, R. E.: The effect of certain steroid hormones on the growth, fattening and carcass quality of heavy lambs. (Abs.) J. Anim. Sci. **9**, 680 (1950).
RAY, D. E., HALE, W. H., MARCHELLO, J. A.: Influence of season, sex and hormonal growth stimulants on feedlot performance of beef cattle. J. Anim. Sci. **29**, 490—495 (1969).
RAY, E. E., BERRY, B. W., REYNOLDS, D. A.: Effects of testosterone-estradiol implants and rations on gains, feed efficiency and carcass traits in lambs. (Abs.) J. Anim. Sci. **25**, 592 (1966).
SLEETH, R. B., PEARSON, A. M., WALLACE, H. D., KROPF, D. H., KOGER, M.: Effects of injection of testosterone, estradiol, and a combination of the two upon growing-fattening swine. J. Anim. Sci. **12**, 322—330 (1953).
STURKIE, P. D.: Avian Physiology. New York: Comstock Publ. Assoc., Cornell Univ., Ithaca 1965.
TICHENOR, D. A., KEMP, J. D., FOX, J. D., MOODY, W. G., DEWEESE, W.: Effect of slaughter weight and castration on ovine adipose fatty acids. J. Anim. Sci. **31**, 671—675 (1970).
THRASHER, G. W., PERRY, T. W., ANDREWS, F. N., BEESON, W. M., STOB, M.: The effect of estrogenic and androgenic compounds upon growth and carcass composition of swine. J. Anim. Sci. **18**, 399—409 (1959).
TURTON, J. D.: The effect of castration on meat production and quality in cattle, sheep and pigs. Anim. Breed. Abstr. **30**, 447—456 (1962).
VOITKEVICH, A. A.: The Feathers and Plumage of Birds. New York: October House 1966.
WHITAKER, M. D., BROWN, H., BARNHART, C. E., KEMP, J. D., VARNEY, W. Y.: Effects of Methylandrostenediol, methyl testosterone and thyroprotein on growth and carcass characteristics of swine. J. Anim. Sci. **18**, 1189—1195 (1959).
WINTERS, L. M., COMSTOCK, R. E., JORDAN, D. F., KISER, O. M.: The effect of sex on the development of the pig. I. Differences in growth between boars and barrows by lines of breeding. J. Anim. Sci. **1**, 41—47 (1942).
WOEHLING, H. L., WILSON, G. D., GRUMMER, R. H., BRAY, R. W., CASIDA, L. E.: Effects of stilbestrol and testosterone pellets implanted into growing-fattening pigs. J. Anim. Sci. **10**, 889—892 (1951).

# VII. Clinical Applications

## A. Effect of Androgens on Bone, Calcium, and Phosphorus Metabolism[1]

H. SPENCER, J. A. FRIEDLAND, and I. LEWIN

### Effect of Androgens on Mineral Metabolism in Animals and Man

#### A. Effect of Androgens in Animals

Androgens have been shown to affect skeletal growth, skeletal maturation, mineralization and collagen formation in animals.

##### 1. Skeletal Growth

Changes in skeletal growth have been observed in androgen-deficient animals. In castrated young male rats, growth was delayed even when these animals were castrated after puberty (ASCHKENASY-LELU and ASCHKENASY, 1959), and this growth inhibition was ascribed to the lack of androgens needed for the stimulation of skeletal growth. The dose of the androgens apparently plays a role in growth and skeletal development (BERGSTRAND, 1950). Variable results have been reported on the effect of androgens on skeletal growth depending on the dose of the androgen. However, other factors, such as the age, sex, and species of the animal, also appear to play a decisive role. Small doses of androgens have been reported to stimulate linear growth in castrated male rats (BERGSTRAND, 1950). The effect of smaller doses of testosterone propionate as well as of methyltestosterone on the acceleration of linear growth were confirmed in female rats (JOSS et al., 1963), and the smaller doses of both of these androgenic compounds did not lead to skeletal maturation. Large doses, however, had the opposite effect and actually inhibited growth (RUBINSTEIN and SOLOMON, 1941). This growth inhibition is probably due to premature closure of the epiphyses induced by the use of androgens (SILBERBERG and SILBERBERG, 1956; SOBEL et al., 1956). In growing female mice, testosterone propionate had an inhibitory effect on the proliferation of the epiphyseal cartilage, while this effect was not as marked in male mice; testosterone propionate also caused degeneration, hyalinization, and sclerosis, as well as calcification of the cartilage (SILBERBERG and SILBERBERG, 1941). The larger doses of testosterone propionate caused skeletal maturation without promoting linear growth in female rats (JOSS et al., 1963). The mechanism of the effect of different doses of testosterone on growth inhibition is not clear and contradictory opinions have been expressed. The effect of large doses of testosterone on growth inhibition may not be a primary effect on bone but rather secondary to suppression of the pituitary function (SILBERBERG and SILBERBERG, 1956). However, in some studies in rats, small doses of testosterone promoted linear growth in the

[1] This study was supported in part by U.S. Public Health Research Grant DE-02486 and in part by the National Dairy Council.

absence of the pituitary gland (TYSLOWITZ, 1943). It has also been postulated that testosterone and growth hormone have a synergistic effect on skeletal growth (SIMPSON et al., 1944). The growth-promoting effect of androgen may in part be due to the nitrogen-retaining effect of testosterone and to the retention of nitrogen in bone, which may stimulate the proliferation and hypertrophy of the cartilage (KOCHAKIAN and MURLIN, 1936).

## 2. Skeletal Maturation

Differences in the effect of different androgenic anabolic agents on skeletal maturation have been observed in various species of animals. The rate of bone maturation in mice was stimulated to a great extent by testosterone and to a somewhat lesser degree by dehydroepiandrosterone (HOWARD, 1962). In young rats, the use of the anabolic agent nandrolone phenylpropionate resulted in an increase of the femur weight as well as in an increase of the calcium and phosphorus content of bone, while the nitrogen content decreased slightly (RONNING et al., 1964). These changes of calcium and phosphorus content in bone were interpreted as early bone maturation. In chicks, maturation of the subepiphyseal bone was stimulated by the combined use of norethandrolone and calcium phosphate (BELANGER and MIGICOVSKY, 1960). However, in rachitic chicks this anabolic agent caused increased bone growth but no further bone maturation (BELANGER and MIGICOVSKY, 1961). The ratio of calcium/nitrogen in bone has been considered an indicator of bone maturation (RONNING et al., 1964). In 2-month-old rats, treatment with the anabolic agent nandrolone phenylpropionate resulted in a slight decrease of the nitrogen content of bone while the calcium content increased, resulting in a high calcium/nitrogen ratio (RONNING et al., 1964).

## 3. Mineralization

Various androgenic compounds have been shown to stimulate bone mineralization in animals. Enchondral ossification was promoted and fracture healing accelerated in castrated rats by the use of the anabolic agent Dianabol (CLEIN and KOWALEWSKI, 1962). In intact young male rats the callus strength of fractured humeri increased significantly by the use of Dianabol or methyltestosterone (WIANCKO and KOWALEWSKI, 1961) and the strength of the callus exceeded that of the control group after 3 weeks of androgen treatment. In a strain of dystrophic mice testosterone propionate prevented the development of osteoporosis (ERSHOFF et al., 1961). In young male and female rats receiving the anabolic agent nandrolone phenylpropionate, the femur weight increased while there was no increase in bone length. This effect was greater in female than in male rats (RONNING et al., 1964). However, the same anabolic agent was ineffective in rabbits with disuse osteoporosis, caused by immobilization of the limbs, in preventing the decrease of the bone weight and the decrease of the amount of cancellous bone, which was determined by planimetric measures (LEVIN et al., 1972). However, in the nonimmobilized controls who also received the anabolic agent nandrolone phenylpropionate, a significant decrease in bone breakability, measured by kinetic energy of the fracturing force, was demonstrated (LEVIN et al., 1972). In an organ culture study of chick embryo frontal bone the accumulation of calcium was doubled by the use of dehydroepiandrosterone sulfate (PUCHE and

ROMANO, 1969), and this effect was dose-dependent. In another study in rats, the retention of radioactive calcium was also increased when the anabolic agents nandrolone phenylpropionate, 19-nortestosterone, and 17β-hexophenylpropionate were administered (BOHR and DAWIDS, 1964).

### 4. Collagen Formation and Cellular Action of Androgens

Several investigators have shown that androgens affect collagen formation in animals. In rachitic chicks the secretion of the organic matrix and the growth of immature bone cells was stimulated by the anabolic agent norethandrolone (BELANGER and MIGICOVSKY, 1961). Subcutaneous injections of the anabolic agent nandrolone phenylpropionate resulted in an increase of the uptake of $^{35}S$ in fractured bone of young male rats (KOWALEWSKI, 1958a), indicating that the androgenic steroid has an anabolic effect on the organic phase of the bone callus. Similar observations were made by the same investigator in cockerels (KOWALEWSKI, 1958b; KOWALEWSKI, 1958c). Although this effect could be obtained in different species of animals with the anabolic agent nandrolone phenylpropionate, it is of interest to note that both methyltestosterone and testosterone propionate had very little effect (KOWALEWSKI, 1958c). Also, the administration of methandrostenolone did not affect the hydroxyproline or hexosamine content of the bone of young rats nor the histological appearance of bone (KOWALEWSKI, 1962). However, several studies reported positive effects on collagen formation when various testosterone derivatives were used. In tissue culture the osteoid formation and calcification of chick embryo frontal bone was accelerated by dehydroepiandrosterone (PUCHE and ROMANO, 1969); the production of femur collagen was stimulated by the use of testosterone propionate in young rats (SMITH and ALLISON, 1965). Both collagen formation and growth of new bone were stimulated in fractured humeri of rats by the use of methandrostenolone (WIANCKO and KOWALEWSKI, 1961). The effect of fluoxymesterone was even more marked, resulting in greater new bone formation, in a larger size of the trabeculae, and in increased calcification (WIANCKO and KOWALEWSKI, 1961).

In an in vitro study, the addition of dehydroepiandrosterone sulfate to the tissue culture medium resulted in increased calcification of chick embryo bone, and this process was associated with lower lactate/glucose ratios and with more glucose-6-phosphate-dehydrogenase (G-6-PD) activity than in the control study (PUCHE and ROMANO, 1970). This finding was interpreted as a shift of the glucose metabolism toward the hexose-monophosphate shunt pathway. These results could also be reproduced in vivo, following treatment of young mice with either dehydroepiandrosterone or testosterone. An effect of these androgens on bone cells could also be demonstrated in this study and a distinct increase of the hypertrophic cell layer of the epiphyseal plate was noted (PUCHE and ROMANO, 1970). In a recent in vitro study it was shown that rat mandibular bone metabolized testosterone (VITTEK et al., 1974). Both androgenic metabolites and to a lesser extent estrogenic metabolites were formed. These studies indicate a direct effect of androgenic activity at the bone level and that androgens may play a significant role in bone metabolism locally. Using $^{45}Ca$ release from bone as an indicator of bone resorption in tissue culture, the PTH-induced bone resorption from fetal rat bone was inhibited by adding testosterone to the tissue culture medium (STERN, 1969).

### 5. Androgen Effects During Corticosteroid Administration

The catabolic effect of corticosteroids on bone has been amply demonstrated and the effect of androgens on bone during the administration of corticosteroids has been studied in several species of animals. For example, maturation of the epiphyseal cartilage matrix in rachitic chicks was inhibited by cortisone and hydrocortisone (BELANGER and MIGICOVSKY, 1960). The decreased binding capacity for $^{35}S$ in bone which was induced by cortisone in growing cockerels was counteracted by the use of the anabolic agent norethandrolone (KOWALEWSKI, 1958b). In another study in cockerels, the effects of cortisone on bone, such as thinning and irregularity of the trabeculae, a decrease of the number of osteoblasts, a decrease of the acid mucopolysaccharides and of the alkaline phosphatase content of bone could be counteracted by the simultaneous use of both cortisone and the anabolic androgenic agent norethandrostenolone (RODIN and KOWALEWSKI, 1963).

The antianabolic effect of corticosteroids on bone can also be counteracted by some androgenic steroids in rats. The decrease of the hydroxyproline and of the hexosamine content of bone induced by the administration of corticosteroids was counteracted by the use of the anabolic agent methandrostenolone (KOWALEWSKI, 1958c). In cortisone-treated rats and guinea pigs the concurrent administration of testosterone and corticosteroids resulted in a significant increase of the mineral content of the femur as compared to the mineral content of animals treated with cortisone alone (SOBEL and BONORRIS, 1963). The anabolic agent norethandrolone counteracted the decreased $^{35}S$ uptake in fractured bone of rats (KOWALEWSKI, 1958a). In another study, it has been shown that the anabolic agent norethandrolone was more effective than methyltestosterone or testosterone propionate in counteracting the decreased binding of $^{35}S$ which was due to the administration of cortisone in rats (KOWALEWSKI, 1958c). Also, the delayed growth rate of young rats receiving cortisone was prevented by the use of norethandrolone (LLAURADO et al., 1959). The decreased skeletal growth of young male adrenalectomized and orchiectomized rats receiving cortisone could be markedly alleviated by the administration of testosterone propionate; however, the combined use of both testosterone and estrogen had a greater effect than the use of testosterone alone during corticosteroid administration (CALDWELL, 1962). The decrease of the tensile strength of healing fractured humeri of castrated male rats receiving cortisone and the inhibitory effect of cortisone on bone healing could be counteracted by the anabolic agent methandrostenolone (CLEIN and KOWALEWSKI, 1962). This type of treatment also promoted enchondral ossification. However, this anabolic agent did not affect the hydroxyproline content of bone. The use of nandrolone phenylpropionate afforded only some protection in rats against the histological effects of corticosteroids in causing osteomalacia (LARON et al., 1963). Also, these rats excreted less calcium in urine and, therefore, the retention of calcium increased (LARON et al., 1963). The growth rate of rats receiving cortisone and of those receiving both cortisone and the anabolic agent 19-norandrostenolone phenylpropionate (Durabolin) did not differ (LARON and BOSS, 1961). The width of the epiphyseal cartilage was markedly reduced and the trabecular bone was altered in the cortisone-treated rats; these changes were not abolished by the concomitant use of cortisone and this anabolic agent. In young mice the accelerated bone maturation induced by dehydroepiandrosterone could be interfered with by the administration of cortisone (HOWARD, 1962). In rabbits the bone resorption induced by corticosteroids could be reversed by the simultaneous use of the corticoid and of several androgenic-anabolic agents (LITTLE, 1970). Also, when testosterone and cortisone were given in an alternating sequence, the anabolic effect of the androgen predominated and the bone resorption rate decreased (LITTLE, 1970).

In a review of the effect of steroids on bone formation, KOWALEWSKI (1962) states that antianabolic steroid hormones inhibit the growth of cellular elements of bone, delay the synthesis of the bone matrix, and produce a negative calcium balance, while anabolic steroids promote bone matrix formation and may counteract the destructive action of corticoids on bone. Additional experimental data from work on young cockerels and rats demonstrated that the inhibitory action of cortisone on bone matrix formation and on mineralization could be protected by the use of the anabolic agent Dianabol (KOWALEWSKI, 1962).

In contrast to the above effects, several investigators could not demonstrate a protective influence of testosterone derivatives against the effects of the corticosteroids on bone or collagen. Methyltestosterone did not counteract the inhibitory effect of corticosteroids on collagen synthesis (BAVETTA et al., 1962). In growing rats, the histologic changes induced by cortisone were affected very little by the use of the anabolic agent nandrolone phenylpropionate (LARON and BOSS, 1961). The simultaneous use of corticosteroids and of the anabolic agent nandrolone phenylpropionate or the combined use of the corticoid and of 19-nortestosterone-17$\beta$-hexophenylpropionate did not alter the decreased retention of radioactive calcium and strontium in rats receiving the corticosteroids (BOHR and DAWIDS, 1964). Also, the administration of androgens alone did not result in increased radiocalcium retention in the control group (BOHR and DAWIDS, 1964).

### 6. Serum Alkaline Phosphatase

Early experiments showed that the enzyme alkaline phosphatase is present in bone and in plasma (ROBISON, 1923). This observation suggested the possibility that this enzyme may play a role in bone formation. Indeed, it has been found that the serum alkaline phosphatase level reflects the activity of osteoblasts. Thus, it should be possible to estimate the effects of androgens on osteoblastic activity by measuring the alkaline phosphatase levels in serum. However, only scant information is available on the changes of the serum levels of this enzyme in animals induced by androgen administration. In the rat, serum alkaline phosphatase levels were increased following the administration of either testosterone propionate or of the anabolic agent nandrolone phenylpropionate (RONNING et al., 1964). In chickens, however, testosterone propionate did not change the level of this enzyme (TANABE and WILCOX, 1961) despite the fact that testosterone exerted a significant growth promoting effect. Tissue culture studies of explants of chick embryo frontal bone did not reveal any difference in alkaline phosphatase activity in the control cultures and in cultures in which dehydroepiandrosterone was added to the tissue culture medium even though mineralization was accelerated (PUCHE and ROMANO, 1969).

## B. Effect of Androgens in Man

The changes in calcium metabolism which androgens induce may be related to the deposition of calcium in bone. However, these changes may also be related to the protein anabolic effect of androgens on the organic bone matrix as the deposition of calcium may be facilitated by the enrichment of the collagenous matrix with protein. The increase of the deposition of calcium may also stabilize and strengthen the bone crystal, and these changes may lead to the inhibition of the process of bone resorption. A recent study has demonstrated the role of protein at the bone level in the process of calcification (OWEN et al., 1973). The administration of one hormone, such as an androgen, may affect the function of other hormones in the body and this interaction may indirectly influence the metabolism of calcium. Although the major effect of testosterone is upon protein metabolism, androgens do affect skeletal growth and development in man.

### 1. Skeletal Growth and Maturation

The effects of gonadal and adrenal androgens on bone growth, skeletal maturation, and mineral metabolism have been observed clinically and experimentally. These effects of androgens in man are similar to those observed in animals, as described in Section A. Skeletal growth in man may be promoted by androgens possibly due to the stimulation of bone matrix formation resulting from the protein anabolic properties of these agents. However, androgens also promote skeletal maturation resulting in early closure of the epiphyses and in arrest of linear growth.

Optimal skeletal growth requires the presence of both androgens and growth hormone and studies in various laboratories suggest a synergistic interaction between androgens and growth hormone in humans. Maximal growth promotion due to androgen in sexually immature boys can occur only in the presence of growth hormone (ZACHMANN and PRADER, 1970). Treatment of hypopituitary dwarfs was more effective using a combination therapy of human growth hormone and androgens rather than human growth hormone alone as a treatment modality (MACGILLIVRAY et al., 1974). It is interesting to note that the administration of testosterone propionate enhanced the secretion of growth hormone in patients with stunted growth (MARTIN et al., 1968). On the other hand, children with dwarfism who have destructive lesions of the pituitary frequently failed to respond to androgen therapy. These observations suggest that the growth-promoting effect of androgens may be mediated by inducing a release of growth hormone (MARTIN et al., 1968).

The dose of the androgen used for stimulating linear growth in children must be carefully determined. Even small doses may lead to early virilization while large doses may cause premature closure of the epiphyses. Therefore, use of the compounds with a high anabolic/androgenic ratio is desirable. Studies of the use of agents with potent anabolic but weak androgenic effects have shown that bone maturation can be delayed while linear growth is enhanced. In young patients with dwarfism, a large dose of methyltestosterone (100 mg/day) increased linear growth but also promoted epiphyseal maturation (MCCULLAGH and ROSSMILLER, 1941). In contrast, when fluoxymesterone was given for 3 months (LARON, 1961) and for 6–7 months (LARON, 1963) in doses of 0.1–0.15 mg/kg body weight to children with growth retardation, an increase in linear growth was observed while there was little, if any, stimulation of skeletal maturation. These studies indicate the importance of using small doses of androgens for maximal growth promotion. Treatment of children with growth retardation using the anabolic agent nandrolone phenylpropionate,

1 g/kg/month, and methandrostenolone, 0.05 mg/kg/day, led to increased linear height and to variable acceleration of skeletal maturation (SCHARER et al., 1960). Increased height without premature closure of epiphyses was achieved when methandrostenolone (30 mg/week) was used intermittently (LEDERER, 1960). Oxandrolone, a potent anabolic agent with weak androgenic properties, was used in several studies. This anabolic agent used in children with growth retardation (DANKOWSKI et al., 1965) and in children with mongolism (RAY et al., 1965) resulted in maximal linear growth without skeletal maturation. In a study of growth-retarded children receiving oxandrolone, an age effect was noted. In children less than 10 years of age, premature closure of the epiphyses was observed while in older children both acceleration of linear growth as well as premature epiphyseal closure occurred (ZANGANAH and STEINER, 1967). Similar age dependent effects were reported in a study using the anabolic agent oxandrolone; however, the differences of the groups were expressed in terms of bone age rather than of chronologic age (BETTMANN et al., 1971).

The use of androgens, such as fluoxymesterone or methyltestosterone, used in normal children of short stature causes an adverse effect on the expected final height by accelerating bone age (MELLMAN et al., 1961). However, in children with a constitutional delay of growth, these androgens did not interfere with achieving the predicted height (KAPLAN et al., 1973). Methyltestosterone given to immature boys of short stature stimulates linear growth as long as the hormone does not cause simultaneously premature closure of the epiphyses. The relative magnitude of these effects depends on the dose and duration of administration of the hormone (BAYLEY et al., 1957). A synthetic androgen, fluoxymesterone, at a dose of 5–10 mg/day given to children of short stature due to various etiologic factors stimulated linear growth more than epiphyseal closure (REILLY and GORDAN, 1961).

Indirect evidence of the influence of androgens on bone maturation and development can be drawn from observations made on castrate and eunuchoid individuals. In these individuals the epiphyseal growth zones are increased because hypertrophy of the cartilage cells and replacement of the provisional cartilage has not occurred (WILKINS, 1953; HOLLO, 1972). Thus, the increased standing height of eunuchs can be explained by a failure of the epiphyses to mature. This indicates that the absence of genital androgens during adolescence does not prevent linear growth, however, the lack of these androgens leads to the development of osteoporosis in eunuchs.

The development of osteoporosis in elderly males has been ascribed to the decrease of production of testicular androgen (REIFENSTEIN and ALBRIGHT, 1947). However, these changes in bone structure cannot be attributed to the lack of androgens alone as the imbalance between the declining production of anabolic sex hormones and of the persistence of antianabolic adrenocortical steroids must be considered (ALBRIGHT and REIFENSTEIN, 1948). In postmenopausal women with osteoporosis, the plasma levels of dehydroepiandrosterone were markedly lower than in normal age-matched controls (HOLLO, 1972). However, the average plasma cortisol level was also significantly lower than in the normal. Another study, however, reported that the blood levels of testosterone and of estrogen did not significantly differ in women with osteoporosis and in an age-matched control group (RIGGS et al., 1973).

## 2. Effect of Androgens on Bone Resorption and Bone Formation

It is well known that the dynamic state of the skeleton depends on the balance between the processes of bone formation and of bone resorption. It has also been recognized that a continuous bone loss occurs during the physiologic process of aging, and that the decrease in bone mass is probably due to an imbalance between bone resorption and bone accretion, the former exceeding the latter. In senile and postmenopausal osteoporosis, the rate of bone resorption has been reported to be increased (LAFFERTY et al., 1964). An extensive review of the factors regulating bone resorption has been published (RAISZ, 1970). Although many publications indicate that bone resorption is increased in osteoporosis (HEANEY, 1962), the evidence is not conclusive (RAISZ, 1970; RIGGS et al., 1969). Androgens have been used for the treatment of osteoporosis for many years (REIFENSTEIN and ALBRIGHT, 1947; ALBRIGHT and REIFENSTEIN, 1948). It has been demonstrated in calcium kinetic studies that testosterone propionate decreased the rate of bone resorption in patients with postmenopausal osteoporosis (LAFFERTY et al., 1964). In a study of a 42-year-old male patient who was treated with testosterone enanthate the bone resorption rate has been reported to decrease from 14.2% to 9.7%; however, the bone formation rate also decreased significantly from 3.9% to 0.2% as determined by microradiography (RIGGS et al., 1969). An increase of the bone resorption rate has also been demonstrated in disuse osteoporosis (HEANEY, 1962). The use of the anabolic agent norethandrolone reduced the increased rate of bone resorption in adults with disuse osteoporosis due to paralysis (HEANEY, 1962). In a study of chronically immobilized, mentally retarded children, the anabolic agent oxymetholone increased the bone density (KEELE and VOSE, 1969). In a subsequent study in immobilized osteoporotic children, these investigators determined the effect of oxymetholone alone as well as in conjunction with the use of fluoride (VOSE and KEELE, 1973). Increased bone growth was observed in the group receiving the anabolic agent and in the group receiving the combination therapy. After 18 months of continuous treatment there was no significant difference between the two groups. However, one year after the discontinuation of the two types of treatment the growth rate of the group receiving the combined treatment of androgens and fluoride continued at approximately the same rate as during the treatment phase, while the growth rate of the group receiving only oxymetholone had decreased slightly. In a study of patients with osteoporosis of unknown etiology the bone formation rate and the exchangeable calcium pool were not affected by treatment with the anabolic agent nandrolone phenylpropionate (DYMLING et al., 1962). In other studies a variety of androgenic compounds did not increase the bone formation rate in patients with osteoporosis due to different etiologies (GORDAN and EISENBERG, 1963; EISENBERG, 1966). These compounds also did not change the miscible calcium pool, however, the rate of the total turnover of the miscible calcium pool was reduced.

## 3. Effect on Serum Calcium, Phosphorus, and Alkaline Phosphatase

### *a) Serum Calcium*

The serum calcium levels did not significantly change during treatment with either testosterone or with anabolic androgenic agents in normocalcemic patients with osteoporosis (REIFENSTEIN and ALBRIGHT, 1947; WHEDON and SHORR, 1957; LAFFERTY et al., 1964; SPENCER et al., 1957). However, in hypercalcemic hyperparathyroid patients, testosterone decreased the serum calcium level (LANDAU and KAPPAS, 1965), indicating that testosterone counteracts bone resorption induced by excess parathyroid hormone. In patients with metastatic bone disease due to carcinoma of the breast, a significant increase of the serum calcium level has been reported (HERRMANN et al., 1949; SWYER et al., 1950; LASZLO et al., 1952; SPENCER and LEWIN, 1963; SPENCER et al., 1967).

*b) Serum Phosphorus*

Almost three decades ago it was demonstrated that the serum phosphorus levels decreased during testosterone treatment in patients with postmenopausal osteoporosis (REIFENSTEIN and ALBRIGHT, 1947; ALBRIGHT and REIFENSTEIN, 1948). This observation was subsequently confirmed by others (LAFFERTY et al., 1964). Similar observations were made in patients with osteoporosis due to immobilization receiving androgen therapy (WHEDON and SHORR, 1957) and in patients with metastatic breast cancer treated with testosterone propionate (GORDAN and EISENBERG, 1963). In a study carried out in this research unit a decrease of the serum phosphorus levels was also observed in patients without bone disease receiving the anabolic agent 17-ethyl-19-nortestosterone (SPENCER et al., 1957) and in those receiving testosterone propionate (SPENCER, unpublished data). Low serum phosphorus levels have also been observed in children with adrenogenital syndrome who have an excess of androgens (EISENBERG, 1966). In contrast, in cases of androgen deficiency induced by castration in young mature males the serum phosphorus level increased gradually and significantly (ALBRIGHT, 1943). It is likely that the decrease of the serum phosphorus levels reported in the studies cited above is due to a direct effect of androgens on bone resulting in decreased bone resorption and in decreased release of phosphorus from bone.

*c) Serum Alkaline Phosphatase*

Several studies have shown that testosterone does not change the serum levels of alkaline phosphatase in normocalcemic patients with osteoporosis (ALBRIGHT and REIFENSTEIN, 1948). Also, the serum alkaline phosphatase level did not increase in patients without bone disease receiving testosterone propionate (SPENCER, unpublished data) and in patients with neoplasia who were free of bone disease (WOODARD, 1953; WOODARD et al., 1954). However, this serum enzyme increased distinctly in patients with metastatic bone disease who responded favorably to androgen therapy (LASZLO et al., 1952; WOODARD et al., 1954). On the other hand, when androgen therapy was associated with complications of hypercalcemia in these patients, an increase of the serum alkaline phosphatase preceded the onset of hypercalcemia (GRIBOFF et al., 1954).

## 4. Effect of Testosterone on Calcium and Phosphorus Balances

ALBRIGHT and REIFENSTEIN have pioneered in studies of the effect of androgens on calcium and phosphorus metabolism in man (REIFENSTEIN and ALBRIGHT, 1947; ALBRIGHT and REIFENSTEIN, 1948). In the early investigations testosterone propionate as well as methyltestosterone were used for the treatment of patients with postmenopausal or senile osteoporosis. Both androgenic compounds resulted in a decrease of the urinary calcium while the decrease in stool calcium was considerably less, indicating that androgens have little effect on the intestinal absorption of calcium. Although the hormone-induced changes were mainly due to a decrease of the calciuria, distinct improvement of the calcium balance occurred indicating a decrease of the calcium loss in these patients.

Another practical application of the effect of androgens on decreasing the urinary calcium has been demonstrated in patients with immobilization (WHEDON and SHORR, 1957). The urinary calcium excretion of these patients is frequently markedly increased and there is a tendency to form urinary tract stones. The administration of testosterone decreased the hypercalciuria significantly and it was shown that this decrease induced by androgens may be dose-dependent. These investigators demonstrated that the calcium-lowering effect was greater when using a dose of 50 mg testosterone propionate rather than one of 25 mg. Also, the observation was made that the decrease of the urinary calcium induced by testosterone propionate was greatest in patients who had hypercalciuria while there was little change of the urinary calcium excretion in patients whose urinary calcium was in the normal range. The smaller or the larger dose of testosterone did not decrease the fecal calcium excretion consistently or significantly in any of the patients studied. Improvement of the calcium balance during the administration of androgens has also been reported by other investigators (HENNEMAN and WALLACH, 1957). The calcium balances became more positive in male patients with senile osteoporosis who received either 25 mg testosterone propionate intramuscularly or 40 mg methyltestosterone orally per day as compared to the calcium balances of these patients in the control studies. The improvement of the calcium balances in these studies was also mainly due to a decrease of the urinary calcium excretion.

In our own studies carried out in patients observed under strictly controlled conditions in the Metabolic Research Unit, only minor overall changes in calcium metabolism have been noted during testosterone propionate administration. The urinary calcium decreased in some cases while there was no change in others. It should, however, be mentioned that all of these patients had a urinary calcium excretion which was in the low normal range prior to the administration of testosterone. Table 1 shows data on the changes in calcium and phosphorus metabolism induced by testosterone propionate of four fully ambulatory male patients. On a low calcium intake averaging 208 mg/day the urinary calcium excretion was low and ranged from 18–44 mg/day. During the administration of testosterone propionate for time periods ranging from 26–34 days, the calcium intake remaining the same (216 mg/day), the urinary calcium excretion decreased slightly in only two of the four patients while there was no change in the remaining two patients. The stool calcium fluctuated irregularly during the administration of testosterone propionate but the overall changes in fecal calcium were minor, indicating that the intestinal absorption of calcium did not improve during testosterone treatment. In view of the minor decrease of the urinary and stool calcium, the changes of the calcium balance were only slight. The calcium balance became either slightly less negative or remained in equilibrium in three of the four patients studied and the calcium balance became slightly more positive in one case (Patient 1).

Table 1. Changes of urinary and fecal calcium excretions and of calcium balance induced by testosterone propionate

| Patient | Study | Days | Calcium, mg/day | | | | Phosphorus, mg/day | | | |
|---|---|---|---|---|---|---|---|---|---|---|
| | | | Intake | Urine | Stool | Balance | Intake | Urine | Stool | Balance |
| 1 | Control | 44 | 216 | 44 | 315 | −143 | 1088 | 514 | 464 | +110 |
| | Testosterone propionate[a] | 34 | 214 | 26 | 282 | − 94 | 1032 | 390 | 342 | +300 |
| 2 | Control | 26 | 219 | 31 | 205 | − 17 | 891 | 627 | 177 | + 87 |
| | Testosterone propionate[a] | 26 | 220 | 20 | 170 | + 30 | 910 | 528 | 165 | +217 |
| 3 | Control | 24 | 200 | 48 | 240 | − 88 | 795 | 572 | 177 | + 46 |
| | Testosterone propionate[a] | 28 | 210 | 41 | 191 | − 22 | 795 | 414 | 214 | +167 |
| 4 | Control | 18 | 196 | 38 | 201 | − 43 | 619 | 640 | 183 | −204 |
| | Testosterone propionate[a] | 32 | 214 | 33 | 188 | − 7 | 687 | 346 | 163 | +178 |

[a] Testosterone propionate, 100 mg, given intramuscularly 3 times per week.

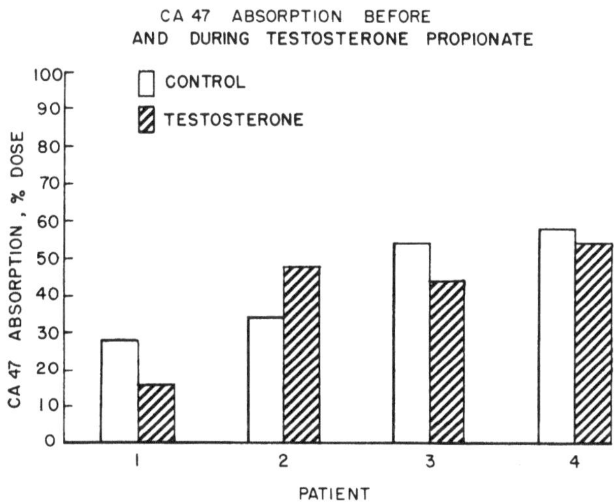


Fig. 1. A tracer dose of $^{47}CaCl_2$ was given orally in the control study and during administration of testosterone propionate; dose of testosterone was 100 mg given 3 times per week for 3–5 weeks

The changes in phosphorus metabolism corresponded to those of the calcium metabolism (Table 1). The dietary phosphorus intake averaged 572 mg/day in the control study and 595 mg/day during testosterone propionate intake. The urinary phosphorus excretion decreased in all patients during hormone therapy and this decrease was greater than the decrease of the urinary calcium excretion. The fecal phosphorus excretion decreased only in one of the four patients studied. The phosphorus balance became more positive in all patients studied primarily due to the decrease of the urinary phosphorus excretion. This decrease reflected the decrease of

Table 2. Effect of testosterone propionate on calcium and nitrogen metabolism in patients with neoplastic bone disease

| Patient | Study | Calcium excretions, mg/day | | Nitrogen balance (mg/day) |
|---|---|---|---|---|
| | | Urine | Stool | |
| 1 | Control | 105 | 140 | + 500 |
| | Testosterone[a] | 95 | 110 | +2300 |
| 2 | Control | 150 | 140 | +1000 |
| | Testosterone[a] | 100 | 180 | +2500 |
| 3 | Control | 100 | 260 | +1500 |
| | Testosterone[a] | 75 | 300 | +2500 |
| 4 | Control | 25 | 520 | − 500 |
| | Testosterone[a] | 20 | 600 | + 400 |

[a] Testosterone propionate, 100 mg, was given intramuscularly 3 times per week.

the urinary nitrogen excretion rather than the decrease of the urinary calcium excretion which was only slight.

The slight effect of androgens on the calcium balance, particularly the lack of change of the stool calcium during the administration of testosterone, was further corroborated by calcium absorption studies using oral tracer doses of the short-lived radioisotope $^{47}$Ca. Figure 1 shows data on the intestinal absorption of calcium of four male patients. These absorption studies were carried out in control studies and during the administration of 100 mg testosterone propionate which was given three times per week for several weeks prior to the oral administration of the $^{47}$Ca tracer. The absorption of calcium determined by this method remained about the same in three of the four patients and increased in one patient.

Results similar to those obtained with testosterone propionate listed in Table 1 were also obtained in patients with metastatic osteolytic bone disease receiving testosterone propionate studied in this Research Unit. Table 2 shows that the fecal calcium excretion did not decrease significantly during the administration of androgens in four patients with osseous metastatic carcinoma. This table also shows that the urinary calcium excretion decreased only slightly during androgen therapy and that the major effect of testosterone propionate was the improvement of the nitrogen balance in each of the four patients studied.

Summarizing the results obtained with testosterone propionate on calcium and phosphorus metabolism in studies carried out in this research unit the following can be stated: testosterone propionate had an inconstant effect on the urinary calcium, little or no effect on stool calcium and on the absorption of calcium, and a more definitive effect on urinary phosphorus excretion. The finding of the minor decrease in stool calcium is in agreement with the results previously reported by others (REIFENSTEIN and ALBRIGHT, 1947; WHEDON and SHORR, 1957; HENNEMAN et al., 1955). Similarly, the findings of the decrease of the urinary calcium as well as of the absorption of calcium are also in agreement with observations reported by others. In a study of male and female patients with osteoporosis various androgenic compounds were reported to decrease the urinary calcium, however, this decrease was

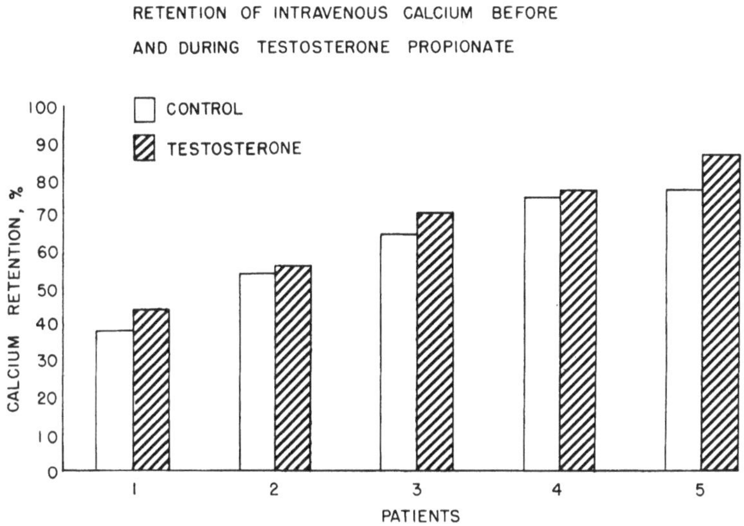


Fig. 2. On the test day 495 mg calcium as the gluconate was infused in 500 ml 5% glucose in water over 4 h. Test was performed in the control study and during administration of 100 mg testosterone propionate, which was given intramuscularly 3 times per week for several weeks

relatively small (EISENBERG, 1966). In a calcium kinetic and calcium balance study of a 75-year-old male with osteoporosis who received testosterone propionate, both the urinary and the fecal calcium decreased slightly (LAFFERTY et al., 1964). Also, the absorption of $^{47}Ca$, determined in this latter study in patients with osteoporosis, did not improve during the administration of testosterone propionate in the majority of the cases. Improvement in calcium metabolism reported by other investigators and to some extent by this group can therefore not be ascribed to improvement of the intestinal absorption of calcium but rather to a decrease of the urinary calcium and/or to the combined decrease of the urinary and fecal calcium. However, a marked decrease of the high urinary calcium excretion has been reported in a patient who was diagnosed as having a parathyroid carcinoma and in patients with primary hyperparathyroidism, indicating that androgens inhibit the active process of bone resorption (LANDAU and KAPPAS, 1965). Similarly, the high urinary calcium of immobilized patients also decreased during testosterone administration (WHEDON and SHORR, 1957).

As androgens result in protein anabolism and improve the nitrogen balance, one may assume that protein is also retained in the bone matrix and that, therefore, additional calcium could be deposited in the newly formed matrix. This aspect was tested in this research unit by infusing calcium intravenously during testosterone administration and by determining the retention of the infused calcium as used in the calcium tolerance test which was developed in this research unit (SCHILLING and LASZLO, 1951; SPENCER et al., 1954; LASZLO and SPENCER, 1956; SPENCER et al., 1964a, b). In this study, 500 mg calcium as the gluconate is infused over 4 h on the day of the calcium tolerance test. The urinary calcium excretion increases to a variable extent on the day of the calcium infusion depending on the calcium metabo-

Table 3. Effects of synthetic androgens on calcium and phosphorus balances

| Study | Days | Calcium, mg/day | | | | Phosphorus (mg/day) | | | |
|---|---|---|---|---|---|---|---|---|---|
| | | Intake | Urine | Stool | Balance | Intake | Urine | Stool | Balance |
| Control | 24 | 153 | 48 | 143 | − 33 | 690 | 503 | 189 | − 2 |
| Hormone[1] | 12 | 147 | 30 | 111 | + 6 | 674 | 223 | 172 | +279 |
| Control | 24 | 114 | 153 | 330 | −369 | 352 | 330 | 152 | −130 |
| Hormone[2] | 12 | 133 | 71 | 162 | −100 | 356 | 213 | 150 | − 7 |
| Control | 58 | 165 | 42 | 181 | − 58 | 631 | 313 | 237 | + 81 |
| Hormone[3] | 18 | 159 | 18 | 120 | + 21 | 634 | 128 | 144 | +362 |
| Control | 18 | 180 | 82 | 96 | + 2 | 715 | 586 | 125 | + 4 |
| Hormone[4] | 10 | 190 | 72 | 79 | + 39 | 785 | 468 | 97 | +220 |
| Control | 18 | 130 | 40 | 177 | − 87 | 474 | 353 | 179 | − 58 |
| Hormone[5] | 20 | 148 | 21 | 175 | − 48 | 522 | 314 | 185 | + 23 |

Hormone 1 = 17-ethyl-19-nortestosterone, 50 mg/day intramuscularly.
Hormone 2 = Dromostanolone propionate, 50 mg/day intramuscularly.
Hormone 3 = estr-4-en-3-one-17β-hydroxy-17-methyl, 10 mg 3 times per day.
Hormone 4 = Fluoxymesterone, 5 mg 4 times daily.
Hormone 5 = 17-methylandrostane-11β,17β-diol, 50 mg 4 times daily.

lism of the individual. The retention of the infused calcium is calculated by determining the difference of the urinary calcium on the day of the calcium infusion and on the day prior to the infusion and by relating this difference to the amount of calcium infused. This study has shown that during testosterone therapy the retention of the intravenously infused calcium was similar to the retention in the control study in the majority of the cases. This is shown in Figure 2. This observation may indicate that most of the nitrogen which is retained during the administration of testosterone is primarily retained in tissues other than bone.

## 5. Effect of Other Synthetic Anabolic Agents on Calcium and Phosphorus Balance

A series of androgenic agents have been synthesized in an attempt to decrease the androgenic effect while preserving or even increasing the anabolic effect of these compounds. These synthetic anabolic androgenic agents were mainly used for the palliative treatment of metastatic carcinoma of the breast. The effect of five synthetic androgenic compounds will be discussed briefly (Table 3). One of the earliest compounds available was 17-ethyl-19-nortestosterone (Nilevar). The metabolic effects of this compound was studied in normocalcemic, normocalciuric patients (SPENCER et al., 1957). A dose response study has shown that the minimally effective dose for nitrogen retention was 50 mg/day. This dose resulted in a small decrease of the urinary calcium excretion, while the fecal calcium also decreased slightly, resulting in a slight to moderate improvement of the calcium balance (Table 3). However, in immobilized hypercalciuric patients who received the anabolic agent norethandrolone the decrease of the urinary calcium excretion was substantial (PLUM and DUNNING, 1958). The average decrease of the urinary calcium excretion in this study was 45%.

Another androgenic synthetic compound (Dromostanolone propionate, Drolban), was also developed primarily for the treatment of advanced carcinoma of the breast. This anabolic agent decreased the urinary and fecal calcium significantly and the calcium balance improved markedly (LEWIN, 1965). Representative data on the effect of this compound on calcium and phosphorus metabolism are shown in Table 3. In a single study of an experimental compound (17β-hydroxy-17-methylestr-4-en-3-one) the urinary and fecal calcium decreased considerably and the calcium balance improved (Table 3). In view of these marked calcium anabolic effects of the latter two compounds, further studies are indicated to confirm these results.

Two other androgenic compounds had either no effect or little effect on calcium metabolism. Table 3 shows that fluoxymesterone had no effect on urinary or stool calcium while it had a distinct nitrogen-retaining effect (not shown in Table 3). The other androgenic compound (17-methylandrostene-11β,17β-diol) had only a moderate and inconstant effect on urinary calcium and no effect on stool calcium.

The phosphorus balances determined during the administration of the five synthetic androgenic compounds (Table 3) show that the urinary phosphorus excretion decreased during the administration of four of these compounds and was very slightly lower during the administration of the fifth compound. The fecal phosphorus excretion remained unchanged during the intake of four of the synthetic androgenic agents and was lower during the intake of one of the synthetic hormones only. All five hormones improved the phosphorus balance, i.e., the phosphorus balance became either more positive or changed from negative to positive values primarily due to the decrease of the urinary phosphorus excretion.

While all five synthetic androgenic compounds had profound and consistent effects on nitrogen retention, the effect on calcium metabolism was variable. It would be tempting to speculate whether the chemical structure of the compound affects the metabolism of protein or of calcium. It is conceivable that the presence or absence of the radical in the 17-position of the sterol ring affects this action. The dissociation of the metabolic effect on nitrogen or calcium may be related to the side chain of the molecule (LEWIN, 1965).

Other examples of the dissociation of the effects of androgenic compounds on calcium and nitrogen metabolism are indicated by the metabolic effects of two additional androgenic compounds, androstane pyrazole and stanolone. Both compounds had an inconstant effect on calcium metabolism while they promoted the retention of nitrogen (WATKIN, 1958; WATKIN et al., 1955; HOWARD and FURMAN, 1962).

The effects of several synthetic anabolic agents on calcium and phosphorus metabolism during corticosteroid administration have been reported (VAN WAYJEN and BUYZE, 1961; VAN WAYJEN, 1962). These studies are described in Section D.3.

## C. Androgen-Estrogen Interaction

While androgens have been observed to have a mineralizing effect (see Section A.3) several reports indicate that androgens may counteract the mineralizing effect of estrogens. Estrogen has been shown to increase the mineral content of bone (WENTWORTH et al., 1940) while large doses of testosterone propionate cause a decrease of the ash content (ARMSTRONG et al., 1945). Furthermore, it has been shown that the effect of estrogen in increasing bone strength can be counteracted by testosterone propionate (GARDNER, 1943). However, this antagonistic effect could not be demonstrated in other studies (URIST et al., 1950). A later study also reported that an increase in bone density due to estrogen can be counteracted by testosterone (BARKER and CROSSLEY, 1962). This effect may be due to inhibition of the suppression of bone resorption induced by estrogen. The possibility that androgen may counteract the action of estrogen on bone is relevant to the treatment of osteoporosis when using a combination of both sex hormones simultaneously.

## D. Therapeutic Uses of Androgens

### 1. Postmenopausal and Senile Osteoporosis

The primary effect of androgens on calcium metabolism is a decrease of urinary calcium, resulting in improvement of the calcium balance while the decrease of the fecal calcium is minor. However, it should be mentioned that the decrease in urinary calcium during androgen therapy reported by other investigators was not impressive when the urinary calcium was normal. This was also the case in the patients studied in our research unit who had a low normal urinary calcium excretion. Usually, patients with postmenopausal osteoporosis have a normal or even a low urinary calcium excretion and any further decrease induced by the use of androgens cannot be expected or is difficult to demonstrate. Studies of the effect of androgens on mineral metabolism carried out in this research unit have also shown that the minor changes in fecal calcium are associated with similar changes of the intestinal absorption of calcium determined independently. It therefore appears that androgens would be most effective in improving the mineral balance of those patients with osteoporosis who have a high urinary calcium excretion (REIFENSTEIN and ALBRIGHT, 1947; ALBRIGHT and REIFENSTEIN, 1948). The observation has been made that the optimum effect of androgens on calcium metabolism is reached after several weeks of treatment, and that this effect persists for several weeks after the discontinuation of the hormone. Androgens have been frequently used in combination with estrogens for the treatment of osteoporosis and this combined treatment has been reported to have a greater effect on calcium metabolism than androgens alone (HENNEMAN and WALLACH, 1957). This combination therapy has also been suggested by other investigators as an effective treatment modality for patients with postmenopausal osteoporosis (COOKE, 1955; HERNBERG, 1960). The mechanism on which this improvement is based is believed to be a decrease of the rate of bone resorption (LAFFERTY et al., 1964).

### 2. Disuse Osteoporosis

Patients with disuse or immobilization osteoporosis due to poliomyelitis or spinal cord injury sustain a marked calcium loss, primarily due to a high urinary calcium excretion. This calcium loss can be decreased by the administration of testosterone propionate (WHEDON and SHORR, 1957) or by the use of the anabolic agent norethandrolone (PLUM and DUNNING, 1958). In disuse osteoporosis, the reduction of the urinary calcium excretion induced by norethandrolone has been attributed to a decrease of the increased rate of bone resorption which is known to be present in immobilization osteoporosis (HEANEY, 1962).

### 3. Corticosteroid-Induced Osteoporosis

Contradictory results have been reported on the effect of androgens on the excretion and retention of calcium during corticosteroid treatment. In a patient with excess production of corticoids (Cushing's syndrome) the high urinary calcium excretion was not affected by the administration of testosterone propionate (ALBRIGHT, 1943). In another study the effect of two anabolic androgenic agents, norethandrolone and nandrolone phenylpropionate, on corticoid induced hypercalciuria could not be adequately demonstrated (MCSWINEY and PRUNTY, 1957; BROCHNER-MORTENSEN et al., 1959). In a calcium kinetic study carried out during long-term corticosteroid therapy the use of the anabolic agent, nandrolone phenylpropionate, did not affect the rate of bone resorption or of bone accretion (DYMLING et al., 1962). However, in other studies, androgens were effective in decreasing the high urinary calcium excretion and in improving the calcium balance in patients in receiving long term corticosteroid therapy (HENNEMAN et al., 1955; REIFENSTEIN, 1958). This effect was obtained in patients with hypercalciuria but not in patients in whom the urinary calcium excretion was in the normal range (HENNEMAN et al., 1955). A decrease of both the urinary and fecal calcium excretion was achieved by the use of methylandrostenediol during corticosteroid therapy and this change resulted in a reversal of the negative calcium balance to a positive balance (FISCHER and HASTRUP, 1954). A similar change from a negative to a positive calcium balance during testosterone propionate was also reported in a patient with multiple myeloma who was treated with ACTH (SKOOG and ADAMS, 1962). Increasing doses of the anabolic agent, methandrostenolone, given during corticosteroid administration, decreased the urinary calcium excretion but the varying doses of this anabolic agent did not have a definitive effect on the fecal calcium excretion (ALMQVIST et al., 1961). However, in another study both the urinary and fecal calcium excretions decreased following the administration of a single dose of either nandrolone phenylpropionate or of nandrolone decanoate (VAN WAYJEN, 1962). Calcium balances were determined during the intake of the anabolic agents norethandrolone, ethylestrenol (Orgabolin), nandrolone decanoate, and methandienone. The first three compounds were reported to improve the calcium balance during corticosteroid administration, while methandienone had a questionable effect (VAN WAYJEN and BUYZE, 1961). The effect of several anabolic androgenic agents on the calcium metabolism of patients receiving long-term corticosteroid therapy has been published by these investigators (VAN WAYJEN and BUYZE, 1961).

### 4. Osteogenesis Imperfecta

Osteogenesis imperfecta is a disease of the connective tissue, which among other clinical manifestations, is primarily characterized by fragile bones. In an attempt to improve the structure and function of the abnormal bone and to reduce the rate of fractures, androgens have been used as a form of treatment in the past (ROPES et al., 1946). In fact, it has been demonstrated that the retention of calcium and phosphorus could be improved in osteogenesis imperfecta with the use of the anabolic agent nandrolone decanoate (MULDOWNEY et al., 1964). However, the efficacy of anabolic agents for the treatment of this disease has been questioned recently (CASTELLS, 1973), based on a report of the lack of effect of these agents (CATTELL and CLAYTON, 1968). With the findings that pyrophosphate metabolism is impaired in osteogenic imperfecta, the administration of magnesium has been suggested for the treatment of this disease (SOLOMONS and STYNER, 1969). Another recent approach to the treatment of osteogenesis imperfecta is the attempt to decrease bone resorption by the use of calcitonin (CASTELLS et al., 1972).

## References

ALBRIGHT, F.: Cushing's Syndrome. In: Harvey Lectures. Lancaster, Penn.: Science Press Printing Comp. 1943.
ALBRIGHT, F., REIFENSTEIN, E. C., JR.: The Parathyroid Glands and Metabolic Bone Disease. Baltimore, Maryland: Williams & Wilkins 1948.
ALMQVIST, S., IKKOS, D., LUFT, R.: Metabolic studies with methandrostenolone (17$\beta$-hydroxy-17$\alpha$-methyl-androsta-1,4-dien-3-one) in human subjects. Acta endocr. (Kbh.) **38**, 413—418 (1961).
ARMSTRONG, W. D., KNOWLTON, M., COUZE, M.: Influence of estradiol and testosterone propionate on skeletal atrophy from disease and on normal bones of mature rats. Endocrinology **36**, 313—322 (1945).
ASCHKENASY-LELU, P., ASCHKENASY, A.: Effects of androgens and oestrogens on the metabolism of proteins and the growth of tissues. In: World Review of Nutrition and Dietetics. Philadelphia: J.B. Lippincott Co. 1959.
BARKER, D. J. P., CROSSLEY, J. N.: Effect of testosterone on estrogen-induced bone formation in mice. Nature (Lond.) **194**, 1088—1089 (1962).
BAVETTA, L. A., BEKHOR, I., NIMNI, M. E.: Effects of hormone administration on collagen biosynthesis in the rat. Proc. Soc. exp. Biol. (N.Y.) **110**, 294—297 (1962).
BAYLEY, N., GORDAN, G. S., LISSER, H.: Long-term experiences with methyltestosterone as a growth stimulant in short immature boys. In: Pediatric Clinics of North America. Philadelphia: W.B. Saunders Company 1957.
BELANGER, L. F., MIGICOVSKY, B. B.: Comparative effects of vitamin D, calcium, cortisone, hydrocortisone, and norethandrolone on the epiphyseal cartilage and bone of rachitic chicks. Develop. Biol. **2**, 329—342 (1960).
BELANGER, L. F., MIGICOVSKY, B. B.: Radiographic and histochemical observations on the effects of norethandrolone and calcionorethandrolone on the bones of rachitic chicks. Minerva Med. **52**, 2116—2118 (1961).
BERGSTRAND, C. G.: The influence of testosterone propionate on skeletal development in the immature rabbit. Acta endocr. (Kbh.) **4**, 91—101 (1950).
BETTMANN, H. K., GOLDMAN, H. S., ABRAMOWICZ, M., SOBEL, E. H.: Oxandrolone treatment of short stature: effect of predicted mature height. J. Pediat. **79**, 1018—1023 (1971).
BOHR, H. H., DAWIDS, S. G.: The effect of cortisone and anabolic steroids on the retention of radioactive calcium and strontium in rats. Acta endocr. (Kbh.) **47**, 223—230 (1964).
BRØCHNER-MORTENSEN, K., GJØRUP, S., THAYSEN, J. H.: The metabolic effect of new anabolic 19-nor-steroids. Acta med. scand. **165**, 197—205 (1959).
CALDWELL, R. A.: The effects of sex hormones in experimental osteoporosis in rats. Brit. J. exp. Path. **43**, 103—115 (1962).
CASTELLS, S.: New approaches to treatment of osteogenesis imperfecta. Clin. Orthop. **93**, 239—249 (1973).
CASTELLS, S., INAMDAR, S., BAKER, R. K., WALLACH, S.: Effects of porcine calcitonin in osteogenesis imperfecta tarda. J. Pediat. **80**, 757—762 (1972).
CATTELL, H. S., CLAYTON, B.: Failure of anabolic steroids in the therapy of osteogenesis imperfecta. J. Bone Jt. Surg. **50 A**, 123—141 (1968).
CLEIN, L. J., KOWALEWSKI, K.: Some effects of cortisone and an anabolic steroid on healing of experimental fractures. Canad. J. Surg. **5**, 108—117 (1962).
COOKE, A. M.: Osteoporosis. Lancet **268**, 877—882, 929—937 (1955).
DANKOWSKI, T. S., LEE, R. A., COHN, R. E., D'AMBROSIA, R. D., LIMAYE, N. R.: Oxandrolone therapy of growth retardation. Amer. J. Dis. Child. **109**, 526—532 (1965).
DYMLING, J. F., ISAKSSON, B., SJORGREN, B.: Anabolic steroids in the treatment of osteopenia. In: Protein Metabolism. Berlin-Heidelberg-New York: Springer 1962.

EISENBERG, E.: Effects of androgens, estrogens and corticoids on strontium kinetics in man. J. clin. Endocr. **26**, 566—572 (1966).

ERSHOFF, B. H., ALFIN-SLATER, R. B., BERNICK, S.: Osteoporosis in dystrophic mice. Arch. Path. **72**, 599—601 (1961).

FISCHER, F., HASTRUP, B.: Cortisone and calcium balance (effect of calcium, vitamin D and methylandrostenediol). Acta endocr. (Kbh.) **16**, 141—148 (1954).

GARDNER, W. U.: Influence of sex and sex hormones on the breaking strength of bones of mice. Endocrinology **32**, 149—160 (1943).

GORDAN, G. S., EISENBERG, E.: The effect of estrogens, androgens and corticoids on skeletal kinetics in man. Proc. roy. Soc. Med. **56**, 1027—1029 (1963).

GRIBOFF, S. L., HERRMANN, J. B., SMELLIN, A., MOSS, J.: Hypercalcemia secondary to bone metastases from carcinoma of the breast. J. clin. Endocr. **14**, 378—388 (1954).

HEANEY, R. P.: Radiocalcium metabolism in disuse osteoporosis in man. Amer. J. Med. **33**, 188—200 (1962).

HENNEMAN, P. H., IRWIN, J. W., WANG, D. M. K., BURRAGE, W. S.: The effect of estrogen and androgen on calcium metabolism during cortisone administration. J. clin. Endocr. **15**, 858 (1955).

HENNEMAN, P. H., WALLACH, S.: A review of prolonged use of estrogens and androgens in postmenopausal and senile osteoporosis. Arch. intern. Med. **100**, 715—723 (1957).

HERNBERG, C. A.: Treatment of postmenopausal osteoporosis with oestrogens and androgens. Acta endocr. (Kbh.) **34**, 51—59 (1960).

HERRMANN, J. B., KIRSTEN, E., KRAKAUER, J. S.: Hypercalcemic syndrome associated with androgenic and estrogenic therapy. J. clin. Endocr. **9**, 1—12 (1949).

HOLLO, I.: Postmenopausal osteoporosis. Pathogenesis and the mode of action of hormone therapy. Ther. hung. **20**, 123—126 (1972).

HOWARD, E.: Steroids and bone maturation in infant mice: relative activities of dehydroepiandrosterone and testosterone. Endocrinology **70**, 131—141 (1962).

HOWARD, R. P., FURMAN, R. H.: Metabolic and serum lipid effects of methylandrostane and methylandrostane pyrazoles. J. clin. Endocr. **22**, 43—51 (1962).

JOSS, E. E., ZUPPINGER, K. A., SOBEL, E. H.: Effect of testosterone propionate and methyl testosterone on growth and skeletal maturation in rats. Endocrinology **72**, 123—130 (1963).

KAPLAN, J. G., MOSHANG, T., JR., BERNSTEIN, R., PARKS, J. S., BONGIOVANNI, A. M.: Constitutional delay of growth and development. Effects of treatment with androgens. J. Pediat. **82**, 38—44 (1973).

KEELE, D. K., VOSE, G. P.: A study of bone density. Amer. J. Dis. Child. **118**, 759—764 (1969).

KOCHAKIAN, C. D., MURLIN, J. R.: The relationship of the synthetic male hormone, androstendion, to the protein and energy metabolism of castrate dogs, and the protein metabolism of a normal dog. Amer. J. Physiol. **117**, 642—657 (1936).

KOWALEWSKI, K.: Comparison of the effects of cortisone and certain anabolic-androgenic steroids on the uptake of radiosulfur in a healing fractured bone. Endocrinology **62**, 493—497 (1958a).

KOWALEWSKI, K.: Uptake of radiosulphur in growing bones of cockerels treated with cortisone and 17-ethyl-19-nortestosterone. Proc. Soc. exp. Biol. (N.Y.) **97**, 432—434 (1958b).

KOWALEWSKI, K.: Uptake of radiosulfate in growing bones of cockerels treated with cortisone and certain anabolic-androgenic steroids. Endocrinology **63**, 759—764 (1958c).

KOWALEWSKI, K.: Effect of steroids on bone formation. In: Protein Metabolism. Berlin-Heidelberg-New York: Springer 1962.

LAFFERTY, F. W., SPENCER, G. E., PEARSON, O. H.: Effects of androgens, estrogens and high calcium intakes on bone formation and resorption in osteoporosis. Amer. J. Med. **36**, 514—528 (1964).

LANDAU, R. L., KAPPAS, A.: Anabolic hormones in hyperparathyroidism. Ann. intern. Med. **62**, 1223—1233 (1965).

LARON, Z.: Effectiveness of fluoxymesterone on linear growth and weight in children with growth retardation and underweight. Acta endocr. (Kbh.) **36**, 541—548 (1961).

LARON, Z.: Use of fluoxymesterone in the treatment of growth retardation. Acta Paediat. **52**, 465—472 (1963).

LARON, Z., BOSS, J. H.: Failure of 19-norandrostenolone phenylpropionate (Durabolin) to prevent the alterations produced by cortisone on the growing bone in rats. Endocrinology **69**, 608—612 (1961).

LARON, Z., KOWADLO, A., ARIE, B. Z., KALISH, I., KENDE, S.: Interaction between 19 nor-androstenolone phenylpropionate (Durabolin) and 6-methyl-prednisolone (Medrol) on the bones of rats fed a low calcium diet. Israel J. exp. Med. **11**, 41—47 (1963).

LASZLO, D., SCHILLING, A., BELLIN, J., GOTTESMAN, E. D., SCHULMAN, C. A.: Effect of testosterone on patients with bone metastases. J. Amer. med. Ass. **148**, 1502—1507 (1952).

LASZLO, D., SPENCER, H.: Newer techniques in the study of calcium metabolism in man and effects of hormones thereon. In: Hormones and the Aging Process. New York City: Academic Press 1956.

LEDERER, J.: Etude clinique d'un nouval anabolisant des proteines. Schweiz. med. Wschr. **90**, 1379—1383 (1960).

LEVIN, S. M., SIMKIN, A., NAOR, E., ROBIN, G. C., GOLDENBERG, A., DORFMAN, H.: A comparison of assay technics in experimental disuse osteoporosis: the effect of nandrolone phenylpropionate. Clin. Orthopaed. Related Res. **82**, 246—252 (1972).

LEWIN, I.: Metabolic effects of structurally related steroid hormones. Fed. Proc. **24**, 638 (1965).

LITTLE, K.: Interaction between catabolic and anabolic steroids. Curr. Ther. Res. **12** (10), 658—676 (1970).

LLAURADO, J. G., TRUNNELL, J. B., CLAUS, J. L.: Some effects of simultaneous administration of norethandrolone and cortisone in the rat. Acta endocr. (Kbh.) **32**, 536—544 (1959).

MACGILLIVRAY, M. H., KOLOTKIN, M., MUNSCHAUER, R. W.: Enhanced linear growth responses in hypopituitary dwarfs treated with growth hormone plus androgen versus growth hormone alone. Pediat. Res. **8**, 103—108 (1974).

MARTIN, L. G., CLARK, J. W., CONNOR, T. B.: Growth hormone secretion enhanced by androgens. J. clin. Endocr. **28**, 425—428 (1968).

MCCULLAGH, E. P., ROSSMILLER, H. R.: Methyl testosterone. III. Effect upon body weight and growth. J. clin. Endocr. **1**, 507—510 (1941).

MCSWINEY, R. R., PRUNTY, F. T. O.: Metabolic effects of three testosterone derivatives including 17$\alpha$-ethyl-19-nortestosterone. J. Endocr. **16**, 28—40 (1957).

MELLMAN, W. J., BONGIOVANNI, A. M., GARRISON, M., STEIKAR, D. D.: Comparison of fluoxymesterone and methyltestosterone as growth stimulants. Pediatrics **28**, 525—530 (1961).

MULDOWNEY, F. P., O'DONOVAN, D. K., GALLAGHER, J. E., CANTWELL, D. F., ABRAHAMSON, M., FREANEY, R.: Bone formation in osteogenesis imperfecta: The effect of anabolic steroid therapy. J. Irish med. Ass. **54**, 132—135 (1964).

OWEN, M., TRIFFITT, J. T., MELICK, R. A.: Albumin in bone. In: Hard Tissue Growth, Repair and Remineralization. Ciba Foundation Symposium 11. London: Ass. Sci. Pub. 1973.

PLUM, F., DUNNING, M. F.: Amelioration of hypercalciuria following poliomyelitis by 17-ethyl-19-nortestosterone (Nilevar). J. clin. Endocr. **18**, 860—871 (1958).

PUCHE, R. C., ROMANO, M. C.: Effect of dehydroepiandrosterone sulfate on the mineral accretion of chick embryo frontal bones cultivated *in vitro*. Calc. Tis. Res. **4**, 39—47 (1969).

PUCHE, R. C., ROMANO, M. C.: The effect of androgens on the carbohydrate metabolism of bone. Calc. Tis. Res. **6(2)**, 133—142 (1970).

RAISZ, L. G.: Physiologic and pharmacologic regulation of bone resorption. New Engl. J. Med. **282**, 909—916 (1970).

RAY, C. G., KIRSCHVINK, J. F., WAXMAN, S. H., KELLEY, V. C.: Studies of anabolic steroids. III. The effect of oxandrolone on height and skeletal maturation in mongoloid children. Amer. J. Dis. Child. **110**, 618—623 (1965).

REIFENSTEIN, E. C., JR.: Control of corticoid-induced protein depletion and osteoporosis by anabolic steroid therapy. Metabolism **7**, 78—89 (1958).

REIFENSTEIN, E. C., JR., ALBRIGHT, F.: The metabolic effects of steroid hormones in osteoporosis. J. clin. Invest. **26**, 24—56 (1947).

REILLY, W. A., GORDAN, G. S.: Dissociation of growth-stimulating and skeleton-maturing actions of the synthetic androgen, fluoxymesterone. J. Pediat. **59**, 188—193 (1961).

RIGGS, B. L., JOWSEY, J., KELLY, P. J., JONES, J. D., MAHER, F. T.: Effect of sex hormones on bone in primary osteoporosis. J. clin. Invest. **48**, 1065—1072 (1969).

RIGGS, B. L., RYAN, R. J., WAHNER, H. W., JIANG, N.-S., MATTOX, V. R.: Serum concentrations of estrogen, testosterone and gonadotropins in osteoporotic and nonosteoporotic postmenopausal women. J. clin. Endocr. **36**, 1097—1099 (1973).

ROBISON, R.: Possible significance of hexosephosphoric esters in ossification. Biochem. J. **17**, 286—293 (1923).

RODIN, A. E., KOWALEWSKI, K.: Histological and histochemical effects of cortisone and an anabolic androgen on long bones of young cockerels and rats. Canad. J. Surg. **6**, 229—236 (1963).

RONNING, O., MAKINEN, E., LAHTINEN, K., FAIJOLA, E.: Effect of nandrolone phenylpropionate on the bone of young rats. Endocrinology **75**, 477—484 (1964).

ROPES, M. W., ROSSMEISL, E. C., BAUER, W.: The effect of estrin in osteogenesis imperfecta. J. clin. Invest. **25**, 932—933 (1946).

RUBINSTEIN, H. S., SOLOMON, M. L.: The growth depressing effect of large doses of testosterone propionate in the castrate albino rat. Endocrinology **28**, 112—114 (1941).

SCHARER, V. K., HABICH, H., PRADER, A.: Wachstumsförderung mit neuen anabolen Steroiden. Helv. med. Acta **27**, 530—534 (1960).

SCHILLING, A., LASZLO, D.: Rate of urinary calcium excretion following intravenous administration as an indicator of bone metabolism. Proc. Soc. exp. Biol. (N.Y.) **78**, 286—289 (1951).

SILBERBERG, M., SILBERBERG, R.: Response of cartilage and bone of growing mice to testosterone propionate. Arch. Path. **32**, 85—95 (1941).

SILBERBERG, M., SILBERBERG, R.: Steroid hormones and bone. In: The Biochemistry and Physiology of Bone. New York City: Academic Press 1956.

SIMPSON, M. E., MARX, W., BECKS, H., EVANS, H. M.: Effect of testosterone propionate on the body weight and skeletal system of hypophysectomized rats. Synergism with pituitary growth hormone. Endocrinology **35**, 309—316 (1944).

SKOOG, W. A., ADAMS, W. S.: Metabolic balance studies of a patient with multiple myeloma given combined ACTH-androgen therapy: Comparisons of three androgens administered in varying dosage on exchanges of calcium, phosphorus and nitrogen. Metabolism **2**, 614—625 (1962).

SMITH, Q. T., ALLISON, D. J.: Cutaneous collagen and hexosamine and femur collagen of testosterone propionate-treated rats of various ages. Biochem. Pharmacol. **14**, 709—720 (1965).

SOBEL, H., BONORRIS, G.: Effect of hormones on femurs and carcass nitrogen of animals treated with antianabolic agents. Metabolism **12**, 246—251 (1963).

SOBEL, E. H., RAYMOND, S., QUINN, K. V., TALBOT, N. B.: The use of methyltestosterone to stimulate growth: Relative influence on skeletal maturation and linear growth. J. clin. Endocr. **16**, 241—248 (1956).

SOLOMONS, C. C., STYNER, J.: Osteogenesis imperfecta: Effect of magnesium administration on pyrophosphate metabolism. Calc. Tis. Res. **3**, 318—326 (1969).

SPENCER, H.: unpublished data.

SPENCER, H., BERGER, E., CHARLES, M. L., GOTTESMAN, E. D., LASZLO, D.: Metabolic effects of 17-ethyl-19-nortestosterone in man. J. clin. Endocr. **17**, 975—984 (1957).

SPENCER, H., HAUSINGER, A., LASZLO, D.: The calcium tolerance test in senile osteoporosis. J. Amer. Geriat. **2**, 19—25 (1954).

SPENCER, H., LEWIN, I.: Derangements of calcium metabolism in patients with neoplastic bone involvement. J. chron. Dis. **16**, 713—726 (1963).

SPENCER, H., LEWIN, I., FRIEDLAND, J. A.: Changes in calcium metabolism in metastatic bone disease. Israel J. med. Sci. **3**, 643—656 (1967).

SPENCER, H., MENCZEL, J., LEWIN, I.: Metabolic and radioisotope studies in osteoporosis. Clin. Orthopaed. **35**, 202—219 (1964a).

SPENCER, H., MENCZEL, J., LEWIN, I., SAMACHSON, J.: Absorption of calcium in osteoporosis. Amer. J. Med. **37**, 223—234 (1964b).

STERN, P. H.: Inhibition by steroids of parathyroid hormone-induced $Ca^{45}$ release from embryonic rat bone in vitro. J. Pharmacol. exp. Ther. **168**, 211—217 (1969).

SWYER, A. J., BERGER, J. S., GORDON, H. M., LASZLO, D.: Hypercalcemia in osteolytic metastatic cancer of the breast. Amer. J. Med. **8**, 724—732 (1950).

TANABE, Y., WILCOX, F. H.: Endocrine control of serum alkaline phosphatase activity in the chicken. Poultry Sci. **49**, 411—416 (1961).

Tyslowitz, R. F.: Trans. 4th Josiah Macy, Jr. Conf. on Metabolic Aspects Convalescence, p. 123, 1943.

Urist, M. R., Budy, A. M., McLean, F. C.: Endosteal-bone formation in estrogen-treated mice. Amer. J. Bone Jt. Surg. **32**, 143—162 (1950).

Van Wayjen, R. G. A.: Long term balance studies on certain anabolic agents. Symposium on Anabolic Therapy, Michigan and Wayne County Academies of General Practice, Detroit, Michigan, March 21, 1962.

Van Wayjen, R. G. A., Buyze, G.: Clinical-pharmacological evaluation of certain anabolic steroids. Acta endocr. (Kbh.) (Suppl. 63) **39**, 18—36 (1961).

Vittek, J., Altman, K., Gordon, G. G., Southren, A. L.: The metabolism of $7\alpha$-$^{3}H$-testosterone by rat mandibular bone. Endocrinology **94**, 325—329 (1974).

Vose, G. P., Keele, D. K.: Bone growth in nonambulatory children on oxymetholone and fluoride therapies. Growth **37**, 213—221 (1973).

Watkin, D. M.: The assessment of protein nutrition in aged man. Ann. N.Y. Acad. Sci. **69**, 902—913 (1958).

Watkin, D. M., Parsons, J. M., Yiengst, M. J., Shock, N. W.: Metabolism in the aged: The effect of stanolone on the retention of nitrogen, potassium, phosphorus, and calcium and on the urinary excretion of 17-keto, 11-oxy, and 17-hydroxy steroids in eight elderly men on high and low protein diets. J. Gerontol. **10**, 268—287 (1955).

Wentworth, J. H., Smith, P. K., Gardner, W. U.: The composition of bones of mice receiving estrogens and androgens. Endocrinology **26**, 61—67 (1940).

Whedon, G. D., Shorr, E.: Metabolic studies in paralytic acute anterior poliomyelitis. J. clin. Invest. **36**, 941—1033 (1957).

Wiancko, K. B., Kowalewski, K.: Strength of callus in fractured humerus of rat treated with anti-anabolic and anabolic compounds. Acta Endocr. (Kbh.) **36**, 310—318 (1961).

Wilkins, L.: Disturbances in growth. Bull. N.Y. Acad. Med. **29**, 280—294 (1953).

Woodard, H. Q.: Changes in blood chemistry associated with carcinoma metastatic to bone. Cancer **6**, 1219—1227 (1953).

Woodard, H. Q., Escher, G. C., Farrow, J. H.: Changes in the blood chemistry of patients with disseminated carcinoma of the breast during endocrine therapy. Cancer **7**, 744—757 (1954).

Zachmann, M., Prader, A.: Anabolic and androgenic effect of testosterone in sexually immature boys and its dependency on growth hormone. J. clin. Endocr. **30**, 85—95 (1970).

Zanganah, F., Steiner, M. M.: Oxandrolone therapy in growth retardation of children. Amer. J. Dis. Child. **113**, 233—241 (1967).

# B. The Use of Anabolic-Androgenic Steroids in Childhood and Adolescence

W. A. DANIEL, JR. and D. L. BENNETT

## I. Introduction

Childhood and adolescence are dynamic periods of life. Adult stature and optimal function can be achieved only if growth and maturation proceed normally. When these processes are disturbed, however, the implications for the individual, both physically and psychologically, are serious, and the desire on the part of the clinician to intervene is understandable.

Ever since the nitrogen-retaining effect of androgens was first demonstrated in 1935 (KOCHAKIAN and MURLIN, 1935), attempts have been made to utilize these hormones clinically. The term "anabolic" generally refers to the conservation of nitrogen with consequent favorable effect on the building of tissue protein. Steroid chemists have succeeded in synthesizing drugs that are more anabolic and less virilizing than testosterone, but none is entirely without problems. In Pediatrics and Adolescent Medicine, the aims of therapy with anabolic-androgenic agents and concern about their adverse effects vary with the underlying condition.

## II. Growth and Development

### A. Growth

Growth, here defined as an increase in size with time, is a complex process requiring the harmonious interaction of diverse factors, is common to all multicellular systems, and occurs by both cell division and cell enlargement. With modern chemical methods, these events can be accurately quantitated rather than merely being observed microscopically (ROOT, 1973). Early growth appears to be due primarily to increase in cell number, which occurs in geometric progression (COOKE, 1970). Subsequently, there is a greater tendency for accretion of protein. Growth hormone is an important regulator of cell division, a process that may occur in spurts rather than as a continuous process. Thyroid hormone is concerned with the control of cell size, which also varies markedly from one tissue to another and from one period of life to another.

In the intact animal, total body mass can be estimated as can the mass of a single isolated muscle cell. The amount of muscle mass tends to approximate overall growth changes. Between infancy and adulthood in males, for example, there is a 14-fold increase in the number of muscle cells. Under the influence of androgens (CHEEK, 1968; TANNER, 1969), muscle cell size continues to increase until the end of the third decade. The mature male has 50% more muscle cells than a mature female

Table 1. Hormones and their effects in bone growth and bone maturation

| Hormone | Linear growth | Bone maturation |
|---|---|---|
| Growth hormone | ++ | + |
| Thyroid hormone | + | ++ |
| Androgens and estrogens | + | ++ |
| Glucocorticoids | − | + |
| + indicates increase | − indicates decrease | |

Adapted from table in HALL et al.,: Clinical Endocrinology, ed. 2, p. 494, Philadelphia: J. B. Lippincott, 1974.

(CHEEK, 1968; CHEEK and HILL, 1970), and they are larger, thus accounting for the superior muscle strength in males.

Growth as it occurs in long bones requires new cartilage formation (chondrogenesis), while skeletal maturation occurs by replacement of cartilage with calcified bone (osteogenesis) (HUBBLE, 1969). Chondrogenesis is influenced by pituitary growth hormone, while thyroid hormone is important in the process of osteogenesis. Gonadal steroids influence both chondrogenesis and osteogenesis (ROOT et al., 1971). In essence, therefore, before puberty, skeletal maturation is under the influence of thyroid and growth hormones and proceeds normally in the absence of androgen. After puberty, the anabolic action of the androgens is important in developing the final skeleton. A summary of these hormonal effects on growth and bone maturation is given in Table 1.

In the control of these complex processes, there is obviously interplay of genetic and environmental factors, many of which are not fully understood. Concern that all may not be proceeding normally is most likely to occur at the time of puberty.

## B. Puberty

Puberty is that period of human development intervening between the time of the appearance of secondary sex characteristics and the attainment of sexual reproduction and completion of somatic growth. It is associated with changes in psychologic perspective, particularly in relation to body image and personal identity.

The age of onset of puberty, the velocity of change, and the age of completion of somatic growth are extremely variable. The physiologic and endocrinologic changes occurring during puberty are more accurately related to sexual characteristics than to chronologic age, and these changes occur sequentially until adult sexual characteristics are attained (TANNER, 1969). The use of secondary sex characteristics as a means of rating maturity has found wide usage, and these stages for boys and girls are presented in Table 2.

In English boys (MARSHALL and TANNER, 1970), the peak height velocity of growth was attained at 14.06 (S.D. $\pm 0.92$) years during genital stage 4 and pubic hair stages 3 and 4. Peak height velocity occurs later in males than in females in relation to chronologic age or stage of sexual maturation. Pubic hair is said to appear between 11.8 and 12.2 years and the onset of adolescent genital growth between 11.5 and 11.8 years in American boys. The peak height velocity in British girls was

Table 2. Classification of genitalia maturity stages in boys and girls

| Stage | Pubic hair | Penis | Testes |
|---|---|---|---|
| 1 | None | Preadolescent | |
| 2 | Slight, long, slightly pigmented | Slight enlargement | Enlarged scrotum, pink, texture altered |
| 3 | Darker, starts to curl, small amount | Penis longer | Larger |
| 4 | Resembles adult type, but less in quantity, coarse, curly | Larger, glans and breadth increase in size | Larger, scrotum dark |
| 5 | Adult distribution, spread to medial surface of thighs | Adult | Adult |

| Stage | Pubic hair | Breasts |
|---|---|---|
| 1 | Preadolescent | Preadolescent |
| 2 | Sparse, lightly pigmented, straight, medial border of labia | Breast and papilla elevated as small mound; areolar diameter increased |
| 3 | Darker, beginning to curl, increased amount | Breast and areola enlarged, no contour separation |
| 4 | Coarse, curly, abundant but amount less than in adult | Areola and papilla form secondary mound |
| 5 | Adult feminine triangle, spread to medial surface of thighs | Mature; nipple projects, areola part of general breast contour |

Adapted from TANNER, J.M.: Growth at Adolescence, ed. 2, pp. 32—37, Oxford, England: Blackwell Scientific Publications, 1962.

reached at 12.14 (S.D. ±0.8) years and occurred during breast and pubic hair stages 2 and 3, preceding menarche in all subjects (MARSHALL and TANNER, 1969). The age of menarche in low to middle income American girls in the southern United States was 12.25 (S.D. ±0.3 mo) years (DANIEL, unpublished).

## 1. Gonadal Function During Puberty

It is generally assumed that there is a functionally integrated increase in the activity of the hypothalamic-pituitary-gonadal axis during adolescence. Radioimmunoassays and competitive protein binding assays have increased the accuracy of hormonal measurements during recent years, but interpretation of data from these methods requires a high degree of caution. In certain circumstances bioactivity and immunoactivity have been dissociated (ROSS et al., 1972), and single serum specimens may be higher or lower than appropriate functional levels because of pulsatile patterns of hormonal secretion (ROOT et al., 1972).

### *a) Females*

The exact role of estrogens in the growth spurt of adolescent females is obscure. It has been proposed that estrogens inhibit the peripheral action of growth hormone while promoting increased growth hormone release in response to various stimuli (ROOT, 1972). Urinary excretions of estrone, estradiol, and estriol increase with age and sexual maturation in females (PENNINGTON and DEWHURST, 1969), and plasma estradiol concentrations correlate significantly with chronologic age, skeletal maturation, and stage of female adolescent development (JENNER et al., 1972).

Adrenal androgens stimulate the growth of pubic and axillary hair (adrenarche), increase linear growth rate, and advance epiphyseal maturation. The excretion of 17-ketosteroids, an index of androgen production, increases slowly with age and stage of sexual maturation, correlating significantly with height (SIZONENKO et al., 1972). Androsterone and etiocholanolone, metabolites of dehydroepiandrosterone (DHA), androstenedione, and other androgens are detectable in the urine of children by 4 years of age (PAULSEN et al., 1966). DHA is first found in the urine at 6–9 years with its excretion increasing rapidly after 10 years of age (TANNER and GUPTA, 1968; VESTERGAARD, 1965). In girls, plasma testosterone levels increase slightly while plasma concentrations of DHA, DHA-sulfate, androstenedione, and androstenediol increase several fold during puberty and are paralleled by changes in the values of estrone (SAEZ et al., 1972; FRASIER and HORTON, 1966; GANDY and PETERSON, 1968; FOREST and MIGEON, 1970; ROSENFIELD et al., 1972). DHA-sulfate is moderately androgenic and may be the most important adrenarchal androgen of the female (ROSENFIELD et al., 1972).

The mechanism by which increased production of adrenal androgens (adrenarche) occurs at puberty is unknown. It has been suggested that either an unidentified pituitary hormone with specific adrenal androgen-stimulating effects is secreted at puberty or that a gonadotropin directly stimulates adrenal androgen synthesis. Evidence in support of either hypothesis is lacking.

### *b) Males*

The physical changes of adolescence in the male are due to the secretion of testicular and, perhaps to a lesser extent, adrenal androgens. Androgens stimulate growth of the phallus and scrotum, and pubic, axillary, facial, and body hair. They also accelerate height velocity and skeletal maturation and increase the number and size of muscle cells. The testes secrete testosterone and dihydrotestosterone, the 5-$\alpha$-reduced product of testosterone that is metabolically active at cellular and nuclear levels. There is also a significant increase in adrenal androgen synthesis during male adolescence. Plasma concentrations of testosterone, DHA, DHA-sulfate, androstenedione, and androstenediol increase progressively from prepubertal to adult levels during sexual maturation (WINTER and FAIMAN, 1972a; GANDY and PETERSON, 1968).

The urinary excretion of total and individual 17-ketosteroids increases with age and stage of sexual maturation (WINTER and FAIMAN, 1972a; SIZONENKO et al., 1972; TANNER and GUPTA, 1968; VESTERGAARD, 1965; PAULSEN et al., 1966). The urinary excretion of testosterone derived from testicular and adrenal secretion and from peripheral conversion of precursors increases markedly from 12 years of age (GUPTA and BUTLER, 1969). Also with increasing age, there is a gradual increase in the urinary excretion of epitestosterone.

Urinary excretion of estrogens also increases progressively during maturation (WINTER and FAIMAN, 1972a), due either to secretion by the adrenals or testes or to the peripheral conversion of androgens to estrogens. Recent studies of spermatic vein blood indicate that the adult testes secrete estradiol and estrone, as well as testosterone (KELCH et al., 1972a).

## 2. The Onset of Puberty

The mechanisms that determine the time at which pubertal development begins remain obscure. Various data (KELCH et al., 1972b; PENNY et al., 1970; DONOVAN and VAN DER WERFF TEN BOSCH, 1965; GUYDA et al., 1969; WINTER and FAIMAN, 1972b; RAMIREZ and MCCANN, 1963; KULIN et al., 1969, 1972; NANKIN et al., 1971; KELCH et al., 1972c, d) suggest that puberty does not represent sudden activation of a previously dormant system, but the gradually increasing function of a system that has been continuously active from the in utero stage of development.

One hypothesis maintains that the sensitivity of the hypothalamic receptor sites that mediate the inhibitory (suppressive) effects of the gonadal sex steroids on gonadotropin release declines as the subject matures (ROOT, 1973). Consequently, larger amounts of gonadotropins are secreted until increased gonadal function reestablishes an equilibrium between pituitary and gonadal activities. Readjustment continues until the quantities of gonadal steroids required for hypothalamic suppression exceed the threshold sensitivity of the peripheral tissues to these materials, at which point the physical signs of puberty appear. Re-equilibration continues until the adult relationship between hypothalamic-pituitary and gonadal activities is established.

## 3. Pituitary Gonadotropin Function During Puberty

Several mechanisms are involved in regulating the secretion of luteinizing hormone (LH) and follicle-stimulating hormone (FSH) (ROOT, 1972). Testosterone, in males, inhibits the secretion of LH (FRANCHIMONT, 1971). One or more testicular factors are important in FSH secretion, although the mechanism is unclear. Serum levels of LH and FSH with normal LH values have been found in men with damaged seminiferous tubules (ROSEN and WEINTRAUB, 1971; VAN THIEL et al., 1972). Testosterone can depress serum FSH in males (LEE et al., 1972), but it is unknown whether this is a direct effect of testosterone or due to its conversion to estrogen. In addition to various hormonal relationships, physical or emotional stress, representing the cerebral cortex, is known to play a part in the control of gonadotropin secretion.

In males, FSH affects the late stages of spermatogenesis (STEINBERGER, 1971) and, earlier, may increase seminiferous tubular binding of testosterone (SEILICOVICH, 1972). Gonadotropins have been detected in the urine of very young boys (KULIN et al., 1967; FITSCHEN and CLAYTON, 1965), and concentrations increase with age. Maturation of the interstitial (Leydig) cells of the testes and secretion of testosterone are caused by LH. The onset of spermatogenesis may be due to LH, which leads to high intratesticular testosterone concentrations (STEINBERGER, 1971). Apparently LH may be present in male infants in detectable quantities that soon decline and remain low until about 6 years of age, then increase remarkably during puberty (BLIZZARD et al., 1970; RIFKIND et al., 1970; BUCKLER and CLAYTON, 1970). Other investigations have shown that serum concentrations of LH rise slowly from birth to about 8 years of age and then increase dramatically (BLIZZARD et al., 1970; WINTER and FAIMAN, 1972a; WEILAND et al., 1970; AUGUST et al., 1972; FAIMAN and WINTER, 1971; PENNY et al., 1970; LEE et al., 1970; YEN et al., 1969; YEN and VICIC, 1970; ROOT et al., 1970; BLIZZARD et al., 1972). In adult males, and perhaps also during adolescent maturation, tonic secretion of gonadotropins occurs in contrast to cyclical secretion in females (FLERKÓ, 1970; DONOVAN and VAN DER WERFF TEN BOSCH, 1965).

## III. Some Special Considerations in Children and Adolescents

In children, the effects of anabolic-androgenic steroids differ in several ways from those in adults. In addition to excessive skeletal maturation, which is discussed in a later section, the problems of virilization and hepatic dysfunction necessitate particular caution in the use of these agents in children and adolescents.

### A. Rate of Growth

Animal experiments have shown repeatedly that the greatest effects of androgens or anabolic steroids are produced in young animals, if the dosage-activity relationship is compared for different ages (KRUSKEMPER, 1968). In contrast to the anabolic effect in human adults leading to hypertrophy of extragenital organs and tissue, which is due to an increase in the protein content of the organism without raising the cell count, in children, anabolic steroids stimulate growth in the same fashion as the natural androgens do. The protein content of the individual cells is not increased in children, but growth is accelerated and most organs are involved in physiologic proportion. The only organ suffering considerable involution after administration of anabolic steroids is the thymus gland.

Acceleration of growth and additional nitrogen retention in response to androgens or anabolic steroids occurs even in healthy children who are in a state of anabolism corresponding to individual age. This is an important difference between children and adults, because in a healthy adult before the age of involution, it is hardly possible to force net protein gain (KRUSKEMPER, 1968).

Rate of growth has been one of the parameters used to assess the effects of anabolic steroids. While these hormones are known to bring about an increased rate of growth when used in children, there are certain situations in which the response is generally poor. Growth rate may not be increased at all in some patients, even if they respond with virilism and accelerated skeletal maturation (MELLMAN et al., 1961; REILLY and GORDON, 1961; HUBBLE and MACMILLAN, 1962; PRADER, 1961). Failure of therapy is most likely to occur in children with skeletal problems, primordial dwarfism, or profound hypopituitarism (PRADER, 1961; HUBBLE and MACMILLAN, 1962). This latter finding has been explained in recent years. Androgens are potent stimuli to the release of growth hormone in males with delayed sexual development (MARTIN et al., 1968; ZACHMANN and PRADER, 1970; ILLIG and PRADER, 1970). It appears that testosterone requires the presence of growth hormone in order to exert its full growth-promoting effect. Therefore, some patients in whom anabolic steroids have had no effect, either on growth or on bone maturation, may, in fact, lack growth hormone (GREENE and CARSTAIRS, 1973). Growth rate cannot be significantly accelerated also during the growth spurt of spontaneous puberty (MARTI-HENNEBERG et al., 1975).

### B. Virilization

Children are more sensitive than adults to the virilizing effects of anabolic steroids (PRADER, 1961; WYNN, 1968), an effect that is undesirable in boys before the age of puberty, and of course, in girls of any age.

It was noted in 1961 (BIERICH, 1961) that virilizing symptoms generally appear after 3–4 months if one uses nortestosterone phenylpropionate over long periods of time. The most regularly occurring signs are the growth of pubic hair as well as change in voice pitch. The voice becomes deeper and rougher. If steroid therapy is discontinued, the voice will usually regain its original pitch but if continued, the voice will break, eventually assuming an even, low quality. Other virilizing phenomena, which become disturbing in girls and indeed may force one to discontinue therapy, are an increase in the breadth of the thorax and shoulders, hirsutism, as well as coarsening of the face and the occurrence of acne. More marked virilization of the genitalia in girls is less common, although increased size of the clitoris has been associated with high doses (THOMAS and MAWWHINNEY, 1973). The possibility of the earlier appearance of spontaneous puberty must be considered when evaluating these changes, particularly in girls. Since the onset of pubertal changes is correlated with specific stages of bone maturation (bone age of 11 in girls and 13 in boys) (PRADER and ILLIG, 1962), if bone age is accelerated by anabolic steroids in prepubertal children, the onset of puberty may be an indirect effect (THOMAS and MAWWHINNEY, 1973). This should not be confused with the direct virilizing effects noted above.

In prepubertal boys, gynecomastia has been seen during anabolic-androgenic administration due to conversion of these agents to estrogens in the body (LARON, 1962). Increased frequency of penile erection has been reported (PRADER and ILLIG, 1962), and there have been some instances of priapism (THOMAS and MAWWHINNEY, 1973). Of greater concern in prepubertal boys, however, is decrease in testicular size, which may be the only evidence of virilism (SOBEL, 1968). Gonadal function is definitely depressed by high doses of anabolic steroids (PRADER, 1961). This effect, as well as decreased testicular size, was noted recently in children treated with oxandrolone (MARTI-HENNEBERG et al., 1975). Testicular volume was assessed by the testicular volume index (BURR et al., 1970), which decreased as long as steroid treatment was continued. Long-term follow-up after discontinuance of therapy has not been reported.

Adult sexual function may be altered when these agents are administered to postpubertal boys (FRASIER, 1973). Anabolic-androgen steroids inhibited pituitary LH secretion in a manner analogous to testosterone when used in animal experiments (BORIS and STEVENSON, 1970). Antigonadotropic effects have also been demonstrated in man. Testicular biopsies following treatment of seven volunteer male prisoners for 8–25 weeks with large doses of norethandrolone showed persistent abnormalities such as disorganization of the seminiferous tubules and peritubular fibrosis (HELLER et al., 1959). Since adverse effects on reproductive function may result from excessive administration of anabolic-androgenic agents in mature men, they may theoretically also interfere with normal testicular maturation in prepubertal and pubertal boys.

Although prepubertal children in general are more sensitive than adults to the virilizing effects of anabolic steroids, some children appear to be particularly predisposed to these problems, namely mentally retarded children and those with profound hypothyroidism (PRADER, 1961). Increased androgenic effects have been noted in hypogonadal subjects and also when glucocorticoid steroids are administered concurrently (WYNN, 1968). For reasons that are unknown, however, some normal individuals also are very sensitive to the androgenic effects of these compounds.

Table 3. Preparations used for their anabolic effect

| Generic name | Chemical name | Trade name | [a]Dose in children | Mode of Administration |
|---|---|---|---|---|
| Methan-drostenelone | $\Delta'$-17α-methyltestos-terone; methandienone | DIANABOL | 0.04 mg/kg/day | Oral |
| Nandrolone phenpro-pionate | 17β-phenylpropionate-estr-4-en-3-one | DURABOLIN | Infants: 12.5 mg every 2–4 weeks Children: 25 mg every 2–4 weeks | i.m. only |
| Stanozolol | 17β-hydroxy-17α-methylandrostano (3,2-c)-pyrazole | WINSTROL | 0.1 mg/kg/day | Oral |
| Oxymetholone | 17α-methyl-2-hydroxymethylene dihydrotestosterone | ADROYD ANADROL | 0.1–0.3 mg/kg/day | Oral |
| Oxandrolone | 17β-hydroxy-17-methyl-2-oxa-5α-androstan-3-one | ANAVAR | 0.05–0.1 mg/kg/day | Oral |

[a] Doses recommended are from KEMPE, C.H. et al.: Current Pediatric Diagnosis and Treatment. Los Altos, California: Lange Medical Publications, 1974.

A very important factor is that the virilizing effect of these hormones is dosage- and time-dependent (PRADER, 1961; PRADER and ILLIG, 1962). Consequently there seems to be a range of low doses, which in the majority of cases, still stimulates growth without undue virilism, which has determined the dosages recommended (Table 3).

A large number of steroids have been developed in an attempt to separate anabolic activity from androgenicity. Although many such agents have a relatively high anabolic-androgenic ratio, it has not been possible to completely separate these two activities in any of them (BORIS and STEVENSON, 1970; MOLDAWER, 1968).

## C. Liver Problems

Abnormalities of liver function may occur with oral anabolic agents, particularly those with 17-alpha-alkyl side groups on the testosterone molecule, and these occur at any age. For example, the serum levels of serum glutamic-oxaloacetic transaminase (SGOT), serum glutamic-pyruvic transaminase (SGPT), and lactic acid dehydrogenase are raised, and the retention of sulfobromophthalein is increased (FRASIER, 1973). Methyltestosterone, one of the first orally effective testosterone derivatives, may produce cholestatic jaundice.

Of very real concern to physicians using anabolic-androgenic steroids in children and adolescents, however, are reports of the development of hepatic carcinoma. This association was hinted at in 1971 (BERNSTEIN et al., 1971) when oxymetholone was thought to have contributed to the development of hepatocellular carcinoma in a patient with Fanconi's anemia. Twelve cases of primary liver neoplasia have now been reported in patients under treatment with androgens (GUY and AUSLANDER, 1973; ZIEGENFUSS and CARABASI, 1973; HENDERSON et al., 1973; MEADOWS et al., 1974), the majority being patients with aplastic anemia.

For this reason, one must be particularly concerned about the widespread use of these drugs by athletes, although liver neoplasia has not yet been reported in this specific group (FRASIER, 1973). The importance of careful abdominal palpation at each examination for the earliest possible detection of hepatic masses must be stressed. In several cases reported to date (JOHNSON et al., 1972; FARRELL et al., 1975), the tumors regressed when administration of the drug was discontinued, thus adding weight to the suggestion that tumors are causally related to androgen administration.

Other untoward effects that have been noted with the use of anabolic-androgenic steroids include retention of fluid or edema, and precordial pain (THOMAS and MAWWHINNEY, 1973). Some of the modified androgens can produce creatinuria.

## IV. Clinical Uses

### Growth Retardation

Children and adolescents with growth retardation probably constitute the most important group in which anabolic-androgenic steroids are used. The aims of therapy in this situation are to increase growth and increase predicted height. In most instances, apart from short stature, the patient will be otherwise normal, so there is greater concern on the part of the clinician that untoward effects, such as virilization and bone maturation, be kept to a minimum in comparison to certain other conditions.

#### 1. Premature Dystrophic Infants

Low-birth-weight neonates constitute a high risk group, and methods for reducing mortality or morbidity and increasing growth, have received much consideration. Early investigations sought hormones that might stimulate growth of these, or early-for-date, infants, but poor results were obtained in trials of administering chorionic gonadotropin (GIUFFRIDA, 1936), thyroid products (MONCRIEFF, 1938), or estrogens (MARTIN, 1929; POTTER, 1937; EINHORN, 1940). Testosterone compounds presented more promising possibilities in several early investigations (SHELTON and VARDEN, 1946; SHELTON et al., 1947; EARLE, 1950), but later studies showed this steroid to be of no value in promoting growth in premature infants (HARDY and WILKINS, 1949; SEITCHIK and AGERTY, 1950; REILLY and EARLE, 1951).

Anabolic steroids were then used in the search for growth and weight stimulants, and results have differed. Norethandrolone was used in 13 premature infants and compared with 8 control infants, but there was no significant influence on weight (MEADOWS et al., 1960). Stanozolol was used in a study of 33 premature infants with 15 control infants (YADAO et al., 1965), but all of these subjects had passed through the first critical month of weight loss and possible infection and had already begun to gain weight. In this investigation, all infants gained weight.

The use of weight gain as the sole criterion of efficacy in assessment of these medications has been criticized, because weight is only one parameter of the physical growth of low-birth-weight infants. Weight gain may be fallacious, for in one study (SNYDERMAN and HOLT, 1961), linear growth of the subjects was not affected. As measurement of body length is not a reliable and sensitive procedure in premature infants, one group (LAMBRECHTS and JADOUL, 1961) used the radiographic technique of DAY and SILVERMAN (1952) to measure the length of the fibula under uniform conditions. Nandrolone phenylproprionate was used in this investigation of 19 infants with birth weights of 1300–2100 g, and both weight and linear growth of the fibulas were increased, leading to the conclusion that this substance was a stimulant for growth. The effect was more pronounced in low-birth-weight infants and was absent when the weight was 2000 g or more.

Current treatment of low-birth-weight infants is based upon adequate nutrition and not the use of anabolic steroids (BRANS et al., 1974).

## 2. Growth Retardation in Older Children

Growth retardation is present in an infant or child if the height is more than two standard deviations below the mean for chronologic age, or a marked decrease in growth velocity is present (ROOT et al., 1971). Physicians are frequently called upon to evaluate real or imagined growth retardation during childhood and/or adolescence.

An extremely sensitive indicator of a child's health is his linear growth, for it is affected by many conditions (ROOT et al., 1971; JOSS and ZUPPINGER, 1973). Preliminary evaluation of such a patient requires a careful history and physical examination. Records of previous height and weight, as well as roentgenographic studies of skeletal maturation, are often of great value. If growth velocity has been slow and bone age is significantly retarded, it is necessary to determine if endocrinologic abnormalities exist or if chronic systemic illness is present (ROOT et al., 1971).

In the evaluation of patients who exhibit growth retardation, although a remediable cause may be found, most often no specific etiology can be determined and the children are classified as having constitutional delayed growth and/or adolescence (ILLIG, 1974), or primordial dwarfism (WILKINS, 1965). The former condition should be regarded as a variant of the normal pattern of growth and development, in which there may be inborn constitutional differences in the activities or rates of maturation of the hypothalamic-pituitary-gonadal axis. These children ultimately attain normal adult height. Primordial dwarfism implies that there are unknown factors causing poor growth.

The literature on the treatment of growth retardation with anabolic-androgenic steroids is voluminous, and sorting out the mass of data is difficult because of several factors. As so frequently happens, initial trials may seem to yield favorable results only to be unsubstantiated by later investigations or a longer period of observation of the same group of patients. Often it is impossible to compare statistically patients, doses of medication, analytic criteria, or the experimental designs (SOBEL, 1968).

### *a) Assessment of Skeletal Maturation*

The precision of the method of GREULICH and PYLE (1959) for assessing skeletal maturation has been questioned (SCHOEN et al., 1970), and it is thought that, at best, an error of 3 to 6 months skeletal age is to be expected (MARSHALL, 1966). Despite these claims, this imprecise method remains one of the best currently available. Since there is sometimes considerable variation in the bone age in different epiphyseal centers, however, other investigators believe it may be preferable to obtain roentgenograms of all centers rather than to make an evaluation from a single area such as the hand and wrist. Other tables have been compiled for estimating the bone age on the basis of the time of ossification of various cartilages (HOERR et al., 1962; PYLE and HOERR, 1955;TODD, 1937).

*b) Prediction of Mature Height*

The ultimate height attained by an individual depends not only upon the rate of growth, but also upon the duration of growth as determined by the time of epiphyseal fusions. The cause of the fusion and the relative roles of the various hormones are not clearly understood (SIMPSON et al., 1950; WILKINS, 1965).

Predicted mature height, based on the tables of BAYLEY and PINNEAU (1952), has frequently been used to assess the effect of anabolic steroids. These tables were based on 192 normal school children who were measured and x-rayed every 6 months, but the tables may not be adequate for marked deviations of bone age (GREENE and CARSTAIRS, 1973). Prediction becomes more reliable as age advances, even for normal children (BAYER and BAYLEY, 1963), but because prediction is affected by the skeletal age of the patient as well as by differences between skeletal and chronologic ages, the imprecision of the former in growth-retarded children causes reliability in this group to be suspect. It is even more difficult to evaluate the effects of genetic and environmental factors on the precision of mathematical predications of mature height (SOBEL, 1968). Recently, a follow-up study demonstrated conclusively that height predictions overestimated actual achieved final heights (KAPLAN et al., 1973).

*c) Experimental Design*

Differences in experimental design account for some discrepancies in published results. For example, increase in predicted mature height is more often reported when the estimate is made at the end of therapy (REILLY and GORDON, 1961; GELLER, 1964; FALLIERS et al., 1963) than after a follow-up period, when it may actually be decreased (GELLER, 1964; FALLIERS et al., 1963). Favorable reports of accelerated growth without increased skeletal maturation are usually based upon results at the end of a course of treatment (MELLMAN et al., 1961; DANOWSKI et al., 1965), but continued observation following cessation of treatment has often required revision of the earlier conclusions (MELLMAN et al., 1961; GELLER, 1964; RAY et al., 1965). The explanation for these differing results is that the full effect of anabolic steroids on skeletal maturation may not be evident until several months after termination of treatment (DANOWSKI et al., 1965; RAY et al., 1965), for the effect on bone usually continues for several months (PRADER, 1961; PRADER and ILLIG, 1962).

*d) Linear Growth Versus Skeletal Maturation*

Excessive skeletal maturation has been encountered in children with endogenous overproduction of androgenic steroids, such as adrenogenital syndrome and other types of male precocity. Closure of the epiphyses occurs prematurely in these conditions with ultimate stunting of growth.

As early as 1941, skeletal maturation was noted to advance rapidly in a boy whose rate of growth was accelerated by methyltestosterone (McCULLAGH and ROSSMILLER, 1941), and this phenomenon was later observed in children who received much smaller doses (DEAMER, 1948). It was subsequently found, however, that use of a combination of testosterone and thyroid hormone increased growth without undue advance in bone age (ESCAMILLA and BENNETT, 1951). At that time, therefore, a debate concerning linear growth versus skeletal maturation was well underway. A critical investigation tackling this fundamental problem showed that methyltestosterone stimulated both linear growth and skeletal maturation and that larger doses produced more rapid epiphyseal fusion, but no greater growth, than did 5 mg per day (SOBEL et al., 1956). In other studies, dangers have been emphasized by some workers (SOBEL et al., 1956; PRADER, 1961), with favorable responses being reported by others (BAYLEY et al., 1957; HORTLING and WAHFORS, 1959).

A more recent follow-up study on treated and untreated patients has been enlightening (KAPLAN et al., 1973). The effects of three androgens on final adult stature was studied in 19 male children with constitutional delay in growth and/or development and compared with 21 untreated control subjects. In all patients, the bone age was retarded. Methyltestosterone, fluoxymesterone, or testosterone enanthate was administered in the treatment groups. Low doses of all three androgens were used, and the average duration of treatment was 5.3 months. Predictions of final height overestimated actual achieved final heights by an average of 1.3 inches in the control group. Androgen therapy did not affect ultimate height adversely and, in fact, treated patients achieved an average final height much closer to the average predicted height than the control group.

Differences of opinion have been expressed also concerning the newer anabolic steroids. Reports concerning fluoxymesterone have been both favorable (REILLY and GORDON, 1961) and critical (MELLMAN et al., 1961), but in these studies, different doses and study designs were employed. Methandienone has been used in several investigations, again with conflicting results (FOSS, 1960; HUBBLE and MACMILLAN, 1962; GREENE and CARSTAIRS, 1973). Similarly, oxandrolone has been subjected to evaluation in cases of growth retardation (RAY et al., 1965; VISSER and CROUGHS, 1966; ZANGENEH and STEINER, 1967; GELLER, 1968; LIMBECK et al., 1971; BETTMAN et al., 1971; MARTI-HENNEBERG et al., 1975). This drug, more often than not, has been observed to favor growth over epiphyseal maturation.

### *e) Recommendations for Treatment*

In the treatment of patients with constitutional delay of growth and/or development, the following guidelines are generally accepted: Treatment should not begin until the child is approaching a physiologic age appropriate for puberty (KAPLAN et al., 1973). Those patients with bone ages greater than 9 years seemed to fare better in an earlier study, although the post-treatment follow-up was only 6–12 months and not final stature (BETTMAN et al., 1971). It is recommended that therapy should not be commenced in patients at a chronologic age less than 12 years or at a bone age less than 10 years.

Minimal doses should be given over a short period of time (Table 3). In a large series (FOSS, 1965), the ultimate height of children with both retarded and normal bone ages was not affected by moderate doses (less than 30 mg per day) of methyltestosterone used over a short period of time. A third group treated intensively with larger doses for a prolonged period, however, did have a loss of ultimate height. The recommendation is, therefore, that minimal doses be used for less than 6 months, with no subsequent treatment for at least 6 months after therapy has been discontinued. Treatment can be recommended after reevaluation of growth progress and skeletal maturation (KAPLAN et al., 1973).

In general, there appears to be little to choose between the individual drugs. None of the hormones used in Kaplan's study affected ultimate height adversely. If the above recommendations are followed, excessive virilization should not occur with any of the commonly used anabolic agents.

The patient's emotional state will often determine the decision to use pharmacologic intervention. One must not forget that constitutional delay of growth and development by definition indicates eventual normal maturity and height. Consequently, a more appropriate form of intervention may be formal psychotherapy and family counseling, and use of anabolic-androgenic steroids should be reserved for those patients with severe psychologic difficulties related to their short stature or delayed development.

### 3. Hypopituitarism

Although hypopituitarism is a relatively rare cause of growth retardation, it is being recognized with increasing frequency because specific diagnostic procedures and therapy are now available (WILKINS, 1965; COSTELLO et al., 1968). The plasma growth hormone response to insulin-induced hypoglycemia and to infusion of certain amino acids can be measured by sensitive radioimmunoassay techniques.

Anabolic-androgenic steroids have been useful in hypopituitarism, and certainly this will remain so as long as human growth hormone is not readily available. Optimal therapy for this condition consists of a combination of anabolic steroids, human growth hormone, thyroxine, and adrenocorticoid hormones. It would, of course, be impossible to achieve normal adult size by treatment with androgens or anabolic steroids alone, as only one of the missing factors would be replaced (KRUSKEMPER, 1968).

A general rule is to prescribe anabolic steroids in minimally active dosage for male patients before their 15th year, and after age 15, to prescribe testosterone in increasing virilizing dosages until all secondary sexual characteristics have appeared (LYSTER and DUNCAN, 1963). The rationale for not using androgens until the age of puberty is that by so doing, the time of sexual development will be appropriate to age, and one may also achieve a maximal increase in height (ASTWOOD, 1970). Sometimes, use of androgens is deferred even longer with this aim in view.

In boys, the same schedule of treatment as that for hypogonadism has been used, and acceleration of growth has been recorded that is as rapid as in individuals with normal pituitary glands. Male patients may be given testosterone enanthate, 200–300 mg intramuscularly, once monthly, or methyltestosterone, 10 mg by mouth 2–3 times daily (BRASEL and BLIZZARD, 1968) (Table 4).

In girls, large doses of androgen cannot be used, and treatment with estrogen induces only a slight acceleration of growth. A substantial spurt in growth was

Table 4. Some preparations of androgenic steroids and suggested doses for replacement therapy in hypogonadism

| Compound | Average replacement therapy[a] | | Mode of administration |
|---|---|---|---|
| | Dose (mg) | Frequency of administration | |
| Methyltestosterone | 25 | Daily | Buccal |
| Fluoxymesterone | 2 | Daily | Oral |
| Testosterone propionate in oil | 25 | Twice/week | Intramuscular injection |
| Testosterone enanthate | 200 | Once/month | Intramuscular injection |
| Testosterone pellets | 400 | Once/6 months | Subcutaneous implantation |

Adapted from table in DILLON, R. S.: Handbook of Endocrinology-Diagnosis and Management of Endocrine and Metabolic Disorders, p. 487, Philadelphia: Lea and Febiger, 1973.
[a] At the beginning of treatment and in other conditions, larger doses may be required.

achieved (MARTIN and WILKINS, 1958) by using 10–20 mg of methyltestosterone with 1 mg of diethylstilbestrol per day. Virilization did not occur and feminine sexual development was reported to be normal, including normal development of pubic and axillary hair. Fluoxymesterone, 2.5 mg orally once or twice daily, has also been used (BRASEL and BLIZZARD, 1968) and may be preferred in that it is more anabolic and less virilizing than methyltestosterone.

## V. Problems of Sexual Maturation

### A. Delayed Sexual Maturation

During the preadolescent years it is impossible to predict which members of a group of children will mature sexually at the average time, which will have a delayed but normal maturation, or to identify those who may remain sexually immature as adults. Delayed adolescence occurs in about 25 of 1000 children (ILLIG, 1974). In boys, particularly, there is frequent complaint of lack of growth and sexual development if they are left behind by contemporaries who have begun the pubertal growth spurt. Often parents' concern over a son's developmental lag evidenced by the short height, lack of muscular development, and continued psychic immaturity, cause them to seek help in coping with the delay. Most of these boys eventually mature in a normal manner, although later than usual. This fact in itself is evidence that the process represents delayed activation or continuing inhibition of the hypothalamic-pituitary-gonadal-adrenal system. Reasons for the delay in pubertal change are unclear.

Delayed sexual maturation may be assumed to be present in a boy if testicular enlargement, the first clinical indication of the onset of puberty, has not begun by 13.5 years or if 5 years have elapsed after the onset and before completion of genital growth (MARSHALL and TANNER, 1970). In girls, the first signs of breast enlargement should occur by age 13 and delayed sexual maturation is present if 5 years have passed between the onset of breast development and menarche (MARSHALL and TANNER, 1969).

One must seek to exclude organic causes of hypogonadism. Primary (hypergonadotropic) hypogonadism refers to developmental abnormalities such as anorchia or

Klinefelter's syndrome, and includes orchitis, trauma, and neoplasm. In secondary (hypogonadotropic) hypogonadism there may rarely be an isolated gonadotropin deficiency, but more commonly it results from failure of pituitary elaboration of the gonadotropic hormones, specifically interstitial cell-stimulating hormone and FSH.

The diagnosis is arrived at via a workup similar to that for short stature per se (ILLIG, 1974). A carefully taken family and personal history, a thorough examination of the patient, the sex maturity rating of puberty (TANNER, 1962), analysis of the patient's growth curve, and roentgenograms of the hand and wrist for bone age, usually determine the need for other investigative procedures to exclude known metabolic, organic, or endocrine causes of delayed puberty and/or small stature. In some patients with extreme growth retardation or suggestive findings of hypopituitarism, growth hormone determinations may be necessary.

The determination of plasma luteinizing hormone (LH) and follicle-stimulating hormone (FSH) after stimulation with luteinizing hormone-releasing hormone (LH-RH) has proved to be of value in differentiating the various causes of delayed puberty (ILLIG, 1974; KULIN and REITER, 1972). Patients with isolated gonadotropin deficiency evince no release of LH and FSH under stimulation. Children with delayed adolescence show a normal rise of LH and FSH, while patients with primary gonadal failure such as anorchia or Klinefelter's syndrome demonstrate an excessive increase of LH and FSH in response to LH-RH (ILLIG, 1974). Even without stimulation, serum gonadotropin concentrations are almost always above normal values in children with primary hypogonadism by 10 years of age (ANDERSON, 1953; WINTER and FAIMAN, 1972b).

As soon as the diagnosis of delayed adolescence is established, regardless of classification, the question of hormone treatment arises. Since the inadequate development of secondary sex characteristics and the small stature of patients with constitutionally delayed puberty are associated with the extremely late onset and slow course of their normal maturation, there is no somatic need for a hormonal induction of virilization and the pubertal growth spurt at this time. Usually reassurances and frequent examinations suffice until anticipated sexual changes appear. However, occasionally peer reaction to a boy's childlike somatotype will cause psychological difficulties and he, his parents and the physician may agree that developmental stimulus is of value.

In this situation, or after having diagnosed a primary gonadal disorder in a boy, androgen replacement may be started at the time of estimated onset of puberty. Methyltestosterone, administered sublingually, may be given 10–20 mg daily for 3–6 months. Fluoxymesterone in a 10–20 mg dose daily has also been used. It should be noted that some experienced observers have found that the androgenic response is limited and that even larger doses may fail to bring about complete sexual development (MCCULLAGH and SCHIFFENBERG, 1954; WILKINS, 1965; PAULSEN, 1968). Where there is complete testicular failure and puberty cannot occur, vigorous and prolonged therapy is required. One of the long-acting esters of testosterone, such as testosterone enanthate, may be given intramuscularly, in a dose of 200 mg at monthly intervals for 3–6 months (Table 4). The action of testosterone when administered orally or by injection is only short-lived. It can, however, be made effective for 6 months by inserting pellets under the skin, obviously with strict aseptic technique.

The dangers of testosterone administration should be appreciated. Frequent physical examination to ascertain genital maturation must be performed, and when there has been a significant increase in size of the testes, androgen should be discontinued, for this change usually indicates functioning of the patient's hypothalamic-pituitary-gonadal axis. Because of the possibility of greater rate of skeletal maturation than increase in linear growth velocity, testosterone should not be prescribed for boys with limited growth potential. A period of time without androgens equal to that of the duration of therapy should be allowed to elapse before evaluating the effects of testosterone administration. The danger of developing hepatocellular carcinoma while receiving androgens has been alluded to, and must be considered a possibility in these patients.

### 1. Klinefelter's Syndrome

In 1942 KLINEFELTER et al. described a syndrome characterized by small testes, azoospermia, testicular hyalinization with fibrosis, and gynecomastia. Since that time various reports have been concerned with this diagnosis, psychologic problems of patients with the syndrome, and possible use of androgens in the management of such patients.

The incidence of Klinefelter's syndrome in the general population is uncertain but is reported to be approximately 0.25% (SOHVAL, 1969). Definitive diagnosis rests upon chromosomal patterns. Because the syndrome is often associated with mental retardation, investigations have been made in institutions, and the reported incidence varies from 0.4–2.38 (PAULSEN et al., 1964). Evaluation of patients on a neurological ward showed 1% to have Klinefelter's syndrome (NIELSEN, 1965; NIELSEN and FISCHER, 1965), whereas 8.5% of all males attending fertility clinics were reported to have positive chromosomal findings. Undoubtedly the incidence is higher in the general population than suspected.

Since Klinefelter's original description, many other clinical features have been described (MYHRE et al., 1970; KIVOWITZ and CORCORAN, 1971; GIORGI and SOMMERVILLE, 1963; JOHNSON et al., 1970; NIELSEN et al., 1970; NIELSEN, 1970), and patients have had extensive psychologic evaluation. The most consistent clinical finding is the presence of small, atrophic testes usually reported to be firm to palpation but which, in young adolescents, have been soft, at least in our experience. Phallic development is variable from hypogenitalism (GIORGI and SOMMERVILLE, 1963) to normal adult size. Although typically many patients with Klinefelter's syndrome have eunuchoid somatotype, that is, tall and with a female distribution of fat, the condition does occur in boys who are shorter than average and who are lean. Many of these patients are mentally retarded, often mildly so, whereas others do not show intellectual deficit.

The most common cytologic finding in Klinefelter's syndrome is 47 chromosomes with an XXY pattern (SOHVAL, 1969), but as many as 24 chromosomal varieties have been found. A constant feature is a positive sex chromatin pattern, the Barr body, as determined by buccal smear. Males having XXY pattern are sterile as are those with almost all of the other varieties, although rare instances have been reported in which patients had mature sperm.

Diagnosis of the syndrome is usually suggested by the small soft testes in adolescents regardless of height, fat distribution of a female type, or gynecomastia, which are frequent characteristics of Klinefelter's syndrome, and in this situation a buccal smear should be examined for Barr bodies. If this test is positive, then chromosomal typing should be performed for definitive classification.

Few hormone assay studies have been performed on males having Klinefelter's syndrome, and almost all available data represent studies involving adult men rather than pre- or adolescent boys. Urinary gonadotropin levels were typically high (GIORGI and SOMMERVILLE, 1963; KLINEFELTER, 1958) and urinary neutral 17-ketosteroids were low, as were pregnanediol and estrone (GIORGI and SOMMERVILLE, 1963). The cause of the gynecomastia that often occurs is uncertain, but this is not thought to be due to production of estrogen (WILKINS, 1965).

Psychologically, boys with Klinefelter's syndrome are said to be more passive and more compliant as infants. As they develop, they are often regarded as being different from siblings, more dependent and immature, and as having difficulties in interpersonal relationships, particularly at school (NIELSEN et al., 1970). These psychological symptoms usually persist into and perhaps exacerbate with adolescence. Often there is antisocial behavior, weak sexual drive, or psychosexual confusion in adolescents having Klinefelter's syndrome (COURT-BROWN, 1962). Pubertal development is frequently not consistent with peers, and this often leads to difficulty.

The management of these boys, child or adolescent, has two significant parts: emotional and physical. In the preadolescent boy, psychological management is predominant and hopefully can alter some of the abnormal self-concepts, difficulties in interpersonal relationships, and relieve much of the dependent-passivity. As adolescence approaches, assuming the diagnosis has been made previous to or at the beginning of puberty, not only psychological therapy, but also hormonal treatment must be considered.

Five institutionalized males with Klinefelter's syndrome, aged 9–24 years, were studied and treated with 200 mg of testosterone enanthate given once every 4 weeks for a period of 7 months (MYHRE et al., 1970). The treatment period was followed by placebo injections for 5 more months. The values of urinary 17-hydroxycorticosteroids, 17-ketosteroids, and pregnanetriol were normal throughout the study. There was, however, significant reduction of plasma 17-hydroxycorticosteroids and 17-hydroxycorticosteroid response to adrenocorticotropic hormone (ACTH) at the end of the treatment. Bone age advanced in all subjects. Physical changes associated with testosterone therapy were described as a more masculine body contour due to increase in muscle mass and redistribution of subcutaneous fat, advancement of secondary sexual characteristics, particularly an increase in phallic size, and the development of a more masculine body hair distribution. Gynecomastia is said to have decreased in two of the three subjects who had the condition prior to treatment. Following treatment, the physical changes were retained for at least 5 months, although one of the three patients had a recurrence of gynecomastia.

During adolescence, with its many and varied physical changes at the beginning, which are often confusing to a boy, an idealized body image is usually present. With further growth, association with peers, and self-evaluation, a realistic body image emerges. Following this, there is development of a concept of self that provides the gyroscopic stability necessary for attaining normal psychological development of adolescence. Because of the frequency of atypical male appearance in adolescents having Klinefelter's syndrome, there is often a disturbance of body image with the lack of development of a realistic concept. The small group of patients referred to above was evaluated psychologically before, during, and after testosterone therapy (JOHNSON et al., 1970) with the Draw-A-Person test (DAP). The first figures drawn showed a sexually undifferentiated, sexually immature, or feminine body image, whereas after 7 months of treatment with testosterone enanthate, the DAP showed an unexpected rapid shift in body image and the figures were masculine or more masculine. A final DAP test given 5 months after discontinuance of hormonal therapy showed that the change in body image was not maintained in the majority of subjects, although it had been in one patient. Prior to receiving testosterone these mildly retarded patients were said to have been cooperative, friendly, and passive, but during therapy, they displayed a marked change in their personalities and became more assertive, alert, and outgoing. They also were reported to have shown an increase in goal-directed behavior and, following treatment, approximately 50% of them were said to be more interested in females.

In one of our patients with Klinefelter's syndrome, the original complaints were that he had no real friends, felt shy, and often felt afraid of other boys at school. This 16-year-old patient's height was 190 cm (6 ft 4 in) and he weighed 82 kg (180 lbs). He was not mentally retarded and his scholastic performance was average. He stated that he felt lonely, that "everyone tells me what to do and nobody likes me." He originally received 25 mg of methyltestosterone orally each day for 1 month, and then the dose was decreased to 10 mg daily. During the first 3 months of therapy he became much more aggressive and irritable, his sex drive increased as did frequency of masturbation. He had fantasies concerning girls and a desire to exhibit his genitalia. He received psychological counseling at weekly intervals and developed a more appropriate appreciation of his new feelings and reactions. Medication was continued for 6 months at the end of which time his physical appearance was more masculine. He was interviewed by the psychologist at 3- and 6-month periods after cessation of therapy, and he continued to be interested in girls, with lesser sexual drive.

Although little data concerning the use of testosterone preparations in adolescent patients with Klinefelter's syndrome are available, it would seem that individualization is required before therapy is begun. From the standpoint of growth and psychological development, treatment should be instituted at an appropriate age for pubertal development to be in keeping with peer development; also, it is likely that the response would be of greater benefit than in an adult patient. Whether testosterone should be given orally or by injection cannot be evaluated by available data and no particular dose can be recommended at this time. Probably the response of the patient to the dose, both physically and psychologically, should be used as a guide. In preadolescent patients, psychological attempts should be made to decrease passivity, to develop better interpersonal relationships, and to increase understanding in parents to bring about a more normal development. There are often difficulties at school relating to other children, and frequently young boys with Klinefelter's syndrome are referred to psychiatrists or psychologists. It is hoped that early diagnosis and appropriate preventive psychological treatment may be of value and, at the beginning of adolescence, that testosterone administration would be considered.

### 2. Gonadal Dysgenesis (Turner's Syndrome)

Turner's syndrome is a special form of congenital ovarian hypoplasia in which usually a fibrotic thread or cord in the broad ligament represents the ovary. Many clinical findings have been reported including short stature, high, arched palate, small mandible, prominent ears, webbed and/or short neck, low posterior hairline, broad chest with apparently widely spaced nipples, cubitus valgus, short metacarpals, congenital peripheral lymphedema, renal anomalies, coarctation of the aorta, hypertension, pigmented nevi, and osteoporosis. Rarely do patients have such a collection of findings, and usually short stature in the second decade causes one to suspect the condition and seek diagnostic tests. Axillary and pubic hair may be present, even in adult quantity, although their development is usually delayed.

The retarded growth in these individuals is anomalous. This is evident from the fact that removal or destruction of the gonads in a prepubertal child will invariably result in a considerable delay in epiphyseal closure. Consequently, when there is a decrease or absence of ovarian function, there will be an increase in growth of the long bones resulting in a tall eunuchoidal stature due in part to the delay in epiphyseal fusion (KUPPERMAN, 1963). Patients with gonadal dysgenesis, however, invariably have a normal bone age and fail to show any eunuchoidal characteristics or marked delay in epiphyseal fusion.

Examination of nuclei in smears of the buccal mucosa, skin biopsies, or leukocytes have shown chromatin-negative (male pattern) nuclear sex in approximately 80% of the patients. Only 45 chromosomes and XO sex chromosome pattern compose the most common karyotype, although mosaic compositions have been described in a significant minority of patients.

Although ovarian dysgenesis may be diagnosed in the neonatal period, it is usually discovered in late childhood or adolescence. The short stature often leads to diagnostic procedures, but frequently the condition is unrecognized until the short female fails to develop secondary sexual characteristics or attain menarche. Obviously any girl of pubertal age who is less than 150 cm (5 ft) in height and has delayed menarche, should be suspected of having Turner's syndrome, and sex chromatin determination from buccal or vaginal smear is indicated (FINLEY and FINLEY, 1970).

Sex hormones are known to affect somatic growth and influence serum concentration of gonadotropins and specific hormone-binding proteins, in addition to their influences on secondary sexual characteristics. Elevated gonadotropin levels are present during the pubertal years in these patients due to the gonadal defect (CUNNINGHAM, 1970). Response to single injections of depot-estradiol or depot-testosterone administered to hypogonadal female teenagers has been measured (ROSENFIELD et al., 1973). This study indicated that a physiologic level of plasma estradiol promptly decreased gonadotropin release without abolishing the variations in LH levels, exerted greater suppressive effect on FSH than upon LH release, and caused persistent inhibition of FSH and LH secretion after the estradiol had fallen from suppressive levels. Slightly increased testosterone-estradiol-binding globulin capacity without inducing a change in thyroxine-binding globulin capacity also occurred, and somatic as well as sexual pubertal growth was produced. This investigation also showed that plasma concentrations of testosterone produced by a single parenteral administration (10 mg of testosterone cyclopentylpropionate) of depot-testosterone did not consistently affect FSH and LH levels and caused very slight depression of testosterone-estradiol-binding globulin.

Objectives of treatment of gonadal dysgenesis are increased height, sexual maturation, and feminization. The sterility of the patient cannot be corrected. Various estrogen preparations have been used to bring about development of the breasts, increased pubic hair, and menstruation, and the recommended time of commencement should be when the last normal classmates start their development (PERHEENTUPA et al., 1974).

Administration of 1.5–2.0 mg of estradiol cyclopentylpropionate (depot-estradiol) each month brings about menstrual bleeding within 3–6 months with increased feminine secondary sexual characteristics (ROSENFIELD et al., 1973).

There is still disagreement as to the value of anabolic-androgenic agents in the therapeutic regimen. Poor responsiveness to anabolic steroids has been reported by some workers (GREENBLATT et al., 1970; PRADER and ILLIG, 1962). Others have found, however, that anabolic steroids have proved to be an effective means of achieving enchanced growth in some of these children (KUPPERMAN, 1963; PERHEENTUPA et al., 1974).

A study of 26 girls having ovarian dysgenesis who received anabolic therapy (fluoxymesterone, 0.1 mg/kg daily for 0.7–3 years) in addition to estrogen indicated no effect on the growth velocity (PERHEENTUPA et al., 1974). Acceleration of growth during puberty is essential for adolescent psychological development and, in many of the girls in this investigation, it was believed that acceleration of growth was achieved with fluoxymesterone. Anabolic treatment was thought to have had a distinctly beneficial psychological influence due to the experience of growing at a time when the patients believed they would not grow any further, and this led to an integration of body image and increased sexual identity. These aspects of treatment, coupled with estrogen substitution and the appearance of menstrual bleeding, brought about rapid emotional improvement for older girls; in younger girls, psychological difference from peers gradually diminished and a normal self-consciousness appeared.

In two cases of Turner's syndrome treated with methandienone (HUBBLE and MACMILLAN, 1962), marked virilization occurred despite the small doses used, and greatly increased skeletal maturation accompanied growth stimulation in both cases. It is noteworthy that many of these patients show evidence of clitoral enlargement as part of their syndrome (KUPPERMAN, 1963). This structure should be checked therefore at regular intervals to establish, if possible, that the development of the clitoris is not due to the therapy, but is merely an expression of the congenital nature of the syndrome itself.

## B. Precocious Puberty

A child is said to be sexually precocious if pubertal sex characteristics appear before 9.5 years of age in a boy (MARSHALL and TANNER, 1970) or before age 8 years in a girl (MARSHALL and TANNER, 1969). Isosexual precocity refers to sexual development appropriate for the phenotype of the child, and heterosexual precocity for instances in which sexual development is inappropriate for the child's phenotype.

Although there is premature functioning of the hypothalamic-pituitary mechanism in isosexual precocious puberty, the reason for such activation is unclear.

There are many classifications of precocious sexual development but, in general, one may consider the changes as being due to early secretion of gonadotropin or to secretion of sex hormones by the gonads or adrenal glands as constituting two broad functional groups. Clinical classification may be based on whether the precocious sexual development is appropriate, that is, isosexual, for the phenotype of the child or inappropriate, that is, heterosexual.

Premature activation of the hypothalamic-pituitary axis is assumed to bring about isosexual precocity. The incidence of this type of precocious development varies greatly in reported investigations, depending upon the particular population evaluated, but it is estimated that 85% of girls and 66% of boys have an unknown etiology (WILKINS, 1965). True sexual precocity may be associated with premature hypothalamic function due to congenital anomalies or damage to the central nervous system (MYHRE et al., 1970), but neurologic abnormalities have not been found in boys with familial isosexual precocity (BEAS et al., 1962). In boys with nonfamilial isosexual precocity one should exclude the presence of a lesion of the central nervous system by examination over a period of years because of the notoriously slow growth of such lesions.

Treatment is rarely indicated or effective in boys who are found to have constitutional isosexual precocity. Psychological problems are rare if parents are wisely counselled, for the psyche of the child is appropriate for chronologic rather than biologic age. Medroxyprogesterone has been given to such boys because it will depress the plasma testosterone concentration and testosterone production rate. However, such treatment does not inhibit accelerated rates of linear growth and skeletal maturation (SCHOEN, 1966; CAMACHO et al., 1972).

Puberty associated with lesions unrelated to mechanisms controlling the hypothalamic-pituitary-gonadal activity are often termed pseudoprecocious puberty. Such abnormalities may elaborate material similar to human chorionic gonadotropin (HCG), which stimulates ovarian and testicular secretion of estrogens and testosterone (ROOT et al., 1968), and these tumors, such as chorionepithelioma of the ovary or testis, convert androgens to estrogens. Testicular neoplasms, the Leydig cell tumor, for example, may produce sex hormones autonomously (ZANGENEH and KELLY, 1968; ROOT et al., 1972b). Other tumors, such as hepatoblastoma in a male, may secrete a gonadotropin to bring about pubertal changes (ROOT et al., 1968). The most common site of extratesticular androgen synthesis in boys is the adrenal gland. Congenital adrenal hyperplasia and virilizing adrenal adenomas or carcinomas are associated with excessive androgen production (KELLER, 1972). Administration of androgens and HCG in the wish to stimulate growth has been responsible for precocious pubertal changes (PLANTIER, 1964) as has unintentional exposure to androgens.

In males with heterosexual precocity, feminizing tumors of the adrenal, teratomas secreting estrogen, neurofibromatosis, and administration of estrogens, intentional or accidental, have been implied as etiologic agents (MYHRE et al., 1970). In the evaluation of these male patients, testicular size should be accurately determined, for those having extratesticular sources of androgens have testicles much smaller than normal for the degree of virilization but appropriate in size for chronologic age. A Leydig cell tumor may often be palpated, at times as an intratesticular mass, and one testicle is larger than the other with the smaller testicle being appropriate for the age of the patient. Bone age is markedly advanced in patients with true, complete sexual precocity, with ectopic gonadotropin production or with hypersecretion of gonadal or adrenal hormones. Causes of precocious puberty inappropriate for a female phenotype are usually ovarian or adrenal tumors, androgen-producing teratomas, or the administration of androgens. In either sex, such a tumor should be removed. If correction of adrenal hyperplasia is carried out, continuous treatment with cortisol will be required in order to suppress the hypothalamic-pituitary-adrenal axis.

The urinary excretion rate of 17-ketosteroids is increased in males with virilizing testicular and adrenal tumors and with congenital hyperplasia; urinary pregnanetriol excretion is elevated in the latter condition but may also be slightly increased in patients with Leydig cell tumors of the testes (ROOT et al., 1972b). If urinary 17-ketosteroid and pregnanetriol excretion rates are elevated, suppression with cortisol should be undertaken in order to confirm a diagnosis of congenital adrenal hyperplasia. Plasma and urinary levels of testosterone are increased in patients with isosexual precocity from many causes. The diagnosis and therapy may not be accomplished until skeletal maturity has reached that typical for 12 to 14 years of age. Medroxyprogesterone is not indicated for these patients (BONGIOVANNI et al., 1972).

Premature adrenarche, like premature thelarche, is usually an isolated clinical finding without generalized systemic hormonal production. The etiology of the condition is not clear. In premature adrenarche, plasma levels of DHA, DHA-sulfate, and androstenediol may be at adolescent concentrations with slightly advanced skeletal maturation (ROSENFIELD et al., 1972; ROSENFIELD, 1971), but this is not always true. Serum gonadotropin levels may be within normal levels or slightly elevated (PENNY et al., 1970; ROOT et al., 1970; GUYDA et al., 1969; KENNY et al., 1969).

## C. Blood Diseases

In this group of indications, the aims of therapy with anabolic-androgenic steroids are different, and there is often greater concern for the patient's life than in the previous categories. Consequently, the clinician may be prepared to accept more of the adverse effects of these agents than he perhaps would otherwise.

### 1. Aplastic Anemia

It has been firmly established that androgens stimulate erythropoiesis (DEL ZOPPO, 1937; VOLLMER and GORDON, 1941; VAN DYKE et al., 1954; FINKELSTEIN et al., 1944; TABER et al., 1943; STEIN and JACOBSEN, 1944; STEINGLASS et al., 1941; CRAFTS, 1946). Adolescent males have higher hematocrit values than adolescent females, and the rise in hematocrits of pubertal boys has a direct relationship with sex maturity ratings (DANIEL, 1973). Such differences are not present before puberty unless there is an androgen-producing tumor bringing about early sexual maturity at an inappropriate age. Conversely, there are fewer circulating erythrocytes in males with hypogonadism, and this condition is corrected after administration of androgens, a finding that was first demonstrated in 1941 (MCCULLAGH and JONES, 1941). When it was later noted that large doses of androgen sometimes caused excessive erythropoiesis, leading to moderate polycythemia, advantage was taken of this effect in the treatment of hypoplastic anemia and the anemia of neoplastic disease (GARDNER and PRINGLE, 1961).

Current therapy for acquired and constitutional aplastic anemia involves the use of androgens and/or corticosteroids. Approximately half of the patients with acquired aplastic anemia treated with androgens undergo partial remission, as do the majority of patients having the constitutional type. In one recent study of 92 children with idiopathic or acquired aplastic anemia, of whom the 5-year overall survival rate was 25%, 61 children were treated with androgens and 31 with corticosteroids (HUGHES, 1973). There was a higher initial remission in the androgen-treated group, but nearly half of these patients relapsed, and the crude survival rate after 5 years did not differ significantly in the two therapeutic categories. One of the largest recent series reported the median survival of 66 androgen-treated patients to be 26 months, whereas it was only 13 months in 46 similar patients who received the same supportive treatment without androgens (PERUGINI et al., 1970). The mortality rate in this study was also lower in the androgen-treated group (53%) when compared to the nontreated patients (83%). Many patients are dependent on continuous androgen administration to maintain satisfactory hemopoiesis, and withdrawal of androgen therapy is often associated with a decrease in hemoglobin to pretreatment levels, necessitating transfusions (SHAHIDI and DIAMOND, 1961).

An increase in the concentration of erythrocyte 2,3-diphosphoglycerate (DPG) induced by androgens has been reported (PARKER et al., 1972). Although regeneration of erythropoiesis is the primary effect of androgen therapy in aplastic anemia, there are usually increases also in the polymorphonuclear leukocytes. Thrombopoiesis in general occurs later and is less pronounced than regeneration of polymorphonuclear leukocytes, although thrombocytes are usually present in sufficient numbers to control bleeding tendencies if the hemoglobin concentration reaches 12 or 13 g per 100 ml (SHAHIDI, 1973).

There seems to be general agreement that children and women with acquired aplastic anemia show a more satisfactory response than do adult males. Although there are no available data relating response to stages of maturity in adolescent boys, it seems likely that the degree of satisfactory change would be related to the stage of maturity and not to chronologic age. Factors governing the response to androgen treatment are not clear, but intercurrent infection or severe hemorrhagic manifestations requiring multiple transfusions are major impediments to successful therapy.

The initial status of the bone marrow in patients with acquired aplastic anemia is an important determining factor in the success of therapy. In general, the majority of patients having remissions associated with androgen therapy show some degree of erythropoietic and myelopoietic activity in the bone marrow at the time of initiation of treatment (SHAHIDI and DIAMOND, 1961; KILLANDER et al., 1969). This is relevant also to the rate of recovery, which is usually more rapid in patients having some erythroid and myeloid elements in the bone marrow at onset of androgen therapy. Patients having severe bone-marrow aplasia have very slowly progressive improvement, and reticulocytosis does not appear in the peripheral blood until after several months of treatment. Although initial response is slow, once a response is established, improvement is usually sustained.

The erythropoietic response depends also to some extent on the etiology. In patients with acquired aplastic anemia, there is frequently a more rapid regeneration and an earlier, higher reticulocyte peak (SHAHIDI and DIAMOND, 1961) than in those with constitutional aplastic anemia.

### 2. Hemolytic Anemias

There are reports of variable success from androgen therapy in patients with chronic hemolytic anemias. Several reports have advocated the use of androgens in the treatment of sickle cell anemia (MENTZER et al., 1969; LUNDH and GARDNER, 1970; ISSACS et al., 1972). As noted previously, androgens have been reported to increase the concentration of erythrocyte 2,3-DPG (PARKER et al., 1972). Theoretically, an elevation of 2,3-DPG should be of benefit in anemia since it results in reduced affinity of hemoglobin for oxygen, and by implication, improved tissue oxygenation. A note of caution has been expressed concerning sickle cell disease, however (FEIG and SEGEL, 1972), in that in this situation further elevation of 2,3-DPG would certainly decrease oxygen saturation of sickle cells and thus might increase sickling in vivo. However, increased red cell mass (LUNDH and GARDNER, 1970) has been noted with androgen use without alteration of the sickling tendency of the erythrocytes, although there was a decrease in the plasma volume, thus accentuating the rise in hemoglobin and hematocrit (LUNDH and GARDNER, 1970).

### 3. Anemia of Renal Failure

Patients with bilateral nephrectomy continue to produce erythrocytes, although at an insufficient rate (NATHAN et al., 1964; ERSLEY et al., 1968), suggesting that erythrocytosis is under hormonal control. Erythropoietin has been detected in the plasma of anephric patients (NAETS and WITTEK, 1968; MIRAND and MURPHY, 1969) and is said to increase after androgen therapy in such patients (MIRAND et al., 1969; DEGOWIN et al., 1970). Because of these findings, patients with chronic renal failure have been treated with androgens (DEGOWIN et al., 1970; RICHARDSON and WEINSTEIN, 1970; SHALDON et al., 1971; PARKER et al., 1972) and an increase in red cell mass obtained (RICHARDSON and WEINSTEIN, 1970). Another group of patients receiving long-term hemodialysis showed a definite rise of hematocrit levels associated with androgen therapy, which declined steadily when the androgens were discontinued (SHALDON et al., 1971).

Of major benefit was a decreased frequency of maintenance transfusions as well as the protein anabolic effect resulting in nitrogen retention, thereby decreasing possible protein breakdown. The latter factor may be extremely important in that sudden withdrawal of hormone therapy in uremic patients may increase the azotemia (SHAHIDI, 1973).

Although all data concerning the use of androgens in chronic renal failure have been related to adults, use of hemodialysis and also renal transplant procedures are now being used more often in the pediatric age group of the population. Whether growth and sexual maturation will respond well to androgen administration is not known at this time, but such therapy may be of particular value at the time of anticipated pubertal change.

## D. Chronic Wasting Diseases

One of the most controversial uses of the anabolic steroids is in the treatment of patients with chronic debilitating and wasting conditions. The aims of such treatment have been to hasten recovery by promoting anabolism of protein with subsequent gain in weight, and to provide symptomatic improvement.

### 1. Malnutrition and Protein Loss

Since depletion of tissue protein is common to all wasting diseases, there has been particular interest in using anabolic steroids in cases of malnutrition and conditions that cause loss of protein. Several reports quoted favorable responses when these agents were used in cases of malnutrition and kwashiorkor (ROY, 1966; DAS, 1968), particularly in conjunction with routine therapy. Some other factors are in action here. Although kwashiorkor is produced in children and in animals by severe protein deficiency, it may, in some instances, be secondary to severe energy restriction (NAISMITH, 1973). Maramus results from grossly restricted intake of all nutrients and energy (MCLAREN, 1974), but dietary factors may be of second-line order of importance. In a multifactorial etiology, poverty, ignorance, bad housing, poor hygiene, and lack of family planning all conspire to produce this condition (CRAVIOTO et al., 1967; KANOWATI and MCLAREN, 1973).

Furthermore, food-consumption data (SUKHATME, 1970) and dietary surveys incriminate energy rather than protein deficit, and increasing the energy intake and not that of protein has produced catch-up growth in some undernourished children (GOPALAN et al., 1973). Obviously with these new insights, anabolic steroids will probably not play an important role in the management of these problems.

Anabolic steroids have been used to increase appetite and cause gain in weight in some cases of anorexia nervosa, and favorable responses have been reported (ABRITTA, 1961; ROSADINI, 1959). For reasons of unreliability and the possibility of hepatic dysfunction in these patients due to their relative starvation state associated with avitaminosis and inadequate protein intake, parenteral anabolic steroids have been recommended (KUPPERMAN, 1963). It was emphasized, however, that these drugs constitute merely symptomatic treatment, and if used, should be accompanied by the appropriate psychotherapy.

In cases of protein-losing enteropathy, anabolic steroids have been tried. It was shown in cases of sprue, for example, that nortestosterone phenylpropionate successfully complimented the usual therapy (KUHLMANN, 1961). The improved intestinal absorption was attributed to the myotrophic effect of the steroid on the hypotonic musculature of the small intestines.

An important cause of protein loss in children is the nephrotic syndrome in which 5–40 g of serum proteins may be lost daily in the urine. In a few cases reported, high doses of depot-testosterone decreased proteinuria (MOENCH and SARTORIUS, 1954). Anabolic drugs alone, however, are rarely sufficiently helpful in decreasing the urinary protein loss. It has been suggested that a diet rich in protein together with plasma transfusion, corticosteroids, and anabolic steroids may produce the best results (KRUSKEMPER, 1968).

Although many studies favor the view that anabolic agents are beneficial under these circumstances (WEWALKA, 1968), others believe that the reported rates of gaining weight are not impressive (FRUEHAN and FRAWLEY, 1963). Also, the significance of observed increase in weight is questionable and open to interpretive error, because the composition of weight gained (water, muscle, or fat) cannot be derived from measurement of weight alone. It has been pointed out that gains or losses of water may counterbalance losses or gains in tissue mass (WATERLOW et al., 1960). It would seem reasonable to state that in the majority of such patients, therefore, the homeostatic mechanisms of the body would of themselves compensate for a deficiency of body protein as long as the diet provided adequate calories and protein intake.

## 2. With Corticosteroids

Even more controversial is the use of anabolic steroids in conjunction with corticosteroids. The use of corticosteroids in the treatment of rheumatic, allergic, and related conditions is attended by a number of complications that have been attributed to catabolic effects. Under conditions of constant intake, compounds such as norethandrolone and nandrolone phenylpropionate have been shown to reverse negative nitrogen balance resulting from administration of 35 mg of cortisone daily without affecting the beneficial effects of the corticosteroid (BRØCHNER-MORTENSEN et al., 1959). The addition of 5 mg daily of methandrostenolone appeared to limit the depletion of tissue protein resulting from moderately large doses of glucocorticoid hormones being used in lengthy courses for conditions such as nephrotic syndrome or ulcerative colitis (EZRIN, 1965). In nephrotic syndrome a negative nitrogen balance became positive when anabolic steroids were used in conjunction with corticosteroids in one study (KLUTHE and RICHTER, 1963). In another study, it was noted that the effects of triamcinolone were mitigated by the addition of stanozolol to the therapeutic regime (ALBANESE, 1968).

Taking an opposing view, another author recommended caution and claimed that it was gross oversimplification to suggest anabolic steroids as a logical treatment for the unwanted metabolic effects of glucocorticoids on the basis of short-term balance experiments (WYNN, 1968). He pointed out that the modes of action of these steroids are different and that the beneficial effects of anabolic steroids, that is, the reversal of the negative nitrogen and calcium balance induced by glucocorticoids shown in short-term metabolic studies, had not persisted. He claimed, further, that this combination may result in an increase of glucocorticoid undesirable effects such as plethora, hypertension, acne, or hirsutism.

Clearly a final answer to this controversy is not yet at hand. In clinical practice, however, it appears that the androgens and the newer derivatives are being used during therapy with corticosteroids in the hope of preventing thinning of the skin, increased fragility of blood vessels, muscle wasting and weakness, peptic ulceration, and demineralization of the skeleton (ASTWOOD, 1970).

## 3. Diabetes Mellitus

Children with inadequately controlled diabetes are usually poor physical specimens who fail to gain weight due to the excessive glucose spillage. Furthermore, the brittleness of the diabetes makes it difficult to administer an adequate dose of insulin to achieve a balance between glycosuria and hypoglycemic attacks.

The hypoglycemic action of anabolic steroids has been noted by several workers (HOUTSMULLER, 1961; CAVALIERI et al., 1962; DARDENNE, 1961; LANDON et al., 1962). There is good documentation in these studies, therefore, that synthetic anabolic steroids have a direct action on the diabetic state by reducing blood sugar and lowering insulin requirements, and they appear also to restore reactivity in the insulin-resistant patient (TAINTER et al., 1964).

Other aspects of uncontrolled diabetes that have been of concern are the loss of body weight, depletion of tissue protein, and increase in the rate of nitrogen excretion. The tissue wastage of diabetes can be overcome by insulin, although this hormone has little effect on the nitrogen balance of nondiabetic animals or patients (KORNER and MANCHESTER, 1960). In experimentally produced insulin-deficient animals, it was believed that the exaggerated loss of nitrogen was caused by the secretion of corticosteroids, which were not balanced by the anabolic action of the deficient insulin. The suggested contribution of anabolic steroids, therefore, was considered to be their ability to antagonize the catabolic actions of corticosteroids. In practice, the use of these agents has been shown to improve the general condition of poorly controlled diabetics through better appetite, weight gains, and increased sense of well-being (TAINTER et al., 1964). Several authors have considered, therefore, that there should be greater use of these preparations in diabetic children (KUPPERMAN, 1962; TAINTER et al., 1964).

Very little work appears to have been done during the past decade, however, and we believe a word of caution is appropriate. This concerns the matter of compliance, a factor that is particularly relevant to adolescents with diabetes (BENNETT and WARD, 1975). Problems in accepting the disease and learning to manage it are very real to the teenager. During this period of life, the insulin requirements are likely to be high and the diabetes more difficult to control. It is vitally important to investigate the possibility of sabotage of the medical regime before resorting to added medication. It may be that some psychological factors need to be dealt with rather than prematurely complicating the treatment program.

## 4. Hyperthyroidism

Anabolic steroids have been used successfully with antithyroid therapy to treat the catabolic state caused by thyrotoxicosis (WEWALKA, 1968). This condition is not infrequently encountered, particularly in adolescents (CUNNINGHAM, 1970; HUNG, 1972), and some of these patients may suffer lack of appetite and great weight loss.

There are specific circumstances in which the use of anabolic steroids may be indicated. The combination of antithyroid and anabolic therapy may be particularly helpful, for example, in thyrotoxic myopathy, although this is rare in children and adolescents. One special indication for anabolic steroids is hypercalcemia with a negative calcium balance (KRUSKEMPER, 1968), and anabolic steroids may be beneficial also during the first weeks of antithyroid therapy or radioiodine therapy, as well as in pre- and postsurgical stages.

Although protein-bound iodine will be lowered during treatment with anabolic steroids (ROSENBERG et al., 1962), this is due to the lowering of thyroid-binding-globulin, and thyroid hormone output is, in fact, unchanged. Likewise the basal metabolism remains unchanged. It should be stressed, therefore, that this is merely an antithyrotoxic auxillary treatment that aims at metabolic symptoms of the disease (KRUSKEMPER, 1968).

## 5. Pulmonary Diseases

### *a) Pulmonary Tuberculosis*

In patients with chronic pulmonary tuberculosis, anabolic steroids may help overcome the protein deficiency in the tissues (KRUSKEMPER, 1968). In several studies, along with antituberculosis drugs, these drugs produced appreciable weight gains (SCHIFFRIN et al., 1961), which have been found to be independent of the manner of administration of the anabolic steroid. Particularly apparent have been the rise of appetite and the improved sense of well-being of patients thus treated. Follow-up observations have shown that after treatment was stopped, the body weight remained at the same high level for a longer period of time and that, generally, there was no rebound phenomenon such as that seen in healthy experimental subjects. Anabolic steroids effect no clear-cut changes in the lungs as revealed by x-ray films, however, and it is not known whether or in what manner the course of pulmonary tuberculosis is influenced by such medications. These agents have been found to be useful as an adjunct to the antituberculous regime for tuberculous lymphadenopathy.

### *b) Cystic Fibrosis*

Fibrocystic disease is a serious disorder of infancy and childhood seen also in adolescents and young adults (GRACEY and ANDERSON, 1969) for which the cause is obscure. Several studies have been conducted concerning the effects of anabolic steroids in this condition as an addition to routine therapies. A review of the literature on this subject revealed that the claimed benefits may have been more apparent than real (SOBEL, 1969). When used in cystic fibrosis patients with growth failure, anabolic steroids have been said to cause weight gain (DENNIS and PANOS, 1965; DOOLEY et al., 1969). Patients have shown subjective improvement, for example, with the use of large doses of methylandrostenolone (GOOD and BESSMAN, 1956). None of these studies, however, was perfect in design, and it was suggested that they should not be the basis for unbridled enthusiasm for the use of anabolic steroids in this condition (SOBEL, 1969).

It is apparent that further research is needed with appropriate experimental design including double-blind protocol and objective analyses of results in terms most relevant to the patient's problem. As yet there is no answer to the question raised by these studies as to whether or not anabolic steroids may affect the oxygenation of patients with severe pulmonary disease, and this may be a worthwhile avenue for further research in this condition. Meanwhile management continues to include time-measured pancreatic extract together with supportive measures (GUPTE, 1974).

## 6. Muscular Dystrophy

Anabolic steroids have been used in muscular dystrophy, an inherited group of wasting diseases becoming manifest in childhood. The most common form of dystrophy is the pseudohypertrophic progressive type first described by DUCHENNE (1868). This condition is seen almost exclusively in males and occasionally in females who are carriers; it follows a predictable, savage course of relentless progression leading to disability and loss of ambulation within 8–10 years after onset of symptoms, and, by the end of the second decade, death from intercurrent infection or respiratory failure. Concern for finding optimal forms of therapy is understandable.

The manifold attempts of therapy with vitamins, inositol, amino acids, cytochrome-C, adenosine triphosphate, prostigmin, malarial induction, or pilocarpine-epinephrine treatment have not yielded satisfactory results with progressive muscular dystrophy (VAN METER, 1953; WALTON, 1963; KRUSKEMPER, 1968). A number of reports, however, describe favorable effects of testosterone propionate and methyltestosterone upon progressive muscular dystrophy (HESSER et al., 1940; WARING et al., 1940; PERLSTEIN and GUTTERMAN, 1950; QUINN and WORCESTER, 1951). As these compounds proved to be too androgenic, protracted treatment was precluded (KRUSKEMPER, 1968), and subsequently, anabolic steroids were tried in an attempt to gain improvement. With the stimulus of prolongation of the lives of Bar Harbor mice suffering from hereditary muscular dystrophy by the use of 17-ethyl-19-noretestosterone (DAWBEN, 1959), many studies were undertaken in the late 1950s with several anabolic steroids (DETONI, 1961; PATEISKY et al., 1961; DAWBEN, 1963). Generally speaking, in about 20–50% of the cases, a clear-cut influence on body weight and on the growth rate in children was achieved, as well as considerable objective improvement of mobility and muscle power associated with the suggestive report of increased work capacity. After a longer period of time, however, these results have not been too convincing. In one double-blind control trial, for example, patients with progressive muscular dystrophy were treated with methandrostenolone and nandrolone decanoate, and no significant subjective or objective improvement could be found in either group, except for increase in appetite and weight gain. Muscle power and functional capacity were not significantly improved (BARWICK et al., 1963).

In one published report a regime of digitoxin and an anabolic steroid, together with physical therapy, was said to be effective in the management of muscular dystrophy (DAWBEN, 1963). Of 37 patients treated for periods of up to 19 months, 22% were said to have shown appreciable improvement and 70% were essentially unchanged. It is not known, unfortunately, how much of the effect noted was due to anabolic steroids, how much to digitoxin, and how much to physical therapy. This identical regime was tested subsequently, and in the patients investigated, one year after the initiation of treatment, all were significantly worse. None showed any change that could be interpreted as representing any improvement (CHARASH, 1965). Physicians at eight other centers with experience in trials with identical therapy in approximately 300 patients, reported no evidence of any objective benefit from the use of this steroid therapy in muscular dystrophy (NEWS CONFERENCE, 1964).

At this time there appears to be little enthusiasm for the use of anabolic steroids in muscular dystrophy. In most of the studies quoted, hopes for the absence of androgenic effects have been in vain. In the majority of cases steroids caused masculinizing phenomena and occasionally precocious closure of epiphyses. There seems little to commend them for routine use in this group of insidiously progressive diseases (KRUSKEMPER, 1968; FALUDI et al., 1968).

## E. Miscellaneous Conditions

There are several other situations in childhood and adolescence in which anabolic steroids have been used.

## 1. Athletics

The self-administration of anabolic steroids is a problem in athletics. Apart from being unethical and hazardous, this practice appears to be quite ineffectual. In one study, the effects of androstenolone, exercise, and androstenolone combined with exercise were studied on muscle strength, physical working capacity, and motor performance, and various anthropometric measures in untrained college students and trained college rugby players over a 16-week period (FOWLER et al., 1965). No significant changes in muscle strength were found in response to this anabolic agent. While exercise increased physical working capacity and motor performance, androstenolone had no effect. The administration of anabolic steroids in combination with exercise did not increase physical working capacity or motor performance beyond the effect of exercise alone.

In a more recent study, untrained University of Southern California College students participated in a weight-lifting program for 28 days (MUNSON, 1970). One-half of the subjects received oxandrolone in addition to a constant weight-lifting regime. Although subjects receiving the anabolic agent showed a significantly greater increase in body weight than placebo-treated controls, there were no significant effects on muscle size, muscle strength, or physical fitness as measured by oxygen consumption during exercise in the oxandrolone-treated group beyond the changes induced by exercise alone. A significant proportion of the increase in body weight was felt to be due to fluid retention.

On the basis of studies such as these, and because of the well-recognized hazardous effects of these steroids, especially when used in large doses over long periods of time, it must be concluded that anabolic-androgenic steroids have no justifiable use in athletics and should be condemned in this situation (FRASIER, 1973).

## 2. Osteogenesis Imperfecta

Osteogenesis imperfecta is considered a disease of the connective tissue (LANGNESS and BEHNKE, 1971; MCKUSICK, 1960; STEVENSON et al., 1970). In the skeleton, the basic abnormality produces a deficiency in the formation of bone and an acceleration in its breakdown, although cartilage proliferation remains normal. The result is a generalized osteoporosis, which is manifested clinically by bone fragility, often leading to severe fractures and skeletal deformities. Lax ligaments, deafness, and blue sclerae are other clinical features frequently found in osteogenesis imperfecta.

The latest information about treatment of osteogenesis imperfecta has been summarized by CASTELLS (1973). Several forms of therapy have been used in the past and were found to be of no beneficial value, such as treatment with thyroid hormone (SCHABAD, 1914), parathormone (HANSEN, 1935), parathyroidectomy (LEFORT, 1938), and thymus extract administration (RYAN, 1932; SECORD et al., 1936). More recent therapies which are reviewed include anabolic steroids, fluoride therapy, magnesium oxide, and calcitonin.

The role of sex hormones in the treatment of osteogenesis imperfecta has been based on the clinical observation that in this condition there is a decrease in the incidence of fractures at puberty and an increase after the menopause. In one study in which the effects of anabolic steroids were studied in children with osteogenesis imperfecta, there was no evidence that anabolic hormones produced increased retention of calcium by the skeleton (CATTELL and CLAYTON, 1968). Although a reduced incidence of fractures was observed in some patients, the drugs produced no dramatic alleviation of the disease, and severe adverse effects were noted.

The lack of effective therapy for children with osteogenesis imperfecta stems from the lack of knowledge of the pathophysiology (CASTELLS, 1973). At the present time the most promising development is the finding that bone turnover is increased in osteogenesis imperfecta and that calcitonin might be able to suppress it. Administration of calcitonin over a prolonged period may thus lead to therapeutic possibilities.

### 3. Inborn Errors of Metabolism

This set of indications for anabolic steroids has only recently appeared in pediatrics. Two patients with the Fanconi (de Toni-Debré-) syndrome were treated for more than 8 months with 17β-hydroxy-17-ethyl-19-norandrost-4-en-3-one (10–20 mg daily) (BERCZELLER and KUPPERMAN, 1960). Linear growth was accelerated, muscle power increased, and the interest of the patient in his environment rose in clear contrast to the preceding lethargic state. Especially noteworthy was the decrease of the cystinuria to the normal value. In another report, the successful treatment of severe cases of cystinosis was reported with 17β-hydroxy-1-methyl-5α-androst-1-en-3-one heptanoate (WEBER and HAGGE, 1963). In this report, the child, who, at the beginning of therapy was dystrophic, thrived well. Fever spikes, glycosuria, and acidosis disappeared; the excretion of amino acids dropped considerably, while the concentration of free amino acids in serum rose. On the basis of this limited experience, further experiments should be conducted to study therapy with anabolic steroids in other cases of inborn errors of metabolism.

### 4. Other Conditions

The use of anabolic steroids in epilepsy was ineffective in 26 patients, compared to response in a similar placebo group (ORLANSKY and OSTERHELD, 1963).

Androgens were formerly used in the treatment of a variety of menstrual disorders such as dysmenorrhea or menorrhagia, but they have been supplanted by the more effective, orally active progestogen compounds (ASTWOOD, 1970).

## VI. Conclusions

Despite the synthesis of drugs with less androgenic effects than testosterone, all anabolic agents are virilizing with adequate dosage and duration of therapy. This factor, together with the other adverse effects discussed earlier, necessitate caution when these hormones are used clinically in children and adolescents. There are, however, legitimate indications, and acceptance of untoward effects obviously depends on the severity and potential dangers of the underlying condition.

In children, androgens have undisputed value in the treatment of male hypogonadism, in which situation they are used long-term as hormonal replacement therapy. Testosterone, or a close derivative, is administered to induce puberty and subsequently to maintain secondary sex characteristics.

These steroids are capable of stimulating erythropoiesis and the formation of other cellular elements of the blood. Consequently they have been of use in the pediatric and adolescent age groups in cases of aplastic anemia, sickle cell disease, and the anemia of chronic renal failure.

In growth retardation, their use is not as clear-cut. Recent experience (KAPLAN et al., 1973) has indicated that anabolic-androgenic steroids can be used in patients with constitutional delayed growth and development without adversely affecting ultimate height. Most authorities agree, however, that use of anabolic steroids in this situation should not be undertaken lightly. For those patients with normal predicted adult height who are not severely disturbed by their immediate height, reassurance may be all that is required (GELLER, 1968) and is undoubtedly the best course of action. For those patients with poor predicted adult height and those with severe psychological problems related to their current stature, regardless of predicted height, anabolic therapy should probably be undertaken. By the use of lower dose schedules, the problem of accelerated bone maturation with premature epiphyseal closure can usually be avoided.

The use of these hormones in wasting conditions during childhood is even more controversial. When promotion of anabolism is the aim of therapy, one should assess the existing state of anabolism before administering the steroids (WYNN, 1968), but there are many uncertainties. The effects on nitrogen balance are so intimately related to the intake of protein and the total calories consumed that, depending upon the conditions, a small increase in either one of these may have a greater effect in practice than the administration of the anabolic agent.

Obviously it is still not possible to draw hard and fast rules. Roles for the clinical use of these steroids in children and adolescents with a variety of disorders need to be further defined. Initial enthusiasm for each anabolic agent has been progressively tempered with reduction in recommended dosage over time and growing awareness of adverse effects. Physicians caring for young people who are afflicted with serious ailments should not indiscriminately try a new panacea that may cause real harm. As with all medications, the benefits must be carefully weighed against the disadvantages.

## References

ABRITTA, J.: Casuistica anabolica en la practica clinica. Sem. med. (Madr.) **119**, 983—990 (1961).

ALABANESE, A. A.: Anticatabolic application of newer anabolic steroids. Med. Tms (Lond.) **96**, 871—881 (1968).

ANDERSON, W. A. D.: Pathology, 2nd Ed., p. 990. St. Louis: C. V. Mosby Comp. 1953.

ASTWOOD, E. B.: Androgens and anabolic steroids. In: GOODMAN, S., GILMAN, A. (Eds.): The Pharmacological Basis of Therapeutics, 4th Ed., pp. 1566—1580. New York: MacMillan Co. 1970.

AUGUST, G. P., GRUMBACH, M. M., KAPLAN, S. L.: Hormonal changes in puberty. III. Correlation of plasma testosterone, LH, FSH, testicular size and bone age with male pubertal development. J. clin. Endocr. **34**, 319—326 (1972).

BARWICK, D. D., NEWELL, D. J., WALTON, J. N.: Methandrostenolone and nandrolone decanoate in muscular dystrophy. Neurology **13**, 12—23 (1963).

BAYER, L. M., BAYLEY, N.: Growth pattern shifts in healthy children: spontaneous and induced. J. Pediat. **62**, 631—645 (1963).

BAYLEY, N., GORDAN, G. S., LISSER, H.: Long-term experience with methyltestosterone as a growth stimulant in short immature boys. Pediat. Clin. Amer. **4**, 819—825 (1957).

BAYLEY, N., PINNEAU, S. R.: Tables for predicting adult height from skeletal age: Revised for use with the Greulich Pyle hand standards. J. Pediat. **40**, 423—441 (1952).

BEAS, F., ZURBRÜGG, R., LEIBOW, S. G., PATTON, R. G., GARDNER, L. I.: Familial male sexual precocity: report on the eleventh kindred found, with observations on blood group linkage and urinary CA-steroid excretion. J. clin. Endocr. **22**, 1095—1102 (1962).

BENNETT, D. L., WARD, M. S.: Diabetic ketoacidosis in adolescents: A guide to management. J. med. Ass. Ala **44**, 499—518 (1975).

BERCZELLER, P. H., KUPPERMAN, H. S.: The anabolic steroids. Clin. Pharmacol. Ther. **1**, 464—482 (1960).

BERNSTEIN, M. S., HUNTER, R. L., YACHNIN, S.: Hepatoma and peliosis hepatitis developing in a patient with Fanconi's anemia. New Engl. J. Med. **284**, 1135—1136 (1971).

BETTMANN, H. K., GOLDMAN, H. S., ABRAMOWICZ, M., SOBEL, E. H.: Oxandrolone treatment of short stature: effect on predicted mature height. J. Pediat. **79**, 1018—1023 (1971).

BIERICH, J. R.: Effects and side effects of anabolic steroids in children. Acta endocr. (Kbh.) **39** (Suppl. 63), 89—110 (1961).

BLIZZARD, R. M., JOHANSON, A., GUYDA, H., BAGHDASSARIAN, A., RAITI, S., MIGEON, C. J.: Recent developments in the study of gonadotropin secretion in adolescence. In: HEALD, F. P., HUNG, W. (Eds.): Adolescent Endocrinology, pp. 1—24. New York: Appleton-Century-Crofts 1970.

BLIZZARD, R. M., PENNY, R., FOLEY, T. P., JR., BAGHDASSARIAN, A., JOHANSON, A., YEN, S. S. C.: Pituitary-gonadal interrelationships in relation to puberty. In: SAXENA, B. B., BELING, C. G., GANDY, H. M. (Eds.): Gonadotropins, pp. 502—523. New York: Wiley-Interscience 1972.

BONGIOVANNI, A. M., MOSHANG, T., PARKS, J. S.: Deferment of puberty in "late" treated congenital adrenal hyperplasia. (Abs.). Pediat. Res. **6**, 352 (1972).

BORIS, A., STEVENSON, R. H., TRMAL, T.: Comparative androgenic, myotrophic and anti-gonadotrophic properties of some anabolic steroids. Steroids **15**, 61—71 (1970).

BRANS, Y. W., SUMNERS, J. E., DWECK, H. S., CASSADY, G.: Feeding the low birth weight infant: orally or parenterally? Preliminary results of a comparative study. Pediatrics **54**, 15—22 (1974).

BRASEL, J. A., BLIZZARD, R. M.: The influence of the endocrine glands upon growth and development. In: WILLIAMS, R. H. (Ed.): Textbook of Endocrinology, 4th ed., pp. 1017—1038. Philadelphia: W. B. Saunders Co. 1968.

BRØCHNER-MORTENSEN, K., GJØRUP, S., THAYSEN, J. H.: The metabolic effect of new anabolic 11-nor-steroids. Acta med. scand. **165**, 197—205 (1959).

BUCKLER, J. M. H., CLAYTON, B. E.: Output of luteinizing hormone in the urine of normal children and those with advanced sexual development. Arch. Dis. Childh. **45**, 478—484 (1970).

BURR, I. M., SIZONENKO, P. C., KAPLAN, S. L., GRUMBACH, M. M.: Hormonal changes in puberty. I. Correlation of serum luteinizing hormone and follicle stimulating hormone with stages of puberty, testicular size, and bone age in normal boys. Pediat. Res. **4**, 25—35 (1970).

CAMACHO, A. M., WILLIAMS, D. L., MONTALVO, J. M.: Alterations of testicular histology and chromosomes in patients with constitutional sexual precocity treated with medroxyprogesterone acetate. J. clin. Endocr. **34**, 279—286 (1972).

CASTELLS, S.: New approaches to treatment of osteogenesis imperfecta. Clin. Orthop. **93**, 239—249 (1973).

CATTELL, H. S., CLAYTON, B.: Failure of anabolic steroids in the therapy of osteogenesis imperfecta, J. Bone Jt Surg. **50 A**, 123—141 (1968).

CAVALIERI, U., QUADRI, A., TAMMARO, A. E.: On a hypoglycemic action of anabolic steroids. Presentation of a case: verification and considerations. Minerva med. **53**, 1155—1157 (1962).

CHARASH, L.: Anabolic steroids in the management of muscular dystrophy. Pediatrics **36**, 402—405 (1965).

CHEEK, D. B.: Human Growth, Body Composition, Cell Growth, Energy and Intelligence. Philadelphia: Lea and Febiger 1968.

CHEEK, D. B., HILL, D. E.: Muscle and liver cell growth: role of hormones and nutritional factors. Fed. Proc. **29**, 1503—1509 (1970).

COOKE, R.: Growth and development. In: HARRISON, T. R. (Ed.): Principles of Internal Medicine, 6th Ed., pp. 31—34. New York: McGraw-Hill 1970.
COSTELLO, J. M., IBBENTSON, H. K., SCOTT, D. J.: Experience in a growth and development clinic with particular reference to growth hormone deficiency. N. Z. med. J. **68**, 67—72 (1968).
COURT-BROWN, W. M.: Sex chromosomes and the law. Lancet **1962 II**, 508—509.
CRAFTS, R. C.: Effects of hypophysectomy, castration, and testosterone propionate on hemopoiesis in the adult male rat. Endocrinology **39**, 401—413 (1946).
CRAVIOTO, J., BIRCH, H. G., DE LICARDIE, E. R., ROSALES, L.: The ecology of infant weight gain in a pre-industrial society. Acta pediat. scand. **56**, 71—84 (1967).
CUNNINGHAM, R. D.: Endocrine diseases. In: DANIEL, W. A., JR. (Ed.): The Adolescent Patient, pp. 249—282. St. Louis: C. V. Mosby Co. 1970.
DANIEL, W. A., JR.: Unpublished data.
DANIEL, W. A., JR.: Hematocrit: Maturity relationship in adolescence. Pediatrics **52**, 338—394 (1973).
DANOWSKI, T. S., LEE, F. A., COHN, R. E., D'AMBROSIA, R. D., LIMAYE, N. R.: Oxandrolone therapy of growth retardation. Amer. J. Dis. Child. **109**, 526—532 (1965).
DARDENNE, U.: The therapeutic applications of anabolic steriods in opthalmology. Acta endocr. (Kbh.) **39** (Suppl. 63), 143—153 (1961).
DAS, P. K.: Anabolic steroids. J. Indian Ass. **50**, 79—81 (1968).
DAWBEN, R. M.: Prolonged survival of dystrophic mice treated with 17-alpha-19-nortestosterone. Nature (Lond.) **184**, 1966—1967 (1959).
DAWBEN, R. M.: Treatment of muscular dystrophy with steroids. New Engl. J. Med. **268**, 912—916 (1963).
DAY, R., SILVERMAN, W. A.: Étud. néo-natal. **1**, 111 (1952).
DEAMER, W. C.: Stimulation of growth in boys by sublingual testosterone therapy. Amer. J. Dis. Child. **75**, 850—859 (1948).
DEGOWIN, R. L., LAVENDER, A. R., FORLAND, M., et al.: Erythropoiesis and erythropoietin in patients with chronic renal failure treated with hemodialysis and testosterone. Ann. intern. Med. **72**, 913—918 (1970).
DEL ZOPPO, R.: Modifiche ematiche ed istologiche da somministrazione di estratti testicolari. Clin. Med. Ital. **68**, 113—130 (1937).
DENNIS, J. L., PANOS, T. C.: Growth and bone-age retardation in cystic fibrosis. Response to an anabolic steroid. J. Amer. med. Ass. **194**, 855—858 (1965).
DETONI, E.: Treatment of progressive muscular dystrophy in particular with the anabolic steroids. Acta. endocr. (Kbh.) **39**, (Suppl. 63), 175—184 (1961).
DILLON, R. S.: Handbook of Endocrinology. Diagnosis and Management of Endocrine and Metabolic Disorders, p. 487. Philadelphia: Lea and Febiger 1973.
DONOVAN, B. T., VAN DER WERFF TEN BOSCH, J. J.: Physiology of Puberty, p. 216. Baltimore: Williams & Wilkins Co. 1965.
DOOLEY, R. R., MOSS, A. J., WRIGHT, P. M., HASSAKIS, P. C.: Norethandrolone in cystic fibrosis of the pancreas. J. Pediat. **74**, 95—102 (1969).
DUCHENNE, G. B.: Recherches pour le paralysie musculare pseudohypertrophique ou paralysie myosclerosique. Arch. gén. Méd. **11**, 5, 179, 305, 421, 552 (1868).
EARLE, A. M.: Methyltestosterone and plasma for premature infants. J. Pediat. **36**, 87—90 (1950).
EINHORN, M. B.: Use of estrogenic substance in treatment of premature infant. Arch. Pediat. **57**, 115—118 (1940).
ERSLEV, A. J., MCKENNA, P. J., CAPELLI, J. P., et al.: Rate of red cell production in two nephrectomized patients. Arch. intern. Med. **122**, 230—235 (1968).
ESCAMILLA, R. F., BENNETT, L.: Pituitary infantilism treated with purified growth hormone, thyroid, and sublingual methyltestosterone. A case report. J. clin. Endocr. **11**, 221—228 (1951).
EZRIN, C.: Anabolic agents. Canad. med. Ass. J. **92**, 529—530 (1965).
FAIMAN, C., WINTER, J. S. D.: Sex differences in gonadotropin concentrations in infancy. Nature New Biol. (Lond.) **232**, 130—131 (1971).
FALLIERS, C. J., JORGENSEN, J. R., TAN, L. S., BUKANTZ, S. C.: Anabolic effects of stanozolol. Amer. J. Dis. Child. **106**, 388—391 (1963).
FALUDI, G., DYKYJ, R., CHAYES, Z. W., NODINE, J. H.: Anabolic steroids in muscular, neurologic and hematologic disorders. J. Amer. med. Wom. Ass. **23**, 346—351 (1968).

FARRELL, G. C., UREN, R. F., PERKINS, K. W., JOSHUA, D. E., BAIRD, P. J., KRONENBERG, H.: Androgen-induced hepatoma. Lancet **1975I**, 430—431.

FEIG, S. A., SEGEL, G. B.: Letter: Androgens 2,3-DPG and sickling. New Engl. J. Med. **287**, 1097—1098 (1972).

FINKELSTEIN, G., GORDON, A. S., CHARIPPER, H. A.: The effect of sex hormones on the anemia induced by hemorrhage in the rat. Endocrinology **35**, 267—277 (1944).

FINLEY, W. H., FINLEY, S. C.: Genetic disorders. In: DANIEL, W. A., JR. (Ed.): The Adolescent Patient, pp. 197—218. St. Louis: C. V. Mosby Co. 1970.

FITSCHEN, W., CLAYTON, B. E.: Urinary excretion of gonadotropins with particular reference to children. Arch. Dis. Child. **40**, 16—26 (1965).

FLERKO, B.: Control of follicle-stimulating hormone and luteinizing hormone secretion. In: MARTINI, L., MOTTA, M., FRASCHINI, F. (Eds.): The Hypothalamus, pp. 351—363. New York: Academic Press 1970.

FOREST, M. G., MIGEON, C. J.: Percentage of testosterone, androstenedione, and dehydroisoandrosterone bound to plasma protein in preadolescent children. J. Pediat. **76**, 732—738 (1970).

FOSS, G. L.: Some experiences with a new anabolic steroid (methandrostenolone). Brit. med. J. **1960 I**, 1300—1305.

FOSS, G. L.: The influence of androgen treatment on ultimate height in males. Arch. Dis. Childh. **40**, 66—70 (1965).

FOWLER, W. M., JR., GARDNER, G. W., EGSTROM, G. H.: Effect of an anabolic steroid on physical performance of young men. J. appl. Physiol. **20**, 1038—1040 (1965).

FRANCHIMONT, P.: The regulation of follicle stimulating hormone and luteinizing hormone secretion in humans. In: MARTINI, L., GANONG, W. F. (Eds.): Frontiers in Neuroendocrinology, pp. 331—358. London: Oxford University Press 1971.

FRASIER, S. D.: Androgens and athletes. Amer. J. Dis. Child. **125**, 479—480 (1973).

FRASIER, S. D., HORTON, R.: Androgens in the peripheral plasma of prepubertal children and adults. Steroids **8**, 777—784 (1966).

FRUEHAN, A. E., FRAWLEY, T. H.: Current status of anabolic steroids. J. Amer. med. Ass. **184**, 527—532 (1963).

GANDY, H. M., PETERSON, R. E.: Measurement of testosterone and 17-ketosteroids in plasma by the double isotope dilution derivative technique. J. clin. Endocr. **28**, 949—977 (1968).

GARDNER, F. H., PRINGLE, J. C., JR.: Androgens and erythropoiesis. I. Preliminary clinical observations. Arch. intern. Med. **107**, 846—862 (1961).

GELLER, J.: The effect of 17-alpha-ethyl-19-nortestosterone and 17-alpha-ethyl-19-nortestosterone 3-enol-propionate on growth and endocrine function in normal young males of short stature. Acta. endocr. (Kbh.) **45**, 13—25 (1964).

GELLER, J.: Oxandrolone effect on growth and bone age in idiopathic growth failure. Acta endocr. (Kbh.) **59**, 307—316 (1968).

GIORGI, E. P., SOMMERVILLE, I. F.: Hormone assays in Klinefelter's syndrome. J. clin. Endocr. **23**, 197—202 (1963).

GIUFFRIDA, S.: Azione degli ormoni preipofisaril del sangue di donna gestante sulla capacità vitale e sullo sviluppo dei prematuri. Ann. Obstet. Ginec. **58**, 635—646 (1936).

GOOD, T. A., BESSMAN, S. P.: Anabolic steroids in cystic fibrosis of the pancreas. Amer. J. Dis. Childh. **111**, 272—277 (1956).

GOPALAN, C., SWAMINATHAN, M. C., KUMARI, V. K., RAO, D. H., VIJAYARAGHAVEN, K.: Effect of caloric supplementation on growth of undernourished children. Amer. J. clin. Nutr. **26**, 563—566 (1973).

GRACEY, M., ANDERSON, C. M.: Cystic fibrosis of the pancreas in adolescence and adulthood. Aust. Ann. Med. **18**, 91—101 (1969).

GREENBLATT, R. B., DALLA PRIA, S., BELLANGER, P.: Hormonal control of growth. Medical Tms (Lond.) **97**, 155—170 (1970).

GREENE, R., CARSTAIRS, L. S.: The effect of anabolic hormones on the growth of undersized boys. Brit. J. clin. Pract. **27**, 3—7 (1973).

GREULICH, W. W., PYLE, S. I.: Radiographic Atlas of Skeletal Development of the Hand and Wrist. Stanford, Calif.: Stanford Univ. Press 1959.

GUPTA, D., BUTLER, H.: Excretion of testosterone and epitestosterone glucuronides in pre-school and pre-adolescent and adolescent children. Steroids **14**, 343—367 (1969).

GUPTE, S.: Treatment of cystic fibrosis. Lancet **1974 II**, 251.

GUY, J. T., AUSLANDER, M. O.: Androgenic steroids and hepatocellular carcinoma. Lancet **1973 I**, 148.

GUYDA, H. J., JOHANSON, A. J., MIGEON, C. J., BLIZZARD, R. M.: Determination of serum luteinizing hormone (SLH) by radioimmunoassay in disorders of adolescent sexual development. Pediat. Res. **31**, 538—544 (1969).

HALL, R., ANDERSON, J., SMART, G. A., BESSER, G. M.: Clinical Endocrinology, 2nd Ed., p. 494. Philadelphia: J. B. Lippincott 1974.

HANSEN, A. E.: Influence of viosterol and parathyroid extract on mineral metabolism in osteogenesis imperfecta. Amer. J. Dis. Child. **50**, 132—157 (1935).

HARDY, J., WILKINS, L.: Methyl testosterone in the treatment of premature infants. J. Pediat. **34**, 439—442 (1949).

HELLER, C. G., MOORE, D. J., PAULSEN, C. A., NELSON, W. C., LAIDLAW, W. M.: Effects of progesterone and synthetic progestins on the reproductive physiology of normal men. Fed. Proc. **18**, 1057—1064 (1959).

HENDERSON, J. T., RICHMOND, J., SUMERLING, M. D.: Androgenanabolic steroid therapy and hepatocellular carcinoma. Lancet **1973 I**, 934.

HESSER, F. H., LANGWORTHY, O. R., VEST, S. A.: Muscle strength in myotonia atrophica (dystrophia myotonica) improved by testosterone propionate. Endocrinology **26**, 241—243 (1940).

HOERR, N. L., PYLE, S. E., FRANCIS, C. C.: Radiographic Atlas of Skeletal Development of the Foot and Ankle, a Standard of Reference. Springfield, Ill.: Chas. C. Thomas 1962.

HORTLING, H., WAHLFORS, K.: Long-term treatment of dwarfism with androgens and thyroid hormone. Acta endocr. (Kbh.) **32**, 563—574 (1959).

HOUTSMULLER, A. J.: The therapeutic applications of anabolic steroids in ophthalmology: biochemical results. Acta. endocr. (Kbh.) **39** (Suppl. 63), 154—174 (1961).

HUBBLE, D.: Endocrine control of growth and skeletal maturation. In: HUBBLE, D. (Ed.): Paediatric Endocrinology, pp. 1—33. Philadelphia: F. A. Davis Co. 1969.

HUBBLE, D., MACMILLAN, D. R.: A study of growth promotion in children. Arch. Dis. Childh. **37**, 518—524 (1962).

HUGHES, D. W.: Aplastic anaemia in childhood: A reappraisal. II. Idiopathic and acquired aplastic anaemia. Med. J. Aust. **2**, 361—366 (1973).

HUNG, W.: Juvenile hyperthyroidism. Clin. Proc. Child. Hosp. **28**, 3—12 (1972).

ILLIG, R.: Delayed adolescence. Pediat. Ann. **3**, 17—28 (1974).

ILLIG, R., PRADER, A.: Effect of testosterone on growth hormone secretion in patients with anorchia and delayed puberty. J. clin. Endocr. **30**, 615—718 (1970).

ISAACS, W. A., EFFIONG, C. E., AYENI, O.: Steroid treatment in the prevention of painful episodes in sickle-cell disease. Lancet **1972 I**, 570—571.

JENNER, M. R., KELCH, R. P., KAPLAN, S. L., GRUMBACH, M. M.: Hormonal changes in puberty. IV. Plasma estradiol, LH, and FSH in pre-pubertal children, pubertal females, and in precocious puberty, premature thelarche, hypogonadism, and in a child with a feminizing ovarian tumor. J. clin. Endocr. **34**, 521—530 (1972)

JOHNSON, F. L., FEAGLER, J. R., LERNER, K. G., MAJERUS, P. W., SIEGEL, M., HARTMANN, J. R., THOMAS, E. D.: Association of androgenic-anabolic steroid therapy with development of hepatocellular carcinoma. Lancet **1972 II**, 1273—1276.

JOHNSON, H. R., MYHRE, S. A., RUVALCABA, R. H. A., THULINE, H. C., KELLEY, V. C.: Effects of testosterone on body image and behavior in Klinefelter's syndrome: A pilot study. Dev. med. Child Neurol. **12** (Suppl. 22), 454—460 (1970).

JOSS, E. E., ZUPPINGER, K. A.: Stunted growth: diagnosis and differential diagnosis in the outpatient. Pediatrician **2**, 135—146 (1973).

KANAWATI, A. A., MCLAREN, D. S.: Failure of thrive in Lebanon. II. An investigation of the causes. Acta pediat. scand. **62**, 571—576 (1973).

KAPLAN, J. G., MOSHANG, T., JR., BERNSTEIN, R., PARKS, J. S., BONGIOVANNI, A. M.: Constitutional delay of growth and development: Effects of treatment with androgens. J. Pediat. **82**, 38—44 (1973).

KELCH, R. P., CONTE, F. A., KAPLAN, S. L., GRUMBACH, M. M.: Evidence for the episodic secretion of LH and decreasing sensitivity of the hypothalamic-pituitary "gonadostat" in adolescent patients with gonadal dysgenesis, (Abs.) Pediat. Res. **6**, 349 (1972a)

KELCH, R. P., GRUMBACH, M. M., KAPLAN, S. L.: Studies on the mechanism of puberty in man. In: SAXENA, B. B., BELING, C. G., GANDY, H. M. (Eds.): Gonadotropins, pp. 524—534. New York: Wiley-Interscience 1972b.

KELCH, R. P., KAPLAN, S. L., GRUMBACH, M. M.: Suppression of urinary and plasma gonadotropins by exogenous estrogens in prepubertal and pubertal children. (Abs.) Pediat. Res. **6**, 354 (1972c).

KELCH, R. P., WEINSTEIN, R. L., JENNER, M. R., KAPLAN, S. L., GRUMBACH, M. M.: Estrone secretion by human, simian and canine testes: effects of HCG on estrone and estradiol. International Congress, Series 256, Endocrinology, Washington, D. C., June 18—24, 1972d, (Abs. 542).

KELLER, P. J.: Induction of ovulation by synthetic luteinizing-hormone releasing factor in infertile women. Lancet **1972II**, 570—572.

KEMPE, C. H., SILVER, H. K., O'BRIEN, D., ET AL.: Current Pediatric Diagnosis and Treatment. Los Altos, Calif.: Lange Medical Publ., 1974.

KENNY, F. M., MIDGLEY, A. R., JR., JAFFE, R. B., GARCES, L. Y., VASQUEZ, A., TAYLOR, F. H.: Radioimmunoassayable serum LH and FSH in girls with sexual precocity, premature thelarche and adrenarche. J. clin. Endocr. **29**, 1272—1275 (1969).

KILLANDER, A., LUNDMARK, K. M., SJOLIN, S.: Idiopathic aplastic anemia in children: results of androgen treatment. Acta paediat. scand. **58**, 10—14 (1969)

KIVOWITZ, J., CORCORAN, J.: Theoretical and practical considerations of Klinefelter's syndrome in children: A report of three cases of 47 XXY. J. Amer. Acad. Child. Psychiat. **10**, 700—712 (1971).

KLINEFELTER, H. F., JR.: In: JONES, H. W., SCOTT, W. W. (Eds.): Hermaphroditism, Genital Anomalies and Related Endocrine Disorders, p. 80. Bailliere: Tindall & Cox, London 1958.

KLINEFELTER, H. F., JR., REIFENSTEIN, E. C., JR., ALBRIGHT, F.: Syndrome characterized by gynecomastia, aspermatogenesis without A-Leydigism, and increased excretion of folliclestimulating hormone. J. clin. Endocr. **2**, 615—627 (1942).

KLUTHE, R., RICHTER, F.: Klinische-experimentelle Untersuchungen über die Wirkung eines anabolen Steroids (Methylandrostenolonenanthat) auf die Stickstoffbilanz von Patienten mit Nephrotischem Syndrom. Klin. Wschr. **41**, 244—245 (1963).

KOCHAKIAN, C. D., MURLIN, J. R.: The effect of male hormone on the protein and energy metabolism of castrate dogs. J. Nutr. **10**, 437—459 (1935).

KORNER, A., MANCHESTER, K. L.: Insulin and protein metabolism. Brit. med. Bull. **16**, 233—236 (1960).

KRUSKEMPER, H.-L.: Anabolic Steroids (DOERING, C. H., Transl.) p. 236. New York: Academic Press 1968.

KUHLMANN, F.: Zur Pathogenese und Therapie der Sprue. Med. Klin. **56**, 1659—1665 (1961).

KULIN, H. E., GRUMBACH, M. M., KAPLAN, S. L.: Changing sensitivity of the pubertal gonadal hypothalamic feedback mechanism in man. Science **166**, 1012—1013 (1969).

KULIN, H. E., GRUMBACH, M. M., KAPLAN, S. L.: Gonadal-hypothalamic interaction in prepubertal and pubertal man: Effect of clomiphine citrate on urinary follicle-stimulating hormone and luteinizing hormone and plasma testosterone. Pediat. Res. **6**, 162—171 (1972).

KULIN, H., REITER, E. O.: Delayed sexual maturation, with special emphasis on the occurrence of the syndrome in the male. In: Conference on the Control of the Onset of Puberty, Airlie, Va., Oct. 8—11, 1972, pp. 238—257.

KULIN, H. E., RIFKIND, A. B., ROSS, G. T., ODELL, W. D.: Total gonadotropin activity in the urine of prepubertal children. J. clin. Endocr. **27**, 1123—1128 (1967).

KUPPERMAN, H. S.: Human Endocrinology, pp. 120—150. Philadelphia: F. A. Davis Co. 1963.

LAMBRECHTS, A., JADOUL, R.: Influence of norandrostenolone phenylpropionate on growth and balance experiments in children. Acta endocr. (Kbh.) **39**, (Suppl. 63), 111—123 (1961).

LANDON, J., WYNN, V., COOKE, J. N., KENNEDY, A.: Effects of anabolic steroid, methandienone, on carbohydrate metabolism in man. Metabolism **11**, 501—512 (1962).

LANGNESS, U., BEHNKE, H.: Collagen metabolites in plasma and urine in osteogenesis imperfecta. Metabolism **20**, 456—463 (1971).

LARON, Z.: Breast development induced by methandrostenolone (Dianabol). J. Clin. Endocr. **22**, 450—452 (1962).

Lee, P. A., Jaffe, R. B., Midgley, A. R., Jr., Kohen, F., Niswender, G. D.: Regulation of human gonadotropins. VIII. Suppression of serum LH and FSH in adult males following exogenous testosterone administration. J. clin. Endocr. **35**, 636—641 (1972).
Lee, P. A., Midgley, A. R., Jr., Jaffe, R. B.: Regulation of human gonadotropins. VI. Serum follicle stimulating and luteinizing hormone determinations in children. J. clin. Endocr. **31**, 248—253 (1970).
Lefort, R.: Osteopsatirose. Bull. Soc. belg. Orthop. **10**, 129 (1938).
Limbeck, G. A., Ruvalcaba, R. H. A., Mahoey, C. P., Kelly, V. C.: Studies on anabolic steroids. IV. The effects of oxandrolone on height and skeletal maturation in uncomplicated growth retardation. Clin. Pharmacol. Ther. **12**, 798—805 (1971).
Lundh, B., Gardner, F. H.: The haematological response to androgens in sickle cell anemia. Scand. J. Haematol. **7**, 389—397 (1970).
Lyster, S. C., Duncan, G. W.: Anabolic androgenic and myotropic activities of derivatives of 7-alpha-19-nortestosterone. Acta endocr. (Kbh.) **43**, 399—411 (1963).
Marshall, W. A.: Assessment of skeletal maturity. In: Van der Werff ten Bosch, J. J., Haak, A. (Eds.): Somatic Growth of the Child, p. 31. Springfield, Ill.: Charles C. Thomas 1966.
Marshall, W. A., Tanner, J. M.: Variations in patterns of pubertal changes in girls. Arch. Dis. Childh. **44**, 291—303 (1969).
Marshall, W. A., Tanner, J. M.: Variation in the pattern of pubertal changes in boys. Arch. Dis. Childh. **45**, 13 (1970).
Marti-Henneberg, C., Niiranen, A. K., Rappaport, R.: Oxandrolone treatment of constitutional short stature in boys during adolescence: Effect on linear growth, bone age, pubic hair and testicular development. J. Pediat. **86**, 783—788 (1975).
Martin, E.: „Unden" bei der Aufzucht von Frühgeburten. Wschr. Geburtsh. Gynäk. **82**, 66—69 (1929).
Martin, L. G., Clark, J. W., Connor, T. B.: Preliminary communication growth hormone secretion enhanced by androgens. J. clin. Endocr. **28**, 425—428 (1968).
Martin, M. M., Wilkins, L.: Pituitary dwarfism: diagnosis and treatment. J. clin. Endocr. **18**, 679—693 (1958).
McCullagh, E. P., Jones, R.: A note on the effect of certain androgens upon red blood cell count and upon glucose tolerance. Cleveland Clin. Quart. **8**, 79—84 (1941).
McCullagh, E. P., Rossmiller, H. R.: Methyltestosterone. III. Effect upon body weight and growth. J. clin. Endocr. **1**, 507—510 (1941).
McCullagh, E. P., Schiffenburg, C. A.: The testes. In: Glandular Physiology and Therapy, 5th Ed., pp. 220—257. (Council on Pharmacy and Chemistry, American Medical Assoc.) Philadelphia: J. B. Lippincott Co. 1954.
McKusick, V. A.: Heritable Disorders of Connective Tissue, 2nd Ed., p. 178. St. Louis: C. V. Mosby Co. 1960.
McLaren, D. S.: The great protein fiasco. Lancet **1974II**, 93—96.
Meadows, A. T., Naiman, J. L., Valdesdapena, M.: Hepatoma associated with androgen therapy for aplastic anemia. J. Pediat. **84**, 109—110 (1974).
Meadows, R. W., Hughes, W. T., Walker, L. C., Ettledorf, J. N.: Effects of oral 19-nortestosterone derivative (nilevar) on growth of premature infants. Amer. J. Dis. Child. **99**, 206—211 (1960).
Mellman, W. J., Bongiovanni, A. M., Garrison, M., Steiker, D. D.: Comparison of fluoxymesterone and methyltestosterone as growth stimulants. Pediatrics **28**, 525—530 (1961).
Mentzer, W. C., August, C. S., Nathan, D. G.: The effects of androgen administration in sickle cell anemia. (Abs.) Pediat. Res. **3**, 378 (1969).
Mirand, E. A., Murphy, G. P.: Erthropoietin alterations in patients with uremia, renal allografts, or without kidneys. J. Amer. med. Ass. **209**, 392—398 (1969).
Mirand, E. A., Murphy, G. P., Steeves, R. A., et al.: Erythropoietin activity in anephric, allotransplanted, unilaterally nephrectomized, and intact man. J. Lab. clin. Med. **73**, 121—128 (1969).
Moench, A., Sartorius, H.: Klinische Untersuchungen über Testosteronbehandlung beim Nephrotischen Syndrom. Klin. Wschr. **32**, 329—330 (1954).
Moldawer, M.: Anabolic agents: clinical efficacy versus side effects. J. Amer. med. Wom. Ass. **23**, 352—356, passim, (1968).

MONCRIEFF, A.: Administration of thyroid gland to premature babies. Arch. Dis. Childh. **13**, 57—64 (1938).

MUNSON, A. R.: Some effects of anabolic steroids during weight training. Ph. D. thesis dissertation, Univ. of Mich., University Micro-films, Ann Arbor, Mich.

MYHRE, S. A., RUVALCABA, R. H. A., JOHNSON, H. R., THULINE, C., KELLY, V. C.: The effects of testosterone treatment in Klinefelter's syndrome. J. Pediat. **76**, 267—276 (1970).

NAISMITH, D. J.: Kwashiokor in western Nigeria: a study of traditional weaning foods, with particular reference to energy and linoleic acid. Brit. J. Nutr. **30**, 567—576 (1973).

NAETS, J. P., WITTEK, M.: Erythropoiesis in anephric man. Lancet **1968I**, 941—943.

NANKIN, H. R., YANAIHARA, T., TROEN, P.: Response of gonadotropins and testosterone to clomiphene stimulation in a pubertal boy. J. clin. Endocr. **33**, 360—363 (1971).

NATHAN, D. G., SCHUPAK, E., STOHLMAN, F., JR., et al.: Erythropoiesis in anephric man. J. clin. Invest. **43**, 2158—2165 (1964).

News Conference called by Muscular Dystrophy Association of America, Sept. 2, Phys. Therapy **44**, 1041—1044 (1964).

NIELSEN, J.: Klinefelter's syndrome in a neurological ward. Acta neurol. scand. **41**, 197—214 (1965).

NIELSEN, J.: Criminality among patients with Klinefelter's snydrome and the XXY syndrome. Brit. J. Psychiat. **117**, 365—369 (1970).

NIELSEN, J., BJARNASON, S., FRIEDRICH, U., FRØLAND, A., HANSEN, H., SORENSEN, A.: Klinefelter's syndrome in children. J. Child. Psychol. **11**, 109—119 (1970).

NIELSEN, J., FISCHER, M.: Sex-chromatin and sex-chromosome abnormalities in male hypogonadal mental patients. Brit. J. Psychiat. **111**, 641—647 (1965).

ORLANSKY, G. J., OSTERHELD, R. G.: The use of a synthetic anabolic, methandrostenolone, as an adjuvant in the treatment of epilepsy. J. Neuropsychiat. **4**, 366—368 (1963).

PARKER, J. P., BEIRNE, G. J., DESAI, J. N., et al.: Androgen induced increase in red-cell 2,3-diphosphyglycerate. New Engl. J. Med. **287**, 381—383 (1972).

PATEISKY, K., SCHINKO, H., HABERLER, H.: A study of the treatment of dystrophia musculorum progressiva with an anabolic steroid: nòrandrostenolone decanoate. Acta endocr. (Kbh.) **39** (Supp. 63), 185—198 (1961).

PAULSEN, C. A.: The testes. In: WILLIAMS, R. H. (Ed.): Textbook of Endocrinology, 4th Ed., pp. 405—458. Philadelphia: W. B. Saunders Co. 1968.

PAULSEN, C. A., DESOUZA, A., YOSHIZUMI, T., LEWIS, B. M.: Results of a buccal smear survey in noninstitutionalized adult males. J. clin. Endocr. **24**, 1182—1187 (1964).

PAULSEN, E. P., SOBEL, E. H., SHAFRAN, M. S.: Urinary steroid metabolites in children. I. Individual 17-ketosteroids in children with normal sexual development. J. clin. Endocr. **26**, 329—339 (1966).

PENNINGTON, G. W., DEWHURST, C. J.: Hormone excretion in premenarcheal girls. Arch. Dis. Childh. **44**, 629—636 (1969).

PENNY, R., GUYDA, H. J., BAGHDASSARIAN, A., JOHANSON, J. A., BLIZZARD, R. M.: Correlation of serum follicular stimulating hormone (FSH) and luteinizing hormone (LH) as measured by radioimmunoassay in disorders of sexual development. J. clin. Invest. **49**, 1847—1852 (1970).

PERHEENTUPA, J., LENKO, H.-L., NEVALAINEN, M. N., SODERHOLM, A., TAIPALE, V.: Hormonal treatment of Turner's Syndrome. Presentation, First International Symposium on Adolescent Medicine, Helsinki, Finland, 1974.

PERLSTEIN, M. A., GUTTERMAN, H.: Testosterone in progressive pseudohypertrophic muscular dystrophy. J. Pediat. **37**, 743—749 (1950).

PERUGINI, S., LUSVARGHI, E., VACCARI, G.: Androgenes et corticosteroides associes dans le tritement des aplasies medullaires acquises. Schweiz. med. Wschr. **100**, 1982—1984 (1970).

PLANTIER, H. A.: Precocious pubescence-iatrogenic. New Engl. J. Med. **270**, 141—142 (1964).

POTTER, M. F.: Oral administration of oestrin to premature babies. Brit. med. J. **1937I**, 1201—1202.

PRADER, A.: The influence of anabolic steroids on growth. Acta. endocr. (Kbh.) **39**, (Suppl. 63), 78—88 (1961).

PRADER, A., ILLIG, R.: Use of anabolic agents in disorders of growth. Ciba Foundation. Protein metabolism. Influence of Growth Hormone, Anabolic Steroids, and Nutrition in Health and Disease. An International Symposium. Leyden 25—29, June, 1962, pp. 383—397.

PYLE, S. I., HOERR, N. L.: Radiographic Atlas of Skeletal Development of the Knee; A Standard of Reference. Springfield. Ill.: Charles C. Thomas 1955.
QUINN, E. L., WORCESTER, R. L.: Chronic thyrotoxic myopathy: report of a case. J. clin. Endocr. **11**, 1564—1571 (1951).
RAMIREZ, D. V., MCCANN, S. M.: Comparison of the regulation of luteinizing hormone (LH) secretion in immature and adult rats. Endocrinology **72**, 452—464 (1963).
RAY, C. G., KIRSCHVINK, J. F., WAXMAN, S. H., KELLEY, V. C.: Studies of anabolic steroids. III. The effect of oxandrolone on height and skeletal maturation in mongoloid children. Amer. J. Dis. Child. **110**, 618—623 (1965).
REILLY, W. A., EARLE, A. M.: Methyltestosterone treatment of premature infants. Amer. J. Dis. Child. **82**, 323—324 (1951).
REILLY, W. A., GORDAN, G. S.: Dissociation of growth-stimulating and skeletal maturing actions of the synthetic androgen, fluoxymesterone. J. Pediat. **59**, 188—193 (1961).
RICHARDSON, J. R., JR., WEINSTEIN, M. B.: Erythropoietic response of dialyzed patients to testosterone administration. Ann. intern. Med. **73**, 403—407 (1970).
RIFKIND, A. B., KULIN, H. E., RAYFORD, P. L., CORGILLE, C. M., ROSS, G. T.: 24-hour urinary luteinizing hormone (LH) and follicle stimulating hormone (FSH) excretion in normal children. J. clin. Endocr. **31**, 517—525 (1970).
ROOT, A. W.: Human Pituitary Growth Hormeone, p. 259. Springfield, Ill.: Charles C. Thomas 1972.
ROOT, A. W.: Endocrinology of puberty: I. Normal sexual maturation. J. Pediat. **83**, 1—19 (1973).
ROOT, A. W., BONGIOVANNI, A. M., EBERLIN, W. R.: Heterosexual precocious puberty: congenital adrenal hyperplasia. Clin. Obstet. Gynec. **11**, 815—834 (1968).
ROOT, A. W., BONGIOVANNI, A. M., EBERLIN, W. R.: Diagnosis and management of growth retardation with special reference to the problem of hypopituitarism. J. Pediat. **73**, 737—753 (1971).
ROOT, A., DECHERNEY, A., RUSS, D., DUCKETT, G., GARCIA, C.-R., WALLACE, E.: Episodic secretion of luteinizing and follicle stimulating hormones in agonadal and hypogonadal adolescents and adults. J. clin. Endocr. **35**, 700—704 (1972a).
ROOT, A. W., MOSHANG, T., JR., BONGIOVANNI, A. M., EBERLIN, W. R.: Concentrations of plasma luteinizing hormone in infants, children and adolescents with normal and abnormal gonadal function. Pediat. Res. **4**, 175—186 (1970).
ROOT, A., STEINBERGER, E., SMITH, K., STEINBERGER, A., RUSS, D., SOMERS, L., ROSENFIELD, R.: Isosexual pseudoprecocity in a 6-year-old boy with a testicular cell adenoma. J. Pediat. **80**, 264—268 (1972b).
ROSADINI, I.: Action of 4-chlorotestosterone acetate in anorexia nervosa (It). Minerva med. **501**, 3001—3003 (1959).
ROSEN, S. W., WEINTRAUB, B. D.: Monotropic increase of serum FSH correlated with low sperm count in young men with idiopathic oligospermia and aspermia. J. clin. Endocr. **32**, 410—416 (1971).
ROSENBERG, I. N., AHN, C. S., MITCHELL, M. L.: Effects of anabolic steroids upon circulating thyroid hormones. J. clin. Endocr. **22**, 612—616 (1962).
ROSENFIELD, R. L.: Plasma 17-ketosteroids with 17-beta-hydroxysteroids in girls with premature development of sexual hair. J. Pediat. **79**, 260—266 (1971).
ROSENFIELD, R. L., BARMACH DE NIEPOMNISZCZE, A. I., OTTO, P.: Plasma androstenediol in normal and abnormal development (Abst.) Pediat. Res. **6**, 355 (1972).
ROSENFIELD, R. L., FANG, V. S., DUPON, C., MOON, H. K., REFETOFF, S.: The effects of low doses of depot estradiol and testosterone in teenagers with ovarian failure and Turner's syndrome. J. clin. Endocr. **37**, 574—580 (1973).
ROSS, G. T., VAN HALL, E. V., VAITUKAITIS, J. L., BRAUNSTEIN, G. D., RAYFORD, P. L.: Sialic acid and the immunologic and biologic activity of gonadotropins. In: SAXENA, B. B., BELING, C. G., GANDY, H. M. (Eds.): Gonadotropins, pp. 417—423. New York: Wiley-Interscience 1972.
ROY, S. L.: Anabolic hormones. Indian med. J. **60**, 107—108 (1966).
RYAN, W. J.: Osteogenesis imperfecta. With a suggestion for treatment. J. Bone Jt Surg. **14**, 939—942 (1932).
SAEZ, J. M., MORERA, A. M., BERTRAND, J.: Plasma concentration of estrone ($E_1$) and estradiol ($E_2$) before puberty in humans. Internat. Congress Steries 256. International Congress of Endocrinology Washington, D. C., June 18—24, 1972, (Abs. 306).

SCHABAD, I. A.: Der Stoffwechsel bei angeborener Knochenbrüchigkeit (Osteogenesis imperfecta). Kinderheilkunde **11**, 230—266 (1914).

SCHIFFRIN, D., REPETTI, L., UCERO, G., CANARO, F.: The use of a new anabolic drug in pulmonary tuberculosis: Día méd. **33**, 378—380 (1961).

SCHOEN, E. J.: Treatment of idiopathic precocious puberty in boys. J. clin. Endocr. **26**, 363—370 (1966).

SCHOEN, E. J., SOLOMON, I. L., MILKOVICH, L.: Bone age in short boys and tall girls. Amer. J. Dis. Child. **119**, 252—254 (1970).

SECORD, E. W., WILDER, R. M., HENDERSON, M. S.: Osteogenesis imperfecta tarda treated with thymus extract. Proc. Mayo Clin. **11**, 1—5 (1936).

SEILICOVICH, A.: Effect of FSH on the uptake of testosterone $^3$H by the seminiferous tubule. Internat. Congress Series 256, International Congress of Endocrinology, Washington, D. C. June 18—24, 1972 (Abs. 409).

SEITCHIK, J. N., AGERTY, H. A.: Experimental use of methyltestosterone in premature infants. Pediatrics **5**, 200—203 (1950).

SHAHIDI, N. T.: Androgens and erythropoiesis. New Engl. J. Med. **289**, 72—80 (1973).

SHAHIDI, N. T., DIAMOND, L. K.: Testosterone-induced remission in aplastic anemia of both acquired and congenital types: further observations in 24 cases. New Engl. J. Med. **264**, 953—967 (1961).

SHALDON, S., KOCH, K. M., OPPERMANN, F. et al.: Testosterone therapy for anaemia in maintenance dialysis. Brit. med. J. **1971III**, 212—215.

SHELTON, E. K., VARDEN, A. E.: The use of methyltestosterone in the treatment of premature infants. J. clin. Endocr. **6**, 812—816 (1946).

SHELTON, E. K., VARDEN, A. E., MARK, J. S.: Experimental use of testosterone compounds in premature infants. J. clin. Endocr. **7**, 708—713 (1947).

SIMPSON, M. E., ASLING, C. E., EVANS, H. M.: Some endocrine influences on skeletal growth and differentiation. Yale J. Biol. Med. **23**, 1—27 (1950).

SIZONENKO, P. C., LEWIN, M.: Problemes physiologiques de la puberté. I. Modifications metaboliques. Arch. franç. Pediat. **29**, 169—183 (1972).

SIZONENKO, P. C., LEWIN, M., BURR, I. M.: Problemes physiologiques de la puberté. II. Les secretions endocrines peripheriques. Arch. franç. Pediat. **29**, 185—201 (1972).

SNYDERMAN, S. E., HOLT, L. E., JR.: The effect of high caloric feeding on the growth of premature infants. J. Pediat. **58**, 237—240 (1961).

SOBEL, E. H.: Anabolic steroids. In: ASTWOOD, E. B., CASSIDY, C. E. (Eds.): Clinical Endocrinology, Vol. 2, pp. 789—797. New York: Grune & Stratton, Inc. 1968.

SOBEL, E. H.: Anabolic steroids in cystic fibrosis. J. Pediat. **74**, 157—158 (1969).

SOBEL, E. H., RAYMOND, C. S., QUINN, K. V., TALBOT, N. B.: The use of methyltestosterone to stimulate growth: relative influence on skeletal maturation and linear growth. J. clin. Endocr. **16**, 241—248 (1956).

SOHVAL, A. R.: Klinefelter's syndrome. Medical Aspects Human sexuality **3**, 69—71, passim. (1969).

STEIN, K., JACOBSEN, B.: The effects of castration and cryptorchidization on the blood picture of the golden hamster. Anat. Rec. **88**, 459 (1944).

STEINBERGER, E.: Hormonal control of mammalian spermatogenesis. Physiol. Rev. **51**, 1—22 (1971).

STEINGLASS, P., GORDON, A. S., CHARIPPER, H. A.: Effect of castration and sex hormones on blood of the rat. Proc. Soc. exp. Biol. **48**, 169—177 (1941).

STEVENSON, C. J., BOTTOMS, E., SHUSTER, S.: Skin collagen in osteogenesis imperfecta. Lancet **1970I**, 860—861.

SUKHATME, P. V.: Size and nature of the protein gap. Nutr. Rev. **28**, 223—226 (1970).

TABER, E., DAVIS, D. F., DOMM, L. V.: Effect of sex hormones on erythrocyte number in the blood of the domestic fowl. Amer. J. Physiol. **138**, 479—487 (1943).

TAINTER, M. L., ARNOLD, A., BEYLER, A. L., POTTS, G. O., ROTH, C. H.: Anabolic steroids in the management of the diabetic patient. N. Y. St. J. Med. **64**, 1001—1009 (1964).

TANNER, J. M.: Growth at Adolescence, 2nd Ed., pp. 32—37. Oxford, England: Blackwell Scientific Publ. 1962.

TANNER, J. M.: Growth and endocrinology of the adolescent. In: GARDNER, L. I. (Ed.): Endocrine and Genetic Diseases of Childhood, pp. 19—69. Philadelphia: W. B. Saunders Co. 1969.

TANNER, J. M.: Sequence and tempo in the somatic changes in puberty. In: The Control of the Onset of Puberty, pp. 448—470. Airlie, Va., Oct. 8—11, 1972.
TANNER, J. M., GUPTA, D.: A longitudinal study of the urinary excretion of individual steroids in children from 8 to 12 years old. J. Endocr. **41**, 139—156 (1968).
THOMAS, J. A., MAWWHINNEY, M. G.: Pharmacology of androgens. In: Synopsis of Endocrine Pharmacology, pp. 140—159. Baltimore: University Park Press 1973.
TODD, T. W.: Atlas of Skeletal Maturation. St. Louis: C. V. Mosby Co. 1937.
VAN DYKE, D. C., CONTOPOULOS, A. N., WILLIAMS, B. S., et al.: Hormonal factors influencing erythropoiesis. Acta. haemat. **11**, 203—222 (1954).
VAN METER, J. R.: Progressive muscular dystrophy—a preliminary report on treatment with amino acids, folic acid and vitamins. Calif. Med. **79**, 297—299 (1953).
VAN THIEL, D. H., SHERINS, R. J., MYERS, G. H., JR., DEVITA, V. T., JR.: Evidence for a specific seminiferous tubular factor affecting follicle-stimulating hormone secretion in man. J. clin. Invest. **51**, 1009—1019 (1972).
VESTERGAARD, P.: Urinary excretion of individual 17-keto-steroids in children. Acta endocr. (Kbh.) **49**, 436—442 (1965)
VISSER, H. K. A., CROUGHS, W.: Effects of androgens and anabolic steroids on growth rate and height. In: VAN DER WERFF TEN BOSCH, J. J., HAAK, A. (Eds.): Somatic Growth of the Child, pp. 290—309. Springfield, Ill.: Charles C. Thomas 1966.
VOLLMER, E. P., GORDON, A. S.: Effect of sex and gonadotropic hormones upon the blood picture of the rat. Endocrinology **29**, 828—837 (1941).
WALTON, J. N.: Clinical aspects of human muscular dystrophy. In: BOURNE, G. H. & GOLARZ, M. N. (Eds.): Muscular Dystrophy in Man and Animals, pp. 264—321. Basel: S. Karger 1963.
WARING, J. J., RAVIN, A., WALKER, C. E.: Steroids in dystrophia myotonica. II: Clinical features and treatment. Arch. intern. Med. **65**, 763—799 (1940).
WATERLOW, J. C., CRAVIOTO, J., STEPHEN, J. M. L.: Protein malnutrition in man. Advanc. Protein Chem. **15**, 131—238 (1960).
WEBER, H., HAGGE, W.: Über die erfolgreiche Behandlung der Zystinose mit einem Anabolicum. Arch. Kinderheilk. **168**, 110—128 (1963)
WEILAND, R. G., YEN, S. S. C., POHLMAN, C.: Serum testosterone levels and testosterone binding affinity in prepubertal and adolescent males: correlation with gonadotropins. Amer. J. med. Sci. **259**, 358—360 (1970).
WEWALKA, F. G.: Anabolic steroids in the management of chronic wasting diseases. J. Amer. med. Wom. Ass. **23**, 339—345 (1968).
WILKINS, L.: The Diagnosis and Treatment of Endocrine Disorders in Childhood and Adolescence, 3rd Ed., Springfield, Ill.: Charles C. Thomas 1965.
WINTER, J. S. D., FAIMAN, C.: Pituitary-gonadal relations in male children and adolescents. Pediat. Res. **6**, 126—135 (1972a).
WINTER, J. S. D., FAIMAN, C.: Serum gonadotropin concentrations in agonadal children and adults. J. clin. Endocr. **35**, 561—564 (1972b).
WYNN, V.: The anabolic steroids. Practitioner **200**, 509—518 (1968).
YADAO, J. S., FRANCO, I. F., CASPELLAN, A., VINALAY, P.: Anabolic therapy in children. J. Phillips Med. Ass. **41**, 201—206 (1965).
YEN, S. S. C., VICIC, W. J.: Serum follicle stimulating hormone levels in puberty. Amer. J. Obstet. Gynec. **106**, 134—137 (1970).
YEN, S. S. C., VICIC, W. J., KEARCHNER, D. V.: Gonadotropin levels in puberty. I. Serum luteinizing hormone. J. clin. Endocr. **29**, 382—385 (1969).
ZACHMANN, M., PRADER, A.: Anabolic and androgenic effect of testosterone in sexually immature boys and its dependency on growth hormone. J. clin. Endocr. **30**, 85—95 (1970).
ZANGENEH, F., KELLEY, V. C.: Granulosetheca-cell tumor of the ovary in children. Amer. J. Dis. Child. **115**, 494—508 (1968).
ZANGENEH, F., STEINER, M. M.: Oxandrolone therapy in growth retardation of children. Amer. J. Dis. Child. **113**, 234—241 (1967).
ZIEGENFUSS, J., CARABASI, R.: Androgens and hepatocellular carcinoma. Lancet **1973I**, 262.

# C. The Hematologic Effects of Androgens

C. W. Gurney

## I. Introduction

Many clinical and experimental findings extending over a span of more than four decades indicate that androgens have a noteworthy effect on red cell formation. The earliest clinical reports of the erythropoietic stimulating properties of androgens were hematologic observations in women receiving large doses of testosterone as therapy of metastatic breast carcinoma (Talbot and Escher, 1949; Kennedy et al., 1949).

Following the demonstrations by Shahidi and Diamond in 1959 and Gardner and Pringle in 1961 that androgens have a clear and beneficial therapeutic effect in aplastic pancytopenias of children and adults, the interest of clinicians in this subject was intensified. Basic and clinical investigations have indicated the nature of the mechanisms by which androgens exert their erythropoietic effects. Elucidation of the physiologic control of red cell production and disturbances in this regulation in disease states have led to the extension, on a rational basis, of androgen therapy for some of the secondary anemias, as well as the primary pancytopenias, of man. Finally, data are emerging which suggest androgens may also have effects on leukopoiesis. It is the goal of this chapter to summarize the hematologic effects of androgens.

## II. The Effects of Androgens on Erythropoiesis

### A. Effects of Androgenic Steroids on Peripheral Blood Values

Normal human males have long been known to have higher peripheral red cell values than normal human females (Williamson, 1916; Vahlquist, 1950). This difference does not emerge until after puberty. It persists even after iron deficiency, so common in the young adult female, has been excluded as a contributing factor to sex differences in blood counts. The basis of the sex difference in red cell mass and peripheral red blood counts is now generally accepted to be the consequence primarily of the erythropoietic support afforded by androgens, although inhibition of erythropoiesis by estrogens (Dukes and Goldwasser, 1961) may also contribute to this difference. Castration of male rats leads to a fall in red counts which can be reversed by serial injections of testosterone (Steinglass et al., 1941). The anemia of hypogonadism in the male is responsive to testosterone therapy (McCullagh and Jones, 1941). Investigations in several laboratories have verified the stimulatory effect of androgens on erythropoiesis in chickens and mammals (Taber et al., 1943;

CRAFTS, 1946; VAN DYKE et al., 1954). The importance of such hematopoietic support in males has been attributed, by an anonymous teleologist, to the survival value of an expanded red cell mass, better able to protect against the devastating effects of hemorrhage secondary to combat-induced wounds. Such trauma, it is speculated, occurs with greater frequency in the male because of heightened aggressivity attributed to the consequences of androgen-induced alterations in behavior. The historical record of the Women's Liberation Movement in the last decade has afforded reason for some to doubt, but not yet reject with finality, this hypothesis.

## B. Mechanisms of Action of Androgenic Steroids on Red Cell Formation

In the last two decades, the mechanism of regulation of erythropoiesis has been elucidated in great detail (KRANTZ and JACOBSON, 1970; KRANTZ, 1973). The potential for cellular differentiation into a number of different hematopoietic lines resides in primitive cells in the bone marrow. Much of our information comes from experiments on mice, but there is no reason to suspect that the general conclusions from these studies do not hold for all mammals. Multipotent hematopoietic murine stem cells are able to produce macroscopic colonies in the spleen when injected into irradiated mice (TILL and MCCULLOCH, 1961), and hence are designated as colony-forming units, or CFC. The mechanisms governing the commitment of CFC into any one hematopoietic cell line are currently not known. However primitive cells, once committed to the erythroid series, probably exist in a dormant state until triggered by a glycoprotein, erythropoietin. Although morphologic criteria do not yet exist to allow for identification and specific characterization of colony-forming cells as opposed to erythropoietin-responsive cells, results from a number of experiments (BRUCE and MCCULLOCH, 1964; HANNA, 1967; O'GRADY et al., 1968; PORTEOUS and LAJTHA, 1968; REISSMANN and SAMORAPOOMPICHIT, 1969; STEPHENSON and AXELROD, 1971; REISSMANN and UDUPA, 1972) afford a strong basis for the conclusion that erythropoietin exerts its primary erythropoietic effect by acting on a cell already committed to the erythroid line.

In a very short time after exposure to erythropoietin, a number of biochemical events occur, including RNA synthesis (KRANTZ, 1973; KRANTZ and GOLDWASSER, 1965; GROSS and GOLDWASSER, 1971), RNA-dependent DNA synthesis (GROSS and GOLDWASSER, 1970), and hemoglobin formation (KRANTZ and JACOBSON, 1970; KRANTZ, 1973; GROSS and GOLDWASSER, 1971; NAKAO et al., 1968). Subsequent maturation of erythroblasts, including a number of cell divisions and ultimately the extrusion of their nuclei by the daughter cells, probably can occur in the absence of erythropoietin, although maturation of early erythroblasts appears to be accelerated by the presence of this glycoprotein (GALLAGHER and LANGE, 1962; PAUL et al., 1973). The final liberation of the red cell from the hematopoietic tissue into the blood stream appears to be facilitated directly or indirectly by the action of erythropoietin on the hematopoietic microvasculature (MCCUSKEY et al., 1972; CHAMBERLAIN et al., 1975).

The common denominator of all stimuli leading to the production of erythropoietin is hypoxia. Most erythropoietin is probably produced by the kidney, or, perhaps as suggested by several investigators (CONTRERA et al., 1966; KURATOWSKA, 1967; ZANJANI et al., 1967; FISHER et al., 1971), by the interaction of a renal factor and an inactive erythropoietin precursor in the plasma. The evidence for such a complicated interaction is not yet universally accepted, and many experimental hematologists still believe an active form of erythropoietin is produced directly by the kidney and to a lesser extent, by other organs of the body. It is important to realize erythropoiesis cannot long continue in the absence of primitive erythroid precursors of the hematopoietic system. Another ingredient generally considered to be essential for normal erythropoiesis is erythropoietin. Other substances which are known to initiate or increase erythropoiesis have all been shown to increase the production of erythropoietin. These include the cobaltous ion (GOLDWASSER et al., 1957), cyclic AMP (SCHOOLEY and MAHLMAN, 1971), and prostoglandins (SCHOOLEY and MAHLMAN, 1972). The only circumstances in which erythroid differentiation appears to occur in the absence of erythropoietin are during erythroblastic crises seen in a form of murine leukemia (SASSA et al., 1966; MIRAND, 1967), and in polycythemia vera of man (KRANTZ, 1968). Erythropoiesis in Di Guglielmo's syndrome appears to be at least partially dependent upon erythropoietin (ADAMSON and FINCH, 1970). As will be discussed in detail later in this chapter, it has been proposed that metabolites of androgenic steroids may also stimulate erythroid differentiation in rodents presumed to be devoid of erythropoietin.

From this summary of the physiology of erythropoiesis, it follows that there are at least two different mechanisms by which androgens might be expected to exert their effects on erythropoiesis. A recounting of the experimental evidence supporting two mechanisms will be preceded by a theoretical consideration of possibilities which, it is hoped, will facilitate the reader's appreciation of the significance of the multitude of independent observations relevant to the topic under examination.

First, androgens might increase the production of erythropoietin. If androgenic steroids were to produce hypoxia around or within the cells responsible for the elaboration of erythropoietin, the increased production of the hormone would be an expected logical consequence. It is relevant to note the change in erythrocyte 2,3-diphosphoglycerate (DPG) content after administration of androgens because the concentration of this substance increases markedly when adult red cells are made hypoxic (OSKI et al., 1970). Patients with chronic renal disease who were maintained on a program of biweekly hemodialysis were found to have substantial increases in their erythrocyte 2,3-DPG content after treatment with testosterone enanthate was initiated (PARKER et al., 1972). The administration of 19-nortestosterone decanoate to squirrel monkeys and mice produced an increase of erythrocyte 2,3-DPG, and a "right shift" of the $P_{50}$ was noted (GORSHEIN et al., 1974). The rise in 2,3-DPG reduces the affinity of hemoglobin for oxygen and constitutes an important adaptive mechanism which increases the partial pressure of oxygen for any given level of red cell oxygen content. It is possible that androgens lower the intracellular oxygen tension by some unknown mechanism. If that is true, the increased production of both 2,3-DPG and erythropoietin would be logical consequences of intracellular hypoxia. If, however, subsequent investigations should show the 2,3-DPG elevation represents a direct effect of androgens rather than a response to impaired intracellular oxidation, then the theoretical basis for our current line of reasoning would crumble, since androgens would then be expected to increase the oxygen tension in the environment of erythropoietin-producing cells, and thereby reduce, rather than increase, the stimulus for the production of erythropoietic factor. No information on the effects of androgens on the subcellular metabolic processes within the cells of erythropoietin production is yet available, therefore intracellular hypoxia or the accentuation of some metabolic consequence of hypoxia must be retained as a theoretical possibility if the evidence is convincing that androgens actually do increase erythropoietin production.

Secondly, androgenic steroids might exert their effect by increasing the number of primitive erythroid precursors of the hematopoietic system or by heightening the sensitivity of these cells to erythropoietin. Such an effect upon the erythroid primitive precursor compartment constitutes an attractive hypothesis when one attempts to account for the beneficial effects of androgen therapy in patients with aplastic anemia, since such patients are presumed to be hypoxic as a consequence of their anemia, and they usually already have very high plasma erythropoietin titers to which their hematopoietic tissues are resistant prior to androgen therapy. Such a hypothesis is on less firm ground, however, in affording an explanation for the erythropoietic effect of androgens in normal mammals. Erythropoiesis in the mouse is quite sensitive to small increases in the red cell mass by transfusion (FRIED and GURNEY, 1965b). Elevation of the peripheral hematocrit by as little as 1% leads to a striking reduction in erythropoiesis, measured by $^{59}Fe$ incorporation into newly formed red cells, presumably because increased delivery of oxygen to the kidney by the additional erythrocytes leads to reduction of endogenous erythropoietin production. Unless androgenic steroids or their metabolites can produce erythroid differentiation and maturation in the absence of erythropoietin, it becomes difficult to see how a postulated androgenic effect on erythroid precursor cells alone could ever produce a substantial degree of plethora in normal mammals, since increased delivery of oxygen (2,3-DPG effect) and a small increase in the erythrocyte mass would be expected to turn off erythropoietin production.

On the basis of the theoretical considerations presented to date, it seems reasonable to expect there might be several mechanisms by which androgens stimulate red cell production. In the remainder of this section, the evidence supporting dual mechanisms will be explored.

Evidence that androgenic steroids increase plasma erythropoietin titers is convincing. Several investigators have shown plasma erythropoietin levels rise in rodents following injections of testosterone. A feature that must be noted, however, (NAETS and WITTEK, 1964; FRIED et al., 1964; MIRAND et al., 1965; FRIED and GURNEY, 1965a), is the large amounts of androgens used in most experiments. The earliest studies on the erythropoietic effect of androgens utilized the plethoric rodent, in which erythropoiesis is essentially eliminated because endogenous elaboration of erythropoietin has ceased. When such studies were conducted in mice, rendered only mildly plethoric by transfusions of small, rather than the usually large, volumes of blood, and erythropoietic effect was observed with as little as two daily injections of 0.1 mg of testosterone propionate (FRIED and GURNEY, 1965b). This dose is equivalent to 4 mg kg mouse body weight, and is similar to the large doses necessary in man if a hematopoietic effect is to be observed. It has generally been assumed that the return of erythropoiesis in such animals is indicative of the resumption of erythropoietin elaboration, but the recent challenge to the central and critical role of erythropoietin in erythropoiesis, to be discussed later, makes it essential to demonstrate this factor in the plasma of recipients of androgens before specific conclusions can be reached.

Following one or two injections of androgen, there is a latent period of over 36 h prior to appearance of erythropoietin in the plasma of the recipient rodent (FRIED et al., 1964). Erythropoietin can be detected by bioassay of plasma obtained from rodents following one or two injections of testosterone, and a synergistic effect on erythropoietin production is observed when androgen therapy is coupled with either hypoxia (GURNEY and FRIED, 1965) or administration of the cobaltous ion (JANDA et al., 1965). Further evidence that the erythropoietic factor, demonstrable by bioassay in the plasma of androgen-treated mice, is indeed erythropoietin is afforded by the demonstration that this activity is neutralized when the plasma is incubated with anti-erythropoietin antibody (FRIED et al., 1966). This antibody, when injected into polycythemic mice at appropriate times, has also been found to abolish the ability of testosterone to increase the rate of radioiron incorporation into newly-formed erythrocytes (SCHOOLEY, 1966). The late rise of erythropoietic factor is eliminated if, at an appropriate time after administration of testosterone, both kidneys are removed, a finding consistent with the renal origin of erythropoietin (FRIED and GURNEY, 1968). Of interest is the striking rise of erythropoietin titer, much greater than that in androgen-treated animals, when administration of the androgen is coupled with bilateral ureter ligation (FRIED and GURNEY, 1968). It is not clear at this time whether or not the postulated production of erythropoietin by the kidneys after androgen administration represents a direct physiologic correlation with the anatomic dependence of renal size on the presence of male hormone (SELYE, 1939; KOCHAKIAN, 1941). Some of the dependence of mouse renal cell ribosomes on androgens (KOCHAKIAN et al., 1969) may be relevant to erythropoietin production, and is worthy of additional study.

Nephrectomized animals retain the ability to produce some erythropoietin when subjected to a hypoxic stress (MIRAND and PRENTICE, 1957; ERSLEV, 1958). The best evidence that extrarenal production of erythropoietin is enhanced by androgens comes from clinical investigations which will be reviewed later in this chapter.

Interesting sex differences in erythropoiesis have been observed (FRIED and GURNEY, 1966). Erythropoiesis is more completely suppressed in female mice by transfusion, than it is in males of the same strain. Following administration of testosterone, the level of erythropoietin demonstrable in the plasma of female mice is substantially greater than the titer in the plasma of male mice. Four months after castration, male mice were found to have acquired erythropoietic responsiveness comparable to that of the normal female (FRIED and GURNEY, 1966).

The erythropoietic effect of androgens has been neutralized, or eliminated, in mice by the simultaneous administration of cyproterone acetate, a synthetic steroid with strong antiandrogen properties (MEDLINSKY et al., 1969). The ratio of cyproterone to testosterone required for complete neutralization of the erythropoietic effect of the androgen is approximately the same as that ratio required to suppress the androgenic effect on the levator anus and genital glands of the castrated rat. Cyproterone acetate does not alter the erythropoietic response to erythropoietin in plethoric mice, and it has therefore been postulated that competition at receptor sites on the surface of the cells responsible for erythropoietin production, rather than interference at the level of the primitive erythroid precursor cells, probably accounts for the neutralization of the erythropoietic effect of androgens.

Evidence for increased production of erythropoietin in response to androgen administration in man is convincing. A low rate of urinary excretion of erythropoietin has been demonstrated in volunteers, and normal values have been established in 24 h urine samples (ADAMSON et al., 1966). The urinary excretion of erythropoietin in normal and hypogonadal males (ALEXANIAN, 1969) increases quite substantially in response to fluoxymesterone treatment. Taken collectively, these laboratory and clinical studies prove quite conclusively that the production and excretion of erythropoietin are increased as a consequence of the administration of androgens.

A number of studies also support the hypothesis that androgens or some of their metabolic products exert another major erythropoietic effect by direct action upon the primitive hematopoietic cells. Porphyrin biosynthesis is increased in embryonic chick liver cultures by 5$\beta$-H steroids of the C-19 and C-21 series (GRANICK and KAPPAS, 1967), and hemoglobin synthesis increases two- to threefold in erythroid cells of chick blastoderm (LEVERE et al., 1967). These 5$\beta$-H compounds are metabolic by-products of testosterone usually present in tissues only in minute concentrations. Stimulation of hemoglobin synthesis is thought to arise from enhanced *de novo* formation of $\Delta$-amino levulinic acid synthetase, the enzyme which is rate-limiting in the synthesis of heme. These observations have been extended to cultures of bone marrow obtained from normal young adult volunteers (NECHELES and RAI, 1969).

Since the hemoglobin concentration per red cell is never increased above normal levels in man, and since the experimental and clinical erythropoietic responses of note in rodents and man take the form of an increase in red cell mass and total numbers of erythrocytes, rather than an increase of hemoglobin content per cell, the full biologic significance and clinical relevance of the direct effect of androgens or their metabolites on hematopoietic tissue needed clarification from experiments of a different design. These were afforded by experiments which indicate that 5$\beta$-compounds etiocholanolone and 19-nortestosterone phenylproprionate induce active cycling of resting colony-forming units (stem cells) of mouse marrow (BYRON, 1970). As described earlier, colony-forming units of a donor mouse are cells which can be assayed by passage and subsequent production of visible splenic nodules in an irradiated recipient (TILL and MCCULLOCH, 1961). These colony-forming units are normally not affected by large doses of tritiated thymidine administered to the donor prior to harvesting of cells (PORTEOUS and LAJTHA, 1968). It is therefore assumed that colony-forming units are dormant, since cells that are in active cycle of DNA synthesis and mitosis will take up enough tritiated thymidine to be destroyed by internal irradiation. If, however, the donor mice had been pretreated with either of the 5$\beta$-metabolites prior to receiving suicidal doses of tritiated thymidine, the number of colony-forming units per $10^4$ marrow cells injected into irradiated recipients was reduced by as much as 32%. From these observations it is reasonable to conclude that steroids with 5$\beta$-configurations may exert an hematopoietic effect by triggering resting stem cells into cycle. Since stem cells, or CFC, are precursors of erythropoietin responsive cells, one mechanism by which 5$\beta$-metabolites might exert an effect on erythropoiesis is to expand the pool of target cells upon which erythropoietin exerts its effect. This hypothesis could be strengthened by the availability of additional information which would explain the factors regulating the transformation of CFC to erythropoietin-responsive cells.

Of particular interest is the erythropoietic effect produced in plethoric rodents by the injection of 5$\beta$-steroid metabolites. The heightened erythropoietic response to the subsequent administration of erythropoietin in polycythemic mice first pretreated with 19-nortestosterone decanoate (GORSHEIN et al., 1973) might be achieved if the sensitivity of the primitive compartment of hematopoietic cells to erythropoietin is increased, or if the numbers of erythropoietin-responsive cells are increased by the 5$\beta$-metabolites. These results could also reflect the additive effects of steroid-induced, erythropoietin-independent erythropoiesis and the red cell formation initiated by the erythropoietic factor. The distinction between these two possibilities is profound. A direct erythropoietic effect by 5$\beta$-metabolites, independent of erythropoietin, has been postulated on the basis of results of an experiment in which anti-erythropoietin antibodies, which will neutralize the effects in plethoric mice of exogenously administered erythropoietin, did not abolish the erythropoietic stimulating effects of 11-ketopregnanolone (GORDON et al., 1970). This observation is so important as to warrant verification and thorough exploration, since it indicates that normal as well as malignant erythropoiesis can proceed in the absence of erythropoietin.

Finally, testosterone itself has been reported to produce an increase in the formation of microscopic erythroid colonies in rabbit (MORIYAMA and FISHER, 1974) and human (MORIYAMA and FISHER, 1975) marrow cultures in vitro. Although traces of erythropoietin, in the free state or bound to cell surfaces or inside the primitive cells, may be present in the marrow cultures at the time they are set up, the possibility of additional production of this stimulating factor appears remote, and the results of these experiments therefore suggest testosterone itself, as well as its 5$\beta$-H metabolites, is capable of exerting a direct effect upon the hematopoietic marrow.

## III. The Use of Androgens in Treatment of Anemias

### A. Pharmacologic Considerations

Unfortunately the pharmacologic parameters of importance in rational therapy of hematopoietic disorders of man have not yet been elucidated. The preferred androgen, if indeed one exists, is not yet known. The problem of selecting an appropriate agent for therapy is further complicated by the demonstration that success is often achieved only with very large doses. Furthermore, the response to therapy may be delayed, and an unresponsiveness cannot be accepted until an adequate trial has been pursued for at least 3 months. Indeed, some authors believe four to six months may be required to obtain a therapeutic effect with anabolic-androgenic steroids. In one report, 18 patients in a series of 24 individuals with aplastic anemia failed to respond, and 17 of the 18 did not survive the requisite time to insure an adequate trial, by these rigorous standards (DAVIS and RUBIN, 1972). The agents used most commonly include dromostanolone, fluoxymesterone, methandrostenolone, methyltestosterone, nandrolone, oxymetholone, stanozolol, and testosterone proprionate (EVENS and AMERSON, 1974). Other compounds used with some success include norethandrolone, testosterone enanthate, and testosterone cyclopentylpropionate. Most hematologists employ one or two androgenic compounds, and the variety of preferences probably indicates no preparation is clearly superior. Injectable preparations have the single advantage of assuring adequate therapy over the requisite long period of time. This factor assumes great importance when such large amounts of medication are required and when the hematologic diseases are so serious that patients often become very discouraged. The masculinizing effects of some of the compounds are particularly troublesome to women and constitute an additional, and quite understandable, factor in accounting for erratic adherence to a program of oral therapy. Nandrolone decanoate, testosterone enanthate, and testosterone cyclopentylpropionate have the advantage over other parenteral preparations of requiring injections only once weekly in order to achieve what is generally thought to be maximal hematopoietic stimulation.

Commitment to weekly injections for a minimum of three to six months is not undertaken lightly by many hematologists, especially in patients with leukopenia or thrombocytopenia, and many clinicians prefer oral androgens. Methyltestosterone, when prescribed generically, is least expensive. Fluoxymesterone and oxymetholone are also employed extensively. These three compounds are examples of 17-alkylated androgens whose structure is thought to dispose their recipients to the risk of hepatotoxicity. The most convincing and successful hematologic response observed by this author in response to fluoxymesterone therapy administered to a patient with aplastic anemia terminated in peliosis hepatis and death from acute liver failure. This patient was one of a small series in which this devastating complication is reviewed (BAGHERI and BOYER, 1974). It is unfortunate that the minimal dose, preferred route of administration, optimal schedule to acquire a maximal hematologic response, and most efficacious specific androgen are not yet known with certainty, and the patient is therefore subjected to discomfort, cost, and side effects that may be excessive if he is to hope for benefits that are all too often less than satisfactory.

Although the application of androgenic therapy to anemias of man has been empirical, the cumulated knowledge of possible mechanisms of steroid effect on erythropoiesis coupled with our current understanding of the pathophysiology of anemic states suggests a rational, if tentative, approach to a consideration of the indications for androgenic therapy of anemias. If androgens have two mechanisms of action—increasing the receptivity of the bone marrow to the stimulus for erythroid differentiation and increasing the magnitude of the stimulus for erythroid differentiation—those anemias characterized by abnormalities of stem cells or erythropoietin responsive cells, and those characterized by inadequate erythropoietin elaboration are the logical anemias to consider for androgen therapy.

## B. Treatment of Hypoplastic Anemias

The first group of diseases, characterized by marrow defects of primitive precursor hematopoietic cells, are among the most serious hematologic abnormalities. These are usually primary pancytopenias of man, although they are commonly designated as aplastic anemias. They are of greatest concern because the patient's dependence upon blood transfusions for survival presents such serious management problems, and because they are frequently fatal even when large doses of androgens are employed. These clinical states are so serious because the defects in hematopoietic differentiation and maturation are usually not limited to the erythroid series. Infection secondary to leukopenia or hemorrhage secondary to thrombocytopenias all too often lead to death of patients whose mechanisms of replication or differentiation of the primitive hematopoietic cells have been damaged.

The use of androgens in aplastic anemias, usually in conjunction with corticosteroid therapy, has attracted most widespread attention. About half of the patients with acquired aplastic anemia and most of the patients with constitutional aplastic anemia experience partial or total remission when treated in this manner (KILLANDER et al., 1969; SHAHIDI, 1973). A beneficial effect is most likely to occur, and recovery is most rapid, if some residual foci of hematopoiesis can be demonstrated in the bone marrow prior to the onset of therapy, and conversely, therapy is usually unrewarding if a more severe degree of aplasia is present, as manifested by severe marrow hypocellularity and greater depression of peripheral blood counts (KILLANDER et al., 1969; SHAHIDI, 1973).

A recent report of the lack of success when androgens were used as therapy of patients with aplastic anemias in a double blind study (BRANDA et al., 1974) accentuates the variability of experience of different hematologists in the evaluation of the effectiveness of androgens. Much depends upon patient selection. Clearly the overall rate of response is better in children with aplastic anemia than it is in adults. The intensity of the hypoplasia is also of great importance and in any individual small series, impressions may be distorted favorably or unfavorably by a few patients whose disease is inordinately mild or severe. All of the known and as yet unknown etiologic agents responsible for aplastic anemia may not produce the same molecular damage in primitive hematopoietic cells. Fundamentally different lesions, with correspondingly different responsiveness to androgens, may be grouped together within the single broad category we now recognize as aplastic anemia.

When we consider our current lack of understanding of the etiologies and pathogenesis of multiple different stem cell defects that are theoretically possible, and when we recall that the most appropriate androgenic preparations and the most effective doses to employ have yet to be established, it is safe to anticipate that the modest, and sometimes very unsatisfactory, results currently reported may yet be improved as our basic and clinical knowledge increase.

Congenital aplasias and hypoplasias of the marrow and the pure red cell anemias of children and the role of androgens in their therapy are sufficiently unique to warrant separate attention, and will be considered in Chapter VII.B and C of this book.

The use of androgens in other forms of aplasia or functional hypoplasia is worthy of comment. Radiation hypoplasia is an example of iatrogenic hypoplasia in which recovery has been reported to be facilitated by anabolic androgenic steroids (EVANS and ELIAS, 1972). The fall in hemoglobin and red cell counts in patients receiving irradiation for carcinoma of the lung was less marked in patients receiving an initial primary dose of 100 mg of nandrolone phenylpropionate intramuscularly followed by weekly 50 mg doses than was the fall in control patients.

Androgens have sometimes been found to be useful in the therapy of the anemia occurring secondary to myelofibrosis. In this unusual disease, the spaces within the bones normally occupied by marrow are replaced by fibroblasts and other connective tissue elements. Such therapy is not always beneficial, and ferrokinetic studies are reported to be helpful in indicating which patients are likely to respond by an increase in effective erythropoiesis (GARDNER and NATHAN, 1966).

## C. Treatment of Secondary Anemias

An argument can be presented that a number of the secondary anemias are characterized by insufficient production of erythropoietin. These include the anemia often observed in patients with infection or inflammation, chronic renal disease, and neoplastic disease uncomplicated by blood loss. Bioassay of plasma samples from such patients seldom shows elevated erythropoietin titers. Unfortunately, the assay systems used have, until recently, been insensitive to normal levels of plasma erythropoietin. Therefore the observation that the elevations of plasma erythropoietin titer customarily demonstrable in the plasma of patients with marrow failure are not found when the donors suffer from secondary anemia of comparable magnitude (KRANTZ and JACOBSON, 1970; GALLAGHER et al., 1959) only suggests, but does not prove that insufficient erythropoietin contributes to the pathogenesis of the disease. Some support for the causal role of erythropoietin deficiency in the pathogenesis of secondary anemias resides in examples of responses of selected patients to therapy with cobaltous chloride (GARDNER, 1953), a substance known to exert its erythropoietic effect in rodents by increasing markedly the elaboration of erythropoietin (GOLDWASSER et al., 1957). Using recently developed marrow culture assay systems of a more sensitive nature, direct evidence supporting relative erythropoietin deficiency as an important cause of anemia secondary to infection, inflammation, and malignancy has been presented (WARD et al., 1971).

To date, the only secondary anemia that has received extensive therapeutic evaluation by androgens is the anemia of chronic renal disease (DEGOWIN et al., 1970; BLUMBERG and KELLER, 1971; SHALDON et al., 1971; ESCHBACH and ADAMSON, 1973; FRIED et al., 1973; KEYSSNER et al., 1974). The other secondary anemias are generally not sufficiently severe to justify the cost or the risks entailed by such therapy. Although anephric rodents and man retain the ability to produce small amounts of erythropoietin when a hypoxic stimulus of sufficient intensity is present, the plasma of anemic uremic patients seldom contains demonstrable titers of erythropoietin when subjected to bioassay (FRIED et al., 1973). Plasma obtained from patients with chronic renal insufficiency maintained on hemodialysis demonstrates erythropoietic activity after androgenic therapy (ESCHBACH and ADAMSON, 1973; FRIED et al., 1973). The presence of the diseased kidneys is not necessary for an erythropoietic effect to be obtained. In one study, anephric patients did not respond to testosterone propionate 150 mg twice weekly, and the authors suggest nephrectomy should therefore be delayed as long as possible so that the major erythropoietin-producing organ can be retained (FRIED et al., 1973). Other investigators have shown, however, that testosterone in a dosage of 500 mg weekly i.m. (BLUMBERG and KELLER, 1971) or fluoxymesterone administered orally in doses of 10 mg per day to women and 30 mg per day to men (ESCHBACH and ADAMSON, 1973) was capable of eliciting an erythropoietic response in anephric patients. Nonetheless, the rate of erythropoiesis was reduced markedly after bilateral nephrectomy. Therefore in the absence of intractable hypertension, renal infection, or abnormal immunologic responses, early bilateral nephrectomy in potential recipients of renal allografts is discouraged (SHALDON et al., 1971; ESCHBACH and ADAMSON, 1973).

## IV. The Effects of Androgens on Leukopoiesis and the Use of Androgens in Leukopenia

Surprisingly little attention has been directed to the effects of androgenic steroids on white cell and platelet production, particularly since many of the children with aplastic anemia treated by SHAHIDI and DIAMOND generally demonstrated a granulocyte increase and often showed some elevation of the low platelet counts as well (SHAHIDI and DIAMOND, 1961). In an excellent review of the regulation of granulopoiesis (GOLDE and CLINE, 1974), the possible effects of androgens on white cell formation were not considered. If androgens were to exert any granulocyte-stimulating properties, a higher neutrophil count would be expected in normal men than is present in women. In reality, however, neutrophil counts are slightly higher in normal women (BAIN, 1975).

Only recently have the leukopoietic effects of androgens in experimental animals received attention. A testosterone-induced leukocytosis has been observed in female mice on the day following the last of two daily injections of testosterone propionate (HOGAN and CODY, 1972), but this elevation disappears in the subsequent two days, and therefore probably represents release of white cells from reserve pools rather than any alteration of leukocyte production. Rats appear to have less depression of white counts from nitrogen mustard if administration of this cytotoxic compound is accompanied by administrations of androgens, but the leukocytic reserves, as measured by endotoxin challenge, are not altered by the addition of male hormone (HORN, 1971a). Anabolic and androgenic hormones also have been reported to have a facilitory effect on leukocyte recovery after X-ray exposure (HORN and ROBINSON, 1971b; HORN, 1971c; HORN, 1972; HORN and PRICE, 1972). Testosterone propionate accelerates recovery of the depression of the polymorphonuclear leukocyte count, particularly if injections are initiated prior to the onset of the irradiation and if a hematopoietic area (one leg) is shielded from the radiation (HORN and PRICE, 1972).

Recent studies have suggested mechanisms by which androgens may exert their stimulatory effect on leukopoiesis. The number of granulocyte colonies in cultures of human marrow are increased when testosterone is added, presumably as a consequence of proliferation of committed granulocyte precursors (ROSENBLUM and CARBONE, 1974). Oxymetholone, nortestosterone, and testosterone are all capable of enhancing granulopoiesis in mice made neutropenic by BCNU, cyclophosphamide, or vinblastine (UDUPA and REISSMANN, 1974). This in vivo effect is neutralized by busulfan, an agent which eradicates primitive hematopoietic cells. It has therefore been suggested that androgenic steroids exert their granulopoietic effect by stimulating the proliferation of colony-forming cells already committed to the granulocytic series. This mechanism is noteworthy in its similarity to the reports of emergence of increased numbers of erythroid colonies observed in cultures of rabbit and human marrow upon the addition of testosterone.

Several reports indicate androgens may have a role in the therapy of some of the neutropenias of man. A reduction in the anticipated granulocyte depression was reported in three cancer patients if, prior to chemotherapy, testosterone was administered (BRODSKY et al., 1964). The 31.9 days of expected leukopenia in cancer patients treated with chemotherapy or irradiation was reduced to 9.8 days in similarly treated patients who also received testosterone propionate intramuscularly in doses of 100 mg 3–7 times a week (HORN et al., 1968). A mild leukopoietic effect was also observed in women with metastatic breast cancer who were treated with $7\beta,17\alpha$-dimethyltestosterone, 200 mg daily orally (HORN et al., 1973). The maximal leukocyte counts were observed in the 30th week of therapy, at which time they were 124.5% of the initial pre-therapy levels. Such effects are mild, but were found to be statistically significant. Nandrolone decanoate and oxymetholone have also been employed in an effort to reduce the hematologic depression produced by methotrexate chemotherapy (RAWBONE and BAGSHAW, 1972). Although the duration of the leukopenia was shortened in patients receiving the anabolic steroids, the importance of the observations was questioned, and the authors expressed the opinion that their study did not "lend much support for the use of anabolic steroids in an attempt to minimize marrow toxicity during cytotoxic chemotherapy." Finally, large doses of testosterone enanthate have been observed to produce both subjective improvement and elevation of the blood granulocyte count in a patient afflicted with cyclic neutropenia (BRODSKY et al., 1965).

## V. Summary

Androgens and their metabolites exert striking stimulatory effects on erythropoiesis, presumably both by increasing erythropoietin production and by increasing the responsiveness of the erythropoietic tissues to erythropoietin. Until a few years ago, it was generally accepted that erythropoietin was essential for normal regulated erythropoiesis in mammals, and only malignant erythropoiesis, seen in polycythemia vera of man and murine viral reticuloendotheliosis, could occur in the absence of this erythropoietic factor. Recent observations suggest the possibility, however, that androgens or their metabolites may, in addition to altering the responsiveness of the hematopoietic tissues, also be capable of inducing erythroid differentiation in the absence of erythropoietin. Because the 5$\beta$-androgenic compounds have such striking effects on the primitive precursor cells of the hematopoietic tissues, increasing the sensitivity of the marrow to erythropoietin, further studies that exclude this erythropoietic factor from the system under observation will be needed before an initiating role of androgens in erythroid differentiation can be accepted.

Androgens also exert a leukopoietic effect in man and rodents, probably by stimulating the proliferation of granulocytic colony-forming cells. A detailed investigation of the effect of androgens on thrombocytopoiesis has yet to be undertaken.

The role of androgens in the therapy of some anemias of man is established. As our knowledge of hematopoietic mechanisms and processes grows, it is reasonable to anticipate an expanded role for androgens in the treatment of anemias and leukopenias of man.

## References

ADAMSON, J. W., ALEXANIAN, R., MARTINEZ, Z., FINCH, C. A.: Erythropoietin excretion in normal man. Blood **28**, 354—364 (1966).

ADAMSON, J. W., FINCH, C. A.: Erythropoietin and the regulation of erythropoiesis in Di Guglielmo's syndrome. Blood **36**, 591—597 (1970).

ALEXANIAN, R.: Erythropoietin and erythropoiesis in anemic man following androgens. Blood **33**, 564—572 (1969).

BAGHERI, S. A., BOYER, J. L.: Peliosis hepatis associated with androgenic-anabolic steroid therapy. Ann. intern. Med. **81**, 610—618 (1974).

BAIN, B. J.: Normal haematological values, sex difference in neutrophil count. Brit. med. J. **1975**, 306—309.

BLUMBERG, A., KELLER, H.: Erythropoiesis in anephric patients. Schweiz. med. Wschr. **101**, 1887—1893 (1971).

BRANDA, R. F., AMSDEN, T. W., JACOB, H. S.: Randomized study of nandrolone therapy for refractory anemia. (Abs.) Clin. Res. **22**, 607A (1974).

BRODSKY, I., DENNIS, L. H., KAHN, S. B.: Testosterone enanthate as a bone marrow stimulant during cancer chemotherapy: Preliminary report, Cancer Chemother. Rep. **34**, 59—63 (1964).

BRODSKY, I., REIMANN, H. A., DENNIS, L. H.: Treatment of cyclic neutropenia with testosterone. Amer. J. Med. **38**, 802—806 (1965)

BRUCE, W. R., MCCULLOCH, E. A.: The effect of erythropoietic stimulation on the hematopoietic colony-forming cells of mice. Blood **23**, 216—232 (1964).

BYRON, J. W.: Effect of steroids on the cycling of hematopoietic stem cells. Nature (Lond.) **228**, 1204 (1970).

CHAMBERLAIN, J. K., LEBLOND, P. F., WEED, R. I.: Reduction of adventitial cell cover, an early direct effect of erythropoietin on bone marrow ultrastructure. (Abs.) Clin. Research **23**, 401A (1975).

CONTRERA, J. F., GORDON, A. S., WEINTRAUB, A. H.: Extraction of an erythropoietin-producing factor from a particulate fraction of rat kidney. Blood **28**, 330—343 (1966).

CRAFTS, R. C.: Effects of hypophysectomy, castration, and testosterone propionate on hemopoiesis in the adult male rat. Endocrinology **39**, 401—413 (1946).

DAVIS, S., RUBIN, A. D.: Treatment and prognosis in aplastic anemia. Lancet **1972**, 871—873.

DEGOWIN, R. L., LAVENDER, A. R., FORLAND, M., CHARLESTON, D., GOTTSCHALK, A.: Erythropoiesis and erythropoietin in patients with chronic renal failure treated with hemodialysis and testosterone. Ann. intern. Med. **72**, 913—918 (1970).

DUKES, P. P., GOLDWASSER, E.: Inhibition of erythropoiesis by estrogens. Endocrinology **69**, 21—29 (1961).

ERSLEV, A. J.: Erythropoietin function in uremic rabbits. Arch. intern. Med. **101**, 407—417 (1958)

ESCHBACH, J. W., ADAMSON, J. W.: Improvement in the anemia of chronic renal failure with fluoxymesterone. Ann. intern. Med. **78**, 527—532 (1973).

EVANS, J. T., ELIAS, E. G.: The erythropoietic response to anabolic therapy in patients receiving radiotherapy. J. clin. Pharm. **12**, 101—104 (1972).

EVENS, R. P., AMERSON, A. B.: Androgens and erythropoiesis. J. clin. Pharm. **14**, 94—101 (1974).

FISHER, J. W., SAMUELS, A. I., MALGOR, L. A.: Androgens and erythropoiesis. Israel J. Med. Sci. **7**, 892—900 (1971).

FRIED, W., DEGOWIN, R., GURNEY, C. W.: Erythropoietic effect of testosterone in the polycythemic mouse. Proc. Soc. exp. Biol. (N.Y.) **117**, 839—842 (1964).

FRIED, W., GURNEY, C. W.: Erythropoietic effect of plasma from mice receiving testosterone. Nature (Lond.) **206**, 1160—1161 (1965a).

FRIED, W., GURNEY, C. W.: Use of mild plethora to demonstrate an erythropoietic effect from small amounts of androgens. Proc. Soc. Exp. Biol. (N.Y.) **120**, 519—521 (1965b).

FRIED, W., GURNEY, C. W.: The erythropoietic response to testosterone in male and female mice. J. Lab. clin. Med. **67**, 420—426 (1966).

FRIED, W., GURNEY, C. W.: The erythropoietic-stimulating effects of androgens. Ann. N.Y. Acad. Sci. **149**, 356—365 (1968).

FRIED, W., JONASSON, O., LANG, G., SCHWARTZ, F.: The hematologic effect of androgen in uremic patients. Ann. intern. Med. **79**, 823—827 (1973).

FRIED, W., MARVER, D., LANGE, R. D., GURNEY, C. W.: Studies on the erythropoietic stimulating factor in the plasma of mice after receiving testosterone. J. Lab. clin. Med. **68**, 947—951 (1966).

GALLAGHER, N. I., LANGE, R. D.: Response to erythropoietin. Proc. Soc. exp. Biol. (N.Y.) **110**, 422—426 (1962).

GALLAGHER, N. I., MCCARTHY, M., HART, K. T., LANGE, R. D.: Evaluation of plasma erythropoietic-stimulating factors in anemic uremic patients. Blood **14**, 662—667 (1959).

GARDNER, F. H.: The use of cobaltous chloride in the anemia associated with chronic renal disease. J. Lab. clin. Med. **41**, 56—64 (1953).

GARDNER, F. H., NATHAN, D. G.: Androgens and erythropoiesis. III. Further evaluation of testosterone treatment of myelofibrosis. New Engl. J. Med. **274**, 420—426 (1966).

GARDNER, F. H., PRINGLE, J. C. JR.: Androgens and erythropoiesis. New Engl. J. Med. **264**, 103—111 (1961).

GOLDE, D. W., CLINE, M. J.: Regulation of granulopoiesis. New Engl. J. Med. **291**, 1388—1395 (1974).

GOLDWASSER, E., JACOBSON, L. O., FRIED, W., PLZAK, L.: Mechanism of the erythropoietic effect of cobalt. Science **125**, 1085—1086 (1957).

GORDON, A. S., ZANJANI, E. D., LEVERE, R. D., KAPPAS, A.: Stimulation of mammalian erythropoiesis by 5β-H steroid metabolites. Proc. nat. Acad. Sci. **65**, 919—924 (1970).

GORSHEIN, D., HAIT, W. N., BESA, E. C., JEPSON, J. H., GARDNER, F. H.: Increased stem cell response to erythropoietin induced by androgens. Endocrinology **93**, 777—780 (1973).

GORSHEIN, D., OSKI, F. A., DELIVORIA-PAPADOPOULOS, M.: Effect of androgens on the red cell 2,3-diphosphoglycerate hemoglobin oxygen affinity and red cell mass in mammals. Proc. Soc. exp. Biol. (N.Y.) **147**, 616—618 (1974)

GRANICK, S., KAPPAS, A.: Steroid control of porphyrin and heme biosynthesis: A new biological function of steroid hormone metabolites. Proc. nat. Acad. Sci. (Wash.) **57**, 1463—1467 (1967).

GROSS, M., GOLDWASSER, E.: On the mechanism of erythropoietin-induced differentiation. VII. The relationship between stimulated deoxyribonucleic acid synthesis and ribonucleic acid synthesis. J. biol. Chem. **245**, 1632—1636 (1970).

GROSS, M., GOLDWASSER, E.: On the mechanism of erythropoietin-induced differentiation. IX. Induced synthesis of 9 S ribonucleic acid and of hemoglobin. J. biol. Chem. **246**, 2480—2486 (1971).

GURNEY, C. W., FRIED, W.: Further studies on the erythropoietic effect of androgens. J. Lab. clin. Med. **65**, 775—782 (1965).

HANNA, I. R.: Response of early erythroid precursors to bleeding. Nature (Lond.) **214**, 355—357 (1967).

HOGAN, G. R., CODY, A. E.: Testosterone-induced leukocytosis in female mice. J. Endocrinol. **53**, 175—176 (1972).

HORN, Y.: The effect of andogenic hormones on bone marrow of rats receiving chemotherapy. Oncology **25**, 512—519 (1971a).

HORN, Y.: Leukopoietic properties of four different androgenic hormones in irradiated rats. Scand. J. Haemat. **8**, 470—475 (1971c).

HORN, Y.: A comparison of two androgenic hormones as leukopoietic agents in irradiated rats. Oncology **26**, 16—24 (1972a).

HORN, Y., HALDEN, A., GORDAN, G. S.: Leukopoietic effect of calusterone (7β, 17α-dimethyltestosterone) in women with advanced breast cancer. Scand. J. Haemat. **10**, 177—180 (1973).

HORN, Y., PRICE, D. C.: The effect of androgens on erythroid and granulocytic marrow recovery in the irradiated rat. Acta Haemat. **48**, 300—306 (1972b).

HORN, Y., ROBINSON, E.: Testosterone as a bone marrow stimulant in irradiated rats. Radiol. clin. Biol. **40**, 133—141 (1971b).

HORN, Y., ROBINSON, E., HOCHMAN, A.: Leukopoietic stimulation by testosterone in cancer patients treated by chemotherapy and irradiation. Radiol. clin. Biol. **37**, 29—35 (1968).

JANDA, W. E., FRIED, W., GURNEY, C. W.: Combined effect of cobalt and testosterone on erythropoiesis. Proc. Soc. exp. Biol. (N.Y.) **120**, 443—446 (1965).

KENNEDY, B. J., TIBBETTS, D. M., NATHANSON, I. T., AUB, J. C.: Chemical and systemic changes in patients with advanced carcinoma of the breast following hormonal therapy and castration. (Abs.) Cancer Res. **9**, 551—552 (1949).

KEYSSNER, J., HAUSWALDT, C., UHL, N., HUNSTEIN, W.: Testosteronbehandlung renaler Anämie. Schweiz. med. Wschr. **104**, 1938—1941 (1974).

KILLANDER, A., LUNDMARK, K., SJOLIN, S.: Idiopathic aplastic anemia in children: Results of androgen treatment. Acta. paediat. scand. **58**, 10—14 (1969).

KOCHAKIAN, C. D., The rate of absorption and effects of testosterone propionate pellets on mice. Endocrinology **28**, 478—484 (1941).

KOCHAKIAN, C. D., NISHIDA, M., HIRONE, T.: Ribosomes of mouse kidney: Regulation by androgens. Amer. J. Physiol. **217**, 383—391 (1969).

KRANTZ, S. B.: Response of polycythemia vera marrow to erythropoietin in vitro. J. Lab. clin. Med. **71**, 999—1012 (1968).

KRANTZ, S. B.: Recent contributions to the mechanism of action and clinical relevance of erythropoietin. J. Lab. clin. Med. **82**, 847—857 (1973).

KRANTZ, S. B., GOLDWASSER, E.: On the mechanism of erythropoietin-induced differentiation. II. The effect on RNA synthesis. Biochem. biophys. Acta **103**, 325—332 (1965).

KRANTZ, S. B., JACOBSON, L. O.: Erythropoietin and the Regulation of Erythropoiesis. Chicago: University of Chicago Press 1970.

KURATOWSKA, Z.: La nature et le mode d'action de l'erythropoietine. Path. Biol. Sem. Hop. **15**, 671—675 (1967).

LEVERE, R. D., KAPPAS, A., GRANICK, S.: Stimulation of hemoglobin synthesis in chick blastoderms by certain 5β androstane and 5β pregnane steroids. Proc. nat. Acad. Sci. (Wash.) **58**, 985—900 (1967).

McCULLAGH, E. P., JONES, R.: A note on the effect of certain androgens upon red blood cells count and upon glucose tolerance. Cleve. Clin. Q. **8**, 79—84 (1941).

McCUSKEY, R. S., MEINEKE, H. A., KAPLAN, S. M.: Studies of the hemopoietic microenvironment. II. Effect of erythropoietin on the splenic microvasculature of polycythemic $CF_1$ mice. Blood **39**, 809—813 (1972).

MEDLINSKY, J. T., NAPIER, C. D., GURNEY, C. W.: The use of an antiandrogen to further investigate the erythropoietic effects of androgens. J. Lab. clin. Med. **74**, 85—92 (1969).

MIRAND, E. A.: Virus-induced erythropoiesis in hypertransfused polycythemic mice. Science **156**, 832—833 (1967).

MIRAND, E. A., GORDON, A. S., WENIG, J.: Mechanism of testosterone action in erythropoiesis. Nature (Lond.) **206**, 270—272 (1965).

MIRAND, E. A., PRENTICE, T. C.: Presence of plasma erythropoietin in hypoxic rats with or without kidney(s) and/or spleen. Proc. Soc. exp. Biol. (N.Y.) **96**, 49—51 (1957).

MORIYANA, Y., FISHER, J. W.: Effects of testosterone and erythropoietin on erythroid colony formation in rabbit bone marrow cultures. Life Sci. **15**, 1181—1188 (1974).

MORIYAMA, Y., FISHER, J. W.: Effects of testosterone on erythroid colony formation in human bone marrow cultures. Blood **45**, 665—670 (1975).

NAETS, J. P., WITTEK, M.: Etude du mecanisme d'action des androgenes sur l'erythropoiese. C. R. Acad. Sci. (Paris) **259**, 3371—3374 (1964).

NAKAO, K., SASSA, S., WADA, O., TAKAKU, F.: Enzymatic studies on erythroid differentiation and proliferation. Ann. N. Y. Acad. Sci. **149**, 224—228 (1968).

NECHELES, T. F., RAI, U. S.: Studies on the control of hemoglobin synthesis: The in vitro stimulating effect of a 5β-H steroid metabolite on heme formation in human bone marrow cells. Blood **34**, 380—384 (1969).

O'GRADY, L. F., LEWIS, J. P., TROBAUGH, F. E., JR.: The effect of erythropoietin on differentiated erythroid precursors. J. Lab. clin. Med. **71**, 693—703 (1968).

OSKI, F. A., GOTTLIEB, A. J., MILLER, W. W.: The effects of deoxygenation of adult and fetal hemoglobin on the synthesis of red cell 2,3-diphosphoglycerate and its in vivo consequences. J. clin. Invest. **49**, 400—407 (1970).

PARKER, J. P., BEIRNE, G. J., DESAI, J. N., RAICH, P. C., SHAHIDI, N. T.: Androgen-induced increase in red-cell 2,3-diphosphoglycerate. New Engl. J. Med. **287**, 381—383 (1972).

PAUL, J., CONKIE, D., BURGOS, H.: Effects of erythropoietin on cell populations and macromolecular synthesis in foetal mouse erythroid cells. J. Embryol. exp. Morphol. **29**, 453—472 (1973).

PORTEOUS, D. D., LAJTHA, L. G.: Restoration of stem-cell function after irradiation. Ann. N. Y. Acad. Sci. **149**, 151—155 (1968).

RAWBONE, R. B., BAGSHAW, K. D.: Anabolic steroids and bone marrow toxicity during treatment with methotrexate. Brit. J. Cancer **26**, 395—401 (1972).

REISSMANN, K. R., SAMORAPOOMPICHIT, S.: Effect of erythropoietin on regeneration of hematopoietic stem cells after 5-fluorouracil administration. J. Lab. clin. Med. **73**, 544—550 (1969).

REISSMANN, K. R., UDUPA, K. B.: Effect of erythropoietin on proliferation of erythropoietin-responsive cells. Cell Tissue Kinet. **5**, 481—489 (1972).

ROSENBLUM, A. L., CARBONE, P. P.: Androgenic hormones and human granulopoiesis in vitro. Blood **43**, 351—356 (1974).

SASSA, S., TAKAKU, F., NAKAO, K.: Regulation of erythropoiesis in Friend leukemia of mice. Med. Biol. (Tokyo) **72**, 12—15 (1966).

SCHOOLEY, J. C.: Inhibition of erythropoietic stimulation by testosterone in mice receiving anti-erythropoietin. Proc. Soc. exp. Biol. (N.Y.) **122**, 402—403 (1966).

SCHOOLEY, J. C., MAHLMAN, C. J.: Stimulation of erythropoiesis in the plethoric mouse by cyclic-AMP and its inhibition by anti-erythropoietin. Proc. Soc. exp. Biol. (N.Y.) **137**, 1289—1292 (1971).

SCHOOLEY, J. C., MAHLMAN, C. J.: Studies with anti-erythropoietin. In: GORDON, A. S., CONDORELLI, M. (Eds.): Regulation of Erythropoiesis, pp. 167—176. Milan: I. Ponte 1972.

SELYE, H.: The effect of testosterone on the kidney. J. Urol. **42**, 637—641 (1939).

SHAHIDI, N. T.: Androgens and erythropoiesis. New Engl. J. Med. **289**, 72—80 (1973).

SHAHIDI, N. T., DIAMOND, L. K.: Testosterone-induced remission in aplastic anemia. Amer. J. Dis. Child. **98**, 293—302 (1959).

SHAHIDI, N. T., DIAMOND, L. K.: Testosterone-induced remission in aplastic anemia of both acquired and congenital types: Further observations in 24 cases. New Engl. J. Med. **264**, 953—967 (1961).

SHALDON, S., PATYMA, W. D., KALTWASSER, P., WERNER, E., KOCH, K. M., SCHOEPPE, W. E.: The use of testosterone in bilateral nephrectomized dialysis patients. Trans. Amer. Soc. Artif. Int. Organs **17**, 104—107 (1971).

STEINGLASS, P., GORDON, A. S., CHARIPPER, H. A.: Effect of castration and sex hormones on blood of the rat. Proc. Soc. exp. Biol. (N.Y.) **48**, 169—177 (1941).

STEPHENSON, J. R., AXELROD, A. A.: Separation of erythropoietin-sensitive cells from hemopoietic spleen colony-forming cells of mouse fetal liver by gravity sedimentation. Blood **37**, 417—427 (1971)

TABER, E., DAVIS, D. E., DOMM, L. V.: Effect of sex hormones on erythrocyte number in the blood of the domestic fowl. Amer. J. Physiol. **138**, 479—487 (1943).

TALBOT, T. R., JR., ESCHER, G. C.: Effects of testosterone propionate on peripheral blood and bone marrow of women with advanced inoperable carcinoma of breast. (Abs.) J. clin. Endocr. **9**, 666 (1949)

TILL, J. E., MCCULLOCH, E. A.: A direct measurement of the radiation sensitivity of normal mouse bone marrow cells. Radiation Res. **14**, 213—222 (1961).

UDUPA, K. B., REISSMANN, K. R.: Acceleration of granulopoietic recovery by androgenic steroids in mice made neutropenic by cytoxic drugs. Cancer Res. **34**, 2517—2520 (1974).

VAHLQUIST, B.: The cause of the sexual differences in erythrocyte hemoglobin and serum iron levels in human adults. Blood **5**, 874—875 (1950).

VAN DYKE, D. C., CONTOPOULOS, A. N., WILLIAMS, B. S., SIMPSON, M. E., LAWRENCE, J. H., EVANS, H. M.: Hormonal factors influencing erythropoiesis. Acta Haematol. **11**, 203—222 (1954).

WARD, H. P., KURNICK, J. E., PISARCZYK, M. J.: Serum level of erythropoietin in anemias associated with chronic infection, malignancy, and primary hematopoietic disease. J. Lab. clin. Med. **50**, 332—335 (1971).

WILLIAMSON, C. S.: Influence of age and sex on hemoglobin: A spediophotometric analysis of nine hundred and nineteen cases. Arch. intern. Med. **18**, 505—528 (1916).

ZANJANI, E. D., CONTRERA, J. F., GORDON, A. S., COOPER, K., WONG, K., KATZ, R.: The renal erythropoietic factor (REF) III. Enzymatic role in erythropoietin production. Proc. Soc. exp. Biol. (N.Y.) **125**, 505—508 (1967).

# D. Cancer in Man

G. S. GORDAN

## I. Introduction

Many steroid hormones and their derivatives are of therapeutic value in specific human cancers. Estrogens, antiestrogens (COLE et al., 1971; O'HALLORAN and MADDOCK, 1974; BLOOM and BOESON, 1974), androgens, and chemical derivatives with less or even no hormonal activity (e.g. $\Delta^1$-testololactone) are capable of producing long, worthwhile regressions in disseminated human breast cancer. Progestational compounds have proved their worth in the treatment of inoperable, recurrent, and metastatic endometrial cancer (KELLEY and BAKER, 1961; MALKASIAN et al., 1974). They are also reported effective in some cases of renal cell cancer on the kidney (BLOOM, 1971). In prostate cancer, by contrast, androgens are generally considered contraindicated, and estrogens are the steroid hormones of greatest therapeutic value. One modified androgen with potent anabolic activity, Nilevar (17α-ethyl-19-nortestosterone), has however been reported of subjective therapeutic benefit in some cases of disseminated prostatic cancer no longer responding to estrogens and/or castration (BRENDLER and WINKLER, 1959). Anabolic-androgenic steroids are also of nonspecific therapeutic value in certain cancers and marrow diseases, neoplastic and nonneoplastic, because of their hematopoietic activity (cf. Chapter VII.C). Less well popularized but perhaps of even greater significance is the fact that these compounds also stimulate white blood cell and platelet production. These properties have been applied only sporadically to clinical conditions but may be of particularly great importance to cancer patients because of the marrow-damaging effects of radiation and cytotoxic chemotherapeutic agents. It was at one time hoped that bronchogenic carcinoma, a common fatal cancer with a very poor cure rate, might be responsive to various steroid hormones. A large, well-controlled study in the United States, however, showed no benefit from testosterone propionate, stilbestrol, progesterone, or cortisone (WOLF et al., 1960).

## II. Endocrine-Responsive Tumors

The three disseminated human cancers with the longest and best established history of response to endocrine therapeutic maneuvers are cancers of the breast, prostate, and endometrium. BEATSON (1896) showed in 1896 that breast cancer responds to oöphorectomy, and HUGGINS and HODGES (1941) similarly reported in 1941 regressions of prostatic cancer induced by castration. Evaluation of efficacy of therapy in these two conditions presents quite different problems. In most cases, breast cancer produces measurable metastases in skin, lymph nodes or visceral organs that can be shown by caliper measurement to shrink if a regression is induced. Most, but not all, osseous metastases are lytic, so these too can be measured, and healing can be observed when recalcification occurs. Prostatic cancer, on the other hand, rarely produces easily measurable lesions, and untreated bone metastases rapidly change from a lytic to a blastic appearance, so that evaluation of a therapeutic effect is most difficult. The one measurable endpoint in prostatic cancer is the serum acid phosphatase level. This raises another problem since a therapeutic maneuver or agent might turn off secretion of the enzyme without necessarily causing a regression of the cancer. This phenomenon is known to occur with amphenone, which turns off production of steroids without affecting the growth of functioning carcinoma of the adrenal cortex. Carcinoma of the breast responds to estrogens, androgens, and their chemical derivatives; prostatic cancer, to estrogens and/or castration. The third endocrine-sensitive cancer, that of the endometrium, responds well to progestational agents (KELLEY and BAKER, 1961). Progestational agents have also been reported effective in the treatment of metastatic renal cancer (BLOOM, 1971).

It should be noted that all of these tumors also respond to cytotoxic chemotherapeutic agents. At the outset, it should be emphasized that the toxicity of cytotoxic agents is not at all comparable to the relatively modest toxicity—really the hormonal effects—of hormonal steroids. As will also be noted later, early use of cytotoxic chemotherapy prejudices against subsequent response to hormonal agents.

## III. Breast Cancer

### A. Attempts to Predict Response

Everyone treating patients must realize that no present treatment, no matter how drastic, whether cytotoxic, endocrine, or endocrine-ablative (hypophysectomy, adrenalectomy) can sterilize the patient of disseminated cancer of the breast, prostate, endometrium, or kidney. This concept is of greatest importance today when increasing numbers of regressions are being reported as a result of combining multiple, highly toxic, expensive cytotoxic chemotherapeutic agents. This combination does indeed have its proper place and time in the treatment of most patients, but if it is used early, those patients who do respond—less than half of all patients—obtain regressions averaging approximately 5 months, have considerable drug-induced morbidity and even some mortality, and—worst of all—have a much reduced chance of responding to subsequent therapy (DAVIS et al., 1974). It therefore follows that in most cases of endocrine-responsive tumors, hormonal therapy should precede rather than follow cytotoxic chemotherapy. Because no treatment produces regressions in all patients, a great effort has gone into means of identifying those breast cancer patients who will respond to any given treatment. An incredibly large literature exists on the use of various clinical, chemical, and highly sophisticated dynamic techniques to identify those patients whose tumors will respond to any given treatment. It would be pointless to review all of the now outmoded proposed prognostic indicators, including urinary calcium excretion, estrogen excretion, conversion of estradiol to estrone, androgen excretion, ratio of etiocholanolone to corticoids in urine (but not in blood) and, more recently, serum prolactin levels. The current vogue is measurement of receptors for estrogen or for prolactin in tumors and metastases. While a great deal of interest is currently generated in this worthwhile endeavor and while many current workers are optimistic, it seems at present premature to think we can now recognize with confidence those patients who will or will not respond to any mode of therapy. The facts that estrogen receptors are present in more than half of primary breast cancers, that there is a great discrepancy between the incidence of receptors in primary tumors and in metastases, and that there is already a tempering of initial enthusiasm in reported studies makes one cautious about accepting this solution for a difficult and important problem. It should also be noted that discriminants based upon urinary excretion of steroid metabolites or other substances in patients with advanced cancer should be viewed with great skepticism in view of the important but little publicized fact that glomerular filtration falls with advance of cancer (GRAHAM et al., 1964). Consequently, urinary excretion of any substance may reflect the state of the kidney rather than of the cancer's responsiveness to a given treatment.

## B. Empirical Response Rates

Therefore, if one accepts the premise that we cannot yet predict which patient will respond to which treatment, it becomes clear that one must rely upon empirical data, that is response rates obtained in patients with disseminated human cancer according to the various, presently accepted variables. In general, it is recognized that most previously available hormonal treatments were less effective in young women than in those progressively farther along beyond the menopause and that patients with metastases in the vital organs (brain, liver, and lung) respond less frequently than those with metastases in less vital areas (skin, lymph, nodes, bones). It is also now well established that oöphorectomy is most effective (25–30% regressions) in women who are still menstruating, that the response rate drops in the first year after the last menstrual period, and that responses more than 1 year after the menopause are extremely rare. While regression rates for oöphorectomy are less than those now obtainable with calusterone or with cytotoxic chemotherapy, oöphorectomy remains the first line of attack—in women of suitable age, of course—because of the much longer duration of regressions, averaging 33 months and ranging up to 11 years (GORDAN, 1974).

## C. Evaluation of Efficacy of Treatment

In trying to choose therapy on the basis of objective regression rates, it is unfortunate that widely disparate criteria have been used in various studies. In addition, the types of subjects for investigational therapy have varied widely from series to series. It is therefore possible to find reports showing unbelievably high regression rates in selected patients evaluated by incomplete or doubtful criteria. Similarly, it is easy to prove that a good compound is useless by evaluating its efficacy in patients who have already undergone cytotoxic chemotherapy, who have a high incidence of visceral metastasis, who are mostly young women who have failed to respond to castration, or by using inappropriate doses. Consequently, the literature is, to say the least, chaotic and confusing, and reported rates are often meaningless. In my opinion, the best data were those obtained by the Cooperative Breast Cancer Group of the

National Cancer Institute up to 1964. From that date on, unfortunately, many patients were included who had already undergone cytotoxic chemotherapy. The fact that this type of treatment compromises against subsequent response to steroid hormone therapy was first made clear while reviewing the cases of F. J. Ansfield at Madison, Wisconsin. He pointed out that their usual procedure included the treatment of patients who had previously been treated with 5-fluorouracil, as permitted by the protocols of the Cooperative Breast Cancer Group. While the rest of the group observed a 21.5% regression rate with testosterone propionate (Cooperative Breast Cancer Group, 1964a), Ansfield observed that only 8% of patients previously treated with 5-fluorouracil responded to this steroid. With the more wide-spread use of cytotoxic chemotherapy, the group regression rate with testosterone propionate dropped from 21.5% to 10% (Cooperative Breast Cancer Group, 1964b). That the drop is actually due to cytotoxic chemotherapy has been demonstrated in our crossover study in which calusterone produced 51% objective regressions as primary or secondary hormonal therapy (46 of 90 cases), but only 19% in patients with previous cytotoxic chemotherapy (GORDAN et al., 1973).

## 1. Established Clinical Variables

In order to obtain comparable groups, it is essential that the patients receiving the experimental treatment being tested be similar in menopausal age and in site of metastasis, to those receiving a reference standard. Obviously, it is not ethical to have an untreated group. Naturally, patients who are receiving the treatment as primary therapy for disseminated disease should not be compared with those who have had previous endocrine treatments, and patients who have had previous cytotoxic chemotherapy must also be considered separately. If groups are of sufficient size, e.g., 20 or more patients, one can expect that the free interval from mastectomy to first known metastasis will probably be approximately comparable in the investigational and reference groups. A long free interval, in general, indicates slow growth, better prognosis, and a better chance of responding to hormonal therapy, administrative or ablative.

## 2. Measurement of Responses

Evaluation requires periodic physical examination with measurement of all palpable or visible metastases, lymph nodes and palpable viscera (especially the liver). We usually examine patients for evaluation at intervals of 6 weeks. If they have been in regression for more than 6 months, as is usual for responders to hormonal therapy, this period may be extended to 3 months. In very rare cases, for example, patients who have been in regression for 5–10 years since oöphorectomy, examination every 6 months may be appropriate. Before treatment and at times of evaluation, patients also undergo an x-ray survey to detect and measure lesions in the skull, cervical, thoracic and lumbar spine, ribs, pelvis, and chest. Comparable views must be taken at each examination, for example, left lateral skull, left lateral spine films. It is desirable but not always possible to obtain similar exposure, kilovoltage, and milliamperage. If technical factors are seriously different, this must be taken into account in interpreting films.

### 3. Other Parameters

Not used in evaluation of efficacy of treatment but of importance for the patients' well-being and for evaluation of other effects of the test treatment are a complete blood count, routine blood chemistry (serum creatinine or BUN, calcium, phosphate, alkaline phosphatase, bilirubin, SGOT and, in selected cases, bromsulfalein for effects of the tumor or the treatment on liver function). Definition of a regression requires that: (1) measurable lesions shrink, or (2) osseous lesions calcify, (3) and that no new lesions appear. Perhaps the most important criterion introduced by the Cooperative Breast Cancer Group, and unfortunately not used universally by other groups or individual investigators, is that all regressions must be confirmed by two extramural examiners who see only photographs, measurements and X-ray films without knowing which patient received which treatment. In my experience, reviewers rarely change the investigator's evaluation, probably because the investigator, knowing he will be reviewed, is extremely careful about what he calls a regression. By the same token, regression rates obtained in reviewed series are considerably lower, and in my opinion, more meaningful than those in which no review is obtained. It should be noted that anabolic-androgenic agents often raise hematocrit, white count and platelet count; these are not considered tumor responses but hormonal effects.

## D. Antitumor Efficacy of Steroidal Agents

### 1. Primary Hormonal Therapy

A list of compounds that have been studied in women with disseminated breast cancer without previous chemotherapy is appended (Tables 1–3). The clinically most important compounds in the treatment of advanced breast cancer are listed in Table 2.

The primary reference standard used by the Cooperative Breast Group for many years was testosterone propionate. As noted, this is an effective agent producing regressions in 21.5% of patients. It is also the best studied of all agents, having been evaluated in 521 patients without previous chemotherapy. Statistical analysis of this series (Cooperative Breast Cancer Group, 1964a) showed that patients who responded to this therapy had significantly better survival than those who did not. Nonetheless, I have not used this therapy nor recommended it in the last 10 years. The reason is that this therapy produces intense virilization in old women, already partially defeminized by mastectomy and often by oöphorectomy, and doomed to die of cancer. In my view it is cruel to induce acne, hirsutism, alopecia, masculine deepening of the voice and, a driving, intense libido that is most unwelcome in older women, many of whom are widows. Reduction of androgenicity without loss of antitumor effect against breast cancer by chemical alteration of the testosterone molecule was first achieved by BLACKBURN et al. (1959). The compound dromostanolone propionate (2α-methyl-dihydrotestosterone propionate) was the first established example of an effective antitumor testosterone derivative with less androgenicity. It also had another unique property. Unlike other steroid hormones, it was as effective when given as secondary *hormonal* therapy (*not* after cytotoxic chemotherapy) as in primary treatment (THOMAS et al., 1962; WITT et al., 1963). It was formerly believed—erroneously—that after a patient had ceased responding to an androgen,

Table 1. Steroids vs. breast cancer (without prior cytotoxic therapy)

| | |
|---|---|
| a) Steroids with no antitumor efficacy | Regression/cases tested |
| 17α-ethynyltestosterone | 0/22 |
| Progesterone | 0/26 |
| Testololactone | 0/22 |
| 17α-methyl-5α-androstan-3β,11β,17β-triol | 0/24 |
| Delalutin | 0/20 |
| 17αmethyl-6-dehydrotestosterone | 0/19 |
| 6α-chloro-17-hydroxyprogesterone acetate | 0/19 |
| 17α-methyl-4β-hydroxytestosterone | 0/22 |
| 17α-methyl-dihydrotestosterone | 0/22 |
| b) Steroids yielding <10% regressions | Regressions/cases treated (%) |
| 16α-estradiol dipropionate | 1/27 (4%) |
| Estradiol benzoate | 1/25 (4%) |
| 5α-androstan-17β-ol propionate | 1/25 (4%) |
| 16α-hydroxytestosterone | 1/25 (4%) |
| 2α-hydroxytestosterone | 1/24 (4%) |
| 11β-hydroxytestosterone | 1/24 (4%) |
| Androsterone | 1/22 (5%) |
| Estrololactone | 1/21 (5%) |
| 2α-methyldihydrotestololactone | 1/21 (5%) |
| Prednisone | 1/21 (5%) |
| 2α,17α-dimethyltestosterone | 2/26 (8%) |
| 17α-methyl-5α-androstan-11β, 17β diol | 2/26 (8%) |
| 3β-fluoro-5α-androstan-17β-ol | 2/24 (8%) |
| 17α-methyl-11β-hydroxytestosterone | 2/23 (9%) |
| 5α-androstan-3α,17β-diol dipropionate | 2/23 (9%) |
| 11β-hydroxydianabol | 2/23 (9%) |
| c) Steroids yielding 10–16% regressions | Regressions/cases (%) |
| 4-androstene-3,17-dione | 2/21 (10%) |
| 2α-fluorotestosterone propionate | 2/21 (10%) |
| 6-dibromomethylene testosterone propionate | 3/24 (13%) |
| 2α-methyl-19-nortestosterone | 3/22 (14%) |
| Dimethylstilbestrol | 3/22 (14%) |
| 5α-androstan-3β,17β-diol diacetate | 3/21 (14%) |
| 17β-hydroxy-17α-methyl-5α-androstan-3,11-dione | 4/28 (14%) |
| 17β-hydroxy-6β-methyl-5α-androstan-3-one propionate | 3/21 (14%) |
| Fluoxymesterone | 30/203 (15%) |
| 4,4-dimethyl-17β-hydroxy-5-androsten-3-one | 3/19 (16%) |
| 7α-methyltestosterone propionate | 7/44 (16%) |
| Anvene | 4/25 (16%) |
| Hexestrol | 4/23 (17%) |
| 2(hydroxymethylene)-5α-dihydrotestosterone | 4/24 (17%) |
| d) Steroids yielding >17% regression | Regressions/cases treated (%) |
| Testosterone propionate | 191/1055 (18%) |
| Dianabol | 4/22 (18%) |
| Dihydrofluoxymesterone | 4/21 (19%) |
| 4-androsten-3α,17α-diol diacetate | 4/21 (19%) |
| Norlutin | 5/26 (19%) |
| [a]Dromostanolone | 14/72 (19%) |
| Dromostanolone propionate | 38/194 (20%) |
| 11-ketofluoxymesterone | 4/20 (20%) |
| 17α-ethynyl-17-hydroxy-$\Delta^{5-10}$estren-3-one | 4/20 (20%) |

Table 1 (continued)

| | |
|---|---|
| Braxorone | 5/25 (20%) |
| D-homotestosterone propionate | 4/20 (20%) |
| 7α-methyl-19-nortestosterone acetate | 5/24 (21%) |
| Stilbestrol | 56/268 (21%) |
| 7α,17α-dimethyltestosterone | 4/19 (21%) |
| 4-androsten-3β,17β-diol-diacetate | 5/23 (22%) |
| ICI 46474 | 10/46 (22%) |
| 17α-methyl-19-nortestosterone | 5/21 (24%) |
| Medroxyprogesterone | 6/24 (25%) |
| Oxylone acetate | 6/23 (26%) |
| A-nortestosterone propionate | 7/26 (27%) |
| $\Delta^1$-testololactone[a] | 8/25 (32%) |
| Nafoxidine | 18/52 (37%) |
| Calusterone | 46/90 (51%) |

[a] nonandrogenic (?).

Table 2. Efficacy of endocrine treatment of disseminated breast cancer: I. Primary hormonal therapy

| Preparation | Dose | Route | Regressions/(%) no. treated |
|---|---|---|---|
| Calusterone | 200 mg/day | oral | 21/41 (51%) |
| $\Delta^1$-testololactone | 1000 mg/day | oral | 8/25 (32%) |
| Testosterone propionate | 100 mg 3x/wk | I.M. | 112/521 (21%) |
| Dromostanolone propionate | 100 mg 3x/wk | I.M. | 39/193 (20%) |
| Fluoxymesterone | 20 mg/day | oral | 24/160 (15%) |
| Stilbestrol | 15 mg/day | oral | 48/261 (18%) |

she should be given a trial with an estrogen and, conversely, after an estrogen, an androgen. It was soon found that testosterone propionate and stilbestrol are very poor agents when applied as secondary hormonal therapy (WITT et al., 1963). Dromostanolone propionate was the first steroidal agent to prove as effective for secondary as for primary hormonal therapy. As will be seen later, another steroidal agent, calusterone, also has this unusual property, which can be exploited by the physician to obtain further regressions in the serial therapy of the woman with disseminated breast cancer. The free alcohol of dromostanolone, unlike the propionate, is effective by mouth. It has not been studied as extensively as the propionate, but the data appear to indicate that it is not a potent androgen when given by this route and that it has antitumor efficacy similar to that of testosterone propionate or of dromostanolone propionate. It is most widely used for its hematopoietic effect.

Fluoxymesterone was also found somewhat effective against advanced breast cancer. Clinically the androgenicity of this agent is hard to delineate. In animal studies it is a very potent androgen. In hypogonadal or juvenile human subjects it causes a disproportionate enlargement of the phallus (penis or clitoris) but little other androgenicity. Strangely, it does not increase libido and rarely causes significant acne. It is poor replacement therapy for hypogonadal men. On the other hand this lack of visible androgenicity is desirable in the treatment of women with advanced breast cancer. Unfortunately, early claims that it produced regressions in 53% of patients (KENNEDY, 1957a, b, 1958) were not realized. In a protocol study of previously untreated patients randomized against testosterone propionate and evaluated by the criteria of the Cooperative Breast Cancer Group, we found that fluoxymesterone produced regressions in only 14% of our patients (LOWE et al., 1961). A group study organized shortly after, including 160 patients, resulted in a group regression rate of 15% (Cooperative Breast Cancer Group, 1961). Thus, this agent is slightly less effective than testosterone propionate or dromostanolone propionate but has the double advantage that it is effective when taken by mouth and that it produces little visible androgenicity. It is, however, a 17α-alkylated steroid and, as such, carries a cholestatic hazard (DE LORIMIER et al., 1965).

One of the most important and interesting steroidal compounds in the treatment of advanced breast cancer is $\Delta^1$-testololactone. It was introduced by SEGALOFF et al. in 1960. When first studied, it produced objective regressions in 32% of cases (SEGALOFF et al., 1962). From 1962 to 1969, however, results of treatment with this compound were poor (Cooperative Breast Cancer Group, 1964b). Attempts to elucidate reasons for the difference by infrared or ultraviolet spectrophotometry or by gas liquid chromatography showed no chemical abnormality of the steroid (HALDEN et al., 1973). The tablets, however, varied greatly in their in vitro dissolution rate as determined by the N.F. procedure. Most preparations of compressed tablets from 1964a to 1969 were virtually indissoluble. Subsequently the Upjohn Company prepared for us compressed tablets of $\Delta^1$-testololactone with a $T_{50}$ of 4.5–8.5 minutes. When tested for antitumor efficacy against advanced breast cancer in women without previous hormonal or cytotoxic chemotherapy, using a dose of 1000 mg a day by mouth, the Upjohn preparation produced regressions in 8 of 25 women (32%) (GORDAN et al., 1973). In this study, $\Delta^1$-testololactone was double-blinded against calusterone in a dose of 200 mg a day by mouth, which produced 12 regressions in 23 women (52%). $\Delta^1$-Testololactone is of particular interest because it has no known hormonal or other effects; its only established action is to produce regressions in advanced breast cancer. Because it is effective in 32% of women with advanced breast cancer and has no known undesirable actions, it is probably the primary treatment of choice in most cases, at least in women 5 or more years past the menopause. In this age group, in favorable cases, primary hormonal therapy can probably best be carried out by administration of either $\Delta^1$-testololactone 1000 mg a day or stilbestrol 15 mg a day without compromising the patient's chances of subsequent response to another relatively innocuous treatment such as calusterone or dromostanolone. Objective regressions produced by primary steroidal therapy almost always last more than 6 months and, in my experience, average 11.5 months in duration.

## 2. Secondary Hormonal Therapy

As already noted, when patients no longer respond to androgens, estrogens are poor secondary hormonal therapy. Similarly, androgens are poor agents for patients who are no longer responding to estrogens (Table 3). By far the most potent antitumor steroid against disseminated human breast cancer thus far identified is calusterone

Table 3. Efficacy of endocrine treatment of disseminated breast cancer: II. Secondary hormonal therapy

| Preparation | Dose | Route | Regressions/(%) no. treated |
|---|---|---|---|
| Calusterone | 200 mg/day | oral | 25/49 (51%) |
| $\Delta^1$-testololactone | 1000 mg/day | oral | 17/40 (42%) |
| + 5-fluorouracil | 500 mg/wk | I.V. | |
| Dromostanolone propionate | 100 mg 3x/wk | I.M. | 5/26 (19%) |
| Prednisone | 30 mg/day | oral | 11/46 (24%) |
| + liothyronine | 50 µg/day | oral | |
| Hydrocortisone | 30 mg/day | oral | 5/23 (22%) |
| + liothyronine | 50 µg/day | oral | |
| Prednisone | 30 mg/day | oral | 1/21 (5%) |
| Testosterone propionate | 100 mg 3x/wk | I.M. | 2/22 (9%) |
| Stilbestrol | 15 mg/day | oral | 2/46 (4%) |
| Hypophysectomy | | transfrontal | 3/27 (11%) |

(7β-17α-dimethyltestosterone). This agent is a weak androgen but, like other 17α-methylated compounds, in doses of 200 mg a day or more, it often produces transient nausea (in 28% of cases) and can cause bromsulfalein retention. In daily doses of 150–200 mg, it has not as yet caused jaundice or hepatocellular damage. Nonetheless, any 17α-alkylated steroid should be administered with great caution in patients who have compromised liver function, for example, in cirrhosis or massive hepatomegaly due to metastatic breast cancer. Calusterone is as effective when given as secondary hormonal therapy as when given as primary hormonal therapy. Unlike *all* other previously tested steroids, it is as effective in women less than 1 year after castration as in old women, and in women with visceral metastases as in those with local disease. It produces objective regressions in 51% of patients including unfavorable cases less than 1 year past the menopause (castration failures) and women with visceral metastases (GORDAN et al., 1973). The only other hormonal steroid of proved value when given alone as secondary hormonal therapy in disseminated breast cancer is dromostanolone propionate. As noted, it produces objective regressions in 19% of patients when given as secondary therapy. Testosterone propionate, methyltestosterone, prednisone, and stilbestrol are of little value at this stage. Curiously, when prednisone 30 mg a day is supplemented with triiodothyronine 50 µg a day, the combination has potent antitumor activity, producing regressions in 24% of patients (GARDNER et al., 1962). Predictably, this dose of corticoid invaribly produces severe Cushing's syndrome. It was therefore a pleasant surprise to find that hydrocortisone 30 mg combined with triiodothyronine 50 µg a day is as effective as the combination of prednisone and triiodothyronine (WITT et al., 1963). This dose of hydrocortisone, of course, is merely replacement therapy and is innocuous. The combination produces regressions in 22% of patients when administered as secondary hormonal therapy. It was discovered during trials to ascertain whether hypophysectomy or the replacement therapy necessitated by hypophysectomy is the effective agent in producing regressions of disseminated breast cancer. At that time, hypophysectomy was performed by the transfrontal route with considerable morbidity and some mortality. Presently, of course, cryohypophysectomy is much easier on the patient and produces regressions in approximately 30% of patients (GORDAN, 1974). It should be noted that corticoids, like cytotoxic chemotherapeutic agents, interfere with bone healing. It is extremely rare to see recalcification of osteolytic lesions in corticoid-treated patients. It seems likely that corticoids and cytotoxic agents each inhibit osteocytic activity. This effect may also explain their ability to reduce the hypercalcemia of malignancy since it is now believed that the blood calcium level is maintained by osteocytic osteolysis. Objective regressions produced by secondary hormonal therapy usually exceed 6 months in duration and average 9 months (WITT et al., 1963).

### 3. Combination of Steroids and Cytotoxic Chemotherapy

Little information is available about the combination of steroidal treatment with cytotoxic chemotherapy. In one report (FRECKMAN et al., 1964), chlorambucil 14–20 mg daily was combined with prednisolone 30 mg a day for 14 days, then tapered to 10 mg twice daily for 7 days and then to 10 mg once daily as maintenance therapy. It was reported that this treatment produced objective regressions in 34% of patients and that these doses caused no serious, disagreeable or toxic effects. The experimental design did not include extramural evaluation or simultaneous control, so it is difficult to know whether these figures are significantly greater than those obtained with chlorambucil alone. GOLDENBERG et al. (1973), using the rigorous criteria of the Cooperative Breast Cancer Group, reported 21% objective regressions for the combination of chlorambucil and prednisone. In another report (HALDEN et al., 1973), also evaluated by the criteria of the Cooperative Breast Cancer Group, including extramural review, the combination of $\Delta^1$-testololactone 1000 mg a day plus 5–fluorouracil 500 mg intravenously once a week produced objective regressions in 17 of 40 patients (42%). White blood cell count and platelet count fell slightly, but sufficiently to cause therapy to be omitted temporarily in seven cases and to terminate this treatment in two others. Bone healing did not occur. The regression rate is higher than has previously been obtained with these doses of either agent alone, but lack of simultaneous controls precludes firm conclusions. Fragmentary data from these and similar studies suggest the possibility of additive effects of the two types of treatment and may indicate different modes of action.

After these types of primary and secondary hormonal therapy, one may then consider the possibility of using a single or multiple cytotoxic agents. The present vogue is for the simultaneous use of 5 cytotoxic chemotherapeutic agents. This is, of course, both toxic and expensive, but does produce regressions in approximately 42% of cases (DAVIS et al., 1974). In this important study by highly experienced chemotherapists, 42% of patients obtained regressions. Although there was no report of an extramural review, the criteria used by these authors are otherwise identical with those of the Cooperative Breast Cancer Group. The median length of response was 5 months. Recalcification of lytic bone lesions occurred in only one patient. Toxicity was thought to have contributed to death in 7 of the 74 women studied, even in the hands of these experienced chemotherapists. A favorable response did *not* lengthen survival, suggesting that the treatment poisoned the host as well as the tumor. After this treatment, only 5 of 28 patients (18%) responded briefly to a variety of therapeutic measures. Nonetheless, at this stage of disease, and after exhausting innocuous treatment, this therapy can still produce worthwhile regression in some patients. After polychemotherapy, one final step still remains, namely the use of adriamycin (BLUM and CARTER, 1974). This agent has the usual toxicity of cytotoxic agents plus cardiotoxicity, limiting its dose severely (RINEHART et al., 1974). Nonetheless, when all else has been exhausted, adriamycin can still produce worthwhile regressions in patients with advanced breast cancer.

## E. Hypercalcemia

### I. Etiology and Pathogenesis

The most common single cause of hypercalcemia is malignancy and the single malignancy most commonly responsible for hypercalcemia is disseminated breast cancer. Even in this condition, however, the mechanism by which hypercalcemia occurs may be heterogeneous. In general, hypercalcemia can result from any of four major pathophysiologic mechanisms: (1) increased intestinal absorption of calcium (as in vitamin D poisoning, sarcoidosis, or milk-alkali syndrome); (2) (rarely) failure to transport calcium from blood to bone (as in myxedema); (3) most important, *increased osteolysis* (as in cancer, thyrotoxicosis, or hyperparathyroidism); and (4) (rarely) increased binding of calcium to serum proteins and citrate (as in Addison's disease). Of these four mechanisms only one, increased osteolysis, has been implicated in the hypercalcemia of breast cancer. Increased osteolysis, in tumor, may result from several different mechanisms. The vast majority of patients with hypercalcemia of disseminated breast cancer have wide-spread osteolytic lesions (GRAHAM et al., 1963). In these patients, serum calcium and phosphate levels are both elevated. Urinary calcium excretion is increased until severe impairment of glomerular filtration supervenes. Tubular reabsorption of phosphate (TRP) is normal and serum parathyroid hormone (PTH) levels are suppressed because of the nonparathyroid hypercalcemia. It should be noted that these chemical manifestations differ greatly from those of hyperparathyroidism, in which serum phosphate levels and TRP are low and PTH is measurable in the plasma by radioimmunoassay. These differences are so striking that it has actually been possible to diagnose parathyroid adenomas accurately in patients with breast cancer because of the presence of hypophosphatemia, low TRP, and measurable parathyroid hormone (GORDAN, 1974). While breast cancer does not secrete parathyroid hormone, it does contain potent osteolytic lipids. These are not vitamin D compounds (GORDAN and SCHACHTER, 1963). The most potent osteolytic agents thus far identified in lipid extracts of human breast cancer are stigmasteryl acetate and $\Delta^7$-sitosteryl acetate. These two phytosteryl esters actually have greater calcium-mobilizing capacity than vitamin $D_3$ or 25-hydroxyvitamin $D_3$ (GORDAN et al., 1967). Nonetheless, it is not possible to prove that these potent osteolytic phytosteryl esters are responsible for the hypercalcemia of breast cancer since serum levels in breast cancer patients are not significantly different from those of other subjects (WHITNEY et al., 1973). Thus, the present state of knowledge of etiology is limited to the observations that in breast cancer, hypercalcemia and hyperphosphatemia occur mostly but not always with osteolytic metastases and that human breast cancer extracts contain potent osteolytic lipids, whose exact relationship to clinical hypercalcemia is not clear.

Hypercalcemia was first discovered in patients with breast cancer during balance studies designed to elucidate the effect of sex hormones. This coincidence led to a long-perpetuated myth that estrogens and androgens are responsible for the hypercalcemia of breast cancer. It is indeed true that hypercalcemia may first occur in a patient with disseminated breast cancer if she becomes nauseated, anorectic, dehydrated and partly immobilized, all of which can occur transiently during the initial phases of estrogen or androgen therapy. Actually, however, hypercalcemia occurs much more commonly in breast cancer patients not receiving hormonal therapy (GRAHAM et al., 1963), and the important work of HORTON and OLSON (1964) showed that administration of androgens actually reduces serum calcium levels in patients with the hypercalcemia of breast cancer. This result is in accord with the now well-established ability of estrogens and androgens to restrain osteocytic osteolysis (GARDNER et al., 1963; GORDAN, 1974). It will be recalled that loss of this restraint by gonadal steroids is responsible for the osteoporosis of hypogonadism, including the postmenopausal state; that serum calcium levels of women in the reproductive years are 0.2–0.3 mg/100 ml lower than in the premenarchal or postmenopausal years; and that estrogens can lower serum calcium levels in normal subjects or even in those with hyperparathyroidism (LANDAU and KAPPAS, 1965; GALLAGHER and NORDIN, 1972). Pharmacologically, estrogens, androgens, and calcitonin have the same ability to turn off osteocytic osteolysis and to lower serum calcium levels. It is not clear at this writing whether corticoids, which also interfere with osteocytic functions, reduce hypercalcemia by this mechanism.

## 2. Treatment of Hypercalcemia

Since hypercalcemia damages the patient by interfering with the action of the antidiuretic hormone on the distal renal tubules causing a diabetes-insipiduslike state and severe dehydration, replacement of blood volume is one of the cardinal features of treatment in all cases. When necessary to correct fluid loss by intravenous therapy, sodium-containing fluids are particularly advantageous, as shown by WALSER (1970). Sodium citrate or sodium sulfate chelates calcium in excretable forms, unlike the phosphate, which is retained and can cause metastatic calcification. Calcium-excreting diuretics (for example Lasix) are probably not good treatment: (1) they can augment potentially fatal dehydration; (2) calcium is not efficiently excreted in the urine; and (3) hypokalemia is a real hazard. Hydration is of value not only as therapy for established hypercalcemia but also as prophylaxis against hypercalcemia in patients with disseminated breast cancer, particularly those with wide-spread osteolytic lesions. Routine prophylactic procedures include mobilization to tolerance and abstinance from milk, cheese, calcium preparations, and pharmacologic doses of vitamin D. Hydration alone will correct hypercalcemia in 50% of cases of disseminated breast cancer. When this fails, or in patients with serum calcium levels above 14 mg/100 ml, corticoids are often effective. Effective doses are the equivalents of 120 mg of hydrocortisone or 20 mg of prednisone or 3 mg of dexamethasone daily, continued until the serum calcium level falls to normal or near-normal levels, whereupon they can be tapered off. The combination of hydration, sodium-containing fluids, and corticoids corrects hypercalcemia in approximately 85% of cases. For the remaining 15%, cytotoxic chemotherapeutic agents in low, minimally toxic doses are effective. Currently, the most popular agent for this cause is mithramycin 25 µg/kg intravenously as a single bolus, repeated in 24–48 h if needed. Similar results can be obtained with other agents, for example, methotrexate 5 mg intravenously, actinomycin D 0.7 mg per day, etc. It is likely that hypercalcemia of breast cancer is less common today because of the widespread, and in my opinion, often premature use of cytotoxic chemotherapeutic agents. Therapeutically, however, they are of great value in the treatment of hypercalcemia by virtue of their ability to poison the osteocytes and turn off osteocytic osteolysis. This effect occurs without necessarily producing tumor regression, as first pointed out by MYERS and MAGILL (1956).

It should be emphasized that the hypercalcemia of breast cancer can be corrected in the vast majority of cases and that treatment is usually worthwhile, even in those patients whose tumor does not respond to hormonal or chemotherapeutic maneuvers. Breast cancer often runs a very indolent course. It often responds to relatively innocuous treatment, and it is possible to make the patient's life comfortable for many years. This principle is well illustrated in a case described by THOMAS et al. (1960). The serum calcium level rose to 22 mg/100 ml in a 40-year-old woman with extensive osteolytic metastases. The hypercalcemia did not respond to hydration or corticoids but was quickly controlled by the administration of D-O-N (6-diazo-5-oxo-l-norleucine) (MYERS and MAGILL, 1956). Her wide-spread metastatic lesions did not respond to any of the then available antitumor agents. On two occasions, 17 months and 2 years after the initial episode, she again manifested moderate transient hypercalcemia. Nonetheless, in the 4½ years following heroic treatment for a record-high hypercalcemia, this woman continued to go hunting and to enjoy life. This anecdote exemplifies the fact that when hypercalcemia is caused by an indolent widespread cancer, it may still be well worthwhile treating.

## References


BEATSON, G.T.: On the treatment of inoperable cases of carcinoma of the mamma: Suggestions for a new method of treatment, with illustrated cases. Lancet **1896 II**, 104, 162.

BLACKBURN, C.M., CHILDS, B.S.: Use of 2 alpha-methyl androstan-17 beta-ol,3-one (2 methyl dihydrotestosterone) in the treatment of advanced cancer of the breast. Proc. Mayo Clin. **34**, 113—126 (1959).

BLOOM, H.J.G.: Medroxyprogesterone acetate (provera) in the treatment of metastatic renal cancer. Brit. J. Cancer **25**, 250—265 (1971).

BLOOM, H.J.G., BOESEN, E.: Antioestrogens in treatment of breast cancer: Value of nafoxidine in 52 advanced cases. Brit. med. J. **1974**, 7—10.

BLUM, R.H., CARTER, S.K.: Adriamycin: A new anticancer drug with significant clinical activity. Ann. intern. Med. **80**, 249—259 (1974).

BRENDLER, H., WINKLER, B.S.: Effect of norethandrolone on 17-ketosteroid excretion in prostatic cancer patients. J. clin. Endocr. **19**, 183—192 (1959).

COLE, M.P., JONES, C.T.A., TODD, I.D.H.: A new anti-oestrogenic agent in late breast cancer; an early clinical appraisal of ICI 46474. Brit. J. Cancer **25**, 270—275 (1971).

Cooperative Breast Cancer Group: Progress report 1956—1960. Cancer Chemother. Rep. **11**, 109—141 (1961).

Cooperative Breast Cancer Group: Testosterone propionate therapy in breast cancer. J. Amer. med. Ass. **188**, 1069—1072 (1964a).

Cooperative Breast Cancer Group: Cancer Chemother. Rep. **41**, Suppl. 1 (1964b).

DAVIS, H.L., RAMIREZ, G., ELLERBY, R.A., ANSFIELD, F.J.: Five-drug therapy in advanced breast cancer: factors influencing toxicity and response. Cancer **34**, 239—245 (1974).

DE LORIMIER, A., GORDAN, G. S., LOWE, R. C., CARBONE, J. V.: Methyltestosterone, related steroids, and liver function. Arch. intern. Med. **116**, 289—294 (1965).
FRECKMAN, H. A., FRY, H. L., MENDEZ, F. L., MAURER, E. R.: Chlorambucil-prednisolone therapy for disseminated breast carcinoma. J. Amer. med. Ass. **189**, 23—26 (1964).
GALLAGHER, J. C., NORDIN, B. E. C.: Treatment with oestrogens of primary hyperparathyroidism in postmenopausal women. Lancet **1972**, 503—507.
GARDNER, B., GRAHAM, W. P., GORDAN, G. S., LOKEN, H. F., THOMAS, A. N., TEAL, J. S.: Calcium and phosphate metabolism in patients with disseminated breast cancer: effect of androgens and of prednisone. J. clin. Endocr. **23**, 1115—1124 (1963).
GARDNER, B., THOMAS, A. N., GORDAN, G. S.: Antitumor efficacy of prednisone and sodium liothyronine in advanced breast cancer. Cancer **15**, 334—337 (1962).
GOLDENBERG, I. S., MCMAHAN, C. A., ESCHER, G. C., VOLK, H., ANSFIELD, F. J., OLSON, K. B.: Secondary chemotherapy of advanced breast cancer. Cancer **31**, 660—663 (1973).
GORDAN, G. S.: Endocrine treatment of breast cancer. Medcom Famous Teachings in Modern Medicine. New York: Medcom, Inc. 1974.
GORDAN, G. S., FITZPATRICK, M. E., LUBICH, W. P.: Identification of osteolytic sterols in human breast cancer. Trans. Ass. Amer. Physicians **80**, 183—188 (1967).
GORDAN, G. S., HALDEN, A., HORN, Y., FUERY, J. J., PARSONS, R. J., WALTER, R. M.: Calusterone ($7\beta$-$17\alpha$-dimethyl testosterone) as primary and secondary therapy of advanced breast cancer. Oncology **28**, 138—146 (1973).
GORDAN, G. S., SCHACHTER, D.: Vitamin D activity of normal and neoplastic human breast tissue. Proc. Soc. exp. Biol. (N.Y.) **113**, 760—761 (1963).
GRAHAM, W. P., GARDNER, B., GORDAN, G. S.: Glomerular filtration in disseminated breast cancer. Arch. intern. Med. **113**, 837—839 (1964).
GRAHAM, W. P., GARDNER, B., THOMAS, A. N., GORDAN, G. S., LOKEN, H. F., GOLDMAN, L.: Hypercalcemia in human breast cancer. Surg. Gynec. Obstet. **117**, 709—714 (1963).
HALDEN, A., HORN, Y., WALTER, R. M., GORDAN, G. S., SCHONFELD, E. M.: Combined high-dose $\Delta^1$-testololactone with low-dose 5-fluorouracil in the treatment of metastatic breast cancer. Oncology **27**, 493—501 (1973).
HORTON, J., OLSON, K. B.: Hypercalcemia associated with cancer of the breast and other organs. In: Proc. 1st Europ. Bone and Tooth Symp., pp. 207—216. Oxford-London-New York-Paris: Pergamon Press 1964.
HUGGINS, C., HODGES, C. V.: I. Effect of castration, of estrogen and of androgen injection on serum phosphatases in metastatic carcinoma of prostate. Cancer Res. **1**, 293—297 (1941).
KELLEY, R. M., BAKER, W. H.: Progestational agents in the treatment of carcinoma of the endometrium. New Engl. J. Med. **264**, 216—222 (1961).
KENNEDY, B. J.: Fluoxymesterone in treatment of advanced breast cancer. Clin. Res. Proc. **5**, 219 (1957a).
KENNEDY, B. J.: Fluoxymesterone in treatment of advanced breast cancer. Cancer **10**, 813—818 (1957b).
KENNEDY, B. J.: Fluoxymesterone therapy in treatment of advanced breast cancer. New Engl. J. Med. **259**, 673—675 (1958).
LANDAU, R. L., KAPPAS, A.: Anabolic hormones in hyperparathyroidism: with observations on the general catabolic influence in man. Ann. intern. Med. **62**, 1223—1233 (1965).
LOWE, R., DE LORIMIER, A. A., GORDAN, G. S., GOLDMAN, L.: Antitumor efficacy of fluoxymesterone: use in advanced breast cancer. Arch. intern. Med. **107**, 241—244 (1961).
MALKASIAN, G. D., DECKER, D. G., JORGENSEN, E. O., WEBB, M. J.: Evaluation of $6,17\alpha$-dimethyl-6-dehydroprogesterone for treatment of recurrent and metastatic gynecologic malignancy. Amer. J. Obstet. Gynec. **118**, 461—465 (1974).
MYERS, W. P. L., MAGILL, G. B.: Alterations in calcium metabolism in cancer patients treated with 6-diazo-5-oxo-L-norleucine; Proc. Soc. exp. Biol. **93**, 314—318 (1956).
O'HALLORAN, M. J., MADDOCK, P. G.: ICI 46,474 in breast cancer. J. Irish Med. Ass. **67**, 38—39 (1974).
RINEHART, J. J., LEWIS, R. P., BALCERZAK, S. P.: Adriamycin cardiotoxicity in man. Ann. intern. Med. **81**, 475—478 (1974).

SEGALOFF, A., WEETH, J. B., MEYER, K. K., RONGONE, E. L., CUNNINGHAM, E. G.: Hormonal therapy in cancer of the breast: XIX. The effect of oral administration of $\Delta^1$-testololactone on clinical course and hormonal excretion. Cancer **15**, 633—635 (1962).
SEGALOFF, A., WEETH, J. B., MURISON, T. J., RONGONE, E. L.: $\Delta^1$-testololactone, androsterone and $5\alpha$-androstane-$3\alpha$-$17\beta$-diol-dipropionate in the treatment of advanced breast cancer. In: PINCUS, G., VOLLMER, E. P. (Eds.): Biological Activities of Steroids in Relation to Cancer, pp. 389—392. New York: Academic Press 1960.
THOMAS, A. N., GORDAN, G. S., GOLDMAN, L., LOWE, R.: Antitumor efficacy of $2\alpha$-methyl-dihydrotestosterone propionate in advanced breast cancer. Cancer **15**, 176—178 (1962).
THOMAS, A. N., LOKEN, H. F., GORDAN, G. S., GOLDMAN, L.: Hypercalcemia of metastatic breast cancer. Surgical Forum **11**, 70—71 (1960).
WALSER, M.: Treatment of hypercalcemia. Mod. Treatment **7**, 662—674 (1970).
WHITNEY, J. O., HORN, H., GORDAN, G. S.: Quantitation of phytosteryl acetates in human serum. Proc. Soc. exp. Biol. (N.Y.) **143**, 1000—1002 (1973).
WITT, J. A., GARDNER, B., GORDAN, G. S., GRAHAM, W. P., THOMAS, A. N.: Secondary hormonal therapy of disseminated breast cancer: comparison of hypophysectomy, replacement therapy, estrogens and androgens. Arch. intern. Med. **111**, 557—563 (1963).
WOLF, J., SPEAR, P., YESNER, R., PATNO, M. E.: Nitrogen mustard and steroid hormones in the treatment of inoperable bronchogenic carcinoma. Amer. J. Med. **29**, 1008—1016 (1960).

# E. Athletics

A. J. Ryan

## I. Introduction

There is scarcely any sport in which muscular strength does not play a role. This strength may be expressed statically, as in supporting a weight, dynamically, as in hitting a baseball, or in terms of endurance, as in pedaling a bicycle for 50 miles or more. Strength may be defined as the ability to produce tension in a muscle. It arises as the result of repeated muscular activity, from the process of learning the repeated action (skill), and from adaptation of the nervous system to the demands imposed on it.

Strength is increased by overloading muscle. The overload may be applied in four different ways, according to Lockhart: (1) the speed of the strength-producing activity can be gradually increased in a progressive manner; (2) the total load can be gradually increased; (3) the total time that the resistance is opposed can be progressively increased; and (4) with a constant load, the total number of performances resisting it can be progressively increased. Based on these possibilities a variety of systems of strength training have been developed: (1) making a single repetition against a maximum resistance; (2) making a semiflexible series of repetitions against a load that progresses from a fraction of the maximum to a maximum resistance (the method of De Lorme) or reversing the process (the Oxford method); (3) performing two-thirds of a maximum number of repetitions against a fixed resistance (the method of Morgan and Adamson); and (4) making a maximum number of repetitions against a relatively low and fixed resistance (the method of Walters).

Hettinger and Müller (1953) were the first to describe the performance of maximum static muscle contractions as isometric exercise. The practice of this technique to increase muscle size and strength was certainly not original with them since Charles Atlas (Romano) had been teaching it as "dynamic tension" for many years. In contrast, exercise in which muscles were shortened or lengthened (concentric and eccentric) was distinguished as isotonic.

Although isometric exercise can produce rapid gains in strength, it has the disadvantages that the prolongation of maximum contractions tends to reduce blood supply within the muscle and that it is difficult to monitor increases in the maximum tension and to motivate the subject to increase his effort. Isotonic exercise does not produce consistent gains through an entire range of joint motion due to changing mechanical advantages. This led Perrine (1968) to develop a machine to allow what he termed isokinetic exercise. This machine provides a constant resistance throughout the range of motion by controlling the speed at which the resistance arm can be moved. Because of the limitations imposed by the nature of the apparatus, the use of this device is limited chiefly to the development of strength in the moderate range required for rehabilitative exercise.

Once strength is developed, it may be maintained over an indefinite period by performing exercise at the same level twice or three times weekly. When activity stops completely, strength falls off at the rate of about 10% a week until it reaches the base from which it started. Strength decreases with aging, beginning at about age 40.

The muscular hypertrophy that occurs from overloading is prized by athletes, especially those concerned with overall development of the entire voluntary musculature, for its esthetic appearance. While it is true that this may be carried to an extreme that appears grotesque, a well-developed body is more pleasing to most persons than one that is thin and undeveloped.

Development of muscle size and strength proceeds slowly and gradually during childhood and cannot be greatly accelerated by training prior to the onset of puberty. Beginning at that point, the potential for increase in muscular bulk accelerates sharply, and in the case of boys, remains high until linear growth ceases. It never increases as much in girls and levels off sooner. The difference appears to be related to the amounts of circulating androgen available, although the relationship between human growth hormone and androgens is a complex one in this regard. There is also a relationship between androgens and the type of body physique, which affects the potential for increasing muscle size and strength. In general, taller persons appear to have greater strength potential if they are of the mesomorphic type, but endomorphs have a mechanical advantage in certain types of strength activities such as weight lifting.

Increases in muscle strength and size cannot take place, regardless of the amount of training, unless adequate supplies of protein are available to the body. Daily requirements for normal replacement needs in the adult vary greatly on an individual basis, but an intake of between 0.8 and 1.0 g/kg daily seems reasonable. Children and adolescents, particularly when engaged in vigorous physical activity, appear to need from 1.5–2.0 g/kg daily. Athletes in training, particularly in sports where muscle size and strength are important, probably need from 2.0–2.5 g/kg daily and some take more. The excess over what is needed for replacement and muscle growth is either converted to fat or excreted. Protein is not used as an energy source except in starvation.

When androgens first became available for experimental and therapeutic use, there was a limited interest among scientists in determining whether they could help maintain or restore strength in aging men, who presumably had low androgen levels. During World War II they were administered to starvation victims for their ability to help reverse negative nitrogen balances.

An American physician interested in weight lifting reasoned, incorrectly, that if androgens, and the anabolic steroids that had been developed as less masculinizing substitutes, would reverse negative nitrogen balances, they would somehow increase the utilization of protein to form muscle in those who already enjoyed a positive balance and had unlimited access to protein. He began to experiment on himself, first with an injected anabolic steroid, and later with an oral one. After a while he was satisfied that he had been able to achieve gains in muscle strength and size greater than those that he had obtained with weight lifting and a high protein diet alone. He reported his result in a magazine catering to weight lifters and body builders and started the craze for the use of these substances, which has still not subsided.

Within a relatively short time, football players, weight throwers, and other athletes had picked up the idea from the body builders. Physicians were besieged with requests for prescriptions, since anabolic steroids in the United States are controlled drugs. Many of them wrote the prescriptions quite willingly at first and even participated in recommending their use and administering intramuscular injections. The abuse became so widespread that at one time it was difficult to find a competitive weight lifter or thrower who was not taking these substances regularly. Football players at all levels were taking them on their own initiative but also in some cases at the high school, college, and professional levels under orders from coaches and even team physicians.

As is the case with so many fads that have afflicted sports over the centuries, all of athletes proceeded without there being any solid evidence that what they were taking was doing a particle of good, without any knowledge of the potential danger of the indiscriminate use of these substances, and without the performance of any objective evaluations of the apparent effects on the subjects of their ministrations. If we had not already witnessed in recent years the willingness of young, and indeed many older, persons to take even more toxic drugs in dosages that were often entirely unknown to them and to accept these substances from complete strangers without any idea of their source or nature, the phenomenon of the abuse of anabolic steroids among athletes would seem incredible. As it is, it can be seen as simply another manifestation of the drug culture that afflicts society.

When it comes to training, the great majority of athletes are lazy and would prefer to find a short cut to accomplish what can only be obtained through the most dedicated labor. Since strength is at a premium in so many sports, the idea that it could be obtained without having to work so hard for it is naturally appealing. The will to believe that this can happen is so powerful that it easily overrides any scientific demonstration to indicate that it will not. The desire to have the extra advantage is so great that any admonitions that grave dangers may confront the user of the supposedly magic substance are ignored. Some athletes would willingly sell their souls to the devil if they thought it would give them an edge.

What is the evidence from clinical and laboratory studies to indicate the effects of the administration of anabolic steroids to the apparently healthy athlete?

## II. Clinical and Laboratory Studies

As of January 1, 1975, 23 reports were found in the world literature dealing with experimental use of anabolic steroids in males for the purpose of discovering whether their administration could improve physical performance. There were no studies reported of effects of these substances on women. These 23 reports involved 18 distinct study groups of young men. The drugs used were Dianabol, nandrolone decanoate, mesterolone, Adroyd, oxandrolone, Nilevar, and Maxibolen. Dianabol was used most frequently.

For seven of these study groups, the authors reported that no significant performance improvements were obtained by the use of the steroids. The experimental design (including the use of double-blind technique) was well formulated, and the results were reported and analyzed completely and without error in the calculations. In the remaining eleven, the improvements that the authors claimed could not be justified because of serious flaws or omissions in the study design, analysis or calculation or all three. The other five reports dealt with supplementary or continuing studies utilizing one or more of the preceding groups.

There are three studies (SAMUELS et al., 1942; SIMONSON et al., 1944; HETTINGER, 1961) in which the effects on physical performance of men following the administration of testosterone are reported. In the report of SAMUELS et al. (1942), four young men (21–30 years of age) were tested for hand-grip strength and nitrogen balance. They were then given 30 mg of testosterone daily for 6 weeks and retested at the end of that time. There was no significant improvement in hand-grip strength and no change in nitrogen balance. SIMONSON et al. (1944) administered methyl testosterone to a group of male subjects 38 years and over and found significant gains in strength and working capacity. HETTINGER (1961) trained men aged 65–70 for 9 weeks and then gave them 4 testosterone injections at the rate of one every 2 weeks. The men were divided into active and sedentary groups. Strength increased significantly in the active group during the period of drug administration and then gradually fell off. It also increased to a lesser extent in the sedentary group. Performance capacity as measured on the bicycle ergometer also improved during the experimental period.

These studies are chiefly of historical interest. In the first of the three, the subjects were not athletes, and in the second two, they were older men who may have been experiencing a declining elaboration of male hormone in their bodies. They are not in any way conclusive and do not necessarily demonstrate anything regarding anabolic steroids.

One study (BROWN and PILCH, 1972) deals with the use of Dianabol and physical performance in male rats. Eighteen rats were divided into three groups. During a 6-week experimental period, one group was given injections of sesame oil, one, a low dose (0.5 mg/kg) of the drug, and the third, a high dose (5.0 mg/kg) of the drug. Injections were given daily. One of each treatment group was then combined into an activity group that involved sedentary activity, running on a treadmill, or high jumping. No significant performance improvement was found in either steroid group.

All 23 reports on the use of anabolic steroids will be described briefly and analyzed in the following pages. They will be presented by groups, in the order described above. In each group they will be listed alphabetically by the last name of the person who appears to be the principal author rather than by the dates of publication of the results. The reason, which is partly seen from the references listed for each article, and partly from the way in which the investigations were conducted, is that the authors were in some cases apparently either not aware of work previously reported or chose to ignore it, so that there is no true chronological sequence of investigation to follow.

There are additionally some articles that are either reviews of the original reports or speculative pieces. Their chief influence, unfortunately, has been to lend credence to the reports that appeared to show significant positive effects and to disregard or to down play those that showed that significant improvements did not occur. They will not be referred to any further or listed in the references to this section.

## A. Reports in which no Improvement in Strength or Power is Claimed Following the Administration of Anabolic Steroids

CASNER et al. (1971) studied 27 young apparently healthy college men. Their previous athletic history was not described nor their experience in weight training. The subjects were randomly divided into 4 groups. The first group was to take the steroid but have only normal daily activity, the second group to take the steroid and have progressive weight training, the third group was to have placebo therapy and normal daily activity, and the fourth group to have placebo therapy and progressive weight training. The weight training was carried out for 40 min a day, 3 days a week, for the 56 day experimental period. The steroid was stanozolol (17$\beta$-hydoxy-17$\alpha$-methylandrostano-[3,2-C]-pyrazole), and it was given in a 2 mg dose 3 times a day on days 1–21 and 28–49, while the controls took lactose placebos on the same schedule. The technique of administration was double blind. Measures of specific gravity, total body water, body weight and muscular strength were made on days 1, 28, and 56. The tritium dilution technique was used to measure total body water, and specific gravity was determined by weighing in air and under water. The quadriceps of the thigh, arm flexors, forearm extensors, trunk extensors, and flexors of the hand and wrist were tested on the Elgin Multiple Angle Testing Machine. All subjects' diets were checked by a dietary diary, which they kept for days 1–7 and 28–35.

Although increases in strength occurred in both the steroid and placebo groups, there was no significant difference between groups. There was a moderate decrease in thigh muscle strength in the steroid group. Body weight gain in the steroid groups was on the average double that in the placebo group. Weight training had no effect on the amount gained, but specific gravity increased only half as much in the steroid group, indicating that the difference in weight gain was due to water retention. This was borne out by an average increase of body water of 1.2 l in the steroid group as opposed to a loss of 0.9 l in the placebo group.

FAHEY and BROWN (1973) worked with 28 male college student volunteers. All had had at least one course in weight training but none were serious competitive lifters. It was not mentioned whether any had taken steroids previously. They were randomly assigned to control (15) and experimental (13) groups. Nandrolone decanoate was given to the experimental group by parenteral injection in a dose of 1.0 mg/kg of body weight every 3 weeks. The control group received sterile saline during weeks 2, 5, and 7 of the study period. Administration was by a double-blind technique. Both groups were trained for 9 weeks, 3 times weekly, with bench press, leg press, biceps curl, latissimus pull, and dead lift exercises on a Universal gym and with standard Olympic barbells. The program emphasized maximum resistance.

Strength, body composition, and endurance were measured before and after the 9-week study period. Strength was measured by maximum bench press and dead lift and by maximum torque exerted in flexion and extension movements at 30 and 180 movements per second of the elbow, shoulder, and knee joints on a Cybex Isokinetic Dynamometer. Lean body weight and fat were measured by underwater densitometry. Right and left arms, right and left thighs, chest and waist were measured with a steel tape. Maximum oxygen uptake was determined on a Monark bicycle ergometer with an initial work rate of 900 kg meters/min increased by 180 kg m every 2 min until the subject was unable to continue.

There were no significant differences between the control and experimental groups in the recorded torques, maximum weight lifts, body composition, anthropometric measurements, or endurance. Both groups showed the types of gains that might be expected as the result of training.

FOWLER et al. (1965) studied 47 young men aged 18–25. Ten were rugby players and the remainder were untrained. The rugby players were randomly assigned to the experimental and control groups, 5 in each. The untrained men were randomly assigned to four groups. Eight received a placebo and took no regular exercise, 9 were given androstenolone and took no exercise, 10 were given placebos and a regular exercise program, and 10 were given androstenolone and a regular exercise program. The training period lasted for 16 weeks. Exercise was given 5 days a week for 50 min, and 20 mg of androstenolone were given daily to the experimental subjects. The technique was double blind.

Weight, height, hand grip in both hands, strength of elbow flexors and extensors and shoulder abductors in both upper extremities, and of hip flexors and knee flexors and of extensors and trunk flexors and extensors, vital capacity, working capacity, pull-ups and push-ups and vertical and long jumps in inches were all measured before and immediately after the training period. Each measure was done in triplicate and the best of three determinations used. Working capacity was measured according to the method of SJÖSTRAND and WAHLUND. A cable tensiometer was used to measure muscle strength.

The only increases significant at the 5% level were in the exercise groups. There were no significant differences at this level between the groups receiving placebos and those receiving androstenolone.

GOLDING et al. (1974) studied 40 amateur weight lifters, all volunteers, who were training an average of 4 times a week for 80 min per session. Their diets were studied by daily diary entries for a period of 3 weeks before the start of the study period, and they were asked to outline their daily physical workout schedule, indicating the length and type of workout. They also completed a questionnaire covering their training history, body weight, and previous use of drugs or diet supplements. Workout schedules varied from 60 min to a maximum of 3 h a day and from 3–7 times a week. No schedules or diets were changed.

The subjects were assigned to four groups in such a way that there were no significant differences between the groups in height, weight, protein intake, workout time, or days of working out. Each subject was required to undergo a physical examination. The first group was to be given a protein supplement but no steroid; the second, steroid but no protein; the third, steroid and protein supplement; and the fourth, no steroid or protein supplement. The protein supplement was given on the basis of body weight (0.5 g/kg). Placebos were used for the no steroid and no protein supplement groups. The technique of administration was double blind. All subjects believed they were receiving steroids and that two different forms of protein supplement were being tested. The steroid (methandrostenolone) was given in a dose of 10 mg per day for 12 weeks on a 3-weeks-on and 1-week-off pattern.

Body weight, 7 skin folds on the right side of the body, 10 body girths, 13 skeletal diameters, and body density by underwater weighing were measured before and immediately after the 12-week study period. Six measures of static strength were measured on a table specially constructed to isolate joint movements, and maximum bench press and arm curls were also measured. There were no significant weight changes between the four groups nor any significant change in fat percentage, skeletal or girth measurements. Although small increases in dynamic and static strength occurred in all groups, there were no significant changes between groups.

MUNSON (1970) studied 32 untrained college students who volunteered from a weight training class. Sixteen subjects were given 10 mg of oxandrolone twice daily, and the other 16 were given placebo capsules in a double-blind technique for 28 days. All subjects were put on an isotonic program of weight resistance exercise with progressive overload for a 56-day period. Bench press, leg press, and biceps curl with a barbell were the exercises employed. The exercises were carried out for 35 min 2 days a week. Total body weight, oxygen consumption, arm, chest and thigh girths, arm, chest and back skin folds, and arm and lower extremity strength were measured before the study period, afte 28 days and after 56 days. Significant increase in average body weight and significant decrease in average total skin-fold measurements for the experimental compared to the control group were noted. There was no intergroup difference of significance in muscle size, strength or oxygen consumption. It was concluded that the weight gain was due to water retention.

O'SHEA and WINKLER (1970) studied 15 competitive college swimmers during an 11-week period. These swimmers had been through 3 months of preseason conditioning involving weight training, calisthenics, and interval swimming 4 hours a day 5 days a week immediately preceding the study. Body weights were measured and some blood chemistry studies were made at the beginning of the study and at 3, 6, and 11 weeks. Nitrogen balances were determined for 3 days towards the end of the 3rd, 6th and 11th weeks. Eight swimmers were given 10 mg/day of oxandrolone for the first 6 weeks of the study and were also given a high-protein supplement in the amount of 0.5 g/kg of body weight. Controls were given nothing and no blind technique was used. Swimmers' times over 92.3 and 184.6 meters were compared to their best times of the preceding season.

In those swimmers who received the steroid there were slight but not significant increases in nitrogen balance during the period of steroid administration. The non-steroid group improved their swimming times at both distances more than the steroid group. There were no significant differences between groups.

STRÖMME et al. (1974) studied 21 healthy males aged 21–27 years at the Norwegian College of Physical Education and Sport. The subjects were divided into two approximately equal groups by random assignment. Ten were given 75 mg/day of mesterolone in a dose of 75 mg/day for the first 4 weeks of the study and 150 mg/day for the second 4 weeks. The other 11 were given a placebo. The technique of administration was double blind. All subjects were given a daily protein supplement of 30 gm with vitamins and minerals added during the experimental period. In addition to their daily program of physical education, the subjects were given weight training at near maximum levels 3 times weekly for $1^1/_2$ h each session to increase strength in knee-hip extension, trunk extension, trunk flexion, and elbow flexion.

Blood samples were taken to determine levels of ICSH, FSH, testosterone, and testosterone binding to plasma proteins at the beginning of the study and at 4 and 8 weeks. Body weight, circumferences of the thigh and upper arm, and maximal strength of the four trained muscle groups were determined at the start and at 2-week intervals during the study. Maximum isometric muscle strength was measured with a strain gauge meter, taking the best of three determinations. A standard bicycle ergometer warm-up prior to strength testing was used to determine aerobic power from heart rate response to maximal loads using the Åstrand-Rhyming nomogram.

Although both steroid and placebo groups improved strength and limb circumference measures during the study period, there were no significant differences between the two groups. There were no significant gains in body weight or aerobic power in either group. Plasma testosterone and protein binding of testosterone were significantly reduced in the steroid group, with no change in serum levels of ICSH and FSH. There were no changes in testosterone or protein binding in the control group.

## B. Reports in which Improvement in Strength or Power is Claimed Following Administration of Anabolic Steroids

ARIEL is represented in the literature by five reports (ARIEL and SAVILLE, 1972a, b, c; ARIEL, 1972, 1973a) on the effects of administration of anabolic steroids to weight lifters. Two of these reports (ARIEL and SAVILLE, 1972a, b) are virtually identical. These reports and two of the others (ARIEL, 1972, 1973) deal with the same group of subjects, while the fifth (ARIEL and SAVILLE, 1972c) appears to deal with a different group. A review of ARIEL's doctoral thesis (ARIEL, 1973b) indicates that the subjects for the first four reports came from this study, the only difference being that ten subjects were studied in the thesis and only six are reported in the papers that grew out of it.

The six male subjects of the first four papers were varsity athletes with 2 years of weight training experience. Their ages ranged from 18–22 years. For a period of 4 months prior to the study, all subjects performed seated, military, and bench presses and did squat lifts 5 days a week and were tested on the 6th and 7th days. Each test was a maximal lift. From the 8th to the 11th week of the study, all subjects were given a placebo and told that it was 10 mg of methandrostenolone. From the 12th to the 15th week, three of the subjects were given 10 mg of the steroid and the other three continued to take the placebo. A double-blind technique was followed. In addition to the strength measures, total patellar reflex time and reflex latency were obtained on the right limb.

Significant gains in strength at the 5% level were recorded by all subjects during the placebo period for the bench and military press and squat. The military press gain reached the 1% level. During the steroid period the steroid group exceeded the control group only in the seated press and squat exercises. Patellar reflex time was decreased in the steroid but not in the control group, but there was no change in latency in either group.

In the fifth study, 20 male weight-lifter volunteers aged 19–25 were divided into two equal groups. The experimental group received 15 mg of methandrostenolone daily for 4 weeks and the control group received nothing. All subjects lifted weights 5 days a week for 4 weeks prior to drug administration. They were tested on the 7th day for maximum lifts in the bench and military press and the squat exercise. During the 4 week experimental period the subjects worked 5 days a week for 2 h a day in a program designed to overload the major muscle groups in the body. They were tested after the experimental period and then again after a 15 week period of detraining.

The increases, if any, during the experimental drug administration period are not reported in this paper. Average decreases in strength levels from the end of the experimental to the end of the detraining period for all three strength measurements are shown in a table. Decreases occurred in both groups but it is claimed that the losses in the steroid group were not significant and that those of the nonsteroid group were significant at the 1% level.

Bowers and Reardon (1972) studied 18 experienced weight trainers who were matched in experimental and control groups, equal in age, size, and strength. A 6 week training program of bench presses, squats, forearm curls, and triceps extensions was used. All subjects received 30 g of 90% protein daily. The experimental group received 10 mg of Dianabol each of the last 21 days of training, and the control group received a placebo for the same period. The study was not double blind. A comparison of tests before and after the 6 week study period showed significant gains of the experimental group over the controls only for the bench press and squat. Biceps and forearm girths were said to be increased in the experimental group but not significantly over the controls. Body weight was significantly increased in the experimental group. There was no improvement in aerobic capacity.

Johnson et al. (1972) studied 31 volunteer college students who were untrained but enrolled in physical education classes. Their ages ranged from 18–38 years. For the first 4 weeks all subjects were engaged in weight training of a type not specified in the report of the study. Pretests were administered "during this time", but the time was not specified except in the case of a blood sample taken at the end of the 3rd week and a sperm count during the 4th week. Other pretests included max $V_{O_2}$ determined during a treadmill exercise (nature unspecified), dynamic strength as measured by one repetition maximum on the bench press and squat, static strength as measured by cable tensiometry for elbow flexion, shoulder flexion and knee extension, circumferences on the arm, forearm, thigh and leg, skin folds at the triceps, subaxillary and subscapular sites, and gross weight.

At the end of the 4th week, 24 of the subjects (method of selection not specified) were randomly but equally assigned to treatment and placebo groups. The 12 treatment subjects took 10 mg of Dianabol daily for 21 consecutive days. The 12 placebo subjects took blank capsules. The 7 control subjects took nothing. All subjects supplemented their diets with 20 mg of 92% protein powder daily. All subjects trained 3 times weekly for about 1 h each time in a progressive overload program of weight exercises including bench press, squat, and seated dumbbell curls. Posttreatment tests were apparently obtained at the end of the 21 day period, but the day or days on which the tests were done are not specified. The technique of drug administration was double blind.

It is stated that there was no significant change in oxygen uptake, but no figures are given. It is also stated that there was no significant change in skin fold measurements, but no figures are given. It is stated that the weight gain was significantly greater for the treatment than the placebo or control group, but the figures show a significant difference only for the placebo group. It is stated in the "Results" that there was a significant increase in the circumferences at five of eight sites measured, but only four site measurements are recorded in the table, and in these the gain of the controls was greater than that of the treatment group in two instances, nearly equal in a second and barely significantly different in the fourth. It is stated that the dynamic and static strength measurements were significantly increased over placebo and controls in all five measurements, but in fact this was true only in shoulder flexion, bench press, and squat.

JOHNSON and O'SHEA (1969) studied 24 college student volunteers who ranged in age from 19–39 years. Twelve who volunteered to take the steroid were paired with 12 volunteers who were unwilling to take it, matching the pairs as well as possible according to age, weight, approximate strength, and training background. In the first 3 weeks all subjects trained for 1 h 3 times weekly in weight training with progressive overloading in bench presses, squats, and seated dumbbell curls. All supplemented their diets with 15 g of 92% protein daily. Seven measurements were made at the end of the 3 weeks: one maximum repetition in the bench press and squat; static strength in shoulder flexion, elbow flexion, elbow extension, knee flexion, and knee extension; skin folds at six unspecified sites with a Harpenden caliper; circumferences of the right arm, forearm, waist, right thigh, and calf; oxygen uptake by the Åstrand (not further specified) oxygen uptake test; blood chemistry (not further specified); and blood lipids.

In the last 3 weeks the treatment group was given 10 mg of Dianabol daily and the control group nothing. Measurements were made again at the end of the 3-week period. There was a significant increase in gross weight for the treatment group and a significant increase in oxygen uptake. Measurements of skin folds were not given, but it was stated that there was no significant change. Lactic dehydrogenase and serum glutamic-oxalacetic transaminase levels increased, and total bilirubin and alkaline phosphatase levels decreased in treated subjects. Free choloesterol levels decreased significantly in the controls. Figures were not given for any of the blood chemical measurements.

It was stated that dynamic and static strength increased significantly in treated individuals. Significant differences between treatment and control subjects were recorded, however, only in the cases of elbow extension and knee flexion. No measurements were recorded for strength in seated dumbbell curls, and a mistake in subtraction in the table showing the measurements for bench press and squat caused the erroneous conclusion that the treatment versus control difference for the bench press was significant. Changes in girth measurements were recorded in a table but not referred to in the text. The only treatment versus control differences occurred in the biceps and calf measures, where, for unexplained reasons, the circumferences for the control group actually became smaller. In the case of the biceps measurement the difference was incorrectly recorded in the table.

BERG and KEUL (1974) studied 10 male students aged 18–23 years and 15 weight lifters aged 20–38 years. The exact duration of the experimental period was not stated but appears to have been about 6 weeks. They were given nandrolone decanoate intramuscularly in 50 mg doses at intervals of 10–14 days while continuing physical training, which involved three dynamic strength activities, bench presses, squats, and back lifts. The authors' principal interest was in changes in the serum lipid, blood lactate, and pyruvate levels and in blood indices of hepatic function. The lactate and pyruvate levels were lower after exercise after the hormone was given. After steroid treatment, resting amino-N levels were higher than before treatment, but there was no increase during exercise as there was before, and levels fell under the resting level after exercise. Determination of maximum oxygen consumption on a bicycle ergometer test was said to show improvements of 5–6%, but no figures were given and the significance of this change was not calculated.

O'SHEA (1971) reported on a study of the effects of 4 weeks of treatment with an anabolic steroid on the strength performance of 20 weight lifters. The subjects were 20 university volunteers aged 19–26 with a minimum of a year of weight training experience each. They were equally divided into two groups by random selection and then paired for comparison as closely as possible according to strength levels. During the study one member of the placebo group had to withdraw because of injury, so the final comparison involved only nine pairs. Strength levels were established during 1 week before the testing period.

For the test period of 4 weeks, all subjects were placed on identical progressive weight training programs 3 times weekly for 80 min each session. The bench press and full squat with weights were the exercises used. Strength was tested at the end of each week of the study with one repetition maximum for each exercise. A 10-mg dose per day of methandrostenolone was the steroid, and the placebo was calcium lactate. The technique of administration was double blind. All subjects were given a 92% protein supplement in the amount of 0.5 g/kg of body weight daily. Gross body weight was measured before, during, and at the end of the study period.

There was a significantly greater increase in body weight in the steroid than in the placebo group. There was no significantly greater improvement in either the bench press or full squat. The report is unfortunately marred by a failure to transfer the figures correctly from the table of individual performances to the table of comparisons so that the "t" scores for the bench press and squat exercises are incorrect as given.

STEINBACH (1968) studied 125 young men aged 17–19 for a period of $3^1/_2$ months during which some trained and some did not, some were given anabolic steroids, some a placebo and some nothing. They were divided equally into five groups: Group A received steroid, but no training; B received nothing and had no training; C received steroid and had training; D received a placebo and had training; and E received a placebo and had no training. Weights were determined before and at the end of the training period. Strength in a single isometric contraction with one arm and one leg were measured. The steroid administered was Dianabol and 3 mg were given twice daily. The training groups performed an isometric contraction at 40% of maximal isometric strength twice daily for 3 sec 5 times a week. In the two training groups, 13 were trained in leg exercise only and 12 in arm exercise only. Body weight, arm, and leg strength were significantly increased in the two groups given Dianabol as compared to the remaining groups.

STAMFORD and MOFFAT (1974) studied 24 experienced male weight trainers. Each had trained for a minimum of 2 years and was in training at the time of the study. Twelve were prisoners and 12 were from the same area of the state. Many of the prisoners had taken steroids before. The subjects were divided into 4 groups matched on the basis of bench-pressing power and body weight. Groups I and II were inmates and III and IV were those outside the institution. The ages were not given. Diet was not controlled. During the first 4 weeks of the study subjects trained 3 days a week, alternating a day of rest with each training day. Exercise sessions lasted 2 h with emphasis on bench pressing, squatting, deadlifting, and curling with maximal poundages on the barbells. In the second 4 weeks the same schedule was continued but the following regimes were added: Group I received 20 mg of Dianabol 5 days a week and 30 g of a 90% protein supplement 7 days a week; Group II received a placebo and the protein supplement; Group III received only the protein supplement; Group IV received nothing.

Evaluation consisted of measuring maximum bench press, maximum repetitions bench presses with 80% of the maximum previously achieved, static strength index (grip, back and leg, and general upper body strengths measured with dynamometers), flexed arm and thigh measures taken cold and nude body weight. These were done prior to commencement, after the first and after the second 4 weeks. Blood tests of liver and kidney functions were taken of all inmates weekly and blood pressures on Monday, Wednesday, and Friday mornings.

The increase in body weight was the only measurement significantly greater for the steroid over the placebo or control groups at a 1% confidence level. The steroid group showed an increase significant at the 5% confidence level over the placebo group in bench pressing and static strength but showed no improvement at all in the maximum bench press repetitions compared to a small nonsignificant gain for the placebo group and a decrease in biceps measurement with a small nonsignificant gain in the thigh measurement. Approximately the same number in each group reported increased appetite and a feeling of increased strength.

WARD (1970) reported a case study of training one athlete with an anabolic steroid and reported gains in strength as a result. This hardly requires further comment. He later reported a study (1973) of 16 volunteer male college students aged 18–26 years. All had had at least one course in weight training. All were then given 5 weeks of additional training in bench pressing, squatting, inclined press and power lifting using 80% of maximum weights in a progressive overloading program 3 days a week. At the end of 5 weeks they were paired on the basis of strength in bench press and squats and divided into two equal groups.

Strength scores for the bench press and squats were determined before and after the 5 weeks of the experimental period. Body weights in air and underwater were taken before and after, and with residual air determinations, body density and body fat contents were determined. The experimental group was given 10 mg of Dianabol daily and the control group a placebo. The technique was single blind. The steroid and placebo were stopped at the end of the fourth week. Only body density, lean body mass and bench press strength were improved in the experimental over the control group at the 1% level of confidence. Strength in squatting and decrease in fat percentage were significantly greater at the 5% confidence level.

WEISS and MÜLLER (1968) studied 32 volunteer high school students, whom they divided into two groups of 16 each. Weight was determined and arm strength measured with a dynamometer before and after a 17-day experimental period. The circumference of the upper arm was measured at rest and with the muscle contracted. The experimental group was given 10 mg of Dianabol daily and the control group a placebo during the 17 days of the study. The technique was single blind. Although the experimental group gained slightly more weight than the controls, the authors stated that, overall, there were no significant improvements in favor of the steroid group.

## C. Discussion of the Results of the Clinical and Laboratory Studies

In the seven studies that showed no improvement in muscle size or strength of the experimental group taking a steroid over the control group, all except one were double blind. No experimental period was less than 4 weeks long, and one was 16 weeks long. In three, high protein supplements were given to at least one of the groups taking steroids. In two studies, serious efforts were made to determine the exact compositions of the subjects' diets. The number of subjects was substantial, ranging from 15–47, with only one study having fewer than 21 subjects. Five different anabolic steroids were used. The experimental designs were sophisticated. The results were carefully and completely reported with no mistakes observable in calculations and no conclusions drawn that were not justified by the methods employed and the data obtained.

Of the 16 reports that claimed to show greater improvement in muscle size or strength in the experimental group taking the steroid than in the control group, four were reported by one principal investigator with the same group of subjects and represented some duplication and some different aspects of one study period. Drug administration was double blind in this study, but in only two of the other 12. In two others, a single-blind technique was used. The experimental periods varied from 17 days to $3^1/_2$ months, with three study periods less than 4 weeks. In only four of the studies was a high protein supplement used. No attempt was made to determine or control the subjects' diets in any of these reports. The numbers of subjects in these reports varied from 1–125. Four of the reports dealt with the same group of six, who were apparently selected by some means not stated, from an original group of 10. The study that involved 125 persons dealt with extremely limited objectives, and the report of the methods and other circumstances indicated that it did not appear to have a strong design. Dianabol was used in all of these studies except for one, in which nandrolone decanoate was used. The experimental designs of many of these reports appeared to be poorly drawn, facts that might have a significant effect on the interpretation of results were omitted, in two studies calculation errors were found that had a bearing on reported results, and in some, conclusions were drawn that were clearly not justified by the results obtained. Weight gains, for example, were reported often only in terms of gross body weight without determining body density or fat percentages; and where fat percentages were estimated, it was usually only by the relatively inaccurate method of measuring skin-fold thickness.

In brief, a careful review of all these reports leads to the inescapable conclusion that in spite of the claims that have been made by some of the investigators that the use of anabolic steroids by athletes in conjunction with progressive resistance training will afford a greater increase in muscle bulk and greater strength, there is, in fact, no substantial evidence that this is so. On the contrary, there is a substantial body of evidence that, even upon very close scrutiny, indicates that anabolic steroids will not contribute significantly to gains in muscle size or strength in healthy young males.

There is an analogy here to the medieval alchemists search for the philosopher's stone that would transmute base metal into gold. In sports where bulk and strength are important, the gold is found in muscle size and in the ability to generate great power; the base metal is the untrained or partly trained muscle. The gold—the medal or the professional promoter's bank account—can be won by the talented with great effort, but some will stop at nothing to reap it more easily if they can. When anabolic steroids have had their day, there will be something else to replace them.

## III. Control of the Use of Anabolic Steroids

Until 1973 there was no reliable method for the detection of the presence of anabolic steroids in the human body. In that year a radioimmunoassay method was announced by Professor Raymond Brooks of St. Thomas Hospital in London. About the same time, a liquid chromatography-mass spectrometry method was announced by Dr. Cedric Shackleton and his colleagues at Northwick Park Hospital in England.

It was decided to test certain athletes at the British Commonwealth Games in Auekland, New Zealand in February, 1974. Urines were collected and 55 specimens were sent to St. Thomas in London for analysis (1974). Nine urines showed positive tests. In two the structure of the steroid used was identified. The athletes were not identified, and no penalties were assessed since the procedure was experimental in nature.

The International Amateur Athletic Federation Medical Committee met in London, June 28–29, 1974. After a full discussion of the situation, they recommended (HOLDER, 1974) that testing for anabolic steroids be carried out at the European Athletic Championships in Rome in September 1974. No penalties were to be assessed. They confirmed their classification of anabolic steroids as banned drugs in athletic competition.

At the 30th International Congress of Pharmaceutical Sciences in Geneva in 1971. Professor ARNOLD BECKETT (1971), head of the Department of Pharmacy at Chelsea College, London, had said that a licensing system for athletes was under consideration. Under this system regular weight checks of athletes were to be kept. A sudden change of weight (presumably gain) in the month preceding a sports event would entitle officials to request a blood and urine check for anabolic steroids. He also stated that the medical committees of the 26 international sports federations would perform grassroots checks of athletes for the presence of anabolic steroids. There does not appear to be an official corroboration of either of these statements since then.

The technical difficulties involved in the selection of athletes to be tested, securing the urine sample under conditions that guarantee its authenticity, safeguarding it until it can be tested, preserving the sample after testing so that a retest may be performed in the event of a challenge, and other similar problems are no different from those encountered in other drug testing programs. There is no need to recount them here or to explain the elaborate procedures necessary.

Testisg only winners in manifestly unsatisfactory since it is those who are destined to lose who are most likely to feel the need of artificial aids. Testing all athletes is much too expensive and time consuming. Sampling techniques are consequently employed. These may tell you something about the frequency of usage, but when penalties are to be assessed, may let some of the guilty go free.

A fear that has been expressed by the proponents of testing is that if the athlete who has been using anabolic steroids stops their use several weeks before the contest at which he knows he will be tested, no detectable amount may remain in his body, but he still may have the advantage of its use. This assumes that there is some advantage to be gained from using them, but this has not been proved beyond reasonable doubt, as has been pointed out in the preceding section of this chapter.

If any control is to come about, it will not come through testing, although this may make some inroads here and there and scare off some of the fearful. It must come about rather through education of athletes. For this to be successful, scientists must face the facts honestly that there is proof that no advantage comes from taking anabolic steroids. When they are ready to do this as a group, then they can succeed in their educational efforts. As long as some continue to hold out, either through stubbornness or ignorance, athletes will not be convinced.

The process of education must then start through all educational institutions, sports clubs, coaching and training associations, sports physicians, professional sports teams, and especially the lay press. At the same time, all testing must be abandoned, because as long as tests are made, athletes will be persuaded that they are being kept away from something that will really help them. The irrational will continue the use of steroids until a new fad seizes their interest.

## IV. Ethical Considerations

The concept that there is such a thing as ethical behavior is based on an understanding that in actions where there are opposing alternatives, there is a right and a wrong way to behave. If a set of rules governing particular behaviors, as in sports, is laid down by some person or persons who has the power to enforce them, we can say arbitrarily that to obey the rules is right and to violate them is wrong. It is impossible, however, to draw up rules that are so exact that they cover every possible behavior in a given situation or to anticipate circumstances that might require additional rules. Interpretations of rules are then made and many times formulated into new rules themselves. Violations of rules may then be deliberate, or accidental, if the person involved innocently misinterprets the particular rule.

Some rules are quite specific, such as "when the ball goes over the boundary line play will stop and the ball will be returned to the field of play by the side opposed to that whose member touched it last." This may create a problem of observation for an official, but once he has decided who touched the ball last, the enforcement of the rule proceeds simply and automatically. If the law says that "The umpires are the sole judges of fair and unfair play," then a great deal of latitude is given to the official to interpret situations that go beyond the written scope of the rules.

A rule which said, "No player shall take an unfair advantage over an opponent" would be rejected out of hand by both players and officials as being so impossibly broad as to defy effective enforcement. This is where ethical conduct comes in. Both opponents have to have an understanding in the beginning that neither will deliberately take advantage of a situation where the other person is temporarily unable to respond to the attack made on him for reasons not germane to the play of the game and for which he is not responsible. In boxing, for example, "hitting on the break" or hitting after the bell has sounded are unethical practices.

More is involved than the performance of an unfair action, since the idea itself is important. An athlete might take an action that he felt would give him an unfair advantage over his opponent, but in the event, no advantage accrued to him. This would still be ethically wrong, whether or not the opponent or others observing the action were aware that an unfair advantage had been sought. The implication is that the athlete must be the judge of the intent of his own actions. This principle is called "sportsmanship," and the athlete who acts according to it is said to be practicing "fair play."

The effects of any drug on any individual are not entirely predictable. Recommendations for drug therapy are based on the administration of particular drugs to animals and humans under controlled conditions. Certain effects are noted, some of which are wanted and others unwanted. If, on balance, the wanted effects produce results which are therapeutically desirable and which appear consistently, and the unwanted effects are minimal, or at least tolerable, the drug can then be recommended for general use. The practitioner must still keep in mind that there will be an individual variation in dose response, that age and size of the subject must be kept in mind, that specific sensitivities will appear in some subjects, and that accumulation of the drug within the body over a period of time may produce toxic reactions.

From all these considerations and from clinical experience in the administration of a drug, guidelines for dosage are established that indicate roughly how much should be given to produce a certain effect in a typical or average subject. With most drugs there is no satisfactory way of determining whether the optimum quantity of the drug to produce the desired effect is present in the subject for the requisite period of time. With those for whom this is possible, regular monitoring would be time-consuming, expensive, and annoying to the subject, so that this is done with only a few drugs or in special circumstances with some others. Clinical observation of the subject's response is the criterion used as a guide by the practitioner. This is partly objective, partly subjective, on the part of both practitioner and subject. Subjective symptoms may be relieved, and objective improvement in a subject's function can be recorded, in fact, when the drug administered is a "placebo," a substance which has no known pharmacologic basis for producing such reactions.

The purpose of this lengthy preamble is not to educate the reader in what he already knows but to remind him that when an athlete takes a drug, either on his own initiative or at the suggestion of a physician, neither he nor the physician may have a clear idea of what may happen as far as his performance is concerned. His subjective reaction to the taking of the drug may lead him to a performance better than he expected; however, the effects of many other factors that may also be operating cannot be excluded. The drug may exert some reaction that is favorable to his performance, and it may be weak or strong depending on personal and circumstantial factors. The effects, if any, both subjective and objective, will vary from athlete to athlete. With substances that are naturally present in the body, or closely resemble natural substances such as the anabolic steroids, the reactions of particular subjects appear to be more variable than in the case of chemicals not ordinarily found in the body.

To say, therefore, that if two opposing athletes take the same drug, it is "fair play" because each has an equal chance of enhancing his performance, is nonsense. And this is entirely apart from all social, economic, geographic and other considerations that might make it possible for all athletes to have equal access to all drugs. There are also religious or other scruples that might prevent an athlete from taking any drug at all in relation to his practice, competition or otherwise.

The athlete who is sincere in his desire to practice sportsmanship, the ideal on which all true sports that are not merely entertainments is based, has an ethical obligation to refrain entirely from taking drugs whose sole function would be to improve his performance. The use of drugs to treat any acute of chronic illness, an injury or to correct a natural deficiency is a matter that should be decided solely between him and his physician and should not create an ethical problem in sports. The use of any drug that he has any reason to believe might prove harmful to his body creates, however, an ethical problem for him personally, which does not necessarily affect others.

## V. Summary

Strength beyond that degree ordinarily developed from the activities of daily living is desirable in most sports and absolutely necessary for any reasonable achievement in some sports. It is therefore highly prized by the athletes in these sports, and most of them are highly motivated to achieve it by whatever means are possible. Exercising muscles under conditions of overloading in a progressive fashion is the normal means of increasing strength.

The presence of an excess of male hormone appears to be the principal reason for the greater ability of post-pubertal males to develop greater muscle size and strength than postpubertal females. Synthetic substances that simulate the actions of male hormone (anabolic steroids) have appeared to be potent in restoring muscle size and strength in persons who have become emaciated from starvation for whatever cause. From this, some have reasoned that these substances would enhance growth of muscle bulk and strength in conjunction with progressive resistance exercise to a greater extent than with the exercise alone.

There has been widespread use of anabolic steroids by athletes competing in sports where added strength is a great advantage, such as in football, weight lifting and weight throwing. This has been encouraged by the publication of reports purporting to show that gains in muscle size and strength could be achieved by postpubertal male athletes also working against progressive weight resistance. This in turn has led to testimonial recommendation from athlete to athlete and from coach to athlete. Although there are reports that these substances have been used by female athletes, there are no published reports to indicate what, if any, effects were achieved, so that this subject must remain moot for the present.

A review of the world literature revealed 23 reports of laboratory or clinical studies dealing with the experimental use of anabolic steroids in males for the purpose of discovering whether their administration could improve physical performance. The major concern was increase in muscle size and strength, but in several instances the effect on endurance as measured by change in the maximum capacity to take in and utilize oxygen was also measured. These 23 reports involved 18 distinct study groups of young men. The anabolic steroids used were Dianabol, nandrolone decanoate, mesterolone, Adroyd, oxandrolone, Nilevar, and Maxibolen. Dianabol was used most frequently.

For seven of these study groups, the authors reported no significant performance improvements as the result of taking anabolic steroids, with or without additional high protein supplements, while undergoing progressive resistance training. No experimental period was less than 4 weeks long, and one lasted for 16 weeks. In three, high protein supplements were given to at least one of the groups taking steroids. In all except one the drug was given on a double-blind protocol. In two studies, serious efforts were made to determine the exact compositions of the subjects' diets. The number of subjects in each study was substantial ranging from 15–47, with only one study having less than 21 subjects. Five different anabolic steroids were used. The experimental designs were sophisticated. The results were carefully and completely reported with no mistakes observable in calculations and no conclusions drawn which were not justified by the methods employed and the data obtained.

Of the 16 reports (of 11 study groups) that claimed greater improvement in muscle size and strength in the experimental than in the control group, 4 were reported by one investigator using the same group of subjects, presenting some duplication but also some different aspects of the study. Drug administration was double blind in this study, but in only 2 of the other 12. In two more, a single-blind technique was used. The experimental periods ranged from 17 days to $3^1/_2$ months, with 3 study periods less than 4 weeks. In only 4 of the studies was a high protein supplement used. No attempt was made to determine or control the subjects' diets in any of these reports. The number of subjects in these reports varied from 1–125. Four dealt with the same group of 6 subjects, who were apparently selected by some means not stated, from an original group of 10. Dianabol was the steroid used in all except one, in which nandrolone decanoate was used. The experimental designs in many of these reports appeared to be poorly drawn, facts which might have a significant bearing on the interpretation of results were omitted, in 2 studies errors of calculation were found which had a bearing on reported results, and in some, conclusions were drawn which were clearly not justified by the results obtained.

A careful review of all these reports leads to the inescapable conclusion that in spite of the claims made by some of the investigators that the use of anabolic steroids by athletes in conjunction with progressive resistance training will afford a greater increase in muscle bulk and greater strength, there is, in fact, no substantial evidence that this is so. On the contrary, there is a substantial body of evidence, which even under very close scrutiny, indicates that anabolic steroids given orally or parenterally do not contribute significantly to gains in muscle size and strength in healthy young males.

A technique for the detection of the presence of the anabolic steroids in the bodies of athletes through a urine examination has been developed. Its reliability is not yet fully known and it cannot apparently differentiate between what may have been taken very recently or at some earlier time. This technique has been adopted for use in certain international sports events where sample testing for other drugs is currently being done. The use of anabolic steroids is forbidden for athletes by the official bodies responsible for the conduct of these events.

There are strong ethical considerations that should militate against the use of anabolic steroids, or any drugs, for the purpose of improving sports performance, entirely apart from any consideration as to whether favorable effects can be achieved. Their use denies the concepts of "fair play" and "sportsmanship." Personal ethics are also involved, related to the taking of a substance for a reason not necessary to life or health, which might harm the body.

## References


ARIEL, G.: The effect of anabolic steroid (methandrostenolone) upon selected physiological parameters. Athletic Training **7**, 190—200 (1972).

ARIEL, G.: The effect of anabolic steroid upon skeletal muscle contractile force. J. Sports Med. Phys. Fit. **13**, 187—190 (1973a).

ARIEL, G.: The effect of anabolic steroid (methandrostenolone) upon the voluntary force of skeletal muscle. University of Massachusetts Ph. D. (1973b).

ARIEL, G., SAVILLE, W.: Effect of anabolic steroids on reflex components. J. appl. Physiol. **32**, 795—797 (1972a).

ARIEL, G., SAVILLE, W.: The effect of anabolic steroids on reflex components. Med. Sci. Sports **4**, 120—123 (1972b).

ARIEL, G., SAVILLE, W.: Anabolic steroids: the physiological effects of placebos. Med. Sci. Sports **4**, 124—126 (1972c).

BECKETT, A.: Use of steroids in athletes proving difficult to control. Med. Trib. **12**, 28 Aug. 18 (1971).

BERG, A., KEUL, J.: Der Einfluß von anabolen Substanzen auf das Verhalten der freien Serumaminosäuren von Normalperson und Schwerathleten in Ruhe und bei Körperarbeit. Österr. Z. Sportmedizin **4**, 11—18 (1974).

BOWERS, R., REARDON, J.: Effects of methandrostenolone (Dianabol) on strength development and aerobic capacity. Med. Sci. Sports **4**, 54 (1972).

BROWN, B. S., PILCH, A. H.: The effects of exercise and Dianabol upon selected performances and physiological parameters in the male rat. Med. Sci. Sports **4**, 159—165 (1972).

CASNER, S., EARLY, R., CARLSON, B. R.: Anabolic steroid effects on body composition in normal young men. J. Sports Med. Phys. Fit. **11**, 98—103 (1971).

DELORME, T. L.: Restoration of Muscle Power by Heavy Resistance Exercises. J. Bone J. Surgery **37 A**, 645—667 (1945).

FAHEY, T. D., BROWN, C. H.: The effects of an anabolic steroid on the strength, body composition and endurance of college males when accompanied by a weight training program. Med. Sci. Sports **5**, 272—276 (1973).

FOWLER, W. M., JR., GARDNER, G. W., EGSTROM, G. H.: Effect of an anabolic steroid on the physical performance of young men. J. appl. Physiol. **20**, 1038—1040 (1965).

GOLDING, L. A., FREYDINGER, J. E., FISHEL, S. S.: Weight, size and strength—unchanged by steroids. Physician and Sports-medicine **2**, 39—45 (1974).

HETTINGER, T.: Influence of male sex hormone (testosterone) on muscle strength and circulatory system in men aged 65—70. Int. Z. angew. Physiol. einschl. Arb. physiol. **18**, 357 (1961).

HETTINGER, T., MÜLLER, E. A.: Muskelleistung und Muskeltraining. Arb. physiol. **15**, 111—126 (1953).

HOLDER, F. W.: Press Statement. International Amateur Athletic Federation Medical Committee. July 3, 1974.

HOLDER, F. W.: I.O.C. stands behind steroids ban. The Wisconsin State Journal. Friday, May 10 (1974).

JOHNSON, L. C., FISHER, G., SILVESTER, L. J., HOFHEINS, C. C.: Anabolic steroid: effects on strength, body weight, $O_2$ uptake and spermatogenesis in mature males. Med. Sci. Sports **4**, 43—45 (1972).

JOHNSON, L. C., O'SHEA, J. P.: Anabolic steroid: effects on strength development. Science **164**, 957—959 (1969).

LOCKHART, A.: Department of Physical Education, University of Southern California. Unpublished paper, 1959.

MORGAN, R. E., ADAMSON, G. T.: Circuit Training, p. 88. London: G. Bell and Sons 1957.

MUNSON, A. R.: Some effects of anabolic steroids during weight training. University of Southern California Thesis, Ed. D. (1970).

O'SHEA, J. P.: The effects of an anabolic steroid on dynamic strength levels of weight lifters. Nutr. Rep. Int. **4**, 363—370 (1971).

O'SHEA, J. P., WINKLER, W.: Biochemical and physical effects of an anabolic steroid in competitive swimmers and weight lifters. Nutr. Rep. Int. **2**, 351—362 (1970).

PERRINE, J. J.: Isokinetic exercise and the mechanical energy potentials of muscle. J. Health Phys. Educ. Rec. **39**, 40—44 (1968).
SAMUELS, L. T., HENSCHEL, A. F., KEYS, A.: Influence of methyl testosterone on muscular work and creatine metabolism in normal young men. J. clin. Endocr. **2**, 649—654 (1942).
SIMONSON, E., KEARNS, W. M., ENGER, M.: Effects of methyl testosterone treatment on muscular performance and the central nervous system of older men. J. clin. Endocr. **4**, 528—534 (1944).
SJÖSTRAND, T.: Changes in the respiratory organs of workmen at an ore smelting works. Acta med. Scand. Suppl. **196**, 687—699 (1947).
STAMFORD, B. A., MOFFAT, R.: Anabolic steroid: effectiveness as an ergogenic acid to experienced weight trainers. J. Sports Med. Phys. Fit. **14**, 191—197 (1974).
STEINBACH, M.: Über den Einfluß anaboler Wirkstoffe auf Körpergewicht, Muskelkraft und Muskeltraining. Sportarzt und Sportmedizin **11**, 485—492 (1968).
STRÖMME, S. B., MEEN, H. D., AAKVAAG, A.: Effects of an androgenic-anabolic steroid on strength development and plasma testosterone levels in normal males. Med. Sci. Sports **6**, 203—208 (1974).
WAHLUND, H.: Determination of physical working capacity. Acta med. Scand. Suppl. **132**, 1—78 (1948).
WALTERS, C. E.: The Effects of Therapeutic Agents on Muscular Strength and Endurance. Phys. Ther. Rev. **40**, 266—270 (1960).
WARD, P.: The effects of an anabolic steroid (Dianabol) on selected physiological paraments: a study case. Track Technique **41**, 1312—1314 (1970).
WARD, P.: The effect of an anabolic steroid on strength and lean body mass. Med. Sci. Sports **5**, 277—282 (1973).
WEISS, V., MÜLLER, H.: Zur Frage der Beeinflussung des Krafttrainings durch anabole Hormone. Schweiz. Z. Sportmedizin **16**, 79—89 (1968).

# F. Miscellaneous Uses of Anabolic Steroids

H. KOPERA

## I. Introduction

The value of anabolic steroids (AST) in the treatment of several more or less well-defined indications has been described in other sections of this book. The aim of the present chapter is to list, discuss, and where possible evaluate the remaining available data on the therapeutic use of these preparations.

It must be pointed out right from the start that anabolic compounds have been and still are employed in a larger number of widely divergent pathologic conditions than the majority of drugs (FOSSATI, 1974). This might lend support to the impression that the rationale for the therapeutic use of AST is questionable. Admittedly, the scientific basis for their administration appears to be questionable in certain instances and many of the published reports on the actions of these drugs describe observations made only in uncontrolled clinical trials. However, as will be shown, the amount of data from controlled investigations is appreciable and the amount of published observations collected in noncontrolled studies is too voluminous to be ignored. The decisive value of controlled clinical experiments for the evaluation of drug actions is undisputed. Nevertheless, this fact does not justify the assumption that a particular drug efficacy is nonexistent just because it is claimed mainly or only on the strength of preliminary observations or uncontrolled trials. This is indeed the case for several indications which will be discussed in this chapter.

Therefore an attempt will be made to review the literature as thoroughly as possible and to try to draw conclusions, make recommendations, or at least give an opinion. This is by no means easy. The relevant publications are international in scope and are thus not easily accessible. Hence completeness of the review, despite the best of intentions, cannot be guaranteed either because of the language barrier or the impossibility of obtaining original papers from certain countries. Most of the publications referred to in this review date from the last decade. Earlier published work can be found in the excellent monographs by OVERBEEK (1966) and by KRÜSKEMPER (1968). Another difficulty is that many of the data cannot be compared directly because they were obtained in trials performed under widely differing conditions. A final difficulty is the fact that the pathologic conditions discussed in this chapter, as entities to be treated with AST, are often rather ill-defined. They have been grouped to facilitate description of the relevant data and, in so doing, the fact that separation of one condition from the other is sometimes quite artificial has been disregarded.

The anabolic preparations in medical use and their comparability as far as wanted and unwanted effects are concerned, have been described elsewhere (see

Chapter II). With very few exceptions (e.g. hyperlipidemias, fibrinolysis, or liver diseases) no clinical pharmacologic basis exists for regarding a compound as being particularly suited to a specific indication. Hence, in general no distinction will be made between the various drugs although not every one has been tried for every indication mentioned. The principal and therapeutically desired effects can be produced by nearly every one of the AST on the market, provided the dosage is correct. Moreover, no attempt has been made to recommend dosages and treatment regimens for therapeutic use. The physician should make his own decision for each individual case. When deciding whether an oral or parenteral preparation should be used, he must of course realize that the oral 17-alkylated steroids are short-acting and the steroid esters are usually depot preparations; this means that the anabolic as well as the androgenic actions of the latter will accumulate if frequency of administration or dose are higher than recommended. The physician will also have to be continuously aware of the possibility of unwanted hormonal or nonhormonal effects. Thus, particular caution should be observed to avoid undesired endocrine effects (androgenic, progestagenic, central inhibiting) in children and women. In patients with confirmed liver disease, or where liver disease is suspected, preference will have to be given to preparations such as steroid esters, which hardly ever disturb the excretory liver function, as opposed to 17-alkylated steroids which often affect this function. In other words, though AST may appear to be relatively nonspecific drugs, a basic knowledge of the pharmacology of AST is essential for their correct application. As the pharmacology of AST is discussed extensively in a separate chapter, recommendations and warnings about unwanted effects will not usually be included in the following discussion.

## II. Rationale for the Clinical Use of Anabolic Steroids

### A. Protein Anabolism

The positive effects of anabolic steroids (AST) on protein metabolism in the human—a nitrogen-saving, protein synthesis-promoting and anticatabolic action—have been established beyond doubt by metabolic studies (see Chapter II). Consequently, there is good reason to use AST at least as a supporting adjuvant in the therapy of conditions with a relative or absolute protein deficiency; the underlying cause of the protein depletion is relatively unimportant whether it be reduced synthesis or excessive breakdown caused directly by disease or indirectly by reduced endogenous production of androgenic-anabolic steroids. Clinical experience has amply shown that specific therapeutic actions and adequate dietary measures can be effectively supported by AST to restore the disturbed equilibrium of protein metabolism and to reduce protein deficiency, both of which occur in many pathologic conditions.

### B. Other Effects

The therapeutic actions of AST are by no means confined to stimulation of protein synthesis and reduction of excessive breakdown. Other metabolic effects, actions on bone tissue, erythropoiesis, and carcinoma are described elsewhere in this book. The present chapter will increase the list of therapeutic indications for AST by providing evidence for additional effects which may or may not be related to protein metabolism, such as promotion of fibrinolysis, a hypolipidemic action, an effect on various metabolic diseases, and an influence upon the psyche.

The beneficial effects of AST upon mood, concentration, psychomotor activity, energy, physical work capacity, appetite, and general well-being appear to be of particular practical importance. In certain countries it has become quite popular to use AST as therapeutic agents for the treatment of such vaguely defined conditions as physical and mental fatigue and exhaustion. This misuse of AST as general roborants is rightly criticized and condemned, mainly because the existence of the supposed actions has not yet been proven by convincing trials and the danger of masking an underlying disease and of producing side-effects is always present. Still, it would seem unwise to disregard the possibility of a therapeutic action on the psyche (a very appealing prospect to many physicians and patients alike) without proper investigation. The sooner this is done, the better; either to eliminate this or that "indication" or to arrive at clear-cut directives.

The effects of AST are so numerous that one is inclined to assume that these compounds exert their influence on the most basic processes of the organism. The reported effects of AST on specific enzymes, the trophic effects on various tissues such as bone, bone marrow, skin, liver, kidney, mucosa, and the positive effects of AST on the immune status and on the general resistance to pathologic noxae are compatible with such an assumption. It is evident that much more research is required to support or disclaim ideas of this kind. However, the stage of pure speculation appears to have been passed. It is not unthinkable that future investigations may supply evidence about fundamental effects of AST which would make possible a plausible explanation for many hitherto poorly understood therapeutic actions.

## III. Principles of Therapeutic Investigations

The principle requirements of drug testing and the various methods recommended for the evaluation of drug therapy have been the subject of many detailed descriptions by such experts as HARRIS and FITZGERALD (1970); MELMON and MORRELLI (1972); LAURENCE (1973); HAMILTON (1974); KUEMMERLE et al. (1976); these principles need not be repeated here. It is generally accepted that therapeutic trials should adhere to a well-designed protocol. Comparative studies using the double-blind technique should preferably be employed to prevent bias from influencing the results. Treatment should be randomized and the data subjected to statistical analysis. These are considered prerequisites for producing conclusive results, and one is very reluctant to accept data from clinical trials which do not meet such requirements.

Unfortunately, most of the data available for discussion in this chapter were obtained from clinical observations in uncontrolled trials. The patient samples were often unselected, treated with different AST, and dosages that cannot be compared. One is inclined to wonder whether some of these investigations justify the attempt of an analysis and/or make an evaluation possible at all. Indeed, some authors readily state that AST are of little or no value in a number of indications to be described later due to the fact that convincing proof of their effectiveness is lacking. The author does not think this is the best approach, and prefers to review the reported observations and evaluate them wherever possible.

# IV. Clinical Results

## A. Organic Diseases

### 1. Hepatic Disorders

A specific treatment for acute and chronic hepatitis, for fatty liver, and for cirrhosis of the liver has not yet been found. Instead, a frightening multitude of therapeutic proposals of unproven value is available and is used with variable confidence and frequency by the medical profession. Current therapy in some countries seems to be an example of unreasonable polypragmasy based on a combination of mysticism, science, and uncontrolled clinical observations. The principles of therapy for these disorders are general and physical treatment, diet, and drugs and among the latter are AST.

Animal Pharmacology

The effect of androgens and of AST on the liver, especially on liver function, has been studied extensively in animal experiments and a detailed description of the results would require a separate review. This, together with the fact that their relevance for the human, and in particular for therapeutic purposes, is often far from clear makes it advisable to limit discussion to a brief mention of a few findings.

The extent to which androgenic agents affect the weight and nitrogen content of the liver depends upon dosage and available protein. These agents accelerate regeneration after partial hepatectomy. They cannot prevent liver necrosis if the diet is protein-deficient, but can retard formation of fatty liver produced by protein deficiency, chronic alcohol intoxication, or high doses of oxytetracycline. Moreover, they protect the liver against the hepatotoxic action of numerous substances under certain circumstances. Similar effects have been found with many AST. Damage of the normal liver by otherwise nontoxic doses of AST has never been observed (EGER, 1964; OVERBEEK, 1966; BIELAWSKI, 1968; KRÜSKEMPER, 1968; PETZOLD et al., 1969; KULCSÁR-GERGELY et al., 1975). Details of the effect of AST on protein synthesis are given in Chapters III and IV.

Effects on Liver Function and Serum Enzymes in the Human and in Animals

Following the first report of jaundice due to methyltestosterone (WERNER, 1947) and the subsequent publication of similar observations (FOSS and SIMPSON, 1959) much work was devoted to the investigation of the influence of androgenic-anabolic steroids on liver function and serum enzymes in animals (OLDERSHAUSEN, 1964a, b; SCHNACK, 1964a, b; WERNZE, 1964a; OVERBEEK, 1966; KRÜSKEMPER, 1968) and in humans with normal livers as well as with hepatic diseases (LINDNER, 1964a; OLDERSHAUSEN, 1964c; POHLE, 1964; SCHNACK, 1964b; VERMEULEN and DEMEULENAERE, 1964; WELLER, 1964a; WERNZE, 1964b; WILDHIRT, 1964; LORIMIER et al., 1965; GOLDSTEIN and COMBES, 1966; KRÜSKEMPER, 1968; ROZMAN et al., 1971). The results do not always completely agree but in man the majority of investigators found: sulfobromophthalein retention (BSP); increased serum glutamic-oxaloacetic transaminase (SGOT, serum aspartate aminotransferase), serum glutamic-pyruvic transaminase (SGPT, serum alanine aminotransferase) and alkaline phosphatase activity; changes in the lactate dehydrogenase, aldolase, and sorbitol dehydrogenase activity. These alterations hardly ever occur with 17-ester steroids (HORKY, 1967; KRÜSKEMPER, 1968); they are seen frequently with most of the (mainly) orally active steroids with an alkyl substitution in the C-17 or C-1 position and a 3-keto group; no relation seems to exist between these actions and the hormonal properties of the compounds.

The changes in the liver function and in the activity of the serum enzymes are usually transient and disappear in spite of continued treatment; they do not necessarily indicate concomitant effects on the microscopic structure of liver tissue. However, deviations from the norm were detected by electron microscopy, suggesting primary damage in the canalicular membranes of the liver cells (see ROZMAN et al., 1971). Only rarely are the alterations so severe that intrahepatic cholestasis with icterus results.

Since BSP retention is interpreted by most investigators as a symptom of impaired liver function, the use of drugs which produce this effect is discouraged for the treatment of liver diseases or disorders where the presence of liver damage is known or suspected.

### *a) Cirrhosis*

Cirrhosis of the liver is a chronic disease characterized by necrosis of liver cells, destruction of the architecture of the liver through subsequent fibrosis and liver cell regeneration. Liver function becomes increasingly impaired. Among the many clinical features and complications of cirrhosis which might be target for an anabolic treatment are: anorexia, disturbance of protein synthesis, dysproteinemia, and a negative nitrogen balance. All of these symptoms cause weakness, weight loss, and general malaise. In addition, the steroids may influence hypogonadism, which is found not infrequently in these patients. Finally, AST may exert a hepatoprotective effect, as seen in animal experiments.

#### Clinical Trials

GIROLAMI was the first clinician to study the therapeutic effect of testosterone in liver cirrhosis systematically. In a review of his trials covering more than 2000 cases (GIROLAMI, 1973), long-term treatment (months to years) with excessively high doses of testosterone esters plus vitamin $B_1$ was reported to have been successful in over 70% of the cases. His publications since 1946 persuaded other therapists to use androgens for this purpose, some of whom were equally successful (PORTO, 1968). When the less virilizing AST became available, the number of researchers investigating the use of androgenic-anabolic compounds in the treatment of liver disease increased appreciably. With this new-found interest came proposals for various new treatment regimens. These usually involved the administration of much smaller doses of the steroids and the combination with glucocorticosteroids. At that time, the indispensability of controlled trials became increasingly apparent. Unfortunately, however, performance of controlled studies in a representative number of comparable patients with chronic liver disorders is notoriously difficult. This difficulty is due to such factors as spontaneous and unpredictable changes in the progress of the disease, uncontrollable influences of diet, physical activity, and infections. This is probably the main reason why data that conclusively prove the efficacy of AST in cirrhosis are

Table 1. Controlled therapeutic trials with androgens and AST in patients with cirrhosis, fatty liver, or other chronic hepatic diseases

| Author(s) | Country | No. of patients | Duration of treatment | Therapeutic effects recorded | Remarks |
|---|---|---|---|---|---|
| WELLS (1960) | Singapore | 80 | 14–86 weeks | + | Comparison of prednisolone + corticotrophin and testosterone propionate treatment with controls, little difference between the two "treatment series" |
| FRANKEN et al. (1963) | W. Germany | 41 | 4–43 months | + | Comparison of prednisone with testosterone propionate treatment; both about equally effective |
| MÜTING (1964) | W. Germany | 47 | 6 weeks, 1 year | + | Comparison of treatment with nandrolone decanoate for 6 weeks and 1 year respectively |
| BLAHNIKOVA and HORKY (1972) | Czecho-slovakia | 56 | 2 weeks | + | Study of the effect of nandrolone phenyl-propionate or dimethyl-androstanolone on N-balance in hypo-, normo-, and hyperprotein-fed patients |
| FENSTER (1966) | USA | 27 | 1 month | – | Comparison of testosterone propionate, and methenolone enanthate with placebo; duration of treatment too short? |
| WEWALKA (1968) | Austria | 33 | 20 weeks | – | Comparison of methenolone enanthate with placebo. According to the author result may be explained by the poor comparability of the patient groups |
| EBERHARDT et al. (1975) | W. Germany | 66 | 6 weeks | – | Comparison of B vitamins + clostebol with placebo. No drug-induced difference in effect |

still lacking. Nevertheless, the use of anabolics as an adjuvant to routine therapy has been recommended by many specialists.

The vast majority of published clinical observations are inadequate and do not comply with the basic requirements of scientific drug evaluation. Admittedly, there are wide variations in the quality of the different studies, and in a number quite convincing results were obtained, although controlled trials are few and far between.

In the different trials and even within the same investigation, etiology, duration of liver disease, and stage of the cirrhosis (e.g., compensated, decompensated) varied

Table 2. Uncontrolled clinical trials with AST in patients with cirrhosis or other chronic liver diseases. All authors reported therapeutic effects

| Author(s) | Country | No. of patients receiving an AST | Duration of treatment |
|---|---|---|---|
| Fiegel and Kelling (1959b) (see also Fiegel and Kelling 1959a; Fiegel and Fischer 1959) | W. Germany | 106 | weeks to 15 months |
| Berkowitz (1962) | USA | 8 | 3 months |
| Craig (1962) | USA | 2 | 24 weeks |
| Schnack et al. (1962) | Austria | 54 | 3–6 months |
| Trocme and Roy (1962) (see also Roy, 1961) | France | 22 | 1 month |
| Weller and Voegt (1962) | W. Germany | 15 | not specified |
| Dubarry et al. (1963) | France | 77 | 1 month or longer |
| Gramsch (1963) | W. Germany | 58 | 10 days–20 weeks |
| Kasich (1963) | USA | 5 | 12–15 weeks |
| Wewalka (1963) | Austria | 41 | at least 1 year |
| Benda and Geyer (1964) | Austria | 7 | not specified |
| Gramsch (1964) | W. Germany | 30 | 2–7 weeks |
| Lindner (1964b) | W. Germany | 20 | more than 6 weeks |
| Schnack (1964c) | Austria | 37 | 6–> 12 months |
| Seifert (1964) | W. Germany | 43 | not specified |
| Weller (1964b) | W. Germany | 136 | up to 16 months |
| Dubarry et al. (1965) | France | 12 | 3 weeks |
| Müting (1965) | W. Germany | 24 | 18 months |
| Blahnikova (1967) | Czechoslovakia | 28 | 14 days |
| Horky (1967) | Czechoslovakia | 69[a] | up to 4 months |
| Lindner (1967a, b) | W. Germany | 231 | 12–94 weeks |
| Wewalka (1967) | Austria | 95 | up to 2 years |
| Pezold (1968) | W. Germany | 49 | several weeks —3 months |
| Hirayama (1970) | Japan | 35 | 4 weeks |
| Hecht (1971) | France | 22 | more than 1 month |
| Markov and Piralov (1972) | USSR | 44 | |
| Moinade (1974) | France | 31 | 2 months |

[a] Some of the patients might have been counted twice.

widely; so did the general condition of the patients, the duration and type of general and drug therapy and diet. These variables make it impossible to pool and evaluate collectively the results. However, in controlled and uncontrolled clinical trials (Tables 1 and 2), the following therapeutic effects have usually been observed and ascribed to the androgenic-anabolic steroids: a favorable influence on protein metabolism; reduction of elevated serum amino acid levels and hyperaminoaciduria, followed by an increase in albumin synthesis; increase of nucleic-acid synthesis (Leevy, 1966); filling up of the depleted extravascular albumin pool; correction of protein clotting factor deficits within weeks (Tamburro and Leevy, 1967) and of dysproteinemia after several months of therapy by significant elevation or even normalization of serum albumin and reduction of $\gamma$-globulins; increased retention of phosphorus, potassium, and nitrogen, the latter even at low protein intake levels;

anticatabolic action in patients receiving treatment with glucocorticosteroids; improvement of the detoxifying function of the liver, particularly with respect to phenol metabolism. Improvement or normalization of pathologic BSP retention, transaminases, serum enzymes, and other liver function tests were also observed. General improvement was observed particularly in patients where malnutrition from low food intake or digestive disturbances might have been of importance. Patients felt an improvement in their condition, their appetite increased, and they gained weight.

No impressive histologic changes in the liver were reported which indicates that the underlying pathologic processes of cirrhosis are not affected by AST. Nevertheless, reduction of mortality was reported by WELLS (1960) and WEWALKA (1967); no effect on survival following short-term therapy was seen by FENSTER (1966).

Absence of a therapeutic effect of an AST was published by DEMEULENAERE (1964) who performed an uncontrolled trial in 10 patients with chronic hepatic disease.

There is as yet no final answer to the question as to whether the observed beneficial effects of AST (or glucocorticosteroids for that matter) actually influence the course of liver cirrhosis. It also remains open for further research to show that the drug actions contribute more to the improvement of the patient than the general anabolic effect on appetite and general well-being. Certainly the overall improvement is due to a number of factors including drug actions, dietary correction of undernourishment, prevention of physical exertion, and the prohibition of alcohol.

### b) *Fatty Liver*

Excessive fat intake and, in particular, chronic alcohol intoxication can lead to the pathologic deposition of fat in the liver cells. Androgenic and anabolic steroids have been shown to lessen fat accumulation in the liver in animal experiments (BIELAWSKI, 1968; see also KRÜSKEMPER, 1968). Moreover, deficient production of male sex hormones appears to increase the sensitivity of the liver to fatty degeneration, necrosis, and cirrhosis (LEEVY, 1970).

Consequently AST have also been used as adjuvant to the therapy of fatty liver. Encouraging observations such as rapid hepatic triglyceride mobilization through accelerated lipoprotein release and/or synthesis (MENDENHALL, 1974), clinical improvement, and speedier recovery were reported (LEEVY, 1962, 1967; LEEVY, 1970; FIGUEROA, 1973); in addition, some of the patients treated successfully in the cirrhosis trials (Table 2) had also been suffering from fatty liver. No beneficial effects of AST were seen by GABUZDA et al. (1950); GROSSMAN and YALOW (1965); FENSTER (1966); EBERHARDT et al. (1975).

Although fatty liver normally does not require drug therapy (MARTINI et al., 1969; GEROK et al., 1971), it appears that the concomitant administration of AST should be considered as an adjunct to the treatment of certain cases refractory to dietary measures (LEEVY, 1968).

### c) *Viral Hepatitis*

The administration of AST is not generally recommended in hepatitis. In controlled clinical trials the addition of an AST to the standard treatment was found to be of no benefit by MIRANDA et al. (1970), whereas WELLER (1964b) and JANNUZZI et al.

(1970) observed a moderate to pronounced therapeutic effect; a beneficial influence was also seen in an uncontrolled study by TOLENTINO (1966).

*Conclusion:* AST have no specific effect on hepatic diseases but are a useful adjunct to the treatment of liver cirrhosis (especially in cases with low serum albumin levels), and advanced chronic liver diseases with protein depletion. A positive effect can only be expected when treatment is continued for at least several months. Unwanted side-effects, primarily hormonal, due to high dosage and prolonged treatment, will have to be taken into account.

Anabolic agents may also be of some benefit in the treatment of fatty liver and in the reparative phase of acute liver diseases.

Whether the combination of glucocorticosteroids with AST improves the effect of the individual drugs on hepatic diseases awaits further investigation.

## 2. Renal Disorders

### Animal Pharmacology

Experiments in animals have demonstrated that androgenic-anabolic steroids affect the kidney directly (KOCHAKIAN, 1944; GIBBA and PETTERINO, 1962; CECCARELLI, 1965; SCHIMMELPFENNIG, 1965; OVERBEEK, 1966; KRÜSKEMPER, 1968; VOSS, 1973a). They have a nephrotropic effect causing renal hypertrophy in female and castrated male rodents, increasing the degree of hypertrophy following unilateral nephrectomy, and reducing kidney atrophy after homolateral ligation of the ureter. Moreover, androgenic-anabolic steroids protect the kidney against various nephrotoxic substances (nephroprotective effect). Anabolic-androgenic agents do not seem to influence kidney function to an appreciable extent in the animal and apparently not at all in the human except perhaps in the aged (DONTAS et al., 1967; see below under Geriatrics).

Because of their nephrotropic, nephroprotective, and several extrarenal effects androgens and AST have been tried as an adjunct to the therapy of renal diseases.

### *a) Acute Renal Failure*

In bilateral nephrectomized or subtotally nephrectomized animals the development of uremia can be slowed down, and the survival time prolonged, by the administration of anabolic-androgenic steroids (SELYE, 1940; MASSON et al., 1949; SZOLD, 1959; SZOLD et al., 1959; AL-HUSSAINI et al., 1965; KROTOV, 1966; not confirmed by GERBER and COTTIER, 1961; SCHIMMELPFENNIG, 1965). This could indicate a possible role for anabolic-androgenic steroids in certain clinical situations.

Acute renal failure can be caused by a large variety of factors. It is characterized clinically by severe oliguria or anuria followed by polyuria, and biochemically by the accumulation of nitrogenous waste products, hyperkalemia, and acidosis as a result of protein breakdown. Reduction of protein catabolism is one of the foremost aims of the complex therapy in acute renal failure. Achievement of this aim is attempted by cutting down protein intake to the bare minimum, counteracting catabolism with the help of anticatabolic-anabolic drugs, and by dialysis procedures.

Elevated 17-hydroxycorticosteroid levels in plasma due either to increased production or decreased excretion of these compounds are found in the anuric or

Table 3. Clinical trials with AST in acute renal failure

| Author(s) | Country | No. of patients receiving AST | Therapeutic effects recorded | Trial conditions |
|---|---|---|---|---|
| BLAGG and PARSONS (1960) (see also PARSONS and MCCRACKEN, 1959; BLAGG and PARSONS, 1961) | Great Britain | 14 | + | Controlled |
| GERBER and COTTIER (1961) | Switzerland | 1 | – | Controlled |
| THAYSEN (1962) | Denmark | (232)[a] | + | Controlled |
| SARRE (1964) (see also SARRE, 1963) | W. Germany | 7 | + | Controlled |
| KESSEL (1964) | W. Germany | 8 | + | Controlled |
| GAUTIER and TÖNZ (1960) | Switzerland | 2 | + | Uncontrolled |
| WIGGINS (1963) | USA | 3 | (+) | Uncontrolled |
| ROBSON et al. (1964a, b) | Great Britain | 200 | + | Uncontrolled |

(+) Slight improvement only.
[a] Only some of the patients were treated with an AST.

oliguric phase of acute renal failure (DEROT and BERNIER, 1951; BLAIR et al., 1961; IISALO et al., 1962). AST may reduce or inhibit the antianabolic or catabolic action of glucocorticosteroids as long as catabolism is not excessive. In the reparative period of polyuria when kidney function is resumed, anabolic agents could support restitution of protein levels. Unfortunately, conclusive trials that would demonstrate the effect of AST or any single drug in this potentially fatal situation are almost impossible to conduct. The course of the clinical condition is influenced by too many factors (e.g., infection, gastrointestinal hemorrhage, changing renal function, stress responses) and the clinical situation requires therapeutic actions of such urgency and diversity that the effect of one drug can hardly be differentiated from other variables. Nevertheless, in view of the results of comparative studies as well as a number of uncontrolled clinical observations in more than 200 patients it seems reasonable to ascribe some of the observed effects to the action of AST.

## Clinical Trials

High doses of androgens (MERRILL, 1955; FREEDMAN and SPENCER, 1957) or AST can significantly reduce the rate of progression of azotemia and the development of the clinical symptoms of uremia. In responsive patients these drugs lower the urea-nitrogen production; they appear to allow a liberal diet including protein without the danger of urea-nitrogen and potassium accumulation. This is advantageous because the protein-saving effect of AST is much better with a high-protein diet (WOLTHUIS, 1963); these substances help to restore nitrogen equilibrium; they improve appetite and general well-being; and they reduce the need for supplementary hemodialysis. (For references see Table 3.)

Apparently hypercatabolism cannot be influenced to any great extent by AST since no favorable response was seen in patients with severe infections, extensive internal hemorrhages, or gross tissue necrosis (GJØRUP and THAYSEN, 1960; THAYSEN, 1962).

The possibility that children with acute renal failure might be particularly responsive to anabolic therapy was mentioned by GAUTIER and TÖNZ (1960).

It would be an exaggeration to regard AST as indispensable drugs for the conservative management of acute renal failure. However, they have been shown to be of definite therapeutic value in some of the patients with not too excessive protein catabolism. It would, therefore, appear advisable to include AST in the treatment of acute renal failure. Despite the high doses necessary, side-effects are minimal since treatment can be restricted to one or a few weeks.

### b) *Chronic Renal Failure*

Chronic renal insufficiency with retention of nitrogenous waste products in the blood produces the clinical condition of uremia. The etiology is manifold, the clinical picture varies, progressing from an asymptomatic stage to anorexia, general wasting, cardiovascular, neurologic and gastrointestinal disturbances, progressive anemia, bone demineralization, severe intoxication, and coma. Elevation of blood urea-nitrogen, creatinine, and acidosis are regularly found. Indeed, there is not so rapid a change in the protein catabolism and renal function as in acute renal failure. Hence, drug trials under controlled conditions in stable azotemia can be more easily performed.

The rationale for the use of AST in chronic renal failure is based upon several factors. AST lower the elevated blood level of intoxicating nitrogenous waste products and/or prevent progressive accumulation of these products in rapidly deteriorating renal function or catabolic episodes. In this way the frequency of dialyses and the accompanying protein losses can be reduced. Life is made more tolerable because a more palatable diet containing more protein can be permitted. The general condition of the patient is improved, thereby reducing the danger of harmful complications. Obviously, all these effects of AST are extrarenal. Improvement of kidney function was never observed.

As in acute renal failure, anabolic drugs are no more than an addendum to the complex therapeutic measures necessary for the management of chronic azotemia.

#### Clinical Trials

In controlled clinical trials (Table 4) short-term administration of AST to mildly or moderately azotemic patients produced a decrease in serum urea nitrogen, urinary nitrogen, as well as urea and phosphorus excretion indicating improvement of uremia through reduction of nitrogenous end-product formation. It was furthermore found that increase in dietary protein did not result in a rise in blood urea nitrogen (SZOLD et al., 1959; THAYSEN, 1962; SIGLER and ISSEKUTZ, 1967; not confirmed by ROBSON et al., 1964b).

In several investigations only some of the treated patients responded to therapy, the occurrence of a "wearing-off" phenomenon was reported (i.e., the therapeutic efficacy diminished in the course of several weeks) or treatment was occasionally followed by a sort of rebound phenomenon (THAYSEN, 1962; ROBSON et al., 1964a;

Table 4. Controlled clinical trials with AST in patients with chronic uremia

| Author(s) | Country | No. of patients receiving AST | Therapeutic effect on azotemia |
|---|---|---|---|
| EATON (1959) | Great Britain | 4 | In all patients |
| GERBER and COTTIER (1961) (see also COTTIER, 1962) | Switzerland | 8 | None |
| THAYSEN (1962) | Denmark | 16 | In some patients |
| SIGLER and ISSEKUTZ (1967) | USA | 11 | In all patients |
| ROBSON et al. (1968) | Great Britain | 14 | In some patients |
| GOLDMAN and ZARDAY (1973) | USA | 2 | In both patients |
| WILLIAMS et al. (1974) | USA | 9 | None (but significant effect on renal anemia) |

Table 5. Uncontrolled clinical trials with AST in patients with chronic azotemia. All authors reported therapeutic effects

| Author(s) | Country | No. of patients receiving AST |
|---|---|---|
| SZOLD et al. (1959) | Hungary | 50 |
| EL-ABRASHI et al. (1961) | Egypt | 15 |
| BERGMANN (1962) | Austria | 17 |
| FAM (1963) | Egypt | 109 |
| WIGGINS (1963) | USA | 3 |
| SNYDER and BREST (1966) | USA | 10 |
| PANI (1968) | Italy | 16 |
| COMAR (1969) | Italy | 50 |
| ESPOSITO et al. (1975) | Italy | 4 |

SNYDER and BREST, 1966; KRÜSKEMPER, 1968). No wearing-off effect was observed by SIGLER and ISSEKUTZ (1967) in the course of 6 weeks treatment with an AST.

GOLDMAN and ZARDAY (1973) conclude from one observation that AST are contraindicated in patients with severe hypertension.

Positive effects of AST in patients with chronic azotemia of different degrees and under various dietary and therapeutic regimens were also seen in a number of uncontrolled clinical trials (Table 5).

The published data on the use of AST in chronic renal failure indicate that these compounds undoubtedly improve azotemia in some uremic individuals. Whether this effect can be maintained over a long period is questionable. Anabolics are probably more effective in the treatment of acute decompensation of renal function and in exacerbations of azotemia than in long-term therapy of chronic renal insufficiency. These drugs appear to allow a more liberal protein intake, thus making possible a more palatable diet. Moreover, they promote the general well-being, improve the physical state, and promote weight gain. The patient's resistance increases and complications decrease.

### c) The Nephrotic Syndrome

The cardinal symptoms of the nephrotic syndrome are edema, proteinuria, hypoalbuminemia, and hyperlipidemia. A similar condition can be produced experimentally by treating animals with antikidney serum according to the Masugi method. The development of nephrotic symptoms in this experimental nephritis was inhibited or, when already present, therapeutically influenced by androgenic (MOENCH et al., 1954; DUTZ et al., 1956; MOENCH and VOGT, 1958) and anabolic substances, particularly in combination with glucocorticosteroids (KLUTHE and LAMPE, 1964). This effect was the basis for the claim for the existence of an anti-inflammatory action of androgens (SARRE, 1964).

The therapeutic use of AST in the nephrotic syndrome is based both on the favorable results in the above-mentioned animal experiments and on the expectation that the anabolics would reduce or prevent the protein catabolic effects of glucocorticosteroids, in many cases the treatment of choice for this disorder.

There is no unanimity about the therapeutic efficacy of AST in patients suffering from the nephrotic syndrome (KRÜSKEMPER, 1968). Encouraging results, including improvement of dysproteinemia, proteinuria, hypercholesterolemia, and nitrogen balance were published by MOENCH and SARTORIUS (1954); SERENI et al. (1959); BEIGLBÖCK and BRUMMUND (1960); CHRISTL et al. (1962); DORN (1963); SARRE (1964); WATSCHINGER (1964); CESNIK and FINK (1964); ESPOSITO et al. (1975). Absence of any useful action was reported by MARKIEWICZ (1965). A combination of glucocorticosteroids and AST which apparently produces the best results was proposed by COTTIER (1962); HAGGENMÜLLER (1964); KLUTHE (1964); ROHWEDDER (1964); SCHINDLER and KORN (1965); POSER et al. (1966); SARRE (1966); KRÜSKEMPER (1968).

Current knowledge leads to the conclusion that AST are useful as adjunctive intermittent therapy in nephrotic syndrome patients receiving glucocorticosteroids. AST may also be tried without corticosteroids particularly in cases in which corticotropin or glucocorticosteroids are contraindicated or in patients who do not respond to their administration.

### d) Renal Anemia

The benefit of AST in the treatment of anemia in patients with chronic renal failure is discussed in Chapter VII.C. In this context it should be mentioned that improvement of hemoglobin, packed cell volume, red cell mass, and a decrease in requirements for blood transfusions as a result of AST therapy was clearly demonstrated in uncontrolled (DOANE et al., 1975) and in controlled comparative trials (HENDLER et al., 1974; WILLIAMS et al., 1974); the effect seems to be dependent from the presence of kidney tissue (DOANE et al., 1975; STENZEL et al., 1975). However, only some moderate and transient effectiveness was observed by GOODMANN and BESSMAN (1975) and no effect of an AST on anemia but production of a reversible hypertriglyceridemia was reported in a controlled trial by CHOI et al. (1974).

*Conclusion:* Renal diseases require a vast array of therapeutic measures. AST have proved to be a satisfactory and harmless adjunct to other drug therapy when employed for the management of acute and chronic renal failure or for the treatment of nephrosis. In both acute and chronic renal failure AST help to counteract protein catabolism, to reduce the development or progression of azotemia and uremia, to combat protein depletion, and to reduce the need for hemodialysis. Moreover, in chronic azotemia AST appear to allow the patient to safely follow a more palatable diet with a higher protein content. In addition, AST reduce anemia and the requirements for blood transfusions. Glucocorticosteroid treatment of the nephrotic syndrome can be improved by the adjunctive intermittent administration of AST. The latter can also be tried in patients not receiving corticosteroids.

AST do not rank prominently among the drugs used for the conservative treatment of renal diseases. However, based on past experience their use as an addendum to the routine therapy is justified and can be safely recommended.

### 3. Cardiac Disorders

The interest in AST in the treatment of cardiac disorders has been restricted to ischemic heart diseases and degenerative processes of the heart muscle.

It is unnecessary to emphasize the enormous problems resulting from the increasing incidence of myocardial infarction, one of the leading causes of death today. There is also no necessity to elaborate on the many unsolved problems in prophylaxis or treatment of ischemic heart disease and degenerative myocardial changes. It is obvious then, that every new idea for a better prophylactic and therapeutic approach to cardiovascular disorders deserves the attention of research workers and practising physicians alike.

#### Animal Pharmacology

It has been shown that androgens (BLASIUS et al., 1956) and AST (BLASIUS et al., 1957; NOWY et al., 1963a) increase the contractile protein actomyosin in the myocardium of rabbits.

In castrated male rats, ligation of the aorta causes only a very moderate increase in protein-, RNA-, and phospholipid synthesis during the subsequent cardiac hypertrophy, whereas in noncastrated rats, synthesis of these products is significantly higher. This suggests that male sex hormones exert a protective function upon the myocardium by facilitating compensatory hypertrophy (MEERSON, 1969); the findings are not completely in agreement with those of NOWY et al. (1963b) in the rabbit. An increase of myocardial protein synthesis following anabolic treatment was also observed in rats under experimental hypodynamia (MAKAROV, 1974).

The incorporation of glycine-2 $^{14}C$ (and methionine-$^{35}S$, PRUSS et al., 1967) into the cardiac protein of the periphery and the center of an infarct is very significantly increased in dogs pretreated with an AST. This indicates a higher rate of protein synthesis due to fibroblastic proliferation around the infarcted area and may indicate that the damaged myocardium is strengthened. The thickness of the scar and its resistance to deformation or the development of an aneurysm is clearly increased by long-term treatment with an anabolic agent. Anabolic substances also increase the activity of the hexosemonophosphate shunt enzymes. Histologic examination of infarcts reveals that fibroblastic repair, phagocytosis, and the removal of necrotic muscle are accelerated. From these findings it can be deduced that AST stimulate the reparative process following experimental myocardial infarction and that this is achieved by action upon biosynthetic processes in the infarcted area independent of possible effects upon hemodynamia or coronary blood flow (GUDBJARNASON et al., 1966). Experimental infarction is immediately followed by a depletion of high energy phosphates in the myocardium which affects the contractile function of the heart muscle. This unfavorable process is increased by a protein-free diet. Administration of AST, however, significantly diminishes the reduction of adenosine triphosphate and creatine phosphate in the noninfarcted tissue and stimulates recovery. This indicates a positive effect of anabolic agents upon the synthetic and mechanical functions of the heart muscle after artificial infarction in dogs (RAVENS, 1970; a dissenting opinion was expressed by OLSON, 1969).

The effect of an AST on the enzyme systems, important in maintaining the energy capacity of the heart muscle, was investigated in rabbits subjected to experimental myocardial infarction. The results indicate that an acceleration of scar tissue organization was achieved. The major effects were seen in the necrotic area and in the immediately adjacent perinecrotic areas. Collagen staining was more intense in these areas and the number of necrotic muscle fibers was somewhat reduced, suggesting a positive influence on the scar tissue organization and the process of myocardial recovery. It is conceivable that this effect could be helpful in preventing cardiac insufficiency and postinfarction aneurysm (NIKOLAEVA et al., 1974). Similar effects of an AST were obtained in cats (MITROFANOV et al., 1973).

The myocardium of burned rats was found to be protected from morphologic changes by an AST (TICHY et al., 1963), whereas no protective action was seen against electrocardiographic changes produced by triiodothyronine in the rat (KRÜSKEMPER and REICHERTZ, 1963); this contrasts with the many established antithyrotoxic effects of anabolic agents.

Most of these results from animal experiments suggest that AST might exert beneficial effects upon the sequelae of myocardial infarction as well as upon some other cardiac disorders in man. Consequently some clinicians have studied the therapeutic application of AST in heart diseases and the number of reports has increased within the last few years. Surprisingly enough however, the majority of cardiologists show very little interest in the subject, which may explain why the problem is still insufficiently investigated in the human.

### Clinical Trials

AST have been tried alone or, as in most cases, in combination with cardiac glycosides and/or other therapies for cardiac disorders such as: chronic degenerative changes of the myocardium; coronary heart diseases; myocardial infarction, and valvular heart diseases. They were administered for periods varying between weeks and years. In the respective publications (Table 6) a more or less favorable therapeutic efficacy was attributed to AST in respect of subjective complaints, contractile function of the heart muscle, and improvement of the hemodynamia possibly by enlargement of coronary collaterals. Some investigators believe that AST potentiate the effect of cardiac glycosides (BEI, 1968; POLISCHUK, 1970). A therapeutic influence on blood lipids has also been assumed.

Although the number of published observations is impressive, in many instances the results of these trials are less than convincing. Often inappropriate or unspecified

Table 6. Clinical trials with androgens or AST in patients with cardiac disorders. All authors reported therapeutic effects

| Author(s) | Country | No. of patients receiving AST | Trial conditions |
|---|---|---|---|
| GOROKHOVSKY and KITAEVA (1970) | USSR | 70 | Controlled |
| LINEVA (1971) | USSR | 42 | |
| BEI (1972) | USSR | 124 | |
| JAFFE (1973) | USA | 100 | |
| JAFFE (1974) | USA | 25 | |
| AGAPOVA et al. (1974) | USSR | 120 | |
| FIEGEL (1962a, b) | | | Uncontrolled |
| FIEGEL and BARGHEER (1966) | W. Germany | 2546 | |
| FIEGEL and EINFELDT (1971) | | | |
| KALLIOMÄKI and SEPPÄLÄ (1963) | Finland | 22 | |
| KUTSCHERA-AICHBERGEN (1966) | Austria | 4 | |
| HAMMER (1966) | W. Germany | 28 | |
| LESCHINSKY et al. (1967) | USSR | 48 | |
| BEI (1968) | USSR | 40 | |
| LESCHINSKY and KHARITONOVA (1970) | USSR | 100 | |
| POLISCHUK (1970) | USSR | 93 | |
| LESCHINSKY and KHARITONOVA (1972) | USSR | 160 | |

parameters were used or when results were obtained with accepted methods (e.g., electrocardiography) they were interpreted uncritically. The available material, therefore, does not sufficiently comply with the requirements of drug evaluation and is inadequate to assess the possible value of AST in cardiac disorders.

*Conclusion:* Interesting results in animal experiments with myocardial hypertrophy and experimental infarction suggest a possible use for AST in some cardiac diseases. However, the clinical evidence for the therapeutic efficacy of AST in cardiac patients is as yet insufficient to warrant their general use for the treatment of heart diseases. With respect to the importance of this possibility it is hoped that the problem will soon be subjected to thorough, well-planned, conclusive investigation. If anabolic agents are given, it should be remembered that most of them can cause fluid retention. Only nandrolone esters appear to be an exception (VERWILGHEN et al., 1966; BESA et al., 1974). Furthermore, all AST improve general well-being and consequently promote physical activity to an extent which can be unfavorable for some cardiac patients.

## 4. Vascular Disorders and Fibrinolytic Disturbances

Normal fibrinolysis is assumed to be important for the maintenance of vascular patency. Fibrinolysis has been found to be defective in a great many diseases (MENON et al., 1969; MENON, 1970). Fibrinolytic activity appears to be low in a large proportion of patients with occlusive vascular diseases (CHAKRABARTI et al., 1966; BRUHN et al., 1974). Fibrinolytic activity is also reduced after trauma or postoperative stress (BROWN et al., 1971); these events are not infrequently followed by the development of deep vein thrombosis and pulmonary embolism, particularly in older patients. Although a causal association between decreased fibrinolytic activity and these pathologic conditions has not been firmly established, the problem of treating or preventing fibrin deposition and thrombus formation has become increasingly important since occlusive vascular disease nowadays ranks first among the causes of death. It is, therefore, important that every attempt be made to find better prophylactic and therapeutic approaches to this problem.

Exogenous fibrinolytic agents such as streptokinase (CHRISTENSEN and MACLEOD, 1945) have proven to be of great value as a short-term therapy for acute vascular thrombosis. However, they were found to be unsuitable for the purpose of prophylaxis. This might be possible through pharmacologic enhancement of fibrinolysis. A vast number of substances diminish the inhibitory effect of antiplasmin and thus enhance the fibrinolytic activity in vitro but only a few of these substances increase fibrinolysis in vivo. Some AST belong to this latter group. The interrelation between the effect of AST on fibrinolysis and their anabolic-androgenic activity is obscure, however, there might be a structure-activity relationship, since most effects of AST on fibrinolysis described hitherto were produced by 17α-alkylated compounds.

### Pharmacodynamic Studies

Large doses of testosterone enhance blood fibrinolytic activity (FEARNLEY and CHAKRABARTI, 1962; WINTHER, 1967). Similar effects were found with the oral AST methenolone, ethylestrenol (FEARNLEY and CHAKRABARTI, 1964), and stanazolol (MENON et al., 1970). A significant increase in plasma plasminogen activator levels in men with ischemic heart disease was produced by short-term treatment with the 17α-alkylated AST ethylestrenol, norethandrolone, methandrostenediol, oxymetholone and methandienone, not, however, with methenolone acetate (WALKER et al., 1975a). In the case of testosterone (WINTHER, 1970), methenolone, and ethylestrenol (FEARNLEY and CHAKRABARTI, 1964), as with sulphonyl-ureas and biguanides, resistance develops to the enhancement of fibrinolytic activity within a relatively short time. Apparently no such resistance develops with stanazolol (DAVIDSON et al., 1972a, b), and HEDNER and NILSSON did not find this phenomenon for ethylestrenol at the level of the vessel wall in patients with deficient fibrinolysis suffering from recurrent venous thrombosis (HEDNER et al., 1975a, b). To overcome the phenomenon of the development of resistance to the fibrinolytic effect of AST, a combination of ethylestrenol with the biguanides metformin or phenformin (in timed-release capsules), respectively, was tested and it was found that these combinations produce a substantial and sustained increase in fibrinolytic activity in a very high percentage of patients (FEARNLEY et al., 1967; FEARNLEY et al., 1968; CHAKRABARTI and FEARNLEY, 1972; CHAKRABARTI and FEARNLEY, 1974). Data indicate that these drugs raise the level of plasminogen activator in the blood either by increasing its rate of formation (FEARNLEY et al., 1971) or by decreasing its rate of removal. The data are not compatible with the hypothesis (ROBERTS and BURKAT, 1968) that this combination is not fibrinolytic but only renders fibrin more susceptible to lysis. The increase in fibrinolytic activity is accompanied by an increase in fibrin degradation products in the serum, indicating a break-down of fibrin and/or fibrinogen and giving strength to the assumption that pharmacologic enhancement of fibrinolysis as measured in vitro is indeed present in vivo. This effect was only temporary in healthy persons and in some atherosclerotic patients, but was sustained in patients with rheumatoid arthritis (FEARNLEY et al., 1969). The observed effects of phenformin combined with ethylestrenol on the fibrinolytic activity and on fibrinogen as well as on platelet adhesiveness (see later) have been confirmed (ISACSON and NILSSON, 1970; BRUHN et al., 1974) and supplemented by new findings: the combination also increases plasminogen and the plasminogen activator content in the wall of superficial vessels. This is thought to occur through stimulation of plasminogen activator synthesis in the vein walls (NILSSON and ISACSON, 1972; NILSSON et al., 1975). According to recent findings, ethylestrenol without phenformin is also capable of producing this effect and development of resistance to this substance was not observed (HEDNER et al., 1975a, b). In contrast to the observations of FEARNLEY et al. (1969) no degradation products of fibrin and/or fibrinogen were found. Here, however, different detection methods were employed and increases in concentration smaller than 10 μg/ml could not be excluded (NILSSON and ISACSON, 1972; BRUHN et al., 1974). Similarly no changes in the serum levels of fibrinogen degradation products were observed in men with ischemic heart disease treated during periods of four weeks with one of six different AST (WALKER et al., 1975a).

The effects of seven AST upon three $\alpha$-globulin antiprotease inhibitors of thrombin and plasmin were investigated in men with ischemic heart disease (WALKER et al., 1975b); a single dose of each AST was tested, the equivalence of the doses was not established. Ethylestrenol, methandrostenediol, oxymetholone, stanazolol, and norethandrolone produced a significant increase of the plasma antithrombin III level; treatment with norethandrolone, methylandrostenediol, oxymetholone, methandienone, and stanazolol caused a significant fall in plasma $\alpha_2$-macroglobulin levels which was insignificant with ethylestrenol (also described by NILSSON et al., 1975); norethandrolone, ethylestrenol as well as methylandrostenediol therapy was associated with a rise in plasma $\alpha_1$-antitrypsin. Methenolone acetate, the only tested non-17$\alpha$-alkylated steroid, failed to cause significant changes in the levels of any of the antiproteases. These results support the assumption that the increase in plasma fibrinolytic activity produced by 17$\alpha$-alkylated AST is rather a primary effect than secondary to low-grade disseminated intravascular coagulation.

The combination of either phenformin or metformin with ethylestrenol produces a sustained reduction of plasma fibrinogen particularly in patients with high levels (CHAKRABARTI and FEARNLEY, 1972). However, only the combination of ethylestrenol with phenformin causes a small reduction of the serum cholesterol level (FEARNLEY and CHAKRABARTI, 1968; CHAKRABARTI and FEARNLEY, 1972; see also SHARMA et al., 1975) and—with a lag time of about 3 months—induces a sustained reduction of platelet adhesiveness in the majority of patients (CHAKRABARTI et al., 1967; CHAKRABARTI, 1972). Metformin plus ethylestrenol instead tends to raise blood cholesterol levels and increases platelet adhesiveness (FEARNLEY and CHAKRABARTI, 1968). Consequently, metformin is considered to be inferior to phenformin and is no longer under extensive study.

No statistically significant difference was found between phenformin plus ethylestrenol or phenformin plus stanazolol (MENON et al., 1970) on the one hand, and either phenformin or metformin plus stanazolol on the other (MENON, 1971) as far as enhancement of fibrinolytic activity is concerned. Both combinations were also similarly effective in reducing plasma fibrinogen and cholesterol (CHAKRABARTI et al., 1970; CISWICKA-SZNAJDERMAN et al., 1974; not confirmed by DAVIDSON et al., 1972a). Stanazolol with phenformin failed, however, to maintain the reduced platelet adhesiveness. This is thought to be a qualitative rather than a quantitative difference with ethylestrenol (CHAKRABARTI et al., 1970; CHAKRABARTI, 1972). Data strongly suggest that stanazolol by itself is capable of enhancing fibrinolytic activity in patients with ischemic heart disease. This action of stanazolol is not improved by combining it with phenformin (DAVIDSON et al., 1972a; DAVIDSON et al., 1972b).

The fact that combinations of phenformin and an oral AST increase fibrinolytic activity, lower plasma fibrinogen and serum cholesterol, and, in some instances (as is the case with ethylestrenol), reduce platelet adhesiveness, suggested their use in pathologic conditions in which deposition of platelets and fibrin on the vessel wall would be detrimental (FEARNLEY et al., 1967; FEARNLEY et al., 1968; FEARNLEY et al., 1969; BRUHN et al., 1974; BIELAWIEC et al., 1975; VERSTRAETE, 1975). So far, the clinical trials have been inadequate with respect to number of patients, extent, and methodology and no final conclusion can be drawn, but the results are encouraging enough to be reported. It is hoped that they will stimulate further research in this very interesting field which is of great practical importance.

### Therapeutic Trials (Table 7)

In *occlusive vascular disease* such as myocardial infarction, angina pectoris, intermittent claudication, recurrent idiopathic deep venous thrombosis, superficial thrombophlebitis, retinal venous thrombosis, and obstructive retinopathy in diabetes, treatment with phenformin plus ethylestrenol—in a few trials also with phenformin plus stanazolol—has been reported to have prophylactic and/or therapeutic value.

The following effects were observed. Within 3 months a significant increase of spontaneous fibrinolytic activity and of the capacity to release plasminogen activator from the vein walls on venous occlusion, both activities remaining normal for the rest of the treatment period of 48 months. An increase in initially low plasminogen activator content in the wall of superficial veins after 3 months, and a normalization thereof in almost all patients after 12 months of treatment was observed (NILSSON et al., 1975). Clinically the combined treatment reduced the incidence of venous thrombosis markedly in patients suffering from recurrent idiopathic venous thrombosis (NILSSON et al., 1975), improved coronary heart disease and obliterative arteriosclerosis (BIELAWIEC et al., 1975), and in addition recurrence of nonfatal infarction appeared to be less frequent than in untreated patients (CHAKRABARTI and FEARNLEY, 1972).

A substantial number of patients with a *cutaneous vasculitis* and decreased fibrinolytic activity who failed to react to conventional methods of treatment responded favorably to treatment with phenformin plus ethylestrenol or phenformin plus stanazolol.

Patients with *Behcet's syndrome*, that is, with extensive deep venous thrombosis, and/or multiple ulcerations, responded dramatically to treatment with phenformin plus ethylestrenol. It is assumed that a decrease in platelet adhesiveness and/or an influence on fibrinolysis in the affected tissue might have helped to bring about the

Table 7. Therapeutic trials with ethylestrenol (ee), stanazolol (stan), phenformin (ph), and metformin (mf) in patients with vascular diseases

| Author(s) | No. of patients | Treatment | | Therapeutic result | Diagnosis |
|---|---|---|---|---|---|
| | | drugs | duration | | |
| FEARNLEY et al. (1967, 1968) | 33 | ee + ph or mf | 21 months | + + + | Occlusive vascular disease incl. myocardial infarction |
| BIELAWIEC et al. (1975) | 75 | stan + ph | >6 months | + + + | |
| CHAKRABARTI and FEARNLEY (1972) | 42 | ee + ph | 36 months | + + | Myocardial infarction |
| NILSSON et al. (1972)<br>NILSSON et al. (1975) | 75 | ee + ph | 3–48 months | + + + | Recurrent idiopathic deep venous thrombosis (60); superficial thrombophlebitis (10); retinal venous thrombosis (5) |
| CUNLIFFE (1968)<br>MENON et al. (1969)<br>MENON (1970) | 9 | ee | 10 weeks–10 months | + + + | Vasculitis, venous ulcers |
| MENON (1972) | 3 | ee + ph | 8 months | + + + | Recurrent thrombophlebitis, thromboembolism |
| CUNLIFFE et al. (1973) | 15 | ee or stan + ph | not specified | + + + | Cutaneous vasculitis |
| DODMAN et al. (1973) | 12 | ee or stan + ph | 9 months | + + | |
| TAN and COPEMAN (1974) | 1 | ee + ph | > 3 months | + + + | Leg ulcerations, blood hypercoagulability |
| DELANEY and BLACK (1975) | 1 | ee + ph | apparently 1 year or longer | + + + | Malignant atrophic papulosis (KÖHLMEIER-DEGOS) |

Table 7 (continued)

| Author(s) | No. of patients | Treatment: drugs | Treatment: duration | Therapeutic result | Diagnosis |
|---|---|---|---|---|---|
| CUNLIFFE and MENON (1969) | 1 | ee + ph | 4 months | + + + | |
| KIRK and HANDLEY (1972) | 3 | ee + ph | 4 weeks–12 months | + + + | Behçet's syndrome |
| MENON (1972) | 1 | ee + ph | 6 years | + + + | |
| GILLIAM et al. (1974) | 5 | ee + ph | years | + + + | Vasculitis of "atrophie blanche" |
| FEARNLEY and CHAKRABARTI (1969) | 4 | ee + ph | up to 3 years | + + + | Raynaud's disease |
| MENON and CUNLIFFE (1969) | 22 | ee + ph | 2–12 months | + | (Raynaud's phenomenon) |
| FEARNLEY (1974) | "several" | ee + ph | | + + + | |
| MICHAELSSON et al. (1974) | 7 | ee + ph | | + + | Allergic pupura |
| MANDAL and MCNULTY (1971) | 1 | ee + ph | 84 days | + + + | Hemolytic-uremic syndrome |
| ALLENBY et al. (1973) | not specif. | ee | 3 weeks | — | Postoperative |
| FOSSARD et al. (1974) | 45 | ee + ph | 4 weeks | — | venous thrombosis |
| FEARNLEY et al. (1966) | 20 | ee + ph | 5–14 months | + + | Severe rheumatoid arthritis |

Therapeutic result: + + + = very good; + + = good; + = moderate or in few cases; — = none.

long-term improvement obtained in some patients. No effect was reported in one patient with incomplete Behcet syndrome by NETHERCOTT and LESTER (1974).

In *vasculitis of "atrophie blanche"* (livedo reticularis with summer ulcerations), a condition usually highly resistant to therapy, treatment with phenformin plus ethylestrenol proved to be very successful; patients became and remained asymptomatic during this therapy.

Encouraging therapeutic results were also seen in some patients with primary or secondary *Raynaud's disease*. It is assumed that these drugs cause a therapeutic defibrination and in addition act via the reduction of platelet adhesiveness which implies that a satisfactory therapeutic effect should not be expected before 2 to 3 months of treatment (FEARNLEY and CHAKRABARTI, 1969). No benefit of phenformin plus ethylestrenol was seen in *scleroderma* patients with the typical changes of Raynaud's syndrome (PAOLINO et al., 1972).

A beneficial influence of phenformin and ethylestrenol resulting in decreased purpuric reactions after provocation has been observed in patients suffering from *allergic purpura* (Schönlein-Henoch purpura).

The platelet stabilizing effect and the fibrinolytic effect of this combination might also have been responsible for the impressive improvement of one patient with *hemolytic-uremic syndrome* who failed to respond to heparin therapy.

Assuming that the postoperative reduction of fibrinolytic activity (fibrinolytic "shutdown") might be related to *postoperative venous thrombosis*, the prophylactic effect of phenformin plus ethylestrenol was investigated. The combination was given 6 weeks preoperatively and 1 week postoperatively. This regimen increased the fibrinolytic activity, reduced the postoperative fall in this activity, and induced a significant increase in fibrin degradation products (BROWN et al., 1971). Administration of ethylestrenol alone for 3 weeks before surgery did not prevent the postoperative reduction of fibrinolytic activity and had no influence on the incidence of venous thrombosis. This failure could be due to the development of resistance to the effect of ethylestrenol (FEARNLEY, 1973). However, the combination of ethylestrenol with phenformin given 3 weeks pre- and 1 week postoperatively—despite exerting the expected effect on fibrinolytic activity (although not increasing fibrin degradation products)—also failed to reduce the incidence of postoperative deep venous thrombosis (monitored with the help of the $^{125}$I-fibrinogen test). This was found in a double-blind randomized trial with 95 patients; the disappointing result might be explained by the shortened period of treatment and the fact that a variety of factors are probably responsible for the development of postoperative venous thrombosis (FEARNLEY et al., 1967; GILES, 1974; NILSSON et al., 1975).

In chronic inflammatory conditions such as rheumatoid arthritis in which deposition and persistence of fibrin may contribute to continuing inflammation, the combined treatment was claimed to lead to clinical improvement due to increased fibrinolytic activity and reduction of fibrinogen levels, both processes acting to reduce fibrin at the site of inflammation.

With the exception of one case of lactic acidosis (NILSSON et al., 1975), no serious side effects were recorded when this combined treatment was given. Phenformin caused weight loss and gastrointestinal symptoms (in most cases transient) in some 15% of the patients. However, this necessitates stopping treatment in only about 5% of the patients.

Table 8. Controlled therapeutic trials with AST in patients with gastrointestinal disorders. All authors reported therapeutic effects

| Author(s) | Country | No. of patients receiving AST | Duration of treatment | Disorder(s) | Comparison with |
|---|---|---|---|---|---|
| CORONA (1964) | Mexico | 6 | 30 days | Post-gastrectomy | 9 placebo-treated patients |
| AKZHIGITOV et al. (1971) | USSR | 30 | 7 days | Acute pancreatitis | 40 controls |
| KRUMS (1971) | USSR | 5 | 20 days | Chronic enterocolitis | 7 controls |
| EKISENINA et al. (1972) | USSR | 104 | 1 month | Chronic enterocolitis | 178 controls |
| LESCHINSKY et al. (1972) | USSR | 364 | 21–24 days | Gastric and duodenal ulcer | 20 controls |
| DEINEKO et al. (1973) | USSR | 56 | 20–28 days | Gastric and duodenal ulcer | 53 controls |
| KOMAROV et al. (1973) | USSR | 121 | | Peptic ulcer | |
| VAKHRUSHEV and TRUSOV (1975) | USSR | 98 | 18–40 days | Peptic ulcer | 99 and 100 patients respectively[a] |
| SINITZIN and AGAFONOWA (1975) | USSR | 75 | a few weeks | Peptic ulcer | 74 and 107 patients respectively[a] |

[a] The two groups were treated with different therapies.

Ethylestrenol can induce sodium and fluid retention as well as menstrual irregularities such as amenorrhea.

*Conclusion:* Oral AST either alone or in combination with a biguanide can produce a sustained increase in fibrinolytic activity and a reduction of fibrinogen, cholesterol, and platelet adhesiveness. These properties definitely merit further investigation for their prophylactic and therapeutic usefulness. The results obtained so far in diseases in which disturbed fibrinolysis, vascular damage, and thrombus formation are present are quite encouraging.

## 5. Gastrointestinal Disorders

In several gastrointestinal disorders protein deficiency and debilitation due to persistent weight loss can sometimes become so severe as to result in inanition and cachexia. These conditions can result from anorexia, decreased caloric intake, maldigestion caused by inadequate enzymatic actions and/or dumping, protein-losing enteropathy, and faulty absorption as sequelae to motility disorders. Many therapeutic measures such as diets and drugs have been employed in an effort to manage this situation. The additional administration of various AST has also been tried as supportive palliative therapy. In order to avoid poor absorption, some investigators prefer parenteral preparations.

Table 9. Uncontrolled therapeutic trials with AST in patients with gastrointestinal disorders. All authors reported therapeutic effects

| Author(s) | Country | No. of patients receiving AST | Duration of treatment | Disorder(s) |
|---|---|---|---|---|
| DELMONT (1961) | France | 5 | | Nontropical sprue |
| BERKOWITZ (1962) | USA | 26 | 5–20 weeks | Postgastrectomy (25) Ulcerative colitis (1) |
| KASICH (1963) | USA | 43 | 12–15 weeks | Duodenal ulcer, dumping syndrome, chronic ulcerative colitis, regional ileitis |
| DITTRICH and SEIFERT (1964) | Austria | 75 | 1–12 months | Gastric resection |
| LAMPITELLA (1964) | Italy | 22 | 7–8 weeks | Peptic ulcer |
| LODS (1964) | France | 5 | 3 months | Postgastrectomy |
| DUBARRY et al. (1965) | France | 28 | | Postgastrectomy, peptic ulcer, various other gastro-intestinal diseases |
| PEVCHIKH (1969) | USSR | 156 | 3–4 weeks | Chronic diseases of gastro-intestinal tract |
| VAKHRUSHEV (1971) | USSR | 104 | 18–34 days | Gastric and duodenal ulcer |
| VEREVKIN and PEVCHIKH (1972) | USSR | 130 | 20 days | Peptic ulcer |
| LOMAZOVA (1973) | USSR | 34 | | Gastric and duodenal ulcer |
| TRUSOV and VAKHRUSHEV (1973) | USSR | 64 | | Chronic gastritis |
| TRUSOV et al. (1974) | USSR | 83 | 22–38 days | Gastric and duodenal ulcer |

The published material contains data from more than a thousand patients. These data were compiled from controlled and uncontrolled trials which were frequently performed with a number of different drugs, making it difficult to estimate the contribution of the AST to the overall effect. Undoubtedly, however, some therapeutic effects are a consequence of the protein anabolic and general actions of AST.

AST are reported to have been beneficial additives to the conventional treatment of chronic gastritis, peptic ulcers, postgastrectomy dumping syndrome, acute and chronic pancreatitis, regional ileitis, ulcerative colitis, chronic enterocolitis, and nontropical sprue (Tables 8 and 9). A therapeutic effect on weight loss, hypalbuminemia, anorexia, and the poor general condition of the patients was consistently found. According to DITTRICH and SEIFERT (1964), the earlier postoperative anabolic therapy was begun in patients with gastric resection the greater the likelihood of weight gain; the contrary was found by BÄR and USBECK (1971) in gastrectomized patients, who did not confirm this observation in a later trial, however (BÄR and USBECK, 1974). Several investigators emphasize a rapid pain-relieving action in peptic ulcers (AKHZIGITOV et al., 1971; VAKHRUSHEV, 1971; DEINEKO et al., 1973; TRUSOV et al., 1974; VAKHRUSHEV and TRUSOV, 1975), other investigators found a particularly favorable influence upon appetite and psyche in conditions where psychosomatic factors seemed to have been of importance. The usefulness of AST in preventing "postgastrectomy bone disease", osteomalacia, and/or osteoporosis following gastrectomy, has not yet been extensively studied.

Clearly, most of the described therapeutic effects have also been observed in conditions other than gastrointestinal disorders, yet their utilization for such disorders has been unduly neglected by the majority of investigators.

*Conclusion:* The protein anabolic effects, as well as the general effects of AST on appetite, sense of well-being, and psychic attitude might be used in the conservative symptomatic therapy of gastrointestinal disorders. Particular indications are diseases accompanied by protein deficiency, weight loss, and the inability to gain weight by other means.

## 6. Muscular Disorders

The interest in AST therapy in muscular disorders is restricted to the muscular dystrophies and one form of muscular atrophy. Muscular dystrophies are hereditary myopathies that is disorders of the muscles without evidence of neural disease, and characterized by painless, symmetrical, progressive, unremitting weakness, and atrophy of the muscle. The term muscular atrophy is, however, reserved for muscular weakness and wasting secondary to neural degeneration; among the various forms only infantile spinal progressive muscular atrophy (often familial) is of interest in this context. Polymyositis, a diffuse inflammatory disorder of striated muscle, has been mentioned only occasionally in connection with AST.

The pathogenesis of muscular dystrophies is unknown. Disturbance of the protein, carbohydrate and creatine metabolism, a defective cell membrane, changes of myofibrils, and loss of myoglobin are discussed as being of pathogenetic importance. Increased creatinuria and high serum levels of muscle enzymes, reduced creatininuria, degeneration of muscle fibers, replacement of muscle mass by adipose and fibrous connective tissue, and alterations of the electromyogram are found.

An effective drug therapy for these relentlessly progressive diseases does not exist. Consequently, countless preparations and therapeutic measures were and are still being tried; among these drugs are vitamins C, E, and B, sugars, amino acids, ATP, cytochrome C, liver and pancreas extracts, malaria induction, pilocarpineepinephrine, and excessive doses of insulin. The role of AST, which were also thought to influence muscular dystrophies and atrophies, merits a brief review.

### Therapeutic Trials

Androgens with and without concomitant physiotherapy have been tried since 1940, with claims of moderate success by various investigators (Krüskemper, 1968; Voss, 1973b), but their powerful virilizing action prohibited widespread clinical use. With the introduction of AST, interest in this group of drugs revived and reports of therapeutic trials appeared in rapidly increasing numbers. Oral and parenteral preparations were used, mostly in doses high enough to induce androgenic side effects of variable, and often considerable intensity. This seemed acceptable since therapeutic benefit in the majority of patients, most of whom were not severely ill, was reported (Table 10). There was subjective and/or objective improvement of functional capacity (fatigability, energy, muscular power), a positive effect on creatinuria (see also

Table 10. Uncontrolled trials with AST in patients with muscular dystrophy (md) and infantile spinal progressive muscular atrophy (i. sp. ma). All authors reported therapeutic effects

| Author(s) | Country | No. of patients | Diagnosis | Duration of treatment | Therapeutic effects on | | | | |
|---|---|---|---|---|---|---|---|---|---|
| | | | | | Functional capacity | Creatin(in)-uria, muscle and/or serum enzymes | Protein metabolism | Muscle histology | EMG |
| DE TONI, E. (1959); DE TONI, G. (1959) | Italy | 3 | md | 10 days and more | + | + | | | |
| BEKÉNY et al. (1960) | Hungary | 24 | md | 6 weeks– 6 months | + | | + | | |
| LAURIJSSENS et al. (1960) | Belgium | 1 | md | not specif. | + | | | | |
| STUR (1960) | Austria | 3 | i. sp. ma | 2–6 months | + | | | | + |
| BROWN and JAMES (1961) | USA | 16 | md | 3 months | + | + | | + | |
| HABERLER and PATEISKY (1961); PATEISKY et al. (1961) | Austria | 12 | md | up to 12 weeks | + | + | + | | (+) |
| HUBER (1961) | Austria | 2 | i. sp. ma | not specif. | (+) | | | | |
| KAESER (1961) | Switzerland | 11 | md | 1–12 months | + | | | | |
| PROBST (1961) | Denmark | 10<br>4 | md<br>i. sp. ma | 4 months | + | + | | | |
| DE TONI (1961) | Italy | 10 | md | not specif. | + | + | | | |
| HANTSCHMANN et al. (1962); MATZELT (1964); MERTENS (1964) | W. Germany | 25 | md | 9 months | + | + | + | | |
| WORM (1962) | Denmark | 1 | md | 1 year | + | | | + | |

Table 10 (continued)

| Author(s) | Country | No. of patients | Diagnosis | Duration of treatment | Therapeutic effects on | | | | |
|---|---|---|---|---|---|---|---|---|---|
| | | | | | Functional capacity | Creatin(in)-uria, muscle and/or serum enzymes | Protein metabolism | Muscle histology | EMG |
| BIERICH et al. (1962); BIERICH (1964) | W.Germany | 10 | md | >1 year | + | + | | | |
| BECKMANN (1964; 1965) | W.Germany | 24 | md | several 4 week treatments | (+) | | + | | |
| DOWBEN[a] (1963; 1965) | USA | 51 | md | up to $3^1/_2$ years | + | + | | | |
| VASELLA et al. (1964) | Switzerland | 27 | md | not specified | (+) | | | | |
| ZWOZUKIAK and ZGORZALEWICZ (1967) | Poland | 20 | md | 5 weeks | + | + | | | |
| BADALYN et al.[b] (1975) | USSR | 306 | md i.sp.ma | 30 days and more twice a year | + | | | | |

EMG: Electromyography.
Therapeutic effect: + = marked or frequently observed; (+) = doubtful or occasionally observed.
[a] Additional treatment with digitoxin.
[b] Additional treatment with other drugs.

Bourdakos and Wolf, 1962), muscle and serum enzymes, a reduction of urinary nitrogen loss, and amelioration of hypoproteinemia, positive changes in muscle histology (however, see also Fröhlich et al., 1962; Nyfos, 1962; Mertens, 1964) and alterations of electromyographic signs (not seen by Fröhlich et al., 1962). The increasing enthusiasm for the apparent efficacy of AST was supported by animal experiments. Mice of the Bar Harbor strain suffering from a hereditary malady closely resembling human muscular dystrophy survived significantly longer when treated with AST (Dowben, 1959; Dowben, 1962; Watzman et al., 1962; Borgman, 1963; not confirmed by Rabinowitz et al., 1963). This effect could be increased by exercise. A positive influence of an AST was also seen in mice, which were experimentally made dystrophic (Igić et al., 1969).

Unfortunately, assessment of a therapeutic effect in muscular dystrophy, a pathologic condition with so many appearances and variations, meets with great difficulties. There is frequently a lack of objectivity in the patients and those observing them, e.g., the parents. The emotional response to a new therapy in such a discouraging disease is often considerable and very disturbing. Overestimation and/or misinterpretation of seemingly reliable objective measurements of muscle strength and performance, of biochemical, histologic, and electromyographic findings is apparently not confined to less experienced investigators. Last but not least, grave diagnostic difficulties seem to be common even among specialists (Dowben, 1962; Dowben et al., 1965).

Obviously, evaluation of drugs in muscular dystrophy is an excellent example of the necessity to follow the principles of therapeutic investigation (Kuhn, 1966). This is clearly seen in a number of reports from (mainly) controlled clinical trials with AST in muscular dystrophy (Table 11) all of which ended with discouraging results as far as objective and lasting improvement was concerned, although weight gain (which is not always to the benefit of the patient) and improvement in well-being were observed. In some forms of dystrophy an even more rapid deterioration appears to have occurred during (Barwick et al., 1963) or after discontinuation of therapy (Dowben and Perlstein, 1961; Dowben, 1962).

In the evaluation of different clinical observations one is frequently confronted with the fact that very persuasive clinical observations are not confirmed by controlled trials. It is then difficult to draw any conclusions especially if disease states and trial conditions are not comparable. On the one hand, one is faced with undisputable results of valid comparative studies, and on the other with differing clinical observations. One cannot, of course, discount the results of controlled trials. However, in the case of AST treatment of muscular dystrophy there is an impressive body of clinical observations in uncontrolled trials, many coming from reliable sources. One should not question the value of these observations as such. What is in dispute is the demonstration of the cause-and-effect relationship between them and the administration of the drug. This is exactly what the controlled trial is intended to demonstrate. Among the more important sources of possible error may have been: neglect of the pronounced general effects produced by the high doses of AST used, in particular on subjective improvement; too few patients; employment of unsuitable methods for monitoring objective changes; and extreme diagnostic difficulties. Any of these factors could have contributed to results that were not reproducible in controlled trials.

Table 11. Clinical trials with AST in patients with muscular dystrophy without therapeutic results

| Author(s) | Country | No. of patients | Duration of treatment | Remarks |
|---|---|---|---|---|
| HUBER (1961) | Austria | 2 | Not specified | |
| DOWBEN (1962) | USA | not specified | „For very long periods of time" | |
| FRÖHLICH et al. (1962) | Switzerland | 9 | 3 months | Subjective improvement of functional capacity not confirmed by objective measurement |
| BARWICK et al. (1963) | Great Britain | 38 | 3 months | Double-blind controlled trial; comparison of 2 AST and placebo |
| LUNDBERG and BLOM (1963) | Sweden | 14 | 1–12 months | Treatment with either an oral or a parenteral AST preparation. Subjective improvement was not confirmed by objective measurement |
| DANOWSKI et al. (1965) | USA | 21 | 5 months or longer | Comparison of an oral and a parenteral AST preparation. Both groups also received digitoxin |
| FOWLER et al. (1965) | USA | 64 | 12 months | Comparative treatment with AST (19), digitoxin (7), AST + digitoxin (18), AST + digitoxin + exercise (14), and placebo (6) |
| HEYCK et al. (1965) | W. Germany | 51 | 3–5 months | Comparative treatment with AST (20) and AST + digitoxin (31) |
| MILHORAT (1965) | USA | 365 | Not specified | Performed in 10 US clinics |

The above not only holds true for studies with AST alone, but apparently also where treatment consisted of AST and cardiac glucosides. The latter were added in the hope that they would decrease abnormal permeability of the cell membrane. The positive results reported by DOWBEN (1963, 1965) were not reproduced by DANOWSKI et al. (1965), FOWLER et al. (1965), and a deleterious effect of this combination on the wasting process of the muscle was claimed by HEYCK et al. (1965).

It has yet to be confirmed that patients with polymyositis, resistant to high dose corticosteroid therapy, become sensitive after treatment with AST (DOWBEN et al., 1965).

*Conclusion:* Apart from a general effect AST have not been proven to exert any therapeutic action in patients suffering from muscular dystrophy or infantile spinal progressive muscular atrophy. Consequently, these drugs cannot be recommended for the routine treatment of these conditions, which are apparently only susceptible to physiotherapy, treatment with mechanical aids, and occasionally orthopedic operations. In cases where AST are used for their general effects, administration should preferably be restricted to short periods. Weight gain not correlated with an increase in muscle strength should be avoided.

## 7. Pulmonary Disorders

*Pulmonary tuberculosis* is a disease which requires not only treatment with highly effective specific drugs but also auxiliary therapeutic measures to improve the physical and psychic condition of the severely affected patient. The chronic infectious process invariably leads to debilitation, protein deficiency, and a progressive decrease in the general condition, a situation which must be counteracted by supportive measures. One of these could be the administration of AST because of their effect upon protein synthesis and catabolism as well as other effects.

### Therapeutic Trials

AST were used quite extensively in pulmonary tuberculosis and an appreciable number of observations collected in controlled (Table 12) and uncontrolled clinical trials (Table 13) have been reported. Irrespective of the preparation and form of administration, adequate dosing always produced a rapid and significant weight gain in emaciated patients. In some investigations the gain was less marked after 4–6 weeks of treatment (wearing-off phenomenon), although a positive effect on body weight persisted in most instances during the whole period of treatment. The weight gain was maintained after the end of treatment; in some cases there was even a further gain in weight. This maintenance of weight gain suggests that not water retention but true weight gain through tissue formation had been achieved.

Concomitantly, promotion of nitrogen retention, reduction of the erythrocyte sedimentation rate, and increase in serum protein were seen. Patients who responded to AST with an initial drop in serum protein (FANFANI et al., 1962; SAITO and SAITO, 1966) while nitrogen was retained might indicate that tissue protein was spared at the expense of serum protein. Occasionally the impression was gained that a positive influence on the specific tuberculous process occurred, that is, a better response to specific treatment than in patients without additional anabolic therapy and a possi-

Table 12. Controlled clinical trials with AST in patients with pulmonary tuberculosis

| Author(s) | Country | No. of patients rec. AST | Duration of treatment | Therapeutic effects on | | | | Remarks |
|---|---|---|---|---|---|---|---|---|
| | | | | General condition | Appetite, weight gain | N balance or serum proteins | Erythrocyte sedimentation rate | |
| Saito et al. (1966) | Japan | 20 | 18 weeks | + | + | + | + | Comparison of two AST (10 + 10 patients) with a placebo (5 patients) |
| Wiles and Mackintosh (1962) | S. Africa | 36 | 10 weeks | | + | | | Double-blind comparison of different doses of one AST with placebo (34 controls) |
| Porsio (1966) | Italy | 33 | 2–4 months | + | + | | | Double-blind crossover trial with one AST and a placebo |
| Saito and Saito (1966) | Japan | 25 | 9–18 weeks | + | | + | | Comparison of 2 AST (8 + 8 patients) with a placebo (4 patients); open trial in 9 patients |
| Porsio (1968) | Italy | 20 | 2 months | + | + | | | Double-blind crossover comparison of 2 AST |

Table 13. Uncontrolled therapeutic trials with AST in patients with pulmonary tuberculosis

| Author(s) | Country | No. of patients rec. AST | Duration of treatment | Therapeutic effects on | | | | | Remarks |
|---|---|---|---|---|---|---|---|---|---|
| | | | | General condition | Appetite, weight gain | Serum proteins | Erythrocyte sedimentation rate | Tuberculous lesion | |
| VIALLIER (1958) | France | 37 | 10 days– >4 weeks | | + | + | | | |
| FANTOLI et al. (1959) | Italy | 19 | 4–16 weeks | + | + | + | | | |
| MORIN (1960) | France | 31 | $1^1/_2$–3 months | + | + | + | | + | |
| POZZETTO (1961) | France | 25 | 2–14 weeks | + | + | | | | Different doses and duration of treatment with one AST |
| DE REN et al. (1961) | France | 15 | several months | | + | | | | |
| FANFANI et al. (1962) | Italy | 30 | 8–9 weeks | + | + | | + | | 2 AST were used |
| FOSSATI (1963) | Libya | 10 | 2×15 days | + | + | | | | |
| YAMASE and HATANAKA (1963) | Japan | 16 | | + | + | | | + | |
| KÜHN (1964) | W. Germany | 41 | 4–15 weeks | + | + | | | + | |
| BÜRGI and REGLI (1969) | Switzerland | 49 | 2 months | + | + | | + | | Combination preparation containing an AST, vitamins, and minerals |
| VADASZ (1969) | Hungary | 57 | 1 single injection | + | + | | | + | Open comparison of 48 AST-treated patients with 48 controls |
| RABIER (1971) | Senegal | 25 | 75 days | + | + | | | | Open comparison with 25 controls |
| REBORA TOGNO (1971) | Mexico | 32 | 80 days | + | + | | | | Open comparison with 15 controls (conservative and surgical treatment) |

ble stimulation of the aspecific fibroblastic transformation of productive residual lesions. Last but not least the favorable changes in appetite, mood, and physical vigor added to the overall beneficial effect of this symptomatic treatment.

The use of AST in *other pulmonary diseases* as reported sporadically (MAGGI and BARBI, 1960; SANTALÓ and ECHEVARNE, 1963; BOTTERO and LOPS, 1964; SOKOLOVA et al., 1965; OLIVERI, 1967; BALMES et al., 1969) is likewise based on the known protein anabolic, and other effects of the drugs and not on possible specific effects. The assumption of an antiasthmatic effect (FALLIERS et al., 1963; AEPLI, 1964; KERREBIJN and DELVER, 1969) has still to be confirmed.

*Conclusion:* AST may be used as an adjuvant to the therapeutic measures usually employed in the treatment of chronic pulmonary tuberculosis and other chronic diseases of the lung.

## B. Metabolic Diseases

### 1. Diabetes Mellitus

Diabetes mellitus is a condition characterized by alterations not only in the carbohydrate metabolism but also in the protein and fat metabolism. Due to the availability of adequate treatment, the early phases of this disease are much less dangerous today than they were in the past, more potential diabetics are being born, and life expectancy of diabetic patients has increased enormously (BEAUMONT and HOLLOWS, 1972). As a consequence, many more patients live long enough to develop the late diabetic syndrome which nowadays threatens the diabetic patient with the various manifestations of diabetic angiopathy.

Unfortunately, neither an optimal prophylaxis nor a curative therapy has been found to prevent or treat late diabetic syndrome. In particular, the degenerative changes taking place predominantly in the capillaries, have become a major concern. In its worst form the diabetic capillaropathy causes the fatal intercapillary glomerulosclerosis, Kimmelstiel-Wilson's disease. Diabetic capillaropathy is also the main contributing cause for diabetic neuromyopathy and at least for some forms of diabetic retinopathy. Rigid and consistent clinical control of the metabolic disturbances present in diabetes by means of diet and antidiabetic drug treatment in conjunction with other therapeutic measures are effective in delaying the clinical onset and in slowing down progression of the diabetic microangiopathy; however, they cannot prevent it or induce improvement once the damage has been done.

Diabetic retinopathy is only one of many manifestations of diabetic angiopathy and can be expected in about 60% of all diabetic patients (DARDENNE, 1961) and in more than 70% of those suffering from the disease for more than 15–20 years (OOSTERHUIS, 1960). It is one of the frequent causes of acquired blindness (BEAUMONT and HOLLOWS, 1972) accounting for 10–15% of all cases. Progression of the disease is not steady but is characterized by an inevitable deterioration in the course of years.

Biochemical investigations in patients with diabetic retinopathy have revealed not only alterations in carbohydrate metabolism but also pathologic changes in protein- and lipid metabolism: Hypoproteinemia and hypoalbuminemia, increased levels of the $\alpha_2$, $\beta_1$ and $\gamma$-globulin fractions, glycoproteins, hexosamines, mucopolysaccharides, and $\beta$-lipoproteins, most of the nonprotein bound amino acids, the glutamic-oxaloacetic transaminase, neutral fats, and to some extent also the triglycerides and nonesterified fatty acids (DARDENNE, 1961; HOUTSMULLER, 1961; HOFMANN and ZEPPEK, 1963; HOUTSMULLER and VAN POPPEL, 1963; HOUTSMULLER and HENKES, 1964). In addition, multiple abnormalities of the fibrinolytic system were reported such as reduced spontaneous fibrinolytic activity, reduced fibrinolytic response to venous occlusion and elevated fibrinogen level (BEAUMONT and HOLLOWS, 1972; ALMÉR et al., 1975).

The progressive vascular damage is made obvious by an increasing capillary fragility throughout the body. In the eye, signs of the various forms of retinopathy develop (microaneurysms, hemorrhages, exudates, vitreous hemorrhages, and subsequent fibrous tissue proliferation and new vessel formation). These manifest themselves clinically as disturbances in visual acuity and retinal function, detectable by the sensitive methods of retinal fluorescein angiography, electroretinography, and electrooculography (HOUTSMULLER, 1961; HOFMANN and ZEPPEK, 1963; HOUTSMULLER and VAN POPPEL, 1963; HOUTSMULLER and HENKES, 1964; BEAUMONT and HOLLOWS, 1972).

It is not yet certain whether the biochemical disturbances contribute to the development of the capillary damage or the latter should be regarded as the consequence of a more fundamental disturbance (GOEBEL, 1975). However, it has been demonstrated that the metabolic alterations correlate positively with the degenerative vascular processes (HOUTSMULLER, 1961; HOFMANN and ZEPPEK, 1963; HOUTSMULLER and VAN POPPEL, 1963). These metabolic alterations are therefore used as additional indicators for monitoring drug effects in diabetic retinopathy.

It should be remembered that diabetic angiopathy is currently thought to be responsible for 50–70% of all deaths in diabetic patients. One then appreciates why countless preparations and treatment regimens have been tried in the last decades. Many of these were claimed to be also effective in diabetic retinopathy. Controlled trials and sufficient clinical experience have proven them to be useless for this purpose (DARDENNE, 1959). The use of AST in the treatment of diabetic retinopathy should be reviewed since some fairly extensive trials have been performed, several of them adhering well to the principles of therapeutic investigation (TAINTER, 1962; KOPERA, 1964; BABEL, 1965; KOPERA, 1965; SCHUMACHER, 1967).

## Therapeutic Trials

The first observation of a therapeutic success of an androgen in diabetic retinopathy was reported by GRÖNBERG and SVANTESON (1951). A preliminary investigation with testosterone propionate was performed by American research workers (SASKIN et al., 1951), and the Dutch ophthalmologist VALK (1959) was first to publish encouraging results using an AST with reduced androgenic qualities. These results stimulated a large number of clinical investigations.

AST, mainly nandrolone esters, have been tested—only occasionally together with drugs other than antidiabetic compounds—in uncontrolled as well as in a few controlled double-blind comparative clinical trials. The investigations were conducted over periods varying from a few months to more than 5 years in several hundred patients suffering from diabetes and different stages of retinopathy. The

Table 14. Clinical trials with AST in patients with diabetic retinopathy. All authors reported therapeutic effects

| Author(s) | Country | No. of patients receiving AST | Duration of treatment | Therapeutic effects on | | | | | |
|---|---|---|---|---|---|---|---|---|---|
| | | | | Vision | Fundus | ERG | Carbo-hydrate | Protein and lipid | Capillary fragility |
| | | | | | | | metabolism | | |
| DARDENNE (1961)[a] (see also DARDENNE, 1964) | W. Germany | 48 | $1^1/_2$ years | | + | | + | + | |
| HOUTSMULLER and HENKES (1960); HOUTSMULLER (1961); HOUTSMULLER et al. (1963); HOUTSMULLER and VAN POPPEL (1963); HOUTSMULLER (1965)[a] | Netherlands | 140 | 2–4 years | + | + | + | + | + | + |
| WEISSEL (1962) | Austria | 7 | 1–4 weeks | | + | | + | | |
| HOFMANN and ZEPPEK (1963) | Austria | 50 | 4 years | + | + | | | + | |
| VALK (1963; 1964) | Netherlands | 43 | $>1$ year | | + | | | | |
| WALSH et al. (1963) | USA | 9 | $1^1/_2$–6 months | + | + | | | | |
| BRETAN et al. (1964) | Hungary | 21 | 3 months | | + | | | | |
| FABRYKANT et al. (1964) FABRYKANT (1962) | USA | 46 | 6 months–$5^1/_2$ years | + | + | | | | |
| HOUTSMULLER and HENKES[b] (1964) | Netherlands | 42 | 4 years | + | | | + | + | + |
| PESTALOZZI (1964) | Switzerland | 25 | 3–21 months | | + | | | | |
| ROUHER and SERPIN (1964) | France | 72 | | | + | | | | |
| CICCHITTI et al. (1966) | Argentine | 14 | 2 months | | + | | | | |
| PETROCHILOS (1967) | Greece | 26 | | + | + | | | | |
| KLIACZKO et al. (1970) | Poland | 43 | | + | + | | | | |
| KLYACHKO et al.[c] (1971) | USSR | 45 | | + | + | | + | | |
| HAMERSKI (1974) | Poland | 86 | 2–3 weeks repeatedly | + | + | | | | |

ERG: Electroretinography.
[a] Controlled trial with some of the patients.
[b] Double-blind comparative trial.
[c] Comparison of 4 different treatment groups, two of them receiving also AST.

following effects were reported (Table 14): improved glucose tolerance; reduced insulin requirement (see also FOZ, 1965; LANDON et al., 1965; SOLER-ARGILAGA, 1964; not observed by KINK, 1962; MOLNAR et al., 1965; LITTMANN, 1974); restored responsiveness in insulin-resistant patients (WEISSEL, 1962); normalization or improvement of disturbances of the protein and lipid metabolism (also reported by KINK, 1962; not seen by OJALA, 1964); reduction of the increased capillary fragility, in some cases full restoration of normal conditions; and improvement of defective fibrinolysis (BEAUMONT and HOLLOWS, 1972). Arrest of the otherwise progressive deterioration of the vision, fundus, and retinal function for the duration of treatment was observed in the majority of patients, improvement was found in a smaller proportion of the patients. These positive results were more pronounced in earlier stages and in non-proliferative forms of diabetic retinopathy and correlated well with the improvement of several biochemical alterations. Therapeutic effects on the ophthalmological symptoms were found in only few cases by WEISSEL (1962) and OJALA (1964) while other investigators obtained hardly any or no improvement at all (Table 15). Finally, the benefit of the general anabolic effect was emphasized by some investigators (WEISSEL, 1962; VASIUKOVA et al., 1969).

These results are obviously conflicting. The role played in the pathogenesis of diabetic retinopathy by biochemical alterations and the value of influencing these alterations by treatment in predicting the prognosis are still questionable. There is no general agreement as to the extent and duration of the therapeutic effects of AST on diabetic retinopathy. There is, however, an impressive body of apparently positive data which is contradicted by relatively few reports of disappointing experiences. Consequently, a final opinion concerning the value of AST for the treatment of diabetic retinopathy cannot be formed on the basis of the current literature. It is, therefore, difficult to understand why AST have seldom been proposed or even mentioned as a therapeutic alternative in recommendations for the treatment of diabetic retinopathy during the last years (BRÜCKNER, 1965; BOUR et al., 1967; KUTSCHERA, 1970; HENKES et al., 1971; HONEGGER and BOHLEBER, 1971; SEVIN and CUENDET, 1971; KÖNIG, 1972; GERRITZEN, 1975; RUPRECHT, 1975). Obviously experts have lost confidence in these steroids for this indication, at least in Western countries. Investigators from Eastern Europe do not seem to share this attitude. In that part of the world AST are still, or even increasingly, recommended for the treatment of diabetic retinopathy (VASIUKOVA et al., 1969; EFIMOV, 1970; KLIACZKO et al., 1970; KLYACHKO et al., 1971; EFIMOV and KARABUN, 1972; HAMERSKI, 1974; VASIUKOVA and MARGOLIS, 1975; JEFIMOV, 1975; see also SCUDERI and SBORGIA, 1974).

It is not possible to give a generally acceptable answer to the question as to whether AST have been a disappointment and a failure in the therapy of diabetic retinopathy. Undoubtedly the clinical experiences of Western ophthalmologists seem to point to this conclusion. However, it should be realized that the problem is still not sufficiently investigated with regard to dose, treatment regimen, choice of drug, possible selection of particularly suitable patients or special forms of the retinopathy, and other factors which might be relevant. The suggestion that AST could play a role in the prophylaxis of diabetic retinopathy has been made repeatedly (HOUTSMULLER and VAN POPPEL, 1963; KOPERA, 1964) but this possibility has yet to be investigated. The present state of affairs is still disappointing in view of the current

Table 15. Clinical trials of AST in patients with diabetic retinopathy which gave poor or no therapeutic results

| Author(s) | Country | No. of patients receiving AST | Duration of treatment | Remarks (quotations from the respective publications) |
|---|---|---|---|---|
| OOSTERHUIS and LOEWER-SIEGER (1962a, b); OOSTERHUIS (1963) | Netherlands | 20 | 7–18 months | After treatment of 10 patients: "In one patient considerable improvement [of diabetic retinopathy] in 2 other patients the recurrent vitreous haemorrhages ceased to occur ... more investigations needed for an ultimate evaluation." In further 10 patients no therapeutic results were obtained |
| OJALA (1964) | Finland | 21 | 5–15 months | "Material is too small to allow ... any conclusion. The biochemical examination shows that there is no response ... may have some effect in early stages of diabetic retinopathy" |
| MOLNAR et al. (1965) | France | 13 | Not specified | No therapeutic effect on diabetic retinopathy and metabolic alterations |
| HIEKKALA (1965); FORSIUS (1967) | Finland | 21 | 1–2 years | Comparative trial. "It was very difficult afterwards to compare the two groups." The AST "may be capable of slowing down the development of diabetic retinopathy" |
| HUNTER et al. (1967) | Gt. Britain | 33 eyes [a] | 6–22 months | Double-blind comparison with placebo-treated controls. "No beneficial effect was observed in the (AST-)treated group." The increase of hemorrhages in the group receiving the AST was not statistically significant |
| LANGSCH (1970) | E. Germany | 20 | 18 days | No effect on carbohydrate metabolism. According to previous experience AST would have disappointed in diabetic retinopathy |

[a] Trial in 49 patients, data of 78 eyes analyzed (45 eyes of placebo group and 33 of AST-treated group).

therapeutic situation, which remains rather hopeless despite the additional employment of photocoagulation, pituitary ablation, cytostatics, growth hormone inhibitors, pyridinolcarbamate, and antilipidemic substances, which are all therapies of mainly palliative value only (KÜCHLE, 1967; HENKES et al., 1971; TIRKINA et al., 1974; CHENG et al., 1975).

Very little can be said about a possible effect of AST on other manifestations of diabetic capillaropathy. AST have been mentioned only sporadically (although hopefully) in connection with suggestions for the treatment of diabetic neuropathy (KINK, 1962; PRIKHOZAN, 1968; ROBERTS, 1969) and nephropathy (WEISSEL, 1962; VALK and VERGROESEN, 1969).

*Conclusion:* Interesting results have been achieved with AST in patients suffering from diabetic retinopathy. The clinical relevance of some of the effects of these drugs remains questionable. Therapeutic efficacy is still assumed by some ophthalmologists whereas this view is no longer held by others. Taking into account the appreciable number of encouraging results reported in the literature, and the still desperate therapeutic situation of diabetic retinopathy as well as the small risk involved in the use of AST, it appears advisable to devote some further investigations to this problem (see also VAISRUB, 1975). At this time there are no objections to the use of AST in this condition, as an adjunct to treatments considered to be of value, but a positive recommendation cannot be given on the basis of the published investigations. When AST are administered, readjustment of the antidiabetic drug dosage might become necessary.

## 2. Hyperlipemias

The major stimulus for the increasing interest in hyperlipoproteinemias is the fact that a linear relationship exists between cholesterol levels and coronary heart disease. The main reason for the world-wide interest in lipid-lowering substances is the, as yet, unproven hope that reduction of lipid levels will substantially diminish the risk of coronary heart disease by preventing deterioration of arterial circulation (Lancet Edit., 1975; WHYTE, 1975).

A number of highly effective hypolipidemic compounds have been found and some of them, for example nicotinic acid, clofibrate, d-thyroxine, cholestyramine, and other bile acid sequestrants are in common use as an adjunct to dietary control of hyperlipoproteinemia. The few AST that were shown to exert an influence upon plasma triglycerides, triglyceride clearing enzymes as well as the mechanism of triglyceride clearance, do not yet belong to the generally accepted lipid lowering drugs; these AST are still in an investigational stage and are employed therapeutically only on a rather small scale.

In contrast to most other indications for which AST can be recommended as a pharmacologic entity, by virtue of their protein anabolic and other effects, their action upon lipid metabolism seems to be unrelated to the androgenic or anabolic properties. They may either increase or decrease $\beta$-lipoprotein and cholesterol levels and each of them seems to have a different pattern of action on the various lipid substances (FIEGEL et al., 1966; FIEGEL and KUKWA, 1968; KRÜSKEMPER, 1972; VOSS, 1973d).

### Pharmacodynamic Studies

*Androsterone* was reported to produce a fall in blood cholesterol, phospholipids, and triglycerides when given intramuscularly. This effect could not be obtained after oral administration (HELLMANN et al., 1959; COHEN et al., 1961). This may be one of the reasons why few further studies were devoted to this substance.

*Methandrostenolone* lowers triglycerides in normal individuals (SKIKANTIA et al., 1967) and reduces the total lipids by lowering the triglyceride and cholesterol concentrations and raising plasma post-heparin lipolytic activity (KUO and QUICK, 1970). However, methandrostenolone can also cause an increase in cholesterol levels when these levels are initially borderline high (STONE, 1963, 1964) or hypercholesteremic (JOSE and MITCHELL, 1964).

*Fluoxymesterone*, tested in comparative controlled trials, produced a significant fall only in the triglycerides in healthy individuals, whereas initial effects upon total lipids and cholesterol disappeared after prolonged drug administration (NOVAK, 1971; NOVAK et al., 1974).

Significant lowering of serum lipid concentrations were reported for *Norbolethone* (ALBANESE, 1965), *Norethandrolone* (WARDLE, 1974), *Oxymesterone* (ZENEROLI, 1970), *Ethylestrenol* (IGARI et al., 1966; SAITO et al., 1966; see also SHARMA et al., 1975), *Stanazolol* (FURMAN et al., 1963; KUO and QUICK, 1970; CIŚWICKA-SZNAJDERMAN et al., used together with Phenformin, 1974), *Nandrolone esters* (HOUTSMULLER, 1961; HOFMANN and ZEPPEK, 1963; HOUTSMULLER et al., 1963; HOUTSMULLER and HENKES, 1964) and *Oxymetholone* (BERKOWITZ et al., 1962; SAITO et al., 1966; YOSHIDA, 1965), although a lack of influence of oxymetholone on blood lipids was also described (ANTONINI and VERDI, 1961).

*Oxandrolone* is the AST which was most extensively studied in this respect. It produces a fall in the triglycerides and cholesterol levels in normal subjects (GHERONDACHE et al., 1967) and is highly effective in lowering plasma lipids (particularly triglycerides, less so cholesterol) in hyperlipidemia (BEAUMONT et al., 1970) types III, IV, V (FEARNLEY and CHAKRABARTI, 1964; ALBANESE, 1965; SACHS and WOLFMAN, 1968; ALBANESE et al., 1970; LEBEAU and ROBINSON, 1970; GLUECK, 1971a, b; ZORRILLA et al., 1973; DOYLE et al., 1974; LITTMANN, 1974; OLSSON et al., 1974; EHNHOLM et al., 1975). The onset of action is in some cases delayed for 10–14 days. Oxandrolone can alter abnormal lipid phenotypes to more normal lipid distributions (LEBEAU and ROBINSON, 1970; ROBINSON and LEBEAU, 1970). It can also normalize abnormalities of the carbohydrate content of plasma apoproteins in patients with hyperlipidemia type III (FONTAINE and MALMENDIER, 1975).

The mechanism of action of oxandrolone is still incompletely understood. It has been proposed that it inhibits excess triglyceride synthesis (SACHS and WOLFMAN, 1968). Kinetic studies revealed that it increases triglyceride clearance (GLUECK et al., 1973). Within 48 h after initiation of drug administration, the postheparin lipolytic enzyme activities (post-heparin lipolytic activity, triglyceride lipase, delipidated triglyceride lipase, monoglyceride hydrolase) and the fractional turnover rate of very low density lipoprotein triglycerides increases while the half-life is shortened during oxandrolone treatment (GLUECK et al., 1972; EHNHOLM et al., 1975; FALLAT and GLUECK, 1976). This might suggest that reduction of triglyceride levels is in part the result of an increased peripheral triglyceride hydrolysis (GLUECK et al., 1970a, b; GLUECK, 1971a, b; FALLAT, 1974; GLUECK et al., 1975). This is consistent with the finding that intravenous fat tolerance increases significantly under oxandrolone treatment (OLSSON et al., 1974). In familial hyperchylomicronemia (type I hyperlipoproteinemia) and very low postheparin lipase activity, a normalization of the latter was produced, although oxandrolone appeared unable to correct the deficiency in the chylomicron clearing mechanism and left the hypertriglyceridemia unaffected

(FAERGEMAN and DAMGAARD-FEDERSEN, 1973). On the other hand methandrostenolone as well as stanazolol were reported to be effective in the same condition (KUO and QUICK, 1970).

The significant reduction of cholesterol in very low and low density lipoproteins can be compensated by an increase in high density lipoproteins so that the net effect might result in no perceptible change in cholesterol level as such (ZORRILLA et al., 1973; OLSSON et al., 1974). By prolonged administration, oxandrolone can also increase cholesterol and $\beta$-lipoprotein cholesterol to abnormal levels through release of $\beta$-lipoprotein arising from the rapid disappearance of very low density lipoproteins. It is, therefore, advisable to monitor both triglycerides and cholesterol in patients receiving prolonged oxandrolone treatment. It is further suggested that hyperlipidemias with high cholesterol levels, such as types II B and III, be treated with a combination of oxandrolone and other drugs more effective in lowering cholesterol levels (OLSSON et al., 1974).

Investigations of coagulation parameters in patients with hypertriglyceridemia receiving oxandrolone treatment suggest that the AST reduce rather than potentiate any tendency to "hypercoagulability" in these patients who were initially predisposed to coronary or peripheral vascular thrombosis. The only effects observed on the clotting mechanism were a marked fall of plasma fibrinogen and a decrease in initially elevated levels of factors VII and X (GLUECK and GLUECK, 1970; GLUECK and GLUECK, 1973).

*Conclusion:* Some AST exert a powerful hypolipidemic action in various types of hyperlipidemia. This property seems to be dependent on the structure of the AST rather than on hormonal effects. By far the greatest experience has been obtained with oxandrolone. At present it cannot be said whether any of these compounds have sufficient hypolipidemic efficacy and/or selectivity in action to compare favorably with the accepted antilipidemic drugs. If AST are used, it should be remembered that they are still in an investigational stage for this purpose and the necessary precautions have to be observed.

### 3. Other Metabolic Disorders

The role played by AST in the management of several genetic metabolic anomalies is discussed in Chapter VII.B. In addition, the following entities are of interest:

*Phenylketonuria:* AST were found to be successful in lowering the experimentally elevated phenylalanine levels in blood, liver, and brain of animals (REISS et al., 1966a) and the blood level of phenylalanine in phenylketonuric patients (REISS et al., 1966b; PANIGADA et al., 1968; BRAMBILLA et al., 1975).

*Cystinosis:* Quite encouraging therapeutic results were obtained with AST as supportive treatment (WEBER and HAGGE, 1963; BERGER et al., 1964; HAGGENMÜLLER, 1964; BAUER and ANTENER, 1966; AHRENDS, 1967; WILK and RAUSCHSCHOTT, 1971).

*Marfan's Syndrome:* This inherited connective tissue disorder is possibly due to a defect in collagen maturation; it might react favorably to AST. JONES et al. (1969) described one patient in whom long-term therapy with an anabolic drug markedly reduced hydroxyproline excretion and increased psychologic and physical well-being. A positive effect was also reported by SCHINDLER (1975).

*Gargoylism (Hurler's Syndrome):* Reduction of excessive urinary excretion of chondroitin sulphate and some objective as well as subjective clinical improvement were obtained with an AST in three patients by NANIVADEKAR and NANIVADEKAR (1966); confirmed by SPENCER (1967).

*Variegate Porphyria:* In a report on cutaneous porphyria in an English family (BARNES et al., 1965) it was mentioned that abdominal and skin symptoms increased during treatment with norethandrolone and decreased during stanazolol treatment; the possibility of a causal connection between administration of AST and changes in clinical symptomatology was not discussed.

## C. Deficiency States and Catabolic Conditions

### 1. Emaciation, Malnutrition

The causes of emaciation and malnutrition are manifold. Constitutional or exogenous factors, psychoses, anorexia nervosa, diseases of the central nervous system, malabsorption, metabolic disorders, malignant processes, chronic infectious diseases, chronic organic diseases, and chronic intoxications are perhaps the most important causes of protein deficiency. It is readily understandable that the curative therapy for the underlying diseases is different for these various conditions but that the additional symptomatic treatment to relieve the deficiency state (whatever its origin) is in principle the same. This treatment generally consists of nutritional and enzymatic substitution, appetite stimulants, and anabolic steroids.

Therapeutic Investigations

In a number of controlled (Table 16) as well as uncontrolled (Table 17) clinical trials in various disease conditions, AST were shown to increase abnormally low body weight significantly when food supply was adequate. A negative result was reported by BAKER et al. (1964) who performed a double-blind multicenter trial in 63 geriatric patients comparing 2 months treatment with an AST and a placebo.

Anabolic treatment was also found to improve appetite and strength as well as psychic and physical well-being in patients with weight lag or loss. The positive influence upon psychologic attitude seems to be particularly beneficial in the treatment of debilitated mentally ill patients, ulcerative colitis, and geriatric patients.

Obviously the effect upon protein metabolism (stimulation of synthesis and/or reduction of catabolism) and a complex of the various other actions exerted by AST will be responsible for the net therapeutic result.

*Conclusion:* AST are therapeutically effective and can be recommended as adjuvants to the symptomatic therapy of emaciation and malnutrition.

### 2. Surgery

Patients undergoing surgery are frequently in rather poor physical condition due to the disease process. Postoperatively their protein metabolism goes through a very marked catabolic phase lasting 3–5 days. This period is commonly followed by a distinctly anabolic phase which usually lasts for about 5–10 days. The duration and

Table 16. Controlled clinical trials with AST in (mainly) underweight patients and one in healthy volunteers. All investigators reported positive results on weight gain and/or N balance

| Author(s) | Country | No. of patients receiving AST | Duration of treatment | Diagnosis | Remarks |
|---|---|---|---|---|---|
| SMITH and JOHNSON[a] (1961) | USA | 30 | 4, 8, 12 weeks | Postmenopausal females with chronic brain syndromes or variants of schizophrenia | Double-blind comparison of 3 different durations of treatment with an AST and a placebo |
| JOHNSON (1962)[a] | USA | 52 | up to 6 weeks | Underweight, mental illness (42) | Comparison of various doses and different durations of treatment with 2 AST preparations and a placebo |
| PAGE (1962)[a] | USA | 74 | 12 weeks | Mentally ill tuberculous patients | Double-blind crossover within-subject comparison of two AST versus placebo |
| GÜRTLER (1963) | Switzerland | 20 | 26–40 days | Pulmonary tuberculosis | Comparison of two doses of one AST |
| KOLODNY (1963) | USA | 70 | about 60 days | Weight loss or weight lag | Within-subject comparison of AST-placebo-AST |
| KORST et al. (1963) | USA | 54 | 6 weeks | Underweight | Blind comparison of 2 doses of test drug, a reference preparation and placebo |
| ALBANESE et al. (1965) | USA | 32 | Balance study, periods up to 6 weeks | Neurologic or vascular disorders, fractures, carcinoma | Determination of duration and degree of N retaining effects of 5 AST |
| KRIZEK et al. (1969) | Czechoslovakia | 9 | 3 days | Healthy volunteers | Comparative study in experimental starvation |

[a] Also referred to in Table 20.

Table 17. Uncontrolled clinical trials with AST in underweight patients. All authors reported therapeutic results

| Author(s) | Country | No. of patients receiving AST | Duration of treatment | Diagnosis |
|---|---|---|---|---|
| GONNARD et al. (1957) | France | 14 | 1 month | Various chronic disease conditions |
| PAYNE (1959) | USA | 35 | 1–16 weeks | Various chronic disease conditions |
| ALY (1961) | W.Germany | 140 | 2 weeks to several months | Various chronic disease conditions |
| HOCK (1961) | USA | 50 | $10^1/_2$ weeks | Geriatric patients |
| HUGUET et al. (1961) | France | 120 | 40 days | Malignant diseases |
| CRAIG (1962) | USA | 61 | 2–24 weeks | Various chronic disease conditions |
| PERGOLA (1962) | Argentine | 200 | 10–15 days | Underweight patients |
| WOLF (1962) | Austria | 50 | not specified | Various chronic disease conditions |
| HARN (1963) | USA | 53 | 3–60 days | Various diseases of underweight |
| MACMASTER et al. (1963) | USA | 47 | 29–60 days | Mentally disturbed underweight |
| SNOW (1963) | USA | 72 | 3 weeks to 1 year | Underweight, chronic diseases |
| JUNGCK and GREENBLATT (1964) | USA | 87 | 6–12 weeks | Underweight patients |
| ROSNICK (1964) | USA | 105 | 3 days to 6 months | Various chronic disease conditions |
| TEC (1971; 1974) | USA | 12 | not specified | Anorexia nervosa |

intensity of both postoperative periods depend on the age of the patient, the preoperative condition, the disease, the anesthesia, the extent of the surgical trauma and of blood loss, the postoperative care, nutrition, as well as the presence or absence of complications. All of these factors ultimately determine the length of convalescence and completeness of recovery. Thus improvement of the preoperative condition, keeping operative trauma to a minimum, and optimal postoperative care will shorten convalescene and positively influence recovery. To achieve this, special patient care, appropriate dietary and drug regimens are very helpful. AST have been used with benefit both pre- and postoperatively.

For obvious reasons the effect of AST in patients in a poor physical condition before operation has not been investigated extensively. However, clinical experience is apparently in agreement with the assumption, based on animal experiments, that debilitated, cachectic patients should benefit from the protein anabolic and the other effects of these compounds. At any rate, their presurgical administration has become general practice at many hospitals. Clearly, parenteral preparations may be preferred for this purpose.

Postoperative protein metabolism after different kinds of surgical procedures has been investigated repeatedly by means of nitrogen balance studies, determination of serum- and urinary N concentrations (Table 18), and by protein turnover measurements with the help of $^{14}$C-leucine (O'KEEFE et al., 1974; see also PORTER and KNIGHT, 1970).

In the immediate postoperative or posttraumatic catabolic phase, reduced food intake, exudate formation, the increasing glucocorticosteroid production (MOORE and BALL, 1952; DUDLEY et al., 1959; HARTENBACH, 1962; AMMEDICK et al., 1968), an abnormally high protein breakdown (BRENNAN et al., 1975), and a consistent fall in protein synthesis (O'KEEFE et al., 1974) cause loss of body protein and increased amino acid oxidation leading to excessive urinary nitrogen excretion. The second or anabolic phase is the physiologic reaction of the organism to the catabolic phase. One can successfully support it by adequate nutrition and the administration of AST. Thus anabolic compounds would be expected to be effective in both the catabolic and anabolic period of recovery after operation.

### Therapeutic Trials

A high intake of nutrients, administered either by the oral or intravenous route, can significantly improve, although not prevent, the immediate postoperative negative nitrogen balance. During this phase similar improvements can also be produced by anabolic steroids. In both cases, the best results were obtained in well-nourished patients who underwent minor surgery. In major surgery this effect does not become significant before the 4th postoperative day. The combination of both measures—high-protein diet or use of parenteral nutrients together with administration of an AST—applied post- and/or preoperatively, produces the largest and most rapid improvement in nitrogen balance. This was clearly demonstrated in controlled comparative trials (Table 18); the report by WENZEL and BÜNTE (1968) is not in agreement with these findings. The fact that in patients treated with AST and high protein intake, serum nitrogen concentration was found to be distinctly lower than in patients receiving high-protein diet alone, suggests that the supplied amino acids are incorporated into the cell protein with the help of the anabolic drugs. In addition to the positive influence of AST on postoperative protein catabolism and on the reactive anabolic phase, these compounds also exert a number of other therapeutic effects in patients recovering from surgical procedures. This can be assumed on the basis of reported observations made in mostly uncontrolled clinical trials (Table 19). All investigators recorded an increase of appetite, weight gain, a better psychologic attitude, and spontaneous activity of the patients. Some investigators emphasized the benefit of improvements in erythropoiesis and wound healing (see also the results of animal experiments reported by WATTS et al., 1965; EHRLICH and HUNT, 1969; HERRMANN and WOODWARD, 1972; and described by VOSS, 1973c). By their favorable effect on the composition of serum proteins and perhaps on the general resistance (see below under Immunology), AST contribute to a speedier and more complete convalescence from surgery with a possible reduction in the incidence of complications.

*In orthopedics* use has been made of the favorable effect of AST on bone and mineral metabolism, which is described in detail in Chapter VII.A. Furthermore a most encouraging therapeutic effect of AST, significantly better than the conventional treatment alone, has been reported in patients with chondromalacia patellae (DARRACOTT, 1973).

Table 18. Investigations of postoperative protein metabolism under influence of AST. In all studies therapeutic effects of AST were reported

| Author(s) | Country | Patient(s) | | Parameter(s) followed | Nature of trial |
|---|---|---|---|---|---|
| | | No. receiving AST (No. on trial) | Diagnosis and/or surgical procedure | | |
| OTTER (1960) | Netherlands | 10 | Gastrectomy | N balance | Open comparison, 2 groups receiving AST and high caloric diet with high or low N content |
| BUCHNER (1961) | Austria | 37 (44) | Injuries | N balance, urinary N excretion | Open comparison with controls |
| REHN and SPINNER (1962) | W. Germany | 40 | Bone fractures, major surgery, chronic infections | N balance, serum proteins | Open comparison of 2 AST preparations |
| KONRAD et al. (1967)<br>AMMERDICK et al. (1968) | W. Germany<br>W. Germany | 30 (60) | Lung, gastric, and bone surgery | N balance, serum N concentration | Open comparison, 6 groups fed normal or high-protein diet, 3 receiving AST |
| PATNAIK et al. (1969) | India | 22 (44) | Minor and major surgery | N balance | Open comparison with controls |
| SCHOPPE (1971) | W. Germany | 15 (30) | Lung surgery | N balance, serum N concentration, serum protein | Open comparison, 2 groups with normal diet, one receiving AST |
| BALAKHOVSKY et al. (1972) | USSR | 34 (62) | Orthopedic surgery | Urinary N, P, Na, K excretion | Open comparison with controls |
| TWEEDLE et al. (1972, 1973) | Great Britain | 15 (30) | Vagotomy, pyloroplasty | N balance | Open comparison, 5 groups fed with low or high caloric diet, 3 receiving a single injection of AST |

Table 19. Therapeutic trials with AST as adjuvant in convalescence from surgery and chronic illness

| Author(s) | Country | No. of patients receiving AST | Duration of treatment | Diagnosis | Therapeutic effects on | | | Remarks |
|---|---|---|---|---|---|---|---|---|
| | | | | | Appetite, weight gain, physical and psychic state | Erythropoiesis | Wound healing | |
| ENDLER (1961) | Austria | 250 | 3–18 months | Rheumatoid diseases, osteoporosis | + | | | |
| FIXSON (1962) | W. Germany | 25 | up to 70 days | Gynecologic surgery | + | + | | Open comparison with control group |
| KUZELL (1962) | USA | 320 | 10 days to 33 months | Rheumatoid diseases, osteoporosis | + | + | | Comparison of 2 doses of an AST |
| SILK (1962) | USA | 134 | 4–32 weeks | Major surgery | + | + | + | Open comparison with lipancreatin- or placebo-treated groups |
| FAM (1963) | Egypt | 40 | 4 months | Urologic surgery | + | | + | |
| IOVINE and GIERSON (1963) | USA | 30 | 150 days | Debilitated emphysema patients | + | | | Comparison with control group; weight gain statistically significant |
| MATSUMOTO et al. (1963) | Japan | not specif. | not specified | Bone, joint lesions | + | | + | |
| SIEGMUND (1963) | Austria | 141 | not specified | Rheumatoid diseases | + | | | |
| BUTTENBERG (1964) | W. Germany | 36 | up to 19 months | Carcinoma of ovary | + | | | |
| DAVILA FLORES et al. (1964) | Mexico | 30 | 30 days | Major surgery | + | | | |
| FIXSON (1964) | W. Germany | 15 | not specified | Gynecologic surgery | + | + | | Comparison with control group |

Table 19 (continued)

| Author(s) | Country | No. of patients receiving AST | Duration of treatment | Diagnosis | Therapeutic effects on | | | Remarks |
|---|---|---|---|---|---|---|---|---|
| | | | | | Appetite, weight gain, physical and psychic state | Erythropoiesis | Wound healing | |
| Gokhale et al. (1964) | India | 3 | 3–4 months | Leprosy | + | + | | |
| Stuhlfauth and Frost (1964) | W. Germany | 82 | 5 weeks or longer | Hypotension | + | | | AST plus desoxycorticosterone acetate |
| Adda (1965) | Algeria | 50 | up to 2 years | Immobilizing diseases | + | | | |
| Brügel (1965) | W. Germany | 208 | not specified | Underweight gastrectomies | + | | | Comparative trial; 8 groups receiving uniform diet, 6 groups treated with different AST |
| Louyot et al. (1965) | France | 63 | 2–6 months | Rheumatoid diseases, other chronic illness | + | + | | |
| Fauchet (1966) | France | 45 | 51–111 days | Immobilizing diseases | + | | | |
| Stepan and Bremova (1967) | Czechoslowakia | 70 | | Rheumatoid diseases | + | | | N retention recorded |
| Vermeulen (1968) | Netherlands | 45 | 3–9 months | Various organ diseases | + | + | | |
| Ahlendorf and Kob (1970) | E. Germany | 119 | 5–7 weeks | Gastrectomy | + | | | Weight increase statistically significant |
| Geoffroy et al. (1970) | France | 8 | not specified | Chronic illness | + | | | |
| Gomes et al. (1970) | Brazil | 20 | 4–11 days | Major surgery | | | + | Open comparison with control group |
| Bär and Usbeck (1971) | E. Germany | 60 | 3–6 months | Gastrectomy | + | | | |
| Solis (1971) | Argentine | 17 | 21 days | Urologic surgery | + | | | |
| Bär and Usbeck (1974) | E. Germany | 195 | 3–6 months, sometimes with 2–3 months interval | Gastrectomy | + | | | |

*Conclusion:* Clinical application of AST in major surgery appears to be justified and recommendable. In the preoperative phase they can ameliorate the physical and psychic state of the patient, thereby increasing ability to cope with disease and surgical trauma. In the postoperative period AST together with high caloric, high-protein diet can reduce protein catabolism by stimulating protein synthesis and inhibiting excessive breakdown. Thus they may speed up convalescence, improve restitution, and lower the risk of complications. A beneficial effect might also be obtained in some special indications in orthopedics.

### 3. Convalescence

The use of AST need not be confined to the period immediately after an operation. It can also be extended with benefit to the later phases of restitution after surgery, to convalescence from chronic infectious diseases and chronic illness of other origin, to recovery from traumata, and restitution after long periods of immobilization. These drugs can help to diminish protein loss, increase protein synthesis, promote retention of calcium, potassium, phosphorus, and stimulate tissue build-up. At the same time they may improve erythropoiesis, immunologic status, resistance to infections, mental attitude, and physical activity of the patient, thus creating conditions favorable for recovery. This can be inferred from published data collected in mainly uncontrolled clinical trials (Table 19).

*Conclusion:* The addition of AST to established therapeutic procedures can be recommended in convalescence from illness, injuries, and immobilization.

## D. Other Conditions

### 1. Neurology

There are a great number of neurologic disorders which have many common features such as muscle weakness, paralysis leading to immobilization with consequent protein loss, demineralization of bone tissue, malnutrition, and the concomitant worsening of the general condition of the individual.

Therapeutic Trials

Reports on the use of anabolic-androgenic steroids have been published indicating that the well-being, physical vigor, muscle tone, activity, protein metabolism, and nutritional status of the patient can be improved (COOPER et al., 1951; HAGENBUCHNER, 1962; BOURGEOIS and VINCENT, 1964; BORS and COMARR, 1971). The fact that AST have such beneficial effects in neurologic disorders is not surprising, since one simply makes use of the protein anabolic and other effects of the drugs as in so many other comparable situations (e.g., deficiency states and catabolic conditions).

One interesting observation is perhaps worth special attention, since it is a good example of a circadian rhythm in drug effectiveness. In paraplegic patients the hypercalciuria resulting from the demineralization of bone tissue due to immobilization, a process which is characteristic of the early months of the illness, follows the normal circadian rhythm found in immobilized healthy subjects; most of the calcium lost is excreted around noon. AST, which are known to induce calcium retention, can be used for antidemineralization therapy but short-acting oral preparations must be properly timed in order to be effective. This was dramatically demonstrated by MOORE-EDE and BURR (1973) who obtained a significant reduction of calcium excretion only when the drug was administered at 6 a.m. Similar trials with long-acting steroids which may be particularly suited to this indication, have not yet appeared in the literature. Effects of AST which so far cannot be explained are reported from trials in cases of multiple sclerosis, "early" diabetic neuropathy, and postherpetic neuralgia.

In a double-blind comparison of an AST with a placebo significant improvement of gait, strength in the arms and hands, together with a most welcome elevation of mood and psychic attitude was recorded in patients suffering from multiple sclerosis (MALL et al., 1968). A similar effect was seen in another trial (CENDROWSKI and KURAN, 1972). However, JOHNSON and MCLELLAN (1972) claim that therapeutic effects of AST are absent in this disease.

In euthyroid patients with the clinical symptoms of a diabetic-type neuropathy, the altered neuromuscular function is often detectable only with the help of the timed repetitive ankle jerk test. Treatment with an AST brought about a dose-related improvement or normalization of the prolonged, timed repetitive ankle jerk response and concomitant improvement or normalization of conduction velocity where this was initially reduced. This effect usually preceded clinical remission of severe neuropathic discomfort by one or more weeks (ROBERTS, 1969). Similar effects of an AST on the knee jerk reflex were observed in a double-blind comparative study with a placebo in healthy athletes; in the steroid treated group one obtained a significantly faster motor time component and an increased latency component of the greatly reduced total reflex time (ARIEL and SAVILLE, 1972).

ROBERTS (1965) found an AST effective for the prophylaxis and treatment of postherpetic neuralgia of less than 1 year's duration which was resistant to other therapy.

The available data on the effects of AST in multiple sclerosis, diabetic neuropathy, and postherpetic neuralgia are certainly encouraging but are not yet sufficient to permit a final conclusion to be made. More research, and confirmation of the reported findings, is required before an answer can be given as to the usefulness and mechanism of action of anabolic compounds in these indications. In fact such studies should be encouraged, because effective remedies for these conditions are needed and the risks involved in AST therapy are minimal.

*Conclusion:* The use of AST as part of symptomatic therapy in neurologic diseases might be of benefit in cases of immobilization, severe muscle weakness or paralysis, debilitation, and emaciation. To what extent reproducible effects can be obtained in indications such as multiple sclerosis, diabetic neuropathy, and postherpetic neuralgia must still await further investigation.

## 2. Psychiatry

Medical management of mental patients is often very difficult and progressive deterioration of the physical state is frequently a serious problem. As an adjuvant to proper care and nutrition AST may be useful in improving physical strength, increasing body weight, correcting hypoproteinemia and, to a minor degree, in stimulating mental activity (BURNETT, 1963; TEC, 1963; LEFRANÇOIS, 1966; SANSOY et al., 1971; TEC, 1971; HUSEMAN and JOHANSON, 1975). In addition, a favorable effect in patients with bladder and bowel incontinence can perhaps be produced (VLAVIANOS et al., 1961). Obviously, these actions are the result of the protein anabolic effect and the other known effects of AST and not specific to psychiatry. There is no substantial evidence that AST exert a beneficial influence on psychiatric disorders sufficient to warrant their administration for specific therapy. The various forms of depression can now be treated much more successfully with antidepressants than with androgenic or androgenic-anabolic steroids. An antipsychotic effect has not been demonstrated so far (SEGERS and NUYTS, 1964); a true psychostimulant action seems to be lacking; and whether neurotic symptoms indeed improve with anabolic treatment as reported in premenstrual symptoms of neurotic character (NAVRÁTIL et al., 1972) must still be confirmed.

*Conclusion:* AST may be tried as supportive therapy in the management of debilitated psychiatric patients and in cases of incontinence. Specific effects in mental illness should not be expected.

## 3. Geriatrics

Aging per se is not a pathologic process. However, it is accompanied by degenerative and debilitating changes which are frequently so pronounced as to require medical intervention. Neglect of adequate diet, malfunction in the digestive system, the inability to metabolize the nutrients ingested, a general decline in the metabolic processes, and the failure of anabolism to keep pace with catabolism, which is usually not diminished, result in malnutrition, weight loss, and protein deficiency as well as mineral and vitamin deficiencies. To this physical wasting must be added the dysfunction of the senescent memory, a decline in mental performance, changes in behavior, decrease in social interest, and diminished relational ability, to name but a few of the important interrelating psychologic factors. It is easy to understand that an aged but healthy individual showing a number of the above-mentioned changes will sometimes require medical care. It is also obvious that as soon as he becomes the victim of one of the many mental or physical ailments developing particularly in advanced age, he will need a treatment that differs from that appropriate for a younger person. The geriatric patient differs in many respects from a younger patient.

### Therapeutic Trials

AST have been tried in geriatrics as adjunctive therapy to improve the general physical and mental condition of the patient and in a few instances specifically to counteract degenerative organic changes. Due to their various effects (particularly the protein anabolic and anticatabolic properties) AST successfully support adequate high-protein diet to combat alimentary protein and mineral deficiency. They help to restore a positive nitrogen balance, to retain potassium and phosphorus (PEARSON et al., 1954; WATKIN et al., 1955; KRÜSKEMPER, 1968), and to produce a significant weight gain as demonstrated in a great number of controlled and uncontrolled clinical trials (Tables 20 and 21); in general it was found that the lower the initial body weight the greater the subsequent weight gain. Poor weight gain was reported by KALLIOMÄKI et al. (1961 b); WOLPERT et al. (1967); and WOLPERT et al. (1968). Increased appetite and energy and a favorable influence on asthenia and hypoproteinemia were also regularly recorded. Improvement of muscular strength and certain psychometric reactions attained the level of statistical significance only after repeated administration (KALLIOMÄKI et al., 1961 a, b). Beneficial effects of androgens and AST on behavior, mood, activity, and the sense of well-being have been frequently reported (KALLIOMÄKI et al., 1961 a; BAUER and KRAMMER, 1964; BUCKERT, 1964; HAMILTON et al., 1964; ALBERTO et al., 1968; DEUTSCH et al., 1970). In tests for memory function some evidence was obtained that AST may prevent or slow down further deterioration but that they are not effective in restoring severely impaired memory (KRAL and WIGDOR, 1959) or improving performance in mental tests (PIATTI and POZZI, 1975). An attempt to quantify the effect of an AST on behavioral changes failed (WOLPERT et al., 1967) although this could be attributed to the dose used and to the possibly inappropriate psychometric method employed in this investigation. Androgens and AST are apparently without effect upon psychiatric ailments commonly seen in geriatric patients such as generalized debility or organic brain disease (HAMILTON et al., 1964; BOILLAT et al., 1971).

A favorable influence of AST on presbycusis (KROATH, 1967) was observed. An improvement of renal function under AST treatment has been reported by DONTAS et al. (1967). This effect was said to be due to a long-lasting increase of the maximal secretory rate of paraaminohippurate following an increase in the number of PAH-transporting units. AST also seem to arrest involutional changes of circulatory functions but do not affect the glomerular filtration rate.

Urinary incontinence of senile demented women did not improve under anabolic therapy in a trial performed by SVENDSEN (1959) whereas FASTNER (1975) recorded a therapeutic effect of an AST in geriatric patients.

*Conclusion:* AST might be useful additions to the nutritional and medical care of elderly subjects with degenerative and debilitating changes. AST can certainly help to improve physical strength, promote weight gain, and favorably influence behavioral changes. More investigations are required to elucidate a possible beneficial action of AST on memory and several specific organic functions.

As long as androgenic drugs are regarded as contraindicated in prostatic carcinoma, particular attention will have to be paid to this possibility. Careful dosing to avoid facial hair-growth and other signs of virilization is advisable in senile women, although it has been shown that virilization does not always occur (HERK, 1962).

## 4. Immunology

The possibility of producing antibodies against different steroids and thus immunizing the organism to its own hormones is not the subject of this chapter but rather the fact that compounds such as AST can influence the immune state of an individual in general.

Table 20. Controlled clinical trials with AST in geriatric patients

| Author(s) | Country | No. of patients receiving AST | Duration of treatment | Diagnosis | Therapeutic effects on | | Remarks |
|---|---|---|---|---|---|---|---|
| | | | | | Appetite, weight gain | Physical activity, energy, mood | |
| ANTONINI and VERDI (1961) | Italy | 10 | 60 days | Geriatric patients | + | | Blind comparison with 10 vitamin-treated controls |
| KALLIOMÄKI et al. (1961a) | Finland | 15 | one single injection | Geriatric patients | + | + | Double-blind comparison with 15 placebo-treated controls |
| KALLIOMÄKI et al. (1961b) | Finland | 33 | 8 weeks | Geriatric patients | (+) | + | Double-blind comparison of 2 doses of one AST with 18 placebo-treated controls |
| MORRISON (1961) | USA | 10 | 12 weeks | Geriatric patients | + | + | Comparison with 11 placebo-treated controls |
| SMITH and JOHNSON (1961)[a] | USA | 30 | 4–12 weeks | Chronic brain syndrome, variants of schizophrenia | + | | Double-blind comparison of 3 groups treated with an AST for 4, 8, and 12 weeks respectively with a placebo-treated group |
| JOHNSON (1962)[a] | USA | 52 | 6 weeks | Underweight, mental illness (42) | | | Comparison of 7 groups treated with different doses of 2 AST and 7 placebo-treated controls |
| PAGE (1962)[a] | USA | 67 | 12 weeks | Mentally ill geriatric patients | + | | Double-blind crossover within-subject comparison of 2 AST versus placebo |

Table 20 (continued)

| Author(s) | Country | No. of patients receiving AST | Duration of treatment | Diagnosis | Therapeutic effects on | | Remarks |
|---|---|---|---|---|---|---|---|
| | | | | | Appetite, weight gain | Physical activity, energy, mood | |
| VILANOVA (1962) | USA | 42 | 4–6 weeks | Chronic brain syndrome | + | | Open comparison with 41 controls |
| BAUER and KRAMMER (1964) | Austria | 74 | 4 weeks | Chronic brain syndrome, other diseases | + | + | Sequential trial with two AST |
| HAMILTON et al. (1964) | USA | 18 | 12 weeks | Debilitated psychotic patients | + | + | Double-blind comparison with 8 placebo-treated controls |
| SETTEL (1964) | USA | 28 | 16 weeks | Underweight | + | + | Crossover comparison of an AST with an enzyme-vitamin preparation |
| BEW (1965) | Great Britain | 18 | 4 weeks | Geriatric patients | + | + | Within-subject comparison with a placebo in 2 groups |
| LACHNIT and ZWERINA (1965) | Austria | 60 | 1 year | Vascular sclerosis | + | + | Open comparison with two AST |
| MEZZENA et al. (1972) | Italy | 23 | 1–3 months | Geriatric patients | + | + | Comparison with 16 cobamamide- and 20 placebo-treated controls |
| DUBOS and HUGONAT (1974) | France | 36 | 3 months | Geriatric patients | + | + | Double-blind comparison with 35 controls |
| PIATTI and POZZI (1975) | Italy | 44 | 2–4 weeks | Geriatric patients | + | | Double-blind comparison with placebo-treated controls |
| GRIBBIN and FLAVELL MATTS (1976) | Great Britain | 25 | 6 weeks | Geriatric patients | + | + | Comparison with 25 placebo-treated controls |

[a] Also referred to in Table 16.

Table 21. Uncontrolled clinical trials with anabolic steroids in geriatric patients

| Author(s) | Country | No. of patients receiving AST | Duration of treatment | Diagnosis | Therapeutic effects on | | |
|---|---|---|---|---|---|---|---|
| | | | | | Appetite, weight gain | Physical activity, energy, mood | Hypo-proteinemia |
| DE LA PENA (1960) | Mexico | 17 | 60 days | Geriatric patients | + | | |
| HOCK (1961) | USA | 50 | up to 56 weeks | Functional diseases | + | + | |
| LEYRITZ et al. (1961) | France | 37 | not specified | Geriatric patients | + | | |
| BUCKERT (1964) | W. Germany | 50 | 6 months | Underweight | + | + | |
| DARDENNE and CANTALA (1964) | France | not specified | 10–32 weeks | Geriatric patients | + | + | |
| MARQUIER (1964) | France | 45 | 1–2 months | Various diseases | + | + | |
| DUTRUEL (1965) | France | 30 | 30 days | Various diseases | + | + | |
| SCHWARZ (1967)[a] | W. Germany | 68 | up to 14 months | Geriatric patients | + | + | |
| ALBERTO et al. (1968) | Italy | 74 | 17 days | Psychiatric disturbances | + | + | + |
| MANAI and PICCININI (1971) | Italy | 100 | 1 month | Various diseases | + | + | + |
| PASQUALICHIO et al. (1972)[b] | Italy | 20 | up to 197 days | Deficiency syndrome | + | + | |
| SOECKNICK (1972)[c] | Austria | 54 | 2 months | Geriatric patients | + | + | |
| CHAPUY and CONVERT (1973) | France | 39 | 2–3 months | Various diseases | + | + | + |
| MOLINO et al. (1973) | Italy | 25 | 90 days | Underweight | + | | + |

[a] Combination preparation with vitamins.
[b] Combination preparation with vitamins and amino acids.
[c] Combination preparation with vitamins and ferrous fumarate.

## Animal Pharmacology

The effect of AST on the nonspecific response to allergic or infective stimuli and on specific antibody production following antigen administration has been studied in animal experiments. The results are conflicting.

AST enhance the increase in various antibody titers when given before or simultaneously with (TOLENTINO, 1962)—but not when given after—the basic immunization (TOLENTINO et al., 1964). Stimulation of immunoglobulin production and an interaction between the antigen and information passed to the immunocompetent cells is discussed as a possible mechanism of action. Following basic diphtheria and tetanus immunization of rabbits, antitoxin titers, although not significantly different, were found to be slightly higher than in control animals when the antigen and the AST were administered concomitantly (DIECKHOFF and SCHNEEWEISS, 1965). The same investigators also reported that the booster-effect of a diphtheria, tetanus, or typhoid vaccination was not influenced by AST. A dose-dependent acceleration and enhancement of production of specific antibodies in rabbits was observed by STÖCKL and GEYRHOFER (1965) when the AST was given 7 days prior to the antigen; simultaneous administration with the vaccine had no appreciable effect, whereas antibody formation was inhibited if higher doses of an AST were applied 1 or 2 weeks after basic immunization.

AST were also reported to increase the resistance to certain bacterial infections, to exert some endotoxin-neutralizing effect and to reduce or nullify the infection-promoting action of cortisone in rats and mice (GHIONE, 1958; KUTZSCHE, 1964; SENECA and PEER, 1967a, b). Guinea pigs sensitized to horse serum were not protected by an AST; on the contrary the intensity of the anaphylactic shock was greater (GHIONE, 1958).

## Investigations in the Human

Among factors reflecting the immune status of a patient is the response of lymphocytes to mitogenic stimulation with phytohemagglutinin. Maintenance of an adequate response of T cells is important. This response is depressed by surgery or radiotherapy, particularly in cancer patients. Treatment with AST before an operation can significantly diminish the usual fall of lymphocyte transformation scores, as was shown in a controlled trial in patients with primary breast cancer (PENTYCROSS et al., 1973).

In comparative studies it was found that the increase in specific antibodies to bacterial antigens was significantly higher in subjects receiving AST than in controls or prednisone-treated individuals (TOLENTINO, 1962; LIOTTA et al., 1967).

These findings together with the therapeutic effects of AST recorded in ulcus cruris, decubital ulcers, and viral hepatitis as well as observations that AST enhance the active immunization and apparently increase the general resistance of patients to infections, is of growing interest. These effects are discussed particularly in connection with the basic actions of AST which are responsible for the therapeutic effects. It appears to be more than pure speculation to assume that an explanation for some of the therapeutic actions might be found in the influence AST exert on the immunologic processes and natural protective mechanisms (TOLENTINO, 1962; NASSO et al.,

1964; Gaillard et al., 1967; Seneca, 1968; Jannuzzi et al., 1970; Tolentino, 1971; see also Castro et al., 1973).

*Conclusion:* An influence of AST on specific antibody formation, on the immune status in general, and on resistance to infection can be assumed on the basis of animal experiments and observations in man. To what extent these phenomena can explain some of the therapeutic effects of AST in humans and animals, and to what extent they could enlarge the field of application of AST in medicine, remains to be seen.

## 5. Catatoxic Effects

### Animal Pharmacology

Selye and his school demonstrated that more than 200 natural and synthetic steroids increase the natural resistance of intact animals to the toxic effects of numerous endogenous and exogenous pharmacologically and chemically unrelated agents (Selye, 1969). This protective, resistance-increasing property which is independent, that is has no correlation with hormonal or other pharmacologic activities, has been named the catatoxic effect. The most active catatoxic steroids are antimineralocorticosteroids (such as spironolactone, spiroxasone) and AST. They induce resistance to a great variety of drug actions and intoxications (Selye et al., 1969; Selye, 1970a, b, c, d, e; Selye and Solymoss, 1970; Selye et al., 1970; Selye, 1971; Solymoss et al., 1971; Tache et al., 1975) and there seems to be a mutual antagonism between catatoxic steroids and thyroxine in this respect (Selye, 1970f). It is assumed that the catatoxic actions are mostly the result of induction of drug-metabolizing and drug-detoxifying enzymes in the hepatic microsomes, although this may not be the only mechanism of action (Selye, 1970a; Selye and Solymoss, 1970; Selye, 1971). Undoubtedly, these findings are of great scientific interest. Whether they are also of importance for the human and in particular for therapeutic applications, remains to be seen. Firstly, all experiments performed so far were in rats. Secondly, the doses necessary to produce a catatoxic effect were very much higher than those used for therapeutic purposes; this might be of relevance in man despite the well-known fact that it is hazardous to postulate correlations between effective doses of drugs in animals and in therapeutic practice. Thirdly, a catatoxic effect is apparently only pronounced when the AST is given days before the noxious agent; this could, of course, severely limit the clinical applicability of catatoxic steroids.

### Possible Relevant Clinical Observations

Enhancement of hepatic microsomal hydroxylase activity produced by catatoxic AST diminishes the biological effect of natural steroids by increasing their catabolism. In healthy females of reproductive age, AST were reported to shorten the luteal phase when given after ovulation. It is not clearly understood whether this effect is the result of an increased rate of progesterone destruction due to enzyme induction alone, or the net result of various effects, e.g. on gonadotropin availability and progesterone biosynthesis or action. However, the catatoxic property of AST might play a role in this interesting effect which could possibly become of some importance for fertility control (Bolch and Warren, 1971).

*Conclusion:* In animal experiments AST were shown to exert very potent catatoxic effects. The relevance of this finding for human therapy has not yet been established.

## 6. Radiotherapy

In candidates for radiation therapy the balance of protein metabolism is often already shifted toward the catabolic side as a result of the malignant process, the concomitant treatment with corticosteroids, and perhaps also insufficient endogenous production of anabolic hormones. Radiotherapy increases protein breakdown and nitrogen loss; it can produce hypoproteinemia and adverse hematologic effects such as anemia, leukopenia, thrombocytopenia, and immunosuppression. In addition, exposure to therapeutic doses of ionizing radiation is usually followed by an acute radiation sickness with nausea, vomiting, diarrhea, headache, tachycardia, and malaise of varying severity.

It should be an essential part of the treatment of these patients to improve their general physical and psychic condition as well as their protein metabolism before radiotherapy is begun. No available measures should be neglected to improve tolerance to radiotherapy and to minimize its adverse reactions.

### Clinical Trials

Based on encouraging experiences in animal experiments (e.g. LOEWIT, 1964; ADÁM et al., 1968; DANYSZ et al., 1968; PICHA, 1969; HORN, 1971; PANEK, 1972a, b, c; AMBRUS and MIRAND, 1973) AST have been used as adjuvant medication in radiotherapy (Table 22). They were found to be effective in improving the general condition of the patient by increasing appetite, body weight, and by stimulating protein synthesis. According to some investigators AST reduce the symptoms of radiation sickness by increasing tolerance to radiation. In patients receiving additional corticosteroid therapy they counteract the ensuing catabolic disturbance. Furthermore it has been shown that AST can protect erythropoiesis both through an increased mobilization of an erythropoietic substance and by a direct stimulatory effect on the bone marrow elements (see chapter on Erythropoiesis). Patients receiving AST while undergoing radiotherapy are less prone to develop anemic symptoms such as changes in red cell mass and hemoglobin than patients not receiving AST. In monkeys subjected to whole-body irradiation or nitrogen mustard, injection of an AST accelerated not only the erythropoietic but also the leukopoietic recovery; no effect on thrombocytopoietic regeneration was seen (AMBRUS and MIRAND, 1973; AMBRUS and AMBRUS, 1975). Observations indicating that AST diminish the erythropoietic depression of radiotherapy have also been recorded in various clinical studies.

Some types of malignant tumors react specifically to the administration of anabolic-androgenic steroids (see Chapter on Cancer). If these patients are additionally treated with radiotherapy, the anabolic drugs may serve a double purpose.

*Conclusion:* AST can be recommended as adjunctive drugs to improve the condition of patients with malignancies before, during and after radiotherapy. AST reduce adverse effects of ionizing radiation on protein metabolism and on the erythropoietic system.

Table 22. Clinical trials with AST in carcinoma patients receiving radiotherapy

| Author(s) | Country | No. of patients receiving AST | Duration of treatment | Therapeutic effects on | | | | | Remarks |
|---|---|---|---|---|---|---|---|---|---|
| | | | | General condition | Appetite, weight gain | Erythropoiesis | Serum-proteins | Radiation sickness | |
| KLEIBEL (1962a) | W. Germany | 38 | 3–7 weeks | + | + | + | + | + | Different AST |
| KLEIBEL (1962c) | W. Germany | >200 | 4–6 weeks | + | + | + | + | + | Different AST |
| KLEIBEL (1964a) | W. Germany | | | | | | | | Different AST |
| KLEIBEL (1962b) (see also KLEIBEL (1964b), BECKER et al. 1966) | W. Germany | 66 | 4–7 weeks | + | + | + | + | + | Different AST |
| QUACK and GAEBEL (1962) | W. Germany | 46 | 6 weeks | + | + | | | | |
| FIXSON (1964) | W. Germany | 10 | not specified | | + | + | | + | Open comparison with 10 controls |
| PICHA (1964) | Austria | 194 | 6 weeks to 5 months | + | + | + | | + | |
| QUACK (1964) | W. Germany | 120 | not specified | + | + | | | | |
| KLEIBEL (1969) | W. Germany | 70 | one single injection | + | | + | + | | |
| PICHA (1969) | Austria | 856 | 5–8 weeks or longer | + | + | + | | | 2-year mortality of Ca colli III is significantly reduced by AST treatment (comparison of 102 patients receiving AST with 102 controls) |
| ADÁM et al. (1971) | Hungary | 100 | not specified | | | + | | | Additional treatment with an iron preparation |
| EVANS and ELIAS (1972) | USA | 12 | 7–11 weeks | | | + statist. signif. | | | Comparison with 11 controls |

Table 23. Clinical trials with androgenic-anabolic steroids in combination with cytotoxic agents. All investigators reported favorable results

| Author(s) | Country | Androgens or AST in combination with |
|---|---|---|
| KILL (1961) | W. Germany | |
| BOND and ARTHUR (1966) | Great Britain | |
| ARTHUR (1968) | Great Britain | Nitrogen mustard derivatives |
| BRODSKY et al. (1972) | USA | (e.g., cyclophosphamide, |
| BRODSKY (1973) | USA | melphalan, chlorambucil) |
| KAHN et al. (1974) | USA | |
| RIGBERG and BRODKSY (1975) | USA | |
| SCHILLING and FINKEL (1975) | USA | |
| LANGNICKEL (1963) | W. Germany | |
| BOND and ARTHUR (1966) | Great Britain | Ethyleneimines |
| TURNER (1966) | Great Britain | |
| ARTHUR (1968) | Great Britain | (e.g., triaziquone, thiotepa) |
| BRODSKY et al. (1964) | USA | |
| BRODSKY et al. (1972) | USA | Pyrimidine derivatives |
| BRODSKY (1973) | USA | (e.g., fluorouracil) |
| KAHN et al. (1974) | USA | |
| RIGBERG and BRODSKY (1975) | USA | |
| RIORDAN and BROWNE (1961) | Ireland | |
| BROWNE (1966) | Ireland | Radioisotopes (e.g., $P^{32}$) |
| KLEIBEL (1972) | W. Germany | |

# V. Anabolics with Other Drugs

## A. Combination with Cytostatics

Cytotoxic drugs are indispensable in the treatment of hematologic malignancies and malignant solid tumors. Unfortunately, they often produce bone marrow suppression when given in effective doses. This is all the more disturbing as the bone marrow reserve is often already severely suppressed by preceding high-dose large-field radiotherapy in patients requiring chemotherapy. Thus, the dose-dependent toxic effects severely limit the use of alkylating substances as well as antimetabolites and radioisotopes, either by prohibiting continuation of an effective therapy or by preventing a necessary dose increase. Therefore, methods to control the toxic effects of cytostatics are urgently needed.

### Clinical Trials

Animal pharmacology indicates that androgens or AST protect the bone marrow from toxic effects of cytostatic agents used in chemotherapy of human cancer (e.g. JOHNSTON and BURN, 1966; BRODSKY and KAHN, 1967; BRÜGGER, 1967; FÖLSCH, 1967; FÖLSCH et al., 1975; see also UDUPA and REISSMANN, 1974). Consequently, clinical studies have been performed in order to find out whether these encouraging results could be reproduced and used in man (Table 23). It is understandable but nonetheless unfortunate that most of the results published so far, comprising data on some 400 treated patients, were produced in noncontrolled trials under conditions which do not allow comparison of one study with another. Drawing unequivocal conclusions from the observations available is hardly possible due to the fact that in most instances the actions of cytostatic drugs and the supportive therapy cannot be differentiated with sufficient accuracy. Hence, the contribution of AST to the overall success rate of the treatment often remains obscure. No therapeutic effect of androgens or AST was found in controlled trials performed by FALKSON et al. (1964); COOKE and BURN (1971); see also BURN and COOKE (1974); COLE et al. (1973); and GOLDENBERG et al. (1975). Nevertheless, through experience, many clinicians are apparently convinced that the combination of chemotherapeutics with androgenic-anabolic steroids is superior to the administration of either alone in the treatment of solid tumors and particularly of hematologic malignancies.

Reports have been published of the beneficial effects of a combination of AST with either nitrogen mustard derivatives (e.g., cyclophosphamide, melphalan, and chlorambucil), ethyleneimines (e.g., triaziquone or thiotepa), antimetabolites such as pyrimidine derivatives (e.g., fluorouracil), or with the radioisotope $P^{32}$.

The following actions of AST are thought to contribute to the therapeutic effect of such combinations: stimulation of the erythropoiesis, granulopoiesis, and possibly also thrombopoiesis during treatment with cytostatic drugs and in the subsequent recovery period (see also Chapter on Erythropoiesis); tolerance of higher doses of cytostatic drugs; a direct influence upon the course of some hematologic malignancies and solid tumors (see also chapter on cancer); enhancement of fibrinolytic activity; increase in the nonspecific immune response and resistance of the host to the neoplasm; and last but not least the effects on serum proteins, body weight, and the general condition of the severely ill patient.

It must be emphasized that doses exceeding the usual recommended dose range of AST have been employed in most clinical studies when bone marrow protection was the goal. Consequently, androgenic side effects of considerable intensity—although less than with the previously used androgens—occurred in women and children. However, this disadvantage seemed to be outweighed by the beneficial effects.

*Conclusion:* Experimental and clinical evidence suggests that AST may be valuable adjuncts to cytostatic drugs in the treatment of hematologic malignancies and solid neoplasms. For final recommendations more clinical experience is required, in particular well-planned large scale cooperative clinical trials.

## B. Combination with Glucocorticosteroids

Glucocorticosteroids are very powerful drugs and have become an essential part of the armamentarium of the modern therapist. Unfortunately, not all of their potent actions are to the benefit of the patient and some of these actions can manifest themselves as unwanted effects. The most prominent actions relevant to this discussion are: a negative influence upon protein metabolism; adrenal function; bone tissue formation; secondary inflammatory reactions; skin and wound healing; as well as an impairment of growth in children. These undesirable actions are of practical importance, hence, numerous animal experiments and a great many clinical trials have been performed to investigate the possibility of counteracting or preventing such effects with AST.

The various studies in animals and in man have been summarized, extensively discussed, and evaluated in a recent well-documented and balanced review by LINET (1970). Therefore, references and details of the subject published before 1970 need not be repeated here and use can be made of LINET's conclusions insofar as they are still valid.

The results of animal experiments strongly indicate that anabolic-androgenic steroids prevent or counteract the above-mentioned unwanted effects of glucocorticosteroids to a very satisfactory degree. In human studies distinctly less favorable results were achieved. This could only be in part attributed to differences in dosage. In addition, the validity of these clinical investigations suffers from the fact that in the majority of them short-term effects were studied. In therapeutic practice, however, glucocorticosteroids are usually administered over very long periods, and most of the untoward effects encountered are the consequence only of long-term treatment. Obviously, many more clinical pharmacologic trials are required before anything definite can be said about the various interactions between glucocorticosteroids and AST in therapeutic use. Nevertheless, a few preliminary conclusions are possible on the basis of the data available.

Protein catabolism as well as increased urinary calcium and phosphorus excretion produced by glucocorticosteroids (established by means of balance studies) can be reduced or prevented by anabolic-androgenic steroids (GEYER and JESSERER, 1961; KARJALA and FORD, 1964; JAPA et al., 1966; LINET, 1970). It appears however, that this beneficial effect is short lasting although several observations indicating a longer duration of action have been reported (LINET, 1970). Furthermore, it is argued that the pathogenetic role played by the protein catabolic action of corticosteroids in unwanted reactions such as osteoporosis or muscular lesions is not sufficiently well established to make anabolic treatment mandatory during corticosteroid therapy (GEYER and JESSERER, 1962). Nevertheless, some authors (e.g. KAISER, 1973; see also KRÜSKEMPER, 1968) are in favor of combined or intermittent treatment with AST during continuous glucocorticosteroid therapy for certain diseases.

Some AST decrease the urinary excretion of 17-ketosteroids and 17-hydroxycorticosteroids, increase the biological half-life of cortisol, and reduce cortisol production by the adrenal cortex. According to VERMEULEN (1964), the decrease in cortisol synthesis is most probably the result of a delayed inactivation, and to a lesser degree, due to an inhibition of pituitary ACTH secretion; the role played by a possible synthetic block at the adrenal level is not clear (VERMEULEN, 1964; LINET, 1970). However, adrenal function and the fate of endogenous corticosteroids in the organism do not seem to be influenced by AST to a clinically significant degree (VERMEULEN, 1964; LINET, 1970; see also MARTINS, 1964; KRÜSKEMPER, 1968). The questionable, but postulated, potentiating effect of AST on glucocorticosteroid actions could be an exception to this assumption if it turns out to be the manifestation of a reduced inactivation of endogenous cortisol and similar synthetic steroids (JAMES et al., 1962). Adrenal atrophy produced by glucocorticosteroids in animals can be partially or totally prevented by AST (KRACHT, 1964; LINET, 1970). A similar beneficial effect in clinical medicine has not yet been demonstrated convincingly.

The effect of AST upon corticosteroid-induced osteoporosis which can be monitored by calcium excretion studies seems to be restricted to producing subjective relief from pain, delaying the onset, and slowing down progression of this often iatrogenic disease. This would appear to be sufficient justification for a trial of AST administered concomitantly, or intermittently, during prolonged glucocorticosteroid therapy (KAISER, 1973).

The glucocorticosteroid-induced inhibition of mitosis in the mouse epidermis (BULLOUGH, 1965), of the collagen synthesis in fibroblast cultures (SCHWARZ and KUTZSCHE, 1966) and of the $^{14}C$-proline incorporation into dermal collagen in vivo in the rat (SIM et al., 1972) could be diminished or prevented by an AST (BULLOUGH et al., 1968; SIM et al., 1972; SIM et al., 1975). A stimulatory effect of an AST on collagen synthesis could be demonstrated by experiments in vitro (SCHWARZ and KUTZSCHE, 1966). This might indicate that the simultaneous administration or topical application of a glucocorticosteroid with an anabolic compound could be advantageous. By favoring collagen biosynthesis the AST might possibly prevent the inhibitory action of a corticosteroid on the reparative phase of an inflammatory reaction and on wound healing (EHRLICH and HUNT, 1969). Moreover, it could counteract atrophy of the skin due to local treatment with corticosteroids. However, this subject must be studied more thoroughly before final conclusions about therapy can be drawn (see also VOSS, 1973c).

The assumed potentiating effect of AST on the anti-inflammatory properties of glucocorticosteroids (FALLIERS et al., 1963; AEPLI, 1964; EL-MEHAIRY and EL-TARABISHI, 1966; LINET, 1970) seems unlikely due to contradictory results (WEST and LEWIS-FANING, 1965; HASLOCK and WRIGHT, 1971), although the possibility of an eventual difference between the various anabolic compounds in this respect deserves consideration. This will necessitate further investigation.

The influence of an AST upon the reduced growth of corticosteroid-treated children was investigated in controlled comparative studies (FALLIERS et al., 1963; KERREBIJN and DELVER, 1969). A favorable trend was indicated but did not attain the level of statistical significance. As this might have been a matter of dosage, no final conclusion is possible.

*Conclusion:* The therapeutic usefulness of AST in counteracting or preventing some of the unwanted reactions of glucocorticosteroids is limited. Anabolic-androgenic drugs can be recommended as concomitant or, even better, intermittent treatment in patients undergoing long-term glucocorticosteroid therapy in order to reduce the antianabolic effect of the latter. AST can also be tried to alleviate subjective symptoms and delay the development of glucocorticosteroid-induced osteoporosis. For that purpose the separate use of the individual drugs is definitely preferable to the use of combination preparations.

The influence of AST on the inhibitory effect of glucocorticosteroids on wound healing and the productive phase of inflammation is insufficiently investigated, as is the claim that AST enhance the anti-inflammatory actions of glucocorticosteroids and prevent growth inhibition caused by these drugs.

## C. Interaction with Other Drugs

Two or more drugs administered concomitantly may produce effects quantitatively or qualitatively different from those of the individual drugs or their summation. This interaction can be the result of a pharmaceutic incompatibility, of an altered fate of the drugs in the organism, of competition for receptor sites, or the consequence of pharmacologic synergism or antagonism. In addition, interaction is always dependent on the specific clinical condition of the patient and on factors related to drug administration (HANSTEN, 1973). Therefore predictions of drug interactions are often hazardous, all the more so since the results of animal experiments cannot be readily extrapolated to man and actions of many drugs are very difficult to measure under clinical conditions.

An interaction is often only recognizable when manifested as an exaggerated response or a serious toxic reaction. Not all drug-drug interactions described in the literature are disadvantageous and many of them are of no clinical relevance and are of academic interest only. However, some interactions are severe enough to endanger the life of the patient. A thorough knowledge of these interactions is vital, particularly now that polypharmacy is practised on such a frightening scale with pharmacologically potent drugs (HUSSAR, 1973). A few interactions of AST with other drugs which merit the attention of the practising physician have been discovered (Table 24).

There is sufficient clinical evidence to assume that patients treated with AST can have a decreased anticoagulant tolerance; unexpected hemorrhages and the need to reduce anticoagulant dosage have been found in cases receiving 17α-alkylated AST and oral anticoagulants concurrently. A similar potentiating effect of AST esters has not been observed (EDWARDS and CURTIS, 1971). The mechanism responsible for the increased sensitivity to anticoagulants is still unknown. Proposed but disputed possibilities are: transient liver dysfunction affecting prothrombin and fibrinogen synthesis; increased clotting factor degradation; competition with plasma-binding of the anticoagulant and production of a higher circulating titer of the latter; increased affinity of the receptor site for the anticoagulant; and the fibrinolytic property of the AST. An effect of AST on the rate of metabolism of the coumarins has not been detected. One group of investigators (LONGRIDGE et al., 1971) regards this drug interaction so seriously as to completely abandon the use of 17-alkylated AST in patients treated with anticoagulants. However, most clinicians think that careful monitoring of the patient is sufficient to prevent complications.

The blood glucose level and insulin requirement of some diabetic patients decrease and the reactivity of insulin-resistant patients may be restored by treatment with AST (not seen by KINK, 1962). The hypoglycemic response to tolbutamide is enhanced by methandrostenolone but not by methenolone acetate or nandrolone phenylpropionate (LANDON et al., 1965). Consequently particular attention should be paid to symptoms of hypoglycemia, to the blood sugar level, and to the requirements of antidiabetic drug therapy in diabetic patients receiving AST.

Methandrostenolone increases the plasma level of oxyphenbutazone possibly due to altered plasma/tissue distribution or inhibition of oxyphenbutazone metabolism. This effect is masked by phenylbutazone, which probably displaces oxyphenbutazone from plasma protein binding sites. The clinical significance of this phenomenon is obscure and no toxic manifestations have been observed.

In animal experiments interactions between barbiturates and testosterone were found: a synergistic effect by GESSNER and GESSNER (1973); induction of steroid-metabolizing enzymes by RESAN et al. (1972); and a similar phenobarbital-induced alteration of steroid biotransformation was also described for nandrolone esters

Table 24. Interactions of AST with other substances potentially harmful to the patient

| References | Substances interacting with AST | Mechanism(s) | Clinical | |
|---|---|---|---|---|
| | | | Findings | Management |
| MURAKAMI et al. (1965); O'GRADY (1966); SCHROGIE and SOLOMON (1966); DRESDALE and HAYES (1967); SCHROGIE and SOLOMON (1967); VERE and FEARNLEY (1968); VAN DAM et al. (1969); EDWARDS and CURTIS (1971); EKERT et al. (1971); LONGRIDGE et al. (1971); DE OYA et al. (1971); ROBINSON et al. (1971) | Oral anticoagulants (apparently only valid for 17α-alkylated AST) | Not established, several proposed (see text) | Decreased tolerance to anticoagulants, unexpected hemorrhages | Careful monitoring of prothrombin time, readjustment of anticoagulant dosage during and immediately after therapy with AST |
| DARDENNE (1961); HOUTSMULLER (1961); TAINTER (1962); WEISSEL (1962); ALBANESE (1965); LANDON et al. (1965) | Insulin and oral hypoglycemics | Poorly understood metabolic effects | Reduction of blood glucose, decreased requirement for insulin and oral antidiabetic drugs in some diabetics; hypoglycemia; disappearance of insulin resistance | Careful control of blood glucose, possible readjustment of antidiabetic drug dosage during AST therapy |
| WEINER et al. (1965); WEINER et al. (1967); HVIDBERG et al. (1968) | Oxyphenbutazone (so far only reported for methandrostenolone) | Not established | Increased plasma level of oxyphenbutazone without clinical symptoms | Careful observation of patients treated with oxyphenbutazone and methandrostenolone (or other AST) |
| RESAN et al. (1972); GESSNER and GESSNER (1973); KULCSAR-GERGELY et al. (1975) | Barbiturates | Induction of drug-metabolizing enzymes; synergistic effect | Only results of animal experiments available | |

(KULCSAR-GERGELY et al., 1975). These interactions led SHAHIDI (1972) to suggest the avoidance of barbiturates in patients receiving androgenic steroids for the stimulation of erythropoiesis.

All tetracyclines except doxycycline seem to increase urinary nitrogen excretion (SHILS, 1963; LITTLE and BAILEY, 1970). Whether this effect can and should be counteracted by AST has not been investigated.

The therapeutic utilization of the interactions between AST and cytostatics or glucocorticosteroids has been discussed in previous chapters.

*Conclusion:* AST can increase sensitivity to oral anticoagulants and antidiabetic drugs. This makes special monitoring of patients receiving either oral anticoagulants or antidiabetic treatment concurrently with AST a mandatory procedure. As long as the clinical implications of the effect of methandrostenolone on the plasma level of oxyphenbutazone are not elucidated, caution is advisable if simultaneous treatment with AST and oxyphenbutazone is employed. The significance of the other possible drug interactions mentioned above is not yet established.

## References

ADÁM, G., MILTÉNYI, L., DÉZSI, Z., JÓNA, G.: Die Wirkung von Nandrolone-Dekanoat auf den Eisenumsatz des Organismus nach Bestrahlung. Strahlentherapie **136**, 113—115 (1968).

ADÁM, G., MILTÉNYI, L., MOLNÁR, A.: Die Wirkung von Nandrolon-Dekanoat und Eisenpräparaten auf das Blutbild von bestrahlten Patienten. Strahlentherapie **141**, 696—698 (1971).

ADDA, M.: Etude sur 50 malades immobilisés traités par le stromba pendant deux ans. Sem. Hôp. Paris 75—76 (1965).

AEPLI, R.: Potentialisation de l'action anti-inflammatoire des corticosteroides par le Dianabol dans le traitement au long cours de l'asthme bronchique. Schweiz. med. Wschr. **94**, 53—57 (1964).

AGAPOVA, E. N., TKACHENKO, A. M., KRUTOVSKAYA, O. V.: Use of anabolic preparations in atherosclerotic cardiosclerosis. Klin. Med. (Moskau) **52**, 42—44 (1974).

AHLENDORF, W., KOB, D.: Stationäre Nachbehandlung von Patienten mit reseziertem Magenkarzinom. Arch. Geschwulstforsch. **36**, 64—75 (1970).

AHRENDS, H.: Beitrag zum familiären Vorkommen und zur Therapie der Cystinose. Z. Kinderheilk. **99**, 101 (1967).

AKZHIGITOV, G. N., USATOVA, I. Y., KAPLAN, L. D.: Use of glucocorticoid and anabolic hormones in the treatment of acute pancreatitis. Terap. Arkh. **43**, 52—57 (1971).

ALBANESE, A. A.: Newer methodology in the clinical investigation of anabolic steroids. J. new Drugs **5**, 208—224 (1965).

ALBANESE, A. A., LORENZE, E. J., ORTO, L. A., SMULLYAN, I. L.: Nutritional and metabolic effects of some newer steroids. N.Y. St. J. Med. **65**, 2116—2125 (1965).

ALBANESE, A. A., WOODHULL, M. L., ORTO, L. A., ZAVATTARO, D. N., WEIN, E. H.: Steroid and dietary effects on blood lipids in elderly persons. Nutr. Rep. Intern. **1**, 231—242 (1970).

ALBERTO, S., COLONGO, P., BRUSASCO, L.: Comments on the anti-asthenic and anti-depressive properties of androstanazol in geriatrics. Minerva med. **59**, 2288—2293 (1968).

AL-HUSSAINI, M., MCGILLIS, A., FRANK, I. N., MCDONALS, D. F.: Anabolic effect of norethandrolone and nandrolone phenpropionate in uremic rats. Invest. Urol. **3**, 537—545 (1965).

ALLENBY, F., JEYASINGH, K., CALNAN, J.: Ethyloestrenol and postoperative venous thrombosis. Lancet **1973 II**, 38—39.

ALMÉR, L.-O., PANDOLFI, M., NILSSON, I. M.: Diabetic retinopathy and the fibrinolytic system. Diabetes **24**, 529—534 (1975).

ALY, F. W.: Klinische Erfahrungen mit einem nichtvirilisierenden, oral wirksamen Testosteron-Derivat (Dehydro-17-Methyltestosteron). Med. Welt **47**, 2468—2473, Part 1 (1961); **48**, 2513—2520, Part 2 (1961).

AMBRUS, C. M., AMBRUS, J. L.: Myeloid regeneration after whole body irradiation, autologous bone marrow transplantation, and treatment with an anabolic steroid. J. Med. Exp. Clin. **6**, 27—32 (1975).

AMBRUS, J. L., MIRAND, E. A.: Effect of autologous bone marrow transplantation and an anabolic steroid on erythropoietin production and hemopoietic recovery after whole body irradiation and treatment with alkylating agents. J. Med. **4**, 65—81 (1973).

AMMEDICK, U., KONRAD, R. M., GOTZEN, E.: Über die Wirksamkeit anaboler Steroide in der postoperativen Phase. Med. Ernähr. **9**, 121—123 (1968).

ANTONINI, F. M., VERDI, G.: Investigations concerning a new anabolic synthetic steroid: oxymetholone. Behaviour of some lipid fractions of the serum. G. Geront. **9**, 1051—1065 (1961).

ARIEL, G., SAVILLE, W.: The effect of anabolic steroids on reflex components. Med. Sci. Sports **4**, 120—123 (1972).

ARTHUR, K.: Some aspects of chemotherapy in breast carcinomatosis. Clin. Radiol. **19**, 351—356 (1968).

BABEL, J.: Le traitement de la rétinopathie diabétique par les anabolisants. Méd. et Hyg. **23**, 485—486 (1965).

BADALYN, L. O., DUNAEVSKAYA, G. N., SITNIKOV, V. T.: Treatment of progressive muscular dystrophy. Zh. Nevropatol. Psikhiatr. **75**, 1317—1323 (1975).

BÄR, W., USBECK, W.: Zur oralen Anwendung anaboler Hormone bei magenlosen Patienten. Dtsch. Gesundh.-Wes. **26**, 1007—1009 (1971).

BÄR, W., USBECK, W.: Orale Anabolikatherapie bei magenlosen Patienten. Zbl. Chir. **99**, 1306—1311 (1974).

BAKER, C. D., BAUM, G.: Use of an anabolic drug in malnutrition in the aged. Practitioner **193**, 683—685 (1964).

BALAKHOVSKY, I. S., OKHOTSKY, V. P., FEDOROV, I. V., DORONIN, E. I., YAKOVLEVA, I. P.: Indications and applications of an anabolic steroid preparation in clinical orthopaedics. Klin. Med. (Moskau) **50**, 110—114 (1972).

BALMES, A., DAUVERCHAIN, J., OTHONIEL, J., JUANCHICH, J.: Intérêt du Durabolin (phenylpropionate de 19-norandrosténolone) dans le traitement des pneumopathies aigues. Vie méd. **29**, 3527—3536 (1969).

BARNES, H. D., OVERTON, J., SWEET, R. D.: Familial cutaneous porphyria. Brit. J. Derm. **77**, 130—136 (1965).

BARWICK, D. D., NEWELL, D. J., WALTON, J. N.: Methandrostenolone and nandrolone decanoate in muscular dystrophy. A controlled trial. Neurology **13**, 12—23 (1963).

BAUER, A., KRAMMER, H.: Erfahrungen mit Stromba in der Geriatrie. Ther. Umsch. **21**, 194—207 (1964).

BAUER, B., ANTENER, I.: Eine wirksame diätetische und medikamentöse Cystinosebehandlung. Helv. paediat. Acta **21**, 19—50 (1966).

BEAUMONT, J.-L., CARLSON, L. A., COOPER, G., FEJFAR, Z., FREDRICKSON, D. S., STRASSER, T.: Classification of hyperlipidaemias and hyperlipoproteinaemias. Bull. Wld. Hlth. Org. **43**, 891 (1970).

BEAUMONT, P., HOLLOWS, F. C.: Classification of diabetic retinopathy, with therapeutic implications. Lancet **1972 II**, 419—424.

BECKER, J., KÄRCHER, K. H., KLEIBEL, F.: Die zusätzliche Behandlung und Verlaufsbeobachtung während der Strahlentherapie und cytostatischen Behandlung. Radiologe **6**, 159—175 (1966).

BECKMANN, R.: Erfahrungen mit anabolen Steroiden bei Myopathien. In: Wirkung und Anwendung anaboler Steroide. Berlin: Medicus Verlag 1964.

BECKMANN, R.: Discussion on: DOWBEN, R. M., Anabole Steroide bei Myopathien. Myopathien. Stuttgart: Georg Thieme Verlag 1965.

BEI, P. I.: The use of nerobol in the treatment of patients with rheumatic heart defects and circulatory failure. Klin. Med. (Moskau) **46**, 104—106 (1968).

BEI, P. I.: The effect of nerobol on the contractile capacity of the myocardium. Klin. Med. (Moskau) **50**, 69—74 (1972).

BEIGLBÖCK, W., BRUMMUND, W.: Über das Problem der anabolen Wirkung von Testosteron Derivaten (Bericht über Erfahrungen mit Dianabol). Med. Welt **22**, 1192—1203 (1960).

BEKÉNY, G., KRAFT, F., LANG, S.: Die Anwendung von Durabolin (19-Nor-Androstenolon-phenylpropionat) in der Behandlung der Dystrophia musculorum progressiva. Nervenarzt **31**, 118—123 (1960).

BENDA, L., GEYER, G.: Untersuchungen über den Eiweiß-Stoffwechsel bei Leberkranken unter anabol wirksamen Steroiden. In: Wirkung und Anwendung anaboler Steroide. Berlin: Medicus Verlag 1964.

BERGER, H., ANTENER, J., BRECHBÜHLER, T., STALDER, G.: Zum Problem der Cystinose. Beobachtungen über den Cystinnachweis im Serum und therapeutische Erfahrungen mit Penicillamin und anabolem Hormon. Ann. Paediat. (Basel) **202**, 465—482 (1964).

BERGMANN, M.: Die Wirkung von „Deca-Durabolin" auf den Reststickstoffspiegel im Blut. Med. Welt **10**, 538—543 (1962).

BERKOWITZ, D.: Anabolic agents in gastroenterology. Paper presented at a Symposium on Anabolic Therapy, Detroit, March 1962.

BERKOWITZ, D., SPITZER, J., LIKOFF, W. P.: Practical significance of serum triglycerides and radioactive fat tolerance: their relation to current therapy for hypercholesterolemia. Amer. J. Cardiol. **10**, 198 (1962).

BESA, E. C., GORSHEIN, D., GARDNER, F. H.: Androgens and human blood volume changes. Comparison in normal and various anemic states. Arch. intern. Med. **133**, 418–425 (1974).

BEW, K.: A double-blind trial of an anabolic steroid in eighteen elderly women. Practitioner **194**, 530—534 (1965).

BIELAWIEC, M., PERZANOWSKI, A., MYSLIWIEC, M.: Long term therapy of patients suffering from occlusive arterial disease with combined therapy with phenformin and stanozolol. Thrombos. Diathes. haemorrh. **34**, 916—917 (1975).

BIELAWSKI, W.: The effect of oxyterracin on rat liver cell in the course of gravidity and estradiol and durabolin treatment (2nd communication). Pol. Arch. Med. wewnet. **40**, 15—20 (1968).

BIERICH, J. R.: Erfahrungen mit Anabolika bei der progressiven Muskeldystrophie im Kindesalter. In: Wirkung und Anwendung anaboler Steroide. Berlin: Medicus Verlag 1964.

BIERICH, J. R., GRÜTTNER, R., LANDBECK, G.: Therapeutic effects of anabolic steroids in progressive muscular dystrophy. Paper at the 10th Internat. Congress of Pediatrics, Lisbon, 1962.

BLAGG, C. R., PARSONS, F. M.: Use of norethandrolone in acute renal failure from obstetric causes. Lancet **1960 II**, 577—581.

BLAGG, C. R., PARSONS, F. M.: Early dialysis and anabolic steroids in acute renal failure. Amer. Heart J. **61**, 287—291 (1961).

BLAIR, A. J., MORGEN, J. C., BECK, J. C.: The plasma 17-hydroxycorticosteroid levels in acute and chronic renal failure. Canad. J. Biochem. **39**, 1617—1624 (1961).

BLAHNIKOVA, L.: A propos de l'indication des stéroides anabolisants dans le traitement des lésions hépatiques chroniques. Gastroenterologie **107**, 354—359 (1967).

BLAHNIKOVA, L., HORKY, J.: Einfluß der anabolen Steroide auf den Eiweißstoffwechsel bei verschiedener Proteinzufuhr bei chronisch Leberkranken. Z. Ernährungsw. Suppl. **13**, 105—112 (1972).

BLASIUS, R., KÄFER, K., SEITZ, W.: Untersuchungen über die Wirkung von Testosteron auf die contractilen Strukturproteine des Herzens. Klin. Wschr. **34**, 324—326 (1956).

BLASIUS, R., KÄFER, K., SEITZ, W.: Untersuchungen über die Abhängigkeit von androgener Wirkung und proteinaufbauendem Effekt der Steroid-Hormone auf die contractilen Muskelproteine des Herzens. Klin. Wschr. **35**, 308—310 (1957).

BOILLAT, J. E., SAXENA, B. M., LEHMANN, H. E., BAN, T. A.: Combined administration of thioridazine, nicotinic acid and fluoxymesterone in the treatment of geriatric patients. Curr. ther. Res. **13**, 541—544 (1971).

BOLCH, O. H., WARREN, J. C.: Induction of premature menstruation with catatoxic steroids. Amer. J. Obstet. Gynec. **111**, 1107—1110 (1971).

BOND, W. H., ARTHUR, K.: Some observations on the use of cytotoxic drugs and anabolic steroids in the treatment of breast carcinomatosis. Paper read at a Symposium at the University Dept. of Surgery, Royal Infirmary, Sheffield, Nov. 1966.

Borgman, R. F.: Increased survival-time in dystrophic mice treated with methylandrostenediol dienanthoylacetate. Nature (Lond.) **197**, 1304 (1963).
Bors, E., Comarr, A. E.: Neurological urology. Basel-München-Paris-New York: S. Karger 1971.
Bottero, A., Lops, M.: L'impiego della dimetazina nelle bronchopneumatopatie specie senile. Gazz. med. ital. **123**, 111 (1964).
Bour, H., Tutin, M., Dorf, G., Guy-Grand, B.: La rétinite diabétique: pathogénie et thérapeutique. Ent. Bichat. **1967**, 213—220.
Bourdakos, N., Wolf, S.: Creatine and muscular dystrophy; relationship of creatine-creatinine metabolism to diet and drugs. Arch. Neurol. **6**, 439—450 (1962).
Bourgeois, M., Vincent, D.: L'oxymétholone. Essai thérapeutique sur 50 malades de neuropsychiatrie. J. Méd. Bordeaux **141**, 260–263 (1964).
Brambilla, F., Giardini, M., Russo, R.: Prospects for a pharmacological treatment of phenylketonuria. Dis. nerv. Syst. **36**, 257—260 (1975).
Brennan, M. F., Tweedle, D., Moore, F. D.: Body nitrogen after trauma. Lancet **1975 I**, 38—39.
Bretan, M., Szabo, Z., Jakab, L.: Beeinflussung der diabetischen Retinopathie mit Nor-Androstenolonphenylpropionat und Decanoat. Orv. Hetil. **105**, 1171—1174 (1964).
Brodsky, I.: The role of androgens and anabolic steroids in the treatment of cancer. Semin. Drug Treatm. **3**, No. 1, 15—25 (1973).
Brodsky, I., Dennis, L. H., Kahn, S. B.: Testosterone enanthate (NSC-17591) as a bone marrow stimulant during cancer chemotherapy: Preliminary report. Cancer Chemother. Rep. **34**, 59 (1964).
Brodsky, I., Kahn, S. B.: The effect of androgens on cancer chemotherapy. In: Cancer Chemotherapy: Basic and Clinical Applications, 275—287. New York: Grune & Stratton 1967.
Brodsky, I., Kahn, S. B., Conroy, J. F.: The effects of androgens on cancer chemotherapy. In: Cancer Chemotherapy II, 303—314. New York: Grune & Stratton 1972.
Brown, I. A., James, E. M.: Nandrolone-phenpropionate in progressive muscular dystrophy. A preliminary report. Arch. Pediat. **78**, 421—431 (1961).
Brown, I. K., Downie, R. J., Haggart, B., Littler, J., Murray, G. H., Robb, P. M., Santer, G. J.: Pharmacological stimulation of fibrinolytic activity in the surgical patient. Lancet **1971 I**, 774—776.
Browne, P. A.: The treatment of secondary deposits in bone from carcinoma of the breast with $P^{32}$ and durabolin. Paper read at a Symposium at the University Dept. of Surgery, Royal Infirmary, Sheffield, Nov. 1966.
Brückner, R.: Lentopathie und Retinopathie, Signale im Diabetes und Prädiabetes. Praxis **54**, 746—755 (1965).
Brügel, H.: Klinische Erfahrungen bei der Behandlung untergewichtiger Magenresezierter. Aktuelle Kongreß-Berichte **1**, 16. Ärztlicher Fortbildungskurs Bad Kissingen. Berlin: Medicus Verlag 1965.
Brügger, W.: Zur Beeinflussung der hämatotoxischen Effekte von Cyclophosphamid durch das anabole Steroid Nandrolonphenylpropionat. Dissertation, Heidelberg, 1967.
Bruhn, H. D., Jipp, P., Okoye, S., Oltmann, A.: Hypofibrinolyse beim akuten Myokardinfarkt. Med. Klin. **69**, 1951—1955 (1974).
Buchner, H.: Zur Frage der hormonalen Beeinflussung des posttraumatischen Eiweißverlustes. Wien. med. Wschr. **111**, 576—580 (1961).
Buckert, D.: Dianabol in der Geriatrie. Dtsch. med. J. **15**, 25—27 (1964).
Bürgi, H., Regli, J.: Zur Ergänzung der tuberkulostatischen Behandlung mit einer Vitamin-Mineral-Anabolikum-Kombination. Praxis **58**, 873—874 (1969).
Bullough, W. S.: Mitotic and functional homeostasis—a speculative review. Cancer Res. **25**, 1683—1727 (1965).
Bullough, W. S., Homan, J. D. H., Laurence, E. B., Overbeek, G. A.: In vitro Untersuchung an der Mäuseohrepidermis über die von Dexamethason induzierte und von Nandrolondecanoat aufgehobene Mitosehemmung. J. Endocr. **41**, 453—454 (1968).
Burn, J. I., Cooke, W. M.: Effect of nandrolone phenylpropionate (NSC-23162E) on the bone marrow suppression caused by cyclophosphamide (NSC-26271): A clinical trial. Cancer Chemother. Rep. **58**, 867—870 (1974).
Burnett, P. C.: A steroidal pyrazole as an anabolic agent in the treatment of geriatric mental patients. J. Amer. Geriat. Soc. **11**, 979—982 (1963).

BUTTENBERG, D.: Anabolika bei gynäkologischen Karzinomen. In: Wirkung und Anwendung anaboler Steroide. Berlin: Medicus Verlag 1964.
CASTRO, J. E., MEDAWAR, P. B., HAMILTON, D. N. H.: Orchidectomy as a method of immunopotentiation in mice. Ciba Found. Symp. **18**, 237—258 (1973).
CECCARELLI, G.: Anabolizzante e lesione renali. Arch. Sci. med. **120**, 374—384 (1965).
CENDROWSKI, W., KURAN, W.: Results of combined administration of anabolic steroids in patients with multiple sclerosis. Neurol. Neurochir. Psychiat. pol. **6**, 573—576 (1972).
CESNIK, H., FINK, E.: Klinische Untersuchungen eines neuen anabolen Steroids. Med. Klin. **59**, 1430—1433 (1964).
CHAKRABARTI, R.: Comparative effects of platelet adhesiveness by phenformin combined with ethylestrenol or stanozolol. Acta Univ. Carol. Med. (Monogr.) (Praha) **53**, 261—262 (1972).
CHAKRABARTI, R., EVANS, J. F., FEARNLEY, G. R.: Effects on platelet stickiness and fibrinolysis of phenformin combined with ethyloestrenol or stanozolol. Lancet **1970 I**, 591—593.
CHAKRABARTI, R., FEARNLEY, G. R.: Phenformin plus ethyloestrenol in survivors of myocardial infarction. Three year pilot study. Lancet **1972 II**, 556—559.
CHAKRABARTI, R., FEARNLEY, G. R.: Pharmacological fibrinolysis in diabetes mellitus. Diabetologia **10**, 19—22 (1974).
CHAKRABARTI, R., FEARNLEY, G. R., EVANS, J. F.: Reduction of platelet stickiness by phenformin plus ethyloestrenol. Lancet **1967 II**, 1012—1014.
CHAKRABARTI, R., FEARNLEY, G. R., HOCKING, E. D., DELITHEOS, A., CLARKE, G. M.: Fibrinolytic activity related to age in survivors of myocardial infarction. Lancet **1966 I**, 573.
CHAPUY, P., CONVERT, G.: Clinical study of strombaject in geriatrics. Cah. méd. Lyon **49**, 36 (1973).
CHENG, H., KOHNER, E. M., KEEN, H., BLACH, R. K., HILL, D. W.: Photocoagulation in treatment of diabetic maculopathy. Lancet **1975II**, 1110—1113.
CHOI, E. S. K., CHUNG, T., MORRISON, R., MYERS, C., GREENBERG, M. S.: Hypertriglyceridemia in hemodialysis patients during oral dromostanolone therapy for anemia. Amer. J. clin. Nutr. **27**, 901—904 (1974).
CHRISTENSEN, L. R., MACLEOD, C. M.: A proteolytic enzyme of serum: characterization, activation, and reaction with inhibitors. J. gen. Physiol. **28**, 559—568 (1945).
CHRISTL, H., BROMMER, R., WILDE, W., BORK, U.: Moderne eiweißanabole Therapie der täglichen Praxis (I). Fortschr. Med. **80**, 667—703 (1962).
CICCHITTI, L., LABAT, J. C., INTRONA, L., MINUCHIN, O.: Accion del deca-noato de nandrolona sobre la retinopatia diabetica. Paper read at Tercer Congreso Argentino de Nutricion, Mendoza, Sept. 1966.
CIŚWICKA-SZNAJDERMAN, M., BERENT, H., RYMASZEWSKI, Z.: The effect of combined treatment with phenformin and stanazolol on blood lipids and fibrinolytic activity in patients with hypertriglyceridaemia. Atherosclerosis **19**, 153—159 (1974).
COHEN, W. D., HIGANO, N., ROBINSON, R. W., LEBEAU, R. J.: Changes in serum lipids and urinary ketosteroids during oral and intramuscular administration of androsterone. J. clin. Endocr. **21**, 1208 (1961).
COLE, M. P., TODD, I. D. H., WILKINSON, P. M.: Cyclophosphamide and nandrolone decanoate in the treatment of advanced carcinoma of the breast—results of a comparative controlled trial of the agents used singly and in combination. Brit. J. Cancer **27**, 396—399 (1973).
COMAR, O. B.: Nuove osservazioni sull'impiego degli steroidi anabolizzanti negli iperazotemici cronici. Minerva urol. **21**, 69—74 (1969).
COOKE, W. M., BURN, J. I.: The effect of nandrolone phenylpropionate (Durabolin) on the bone-marrow suppression caused by cyclophosphamide. Brit. J. Surg. **58**, 302—303 (1971).
COOPER, I. S., RYNEARSON, E. H., MCCARTY, C. S.: Testosterone propionate as a nitrogen-sparing agent after spinal cord injury. J. Amer. med. Ass. **145**, 549—553 (1951).
CORONA, R. G.: Oximetolona en gastroectomizados. Rev. Med. Mex. **44**, 193—199 (1964).
COTTIER, P.: Die Anwendung von katabolen und anabolen Hormonen bei Nierenkrankheiten. Dtsch. med. J. **13**, 329—334 (1962).
CRAIG, P. E.: Preliminary report of a long acting anabolic agent, nandrolone decanoate. Clin. Med. **69**, 5, 1154—1156 (1962).
CUNLIFFE, W. J.: An association between cutaneous vasculitis and decreased blood-fibrinolytic activity. Lancet **1968 I**, 1226—1228.

CUNLIFFE, W. J., DODMAN, B., ROBERTS, B. E.: Clinical and laboratory double-blind investigation on the effect of fibrinolytic therapy in patients with cutaneous vasculitis. Ann. rheum. Dis. **32**, 591—592 (1973).
CUNLIFFE, W. J., MENON, I. S.: Treatment of Behçet's syndrome with phenformin and ethyloestrenol. Lancet **1969 I**, 1239—1240.
DAM, F. E., VAN, GRIBNAU OVERKAMP, M. J. H., BENRAAD, T. J., HAANEN, C. A. M.: Over de invloed van enkele geneesmiddelen op het effect en de verwerking van het indirecte anticoagulans ethylbiscoumacetaat. Ned. T. Geneesk. **113**, 55—56 (1969).
DANOWSKI, T. S., SABEH, G., VESTER, J. W., SARVER, M. E., SUNDER, J. H.: Muscular dystrophy. Arch. intern. Med. **115**, 294 (1965).
DANYSZ, A., KOCMIERSKA-GRODZKA, D., PANEK, R., MATZKOVYAK, I.: Antimetabolites in the treatment and prophylaxis of radiation sickness. Med. Radiol. Moskva **13**, 3—9 (1968).
DARDENNE, M. U.: Treatment of retinopathia diabetica with anabolic steroids and light coagulation, and the effect of this therapy on the pathological pattern of free amino acids and protein fractions in serum. Excerpta med. Intern. Congr. Ser. **74**, 177 (1964).
DARDENNE, P., CANTALA, P.: Le traitement de l'involution sénile par un stéroide anabolisant (oenanthate de méthénolone). Toulouse méd. **65**, 787—796 (1964).
DARDENNE, U.: Die Therapie der Retinopathia diabetica. Arzneimittel-Forsch. **9**, 672 (1959).
DARDENNE, U.: The therapeutic applications of anabolic steroids in ophthalmology. Acta endocr. (Kbh). **63**, 143—153 (1961).
DARRACOTT, J.: Treatment of the painful knee fulfilling diagnostic criteria for "chondromalacia patellae". Curr. med. Res. Opin. **1**, 412—420 (1973).
DAVIDSON, J. F., LOCHHEAD, M., MCDONALD, G. A., MCNICOL, G. P.: Fibrinolytic enhancement by stanozolol: a double blind trial. Brit. J. Haemat. **22**, 543—559 (1972a).
DAVIDSON, J. F., MCDONALD, G. A., CONKIE, J. A.: Fibrinolytic enhancement by stanozolol—a two year study. Brit. J. Haemat. **22**, 639—640 (1972b).
DAVILA FLORES, O. E., HERRERA, H., ORTEGA, R.: Effecto anabólico protéinico del decanoato de norandrostenolona. Médico (Méx). **8**, 17—19 (1964).
DEINEKO, N. F., SHUSTVAL, N. F., IKHNENKO, R. I., BAKUMENKO, I. S.: Treatment of peptic ulcer with anabolic hormones. Klin. Med. (Moskau) **51**, 57—63 (1973).
DELANEY, T. J., BLACK, M. M.: Effect of fibrinolytic treatment in malignant atrophic papulosis. Brit. med. J. **1975 III**, 415.
DELMONT, J.: Non-tropical sprue. Modern concepts in connection with five cases. Thesis, Marseille, 1961.
DEMEULENAERE, L.: Erfahrungen mit Methenolon bei Leberzirrhose und chronischer Hepatitis. In: Wirkung und Anwendung anaboler Steroide. Berlin: Medicus Verlag 1964.
DEROT, M, BERNIER, J. J.: "Anuric nephritis": Treatment of nephropathies with testosterone propionate. Letter to Editor. J. Amer. med. Ass. **145**, 47 (1951).
DEUTSCH, M., SAXENA, B. M., LEHMANN, H. E.: Combined administration of thioridazine and fluoxymesterone in the treatment of geriatric patients. Curr. ther. Res. **12**, 805—809 (1970).
DIECKHOFF, J., SCHNEEWEISS, B.: Der Einfluß von katabolen und anabolen Steroiden auf die Antikörperbildung. Dtsch. Gesundh.-Wes. **20**, 189—194 (1965).
DITTRICH, H., SEIFERT, E.: Zur Anwendung anaboler Steroide bei Magenresezierten. Med. Klin. **59**, 821—823 (1964).
DOANE, B. D., FRIED, W., SCHWARTZ, F.: Response of uremic patients to nandrolone decanoate. Arch. intern. Med. **135**, 972—975 (1975).
DODMAN, B., CUNLIFFE, W. J., ROBERTS, B. E., SIBBALD, R.: Clinical and laboratory double-blind investigations on effect of fibrinolytic therapy in patients with cutaneous vasculitis. Brit. med. J. **1973 II**, 82—84.
DONTAS, S., PAPANICOLAOU, N. T., PAPANAYIOTOU, P., MALAMOS, B. K.: Long-term effects of anabolic steroids on renal functions in the aged subject. J. Geront. **22**, No. 3, 268—273 (1967).
DORN, W.: Landarzt **39**, 4 (1963).
DOWBEN, R. M.: Prolonged survival of dystrophic mice treated with 17α-ethyl-19-nortestosterone. Nature (Lond.) **184**, Suppl. 25, 1966 (1959).
DOWBEN, R. M.: Muscular dystrophy. Paper read at a Symposium on Anabolic Therapy, Detroit, Michigan, March 1962.

DOWBEN, R. M.: Treatment of muscular dystrophy with steroids. New Engl. J. Med. **268**, 912 (1963).
DOWBEN, R. M.: Anabole Steroide bei Myopathien. Myopathien. Stuttgart: Georg Thieme Verlag, 1965.
DOWBEN, R. M., PERLSTEIN, M. A.: Muscular dystrophy treated with norethandrolone. Arch. intern. Med. **107**, 245—251 (1961).
DOWBEN, R. M., VAWTER, G. F., BRANDFONBRENNER, A., SNIDERMAN, S. P., KAEGY, R. D.: Polymyositis and other diseases resembling muscular dystrophy. Arch. intern. Med. **115**, 584—594 (1965).
DOYLE, A. E., PINKUS, N. B., GREEN, J.: The use of oxandrolone in hyperlipidaemia. Med. J. Aust. **1**, 127—129 (1974).
DRESDALE, F. C., HAYES, J. C.: Potential dangers in the combined use of methandrostenolone and sodium warfarin. J. med. Soc. N.J. **64**, 609 (1967).
DUBARRY, J. J., AUCHE, Y., BERNARD, J.-P., FONTANGES, X.: Application thérapeutique des médications anabolisantes en gastro-enterologie à propos du Primobolan-dépôt. Gaz. méd. Fr. **72**, 2896—2898 (1965).
DUBARRY, J. J., TOURNERIE, J., MARQUIER, A.: L'androgénothérapie dans le traitement des cirrhoses. Presse méd. **71**, 1311—1313 (1963).
DUBOS, G., HUGONAT, R.: Therapeutic trial. Strombaject in geriatrics. Lyon méd. **23**, 183—186 (1974).
DUDLEY, H. A. F., ROBSON, J. S., SMITH, M., STEWART, C. P.: The permissive role of adrenal cortical hormones after injury in man. Metabolism **8**, 895 (1959).
DUTRUEL, R.: Sur trente cas de 'grands vieillards' malades chroniques traités par l'oxymétholone. Rev. franç. Géront. **11**, 59—62 (1965).
DUTZ, H., VOIGT, K., WENDLER, J.: Über die Beeinflußbarkeit der experimentellen Ratten-Nephritis durch Testosteron, Oestradiol, Kastration und somatotropes Hormon. Z. ges. inn. Med. **11**, 1115—1120 (1956).
EATON, J. C.: Metabolic effects in chronic renal failure. Proc. roy. Soc. Med. **52**, 511—514 (1959).
EBERHARDT, G., MÖHR, J., OEHLERT, W., SCHMIDT, E., SCHMIDT, F. W., VONDRÁŠEK, P.: Kontrollierte Studie zur therapeutischen Wirkung von B-Vitaminen und einem anabolen Steroid bei chronischer Hepatitis. Dtsch. med. Wschr. **100**, 2074—2082 (1975).
EDWARDS, M. S., CURTIS, J. R.: Decreased anticoagulant tolerance with oxymetholone. Lancet **1971 II**, 221.
EFIMOV, A. S.: Efficacy and mechanism of action of anabolic steroids in diabetic angiopathy. Ter. Arkh. **42**, 81—85 (1970).
EFIMOV, A. S., KARABUN, P. M.: Mechanism of action of anabolic steroids in the treatment of diabetes mellitus. Ter. Arkh. **44**, 90—94 (1972).
EGER, W.: Experimentelle Untersuchungen zur Frage der Leberschädigung und Oxymetholon im akuten und chronischen Versuch. In: Wirkung und Anwendung anaboler Steroide. Berlin: Medicus Verlag 1964.
EHNHOLM, G., HUTTUNEN, J. K., KINNUNEN, P., MIETTINEN, T. A., NIKKILA, E. A.: Effect of oxandrolone treatment on the activity of lipoprotein lipase, hepatic lipase and phospholipase $A_1$ of human post-heparin plasma. New Engl. J. Med. **292**, 1314—1317 (1975).
EHRLICH, H. P., HUNT, T. K.: The effects of cortisone and anabolic steroids on the tensile strength of healing wounds. Ann. Surg. **170**, 203—206 (1969).
EKERT, H., MUNTZ, R. H., COLEBATCH, J. H.: Decreased anticoagulant tolerance with oxymetholone. Lancet **1971 II**, 609—610.
EKISENINA, N. I., KRUMS, L. M., LUKASH, L. K., MEDVEDEVA, T. F., TAITS, N. S., FURSOVA, G. A., CHANYSHEVA, R. I.: Nerobol in a complex therapy of chronic enterocolitis. Sovetsk. Med. **35**, 68—72 (1972).
EL-ABRASHI, N., EL-DANASONRY, M.: Trial of anabolic agents in cases of chronic renal failure with uremia. J. Egypt. med. Ass. **44**, 333—339 (1961).
EL-MEHAIRY, M. M., EL-TARABISHI, N.: Study of the effect of combining anabolics with steroids in long-term therapy of bronchial asthma. J. Egypt. med. Ass. **43**, 646—651 (1966).
ENDLER, F.: Praktische Erfahrungen mit einem neuen anabolen Wirkstoff in der Orthopädie. Wien. klin. Wschr. **73**, 481—485 (1961).

Esposito, R., Pluvio, M., Giordano, D.: Anabolic agents in kidney disease: the effect of formebolone on protein synthesis in patients with renal insufficiency or nephrosis. Curr. Med. Res. Opin. **3**, 43—45 (1975).

Evans, J. T., Elias, E. G.: The erythropoietic response to anabolic therapy in patients receiving radiotherapy. J. clin. Pharmacol. **12**, 101—104 (1972).

Fabrykant, M.: Diabetic retinopathy: Preliminary report on the use of anabolic steroids and high-protein low-fat diets. J. Amer. Geriat. Soc. **10**, 743—748 (1962).

Fabrykant, M., Gelfand, M. L., Rosenberg, A. S.: Further experience with anabolic steroids in diabetic retinopathy: Factors precipitating retinal hemorrhages. Amer. J. med. Sci. **248**, 304—316 (1964).

Faergeman, O., Damgaard-Federsen, F.: Increase of post-heparin lipase activity by oxandrolone in familial hyperchylomicronemia. Scand. J. clin. Lab. Invest. **31**, 27—31 (1973).

Falkson, G., Falkson, H. C., Minnaar, P. C.: A clinical evaluation of stanozolol as an adjuvant to cytostatic treatment of cancer. S. Afr. Cancer Bull. **8**, 54—62 (1964).

Fallat, R.: Effects of anabolic and progestational agents upon triglycerides and triglyceride kinetics in normal and hyperlipemic patients. Lipids **9**, 117—120 (1974).

Fallat, R. W., Glueck, C. J.: Familial and acquired type V hyperlipoproteinemia. Atherosclerosis **23**, 41—62 (1976).

Falliers, C. J., Jorgensen, J. R., Tan, L. S., Bukantz, S. C.: Anabolic effects of stanozolol. Amer. J. Dis. Child. **106**, 388—401 (1963).

Fam, A.: The use of anabolizers in urology. J. Egypt. med. Ass. **46**, 100—110 (1963).

Fanfani, A., Ricca, M., Cherici, P.: Esperienze di terapia anabolizzante in vecchi TBC. G. Geront. **10**, 659—671 (1962).

Fantoli, U., Boccitto, G., Lorenzetti, L.: Azione del fenilpropionato di 19-nor-androstenolone sulla disprotidemia in soggetti affetti da tubercolosi polmonare. Ann. Ist. Forlanini **2**, 218—237 (1959).

Fastner, Z.: X. Intern. Congress of Gerontology, Jerusalem (1975).

Fauchet, R.: Utilisation du stanozolol chez les malades immobilisés. Sem. Hôp. Paris 70—74 (1966).

Fearnley, G. R.: Ethyloestrenol and postoperative venous thrombosis. Lancet **1973 II**, 95.

Fearnley, G. R.: Fibrinolysis and deep-vein thrombosis. Lancet **1974 I**, 454—455.

Fearnley, G. R., Chakrabarti, R.: Increase of blood fibrinolytic activity by testosterone. Lancet **1962 II**, 128.

Fearnley, G. R., Chakrabarti, R.: Pharmacological enhancement of fibrinolytic activity of blood. J. clin. Path. **17**, 328 (1964).

Fearnley, G. R., Chakrabarti, R.: Oral fibrinolytic agents. Brit. med. J. **1968 III**, 498.

Fearnley, G. R., Chakrabarti, R.: Phenformin and ethyloestrenol for Raynaud's disease. Lancet **1969 II**, 1203.

Fearnley, G. R., Chakrabarti, R., Evans, J. F.: Fibrinolytic treatment of rheumatoid arthritis with phenformin plus ethyloestrenol. Lancet **1966 II**, 757—761.

Fearnley, G. R., Chakrabarti, R., Evans, J. F.: Effect of ethyloestrenol combined with phenformin or with metformin on platelet stickiness and serum-cholesterol in patients with occlusive vascular disease. Lancet **1968 II**, 1004—1007.

Fearnley, G. R., Chakrabarti, R., Evans, J. F.: Fibrinolytic and defibrinating effect of phenformin plus ethyloestrenol in vivo. Lancet **1969 I**, 910—914.

Fearnley, G. R., Chakrabarti, R., Evans, J. F.: Mode of action of phenformin plus ethylestrenol on fibrinolysis. Lancet **1971 I**, 723—725.

Fearnley, G. R., Chakrabarti, R., Hocking, E. D., Evans, J. F.: Fibrinolytic effects of diguanides plus ethyloestrenol in occlusive vascular disease. Lancet **1967 II**, 1008—1011.

Fenster, L. F.: The nonefficacy of short-term anabolic steroid therapy in alcoholic liver disease. Ann. intern. Med. **65**, 738—744 (1966).

Fiegel, G.: Beeinflussung des Myocardstoffwechsels durch anabole Substanzen, ihre Verwertung in der Rehabilitationsbehandlung chronischer Herzmuskelschäden. Med. Klin. **57**, 519—523 (1962a).

Fiegel, G.: Die reparativen Eigenschaften anaboler Steroide am chronisch geschädigten Myokard. Wien. med. Wschr. **112**, 951—954 (1962b).

FIEGEL, G., BARGHEER, R.: Katamnestische Ergebnisse nach Behandlung Herzkranker mit anabolen Steroiden. Med. Klin. **61**, 1870—1872 (1966).
FIEGEL, G., BARGHEER, R., KUKWA, D.: Das Verhalten des Lipidstoffwechsels unter gleichzeitiger Gabe von Anabolika und essentiellen Phospholipiden. Arzneimittel-Forsch. **16**, 443—446 (1966).
FIEGEL, G., EINFELDT, H.: Die reparativen Eigenschaften anaboler und androgener Steroide in der Rehabilitationsbehandlung von 2546 Herzkranken. Therapiewoche **21**, 4042—4046 (1971).
FIEGEL, G., FISCHER, E. K.: Antitoxisch-anabole Behandlung von Leberschäden. Ärztl. Prax. **11**, 1526—1527 (1959).
FIEGEL, G., KELLING, H.-W.: Das Verhalten von Nor-Androstenolonphenylpropionat bei schweren Leberschädigungen. Vergleichende Untersuchungen zu früheren Ergebnissen mit Testosteron-Propionat. Ärztl. Forsch. **13**, I/158-I/161 (1959a).
FIEGEL, G., KELLING, H.-W.: Die Beeinflussung des schweren Leberparenchymschadens unter Zufuhr hochdosierter Androgenderivate und von Corticosteroiden. Ärztl. Forsch. **13**, I/574—I/578 (1959b).
FIEGEL, G., KUKWA, D.: Experimentelle Studie über das Verhalten der Blutlipide während längerer Einwirkung verschiedener anaboler Steroide. Arzneimittel-Forsch. **18**, 452—456 (1968).
FIGUEROA, R. B.: Mesterolone in steatosis and cirrhosis of the liver. Acta hepato-gastroenterol. **20**, 282—290 (1973).
FIXSON, U.: Methenolonoenanthat (Primobolan-Depot) in der postoperativen Behandlung. Med. Welt (Stuttg.) **19**, 1109—1114 (1962).
FIXSON, U.: Einfluß von Anabolika auf den postoperativen Verlauf. In: Wirkung und Anwendung anaboler Steroide. Berlin: Medicus Verlag 1964.
FIXSON, U.: Zur Wirkung von Anabolika auf den postoperativen Verlauf nach gynäkologischen Karzinomoperationen. In: Wirkung und Anwendung anaboler Steroide. Berlin: Medicus Verlag 1964.
FÖLSCH, E., BRÜGGER, W., IMMICH, H.: Zur Beeinflussung der hämatotoxischen Effekte von Cyclophosphamid durch das anabole Steroid Nortestosteronphenylpropionat. Verh. dtsch. Ges. inn. Med. 73. Kongreß 1967, München: J. F. Bergmann 1967.
FÖLSCH, E., BRÜGGER, E. W., BRANDS, W.: Zur Beeinflussung der hämotoxischen Effekte von Cyclophosphamide durch anabole Steroide. Arzneimittel-Forsch. **25**, 59—66 (1975).
FONTAINE, M., MALMENDIER, C. L.: Carbohydrate content of human VLDL, IDL, LDL and HDL plasma apoproteins from fasting normal and hyperlipemic patients. Clin. chim. Acta **64**, 91—93 (1975).
FORSIUS, H.: The retinopathic initial symptoms of juvenile diabetes. Acta ophthal. (Kbh.) **45**, 255—256 (1967).
FOSS, G. L., SIMPSON, S. L.: Oral methyltestosterone and jaundice. Brit. med. J. **1959 I**, 259—263.
FOSSARD, D. P., FIELD, E. S., KAKKAR, V. V., FRIEND, J. R., CORRIGAN, T. P., FLUTE, P. T.: Fibrinolytic activity and postoperative deep-vein thrombosis. Lancet **1974 I**, 9—11.
FOSSATI, C.: Anabolic steroids in the treatment of chronic pulmonary tuberculosis. Panminerva med. **5**, 381—383 (1963).
FOSSATI, C.: Sull'impiego degli steroidi anabolizzanti in terapia. Clin. ter. **70**, 169—202 (1974).
FOWLER, W. M., PEARSON, C. M., EGSTROM, G. H., GARDNER, G. W.: Ineffective treatment of muscular dystrophy with an anabolic steroid and other measures. New Engl. J. Med. **272**, 875—882 (1965).
FOZ, M.: Esteroides anabolizantes. I—algunas consideraciones sobre esteroides anabolizantes. Ann. Med. (Med.) **50**, 194—202 (1965).
FRANKEN, F. H., DAWECKE, H., GRIES, E. A., SCHWEINITZ, H. A. v., HOLZGREWE, H., FORSTMANN, U.: Langzeitbehandlung chronischer Lebererkrankungen mit Prednison und Methandrostenolon. Dtsch. med. Wschr. **88**, 1979—1985 (1963).
FREEDMAN, P., SPENCER, A. G.: Testosterone propionate in the treatment of renal failure. Clin. Sci. **16**, 11 (1957).
FRÖHLICH, R., MUMENTHALER, M., WIESENDANGER, M.: Ergebnisse der Anwendung von anabolen Hormonen bei myatrophischen Erkrankungen. Praxis **51**, 1296—1300 (1962).
FURMAN, R. H., HOWARD, R. P., BERKOWIT, D.: Idiopathic hyperglyceridaemia treated with methyltestosterone and methandienone. Lancet **1963 I**, 837.

GABUZDA, G. J., JR., ECKHARDT, R. D., DAVIDSON, C. S.: Effect of choline and methionine, testosterone propionate and dietary protein on nitrogen balance in patients with liver disease. J. clin. Invest. **29**, 566 (1950).

GAILLARD, L., DELPHIN, D., FERRAT, Y.: La méthode de Nasso: une méthode imprévue mais efficace de prophylaxie de la varicelle en collectivité d'enfants. Pédiatrie **22**, 353—355 (1967).

GAUTIER, E., TÖNZ, O.: Anaboles Hormon in der Behandlung der Anurie im Kindesalter. Helv. med. Acta **27**, 535—538 (1960).

GEOFFROY, H., COUDOUX, P., MAFFIOLI, C., CHARDON, P., DALLE, M.: Etude de l'effet thérapeutique d'un nouvel anabolisant: le bêta (p-hexyloxyphenyl) propionate de 19 nortestosterone, Anador. Gaz. méd. Fr., **77**, 3842—3846 (1970).

GERBER, W., COTTIER, P.: Zur Wirkung anaboler Hormone auf die Azotämie bei Nierenversagen. Helv. med. Acta **28**, 197—215 (1961).

GEROK, W., SICKINGER, S., HENNEKEUSER, H. H. (Hrsg.): Alkohol und Leber. Stuttgart-New York: Schattauer 1971.

GERRITZEN, F. M.: Diabetische retinopathie. Ned. T. Geneesk. **119**, 964—969 (1975).

GESSNER, P. K., GESSNER, T.: The interaction of barbital and testosterone relative to their hypnotic effects. Arch. int. Pharmacodyn. **201**, 52—58 (1973).

GEYER, G., JESSERER, H.: Kann der gesteigerte Eiweiß-Katabolismus während einer Cortisontherapie durch anabole Steroide ausgeglichen werden? In: Die endokrine Behandlung des Mamma- und Prostatacarcinoms. Endokrine Regulationen des Kohlehydratstoffwechsels. Berlin-Göttingen-Heidelberg: Springer 1961.

GEYER, G., JESSERER, H.: Ergebnisse und Probleme der Therapie mit eiweißanabolen Steroiden, diskutiert an Erfahrungen mit Methandrostenolon (Dianabol/Ciba). Dtsch. Arch. klin. Med. **208**, 252—278 (1962).

GHERONDACHE, C. N., DOWLING, W. J., PINCUS, G.: Metabolic changes induced in elderly patients with an anabolic steroid oxandrolone. J. Geront. **22**, 290 (1967).

GHIONE, M.: Anti-infective action of an anabolic steroid. Proc. Soc. exp. Biol. **97**, 773—775 (1958).

GIBBA, A., PETTERINO, E.: Studio sull'azione renoprotettiva di un nuovo steroide anabolizzante: il 19-nor-androstenolone decanoato. Minerva urol. **14**, 3—15 (1962).

GILES, A. R.: Fibrinolytic activity and deep-vein thrombosis. Lancet **1974 I**, 803.

GILLIAM, J. N., HERNDON, H., PRYSTOWSKY, S. D.: Fibrinolytic therapy for vasculitis of atrophie blanche. Arch. Derm. Symp. (Chic.) **109**, 664—667 (1974).

GIROLAMI, M.: La cirrosi del fegato può guarire clinicamente in una alta percentuale di casi con trattamento a base di testosterone e vitamina $B_1$ ad alte dosi. Minerva med. **64**, 1567—1623 (1973).

GJØRUP, S., THAYSEN, J. H.: The effect of anabolic steroid (Durabolin) in the conservative management of acute renal failure. Acta med. scand. **167**, 227—238 (1960).

GLUECK, C. J.: Progestins, anabolic-androgens, estrogens: effects on triglycerides and lipases. Clin. Res. **17**, 475 (1971 a).

GLUECK, C. J.: Effects of oxandrolone on plasma triglycerides and postheparin lipolytic activity in patients with types III, IV, and V familial hyperlipoproteinemia. Metabolism **20**, 691—702 (1971 b).

GLUECK, C. J., FALLAT, R. W., MENDOZA, S.: Effect of sex hormones on protamine inactivated and resistant post-heparin plasma lipases. Clin. Res. **23**, 498A (1975).

GLUECK, C. J., FORD, S., FALLATT, R., STEINER, P.: Triglyceride removal efficiency and lipases: Effects of oxandrolone. Circulation **46**, Suppl. II, 263 (1972).

GLUECK, C. J., FORD, S., STEINER, P., FALLAT, R.: Triglyceride removal efficiency and lipoprotein lipases—effects of oxandrolone. Metabolism **22**, 807 (1973).

GLUECK, C. J., STEINER, P., SWANSON, P.: Amelioration of endogenous hyperglyceridemia by norethindrone acetate and oxandrolone. Clin. Res. **18**, 454 (1970a).

GLUECK, C. J., SWANSON, F., HUTSELL, T.: Effects of oxandrolone and norethindrone acetate on tissue and plasma lipases, and triglycerides. Circulation **42**, Suppl. 3, 158 (1970b).

GLUECK, H. I., GLUECK, C. J.: Clotting mechanism in patients with hypertriglyceridemia during therapy with anabolic or progestational drugs. Clin. Res. **18**, 611 (1970).

GLUECK, H. I., GLUECK, C. J.: Clotting mechanism in patients with hypertriglyceridemia during therapy with anabolic or progestational drugs. Thrombos. Diathes. haemorrh. (Stuttg.) **29**, 499—509 (1973).

GOEBEL, F.-D.: Diabetes mellitus und Mikroangiopathie. Dtsch. med. Wschr. **100**, 382—385 (1975).

GOKHALE, B. B., KOTNIS, M. G., KASABEDAR, P. V.: A preliminary note on anabolic therapy in leprosy. J. trop. Med. Hyg. **67**, 284—285 (1964).

GOLDENBERG, I. S., SEDRANSK, N., VOLK, H., SEGALOFF, A., KELLEY, R. M., HAINES, C. R.: Combined androgen and antimetabolite therapy of advanced female breast cancer. Cancer **36**, 308—310 (1975).

GOLDMAN, A. G., ZARDAY, Z.: Effects of oxandrolone in advanced renal insufficiency. N.Y. St. J. Med. **73**, 316—319 (1973).

GOLDSTEIN, J., COMBES, B.: The effect of steroids on the activity of the enzyme that catalyzes sulfobromophthalein-glutathione conjugation. J. Lab. clin. Med. **67**, 830—835 (1966).

GOMES, J. M. P., LAGRECA, J. R., MORAES, C. R.: Emprêgo de anabolisante em cirurgia cardiaca. Hospital (Rio de J.) **78**, 139—145 (1970).

GONNARD, P., LEFRANC, G., GOSSE, R., PARIS, E.: Recherches sur les propriétés protéino-anabolisantes de l'androstanolone base. Sem. Hôp. Paris **33**, 1 (1957).

GOODMAN, J., BESSMAN, A. N.: Effect of nortestosterone decanoate on red cell 2,3 diphosphoglycerate and hematocrit in hemodialysis patients. Clin. Pharmacol. Ther. **17**, 167—170 (1975).

GOROKHOVSKY, B. I., KITAEVA, I. T.: The use of anabolic hormones in myocardial infarction. Klin. Med. (Moskau) **48**, 29—34 (1970).

GRAMSCH, H.: Klinischer Beitrag über die Anwendung von Primobolan bei Hepatopathien. Med. Welt (Stuttg.) **1963 IV**, 212—214.

GRAMSCH, H.: Beobachtungen über die orale Methenolontherapie bei Hepatopathien. In: Wirkung und Anwendung anaboler Steroide. Berlin: Medicus Verlag 1964.

GRIBBIN, H. R., FLAVELL MATTS, S. G.: Mode of action and use of anabolic steroids. Brit. J. clin. Pract. **30**, 3—5, 7—9 (1976).

GRÖNBERG, A., SVANTESON, G.: Svenska Laekartidn. **48**, 2005 (1951) cited in W. NORDLÖW, A. GRÖNBERG, G. SVANTESON: Sexualhormonterapi vid retinopathia diabetica. Nord. Med. **49**, 742—744 (1953).

GROSSMAN, J., YALOW, A. A.: Effects of anabolic steroids on albumin metabolism. J. clin. Endocr. **25**, 698—707 (1965).

GUDBJARNASON, S., FENTON, J. C., WOLF, P. L., BING, R. J.: Stimulation of reparative processes following experimental myocardial infarction. Arch. intern. Med. **118**, 33—40 (1966).

GÜRTLER, J.: Untersuchungen über die Möglichkeit der Steigerung des Körpergewichts mit Stromba. Praxis **52**, 900—902 (1963).

HABERLER, H., PATEISKY, K.: Dystrophia musculorum progressiva: Results of clinical and biochemical examinations in the course of treatment with an anabolic substance (norandrostenolondecanoate). Excerpta med. Int. Congr. Ser. **38**, 161—162 (1961).

HAGENBUCHNER, K.: Über den Anwendungsbereich anabol wirksamer Steroide in der Neurologie. Wien. klin. Wschr. **74**, 214—219 (1962).

HAGGENMÜLLER, F.: Anabole Steroide bei Kindern mit nephrotischem Syndrom. In: Wirkung und Anwendung anaboler Steroide. Berlin: Medicus Verlag 1964.

HAGGENMÜLLER, F.: Wirkung der Anabolikumbehandlung bei einem Kind mit Debré-de Toni-Fanconi-Erkrankung und Zystin-Speicherkrankheit. In: Wirkung und Anwendung anaboler Steroide. Berlin: Medicus Verlag 1964.

HAMERSKI, W.: Durabolin in the treatment of diabetic proliferative angioretinopathy. Klin. oczna **44**, 461—465 (1974).

HAMILTON, L. D., BENNETT, J. L., SILVER, J.: Nandrolone phenpropionate in the treatment of geriatric patients with chronic brain damage. J. Amer. Geriat. Soc. **12**, 373—378 (1964).

HAMILTON, M.: Lectures on the Methodology of Clinical Research, 2nd Ed. Edinburgh-London: Churchill Livingstone 1974.

HAMMER, F.: Zur Behandlung von Altersherzerkrankungen mit anabolen Steroiden. Med. Welt. (Stuttg.) **17**, (N.F.) 2814—2817 (1966).

HANSTEN, PH. D.: Drug Interactions, 2nd Ed. Philadelphia: Lea & Febiger 1973.

HANTSCHMANN, N., MATZELT, D., MERTENS, H. G., NOWAKOWSKI, H.: Zur Behandlung von Muskelkrankheiten mit anabolen Steroiden. Dtsch. med. Wschr. **87**, 2619—2634 (1962).

HARN, C. S.: Clinical experiences with Winstrol. M.D. Journal **12**, 154—161 (1963).

HARRIS, E. L., FITZGERALD, J. D.: The Principles and Practice of Clinical Trials. Edinburgh-London: E. & S. Livingstone 1970.

HARTENBACH, W.: Indikationen und Kontraindikationen für die Anwendung kataboler und anaboler Hormone in der Chirurgie. Münch. med. Wschr. **104**, 1567—1574 (1962).

HASLOCK, D. I., WRIGHT, V.: A corticosteroid-sparing agent in rheumatoid arthritis. A trial of nandrolone decanoate (Deca-Durabolin). Clin. Trials J. **4**, 31—34 (1971).

HECHT, Y.: Traitement de l'insuffisance hépatocellulaire des cirrhoses par les anabolisants. Rev. franç. Gastro-Entérol. **65**, 51—61 (1971).

HEDNER, U., NILSSON, I. M., ISACSON, S.: Phenformin and ethyloestrenol in recurrent deep venous thrombosis. Presented at 10. Erfurter Konferenz über Hämostase und Thrombose, Mai 1975. Folia haemat. (Lpz.) In press. (1975a).

HEDNER, U., NILSSON, I. M., ISACSON, S.: Effect of ethyloestrenol on fibrinolysis in the vessel wall. Thrombos. Diathes. haemorrh. (Stuttg.) **34**, 609 (1975b).

HELLMANN, L., BRADLOW, H. L., ZUMOFF, B., FUKUSHIMA, D. K., GALLAGHER, T. F.: Thyroid-androgen interrelations and the hypocholesteremic effect of androsterone. J. clin. Endocr. **19**, 936 (1959).

HENDLER, E. D., GOFFINET, J. A., ROSS, S., LONGNECKER, R. E., BAKOVIC, V.: Controlled study of androgen therapy in anemia of patients on maintenance hemodialysis. New Engl. J. Med. **291**, 1046—1051 (1974).

HENKES, H. E., RIASKOFF, S., HOUTSMULLER, A. J.: Nieuwe ontwikkelingen in de behandeling van de diabetische retinopathie. Ned. T. Geneesk. **115**, 1765—1769 (1971).

HERK, E. J., VAN: An investigation of the effect of the steroid ethylestrenol on facial hairgrowth. Acta endocr. **41**, 407—410 (1962).

HERRMANN, J. B., WOODWARD, S. C.: An experimental study of wound healing accelerators. Amer. Surg. **38**, 26—34 (1972).

HEYCK, H., LAUDAHN, G., LÜDERS, C. J., MÜLLER-STEPHANN, H., SCHMIDT-PETER, P.: Anabolic steroids and digitoxin in the treatment of progressive muscular dystrophy. Acta paediat. scand. **54**, 205—217 (1965).

HIEKKALA, H.: The influence of anabolic hormones on retinopathy in juvenile diabetics. Diabetologia **1**, 134 (1965).

HIRAYAMA, C., KIMURA, N., MASUYA, T.: Anabolic steroid effect on hepatic protein synthesis in patients with liver cirrhosis. Digestion **3**, 41—47 (1970).

HOCK, C. W.: Medical treatment of the functionally ill geriatric patient. Geriatrics **16**, 182—184 (1961).

HOFMANN, H., ZEPPEK, E.: Über die Behandlung der Retinopathia diabetica mit einem anabolen Steroid. Klin. Mbl. Augenheilk. **143**, 821—832 (1963).

HONEGGER, H., BOHLEBER, U.: Klinik und Therapie der Retinopathia diabetica. Med. Mschr. **25**, 298—308 (1971).

HORKY, J.: The use of anabolic steroids in the treatment of liver diseases. T. Gastro-ent. **10**, 109—126 (1967).

HORN, Y.: Leucopoietic properties of four different androgenic hormones in irradiated rats. Scand. J. Haemat. **8**, 470—475 (1971).

HOUTSMULLER, A. J.: The therapeutic applications of anabolic steroids in ophthalmology: biochemical results. Acta endocr. (Kbh.) Suppl. **63**, 154—169 (1961).

HOUTSMULLER, A. J.: Influence of deca-durabolin on the usual course of the glucose tolerance test in diabetes. Diabetologia **1**, 80 (1965).

HOUTSMULLER, A. J., HENKES, H. E.: Klinische und biochemische Effekte von Nandrolon phenylpropionat (Durabolin) und Nandrolondecanoat (Deca-Durabolin) bei diabetischer Retinopathie. Ber. dtsch. ophthal. Ges. **63**, 57—62 (1960).

HOUTSMULLER, A. J., HENKES, H. E.: Clinical investigations of patients with diabetic retinopathy treated with anabolic steroids. Paper presented at the Vth Congress of the International Diabetes Federation, Toronto, July, 1964.

HOUTSMULLER, A. J., HENKES, H. E., VAN POPPEL, A. L. A.: Clinical investigation of patients with diabetic retinopathy treated with anabolic steroids. Paper read at the Joint Congress Irish Ophthalmological Society—British Diabetic Society, Dublin, April 1963.

HOUTSMULLER, A. J., VAN POPPEL, A. L. A.: Treatment of diabetic retinopathy with anabolic steroids. Ophthalmologica (Basel) **145**, 185—206 (1963).

HUBER, E. G.: Therapeutische Erfahrungen mit anabolen Hormonen. Münch. med. Wschr. **103**, 1133 (1961).

HUGUET, F.-J.-D., TOURNIGAND, P., DURAN, G.: Effets de l'androstanolone-base sur 120 malades. Presse méd. **69**, 2433—2434 (1961).

HUNTER, P. R., COTTON, S. G., KELSEY, J. H., BLOOMS, A.: Controlled trial of methandienone in treatment of diabetic retinopathy. Brit. med. J. **1967 III**, 651—653.

HUSEMAN, C., JOHANSON, A.: Growth hormone deficiency in anorexia nervosa. J. Pediat. **87**, 946—948 (1975).

HUSSAR, D. A.: Drug interactions. Amer. J. Pharm. **145**, 65—116 (1973).

HVIDBERG, E. F., DAYTON, P. G., READ, J. M., WILSON, C. H.: Studies of the interaction of phenylbutazone, oxyphenbutazone, and methandrostenolone in man. Proc. Soc. exp. Biol. (N.Y.) **129**, 438—443 (1968).

IGARI, J., HAYASHI, Y., USHIO, M., MIKI, S., IWADO, M.: The effect of long-term administration of anabolic steroid upon total serum cholesterol level. Jap. J. Geriat. **3—4**, 275—282 (1966).

IGIĆ, R., JELICIC, J., NIKULIN, E., STERN, P.: Myotrophischer Effekt des Metenolazetat auf die experimentelle Muskeldystrophie der Ratte. Endokrinologie **54**, 103—106 (1969).

IISALO, E., KASANEN, A., KÄRKI, N., TALA, E.: The effect of an anabolic steroid, nor-testosterone phenyl propionate (durabolin), on the urinary and plasma steroids in patients with normal and decreased renal function. Acta med. scand. (Kbh.) **171**, 183—186 (1962).

IOVINE, G., GIERSON, H. W.: Methandrostenolone as an anabolic agent in emphysema. Curr. ther. Res. **5**, 219—231 (1963).

ISACSON, S., NILSSON, I. M.: Effect of treatment with combined phenformin and ethyloestrenol on the coagulation and fibrinolytic systems. Scand. J. Haemat. **7**, 404—408 (1970).

JAFFE, M. D.: Ethylestrenol, coronary artery disease and collaterals. Clin. Res. **21**, 813 (1973).

JAFFE, M. D.: Improvement in postexercise "ischemic" electrocardiographic changes by testosterone cypionate treatment. Clin. Res. **22**, No. 4, 595 A (1974).

JAMES, V. H. T., LANDON, J., WYNN, V.: Effect of an anabolic steroid (methandienone) on the metabolism of cortisol in the human. J. endocr. (Kbh.) **25**, 211—220 (1962).

JANNUZZI, C., BASSETTI, D., FRIGERIO, G.: Ricerche cliniche controllate sull'effetto di steroidi anabolizzanti nell'epatite virale. Minerva med. **61**, 1902—1911 (1970).

JAPA, J., NARBUTT, B., HANKIEWICZ, J., STAS, J.: The influence of encorton and of dianabol administration on protein metabolism in Addison's disease. Endokr. pol. **17**, 1—10 (1966).

JEFIMOV, A. S.: Die Behandlung der diabetischen Angiopathien. IV. Internationales Donau-Symposium über Diabetes mellitus, Dubrovnik, Abstract 133 (1975).

JOHNSTON, I. D. A., BURN, J. I.: The effect of nandrolone (durabolin) on the white cell counts of rats treated with chlorambucil. Paper read at a Symposium at the University Dept. of Surgery, Royal Infirmary, Sheffield, Nov. 1966.

JOHNSON, P. C., JR.: Weight gain in geriatric patients. Paper presented at a Symposium on Anabolic Therapy, Detroit, Michigan, March 1962.

JOHNSON, R. H., MCLELLAN, D. L.: Multiple sclerosis. Practitioner **209**, 183—190 (1972).

JONES, C. R., VAN ITALLIE, T. B., STEVENSON, S. S., PIGMAN, W. W.: Decrease in hydroxyproline excretion in a patient with Marfan's syndrome after methandrostenolone therapy. J. Amer. med. Ass. **207**, 2085—2086 (1969).

JOSE, A. D., MITCHELL, A. S.: Elevation of serum cholesterol by an anabolic steroid. Lancet **1964 I**, 473—474.

JUNGCK, E. C., GREENBLATT, R. B.: Methandrostenolone, an oral anabolic agent. Sth. med. J. (Bgham, Ala.) **57**, 909—913 (1964).

KAESER, H. E.: Zur Frage der Behandlung der Myopathien mit Sexualhormonen und ihren Derivaten. Nervenarzt **32**, 38—40 (1961).

KAHN, S. B., CONROY, J. F., BRODSKY, I.: Effect of nandrolone phenpropionate on myelopoiesis. J. med. exp. Clin. **5**, 303—309 (1974).

KAISER, H.: Cortisonderivate in Klinik und Praxis. 6th Ed., Stuttgart: Thieme 1973.

KALLIOMÄKI, J. L., MARKKANEN, T. K., PIRILÄ, A. M., RUIKKA, I.: Norandrostenolone decanoate—a new anabolic steroid with prolonged effect. Ann. med. intern. fenn. **50**, 178—184 (1961 a).

KALLIOMÄKI, J. L., PIRILÄ, A. M., RUIKKA, I.: A therapeutic trial with ethyl-estrenol in geriatric patients. Acta endocr. (Kbh.) Suppl. **63**, 124—128 (1961 b).

KALLIOMÄKI, J. L., SEPPÄLÄ, P.: Norandrostenolone decanoate as a cardiac anabolizer studied by means of electrocardiographic changes. Cardiologia **43**, 124—128 (1963).

KARJALA, R. J., FORD, R. V.: Prolonged anticatabolic activity of nandrolone decanoate during corticosteroid therapy. Geriatrics **19**, 511—516 (1964).

KASICH, A. M.: Clinical evaluation of nandrolone phenpropionate in patients with gastroinstestinal disease. Amer. J. Gastroent. **40**, 628—633 (1963).

KERREBIJN, K. F., DELVER, A.: Ethylestrenol (Orgabolin): Effects on asthmatic children during corticosteroid treatment. Scand. J. Resp. Dis. Suppl. **68**, 70—77 (1969).

KESSEL, M.: Untersuchungen über den Einfluß anaboler Substanzen auf die Stickstoff-Retention bei akutem Nierenversagen. In: Wirkung und Anwendung anaboler Steroide. Berlin: Medicus Verlag 1964.

KILL, J. M.: Behandlung einer an Brustkrebs Erkrankten mit Endoxan und anabolem Hormon. Erfahrungsbericht. Münch. med. Wschr. **103**, 1767—1768 (1961).

KINK, R.: Über die Wirkung anaboler Steroide beim Diabetes mellitus. Med. Mschr. **16**, 463—467 (1962).

KIRK, J., HANDLEY, D. A.: Experiences in the use of fibrinolytic agents in Behçet's syndrome. Aust. J. Derm. **13**, 5—10 (1972).

KLEIBEL, F.: Die Wirkungen von Androstan-17$\beta$-ol-3-on auf den Stoffwechsel Karzinomkranker während der Strahlenbehandlung. Strahlentherapie **118**, 446—453 (1962 a).

KLEIBEL, F.: Die anabole Wirkung von Norsteroiden und Ethylestrenol in der Strahlenbehandlung. Münch. med. Wschr. **104**, 2514—2518 (1962 b).

KLEIBEL, F.: Die Behandlung kataboler Zustände in der Radiologie. Med. Welt **47**, 2529 (1962c).

KLEIBEL, F.: Wirkung von Anabolika auf den Stoffwechsel Tumorkranker während der Strahlentherapie. In: Wirkung und Anwendung anaboler Steroide. Berlin: Medicus Verlag 1964a.

KLEIBEL, F.: Klinische Veränderungen am Eiweiß-, Stickstoff-, Nukleoproteid-, Purin-, Phosphor-, Elektrolyt- und Wasserhaushalt während der Strahlentherapie. In: Einführung in die klinisch-experimentelle Radiologie. München-Berlin: Urban & Schwarzenberg 1964 b.

KLEIBEL, F.: Langwirkendes Anabolikum (Clinibolin) als Adjuvans in der Tumortherapie. Münch. med. Wschr. **111**, 2452—2454 (1969).

KLEIBEL, F.: Klinische Ergebnisse der Kombinationsbehandlung von $^{32}$Phosphor und Anabolika bei multiplen Knochenmetastasen. Fortschr. Röntgenstr. **116**, 66—67 (1972).

KLIACZKO, W. R., TIRKINA, T. N., MICHAJLOWA, I. N. A.: Diabetic retinopathy and its complex treatment with particular reference to therapy with anabolic steroids. Endokr. pol. **21**, 383—392 (1970).

KLUTHE, R.: Die N-Bilanz bei nephrotischem Syndrom unter verschiedenen Steroiden. In: Wirkung und Anwendung anaboler Steroide. Berlin: Medicus Verlag 1964.

KLUTHE, R., LAMPE, J.: Experimenteller Beitrag zur Strahlentherapie der Ratten-Masugi-Nephrose. In: Wirkung und Anwendung anaboler Steroide. Berlin: Medicus Verlag 1964.

KLYACHKO, V. R., TIRKINA, T. N., RYABCHUK, O. G., MAZOVETSKY, A. G.: Treatment of diabetic retinopathies. Probl. Endokr. **17**, 25—30 (1971).

KOCHAKIAN, C. D.: A comparison of the renotrophic with the androgenic activities of various steroids. Amer. J. Physiol. **142**, 315—325 (1944).

KÖNIG, H.: Behandlung der Retinopathia diabetica. Dtsch. med. Wschr. **97**, 1196 (1972).

KOLODNY, A. L.: Methandrostenolone (Dianabol) in the clinical management of weight deficit. Med. Tms (N.Y.) **91**, 9—19 (1963).

KOMAROV, F. I., LISOVSKY, V. A., LISOVSKY, I. V., SCHEDRUNOV, V. V.: The effect of nerobol and aminocrovin on the free amino acids content in peptic ulcer. Sovetsk. Med. **36**, 83—87 (1973).

KONRAD, R. M., AMMEDICK, U., HUPFAUER, W., RINGLER, W.: Der Effekt anaboler Steroide auf die Stickstoffbilanz bei Patienten nach Lungenoperationen. Chirurg **38**, 168—171 (1967).

KOPERA, H.: Über therapeutische Versuche mit anabolen Steroiden bei diabetischer Retinopathie. Abhandl. dtsch. Akad. Wiss. Berlin Kl. Med. **3**, 361—404 (1964).

KOPERA, H.: Clinical Experiences with Anabolic Steroids in some Special Indications. Hormonal Steroids, Biochemistry, Pharmacology and Therapeutics, Vol. 2, pp. 195—200. New York: Academic Press 1965.
KORST, D. R., BOWERS, C. Y., FLOKSTRA, J. H., MCMAHON, F. G.: Clinical evaluation of a new anabolic agent 7α,17α-dimethyltestosterone (bolasterone). Clin. Pharmacol. Ther. **4**, 734—739 (1963).
KRACHT, J.: Anabole Hormone und sekundäre Nebennierenrindenatrophie. In: Wirkung und Anwendung anaboler Steroide. Berlin: Medicus Verlag 1964.
KRAL, V. A., WIGDOR, B. T.: Androgen effect on senescent memory function. Geriatrics **14**, 450—456 (1959).
KRIZEK, V., STEPANEK, P., SADILEK, L.: Starvation and anabolic steroids. Metab. Clin. Exp. **18**, 585—592 (1969).
KROATH, F.: Zur Behandlung der Presbyakusis mit Sexualhormonen. Z. Laryng. Rhinol. **46**, 552—555 (1967).
KROTOV, V. P.: The effect of anabolic steroids on the course of uremia. Bull. exp. biol. Med. **62**, 34—37 (1966).
KRÜSKEMPER, H.-L.: Anabolic Steroids. New York-London: Academic Press 1968.
KRÜSKEMPER, H.-L.: Androgene, anabole Steroide und Fettstoffwechsel. Z. Ernährungsw. Suppl. **12**, 16—21 (1972).
KRÜSKEMPER, H.-L., REICHERTZ, P.: Der Einfluß eines anabol wirksamen Steroids (1-Methyl-17β-hydroxy-5α-androst-1-en-3-on) auf Veränderungen des Elektrokardiogramms bei experimenteller Hyperthyreose. Z. ges. exp. Med. **137**, 85 (1963).
KRUMS, L. M.: Dynamic changes of some protein metabolism indices in patients with chronic enterocolitis against the background of nerobol medication. Vop. Pitan. **30**, 44—46 (1971).
KÜCHLE, H. J.: Diabetes und Auge. Med. Welt **18**, 1985—1988 (1967).
KÜHN, W.: Über die Wirkung einer neuartigen anabol wirksamen Substanz (1-Methyl-$\Delta^1$-androstenolönanthat) auf die Ernährungsverhältnisse bzw. den Verlauf tuberkulöser Lungenerkrankungen. Dissertation, Med. Akad. Düsseldorf 1964.
KUEMMERLE, H. P., GARRETT, E. R., SPITZY, K. H.: Klinische Pharmakologie und Pharmakotherapie. 3rd ed., München-Berlin-Wien: Urban & Schwarzenberg 1976.
KUHN, E.: Progressive Muskeldystrophie, Myotonie, Myasthenie: Gespräch über die Therapie der progressiven Muskeldystrophie (Gesprächsleitung: R. BECKMANN), pp. 145—155. Berlin-Heidelberg-New York: Springer 1966.
KULCSAR-GERGELY, J., KULCSAR, A., KISS, A.: The role of sex differences in the effect of anabolics on the liver. Arzneimittel-Forsch. **25**, 417—420 (1975).
KUO, P. T., QUICK, R. E.: Improvement of hypertriglyceridemia by anabolic agent in primary type I hyperlipoproteinemia. Circulation **42**, Suppl. 3, 18 (1970).
KUTSCHERA, E.: Probleme der diabetischen Retinopathie. Wien. klin. Wschr. **82**, 706—710 (1970).
KUTSCHERA-AICHBERGEN, H.: Anabolika bei biochemischen Myokardschäden. Wien. med. Wschr. **116**, 352—354 (1966).
KUTZSCHE, A.: Experimentelle Beeinflussung von Tierinfektionen durch Methenolonester. In: Wirkung und Anwendung anaboler Steroide. Berlin: Medicus Verlag 1964.
KUZELL, W. C.: Methandrostenolone in rheumatic diseases and osteoporosis. Geriatrics **17**, 428—441 (1962).
LACHNIT, K. S., ZWERINA, R.: Zur Frage der Anwendung anaboler Steroide in der Geriatrie. Münch. med. Wschr. **107**, 1125—1128 (1965).
LAMPITELLA, P.: Gli steroidi anabolici nel trattamento dell'ulcera sperimentale alla Shay. Boll. Soc. ital. Biol. sper. **40**, 1584—1587 (1964).
Lancet Editorial: Lipid-lowering drugs after myocardial infarction. Lancet **1975 I**, 501—502.
LANDON, J., WYNN, V., SAMOLS, E.: The effect of anabolic steroids on blood sugar and plasma insulin levels in man. Metabolism **12**, 924 (1965).
LANGNICKEL, D.: Zur Behandlung von Malignomen mit Tris-Äthyleniminobenzochinon, 19-Norandrostenolon-Decanoat und Prednison. Med. Welt **18**, 1015—1022 (1963).
LANGSCH, H.-G.: Zur Frage der Beeinflussung des Kohlehydratstoffwechsels bei Diabetikern durch Oral-Turinabol. Medicamentum **11**, 106—107 (1970).
LAURENCE, D. R.: Clinical Pharmacology. 4th Ed. Edinburgh-London: Churchill Livingstone 1973.

LAURIJSSENS, M., VUYLSTEEK, K., VERMEULEN, A.: Progressive myopathische dystrofie. Belg. T. Geneesk. **16**, 774—784 (1960).

LEBEAU, R. J., ROBINSON, R. W.: Effect of an anabolic steroid on hyperlipemia. Cardiovasc. Res. VI. World Congr. Cardiol. 196 (1970).

LEEVY, C. M.: Fatty liver: A study of 270 patients with biopsy proven fatty liver and a review of the literature. Medicine (Balt.) **41**, 249 (1962).

LEEVY, C. M.: Abnormalities of hepatic DNA synthesis in man. Medicine **45**, 423—433 (1966).

LEEVY, C. M.: Clinical diagnosis, evaluation and treatment of liver disease in alcoholics. Fed. Proc. **26**, 5, 1474—1481 (1967).

LEEVY, C. M.: Cirrhosis in alcoholics. Med. Clin. N. Amer. **52**, 6, 1445—1455 (1968).

LEEVY, C. M.: Therapie der Alkoholfettleber und Hepatitis. Therapiewoche **20**, 2396—2398 (1970).

LEFRANÇOIS, J.-J.: De l'emploi d'un stéroide anabolisant en milieu neuro-psychiatrique. Rev. Neuropsichiat. Ouest **12**, 17—27 (1966).

LESCHINSKY, L. A., KHARITONOVA, V. V.: Experience in the therapeutic employment of nerobol in combination with potassium salts in patients with the syndrome of chronic coronary insufficiency. Klin. Med. **48**, 38—42 (1970).

LESCHINSKY, L. A., KHARITONOVA, V. V.: Combined therapy with nerobol and potassium salts in ischemic disease of the heart. Terap. Arkh. **44**, 40—44 (1972).

LESCHINSKY, L. A., PEVCHIKH, V. V., PETROV, N. M.: The use of methylandrostendiol in myocardial dystrophies and atherosclerotic cardiosclerosis. Ter. Arkh. **39**, 84—87 (1967).

LESCHINSKY, L. A., PEVCHIKH, V. V., TRUSOV, V. V., VAKHRUSHEV, Y. M., VEREVKIN, A. M.: Comparative investigation into the effectiveness of nerobol and quateron as well as their combinations in gastric and duodenal ulcers. Klin. Med. (Moskau) **50**, 68—73 (1972).

LEYRITZ, M. J., CAPRON, C., CAPRON, H., SANCHEZ, F., TIMSIT, M.: Action favorable du P.P.N.A. chez les malades atteintes de démence de la sénilité. Ann. méd.-psychol. **5**, 1—3 (1961).

LINDNER, H.: Zum Verhalten von Bilirubin, alkalischer Phosphatase, Transaminasen (GOT und GPT) und des Bromsulfophthaleintestes bei Patienten mit Leberzirrhose und chronischer Hepatitis unter peroraler Primobolan-Therapie. In: Wirkung und Anwendung anaboler Steroide. Berlin: Medicus Verlag 1964a.

LINDNER, H.: Klinische Beobachtungen über Langzeitbehandlung mit Primobolan. In: Wirkung und Anwendung anaboler Steroide. Berlin: Medicus Verlag 1964b.

LINDNER, H.: Langzeittherapie der Leberzirrhose mit anabolen Steroiden. Med. Klin. **62**, 59—64 (1967a).

LINDNER, H.: Treatment of liver cirrhosis with anabolic steroids. Recent Advanc. Gastroenterol. **3**, 662—664 (1967b).

LINET, O.: Interactions between androgenic-anabolic steroids and glucocorticoids. Fortschr. Arzneimittel-Forsch. **14**, 139—195 (1970).

LINEVA, V. A.: On the correcting effect of anabolic preparations on the protein composition of the blood in patients with acute coronary insufficiency. Kardiologija **11**, 75—78 (1971).

LIOTTA, S., SCANDARIATO, V., DE LAURENZI, A.: Influenza del prednisone e degli steroidi anabolizzanti sulla risposta anticorpale ad antigeni stafilococcici. Progr. med. **23**, 432—435 (1967).

LITTLE, P. J., BAILEY, R. R.: Effect of tetracyclines on level of blood urea in patients with renal failure. Aust. Ann. Med. **19**, 88 (1970).

LITTMANN, L.: Changes of glucose tolerance and plasma immunoreactive insulin level before and after treatment of type IV and type V hyperlipoproteinemia with oxandrolone. Diabetologia **10**, 396 (1974).

LODS, J.-C.: Intérêt d'un nouvel anabolisant à action prolongée en gastro-entérologie. Gaz. méd. Fr. **71**, 3338—3339 (1964).

LOEWIT, K.: Über den Einfluß anaboler Hormone auf die Folgen einer Strahlenschädigung bei der weißen Maus. Strahlentherapie **125**, 281—298 (1964).

LOMAZOVA, M. M.: Methyluracil and nerobolin in the antirecurrence treatment of patients with ulcer disease. Vrach. Delo **1**, 56—57 (1973).

LONGRIDGE, R. G. M., GILLIAM, P. M. S., BARTON, G. M. G.: Decreased anticoagulant tolerance with oxymetholone. Lancet **1971 II**, 90.

LORIMIER, A. A., DE, GORDAN, G. S., LOWE, R. C., CARBONE, J. V.: Methyltestosterone, related steroids and liver function. Arch. intern. Med. **116**, 289—294 (1965).

LOUYOT, P., GAUCHER, A., BENOIT, P.: Etude clinique d'un nouvel anabolisant, le stanozolol. Ann. méd. Nancy **4**, 203—211 (1965).

LUNDBERG, P. O., BLOM, S.: Behandling av muskeldystrofier med anabola steroider. Opusk. med. (Stockh.) **8**, 125—129 (1963).

MACMASTER, D. R., ALAMIN, K.: Treatment of severe weight loss with methandrostenolone—a less virilizing anabolic agent. Amer. J. Psychiat. **120**, 179—180 (1963).

MAGGI, C., BARBI, G.: Lo stato nutritionale proteico del bronchitico cronico. Arch. Sci. med. **110**, 668 (1960).

MAKAROV, G. A.: The mechanism of disturbances of synthetic processes in tissues in prolonged hypodynamia. Patol. Fiziol. Eksp. Ter. **4**, 41—45 (1974).

MALL, G., HEILBRUNN, A., PAARMANN, H. F., JÜNEMANN, H. J., RHEINGANS, W. R.: Die Behandlung der multiplen Sklerose mit anabolen Steroiden. Med. Klin. **63**, 2075—2077 (1968).

MANAI, G., PICCININI, G.: La metazina capronato (2-metil-5-androstan 17-ol-3,3'-azina-17-capronato) in terapia geriatrica. G. Geront. **19**, 357—367 (1971).

MANDAL, B. K., MCNULTY, M.: Treatment of haemolytic-uraemic syndrome with phenformin and ethyloestrenol. Lancet **1971 II**, 1036.

MARKIEWICZ, K.: Durabolin in the management of nephrotic syndrome. Wiad. lek. **18**, 661—664 (1965).

MARKOV, M. I., PIRALOV, G. V.: Use of corticosteroids and anabolic steroids in the treatment of patients with liver cirrhosis. Vrach. Delo **4**, 70—72 (1972).

MARQUIER, A.: L'oxymetholone en gériatrie. Expérience sur quarantecinq cas. Rev. franç. Géront. **10**, 294—296 (1964).

MARTINI, G. A., STROHMEYER, G., DÖLLE, W.: Was ist gesichert in der Therapie von Leberkranken? Internist **10**, 456—464 (1969).

MARTINS, J. K.: Corticosteroid side effects counteracted by anabolic therapy. Gen. Pract. Kansas **27**, 6—9 (1964).

MASSON, G., CORCORAN, A. C., PAGE, I. H.: Dietary and hormonal influences in experimental uremia. J. Lab. clin. Med. **34**, 925—931 (1949).

MATSUMOTO, S., TAKABATAKE, Y., SEIKE, S.: Clinical and experimental study on the effect of protein anabolic hormone, deca-durabolin, on lesions of bones and joints. Excerpta med. (Sect. 3) **17**, 536 (1963).

MATZELT, D.: Beeinflussung von Serum-Enzymaktivitäten durch anabole Steroide bei muskelkranken Patienten. In: Wirkung und Anwendung anaboler Steroide. Berlin: Medicus Verlag 1964.

MEERSON, F. S.: Hyperfunktion, Hypertrophie und Insuffizienz des Herzens. Berlin: VEB Verlag Volk und Gesundheit 1969.

MELMON, K. L., MORRELLI, H. F.: Clinical Pharmacology. New York: MacMillan Company 1972.

MENDENHALL, C. L.: Augmented release of hepatic triglycerides with anabolic steroids in patients with fatty liver. Amer. J. dig. Dis. **19**, 122—126 (1974).

MENON, I. S.: Treatment of cutaneous vasculitis and recurrent superficial phlebitis and thromboembolism. Antiseptic **67**, 776—780 (1970).

MENON, I. S.: Metformin and stanozolol in blood fibrinolytic activity. Brit. med. J. **1971 I**, 289—290.

MENON, I. S.: The role of phenformin and ethyloestrenol or stanozolol in the treatment of decreased fibrinolytic states. Acta Univ. Carol. (Med. Monogr.) **52**, 19—23 (1972).

MENON, I. S., CUNLIFFE, W. J.: Phenformin and ethyloestrenol for Raynaud's disease. Lancet **1969 II**, 1135.

MENON, I. S., CUNLIFFE, W. J., DEWAR, H. A.: Preliminary report of beneficial effect of phenformin in combination with ethyloestrenol in treatment of cutaneous vasculitis and Behçet's syndrome. Postgrad. med. J. (Suppl. May), 62—63 (1969).

MENON, I. S., CUNLIFFE, W. J., WEIGHTMAN, D., DEWAR, H. A.: Phenformin and stanozolol in blood fibrinolytic activity. Brit. med. J. **1970 I**, 428.

MERRILL, J. P.: The Treatment of Renal Failure. New York: Grune & Stratton 1955.

MERTENS, H. G.: Anabole Steroide bei Muskelkrankheiten. In: Wirkung und Anwendung anaboler Steroide. Berlin: Medicus Verlag 1964.

MEZŽENA, R., FORESTI, M., TRISOTTO, M., BRUNI, G.: Studio di farmacologia umana sul quinbolone. G. Geront. **20**, 361—378 (1972).

MICHAELSSON, G., PETTERSON, L., JUHLIN, L.: Purpura caused by food and drug additives. Arch. Derm. **109**, 49—52 (1974).

MILHORAT, A. T.: Discussion on DOWBEN, R. M., Anabole Steroide bei Myopathien. Myopathien. Stuttgart: Georg Thieme Verlag 1965.

MIRANDA, G., JIMÉNEZ, G., LARA, F.: Estudio comparativo en hepatitis viral tratada con anabolicos. Acta méd. costaric. **13**, 155—159 (1970).

MITROFANOV, V. S., SELIVERSTOV, S. A., MARKIN, V. A.: Evaluation of the nerobol and prednisolone effectiveness in experimental myocardial infarction. Farmakol. i. Toksikol. **36**, 316—319 (1973).

MOENCH, A., SARRE, H., SARTORIUS, H.: Klinische und experimentelle Untersuchungen zur Frage der Beeinflussung von Nierenläsionen durch Sexualhormone, fetale Nieren-Trockenzellen und Heparin. Verh. dtsch. Ges. inn. Med. **60**, 527—534 (1954).

MOENCH, A., SARTORIUS, H.: Klinische Untersuchungen über Testosteronbehandlung beim nephrotischen Syndrom. Klin. Wschr. **32**, 329—330 (1954).

MOENCH, A., VOGT, A.: Studien zur Frage der Hormonbehandlung bei experimenteller Nephritis-Nephrose der weißen Ratte. Verh. dtsch. Ges. inn. Med. **64**, 527—531 (1958).

MOINADE, S.: Action du trenbolone dans les dénutritions les cirrhoses et les ostéoporoses, étude à partir de 31 cas. Vie méd. **6**, 605—608 (1974).

MOLINO, N., BELLUARDO, C., ROSSO, B.: Le sindromi disprotidemiche in geriatria e loro trattamento con un nuovo steroide anabolizzante derivato dall'androstano (2-formil-11 alfa-idrossi-delta-metiltestosterone). Nostra esperienza clinica. Clin. Ter. **66**, 167—198 (1973).

MOLNAR, G. D., ROSEVAER, J. W., GASTINEAU, C. F., MOXNESS, K. E.: Effects of anabolic steroids on diabetic instability. Amer. J. med. Sci. **249**, 280—290 (1965).

MOORE, F. D., BALL, R. M.: The Metabolic Response to Surgery. Springfield: Ch. C. Thomas 1952.

MOORE EDE, M. C., BURR, R. G.: Circadian rhythm of therapeutic effectiveness of oxymetholone in paraplegic patients. Clin. Pharmacol. Ther. **14**, 448—454 (1973).

MORIN, J.-P.: Contribution a l'étude des anabolisants dans le traitement de la tuberculose pulmonaire. Thesis, Marseille 1960.

MORRISON, B. O.: Use of analeptic, anti-depressant and anabolic drugs in treatment of the aged. J. Mich. St. med. Soc. **60**, 723—726 (1961).

MÜTING, D.: Die Wirkung einer Langzeitbehandlung mit einer anabolen Substanz (Nandrolon-decanoat) auf Proteinsynthese und -abbau sowie Ausscheidungs- und Entgiftungsfunktion der Leber bei chronischen Leberkrankheiten und Diabetes mellitus. Klin. Wschr. **42**, 843—848 (1964).

MÜTING, D.: Protein metabolism, excretion, and detoxication functions of the liver after long-term treatment of liver cirrhosis with an anabolic substance. Amer. J. dig. Dis. **10**, 790—795 (1965).

MURAKAMI, M., ODAKE, K., MATSUDA, T. et al.: Effects of anabolic steroids on anticoagulant requirements. Jap. Circ. J. **29**, 243 (1965).

NANIVADEKAR, S. A., NANIVADEKAR, A. S.: Gargoylism (Report of 3 cases and their treatment with nandrolone). J. J.J. Group Hosp. Grant med. Coll. **11**, 176—181 (1966).

NASSO, S., PEDRETTI, G., LEVORATO, F.: Tentativo di profilassi della varicella con steroidi anabolizzanti ed antirachitici. Minerva pediat. **16**, 390—394 (1964).

NAVRÁTIL, J., HRBEK, J., DVORÁK, M.: The therapeutic effect of stenolon spofa on female neurotic patients. Activ. nerv. sup. (Praha) **14**, 102—103 (1972).

NETHERCOTT, J., LESTER, R. S.: Azathioprine therapy in incomplete Behçet syndrome. Arch. Derm. **110**, 432—434 (1974).

NIKOLAEVA, L. F., CHERPACHENKO, N. M., VESELOVA, S. A., SOKOLOVA, R. I.: Mechanisms of drug action on recovery processes of cardiac muscle in myocardial infarction. Circulation Res. **34**/**35**, Suppl. III, 202—214 (1974).

NILSSON, I. M., HEDNER, U., ISACSON, S.: Phenformin and ethyloestrenol in recurrent venous thrombosis. Acta med. scand., **198**, 107—113 (1975).

NILSSON, I. M., ISACSON, S.: Thrombolytic therapy: long-term enhancement with oral agents. Effect of treatment with combined phenformin and ethyloestrenol on the coagulation and fibrinolytic systems. J. clin. Path. **25**, 638—639 (1972).

NOVAK, E.: The effect of fluoxymesterone on plasma lipids in men. Int. med. Digest **6**, 17—23 (1971).

NOVAK, E., HENDRIX, J. W., SECKMAN, C. E., PENNER, J. A.: A pharmacologic evaluation of fluoxymesterone in normal men. Curr. ther. Res. **16**, 251—260 (1974).
NOWY, H., FRINGS, H. D., FROST, H.: Effekte von Anabolika auf den Herzmuskel. Arzneimittel-Forsch. **13**, 716—717 (1963b).
NOWY, H., FRINGS, H. D., SEITZ, W., HASHEMI, I.: Effekte von Anabolika auf den Herzmuskel. Arzneimittel-Forsch. **13**, 436—438 (1963a).
NYFOS, L.: Durabolinbehandling af muskeldystrofi. Ugeskr. Laeg. **124**, 1412 (1962).
O'GRADY, B. A.: The influence of drugs and diseases on the prothrombin time. Canad. J. med. Technol. **28**, 140 (1966).
OJALA, L.: Experiences on deca durabolin treatment of diabetic retinopathy. Acta ophthal. (Kbh.) **42**, 519—522 (1964).
O'KEEFE, S. J. D., SENDER, P. M., JAMES, W. P. T.: Catabolic loss of body nitrogen in response to surgery. Lancet **1974 II**, 1035—1038.
OLDERSHAUSEN, H.-F. v.: Verfolgung der Radiojodbengalrosaelimination bei Ratten nach einmaliger sowie langfristiger Verabreichung verschiedener anaboler Steroide. In: Wirkung und Anwendung anaboler Steroide. Berlin: Medicus Verlag 1964a.
OLDERSHAUSEN, H.-F. v.: Verfolgung der Serumaktivitäten an Transaminasen und Aldolase bei Kaninchen und Hunden nach langfristiger Verabreichung von Methenolon. In: Wirkung und Anwendung anaboler Steroide. Berlin: Medicus Verlag 1964b.
OLDERSHAUSEN, H.-F. v.: Zur Abhängigkeit funktioneller Veränderungen der Leber von der chemischen Struktur, Dosis und Applikationsweise anaboler Steroide bei lebergesunden Patienten und Kranken mit Leberzirrhose. In: Wirkung und Anwendung anaboler Steroide. Berlin: Medicus Verlag 1964c.
OLIVERI, M.: La dimetazina quale terapia coadjuvante alla telecobalto-terapia in portatori di neoplasic polmonari. Gazz. med. ital. **126**, 399 (1967).
OLSON, R. E.: Metabolic interventions in the treatment of infarcting myocardium. Circulation **40**, Suppl. 4, IV-195-IV-201 (1969).
OLSSON, A. G., ORÖ, L., RÖSSNER, S.: Effects of oxandrolone on plasma lipoproteins and the intravenous fat tolerance in man. Atherosclerosis **19**, 337—346 (1974).
OOSTERHUIS, J. A.: Retinopathia diabetica. Thesis, Amsterdam 1960.
OOSTERHUIS, J. A.: Discussion to a lecture of L.E.M. Valk. Ned. T. Geneesk. **107**, II, 1775 (1963).
OOSTERHUIS, J. A., LOEWER-SIEGER, D. H.: Behandeling van patienten met diabetische retinopathie met nandrolondecanoaat (Deca-durabolin). Ned. T. Geneesk. **106**, II, 1574 (1962a).
OOSTERHUIS, J. A., LOEWER-SIEGER, D. H.: Treatment of diabetic retinopathy with nandrolone-decanoate (decadurabolin). Ophthalmologica **144**, 346—370 (1962b).
OTTER, G. DEN: The influence of anabolic substances on the nitrogen balance in surgical patients. Arch. chir. neerl. **12**, 496—504 (1960).
OVERBEEK, G. A.: Anabole Steroide, Chemie und Pharmakologie. Berlin-Heidelberg-New York: Springer-Verlag 1966.
OYA, J. C. DE, DEL RIO, A., NOYA, M., VILLANUEVA, A.: Decreased anticoagulant tolerance oxymetholone in paroxysmal nocturnal haemoglobinuria. Lancet **1971 II**, 259.
PAGE, C. W.: The treatment of debility: a controlled experience. Paper presented at a Symposium on Anabolic Therapy, Detroit, Michigan, March 1962.
PANEK, R.: Untersuchungen über den Mechanismus der strahlenschützenden Wirkungen der anabolen Steroide. Strahlentherapie **143**, 211—218 (1972a).
PANEK, R.: Untersuchungen über den Mechanismus der strahlenschützenden Wirkung der anabolen Steroide. Strahlentherapie **143**, 444—453 (1972b).
PANEK, R.: Untersuchungen über die Steigerung des strahlenschützenden Effekts der anabolen Steroide. Strahlentherapie **143**, 559—569 (1972c).
PANI, A.: L'associazione steroidi anabolizzanti ed insulinaglucosio nella terapia degli stati iperazotemici. Clin. ter. **46**, 303—314 (1968).
PANIGADA, G., PERLETTI, L., MAGNI, L., GIOVANNINI, M.: Influenza di uno steroide anabolizzante sulla fenilalaninemia in lattanti distrofici. Minerva pediat. **20**, 491—494 (1968).
PAOLINO, J. S., KAPLAN, D., LAZARUS, R., EMMANUEL, G., DIAMOND, H. S.: Phenformin and ethyloestrenol in scleroderma. Lancet **1972 I**, 1023.
PARSONS, F. M., MCCRACKEN, B. H.: The artificial kidney. Brit. med. J. **1959 I**, 740—751.
PASQUALICCHIO, A., PASOLI, C. A., FORTE, P. L., CORBELLARI, G. C.: Contributo al trattamento delle sindromi carenziali nel vecchio. Minerva med. **63**, 2425—2430 (1972).

Pateisky, K., Schinko, H., Haberler, H.: A study of the treatment of dystrophia musculorum progressiva with an anabolic steroid: nor-androstenolone decanoate. Acta endocr. (Kbh.) **63**, 185—196 (1961).

Patnaik, D. N., Nath, R., Pathak, I. C.: The effect of methandienone (an anabolic steroid) on nitrogen metabolism following surgical trauma. Indian J. med. Res. **57**, 1751—1760 (1969).

Payne, R. W.: Anabolism, appetite, and ambition. Kentucky Acad. gener. Pract. J. **5**, 20—21 (1959).

Pearson, S., Weissberg, J., McGavack, T. H.: Steroid studies. I. Metabolic effects of androstanolone in aged people. J. Amer. Geriat. Soc. **2**, 26 (1954).

Pena, E. de la: Clinical use of oxymetholone in geriatrics. Sem. Méd. (Méx.) **27**, 205—207 (1960).

Pentycross, C. R., Toussis, D., McKinna, J. A., Lawler, S. D., Greening, W. P.: Effect of hormone therapy on mitogenic responses of lymphocytes from patients with cancer of the breast. Lancet **1973 II**, 177—179.

Pergola, I.: Pren. méd. argent. **49**, 274 (1962), cited by Tainter, M. L., Arnold, A., Beyler, A. L., Coulston, F., Potts, S. G. O., Roth, C. H.: A summary of the comparative therapeutic actions of stanozolol, an oral anabolic steroid. Clin. Med. **1964**, 2089—2102.

Pestalozzi, D.: Zur Behandlung der Retinopathia diabetica mit Deca-Durabolin. Ophthalmologica **147**, 125—133 (1964).

Petrochilos, A. M.: Steroides anabolisants sur la rétinopathie diabétique. Nosokom. Chron. **29**, 201—203 (1967).

Petzold, H., Matzkowski, H., Burckhardt, M.: Der Einfluß von Oral-Turinabol auf die „Thioacetamidzirrhose" der Rattenleber. Dtsch. Z. Verdau.- u. Stoffwechselkr. **29**, 3 (1969).

Pevchikh, V. V.: Anabolic steroids in the therapy of patients suffering from some diseases of the gastro-intestinal tract. Ter. Arkh. **41**, 69—72 (1969).

Pezold, F. A.: Anabole Hormone bei chronischer Hepatitis. Orale Therapie mit 1α,7α-Bis(acetylthio)-17α-methyltestosteron. Münch. med. Wschr. **110**, 2663—2668 (1968).

Piatti, N., Pozzi, G.: Sperimentazione con 2-formil-17α-metilandrosta-1,4-dien-11α, 17β-diol-3-one in clinica geriatrica. Clin. Ter. **74**, 145—166 (1975).

Picha, E.: Anabole Steroide bei der Strahlentherapie weiblicher Genitalkarzinome. In: Wirkung und Anwendung anaboler Steroide. Berlin: Medicus Verlag 1964.

Picha, E.: Neue Gesichtspunkte zur Anwendung anaboler Steroide in der gynäkologischen Strahlentherapie. Strahlentherapie **138**, 300—322 (1969).

Pohle, H.: Verlaufsbeobachtungen einzelner Leberfunktionsprüfungen unter oraler Applikation von Methenolon und 1-Phenylpropanol bei Lebergesunden. In: Wirkung und Anwendung anaboler Steroide. Berlin: Medicus Verlag 1964.

Polischuk, M. A.: The clinical assessment of the use of strophanthin and nerobol in cardiac insufficiency. Klin. Med. (Moskau) **48**, 53—56 (1970).

Porsio, A.: Controlled clinical trials of a new anabolic agent in chronic TB patients. Clin. ter. **37**, 502—518 (1966).

Porsio, A.: Ricerche cliniche controllate su due anabolizzanti ad azione protratta. Minerva med. **59**, 824—828 (1968).

Porter, R., Knight, J. (ed.): Energy metabolism in trauma. A Ciba Foundation Symposium. London: J. & A. Churchill 1970.

Porto, A. L.: O tratamento testosterónico da cirrose hepática. Thésis, Coimbra 1968.

Poser, G., Kluthe, R., Adler, P.: Zur Therapie des nephrotischen Syndroms. Med. Klin. **61**, 432—435 (1966).

Pozzetto, M. H.: Un anabolisant, le phenylpropionate de nandrolone, en pneumophtisiologie. Rev. Lyon Méd. **10**, 961—968 (1961).

Prikhozan, V. M.: Pathogenesis of the nervous system affection in diabetes mellitus. Probl. Endokrinol. Gormonterap. **14**, 14—18 (1968).

Probst, J.-H.: Treatment of children suffering from disorders of muscular movement with durabolin and decadurabolin. Proc. VII. Intern. Congr. Neurology **2**, 645—652 (1961).

Pruss, G. M., Markovskaia, G. I., Bosko, A. P.: In: Profilaktika i lechenie koronarnoi i serdechnoi nedostatochnosti. Leningrad 1967.

Quack, G.: Primobolan als Zusatztherapie bei der Strahlenbehandlung. In: Wirkung und Anwendung anaboler Steroide. Berlin: Medicus Verlag 1964.

QUACK, G., GAEBEL, E.: Primobolan-Depot in der Strahlentherapie. Med. Mitt. Schering **23**, 16 (1962)

RABIER, M.: Utilisation d'un anabolisant dans la tuberculose pulmonaire du noir africain. Afr. méd. **95** (1971).

RABINOWITZ, J. L., CHASE, G. D., MYERSON, R. M.: Failure of anabolic steroid, oxymetholone, to influence survival time of dystrophic mice. Amer. J. med. Sci. **246**, 456—458 (1963).

RAVENS, K. G.: Der Einfluß von anabolen Hormonen und von Diät auf den myokardialen Gehalt an energiereichen Phosphaten während der Infarktheilung. Verh. dtsch. Ges. Kreisl.-Forsch. **36**, 314—319, 389 (1970).

REBORA TOGNO, F.: El decanoato de nandrolona como coadyuvante del medico o quirurgico de la tuberculosis pulmonar. Sem. Méd. (Méx.) **67**, 146—150 (1971).

REHN, J., SPINNER, M.-E.: Die Anwendung anabol wirksamer Hormone in der Chirurgie. Med. Welt (Stuttg.) **25**, 1401—1405 (1962).

REISS, M., HILLMAN, J. C., REISS, J., DALEY, N., HAYLOCK, S.: Influence of anabolic hormones on phenylalanine metabolism. I. Studies on chronic phenylketonuric patients. J. ment. Defic. Res. **10**, 20 (1966b).

REISS, M., SIDEMAN, M. B., PLICHTA, E. S.: Influence of anabolic hormones on phenylalanine metabolism. II. Studies in animals. J. ment. Defic. Res. **10**, 130—140 (1966a).

REN, G. DE, MEYER, D., SIMON, G.: Utilisation du phenyl-propionate de nor-androstenolone en phtisiologie. Vie méd. **42**, 739—751 (1961).

RESAN, T. K., SHAHIDI, N. T., KORST, D. R.: Effect of phenobarbital on testosterone-induced erythropoiesis. J. Lab. clin. Med. **79**, 187 (1972).

RIGBERG, S. V., BRODSKY, I.: Potential roles of androgens and the anabolic steroids in the treatment of cancer. A review. J. Med. **6**, 271—290 (1975).

RIORDAN, D. J., BROWNE, P. A.: The treatment of secondary deposits in bone from carcinoma of the breast with radio-active phosphorus and durabolin. J. Irish med. Ass. **99**, 40—44 (1961).

ROBERTS, H. J.: Radiotherapy in herpes zoster. Lancet **1965 II**, 1074.

ROBERTS, H. J.: Timed repetitive ankle jerk (TRAJ) responses in 'early' diabetic neuropathy: Diagnostic and therapeutic observations, with particular reference to the effects of nandrolone. Diabetes **18**, 370—371 (1969).

ROBERTS, P. S., BURKAT, R. K.: Effects of biguanides on reactions of thrombin and on 1-stage prothrombin time of standard human plasma. Ann. N.Y. Acad. Sci. 714 (1968).

ROBINSON, B. H. B., HAWKINS, J. B., ELLIS, J. E.: Decreased anticoagulant tolerance with oxymetholone. Lancet **1971 I**, 1356.

ROBINSON, R. W., LEBEAU, R. J.: Effect of oxandrolone on hyperlipemia. Circulation **42**, Suppl. 3, 55 (1970).

ROBSON, A. M., ASHCROFT, R., CLARKSON, B. A., HORN, D. B., KERR, D. N. S.: The effect of an anabolic steroid (Deca-Durabolin) on urea production. Proc. europ. Dialys. Transpl. Ass. (EDTA Proc.) **1**, 159 (1964a).

ROBSON, A. M., ASHCROFT, R., CLARKSON, B. A., HORN, D. B., KERR, D. N. S.: Diet and anabolic steroids. In: SHELDON, S., COOK, S. (Eds.): Acute Renal Failure. Oxford: Blackwell 1964b.

ROBSON, A. M., ASHCROFT, R., CLARKSON, B. A., HORN, D. B., KERR, D. N. S.: Urea metabolism in chronic uremia: Effect of protein restriction, anabolic steroids and antibiotics. In: BERLYNE, G. M. (Ed.): Nutrition in Renal Disease. Edinburgh: Livingstone 1968.

ROHWEDDER, H. J.: Beobachtungen bei der Anwendung anaboler Steroide bei Kindern mit nephrotischem Syndrom. In: Wirkung und Anwendung anaboler Steroide. Berlin: Medicus Verlag 1964.

ROSNICK, M. J.: Use of anabolic steroid, stanozolol, to promote weight gain in underweight patients. Clin. Med. **71**, 989 (1964).

ROUHER, F., SERPIN, G.: Action du phenyl-propionate de nor-androsténolone dans la rétinopathie diabétique. Presse méd. **72**, 2211—2212 (1964).

ROY, J. A.: Study of the correction of protein disorders seen in the course of alcoholic cirrhosis by androstanolone. Thesis, Bordeaux 1961.

ROZMAN, C., URBANO, A., GALERA, H.: Lebertoxizität der anabolen Steroide. Münch. med. Wschr. **113**, 965—967 (1971).

RUPRECHT, K. W.: Diabetische Retinopathie. Med. Welt **26**, 1822—1825 (1975).

SACHS, B. A., WOLFMAN, L.: Effect of oxandrolone on plasma lipids and lipoproteins of patients with disorders of lipid metabolism. Metabolism **17**, 400—410 (1968).
SAITO, T., SAITO, M.: Studies on the action mechanism of anabolic steroids. II. The effect on the serum glycoproteins. Iryo **20**, 1154—1161 (1966).
SAITO, T., SHIRATA, S., NAKAMURA, R., SAITO, M.: Studies on the action mechanisms of anabolic steroids. I. Clinico-pharmacological studies on their effects upon the far advanced pulmonary tuberculous patients. Iryo **20**, 565—572 (1966).
SANSOY, P. M., NAYLOR ROY, A., SHIELDS, L. M.: Anabolic action and side effect of oxandrolone in 34 mental patients. Geriatrics **26**, 139—143 (1971).
SANTALÓ, M., ECHEVARNE, F.: Untersuchungen des Mongolismus und seine Behandlung mit Norandrostenolonphenylpropionat. Med. Welt **34**, 1683—1686 (1963).
SARRE, H.: Die Therapie des akuten Nierenversagens. Mkurse ärztl. Fortbild. **13**, 298 (1963).
SARRE, H.: Einführung zu „Anabolika bei Nierenerkrankungen“. In: Wirkung und Anwendung anaboler Steroide. Berlin: Medicus Verlag 1964.
SARRE, H.: Diuresetherapie. Med. Klin. **61**, 2009—2013 (1966).
SASKIN, E., WALDMAN, S., PELNER, L.: Diabetic retinopathy; a new approach to therapy with a steroid hormone testosterone propionate. Amer. J. Ophthal. **3**, 613 (1951).
SCHILLING, A., FINKEL, H. E.: Ancillary measures in treatment of myeloma. Use of immune serum globulin, fluoride, or androgen. Arch. intern. Med. **135**, 193—196 (1975).
SCHIMMELPFENNIG, W.: Anabole Steroide und Nierenerkrankungen. Dtsch. Gesundh.-Wes. **21**, 97—100 (1965).
SCHINDLER, H.: Marfan-Syndrom mit Morbus Addison und Akromegalie. Wien. med. Wschr. **125**, 643—644 (1975).
SCHINDLER, H., KORN, A.: Anabole Steroide und ihre hormonellen Nebenwirkungen. Münch. med. Wschr. **107**, 479 (1965).
SCHNACK, H.: Tierexperimentelle Untersuchungen mit anabolen Steroiden. In: Wirkung und Anwendung anaboler Steroide. Berlin: Medicus Verlag 1964a.
SCHNACK, H.: Klinische Untersuchungen mit Steroiden der Primobolan-Reihe im Kurzversuch. In: Wirkung und Anwendung anaboler Steroide. Berlin: Medicus Verlag 1964b.
SCHNACK, H.: Klinische Erfahrungen über Langzeitbehandlung von Leberzirrhosen mit anabolen Steroiden der Primobolan-Reihe. In: Wirkung und Anwendung anaboler Steroide. Berlin: Medicus Verlag 1964c.
SCHNACK, H., SCHOBEL, B., WEWALKA, F.: Therapie der Leberzirrhose mit anabolen Steroiden. Wien. klin. Wschr. **74**, 833—836 (1962).
SCHOPPE, W.-D.: Die Beeinflussung des Eiweißstoffwechsels lungenoperierter Patienten durch anabole Steroide bei postoperativer Zufuhr von kliniksüblicher Kost. Dissertation, Düsseldorf 1971.
SCHROGIE, J. J., SOLOMON, H. M.: Hazards of multiple drug therapy in patients taking coumarin anticoagulants. Circulation **34**, Suppl. III—210—III—211 (1966).
SCHROGIE, J. J., SOLOMON, H. M.: The anticoagulant response to bishydroxycoumarin. II. The effect of D-thyroxine, clofibrate, and norethandrolone. Clin. Pharmacol. Ther. **8**, 70—77 (1967).
SCHUMACHER, H.: Klinik und Pathophysiologie der diabetischen Retinopathie. Med. Welt (Berl.) **18**, 96—100 (1967).
SCHWARZ, F. K. T.: Revitalisierung beim alternden Menschen durch eine multivalente Vitamin-Mineral-Anabolikum-Kombination. Hippokrates **38**, 900—902 (1967).
SCHWARZ, W., KUTZSCHE, A.: Elektronenmikroskopische Untersuchungen über die anabole und katabole Wirkung von Dexamethason und Methenolonester auf Fibroblasten in vitro. Klin. Wschr. **44**, 979—982 (1966).
SCUDERI, G., SBORGIA, G.: Terapia della retinopatia diabetica. Clin. Ter. **70**, Suppl. **3**, 42—61 (1974).
SEGERS, J., NUYTS, J. A.: Clinical trial with oxymetholone in a group of schizophrenic patients. Acta neurol. belg. **64**, 736—745 (1964).
SEIFERT, E.: Methenolonazetat und -önanthat bei chronischer Hepatitis und Leberzirrhose. In: Wirkung und Anwendung anaboler Steroide. Berlin: Medicus Verlag 1964.
SELYE, H.: On the protective action of testosterone against the kidney damaging effect of sublimate. J. Pharmacol. exp. Ther. **68**, 454—457 (1940).

SELYE, H.: Catatoxic steroids. Canad. med. Ass. J. **101**, 51—52 (1969).
SELYE, H.: Prevention of colchicine intoxication by catatoxic steroids. Endocrinologia Exp. **4**, 71—76 (1970a).
SELYE, H.: Resistance to picrotoxin poisoning induced by catatoxic steroids. Agents and Actions **1**, 133—135 (1970b).
SELYE, H.: Resistance to various pesticides induced by catatoxic steroids. J. Amer. med. Ass. **214**, 2072 (1970c).
SELYE, H.: Hormone und Widerstandsfähigkeit. Münch. med. Wschr. **112**, 1401—1407 (1970d).
SELYE, H.: Protection by catatoxic steroids against cycloheximide intoxication. Toxical appl. Pharmacol. **17**, 721—725 (1970e).
SELYE, H.: Conditioning of catatoxic steroid actions by the thyroid. J. Med. **1**, 43—55 (1970f).
SELYE, H.: Protection against LSD by various steroids. Life Sci. **10**, 1135—1140 (1971).
SELYE, H.: Hormones and resistance. Berlin-Heidelberg-New York: Springer-Verlag 1971.
SELYE, H., JELINEK, J., KRAJNY, M.: Prevention of digitoxin poisoning by various steroids. J. Pharm. Sci. **58**, 1055—1059 (1969).
SELYE, H., SOLYMOSS, B.: Protection by catatoxic steroids against meprobamate. Neuropharmacology **9**, 327—332 (1970).
SELYE, H., YEGHIAYAN, E., MANDEVILLE, R.: Protection by catatoxic steroids against dihydrotachysterol intoxication. Atherosclerosis **11**, 321—331 (1970).
SELYE, H., YEGHIAYAN, E., MECS, I.: Prevention of nicotine intoxication by catatoxic steroids. Arch. int. Pharmacodyn. **183**, 235—238 (1970).
SENECA, H.: Endotoxic shock and hyperpyrexia in urological practice. J. Amer. Geriat. Soc. **16**, 733—744 (1968).
SENECA, H., PEER, P.: Effect of estrogens and anabolic steroids on renal infections in normal and immunized/sensitized rats and mice. Bact. Proc. 99 (1967a).
SENECA, H., PEER, P.: Effect of estrogens and anabolic steroids on experimental infections. International Congress of Chemotherapy, Vienna 1967b.
SERENI, F., PICENI-SERENI, L., DE RITIS, L.: Influence of an anabolising steroid on the metabolic balance of nitrogen in the nephrotic syndrome. Minerva pediat. **11**, 1124—1128 (1959).
SETTEL, E.: Protein metabolism in the elderly: A cross-over study of a hormonal anabolic agent (methandrostenolone) and a non-hormonal agent (enzyme-vitamin complex). J. Amer. Geriat. Soc. **12**, 538—547 (1964).
SEVIN, R., CUENDET, J. F.: La rétinopathie diabétique. Son traitement médical. Praxis **60**, 612—614 (1971).
SHAHIDI, N. T.: Barbiturate effect in androgen therapy. New Engl. J. Med. **287**, 309 (1972).
SHARMA, K. K., MISRA, M. B., DIXIT, B. B., SASTRI, P. S. R.: Effect of ethyloestrenol on serum cholesterol in rabbits. Indian Drugs Annual 17—19 (July 1975).
SHILS, M. E.: Renal disease and metabolic effects of tetracycline. Ann. intern. Med. **58**, 389 (1963).
SIEGMUND, G.: Die Anwendung eines anabolen Steroids bei Erkrankungen des rheumatischen Formenkreises. Med. Klin. **58**, 995—998 (1963).
SIGLER, H., ISSEKUTZ, B.: Effects of anabolic steroids in chronic renal failure. Arch. intern. Med. **120**, 408—416 (1967).
SILK, R. E.: Management of protein and fat losses in surgical patients. J. Albert Einstein med. Center **10**, 181—187 (1962).
SIM, A. W., HENDERSON, P. S., VIES, J., VAN DER: The effect of steroids on the biosynthesis of collagen in rat skin. Paper presented at III. European Symposium on Connective Tissue Research, Turku, August 1972.
SIM, A. W., PICTON, W., FOX, P. K., WALKER, G. B.: The effect of topical corticosteroids on the metabolism of dermal collagen. In: Proceedings of Glaxo Symposium on Mechanism of Action of Corticosteroids, Edinburgh-London: Livingstone 1975.
SINITZIN, P. D., AGAFONOWA, G. A.: Clinical evaluation of the effect of anabolic preparations on the effectiveness of treatment in peptic ulcer. Sovetsk. Med. 150—151 (1975).
SKIKANTIA, S. G., JAYARAO, K. S., PRASAD, P. S.: Effect of a C-17, alkylated steroid, methylandrostenolone, on plasma lipids of normal subjects. Amer. J. med. Sci. **254**, 201 (1967).
SMITH, C., JOHNSON, P. C.: Anabolic effect of 19-nortestosterone phenylpropionate in underweight geriatric women. J. Amer. Geriat. Soc. **9**, 304—311 (1961).

SNOW, E. W.: Stanozolol, a new anabolic agent: Clinical observations in seventy-two patients. Med. Tms **91**, 149—157 (1963).

SNYDER, D., BREST, A. B.: Chronic renal insufficiency treated with anabolic steroid. Effect on acid-base balance, protein metabolism and hematopoiesis. J. Amer. Geriat. Soc. **14**, 21 (1966).

SOECKNICK, H.: Fertabolin in der Geriatrie. Prakt. Arzt **299**, 509—526 (1972).

SOKOLOVA, T. S., FEDOTOVA, A. M., BALASHOVA, U. A.: The complex employment of anabolic hormones and antihistamine preparations in treating allergic diseases in early childhood. Pediatriia (Moskwa) **44**, 34—39 (1965).

SOLER-ARGILAGA, C.: Esteroides anabolizantes. II. algunos aspectos de nuestra experiencia clinica sobre esteroides anabolizantes. An. Med. (Med.) **50**, 203—219 (1964).

SOLIS, W. A.: Stanozolol in urological surgery. Dia Médico **43**, 1669—1670 (1971).

SOLYMOSS, B., TOTH, S., VARGA, S., SELYE, H.: Protection by spironolactone and oxandrolone against chronic digitoxin or indomethacin intoxication. Toxical appl. Pharmacol. **18**, 586—592 (1971).

SPENCER, D. A.: Letter to the editor. Med. News, No. 228, 12 (1967).

STENZEL, K. H., CHEIGH, J. S., SULLIVAN, J. F., TAPIA, L., RUGGIO, R. R., RUBIN, A. L.: Clinical effects of bilateral nephrectomy. Amer. J. Med. **58**, 1, 69—75 (1975).

STEPAN, J., BREMOVA, A.: Stromba bei Erkrankungen des Bewegungsapparates. Med. Klin. **62**, 1470—1474 (1967).

STÖCKL, W., GEYRHOFER, K.: Die Beeinflussung der Antikörperbildung durch Sanabolicum. Wien. tierärztl. Mschr. **52**, 94—98 (1965).

STONE, M. C.: Idiopathic hyperglyceridemia treated with methyltestosterone and methandienone. Lancet **1963 I**, 477.

STONE, M. C.: Elevation of serum-cholesterol by an anabolic steroid. Lancet **1964 I**, 474.

STUHLFAUTH, K., FROST, H.: Medikamentöse Rehabilitation und Behandlung der Hypotonie mit Steroidhormonen. Münch. med. Wschr. **106**, 498—502 (1964).

STUR, O.: Durabolinbehandlung bei infantiler spinaler Muskelatrophie. Wien. klin. Wschr. **72**, 54—55 (1960).

SVENDSEN, B. B.: Negative results of treatment with nortestosterone in urinary incontinency in women with senile dementia. Nord. Med. **62**, 1516—1517 (1959).

SZOLD, E.: Die Wirkung von Methylandrostendiol und Norandrostenolon in experimenteller Urämie. (Hungarian). Magy. Sebész. **12**, 368—369 (1959).

SZOLD, E., SZENDRÖI, Z., WEISZ, P., PINTER, I., DESI, I., KADAS, T.: Anabolic steroids in uremia. Lancet **1959 II**, 368—369.

TACHE, Y., TACHE, J., SELYE, H.: Inhibition of the effects of Alfathesin and other steroid anesthetics by catatoxic steroids in rats. Arzneimittel-Forsch. **25**, 1603—1606 (1975).

TAINTER, M. L.: Anabolic steroids in the management of the diabetic patient. N.Y. St. J. Med. **64**, 1001—1009 (1962).

TAMBURRO, C., LEEVY, C. M.: Protein clotting factor synthesis in liver disease. Gastroenterology **52**, 325 (1967).

TAN, R. S. H., COPEMAN, P. W. M.: Blood hypercoagulability causing leg ulcerations. Brit. J. Derm. **91**, 46—48 (1974).

TEC, L.: Durabolin in anorexia nervosa. Amer. J. Psychiat. **120**, 282 (1963).

TEC, L.: Anorexia nervosa—follow-up on a special method of treatment. Amer. J. Psychiat. **127**, 1702 (1971).

TEC, L.: Nandrolone in anorexia nervosa. J. Amer. med. Ass. **229**, 1423 (1974).

THAYSEN, J. H.: Protein metabolism. Internat. Symposion, Leyden. Berlin-Göttingen-Heidelberg: Springer 1962.

TICHY, A., DOLECEK, R., KLABUSAY, L.: Experimental cardiopathy in burned rats. Rozhl. Chir. **42**, 230—236 (1963).

TIRKINA, T. N., ALEKSANDRIVA, L. M., KLYACHKO, V. R., MAZOVETSKY, A. G.: Pyridinolcarbamate (anginin) in the treatment of diabetic angiopathy. Ter. Arkh. **46**, 107—110 (1974).

TOLENTINO, P.: Protein anabolic steroid hormones and antibody formation. Ann. paediat. **199**, 467—471 (1962).

TOLENTINO, P.: Gli steroidi anabolizzanti nella terapia delle epatiti virali. Epatologia **12**, 461—469 (1966).

TOLENTINO, P.: Anabolische Steroide und Immunität. Münch. med. Wschr. **113**, 953—955 (1971).

TOLENTINO, P., TERRAGNA, A., JANNUZZI, C.: Ormoni steroidei ed anticorpopoiesi. Mancata azione di stimolo da parte di steroidi proteino-anabolici nella somministrazione distanziata dall'antigene o nella risposta anamnestica. Ann. Sclavo **6**, 439—444 (1964).

TONI, E. DE: Gli effetti degli steroidi anabolizzanti sul ricambio creatinico nella miodistrofia progressiva. Clin. pediat. **41**, 441—445 (1959).

TONI, E. DE: Treatment of progressive muscular dystrophy in particular with the anabolic steroids. Acta endocr. (Kbh.) **63**, 175—184 (1961).

TONI, G. DE: Sulla possibilitá di limitare o addirittura inibire la creatinuria patologica della miodistrofia progressiva mediante l'uso di forti dosi di steroide anabolizzanti. Boll Soc. ital. Biol. sper. **35**, 110—114 (1959).

TROCME, P., ROY, J.: Essai de correction par l'androstanolone des troubles du métabolisme des protides chez les cirrhotiques. Sem. Hôp. Paris **38**, 22—25 (1962).

TRUSOV, V. V., VAKHRUSHEV, Y. M.: Evaluation of the therapeutic effect of the anabolic steroid agent androstanasol in patients with chronic gastritis. Vrach. Delo **8**, 39—42 (1973).

TRUSOV, V. V., VAKHRUSHEV, Y. M., ORESHKOV, T. M.: Clinical evaluation of the therapeutic action of a new soviet-made preparation—androstanasol in patients with peptic ulcer. Ter. Arkh. **46**, 96—100 (1974).

TURNER, R. L.: Combined androgenic and cytotoxic therapy in carcinoma of the breast. Paper read at a Symposium at the University Dept. of Surgery, Royal Infirmary, Sheffield, Nov. 1966.

TWEEDLE, D. E. F., WALTON, C., JOHNSTON, I. D. A.: The effect of a long-acting anabolic steroid on nitrogen balance during the anabolic phase of recovery from abdominal surgery. Brit. J. Surg. **59**, 300 (1972).

TWEEDLE, D. E. F., WALTON, C., JOHNSTON, I. D. A.: The effect of an anabolic steroid on postoperative nitrogen balance. Brit. J. clin. Pract. **27**, 130—132 (1973).

UDUPA, K. B., REISSMANN, K. R.: Stimulation of granulopoiesis by androgens without concomitant increase in colony-stimulating factor. Clin. Res. **22**, 609 A (1974).

VADÁSZ, I.: Effect of nandrolone decanoate on some forms of pulmonary tuberculosis. Arzneimittel-Forsch. **19**, 100—103 (1969).

VAISRUB, S.: Intravascular factors in diabetic retinopathy. Editorial, J. amer. med. Ass. **233**, 1303 (1975).

VAKHRUSHEV, Y. M.: The experience of the therapeutic use of anabolic steroid of nerobol in combination with quateron in exacerbation of peptic ulcer. Terap. Arkh. **43**, 26—30 (1971).

VAKHRUSHEV, Y. M., TRUSOV, V. V.: Comparative characteristics of therapeutic efficacy of a number of medicinal combinations of patients with peptic ulcer. Terapevt. Arkh. **47**, 108—113 (1975).

VALK, L. E. M.: Influence favorable du phénylpropionate de 19-nor-androsténolone sur rétinopathie diabétique. Bull. Soc. franç. Ophtal. **72**, 596—608 (1959).

VALK, L. E. M.: Successen en tegenslagen bij de behandeling van diabetische retinopathie met anabole steroiden gedurende de laatste vijf jaren. Ned. T. Geneesk. **107**, II, 1773—1776 (1963).

VALK, L. E. M.: Nandrolone decanoate, a useful anabolic steroid in diabetic retinopathy instead of destruction of the hypophysis. Excerpta med. Intern. Congr. Ser. **74**, 36 (1964).

VALK, L. E. M., VERGROESEN, J. A. J.: Een gunstige invloed van nandrolon decanoaat op oog en nier in enige gevallen van de ziekte van Kimmelstiel-Wilson; enige moeilijkheid bij de beoordeling van het nierbiopt. Ned. T. Geneesk. **113**, 99 (1969).

VASIUKOVA, E. A., MARGOLIS, M. G.: Complex treatment of patients with diabetes mellitus with diabetic retinopathy. Klin. Med. (Moskau) **53**, 78—82 (1975).

VASIUKOVA, E. A., ZEFIROVA, F. S., BRISKIN, A. I.: Anabolic steroids in the treatment of endocrine disease. Ter. Arkh. **41**, 9—15 (1969).

VASSELLA, F., ROSSI, E., AEBI, U., RICHTERICH, R.: Anabole Steroide bei progressiver Muskeldystrophie. In: Wirkung und Anwendung anaboler Steroide. Berlin: Medicus Verlag 1964.

VERE, D. W., FEARNLEY, G. R.: Suspected interaction between phenindione and ethyloestrenol. Lancet **1968 II**, 281.

VEREVKIN, A. M., PEVCHIKH, V. V.: Changes in gastric and hepatic function in peptic ulcer during treatment with an anabolic steroid preparation—nerobol. Ter. Arkh. **44**, 36—39 (1972).

VERMEULEN, A.: Influence of anabolic steroids on secretion and metabolism of cortisol. In: Structure and metabolism of corticosteroids. London-New York: Academic Press 1964.
VERMEULEN, A., DEMEULENAERE, L.: Einfluß anaboler Steroide auf Leberenzyme. In: Wirkung und Anwendung anaboler Steroide. Berlin: Medicus Verlag 1964.
VERMEULEN, H. J.: 17β-hydroxy-4-estren-3-one laurate, a new long-acting anabolic steroid. Arzneimittel-Forsch. **18**, 1458—1460 (1968).
VERSTRAETE, M.: The position of long-term stimulation of the endogenous fibrinolytic system: Present achievements and clinical perspectives. Thrombos. Diathes. haemorrh. **34**, 613—622 (1975).
VERWILGHEN, R., LOUWAGIE, A., WAES, J., VANDENBROUCKE, J.: Anabolic agents and relative polycythaemia. Brit. J. Haemat. **12**, 712—716 (1966).
VIALLIER, J.: Action de l'androstanolone base chez les tuberculeux pulmonaires. Lyon méd. **90**, 32 (1958).
VILANOVA, J.: The use of methandrostenolone (Dianabol) in improving nutrition in geriatric patients. J. med. Soc. N. Jersey **59**, 65—68 (1962).
VLAVIANOS, G., SEAMAN, G., VLAVIANOS, M.: A further clinical study of anabolic steroids in incontinence. Amer. J. Psychiat. **118**, 539—542 (1961).
VOSS, H. E.: Androgene und Nieren. In: VOSS, H. E., OERTEL, G.: Androgene I, pp. 551—556. Berlin-Heidelberg-New York: Springer-Verlag 1973a.
VOSS, H. E.: Androgene und Muskulatur. In: VOSS, H. E., OERTEL, G.: Androgene I, pp. 559—566. Berlin-Heidelberg-New York: Springer-Verlag 1973b.
VOSS, H. E.: Wirkung der Androgene auf die Wundheilung: In: VOSS, H. E., OERTEL, G.: Androgene I, pp. 576—577. Berlin-Heidelberg-New York: Springer-Verlag 1973c.
VOSS, H. E.: Die anti-hypercholesterinämische Wirkung der Androgene. In: VOSS, H. E., OERTEL, G.: Androgene I, pp. 604—606. Berlin-Heidelberg-New York: Springer-Verlag 1973d.
WALKER, I. D., DAVIDSON, J. F., YOUNG, P., CONKIE, J. A.: Plasma fibrinolytic activity following oral anabolic steroid therapy. Thrombos. Diathes. haemorrh. **34**, 236—245 (1975a).
WALKER, I. D., DAVIDSON, J. F., YOUNG, P., CONKIE, J. A.: Effect of anabolic steroids on plasma antithrombin III, $\alpha_2$-macroglobulin and $\alpha_1$-antitrypsin levels. Thrombos. Diathes. haemorrh. **34**, 106—114 (1975b).
WALSH, F. M., HARRIS, L. D., SCHIPKE, J. M.: The treatment of diabetic vitreous hemorrhages. Minerva Med. **46**, 641—644 (1963).
WARDLE, E. N.: Study of the lipid-lowering action of choloxin and nilevar in patients with chronic renal failure. Postgrad. med. J. **50**, 737—740 (1974).
WATKIN, D. M., PARSONS, J. M., YIENGST, M. J., SHOCK, N. W.: Metabolism in the aged: the effect of stanolone on the retention of nitrogen, potassium, phosphorus, and calcium and on the urinary excretion of 17-keto, 11-oxy, and 17-hydroxy steroids in eight elderly men on high and low protein diets. J. Gerontol. **10**, 268—287 (1955).
WATSCHINGER, B.: Verlaufsbeobachtungen bei Nephrosekranken. In: Wirkung und Anwendung anaboler Steroide. Berlin: Medicus Verlag 1964.
WATTS, G. T., BADDELEY, M. R., WELLINGS, R., EVANS, J.: The nature of wound healing: Experimental tensile strength studies with deca durabolin and $S^{35}$. Ann. Surg. **162**, 109—112 (1965).
WATZMAN, N., KINNARD, W. J., ACETO, M. D., BUCKLEY, J. P.: Effects of certain compounds on experimental muscular dystrophy. J. pharm. Sci. **51**, 1090—1095 (1962).
WEBER, H., HAGGE, W.: Über die erfolgreiche Behandlung der Zystinose mit einem Anabolicum. Arch. Kinderheilk. **168**, 2 (1963).
WEINER, M., SIDDIQUI, A. A., BOSTANCI, N., DAYTON, P. G.: Drug interactions: The effect of combined administration on the half-life of coumarin and pyrazolone drugs in man. Fed. Proc. **24**, 153 (1965).
WEINER, M., SIDDIQUI, A. A., SHAHANI, R. T., DAYTON, P. G.: Effect of steroids on disposition of oxyphenbutazone in man. Proc. Soc. exp. Biol. Med. **124**, 1170—1173 (1967).
WEISSEL, W.: Anaboles Hormon bei malignem oder kompliziertem Diabetes mellitus. Wien. klin. Wschr. **74**, 234—236 (1962).
WELLER, O.: Die Wirkung von Methenolon auf die gesunde Leber. In: Wirkung und Anwendung anaboler Steroide. Berlin: Medicus Verlag 1964a.
WELLER, O.: Die Verwendung von Methenolon bei akuten und chronischen Leberparenchym-Erkrankungen. In: Wirkung und Anwendung anaboler Steroide. Berlin: Medicus Verlag 1964b.

WELLER, O., VOEGT, H.: Therapie der Leberzirrhose. Therapiewoche **12**, 665—668 (1962).
WELLS, R.: Prednisolone and testosterone propionate in cirrhosis of the liver. A controlled trial. Lancet **1960 II**, 1416—1419.
WENZEL, B., BÜNTE, H.: Die Verwertung parenteral zugeführter Aminosäuren nach chirurgischen Operationen. Med. Ernähr. **9** (1968).
WERNER, S. C.: Clinical syndromes associated with gonadal failure in men. Amer. J. Med. **3**, 52—66 (1947).
WERNZE, H.: Aufnahme und Ausscheidung von Bromsulfophthalein unter dem Einfluß anaboler Hormone durch die Leberzelle der Ratte. In: Wirkung und Anwendung anaboler Steroide. Berlin: Medicus Verlag 1964a.
WERNZE, H.: Die Leberfunktion unter dem Einfluß verschiedener Testosteronabkömmlinge (Untersuchungen von Serum-Fermenten und Bromsulfophthalein-Ausscheidung). In: Wirkung und Anwendung anaboler Steroide. Berlin: Medicus Verlag 1964b.
WEST, H. F., LEWIS-FANING, E.: Controlled trial of methenolone acetate in prednisolone-treated rheumatoid arthritic patients. Ann. rheum. Dis. **24**, 576—580 (1965).
WEWALKA, F.: Anabolika bei chronischen Leberkrankheiten. Z. Gastroenterologie **1**, 40—46 (1963).
WEWALKA, F.: Anabole Steroide und Leberkrankheiten. Wien. med. Wschr. **117**, 293—299 (1967).
WEWALKA, F.: Die Behandlung der chronischen Leberkrankheiten mit anabolen Steroiden. Therapiewoche **18**, 766—770 (1968).
WHYTE, H. M.: Potential effect on coronary-heart-disease morbidity of lowering the blood-cholesterol. Lancet **1975 I**, 906—910.
WIGGINS, R. A.: Observations of the use of 19-nortestosterone in patients with oliguric uremia. Sth. med. J. **56**, 669—672 (1963).
WILDHIRT, E.: Über die Leberverträglichkeit von Anabolika. In: Wirkung und Anwendung anaboler Steroide. Berlin: Medicus Verlag 1964.
WILES, F. J., MACKINTOSH, J.: Effect of durabolin in tuberculosis patients. Med. Proc. **8**, 56—57 (1962).
WILK, F., RAUSCH-SCHOTT, G.: Die Zystinose. Wien. med. Wschr. **121**, 349—353 (1971).
WILLIAMS, J. S., STEIN, J. H., FERRIS, T. F.: Nandrolone decanoate therapy for patients receiving hemodialysis. A controlled study. Arch. intern. Med. **134**, 289—292 (1974).
WINTHER, O.: Testosterone and fibrinolytic activity. Scand. J. clin. Lab. Invest. **99**, 207—210 (1967).
WINTHER, O.: Resistance to testosterone as a stimulation of blood fibrinolytic activity in patients with cancer of the breast. Lancet **1970 I**, 625.
WOLF, H.: Zur Behandlung der Untergewichtigkeit mit einem oral wirksamen Anabolikum. Wien. med. Wschr. **112**, 382—384 (1962).
WOLPERT, A., SHEPPARD, C., MERLIS, S.: Method for evaluation of behavioural changes in aged hospital patients during anabolic steroid therapy. J. amer. Geriat. Soc. **15**, 470—474 (1967).
WOLPERT, A., SHEPPARD, C., MERLIS, S.: Body weight as an index of therapeutic effect in anabolic steroid therapy. J. amer. Geriat. Soc. **16**, 323—327 (1968).
WOLTHUIS, F. H.: Balance studies on protein metabolism; significance for the treatment of renal failure. Voeding **24**, 372—386 (1963).
WORM, J.: Progressiv muskeldystrofi behandlet met durabolin. Ugeskr. Laeg. **124**, 1325—1327 (1962).
YAMASE, F., HATANAKA, H.: Effect of anabolic steroid for treatment of pulmonary tuberculosis. Med. J. Mut. Aid Ass. **12**, 4 (1963).
YOSHIDA, H.: Clinical studies on the decholesterolizing effect of anadrol. Saishin Igaku **20**, 2991—2997 (1965).
ZENEROLI, M. L.: Sulle variazione della colesterolemia durante trattamenti prolungati con metandrostenolone, ossimesterone e nandrolone. Endocr. Sci. Const. **31**, 107—114 (1970).
ZORRILLA, E., MAGOS, C., SERRANO, P. A.: Efecto de la oxandrolona sobre los lipidos sericos en las hiperlipidemias primarias tipo IV y $II_b$. Arch. Inst. Cardiol. Méx. **43**, 270—278 (1973).
ZWOZUKIAK, W., ZGORZALEWICZ, B.: Early and distant results of the treatment with the anabolic steroids of the progressive muscular dystrophy. Neurol. Neurochir. pol. **1**, 345—351 (1967).

# VIII. Appendix

Structures, systematic, trivial, generic and trade names of protein anabolic steroids

| Structure | Systematic name[a] | Trivial and/or generic name | Trade name | Manufacturer |
|---|---|---|---|---|
| *I. Testosterone-derived protein anabolic steroids* | | | | |
| 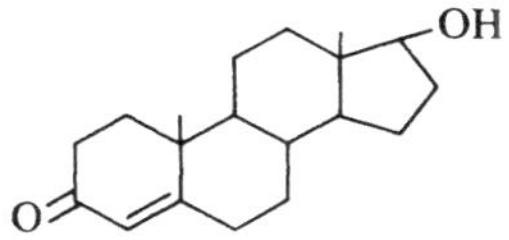 | 17β-Hydroxyandrost-4-en-3-one | Testosterone | Geno-cristaux | Gremy |
| | | | Malestrone | Kirk, N.Y. (Columbia) |
| | | | Orquisteron/Frosst | Schering A.G. |
| | | | Primotest | Schering |
| | | | Oreton | |
| | | -acetate | Aceto-Sterandryl | Roussel |
| | | | Aceto-Testoviron | Schering A.G. |
| | | | Perandrone A | CIBA |
| | | -propionate | Oreton | Schering |
| | | | Perandren | Ciba-Geigy |
| | | | Sterandryl | Roussel |
| | | | Testoviron | Schering |
| | | | Anertan | Boehringer (M) |
| | | | Enarmon | Takeda (Japan) |
| | | -(3-*p*-hexyloxiphenyl)-propionate | Androdurin | Leo, (Helsingborg) |
| | | -heptanoate (enanthate) | Delatestryl | Squibb |
| | | | Androtardyl | Schering-Sepps |
| | | | Testoenant | Geymonod |
| | | -cyclopentyl propionate | Depot testosterone cypionate | Upjohn |
| 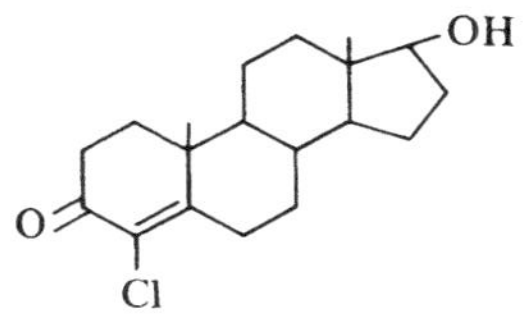 | 4-Chloro-17β-hydroxy-androst-4-en-3-one | Chlorotestosterone | Steranabol | Farmitalia |
| | | -acetate | Macrobin | Teikoku |
| | | | Turinabol | Jenapharm |
| | | | Turinabol inj. | Jenapharm |
| | | -capronate | Macrobin-depot | Teikoku |

Appendix (continued)

| Structure | Systematic name[a] | Trivial and/or generic name | Trade name | Manufacturer |
|---|---|---|---|---|
| 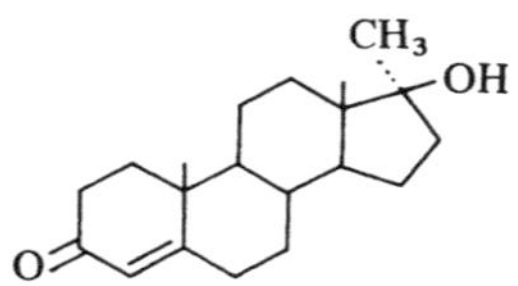 | 17β-Hydroxy-17-methyl-androst-4-en-3-one | Methyltestosterone | Metandren<br>Anertan<br>Oreton<br>Android | Ciba<br>Boehringer (M)<br>Schering<br>Brown |
| 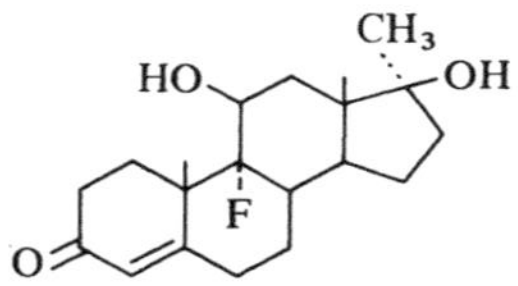 | 9α-Fluoro-11β,17β-dihydroxy-17-methylandrost-4-en-3-one | Fluoxymesterone | Halotestin<br>Ultandren<br>Androfluorene<br>Fluotestin<br>Orateston | Upjohn<br>Ciba<br>Midy<br>Roter<br>Hoechst |
| 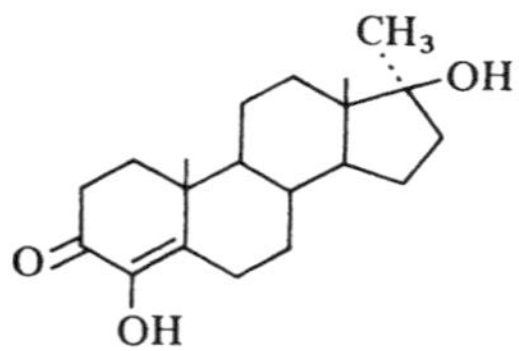 | 4,17β-Dihydroxy-17-methyl-androst-4-en-3-one | Oxymesterone, 4-Hydroxymethyltestosterone | Oranabol<br><br><br><br>Sanaboral<br>Theranabol | Farmitalia<br>Kabi<br>Lipfa<br>May & Baker<br>Ikapharm<br>Theraplix |
| 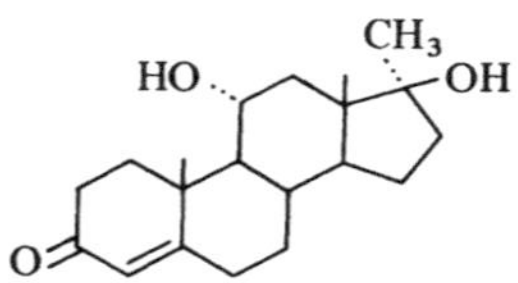 | 11α,17β-Dihydroxy-17-methyl-androst-4-en-3-one | 11α-Hydroxy-methyltestosterone | — | — |
| 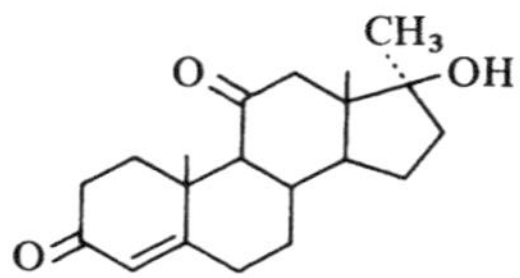 | 17β-Hydroxy-17-methyl-androst-4-ene-3,11-dione | 11-Oxomethyltestosterone | — | — |

Appendix (continued)

| Structure | Systematic name[a] | Trivial and/or generic name | Trade name | Manufacturer |
|---|---|---|---|---|
| 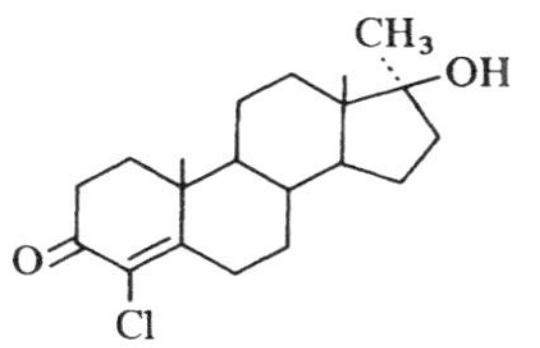 | 4-Chloro-17β-hydroxy-17-methyl-androst-4-en-3-one | Chloromethyltestosterone | Turinabol tabl. | Jenapharm |
| 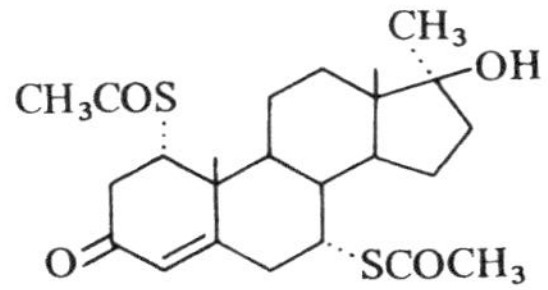 | 1α,7α-Bis(acetylthio)-17β-hydroxy-17-methyl-androst-4-en-3-one | Thiomesterone | Emdabol | Merck A.G. |
| 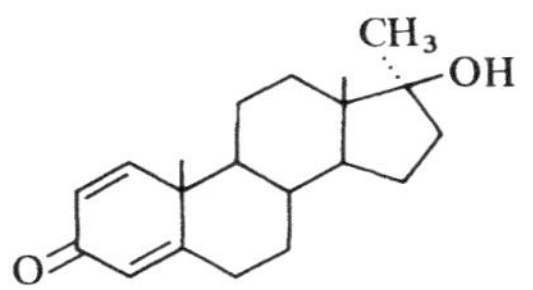 | 17β-Hydroxy-17-methyl-androsta-1,4-dien-3-one | Methandrostenolone<br>Methandienone<br>1-Dehydromethyltestosterone | Dianabol<br>Abirol<br>Geabol<br>Nabolin<br>Nerobil<br>Vanabol | Ciba<br>Takeda<br>Gea<br>Eisai<br>Richter<br>Vitrum |
| | | -cyclopentenyl ether<br>(gen.) Quinbolon | Anabolicus | Vister |
| 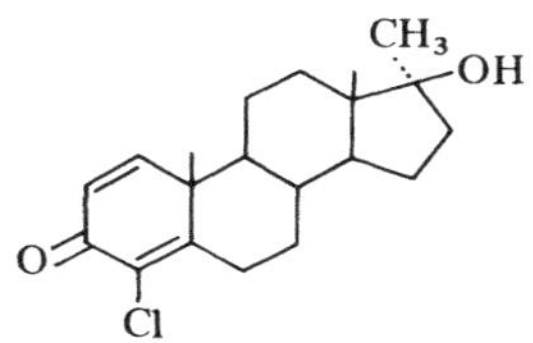 | 4-Chloro-17β-hydroxy-17-methylandrosta-1,4-dien-3-one | Chloro-1-dehydromethyl-testosterone | Oral-Turinabol | Jenapharm |
| 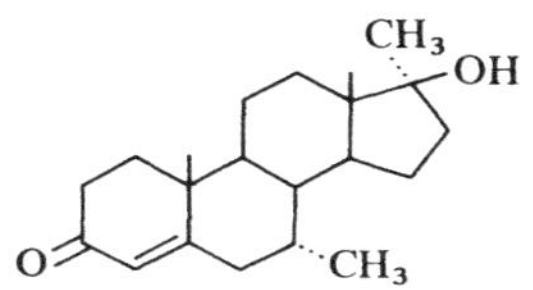 | 17β-Hydroxy-7α,17α-dimethyl-androst-4-en-3-one | Bolasterone,<br>Dimethyltestosterone | Myagen | Upjohn |
| | | Callusterone | Methosarb | Upjohn |

Appendic (continued)

| Structure | Systematic name[a] | Trivial and/or generic name | Trade name | Manufacturer |
|---|---|---|---|---|
| *II. 19-Nortestosterone-derived protein anabolic steroids* | | | | |
| 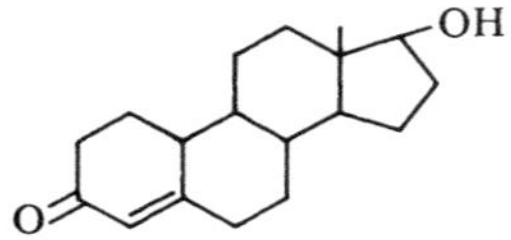 | 17$\beta$-Hydroxyestr-4-en-3-one | 19-Nortestosterone | Nerobolil | Medimpex |
| | | Nandrolone | Nortestonate | Upjohn |
| | | -n-capronate | Methybol-depot | Mepha |
| | | -propionate | Norybol | Serono |
| | | | Anabolicus-Serono | Ausonia |
| | | -phenylpropionate | Activin | Aristegui |
| | | | Durabol | Pharmacia |
| | | | Durabolin | Organon |
| | | | Nerobolil | Richter |
| | | | Neutrosteron | Organon |
| | | -furylpropionate | Demolon | Mochida |
| | | -hexahydrobenzoate | Nor-Durandon | Ferring |
| | | -hexyloxyphenylpropionate | Anadur | Leo-Lundbeck |
| | | -laurate | Laurabolin | Organon (Vet.) |
| | | -undecylate | Dynabolon | Theramex |
| | | -decanoate | Abolon | Benzon |
| | | | Deca-Durabol | Pharmacia |
| | | | Deca-Durabolin | Organon |
| | | | Eubolin | Futerapica |
| | | | Nordecon | IBSA |
| | | | Retabolil | Richter |
| | | -hemisuccinate | Menidrabol | Menarini |
| | | -cyclohexylpropionate | Fherbolico | Fher |
| | | | Sanabolicum | Sanabo |
| | | -cyclopentylpropionate | Sterocrinolo | Orma |
| | | | Depo-Nortestonate | Upjohn |
| | | | Pluropon | Boehringer |
| | | -4-methylbicyclo[2.2.2]oct-2-ene-1-carboxylate | — | Syntex |

Appendix (continued)

| Structure | Systematic name[a] | Trivial and/or generic name | Trade name | Manufacturer |
|---|---|---|---|---|
| 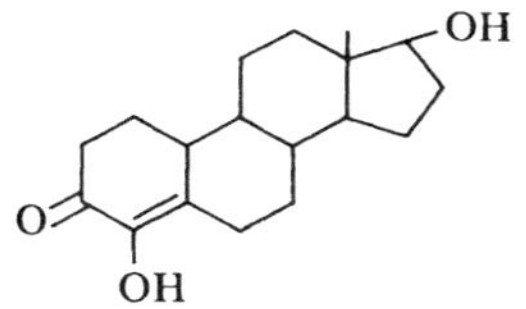 | 4,17β-Dihydroxyestr-4-en-3-one | Oxabolone<br>4-hydroxy-19-nortestosterone<br>-cyclopentylpropionate | Steranabol-depot | Farmitalia |
| 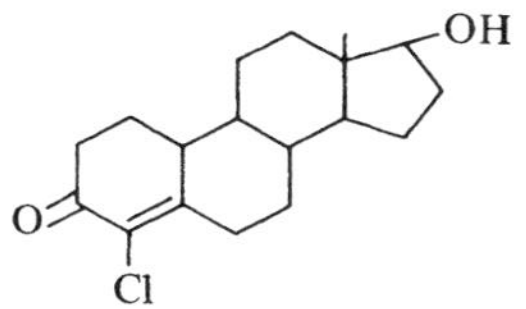 | 4-Chloro-17β-hydroxyestr-4-en-3-one | Chloro-19-nortestosterone<br>-acetate | Steranabol | Farmitalia |
| 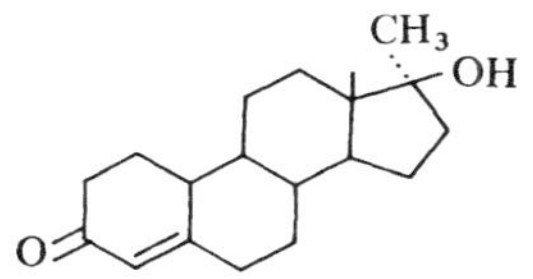 | 17β-Hydroxy-17-methyl-estr-4-en-3-one | Methyl-19-nortestosterone | Methalutin<br>Orgasteron | Syntex<br>Organon |
| 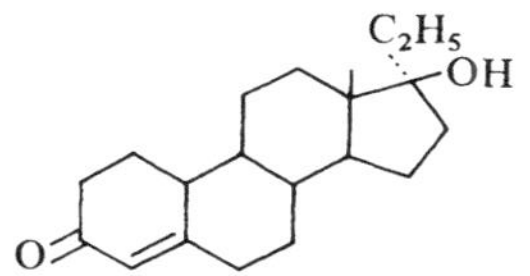 | 17β-Hydroxy-19-nor-pregn-4-en-3-one | Norethandrolone<br>Ethylnortestosterone<br><br>-propionate | Nilevar<br>Nor-Neutrormone<br>Pronabol<br>Solevar | Searle<br>ICI<br>Isis<br>Byla |
| 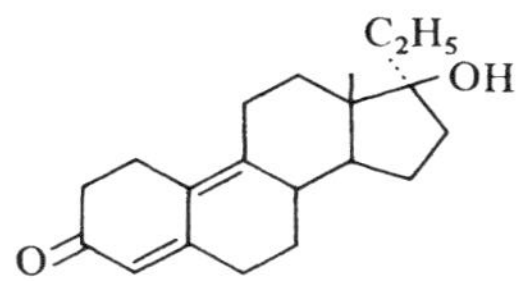 | 17β-Hydroxy-19-nor-pregna-4,9-dien-3-one | Ethyldienolone | — | — |

Appendix (continued)

| Structure | Systematic name[a] | Trivial and/or generic name | Trade name | Manufacturer |
|---|---|---|---|---|
| 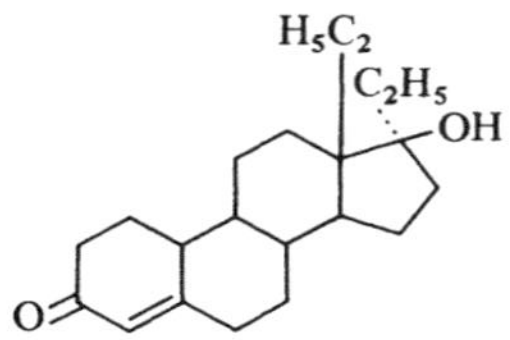 | (*dl*)-17$\beta$-Hydroxy-13$\beta$,17$\alpha$-diethylgon-4-en-3-one ((*dl*)-13-Ethyl-17-hydroxy-18,19-dinor-17$\alpha$-pregn-4-en-3-one) | Norbolethone | Genabol | Wyeth |
| *III. Androstane-derived protein anabolic steroids* | | | | |
| 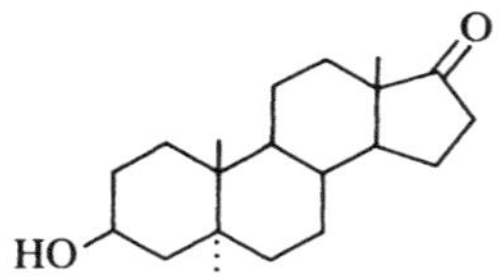 | 3$\beta$-Hydroxy-5$\alpha$-androstan-17-one | Epiandrosterone<br>Isoandrosterone | — | — |
| 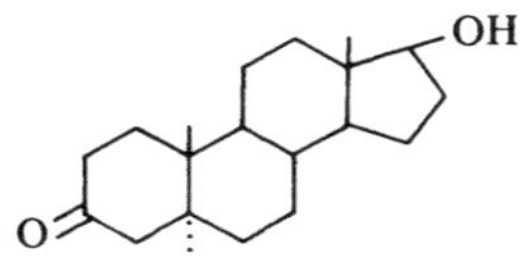 | 17$\beta$-Hydroxy-5$\alpha$-androstan-3-one | Androstanolone<br>Stanolone<br>Dihydrotestosterone | Anaboleen<br><br>Anabolex<br><br>Androlone<br>Proteina<br>Protona<br>Apeton | B.A.G.<br>Uni Chemie<br>Lloyd-Hamol<br>Samil<br>Orma<br>Gremy<br>Gremy<br>Fujisawa |
| | | -valerianate | Apeton depot | Fujisawa |
| 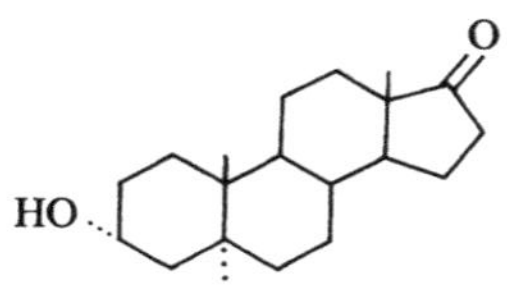 | 3$\alpha$-Hydroxy-5$\alpha$-androstan-17-one | Androsterone | — | — |

| Structure | Systematic name[a] | Trivial and/or generic name | Trade name | Manufacturer |
|---|---|---|---|---|
| 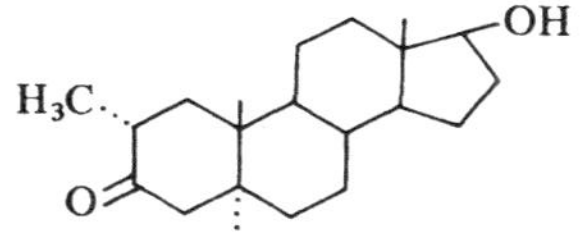 | 17$\beta$-Hydroxy-2$\alpha$-methyl-<br>5$\alpha$-androstan-3-one | Drostanolone<br>2$\alpha$-Methylandrostanolone | Drolban<br>Masterone | Lilly<br>Recovdati<br>Sarva<br>Syntex |
| | | 2$\alpha$-Methyldihydrotestosterone | Metorman | Latino |
| | | -propionate | Mastizol | Shionogi |
| 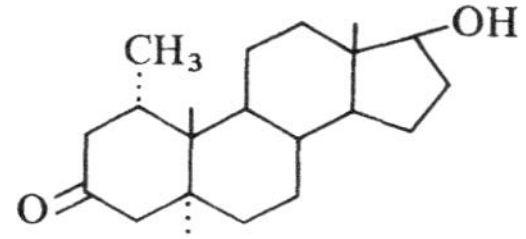 | 17$\beta$-Hydroxy-1$\alpha$-methyl-<br>5$\alpha$-androstan-3-one | Mesterolone | Mestoran | Schering<br>(Denmark) |
| | | | Proviron | Schering A.G. |
| 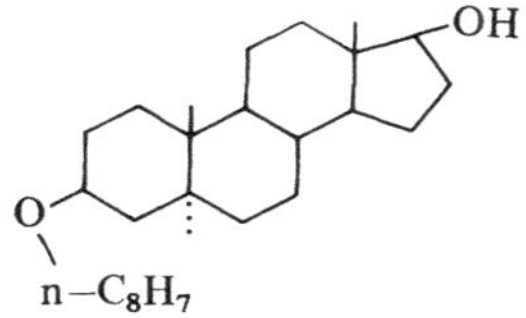 | 5$\alpha$-Androstane-3$\beta$,17$\beta$-diol-<br>3-n-octyl-enol-ether | Androstanediol-<br>3-n-octyl-enol ether | Ectovis<br>Ectovister | Vister<br>Drovyssa |
| 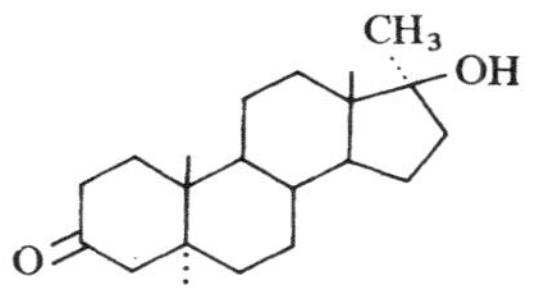 | 17$\beta$-Hydroxy-17-methyl-<br>5$\alpha$-androstan-3-one | Mestanolone<br>Methylandrostanolone | Ermalon<br>Androstalone | Roussel<br>Roussel |
| | | -enanthoyl-acetate | Notandron-depot | Boehringer (M) |
| 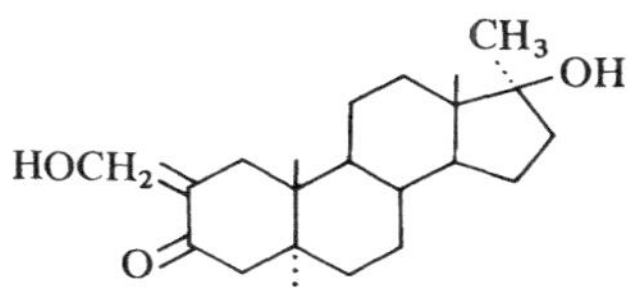 | 17$\beta$-Hydroxy-17-methyl-<br>2-(hydroxy-methylene)-<br>5$\alpha$-androstan-3-one | Oxymetholone | Adroyd | Parke-Davis<br>Sankyo |
| | | | Anadrol | Shionogi<br>Syntex |
| | | | Anadroyd<br>Anapolon<br>Anasterone<br>Anasterona<br>Nastenon<br>Protanabol<br>Synasterobe | Parke-Davis<br>ICI<br>Syntex<br>Latino<br>Cassenne<br>Recordati<br>Sarva |

Appendix (continued)

| Structure | Systematic name[a] | Trivial and/or generic name | Trade name | Manufacturer |
|---|---|---|---|---|
| *IV. Heterocyclic protein anabolic steroids* | | | | |
| 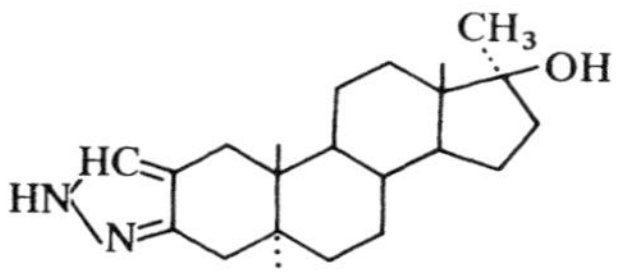 | 17-Methyl-5α-androstano-[3,2-c]-pyrazol-17β-ol | Stanozolol | Winstrol<br><br><br>Stromba<br>Tevabolin | Yamanouchi<br>Winthrop<br>Zambon<br>Winthrop<br>Teva |
| 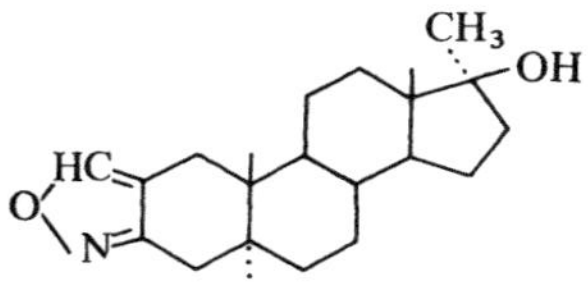 | 17-Methyl-5α-androstano-[3,2-c]-isoxazol-17β-ol | Androisoxazole | Androxan<br>Neo-Ponden<br>Neo-Pondus | Leo-Lundbeck<br>Serono<br>Ausonia |
| 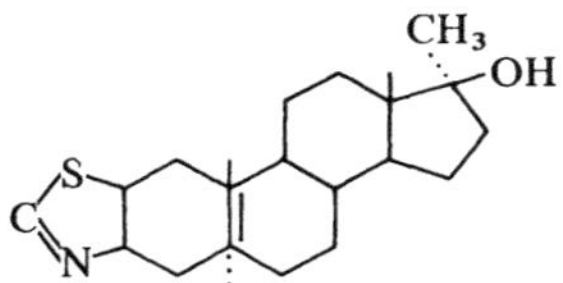 | 2′,17-Dimethylandrost-5(10)-eno[3,2-d]-thiazol-17β-ol | — | — | — |
| 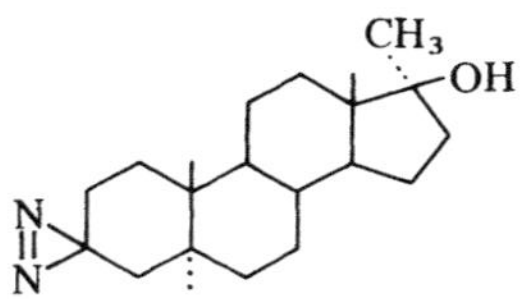 | 3-Azi-17-methyl-5α-androstan-17β-ol | Methyldiazirinol | — | — |
| 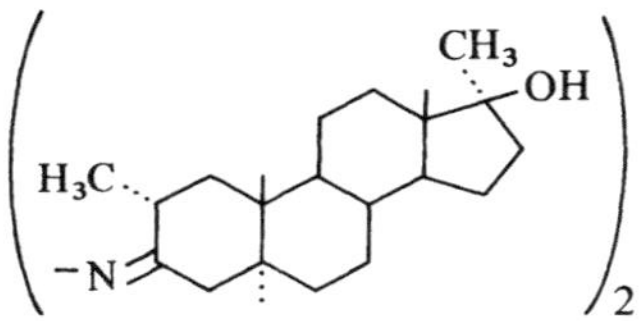 | 17β-Hydroxy-2α,17-dimethyl-5α-androstan-3,3′-azine | Dimethazine | Roxilon<br>Dostalon | Richter<br>Richter |

Appendix (continued)

| Structure | Systematic name[a] | Trivial and/or generic name | Trade name | Manufacturer |
|---|---|---|---|---|
| 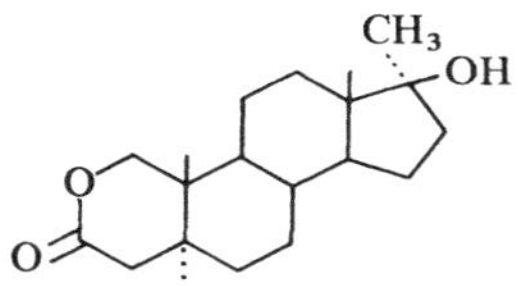 | 17β-Hydroxy-17-methyl-2-oxa-5α-androstan-3-one | Oxandrolone | Anavar | Searle |
| *V. Miscellaneous structure protein anabolic steroids* | | | | |
| 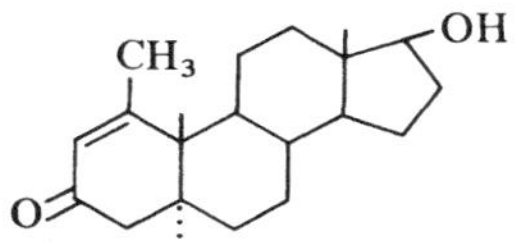 | 17β-Hydroxy-1-methyl-5α-androst-1-en-3-one | Methenolone | Nibol | Squibb |
| | | | Primobolan | Schering A.G. |
| | | -acetate | Primobolan-Acetate | Schering A.G. |
| | | -enanthate | Primobolan-Depot | Schering A.G. |
| 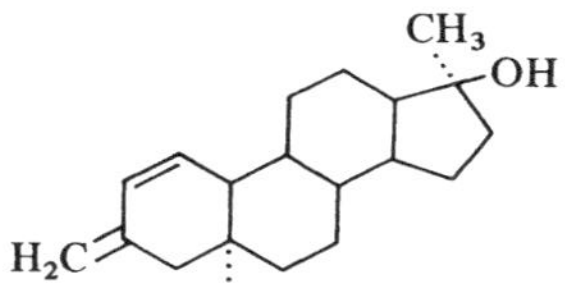 | 17-Methyl-3-methylene-5α-androst-1-en-17β-ol | — | — | — |
| 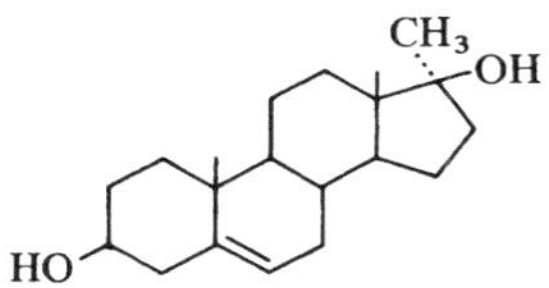 | 17-Methyl-androst-5-ene-3β,17β-diol | Methylandrostenediol | Crestabolic | Nutrition Control Products |
| | | Methandriol | Diandrin | Astra |
| | | | Megabion | Teikoku |
| | | | Metilandrostendiol | Schering A.G. |
| | | | Neosteron | Organon<br>Pharmacia |
| | | | Neutrormone | ISI |
| | | | Notandren | Boehringer (M) |
| | | | Protandren | Ciba |
| | | | Stenediol | Organon |

Appendix (continued)

| Structure | Systematic name[a] | Trivial and/or generic name | Trade name | Manufacturer |
|---|---|---|---|---|
| | | -3-propionate | Metilbisexovis<br>Methilbisexovister<br>Metildiolo | Vister<br>Drovyssa<br>Orma |
| | | -dipropionate | Anabolin<br>Metandiol | Rafa<br>Roussel |
| | | -dienanthoylacetate | Notandren-depot | Boehringer (M) |
| 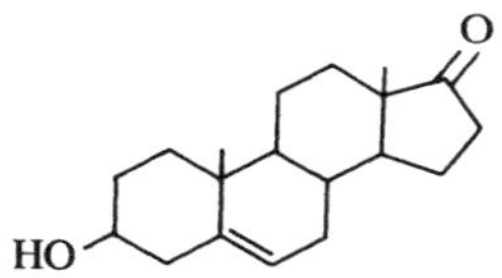 | 3β-Hydroxyandrost-5-en-17-one | Dehydroepiandrosterone | Psicosterone | ICI |
| | | Dehydroisoandrosterone | Diandrone | Organon |
| | | -acetate | 17-Chetovis | Vister |
| 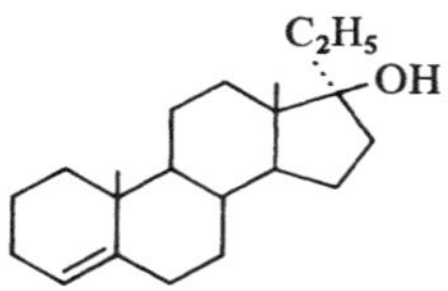 | 19-Norpregn-4-en-17β-ol | Ethylestrenol | Duraboral<br>Maxibolin<br>Orabolin<br>Orgaboral<br>Durabolin-0<br>Orgabolin | Organon<br>Organon<br>Organon<br>Organon<br>Organon<br>Organon |
| 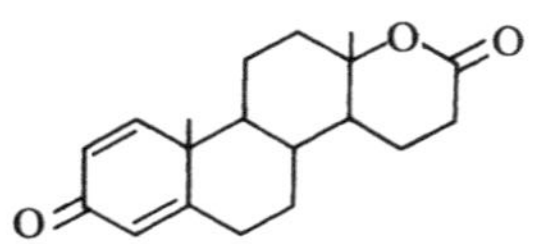 | 17α-Oxo-D-homo-androsta-1,4-diene-3,17-dione (1,2,3,4,4a,4b,7,9,10,10a-decahydro-2-hydroxy-2,4b-dimethyl-7-oxo-1-phenanthrene propionic acid) | Δ′-testololactone | Teslac | Squibb |

[a] Chemical Abstracts Systematic Name, Collective Subject Indices, Chemical Abstract.
e. Prepared by ARNOLD, A., POTTS, G. O., and BEYLER, A. L.

# Author Index

Page numbers in *italics* refer to bibliography

# Subject Index

# Addendum
# VII. E. Athletics

A. J. RYAN

A second study on the effect of the administration of an anabolic steroid to laboratory rats on their physical performances has been published by HICKSON et al. (1976). The activity was high intensity spring training. The study design was good and the report complete. It was well controlled. No improvement in this activity was noted as the result of the administration of Dianabol. There was no significant increase in muscle weights.

A rat study by ROGOZKIN et al. (1975) conducted over a period of 3 years showed that the administration of methandrostenolone and a high-protein diet with 6 min of physical activity daily resulted in transient changes in enzyme activity and increase in RNA synthesis in skeletal muscles. They postulate that this anabolic steroid has a stimulating effect on the transcription process in these muscles. In another series of experiments carried out over 84 days in which methandrostenolone was given by injection 3 times while training was continued, weight gain was observed but body water was not measured. Decreases in the weights of the ventral prostate and seminal vesicles were found which were reversible on the cessation of hormone therapy.

HERVEY (1975) studied 13 men aged 19–25 taking a physical education course in a 12-week double-blind cross-over design while they followed a standard daily weight-training routine and took 100 mg of methandrostenolone daily for 6 weeks and a placebo for 6 weeks. Lean body mass was estimated by the use of $^{40}K$ and a whole-body counter and underwater weighing. There was a significant increase in this measurement. Strength was measured by arm, trunk and leg work, leg and hand dynamometry and a forceplatform jump.There was no significant inter-group difference. Measurement of max $VO_2$ on a bicycle ergometer also showed no intra-group differences. Steroid metabolic studies showed decreased endogenous testosterone production during the period of anabolic steroid administration.

FREED et al. (1975) reported a study of 13 experienced male weight lifters who took part in a double-blind cross-over trial of methandienone 10 or 25 mg a day and who were taking high protein diets. Treatment was for two consecutive 6-week periods with the "low" or "high" dose or the placebo. Strength was measured by six standard weight lifting exercises, lean body weight was recorded and skinfold thickness determined. Ages ranged from 18 to 30. Six took the 10 mg and 7 the 25 mg dose. The defects of this study are so great that the "results" will not be discussed. The men were allowed to report their own weights and all strength measurements. Of the 7 who started on the steroid, 3 dropped out before the 6 weeks were completed. Of the six who started on the placebo, two reported no results to the investigators and two withdrew during the second 6 weeks because of side-effects from the medication. This left only 6 of the original 13.

Win-May and Mya-Tu (1975) reported on an 18-week study of 31 male university students aged 18–23 given 5 mg of methandrostenolone or a placebo in a double-blind design.

Gross body weight, extremity circumferences and skinfold thicknesses were measured. Strength measures recorded were numbers of sit-ups and pull-ups, distance on standing broad jump and hand grip as determined by a dynamometer. Gross body weight was significantly increased in the steroid but not the placebo group. No measure of body water was made. Body fat was significantly increased in the steroid group which could account for some if not all of the difference in body weight. Significant improvements were noted in all four strength measures for the steroid group, but only at the 5% level of confidence.

Hagerman et al. (1975) studied five experienced weight trainers aged 20–23 over a period of 20 weeks. Three of the subjects took 5 mg of methandrostenolone daily on a 21-day-on and 7-day-off schedule during this time. The controls took nothing. Training was vigorous and emphasized power lifting. The days and hours worked were not specified. All subjects showed improvements in performance but no analysis of significance was made. No undersirable effects on liver function were noted over a 12-week period. No consistent patterns of 17-ketosteroid responses were observed and all values fell within the normal ranges for healthy young adults. There were no indications of abnormal testicular or adrenal suppression.

Johnson et al. (1975) studied 21 male college students aged 19–33 who engaged in an interval running program 6 days a week over a 3-week period. Each subject received a high protein diet supplement. A double-blind design was used for the administration of 6 mg of stanozolol per day to members of the treatment group. Maximum oxygen uptake, mile run time, skinfold thickness, body weight and elbow flexion and knee extension strength were measured. Differences between the steroid and control groups were not statistically significant for any of the dependent variables.

Keul et al. (1976) studied 10 non-athletes (average age 22) and 15 weight lifters (average age 27) before and for 3 months after taking nandrolone decanoate. Six of the lifters had been taking this substance for at least 3 years. A control group of 7 weight lifters was also studied. Testing consisted of a maximum test ride on a bicycle ergometer, three strength measures with barbells, observation of heart rates during the ergometric test and a number of measurements of blood elements. The experimental subjects were given 50 mg of nandrolone decanoate by i.m. injection every 12 days for a period of 2 months. Controls were given nothing. No significant performance improvements were noted over the controls for strength measures or endurance. No deleterious side-effects were noted in the subjects from the steroid or in 26 other athletes who had been taking the same substance. Some evidence of liver damage was found in 31 other athletes who had been taking oral anabolic allyl steroids. These were said to have been reversed on stopping the medication.

Hervey et al. (1976) studied 11 male physical education students aged 19–25 years during a course of weight training. After 2 weeks of training without medication, half were given 100 mg methandienone a day for 6 weeks and the other half a placebo. All stopped training and medication for 5 weeks, then resumed training for 1 week and then undertook 6 more weeks of training with the treatments crossed over. Body weight and composition, muscular strength, standing jump performance and maximum oxygen uptake were studied. Plasma homones, urinary cortisol, plasma bilirubin and liver enzymes were also measured. There were no significant differences in physical performance between the control and treated groups.

BROOKS et al. (1975) reported their method of detection of anabolic steroids by radioimmunoassay and WARD et al. (1975) reported their gas chromatographic-mass spectrometric methods. POP et al. (1976a, b) described a method similar to that of BROOKS. In a test of this method on 10 wrestlers taking 20 mg a day of methandrostenolone, they determined that the drug could no longer be detected 2–3 weeks after it was stopped.

## References


BROOKS, R. V., FIRTH, R. G., SUMNER, N. A.: Detection of anabolic steroids by radioimmunoassay. Brit. J. Sports Med. **9**, 89–92 (1975).

FREED, D. L. J., BANKS, A. J., LONGSON, D., BURLEY, D. M.: Anabolic steroids in athletics: Cross-over double-blind trial on weightlifters. Brit. med. J. **1975 II**, 471–473.

HAGERMAN, F. C., JONES-WITTERS, P., RANSON, R.: The effects of anabolic steroid ingestion on serum enzyme and urine 17-ketosteroid levels. J. Sports Med. Phys. Fit. **15**, 287–295 (1975).

HERVEY, G. R.: Are athletes wrong about anabolic steroids? Brit. J. Sports Med. **9**, 74–77 (1975).

HERVEY, G. R., HUTCHINSON, I., KNIBBS, A. V., BURKINSHAW, L., JONES, P. R. M., MORGAN, N. G., LEVELL, M. J.: Anabolic effects of methandienone in men undergoing athletic training. Lancet **1976 II**, 699–702.

HICKSON, R. C., HEUSNER, W. W., VAN HUSS, W. D., JACKSON, D. E., ANDERSON, D. A., JONES, D. A., PSALEDAS, A. T.: Effects of Dianabol and high-intensity spring training on body composition of rats. Med. Sci. in Sports **8**, 191–195 (1976).

JOHNSON, L. C., ROUNDY, E. S., ALLSEN, P. E., FISHER, A. G., SILVESTER, L. J.: Effect of anabolic steroid treatment on endurance. Med. Sci. in Sports **7**, 287–289 (1975).

KEUL, J., DEUS, B., KINDERMANN, W.: Anabole hormone: Schädigung, Leistungsfähigkeit und Stoffwechsel. Wschr. Klin. Prax. **71**, 497–503 (1976).

POP, T., DRAGAN, I., UTA, I., PITCU, E.: A new modern method of making evident anabolic steroids in the urine of high performance sportsmen. Med. dello Sport **29**, 10–12 (1976a).

POP, J., DRAGAN, I., UTA, I., PITCU, E.: Values of the urinary testosterone 2–3 weeks after a treatment with anabolic steroids. Med. dello Sport **29**, 13 (1976b).

ROGOZKIN, V., BAZULKO, A., TCHAIKOVSKY, V., PLISKIN, A., PSHENDIN, A., SILBER, M.: Anabolic and androgenic effects of methandrostenolone during systematic physical activity in rats. Brit. J. Sports Med. **9**, 65–69 (1975).

WARD, R. J., SHACKLETON, C. H. L., LAWSON, A. M.: Gas chromatographic-mass spectrometric methods for the detection and identification of anabolic steroid drugs. Brit. J. Sports Med. **9**, 93–97 (1975).

WIN-MAY, M., MYA-TU, M.: The effect of anabolic steroids on physical fitness. J. Sports Med. Phys. Fit. **15**, 266–271 (1975).

**Handbuch der experimentellen Pharmakologie**
**Handbook of Experimental Pharmacology**

Heffter-Heubner, New Series

---

Vol. 4: **General Pharmacology**

Vol. 10: **Die Pharmakologie anorganischer Anionen**

Vol. 11: **Lobelin und Lobeliaalkaloide**

Vol. 12: **Morphin und morphinähnlich wirkende Verbindungen**

Vol. 14: Part 1: **The Adrenocortical Hormones I**
Part 2: **The Adrenocortical Hormones II**
Part 3: **The Adrenocortical Hormones III**

Vol. 15: **Cholinesterases and Anticholinesterase Agents**

Vol. 16: **Erzeugung von Krankheitszuständen durch das Experiment/Experimental Production of Diseases**
Part 1: **Blut.** In Vorbereitung
Part 2: **Atemwege**
Part 3: **Heart and Circulation**
Part 4: **Niere, Nierenbecken, Blase**
Part 5: **Liver**
Part 6: **Schilddrüse.** In Vorbereitung
Part 7: **Zentralnervensystem**
Part 8: **Stütz- und Hartgewebe**
Part 9: **Infektionen I**
Part 10: **Infektionen II**
Part 11A: **Infektionen III**
Part 11B: **Infektionen IV**
Part 12: **Tumoren I**
Part 13: **Tumoren II**
Part 14: **Tumoren III.** In preparation
Part 15: **Kohlenhydratstoffwechsel, Fieber/Carbohydrate Metabolism, Fever**

Vol. 17: Part 1: **Ions alcalino-terreux I**
Part 2: **Ions alcalino-terreux II**

Vol. 18: Part 1: **Histamine**
Part 2: **Anti-Histaminics.** In preparation

Vol. 19: **5-Hydroxytryptamine and Related Indolealkylamines**

Vol. 20: Part 1: **Pharmacology of Fluorides I**
Part 2: **Pharmacology of Fluorides II**

Vol. 21: **Beryllium**

Vol. 22: Part 1: **Die Gestagene I**
Part 2: **Die Gestagene II**

Vol. 23: **Neurohypophysial Hormones and Similar Polipeptides**

Vol. 24: **Diuretica**

Vol. 25: **Bradykinin, Kallidin and Kallikrein**

Vol. 26: **Vergleichende Pharmakologie von Überträgersubstanzen in tiersystematischer Darstellung**

Vol. 27: **Anticoagulantien**

Vol. 28: Part 1: **Concepts in Biochemical Pharmacology I**
Part 3: **Concepts in Biochemical Pharmacology III**

Vol. 29: **Oral wirksame Antidiabetika**

Vol. 30: **Modern Inhalation Anesthetics**

Vol. 32: Part 1: **Insulin I**
Part 2: **Insulin II**

Vol. 34: **Secretin, Cholecystokinin, Pancreozymin and Gastrin**

Vol. 35: Part 1: **Androgene I**
Part 2: **Androgens II and Antiandrogens/Androgene II und Antiandrogene**

Vol. 36: **Uranium — Plutonium — Transplutonic Elements**

Vol. 37: **Angiotensin**

Vol. 38: Part 1: **Antineoplastic and Immunosuppressive Agents I**
Part 2: **Antineoplastic and Immunosuppressive Agents II**

Vol. 39: **Antihypertensive Agents**

Vol. 40: **Organic Nitrates**

Vol. 41: **Hypolipidemic Agents**

Vol. 42: **Neuromuscular Junction**

**Springer-Verlag Berlin Heidelberg New York**

www.ingramcontent.com/pod-product-compliance
Ingram Content Group UK Ltd.
Pitfield, Milton Keynes, MK11 3LW, UK
UKHW051003220926
475831UK00003B/105
* 9 7 8 3 6 4 2 6 6 3 5 5 0 *